NEUROMETHODS

Series Editor
Wolfgang Walz
University of Saskatchewan
Saskatoon, SK, Canada

For further volumes:
http://www.springer.com/series/7657

Neuromethods publishes cutting-edge methods and protocols in all areas of neuroscience as well as translational neurological and mental research. Each volume in the series offers tested laboratory protocols, step-by-step methods for reproducible lab experiments and addresses methodological controversies and pitfalls in order to aid neuroscientists in experimentation. *Neuromethods* focuses on traditional and emerging topics with wide-ranging implications to brain function, such as electrophysiology, neuroimaging, behavioral analysis, genomics, neurodegeneration, translational research and clinical trials. Neuromethods provides investigators and trainees with highly useful compendiums of key strategies and approaches for successful research in animal and human brain function including translational "bench to bedside" approaches to mental and neurological diseases.

Neurodegenerative Diseases Biomarkers

Towards Translating Research to Clinical Practice

Edited by

Philip V. Peplow

Department of Anatomy, University of Otago, Dunedin, New Zealand

Bridget Martinez

Department of Pharmacology, Department of Medicine, Reno School of Medicine, University of Nevada, Reno, NV, USA

Thomas A. Gennarelli

Department of Neurosurgery, George Washington University, West Chester, PA, USA

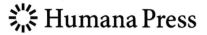 Humana Press

Editors
Philip V. Peplow
Department of Anatomy
University of Otago
Dunedin, New Zealand

Thomas A. Gennarelli
Department of Neurosurgery
George Washington University
West Chester, PA, USA

Bridget Martinez
Department of Pharmacology
Department of Medicine
Reno School of Medicine
University of Nevada
Reno, NV, USA

ISSN 0893-2336   ISSN 1940-6045 (electronic)
Neuromethods
ISBN 978-1-0716-1714-4   ISBN 978-1-0716-1712-0 (eBook)
https://doi.org/10.1007/978-1-0716-1712-0

Cover Caption: This custom piece was created by Dr. Bridget Martinez and Dr. Donald Mario Robert Harker using acrylic paint on canvas with the addition of graphic design. This artistic rendition encapsulates the cognitive abilities that machines seek to recreate – an unparalleled, beautiful assembly - Bridget Martinez, MD, PhD

This Humana imprint is published by the registered company Springer Science+Business Media, LLC part of Springer Nature.
The registered company address is: 1 New York Plaza, New York, NY 10004, U.S.A.

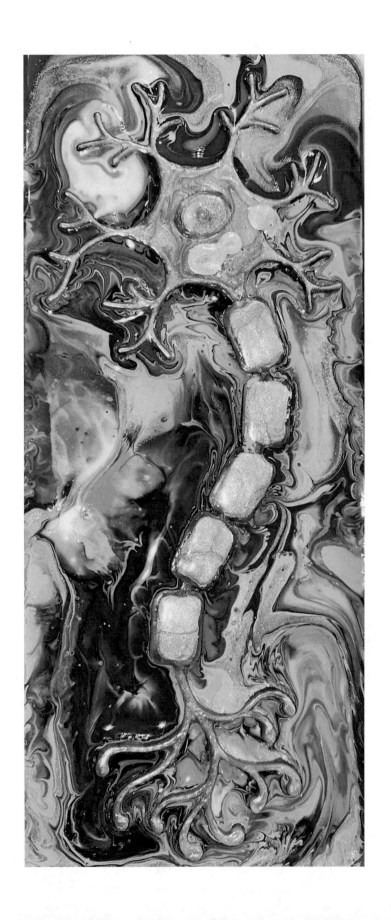

Dedication

The COVID-19 virus has affected so many families in different parts of the world, and for some life will never be the same again. A few countries have done well in containing the virus, but in others the virus is/has been rampant with huge loss of lives. While we may never know the origin of the virus, we must be vigilant to ensure that nothing like this happens again. Many medical researchers, frontline hospital workers, and first responders have lost their lives in fighting the virus, and we pay tribute and dedicate this book to them. We look forward to the world slowly recovering from the effects of the pandemic and remember the words of Captain Sir Thomas Moore who valiantly fundraised for the NHS in Britain "Tomorrow will be a good day."

> *With all its sham, drudgery and broken dreams, it is still a beautiful world. Be cheerful. Strive to be happy.*
>
> (Desiderata, Max Ehrmann)

Philip V. Peplow
Bridget Martinez
Thomas A. Gennarelli
31 May 2021

Preface to the Series

Experimental life sciences have two basic foundations: concepts and tools. The *Neuromethods* series focuses on the tools and techniques unique to the investigation of the nervous system and excitable cells. It will not, however, shortchange the concept side of things as care has been taken to integrate these tools within the context of the concepts and questions under investigation. In this way, the series is unique in that it not only collects protocols but also includes theoretical background information and critiques which led to the methods and their development. Thus it gives the reader a better understanding of the origin of the techniques and their potential future development. The *Neuromethods* publishing program strikes a balance between recent and exciting developments like those concerning new animal models of disease, imaging, *in vivo* methods, and more established techniques, including, for example, immunocytochemistry and electrophysiological technologies. New trainees in neurosciences still need a sound footing in these older methods in order to apply a critical approach to their results.

Under the guidance of its founders, Alan Boulton and Glen Baker, the *Neuromethods* series has been a success since its first volume published through Humana Press in 1985. The series continues to flourish through many changes over the years. It is now published under the umbrella of Springer Protocols. While methods involving brain research have changed a lot since the series started, the publishing environment and technology have changed even more radically. Neuromethods has the distinct layout and style of the Springer Protocols program, designed specifically for readability and ease of reference in a laboratory setting.

The careful application of methods is potentially the most important step in the process of scientific inquiry. In the past, new methodologies led the way in developing new disciplines in the biological and medical sciences. For example, Physiology emerged out of Anatomy in the nineteenth century by harnessing new methods based on the newly discovered phenomenon of electricity. Nowadays, the relationships between disciplines and methods are more complex. Methods are now widely shared between disciplines and research areas. New developments in electronic publishing make it possible for scientists that encounter new methods to quickly find sources of information electronically. The design of individual volumes and chapters in this series takes this new access technology into account. Springer Protocols makes it possible to download single protocols separately. In addition, Springer makes its print-on-demand technology available globally. A print copy can therefore be acquired quickly and for a competitive price anywhere in the world.

Saskatoon, SK, Canada   *Wolfgang Walz*

Preface

Neurodegenerative diseases (NDs) are increasing in incidence in adults, and this increase is expected to continue. The increase is exacerbated by extended life expectancy from improved health care in many developed and developing countries. Considerable effort is being given to developing reliable and validated biomarkers for NDs so as to dramatically accelerate research on the etiology, pathophysiology, and disease progression of a number of prevalent and devastating NDs.

It is the goal of this book to provide a forum for both experimental and clinical international experts in the field of ND research to present recent data on the latest achievements in new and emerging technologies for biomarkers and for innovations in their assessment. Theoretical backgrounds together with tested protocols that reproduce experimental, clinical laboratory, and instrumental methods for educational purposes are presented. It is hoped that the topics covered herein will extend knowledge on the role of novel biomarkers in different types of NDs and that this will lead to a more effective approach to clinical management and ultimately to benefit patient care.

We wish to express our deep appreciation and gratitude to each of the chapter authors for the time and effort spent on writing informative reviews on their respective areas of clinical and research interest. This task was complicated by the national lockdowns in many countries during the global COVID-19 pandemic. Also we wish to thank Professor Wolfgang Walz, Series Editor, Springer *Neuromethods* series, and Anna Rakovsky, Assistant Editor, Springer Protocols, for their help, encouragement, and advice in putting together this book.

Dunedin, New Zealand
Reno, NV, USA
West Chester, PA, USA
27 March 2021

Philip V. Peplow
Bridget Martinez
Thomas A. Gennarelli

Contents

PART I   INTRODUCTION

PART II   RESEARCH METHODS

Contributors

CATARINA M. ABREU • *International Iberian Nanotechnology Laboratory, Braga, Portugal; Swansea University Medical School, Sketty, Swansea, UK*

LEONOR CERDÁ ALBERICH • *Radiology, Neurology and Biomedical Imaging Research Group (GIBI230), La Fe Health Research Institute, La Fe University and Polytechnic Hospital, Valencia, Spain*

FIONA M. BRIGHT • *Department of Biomedical Sciences, Faculty of Medicine Health and Human Sciences, Dementia Research Centre, Macquarie University, Sydney, NSW, Australia*

RUDY J. CASTELLANI • *Department of Pathology, Anatomy, and Laboratory Medicine, West Virginia University, Morgantown, WV, USA; Department of Neuroscience, Rockefeller Neuroscience Institute, West Virginia University, Morgantown, WV, USA*

JUAN FRANCISCO VÁZQUEZ COSTA • *Radiology, Neurology and Biomedical Imaging Research Group (GIBI230), La Fe Health Research Institute, La Fe University and Polytechnic Hospital, Valencia, Spain*

ANDREA CRUZ • *International Iberian Nanotechnology Laboratory, Braga, Portugal; ProChild CoLAB Against Child Poverty and Social Exclusion, Portuguese Foundation for Science and Technology (FCT) Collaborative Laboratory, Guimarães, Portugal*

RUBEN K. DAGDA • *Department of Pharmacology, Reno School of Medicine, University of Nevada, Reno, NV, USA*

EDOARDO ROSARIO DE NATALE • *Neurodegeneration Imaging Group, University of Exeter Medical School, London, UK*

EVA DRAZANOVA • *Department of Pharmacology, Faculty of Medicine, Masaryk University, Brno, Czech Republic; Institute of Scientific Instruments of the Czech Academy of Sciences, Brno, Czech Republic*

PAULO P. FREITAS • *International Iberian Nanotechnology Laboratory, Braga, Portugal*

JENNY N. T. FUNG • *School of Biomedical Sciences, Faculty of Medicine, The University of Queensland, St Lucia, Brisbane, QLD, Australia*

THOMAS A. GENNARELLI • *Department of Neurosurgery, Medical College of Wisconsin, Milwaukee, WI, USA; Department of Neurosurgery, George Washington University, West Chester, PA, USA*

DONALD M. R. HARKER • *School of Medicine, St. Georges Medical University, True Blue, Grenada*

ROBERT D. HENDERSON • *Centre for Clinical Research, The University of Queensland, Brisbane, QLD, Australia; Department of Neurology, Royal Brisbane and Women's Hospital, Brisbane, QLD, Australia*

CORY J. HOLDOM • *Centre for Clinical Research, The University of Queensland, Brisbane, QLD, Australia; Australian Institute for Bioengineering and Nanotechnology, The University of Queensland, Brisbane, QLD, Australia*

JIN-HUI HOR • *Institute of Molecular and Cell Biology, A*STAR Research Entities, Singapore, Singapore; Department of Biological Sciences, National University of Singapore, Singapore, Singapore*

TOMÁŠ HROMÁDKA • *Institute of Neuroimmunology, Slovak Academy of Sciences, Bratislava, Slovakia; Axon Neuroscience R&D Services SE, Bratislava, Slovakia*

MICHAEL KASSIOU • *Faculty of Science, School of Chemistry, The University of Sydney, Sydney, NSW, Australia*

AMIT KHAIRNAR • *Department of Pharmacology and Toxicology, National Institute of Pharmaceutical Education and Research (NIPER), Ahmedabad, Gandhinagar, Gujarat, India*

ANNIE KILLORAN • *Department of Neurology, University of Iowa Hospital and Clinics, Iowa City, IA, USA*

RITUSHREE KUKRETI • *Genomics and Molecular Medicine Unit, CSIR-Institute of Genomics and Integrative Biology (IGIB), Delhi, India*

SHRIKANT KUKRETI • *Nucleic Acids Research Lab, Department of Chemistry, University of Delhi (North Campus), Delhi, India*

WEIDONG LE • *Center for Clinical Research on Neurological Diseases, The First Affiliated Hospital, Dalian Medical University, Dalian, China; Liaoning Provincial Key Laboratory for Research on the Pathogenic Mechanisms of Neurological Diseases, The First Affiliated Hospital, Dalian Medical University, Dalian, China; Institute of Neurology, Sichuan Academy of Medical Science-Sichuan Provincial Hospital, Medical School of UESTC, Chengdu, China*

JOHN D. LEE • *School of Biomedical Sciences, Faculty of Medicine, The University of Queensland, St Lucia, Brisbane, QLD, Australia*

JIATONG LI • *Department of Anatomy, Histology and Embryology, School of Basic Medical Sciences, Fudan University, Shanghai, China*

MARTIN W. LO • *School of Biomedical Sciences, Faculty of Medicine, The University of Queensland, St Lucia, Brisbane, QLD, Australia*

LUIS MARTÍ-BONMATÍ • *Radiology, Neurology and Biomedical Imaging Research Group (GIBI230), La Fe Health Research Institute, La Fe University and Polytechnic Hospital, Valencia, Spain; Radiology Department, La Fe Health Research Institute, La Fe University and Polytechnic Hospital, Valencia, Spain*

BRIDGET MARTINEZ • *Department of Pharmacology, Department of Medicine, Reno School of Medicine, University of Nevada, Reno, NV, USA*

MIGUEL MAZÓN • *Radiology, Neurology and Biomedical Imaging Research Group (GIBI230), La Fe Health Research Institute, La Fe University and Polytechnic Hospital, Valencia, Spain*

PAMELA A. MCCOMBE • *Centre for Clinical Research, The University of Queensland, Brisbane, QLD, Australia; Department of Neurology, Royal Brisbane and Women's Hospital, Brisbane, QLD, Australia*

INÈS MENDES PINTO • *International Iberian Nanotechnology Laboratory, Braga, Portugal*

SHI-YAN NG • *Institute of Molecular and Cell Biology, A*STAR Research Entities, Singapore, Singapore; Yong Loo Lin School of Medicine (Physiology), National University of Singapore, Singapore, Singapore; National Neuroscience Institute, Singapore, Singapore*

SHYUAN T. NGO • *Centre for Clinical Research, The University of Queensland, Brisbane, QLD, Australia; Australian Institute for Bioengineering and Nanotechnology, The University of Queensland, Brisbane, QLD, Australia; Department of Neurology, Royal Brisbane and Women's Hospital, Brisbane, QLD, Australia; Queensland Brain Institute, The University of Queensland, Brisbane, QLD, Australia*

PHILIP V. PEPLOW • *Department of Anatomy, University of Otago, Dunedin, New Zealand*

MARIOS POLITIS • *Neurodegeneration Imaging Group, University of Exeter Medical School, London, UK*

SONG QIN • *Department of Anatomy, Histology and Embryology, School of Basic Medical Sciences, Fudan University, Shanghai, China*

MARY-LOUISE ROGERS • *Department of Human Physiology and Centre for Neuroscience, Flinders University, Adelaide, SA, Australia*

JANA RUDA-KUCEROVA • *Department of Pharmacology, Faculty of Medicine, Masaryk University, Brno, Czech Republic*

ANNA SALAMERO-BOIX • *Georg-Speyer-Haus, Institute for Tumor Biology and Experimental Therapy, Frankfurt am Main, Germany; Biological Sciences, Goethe University Frankfurt, Frankfurt am Main, Germany*

MUNIRAH MOHAMAD SANTOSA • *Institute of Molecular and Cell Biology, A*STAR Research Entities, Singapore, Singapore; Yong Loo Lin School of Medicine (Physiology), National University of Singapore, Singapore, Singapore*

MICHAEL SCHULZ • *Georg-Speyer-Haus, Institute for Tumor Biology and Experimental Therapy, Frankfurt am Main, Germany; Biological Sciences, Goethe University Frankfurt, Frankfurt am Main, Germany*

LISA SEVENICH • *Georg-Speyer-Haus, Institute for Tumor Biology and Experimental Therapy, Frankfurt am Main, Germany; German Cancer Consortium (DKTK, Partner Site Frankfurt/Mainz) and German Cancer Research Center (DKFZ), Heidelberg, Germany; Frankfurt Cancer Institute (FCI), Goethe University Frankfurt, Frankfurt am Main, Germany*

YAPING SHAO • *Center for Clinical Research on Neurological Diseases, The First Affiliated Hospital, Dalian Medical University, Dalian, China; Liaoning Provincial Key Laboratory for Research on the Pathogenic Mechanisms of Neurological Diseases, The First Affiliated Hospital, Dalian Medical University, Dalian, China*

ANJU SINGH • *Department of Chemistry, Ramjas College, University of Delhi, Delhi, India*

FREDERIK J. STEYN • *Centre for Clinical Research, The University of Queensland, Brisbane, QLD, Australia; School of Biomedical Sciences, The University of Queensland, Brisbane, QLD, Australia; Department of Neurology, Royal Brisbane and Women's Hospital, Brisbane, QLD, Australia*

ANDREI SURGUCHOV • *Department of Neurology, Kansas University Medical Center, Kansas City, KS, USA*

NIKOLETTA SZABO • *Department of Neurology, Albert Szent-Györgyi Clinical Center, Faculty of Medicine, University of Szeged, Szeged, Hungary*

AMADEO TEN-ESTEVE • *Radiology, Neurology and Biomedical Imaging Research Group (GIBI230), La Fe Health Research Institute, La Fe University and Polytechnic Hospital, Valencia, Spain*

THOMAS VOGELS • *Sylics (Synaptologics B.V.), Bilthoven, The Netherlands; Department of Psychiatry and Neurochemistry, University of Gothenburg, Gothenburg, Sweden; Department of Neurodegenerative Disease, UCL Queen Square, Institute of Neurology, University College London, London, UK*

NANXING WANG • *Center for Clinical Research on Neurological Diseases, The First Affiliated Hospital, Dalian Medical University, Dalian, China; Liaoning Provincial Key Laboratory for Research on the Pathogenic Mechanisms of Neurological Diseases, The First Affiliated Hospital, Dalian Medical University, Dalian, China*

ERYN L. WERRY • *Faculty of Science, School of Chemistry, The University of Sydney, Sydney, NSW, Australia; Brain and Mind Centre, Faculty of Medicine and Health, The University of Sydney, Sydney, NSW, Australia*

HEATHER WILSON • *Neurodegeneration Imaging Group, University of Exeter Medical School, London, UK*

TRENT M. WOODRUFF • *School of Biomedical Sciences, Faculty of Medicine, The University of Queensland, St Lucia, Brisbane, QLD, Australia; Queensland Brain Institute, The University of Queensland, St Lucia, Brisbane, QLD, Australia*

HUI XI • *College of Science, Harbin Institute of Technology, Shenzhen, China*

GUOWANG XU • *CAS Key Laboratory of Separation Science for Analytical Chemistry, Dalian Institute of Chemical Physics, Chinese Academy of Sciences, Dalian, China*

XIAOJIAO XU • *Center for Clinical Research on Neurological Diseases, The First Affiliated Hospital, Dalian Medical University, Dalian, China; Liaoning Provincial Key Laboratory for Research on the Pathogenic Mechanisms of Neurological Diseases, The First Affiliated Hospital, Dalian Medical University, Dalian, China*

YANG ZHANG • *College of Science, Harbin Institute of Technology, Shenzhen, China*

Part I

Introduction

Chapter 1

Prevalence, Needs, Strategies, and Risk Factors for Neurodegenerative Diseases

Philip V. Peplow, Bridget Martinez, and Thomas A. Gennarelli

Abstract

Effective treatments are not yet available for neurodegenerative diseases, which are rapidly increasing in number as life spans in many countries continue to lengthen. Neurodegenerative diseases place a huge burden on families and on healthcare systems as they are a common cause of morbidity and cognitive impairment in older adults. The cost and societal impact of neurodegenerative disorders are increased by the disability they cause, and the high levels of daily, supervised care and assistance needed. Major unmet needs have been defined and strategies advanced to mitigate neurodegenerative diseases. Novel and multidisciplinary approaches are needed, as well as the strengthening and extension of existing capabilities across basic, clinical, social care and translational research, to achieve significant impact.

Key words Neurodegenerative disease, Prevalence, Health burden, Needs, Strategies, Risk factors

1 Prevalence of Neurodegenerative Diseases

Despite the increasing prevalence of neurodegenerative diseases, due partly to increasing longevity, effective treatments are not yet available. Alzheimer's disease, Parkinson's disease, and amyotrophic lateral sclerosis are three of the major neurodegenerative diseases [1]. The occurrence and incidence of these diseases increases with age; therefore, the number of cases is expected to expand in the foreseeable future as lifespans in many countries continue to lengthen due to decreases in the mortality of cardiovascular disease and cancer. Genetic and environmental factors as contributing causes of neurodegenerative diseases are undefined. The incorporation of molecular techniques, including genomics, proteomics, and measurements of environmental toxic burdens into epidemiological research, may enhance progress on characterizing pathological mechanisms and identifying specific risk factors, especially for sporadic (nonfamilial) forms of these diseases [1].

Philip V. Peplow, Bridget Martinez and Thomas A. Gennarelli (eds.), *Neurodegenerative Diseases Biomarkers: Towards Translating Research to Clinical Practice*, Neuromethods, vol. 173, https://doi.org/10.1007/978-1-0716-1712-0_1,
© Springer Science+Business Media, LLC, part of Springer Nature 2022

In a recent report, the number of cases of Alzheimer dementia (AD) in the United States was estimated to be 5.3 million [2, 3], with an additional 2.2 million having other forms of acquired dementia [3]. As the population ages and overall life expectancy increases, dementia will increase further. Approximately 8.4 million individuals aged >65 years will have AD or another form of dementia such as frontotemporal dementia by 2030 [3]. The population in the USA aged >65 years is projected to be 84 million in 2050, almost twice the estimated population of 43 million in 2012 [4]. The prevalence of dementia worldwide was estimated at 44 million individuals in 2016 [5] and is predicted to double every 20 years [6].

2   Societal and Health System Burden of Neurodegenerative Diseases

Neurodegenerative diseases place a huge burden on families and on healthcare systems as they are a common cause of morbidity and cognitive impairment in older adults. Each of these disorders has varied epidemiology, clinical symptoms, diagnostic criteria, laboratory and neuroimaging features, pathology, treatments, and differential diagnoses [7]. The cost and societal impact of dementia are increased by the disability it causes, and the high levels of daily, supervised care and assistance needed. In the USA, the lower annual cost estimate for AD and other dementias is $243 billion (in 2014 dollars), of which AD accounts for $170 billion, while the total annual cost estimate of neurological diseases is $800 billion (in 2014 dollars) [3]. In 2017, the total costs for health care, long-term care, and hospice for people with AD and other dementia were ~$259 billion [2].

3   Unmet Needs in Neurodegenerative Disease

The EU Joint Programme Neurodegenerative Disease (JPND) Research Group in 2014 agreed on the major unmet needs in the field of neurodegenerative disease that need to be addressed [8]. These needs were as follows.

1. To identify markers for early diagnosis of neurodegenerative disease.

2. To identify tests for a reliable measurement of the progression of the neurodegenerative diseases.

3. To create noninvasive methodologies for the study of brain dysfunctions.

4. To identify targets for treatments aimed at modifying (slowing, arresting, reverting) the progression of neuronal degeneration.

5. To improve symptomatic treatments, with full characterization of the neuronal circuits causative of symptoms, particularly when clinical manifestations are heterogeneous.

6. To identify drugs able to modify/prevent the progression of neurodegenerative disease.

7. To identify environmental factors/gut microbiome alterations which may increase the risk of onset or severity of sporadic forms of neurodegenerative diseases.

4   Strategies for Neurodegenerative Disease Mitigation

Recognizing the rapidly increasing costs of neurological diseases, the following strategies have been proposed [3]:

1. Acceleration of translational research in preventive and disease-modifying therapy. New therapies that are only disease-modifying and not curative can considerably reduce costs. For example, a new treatment that delays the onset of Alzheimer's disease by only 5 years would eliminate 50% of cases, and to delay onset by 10 years would eliminate 75% of patients, with a potential saving of >$175 billion annually (in US dollars) [9].

2. Enhanced outcome and comparative effectiveness research. Funding of research into outcomes and comparative effectiveness of current treatments to determine which are the most effective and cost-efficient is of the upmost importance. Such research has the potential to quickly impact cost, morbidity, and mortality in substantive ways, and funding should be increased to support a sufficient level of research into at least the most common and costliest of neurological diseases.

3. Comprehensive databasing and tracking of neurological disease. A means of national data collection including clinical features, response to therapies, and ongoing burden of neurological disease and its economic impact is essential. This is important for assessment of the status quo, and also to accurately assess the success of new strategies to improve health and lower costs.

4. *Taking advocacy to the next level.* Coordinated advocacy efforts at the individual, institutional, organizational, and local, regional, and national government levels are necessary to promote prioritization, funding, and implementation of these initiatives.

In 2019, JPND released a research and innovation strategy document in which five thematic scientific priorities were identified [10]:

Theme One: The origins and progression of neurodegenerative diseases.

Theme Two: Disease mechanisms and models.

Theme Three: Diagnosis, prognosis, and disease definitions.

Theme Four: Developing therapies, preventive strategies and interventions.

Theme Five: Health and social care.

As noted in the document, novel and multidisciplinary approaches will be needed, as well as the strengthening and extension of existing capabilities across basic, clinical, and social care and translational research, to achieve impact [11].

5   Challenges of Neurodegenerative Diseases

To reduce the incidence and burden of neurodegenerative diseases requires being able to detect the disease at an early stage so that treatments can be initiated to reduce its severity and retard its progression. In addition, in individuals at risk of developing a neurodegenerative disease, being able to detect the disease at the preonset stage (asymptomatic) and begin treatment would also have a considerable impact on disease incidence and health system burden, for example, detecting mild cognitive impairment (MCI) for Alzheimer's disease and rapid eye movement sleep behavioral disorder for Parkinson's disease [12]. Also patients experiencing clinically isolated syndrome (CIS) may transition to multiple sclerosis [13].

Identifying modifiable risk factors of neurodegenerative disorders would also assist in this regard. Functional, molecular, and imaging biomarkers of disease have the potential to advance the diagnosis, prognosis, and response to treatment of the disease.

6   Risk Factors for Neurodegenerative Diseases

1. *Aging.* Aging is the main risk factor for most neurodegenerative diseases, including Alzheimer's disease and Parkinson's disease [14].

2. *Environment.* Living within close distance of a waterbody prone to frequent blooms of cyanobacteria appears to be a significant risk factor for amyotrophic lateral sclerosis in Northern England [15]. Exposure to heavy metals and pesticides has been associated with development of Alzheimer's disease [16].

3. *Vascular risk factors.* Heart disease, stroke, and hypertension may be factors associated with progression of Alzheimer's disease [17].

4. *Diabetes.* Diabetes may predispose to Parkinson-like pathology [18].

5. *Smoking.* Smoking has been associated with progression of Alzheimer's disease [19].

6. *Prediagnosis blood lipid levels.* Cholesterol: total, high-density lipoprotein, low-density lipoprotein [LDL-C], and triglyceride concentrations may be factors associated with progression of Alzheimer's disease [19].

7   Conclusion

A considerable investment in funding and research effort will be required to stem the rising numbers and burden of neurodegenerative diseases. The chapters in this book present new information on the pathophysiology and functional, molecular, and imaging biomarkers of some of the common neurodegenerative diseases. This book hopes to define paths to mitigate neurodegenerative disorders.

References

1. Checkoway H, Lundin JI, Kelada SN (2011) Neurodegenerative diseases. IARC Sci Publ 163:407–419

2. Alzheimer's Association (2017) Alzheimer's disease facts and figures. Alzheimers Dementia 13(4):325–373. https://doi.org/10.1016/j.jalz.2017.02.001

3. Gooch CL, Pracht E, Borenstein AR (2017) The burden of neurological disease in the United States: a summary report and call to action. Ann Neurol 81(4):479–484. https://doi.org/10.1002/ana.24897

4. Ortman JM, Velkoff VA, Hogan H. An aging nation: the older population in the United States. U.S. Department of Commerce Economics and Statistics Administration. https://www.census.gov/prod/2014pubs/p25-1140.pdf

5. Nichols E, Szoeke CE, Vollset SE, Abbasi N, Abd-Allah F, Abdela J et al (2019) Global, regional, and national burden of Alzheimer's diseases and other dementias, 1990-2016: a systematic analysis for the Global Burden of Disease Study 2016. Lancet 18(1):88–106. https://doi.org/10.1016/S1474-4422(18)30403-4

6. Reitz C, Brayne C, Mayeux R (2011) Epidemiology of Alzheimer disease. Nat Rev Neurol 7(3):137–152. https://doi.org/10.1038/nrneurol.2011.2

7. Erkkinen MG, Kim MO, Geschwind MD (2018) Clinical neurology and epidemiology of the major neurodegenerative diseases. Cold Spring Harb Perspect Biol 10(4):a033118. https://doi.org/10.1101/cshperspect.a033118

8. https://www.neurodegenerationresearch.eu/uploads/media/JPND_Exp_Models_Final_report_Jan_2014_-_DM.pdf

9. Brookmeyer R, Gray S, Kawas C (1988) Projections of Alzheimer's disease in the United States and the public health impact of delaying disease onset. Am J Public Health 88(9):1337–1342. https://doi.org/10.2105/ajph.88.9.1337

10. https://www.neurodegenerationresearch.eu/wp-content/uploads/2019/04/Full-JPND-Research-and-Innovation-Strategy-3.04.pdf

11. https://www.scitecheuropa.eu/the-global-challenge-of-neurodegenerative-disorders/97957/

12. Hu CJ, Octave JN (2019) Editorial: Risk factors and outcome predicating biomarker of neurodegenerative diseases. Front Neurol 10:45. https://doi.org/10.3389/fneur.2019.00045

13. https://www.nationalmssociety.org/nationalmssociety/media/msnationalfiles/brochures/brochure-just-the-facts.pdf

14. Hou Y, Dan X, Babber M, Wei Y, Hasselbalch SG, Croteau DL, Bohr VA (2019) Ageing as a risk factor for neurodegenerative disease. Nat Rev Neurol 15(10):565–581. https://doi.org/10.1038/s41582-019-0244-7

15. Henegan P, Butt T, Crothers J, Waters B, Stommel E (2018) Exploring the potential risk factors of neurodegenerative disease in autopsy cases (P1.132). Neurology 90 (15 Suppl):P1.132

16. Chin-Chan M, Navarro-Yepes J, Quintanilla-Vega B (2015) Environmental pollutants as risk factors for neurodegenerative disorders: Alzheimer and Parkinson diseases. Front Cell Neurosci 9:124. https://doi.org/10.3389/fncel.2015.00124

17. Helzner EP, Luchsinger JA, Scarmeas N, Cosentino S, Brickman AM, Glymour MM, Stern Y (2009) Contribution of vascular risk factors to the progression of Alzheimer disease. Arch Neurol 66(3):343–348. https://doi.org/10.1001/archneur.66.3.343

18. Pagano G, Polychronis S, Wilson H, Giordano B, Ferrara N, Niccolini F, Politis M (2018) Diabetes mellitus and Parkinson disease. Neurology 90:e1654–e1662. https://doi.org/10.1212/WNL.0000000000005475

19. Helzner EP, Luchsinger JA, Scarmeas N, Cosentino S, Brickman AM, Glymour MM, Stern Y (2009) Contribution of vascular risk factors to the progression in Alzheimer disease. Arch Neurol 66(3):343–348. https://doi.org/10.1001/archneur.66.3.343

Part II

Research Methods

Chapter 2

Activation of Microglia and Macrophages in Neurodegenerative Diseases

Anna Salamero-Boix, Michael Schulz, and Lisa Sevenich

Abstract

Activation of innate and adaptive immune responses represents a hallmark of neurological disorders. In addition to reactive brain-resident myeloid cells, recruitment of peripheral immune cells to the central nervous system (CNS) critically contributes to disease propagation in neuroinflammatory and neurodegenerative disorders. However, the role of different myeloid subpopulations remains controversial. Recent technological advances including the development of lineage tracing models and unbiased single cell screening approaches significantly contribute to our understanding of myeloid cell heterogeneity underlying lineage-specific disease-associated functions. Such insight provides critical knowledge for the development of myeloid cell–targeted therapies to combat neurological disorders.

Keywords CNS immune landscape, Microglia, Macrophages, Neurological disorders, Neuroinflammation, Neurodegeneration, Lineage tracing models, scRNA-Seq, Myeloid-targeted therapies

1 Introduction

In order to guarantee proper neurological function, the central nervous system (CNS) has to balance between the ability to detect and clear harmful factors but, at the same time, prevent inflammatory responses that damage delicate anatomical structures. As a consequence, the CNS evolved its own monitoring system that significantly differs from immune surveillance and host defense mechanisms of other organs [1]. The contribution of cells from the systemic immune system vs. brain-resident populations in CNS immune surveillance remains controversial. Recent in-depth analyses provide unprecedented insight into cellular players and underlying mechanisms of CNS immunity and functions of different cell populations during disease progression. Given the exclusion of peripheral immune cells from the brain parenchyma, resident cell

Anna Salamero-Boix and Michael Schulz contributed equally with all other contributors.

Philip V. Peplow, Bridget Martinez and Thomas A. Gennarelli (eds.), *Neurodegenerative Diseases Biomarkers: Towards Translating Research to Clinical Practice*, Neuromethods, vol. 173, https://doi.org/10.1007/978-1-0716-1712-0_2,
© Springer Science+Business Media, LLC, part of Springer Nature 2022

types exert important immune functions within the brain parenchyma (Fig. 1). In addition to microglia (MG), macrophages that reside within border-associated regions of the CNS represent critical safeguards at the interface between the immunologically privileged parenchyma and the periphery [2, 3]. Together with MG, border-associated macrophages (BAM; also known as CNS-associated macrophages (CAM)—here collectively denoted as AMs) constitute the major immune cell types in the CNS [4, 5]. A recent single cell RNA sequencing (scRNA-Seq) study sheds light onto the origin of MG in humans and demonstrates high similarity between the developmental features of mice and humans [6]. Both MG and AM are derived from embryonic yolk sac progenitors as revealed by fate mapping studies (MG: Ginhoux et al. [7]) in combination with scRNA-Seq (AMs: Utz et al. [8]). They populate the developing CNS early during embryogenesis and contribute to regional immune cell heterogeneity with tissue-specific transcriptional signatures and functional adaptation [8, 9]. Several independent in-depth analyses have identified distinct subclasses of AMs in perivascular areas (pv), the choroid plexus (cp), and the meninges (m) using mass cytometry or scRNA-Seq in combination with fate-mapping models [5, 8, 9]. In line with their localization at the interface between the CNS and the periphery, AMs show higher expression of CD38 and major histocompatibility complex (MHC) class II indicating an important role in antigen presentation and T cell activation [5].

Although present at small numbers, blood-borne monocyte-derived macrophages (MDM), dendritic cells and different lymphocyte populations represent the other important group of immune cells that patrol border-associated regions of the CNS [10, 11]. Different routes have previously been described through which peripheral immune cells infiltrate into the CNS. Systemic immune cells can access the CNS, for example, via nonfenestrated vascularized stroma of the blood–cerebral spinal fluid (CSF) barrier, the perivascular space, or postcapillary venules that penetrate the parenchyma [12]. Moreover, it was previously described that myeloid cells can migrate through channels that connect the skull bone marrow with the meninges [13]. Different to the traditional view of the CNS as an immune-privileged organ it is now increasingly appreciated that a wide variety of brain-resident and blood-borne cell types constitute the CNS immune landscape under steady state conditions. Seminal work resulted in the molecular characterization (i.e., transcriptomic data) on the single cell level of different cell types in the mouse CNS [14, 15] and human brain [16]. Cytometry approaches (e.g., cytometry by time of flight (CyTOF) or flow cytometry) reveal a profound contribution of immune cell types to the CNS immune cell compartment, however with distinct distributions [4, 5]. Several classes of cells of lymphoid (e.g., CD3$^+$ T cells, NK cells, B cells) and myeloid origin (e.g., dendritic cells,

CNS myeloid cells under steady state

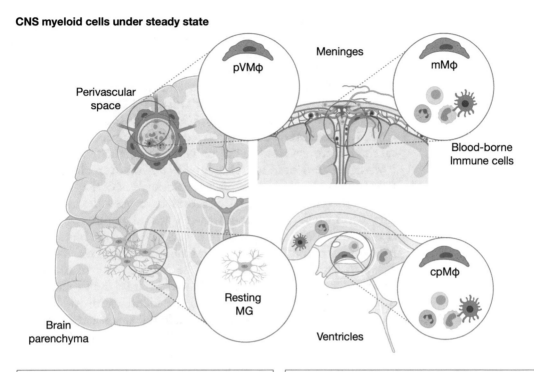

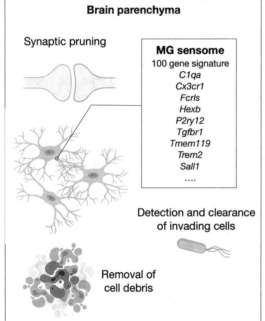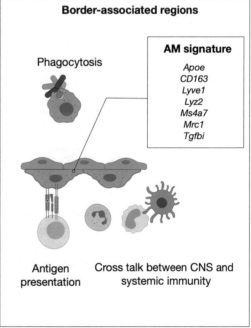

Fig. 1 CNS myeloid cells under steady state. The CNS is populated by different myeloid populations. Microglia (MG) represent the most abundant cell type in the brain parenchyma. As the brain-resident representative of the innate immune system, MG exert critical functions in tissue homeostasis, immune surveillance and host defense. Other CNS macrophages are found in areas at the interface between the brain and the periphery. Macrophage populations are denoted based on their localization as meningeal (m), choroid plexus (cp), and perivascular (pv) macrophages. Macrophages at the CNS interface show higher phagocytic and antigen

granulocytes, monocytes, and macrophages) have been identified, most of them residing in the meninges and choroid plexus [4].

Microgliosis represents a common feature of neurological pathology including immunologically driven diseases such multiple sclerosis (MS) as well as neurodegenerative disorders such as Alzheimer's disease (AD) or Parkinson's disease (PD) [17, 18]. Microgliosis describes the activation of MG in response to damage in the CNS. Reactive MG are characterized by pronounced morphological changes from a ramified morphology with branched protrusions towards an amoeboid morphology [19]. Moreover, alterations in transcriptional programs of disease-associated MG that are shared across different neurological disorders have been described [20] (Fig. 2). Analysis of disease-associated gene signatures do not discriminate distinct subpopulations, such as AMs or blood-borne MDM in neuroinflammatory and neurodegenerative disorders. This can largely be attributed to the previous lack of tools that reliably distinguish MG from AMs and MDM. The development of genetic lineage tracing models, the identification of lineage-specific markers and the advent of single cell approaches allows to systematically query transcriptional and translational programs of myeloid subpopulations in multiple neurological disorders and at distinct stages of disease progression. In this chapter, we discuss recent insights on the key myeloid cell types associated with neurological disorders. We highlight different approaches to discriminate myeloid subpopulations and discuss advantages and disadvantages of each strategy. Furthermore, we summarize recent insights into transcriptional programs to highlight knowledge on disease-associated phenotypes and discuss approaches for therapeutic targeting of myeloid populations in neurodegenerative disorders.

2 Tools to Study Brain-Resident and Recruited Myeloid Cells in Neurological Disorders

Although brain-resident and infiltrating monocyte-derived macrophages stem from different ontogenetic origins and display lineage-restricted transcriptomic programs, it remains challenging to discriminate individual populations upon infiltration of peripheral myeloid cells into the CNS under disease conditions. Given the phenotypic assimilation of disease-associated brain-resident and recruited macrophages, sophisticated tools have to be employed to reliably distinguish individual populations.

Fig. 1 (continued) presentation capacity than MG and are implicated in the crosstalk with patrolling peripheral immune cells that can be found in the cerebrospinal fluid in the meninges and choroid plexus. Under physiological conditions, the blood brain barrier (BBB) restricts the entry of peripheral cells into the brain parenchyma. Figure was generated with Biorender.

CNS myeloid cells in neurological disease

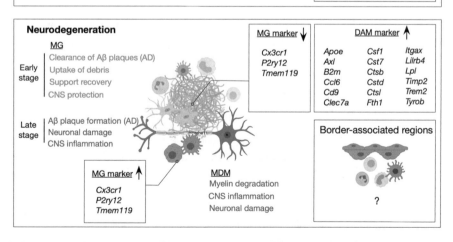

Recruitment of blood-borne
lymphoid and myeloid cells

Recruitment and activation
of microglia = Microgliosis

Inflammation-driven disorders

MG
Uptake of debris
CNS protection

Myelin degradation
CNS inflammation

No induction
of MG-like
markers in EAE
models

MDM
Myelin degradation
CNS inflammation
Neuronal damage

MG marker ↓
Egr1
Mafb
Mef2a
P2ry12
Sall1
Tmem119
Tgfbr1

DAM marker ↑	
Axl	Csf1
Apoe	Itgax
Ccl2	Lilrb4
Clec7a	Spp1

Border-associated regions

Activation / expansion
of autoreactive T cells

Neurodegeneration

MG
Clearance of Aβ plaques (AD)
Early stage — Uptake of debris
Support recovery
CNS protection

Late stage — Aβ plaque formation (AD)
Neuronal damage
CNS inflammation

MG marker ↑
Cx3cr1
P2ry12
Tmem119

MDM
Myelin degradation
CNS inflammation
Neuronal damage

MG marker ↓
Cx3cr1
P2ry12
Tmem119

DAM marker ↑		
Apoe	Csf1	Itgax
Axl	Cst7	Lilrb4
B2m	Ctsb	Lpl
Ccl6	Cstd	Timp2
Cd9	Ctsl	Trem2
Clec7a	Fth1	Tyrob

Border-associated regions

?

Fig. 2 CNS myeloid cells in neurological disease. The development of brain disorders has profound consequences on the CNS immune landscape. Loss of BBB integrity revokes the exclusion of peripheral immune cells. Myeloid and lymphoid cells that patrol the border-associated areas are recruited to pathologic lesions. Danger signals associated with pathological insult lead to the recruitment and activation of MG inducing the formation of microgliosis. MG lose expression of homeostatic markers and upregulate disease-associated transcriptional programs. In neurodegeneration, induction of MG-like gene expression in recruited

2.1 Morphological Characteristics

Before the development of lineage tracing models, lineage-specific marker combinations or unbiased single cell "omic" approaches (*see* below), discrimination of MG from other macrophage subpopulations in the CNS was achieved by morphological criteria. Resting MG can be identified based on their characteristic morphology with long and branched protrusions that survey their surroundings to detect potential damage or invasion of pathogens [21]. Reactive MG gradually retract their protrusions and display enlarged cell bodies [22]. Three-dimensional reconstruction of MG morphology is instrumental to investigate MG reactivity in response to different pathological stimuli and during disease progression [23]. Yamasaki et al. employed the $Ccr2^{RFP/+}Cx3cr1^{GFP/+}$ mouse model to label Ccr2-expressing monocytes in red and Cx3cr1-expressing cells (i.e., MG) in green to distinguish MDM from MG and study their morphological changes during distinct stages of experimental autoimmune encephalitis (EAE), a model for MS [24]. In addition to the identification based on reporter gene expression, the authors performed serial block-face electron microscopy to interrogate morphological characteristics of each population in alignment with reporter gene expression based on criteria involving cell volume and primary processes. This analysis confirmed that MG and MDM display different morphology at disease onset of EAE. MG are larger and harbor multiple ramifications compared with macrophages, whereas MDM contain irregular nuclei, granules, and microvilli [24]. So far, no other reports have interrogated the ultrastructural differences between MG, AM, and MDM.

2.2 Bone Marrow Transplantation and Parabiosis

Traditionally, full-body irradiation and bone marrow transplantation experiments have been employed to decipher the origin of the brain myeloid compartment. In such settings, donor cells are distinguishable from cells of the recipient. The most commonly used strategy takes advantage of mouse models that express fluorescent proteins under a ubiquitously expressed promoter. Alternatively, different MHC class II haplotypes, presence of Y chromosome or CD45.1 vs. CD45.2 can be used to discriminate donor from recipient cells. Bone marrow from donor mice is then transferred into myeloablated recipients, in which the kinetic and tropism of engraftment can be monitored [25]. While this approach has

Fig. 2 (continued) monocyte-derived macrophages (MDM) was observed. Opposing roles for MG and MDM have been shown. While MG are often involved in CNS protection at early disease stages, MDMs are involved in myelin degradation, CNS inflammation and neurotoxicity. At later stages, protective MG functions are converted into disease-promoting functions that enhance CNS inflammation and tissue damage. In inflammation-driven disorders such as EAE or MS, macrophages in border-associated regions of the CNS activate autoreactive T cells and promote their infiltration into the parenchyma. A disease-associated role for mMΦ, cpMΦ, or pvMΦ has not yet been described. Figure was generated with Biorender

been instrumental in addressing the question on the contribution of brain-resident vs. recruited myeloid cells in the CNS under homeostatic and disease conditions, a number of experimental caveats complicates the applicability of bone marrow chimera. For example, intravenous injection of donor cells results in the presence of hematopoietic progenitors in the circulation that are normally not released into the bloodstream [26]. Moreover, myeloablation is often achieved through the application of lethal full-body irradiation. Ionizing radiation is known to disrupt the blood brain barrier (BBB), which results in an artificial influx of bone marrow–derived cells into the brain parenchyma. The use of head shields during irradiation prevents this experimental caveat. Indeed, direct comparison of CNS engraftment of bone marrow–derived myeloid cells confirms that the use of full body irradiation artificially enhances infiltration of bone marrow–derived myeloid cells into the CNS [27]. Alternatively, myeloablation can be achieved by different chemotherapeutic agents such as busulfan or treosulfan. In line with results obtained from experiments with head shields to protect the CNS from lethal irradiation, low CNS engraftment of bone marrow–derived myeloid cells is observed after preconditioning with busulfan and treosulfan [28]. While bone marrow transplantation requires myeloablation for engraftment of donor cells, parabiosis, the surgical joining of two syngeneic mice, enables tracking of infiltrating myeloid cells under homeostatic conditions or in response to different pathological insults [29]. However, several drawbacks are associated with this technique including low rates of chimerism of approximately 30% for Ly6C^{hi} monocytes, high stress levels for parabionts and the technical challenges of the procedure. Given the profound ethical concerns, approval for parabiosis is not granted in many countries.

2.3 In Vivo Transduction

Until now, the use of macrophage-specific promoters for targeted transduction with lentiviral particles has been employed in only two studies. Intracerebroventricular CD11b-virus administration targeted more than 80% microglia and perivascular macrophages [30]. Administration of microRNA (miR)-9 targeted vectors by the intracranial route marked MG around the injection area [31]. The recent discovery of specific markers associated with MG and blood-derived cells (*see* below) allows more specific targeting of different cell populations. Apart from the copy-number effects and random integration that is associated with the application of lentiviral vectors, this procedure is also technically challenging, highly invasive and immunogenic.

2.4 Lineage Tracing Models

Genetic lineage tracing models allow the labeling of a subset of cells to trace their evolution with minimal disturbance of their physiological function. Such strategies usually employ the expression of recombinase enzymes in a lineage-specific manner to modulate the expression of a conditional reporter gene.

The widely used knockin $Cx3cr1^{Cre}$ mouse line expresses Cre recombinase in $Cx3cr1^+$ cells, which include macrophages, circulating monocytes and dendritic cells [32]. When $Cx3cr1^{Cre}$ mice are crossed to loxP-flanked mice, Cre-mediated recombination deletes or reverts the floxed sequences depending on the orientation of the loxP sites. Similar to this system, the $Cx3cr1^{CreER}$ model expresses the Cre^{ERT2} fusion protein and enhanced yellow fluorescent protein (EYFP) in $Cx3cr1^+$ cells. In this inducible model, a tamoxifen-pulse is required to mediate Cre-recombination [33], which allows more precise temporal control of genetic modification. In contrast to peripheral cells, long-lived and self-renewing $Cx3cr1^+$ CNS macrophages retain the label [34]. However, concerns have been raised due to leakage of the system into neurons [35]. A further modification of this model employs the crossing of $Cx3xr1^{CreER}$ mice with Confetti mice [36]. The resulting Microfetti mice ($Cx3cr1^{creER/+}$ $R26R^{Confetti/+}$) is a reporter strain in which $Cx3cr1^+$ cells can randomly express green, yellow, red, or cyan fluorescent protein (GFP, YFP, RFP, or CFP). Thus, this line can be utilized to map individual microglia [37].

In the dual $Cx3cr1^{GFP/+}$:$Ccr2^{RFP/+}$ model, GFP is expressed in $Cx3cr1^+$ cells such as MG and other CNS-resident macrophages whereas RFP is expressed in monocytes, T cells, and NK cells [32, 38]. However, an increase in $Cx3cr1$ levels and a downregulation of $Ccr2$ expression is observed in MDM upon recruitment to CNS [39, 40]. Similar to the $Ccr2^{RFP}$ model, the $Flt3^{Cre}$ model [41] allows the tracing of monocytes and granulocyte-monocyte progenitors and thus its combination with other macrophage-labeling approaches has been used in brain disease models [42, 43]. Of note, higher $Flt3$ gene recombination is found in choroid plexus macrophages (cpMΦ), whereas lower levels are found in MG, meningeal (mMΦ), and perivascular macrophages (pvMΦ) [34]. Cre transmission in the $Flt3^{Cre}$ model is restricted to males which has to considered when planning mating strategies.

To better understand monocyte dynamics, lineage tracing of $Cxcr4^{CreER}$ differentiates hematopoietic stem cell–derived monocytes from MG, pvMΦ and mMΦ. Importantly, this discrimination strategy remains valid upon stroke [44]. Recently, the $Pf4^{Cre}$ mouse line (Cxcl4) was shown to label CD206 + Lyve1+ pvMΦ as well as mMΦ and to a lesser extent cpMΦ [45, 46]. Around 5% of gene recombination was found in MG, which appeared as a group of cells distributed along the parenchyma, in neurons and in platelets. More insights into the AM population can be achieved by using the $CD206^{CreER}$ and the $Lyve1^{Cre}$ mouse strains, which are differentially expressed in steady-state AMs and MG. However, CD206 is also highly expressed in M2-polarized macrophages and Lyve1 expression is predominantly found in lymphatic endothelium [47, 48].

Other mouse models are available to label MG. For example, Sall1 is expressed in MG and approximately 5% of mMΦ and pvMΦ [49] as well as cpMΦ [9]. In the Sall1^{CreER} mouse model, tamoxifen-regulated expression of Cre recombinase occurs in Sall1-expressing cells [50]. Sall1-driven GFP reporter gene expression allows for the evaluation of population dynamics of Sall1+ cells [51]. The downside of these models is the unspecific targeting of non-myeloid CNS cells and non-hematopoietic cells in liver, kidney, and heart. Moreover, Sall1 expression is downregulated in disease-associated MG [52, 53]. Recently, new mouse strains were designed to trace MG by targeting P2ry12 [45], Tmem119 [54, 55], or Hexb [56]. By means of the CRISPR-facilitated homologous recombination, the recently generated P2ry12^{CreER} mouse line preserves P2ry12 function while allowing for MG tracing. Although decreased transcript levels have been described, MG morphology and P2ry12 protein expression remain unaffected. In addition to MG, approximately 25% of cpMΦ and mMΦ are targeted by this strategy. No recombination was observed in circulating myeloid cells in adult mice, although a small percentage of splenic monocyte/macrophage progenitors show tdTomato expression. The P2ry12^{CreER} model has been employed to genetically label MG during development, ischemic injury and EAE [45]. The Tmem119^{EGFP} and Tmem119^{tdTomato} knockin models also label MG [54, 55]. The Tmem119^{CreERT2} allows inducible control of gene expression in MG. Under healthy conditions, *Tmem119* is also expressed in a small population of cpMΦ but its expression is absent in mMΦ and pvMΦ. Furthermore, reactive MG downregulate this marker and MDM increase its expression [42, 57]. CRISPR-Cas9 genome editing was employed to generate an inducible model under the control of the Hexb promoter. In this model, MG as well as a small fraction of pvMΦ and mMΦ are labeled [56]. Hexb expression remains stable under disease conditions. Therefore, Hexb-driven models are advantageous compared to the use of other promoters such as Sall1, P2ry12 or Tmem119 in the context of CNS disease [58].

Lineage tracing models are considered as the method of choice in studies that aim to investigate the ontogenetic origin of certain cell populations or reliably distinguish between lineages of different origin. However, the use of lineage-tracing models requires time-consuming and cost-intense crossing of mouse lines. Furthermore, knockin strategies may lead to haploinsufficiency and functional deficiencies of the corresponding genes with possible effects on the respective cell population [59].

2.5 Lineage-Specific Markers

Flow cytometry has been commonly used to discriminate MG from other macrophages. Traditionally, MG are described as CD45^{int}CD11b$^+$, whereas other CNS macrophages are identified as CD45^{high}CD11b$^+$. However, under inflammatory conditions, MG upregulate CD45 and tumor-associated MDM downregulate it [42, 60]. Moreover, discrimination with CD45 levels is not applicable for human samples [42]. Inflammatory monocytes are distinguished based on Ly6C^{high} expression from Ly6C^{low}-expressing cells such as MG and macrophages derived from the periphery. However, once in the brain, monocytes downregulate the typical markers Ccr2 and Ly6C to the levels present on CNS myeloid cells [40, 61]. Therefore, the use of novel markers such as CD49d and CD49e is required to identify macrophages and monocytes in brain disorders and neurodegeneration models. However, CD49d expression can be found on other infiltrating immune cells (e.g., lymphocytes, dendritic cells, and AMs) [42, 62]. The AM core expression profile is downregulated in neuroinflammatory diseases. Interestingly, *Ms4a7* expression remains stable and thus enables AMs distinction from MG [63]. As mentioned earlier, homeostatic MG markers such as P2ry12, Sall1, and Tmem119 are downregulated in pathologic conditions [64, 65]. Importantly, Hexb and Siglech are part of the healthy MG signature and most remarkably their expression is stable upon different stimuli [56, 66], and can be used to differentiate between microglia and AMs/MDCs [67].

2.6 Large-Scale "omic" Approaches

The use of the approaches mentioned above has significantly contributed to our understanding of the biology of brain-resident and blood-borne myeloid cells in health and disease. However, each technique is associated with technological caveats that bias the results. Moreover, mechanistic analyses of disease-associated functions of immune cells in neuroinflammation and neurodegeneration often reveal conflicting data indicating a high degree of context dependency, cellular heterogeneity, and functional plasticity in response to different pathological stimuli. Given the high complexity of immune functions in disease progression, recently developed large-scale screening approaches allow an unbiased view on the immune landscape in the CNS under different conditions. The "omics" field together with other in-depth and high-throughput techniques is rapidly evolving. Popular gene expression approaches (e.g., microarrays) have been substituted and/or extended with advanced techniques including single cell RNA sequencing (scRNA-Seq), single-nucleus RNA sequencing (snRNA-Seq), multispectral imaging technologies (e.g., Phenoptics™), in situ RNA staining (different in situ hybridization (ISH) approaches), fluorescence, and mass cytometry (e.g., CyTOF). Together these elegant methodological advances led to a tremendous increase of our understanding of "Next-Generation Neuroimmunology" [68].

3 Cellular and Molecular Identity of CNS-Associated Immune Cells at Steady State

Unbiased screening approaches are employed to analyze the immune landscape in the CNS including the brain parenchyma and border-associated regions. Those analyses focus on population heterogeneity of CNS myeloid cells as well as the identity of lymphoid and myeloid cells that patrol border-associated regions of the brain as part of their role in immune surveillance and host defense. The presence of different MG subpopulations was proposed 30 years ago based on different morphological features [69]. Transcriptomic analyses with microarrays [70] and scRNA-Seq [71, 72] confirm the presence of distinct myeloid cell subpopulations during development and adulthood in the mouse brain. Interestingly, the heterogeneity of MG is most pronounced during embryonic development [71, 72]. All subclusters are equipped with a core marker gene set consisting of *C1qa, Cx3cr1, Fcrls, P2ry12, and Trem2*, which enables the cells to quickly respond to danger signals caused by damage, invasion of cells from the periphery, or other pathological stimuli associated with neurological diseases. Together with *Hexb, Tmem119*, and *Tgfbr1* the aforementioned genes form a consensus profile highly enriched in MGs compared to peripheral monocytes, or other brain-resident cell types [57, 73]. Those genes belong to a set of 100 genes that are characterized as the MG "sensome" [73] (Fig. 1). Members of this set of genes belong to purinergic receptors (e.g., *P2ry12, P2ry13*), chemokine receptors (e.g., *Cx3cr1*), and innate immune and Toll-like receptors. The function of other genes such as *Tmem119* remains unknown. The same microglial homeostatic genes (e.g., *Cx3cr1, P2ry12, Tmem119*) as well as overlapping phenotypic features can be found in other species including human as recently described in an RNA-Seq meta-analysis [74]. Interestingly, this massive bulk and scRNA-seq approach concludes that human MG represent a highly heterogeneous population, whereas mouse MG are more uniform [74]. Spatiotemporal heterogeneity in human microglia has been shown before using bulk [75], or scRNA-seq [72] as well as CyTOF [76]. For example, Böttcher et al. determined the expression levels of almost 60 markers from several patients in different brain regions. Unsupervised clustering confirmed the unique profile of MG, which were clearly segregated from control cells (e.g., cells from the CSF and PBMCs). Moreover, this study validated high abundance of MG homeostatic marker on protein level, for example, CX3CR1, P2RY12, TMEM119, and TREM2. However, differences of protein abundance between MG obtained from fresh vs. post-mortem biopsies are observed. Furthermore, a combination of eight markers (CD11c, CD206, CD45, CD64, CD68, CX3CR1, HLA-DR and IRF8) has been defined as sufficient to differentiate between distinct subpopulations [76]. Interestingly, CD206 or MRC1 (macrophage mannose receptor C-type 1), is found in a small subset of human MG, although traditionally associated with M2-polarization of macrophages and previously described as a marker for distinct AMs in mouse brain. Several markers have been validated to distinguish AMs from MG under steady-state conditions including *Apoe, CD163, Lyve1, Lyz2,*

Ms4a7, *Mrc1*, and *Tgfbi* [8, 9, 63]. Remarkably, a small subset of cpMΦ exhibited similarities to MGs, including the expression of *Sall1* [9], which together with colony stimulating factor 1 (CSF1) and transforming growth factor-beta (TGF-ß), is crucial for MG identity and function [49] during development [8] and in adulthood [57]. In line with differential marker expression, it was proposed that MG are predominantly involved in CNS tissue homeostasis including synaptic pruning, removal of cellular debris and detection and clearance of invading bacteria or viruses to prevent infection of the CNS [77]. By comparison, AMs exhibit higher expression of genes involved in phagocytosis and antigen presentation suggesting a critical role in the crosstalk with peripheral immune and inflammatory cells at CNS interfaces [5].

4   Disease-Associated Immune Cells in CNS Disorders

Neuroinflammation, characterized by microgliosis and infiltration of systemic immune cells to pathologic lesions, represents a hallmark of a broad range of CNS disorders including inflammation-driven autoimmune diseases such as MS as well as neurodegenerative disorders such as AD and PD (Fig. 2).

The extent of neuroinflammation differs between individual neurological disorders [62].This is particularly evident for the contribution of T cells to disease progression. In autoimmune diseases like MS, attack of the central nervous system is mediated by infiltrating immune cells, i.e., autoreactive T cells [78]. Interestingly, it was proposed that AMs in the meninges and perivascular space are involved in activating autoreactive T cells to undergo expansion and enter the parenchyma in MS [79–82]. In contrast, T cells only play subordinate roles in disease initiation in neurodegenerative disorders. Disease-associated roles of myeloid cells in CNS disorders remain controversial. For example, it was previously proposed that the crosstalk between myeloid cells and autoreactive, myelin-degrading T cells contributes to the formation of pathologic lesions in MS [83, 84]. However, other studies rather indicate that myeloid cells contribute to CNS recovery by supporting remyelination [85], secretion of neuroprotective cytokines [86], and clearance of myelin debris [87]. Similar observations of opposing effects are also reported for neurodegenerative disorders [24]. A possible explanation for opposing roles of myeloid cells could be different disease-associated functions of brain-resident and recruited myeloid cell types. Indeed, functional studies suggest that brain-resident MG are rather implicated in clearance of cell debris, CNS protection and repair, whereas MDM mediate demyelination (Fig. 2). However, there are also reports on disease-promoting functions of MG. For example, it was demonstrated that microglia promote tau accumulation in neurons [88]. Moreover, neuronal death mediated by

A1-astrocytes is induced by activated MG via IL-1α, TNFα, and C1q [89]. A1 and A2 astrocytes describe different functional states of reactive astrocytes. A1 astrocytes are associated with neurotoxicity, whereas A2 astrocytes promote tissue repair and neuronal survival [89].

The introduction of high-throughput screening approaches allows for an unbiased view on the immune landscape in different neurological disorders at distinct stages of disease progression. Recent large-scale single cell "omic" approaches together with functional studies shed light onto the heterogeneity of myeloid subpopulations. These studies provide important insight into lineage-specific transcriptional programs at distinct stages of disease progression and thus provide rational explanations for previously conflicting data.

Krasemann et al. performed a comprehensive study of transcriptomic changes of disease-associated myeloid cells in different neurodegenerative (e.g., superoxide dismutase (SOD)-1 model for amyotrophic lateral sclerosis (ALS), APP-PS1 model for AD harboring mutations in the amyloid precursor protein (APP) and presenilin 1 (PSEN1) genes) and neuroinflammatory experimental autoimmune encephalomyelitis (EAE) models for MS [52]. Strikingly, all sequencing data clustered according to two conditions: controls vs. disease-associated cells. This suggests the induction of profound transcriptional changes in response to pathological stimuli with high similarities shared across different neurological diseases. Disease-associated cells reversibly downregulated the expression of distinct homeostatic markers over time (e.g., *Sall1*, *Tgfbr1*). Among the overlapping genes (e.g., *Axl*, *Clec7a*, *Csf1*, *Itgax*, *Spp1*) *Apoe* is the only gene that shows inverse correlation to disease progression in each model. Pathway analysis and subsequent functional validation with *Apoe* knockout mice revealed a central role in rapidly mediating the transition towards a reactive MG phenotype (denoted here MGnD), which was Trem2-dependent [52]. In line with those findings, Shi et al. demonstrated that Apoe-dependent depletion of MGs during a critical phase of disease progression in the P301S model for tauopathy, completely blocked neurodegeneration [90]. The upregulation of Trem2 in MG most likely also directly influences the downregulation of homeostatic genes, since its accessory protein (Dap12 or Tyrobp) is directly involved in regulating 40 out of 100 sensome genes [73].

Two other seminal studies examined the myeloid landscape of AD-like mice models in more detail on the single cell level (5XFAD model [91] that co-expresses five genes of familial AD) [58] and the inducible transgenic CK-p25 model [92] expressing cyclin-dependent kinase 5 and its regulatory, truncated subunit p25 [66]). Together both studies provide novel high-dimensional insight, showing the existence of distinct MG subpopulations which dynamically develop during disease progression.

Remarkably, Keren-Shaul and colleagues identified a specific "danger/disease-associated microglia" population (referred to as DAM) which is specifically associated to AD plaques preferentially in the cortex of mice and patients [58]. Furthermore, the authors were able to assign the transition from normal to DAM MG in a consecutive manner: initially Trem2-independent and later Trem2-dependent, again underlining the importance of this receptor for disease-associated MGs. The main markers found to be upregulated in DAMs are *Apoe, Axl, B2m, Ccl6, Cd9, Clec7a, Csf1, Cst7, Ctsb, Ctsd, Ctsl, Fth1, Itgax, Lilrb4, Lpl, Timp2, Trem2, Tyrobp*. At the same time, gene expression of MG homeostatic markers (*Cx3cr1, P2ry12, Tmem119*) is downregulated [58]. Functional annotation of gene expression changes during disease progression identified two main MG responses: in early stages several proliferation markers are upregulated, whereas late responses are driven by distinct immune active genes (e.g., cathepsins, chemokines, and distinct complement members) [66].

A comprehensive overview of changes in the immune composition upon different pathologic insults is described in a study by Ajami et al. using CyTOF data in an unsupervised manner [62]. The authors identified three CNS-resident myeloid cell clusters present in all models examined (EAE, mSOD1, R6/2 (Huntington disease (HD) model), although with stage and disease-specific distributions. In contrast, monocyte-derived cells (MDCs) are distributed across five different clusters with various cytokine profiles, but the HD model completely lacked infiltrating MDCs, similar to healthy controls and recovered EAE mice [62]. Most importantly, CNS-resident and MDC from the periphery show distinct signaling phenotypes. Blockade of recruitment of one MDC cluster using an antibody against CD49e resulted in significantly improved EAE scores. Another scRNA-Seq study focused on the characterization of mononuclear phagocytes infiltrating into the spinal cord in EAE over time [93]. Interestingly the authors identified one monocyte subset (Cxcl10$^+$), which significantly contributed to pathology, since blocking its recruitment improved the course of EAE. Contrary to that, noninflammatory lysophosphatidylcholine (LPC)-induced demyelination within the spinal cord induces activation of certain MG subpopulations, which however outnumber and regulate the infiltration and dispersion of MDCs [94]. Another recent study combined scRNA-Seq with identification of oxidant stress in the spinal cord of an EAE model and identified a core oxidative stress gene signature coupled to coagulation and glutathione-pathway genes shared between various clusters of resident MGs (three clusters) and infiltrating MDCs (seven clusters). Based on these findings the authors propose druggable pathways to suppress neurotoxic innate immune responses [95]. Although with comparatively lower cell numbers, another study demonstrated the changing immune landscape over time in

EAE [63]. The authors dissected the lymphoid and myeloid cell population in a spatiotemporal manner. Important results are the contribution of CNS resident AMs to disease progression which show a clear upregulation of MHC II molecules, further confirming a central role for antigen presentation towards infiltrating T cells. Similarly, different subsets of MDCs are recruited most likely through the leptomeningeal and perivascular routes, showing also high antigen presentation capabilities. However, the cells are not constantly integrated into the CNS myeloid niche as previously reported [62]. Similar to other studies, CNS-resident MGs are found to consist of four distinct disease-associated subsets, showing expansion by clonal proliferation. However, their role remains controversial.

Comparing these findings to the human situation, two recent scRNA-Seq studies identified distinct myeloid subsets in brains of MS patients. MG are present in different activation states and exhibited clear pathology-associated subclustering, with an enrichment of e.g., distinct complement factors, and simultaneous downregulation of homeostatic markers (e.g., *P2RY12* and *TMEM119*) [72, 96]. Since several recent findings highlight the importance of the Trem2-Apoe signaling axis which is central for sensing tissue damage and restricting its spread [97, 98], Zhou et al. performed in-depth analysis of TREM2 functions for AD pathology in mice and humans using snRNA-Seq [99]. Overall, 5XFAD mice deficient for Trem2 exhibit fewer MG and reduced levels of DAM signatures, independently confirming previous results [58]. Surprisingly, overall transcriptomic signatures of seven distinct human MG show remarkable differences, like the upregulation of distinct homeostatic markers (e.g., *P2RY12* and *TMEM119*). However, one of the subsets shows a high overlap with mouse DAM signatures, presenting an upregulation of *HLA-DR, TREM2* and *AIF1* in disease-associated and aged MGs. The transcriptomes of MGs from normal or mutant (R62H) TREM2 variant carriers show only few changes, and MG numbers were similar. In contrast, another form of TREM2 (R47H) which is associated with an even higher risk of developing AD has a greater impact on MGs transcriptomes. Translating these findings back into a mouse model, it was demonstrated that antibody-mediated blocking of this human mutant TREM2 receptor variant in mice results in upregulation of markers related to metabolic activation and proliferation, as examined by scRNA-Seq. These clusters of MG subsets are distinct from DAM and improved overall disease outcome [100].

Integrating several of the studies mentioned above, a meta-analysis examined gene expression levels across a variety of neuropathologies including ischemia, infection, inflammation, demyelination, cancer, and neurodegeneration, and by application of meta-cell clustering evaluated similarities and differences [20]. Three modules emerge as particularly important: genes belonging to proliferation, inflammation, or neurodegeneration, whereas the latter module consisting of 134 genes shows upregulation in almost all analyzed datasets. Focusing on Trem2, this meta-analysis further highlights its importance and that complete transcriptional responses to ß-amyloid (Aβ) pathologies can only be achieved by full function, although the outcome is context-dependent.

4.1 Myeloid Cell–Based Therapies for Neuroinflammation and Neurodegenerative Diseases

Given the key functions in mediating disease progression, myeloid cells associated with CNS disorders are emerging as promising therapeutic targets in neuroinflammation and neurodegenerative disorders. Large scale single cell-based studies help to elucidate disease-associated functions of different populations of CNS myeloid cells including MG, AMs and blood-borne MDM. Detailed analysis of transcriptional programs together with knowledge on functional changes in disease-associated myeloid cell populations are necessary to develop myeloid cell–targeted therapies against neurodegenerative and neuroinflammatory disorders. However, it remains challenging to specifically target disease-associated subpopulations and at the same time spare those fulfilling physiological functions. In the following section, we will discuss different strategies to modulate the myeloid compartment including myeloid cell depletion, prevention of the recruitment of blood-borne myeloid cells to the brain and reprogramming of disease-associated functions.

4.1.1 CSF1R Inhibition

Reactive myeloid cells are an integral part of neuroinflammatory and neurodegenerative disorders. Hence, myeloid cell depletion represents a promising strategy to remove disease-promoting effects. Given the rapid repopulation of MG pools after depletion, it was furthermore proposed to exchange disease-associated cell pools with newly generated, healthy ones. One of the most commonly used strategies for macrophage depletion in preclinical models relies on the inhibition of the colony stimulation factor 1 receptor (CSF1R) using neutralizing antibodies (e.g., RG7115) or small molecule inhibitors (e.g., PLX3397, PLX5562, JNJ-527, or BLZ945). Blockade of CSF1R in myeloid cells leads to inhibition of one of the central macrophage differentiation and survival pathways and thus results in depletion of the vast majority of macrophages. A series of preclinical studies evaluated the consequences of CSF1R inhibition in different neurological diseases including traumatic brain injury (TBI), PD, AD, and MS. Results obtained from neurological diseases reveal context depending responses upon CSF1R inhibition. Using the 5XFAD mouse model, it was demonstrated that MG depletion with the CSF1R inhibitor PLX3397 leads to reduced neuroinflammation and prevented neuronal loss. However, no effects on Aβ levels or plaque load is observed when mice are treated at advanced disease stages

[101]. Treatment of 5XFAD mice with PLX3397 at early disease stages results in reduced accumulation of plaque deposition [102]. Using the more selective CSF1R inhibitor PLX5562, Spangenberg et al. demonstrated impaired Aβ plaque formation in response to MG depletion in advanced stage AD pathology [103]. A direct comparison of PLX5562 doses that deplete MG and CSF1R⁺ myeloid cells in the periphery vs. concentrations that only target cells in the periphery demonstrated diminished plaque formation only in response to higher doses that deplete MG pools. This suggests that MG play a key role in plaque formation. A series of studies evaluated the efficacy of CSF1R inhibition at concentrations that inhibit MG proliferation but does not lead to MG depletion. In contrast to the results obtained by Spangenberg et al., studies using multiple AD-like or tau pathology models demonstrate efficacy of the CSF1R inhibition at concentration that inhibit MG proliferation without depletion of the cell population. It was recently demonstrated that MG depletion by CSF1R inhibition followed by inhibitor withdrawal leads to rapid repopulation of brain from a persisting pool of cells in a spatiotemporal manner [104, 105]. Interestingly, repopulated cells show high overlap of their transcriptomes with normal MG as further highlighted by their potential to respond similarly to LPS challenge [106]. It was therefore proposed to employ MG repopulation strategies to exchange the disease-associated MG pool with a fresh cell pool. Testing this strategy in the 5XFAD model revealed that MG repopulation is accompanied with reappearance of plaque pathology. A similar strategy was employed in a model of TBI. The authors demonstrated that short-term elimination of MG during the chronic phase of TBI followed by repopulation results in long-term improvement in neurological function, suppression of neuroinflammation and a reduction in persistent neurodegenerative processes [107]. Although repopulating MG resemble characteristics of normal MG in the absence of pathological stimuli or after complete resolution of damage, the results obtained by Spangenberg et al. suggest that repopulating MG populations in disease models rapidly acquire disease-promoting functions.

CSF1R inhibition was also tested in EAE models. Similar to findings in AD models, MG depletion during the symptomatic phase of EAE reduces neuroinflammation and supports remyelination and recovery [108]. Similar effects are observed in a cuprizone model in which MG depletion by the inhibitor BLZ945 results in region-specific enhancement of remyelination [109]. In contrast, data from PD models reveal that MG depletion with PLX3397 exacerbates the impairment of locomotor activities and promoted neuronal loss. Worsening of disease is accompanied with production of inflammatory mediators and increased infiltration of leukocytes. These data suggest a protective role of MG against neuroinflammation and neurotoxicity in PD [110].

In summary, CSF1R inhibition in neuroinflammation and neurodegenerative disorders shows a broad spectrum of treatment responses. Treatment efficacy depends among others on the disease entity, the disease stage as well as the concentration and selectivity of the CSF1R inhibitor. Although preclinical testing of CSF1R inhibition reveals therapeutic efficacy without overt side effects of MG depletion, it remains unclear whether human patients would equally well tolerate CSF1R inhibition, given its central function of macrophage biology. PLX3397 was previously administered to glioblastoma patients in a phase 2 clinical study. Although this study did not show therapeutic efficacy against glioblastoma in this patient cohort, data indicate that a 2-week treatment period is well tolerated [111]. CSF1R inhibition can be viewed as a promising treatment strategy for individual CNS disorders. However, this strategy targets all CSF1R⁺ myeloid cells regardless of their disease-associated function or localization. Given the high heterogeneity of CNS myeloid cell populations with potentially opposing roles emphasizes the need for the development of more selective strategies that target disease-promoting phenotypes but spare physiologically important functions.

4.1.2 Emerging Myeloid Cell Targets in Neurological Disorders

Recent screening approaches and functional studies uncovered a broad range of candidate factors critical for the propagation of disease-associated phenotypes in individual myeloid subpopulations. Pharmacological inhibition of those factors could therefore open new avenues for myeloid cell–targeted therapies against neurological disorders. In this context, signaling components of the Apoe–Trem2 axis represent promising targets to modulate MG phenotypes. Employing genetic silencing approaches, it was demonstrated that Apoe depletion reduced clinical severity of EAE likely through limiting neuroinflammation [112]. Krasemann et al. demonstrated that genetic silencing of Trem2 suppresses the Apoe pathway leading to restoration of homeostatic MG in AD models and reduced disease severity [52]. In contrast, it was shown that Trem2 deficiency augments Aβ accumulation and neuronal loss in AD models [113]. Moreover, soluble Trem2 (sTrem2) reduces Aβ plaque formation in AD models. This effect is mediated through increased microglia proliferation and clearance of Aβ plaques, while MG depletion abrogates the neuroprotective effects of sTrem2. This study suggests that elevated levels of sTrem2 could be explored for AD therapy [114]. A newly developed Trem2 activating antibody, AL200c, was recently tested in an AD mouse model. Administration of AL200c induces MG proliferation and impacts their response to Aβ. Although AL200c treatment has only marginal effects on Aβ load, the authors report reduced filamentous plaques and neurite dystrophy as well as improvement of behavioral abnormalities associated with AD [100]. Importantly, AL200c was well tolerated in a first clinical trial with no drug-related severe adverse events or dose-limiting toxicities [100].

Targeting the Trem2-Apoe axis represents a promising approach to modulate MG activation states. Given the accumulating evidence that blood-borne myeloid cells represent critical drivers of disease progression, selective targeting of disease-associated MDMs has also been considered as promising strategies. Infiltration of Ccr2+ monocytes promotes CNS inflammation and has been associated with neuronal damage [115]. Blockade of Ccr2+ monocyte recruitment ameliorates the disease status. Getts et al. employed immune-modulating microparticles to target Ccr2+ inflammatory monocytes. Treatment with immune modulating particles inhibits inflammatory monocytes in an EAE model leading to reduced inflammation and reduced EAE scores [116]. Likewise, selective depletion of Ccr2+ inflammatory monocytes by genetic Ccr2 silencing strongly reduces CNS autoimmunity [117].

Here we discussed a selection of potential targets and therapeutic strategies that could be exploited for myeloid targeted therapies. In particular in-depth single cell sequencing analysis generated a comprehensive list of potential candidate factors that are part of core signatures of disease-associated myeloid populations. Functional studies employing genetic or pharmacological inhibition of candidate factors are now urgently needed to gain further mechanistic insight and to identify druggable targets for immune modulatory strategies.

5   Conclusion

The critical role of myeloid cells in CNS disorders is increasingly recognized. Evaluation of disease-associated roles of CNS myeloid cells often results in conflicting data and the identification of opposing roles during disease progression. To date, we can appreciate the high functional plasticity and population heterogeneity of myeloid populations in neurological disorders originating from brain-resident and peripheral cell pools within a spatial-temporal continuum. Combined knowledge from the recent literature represents a valuable resource to study the function of myeloid cells in distinct neurological disorders. Although individual disorders are initiated by distinct pathological stimuli, transcriptional programs of myeloid subpopulations display striking similarities across different CNS diseases including neuroinflammatory diseases and neurodegenerative disorders as discussed in this chapter. Hence, combined insight from individual disciplines will significantly contribute to evaluation of lineage-specific roles of myeloid cells in CNS diseases and pave the way for the rational design of myeloid cell targeted therapies against neurological disorders.

References

1. Aloisi F (1999) The role of microglia and astrocytes in CNS immune surveillance and immunopathology. Adv Exp Med Biol 468:123–133. https://doi.org/10.1007/978-1-4615-4685-6_10

2. Prinz M, Erny D, Hagemeyer N (2017) Ontogeny and homeostasis of CNS myeloid cells. Nat Immunol 18(4):385–392. https://doi.org/10.1038/ni.3703

3. Herz J, Filiano AJ, Smith A, Yogev N, Kipnis J (2017) Myeloid cells in the central nervous system. Immunity 46(6):943–956. https://doi.org/10.1016/j.immuni.2017.06.007

4. Korin B, Ben-Shaanan TL, Schiller M, Dubovik T, Azulay-Debby H, Boshnak NT, Koren T, Rolls A (2017) High-dimensional, single-cell characterization of the brain's immune compartment. Nat Neurosci 20 (9):1300–1309. https://doi.org/10.1038/nn.4610

5. Mrdjen D, Pavlovic A, Hartmann FJ, Schreiner B, Utz SG, Leung BP, Lelios I, Heppner FL, Kipnis J, Merkler D, Greter M, Becher B (2018) High-dimensional single-cell mapping of central nervous system immune cells reveals distinct myeloid subsets in health, aging, and disease. Immunity 48 (2):380–395.e6. https://doi.org/10.1016/j.immuni.2018.01.011

6. Bian Z, Gong Y, Huang T, Lee CZW, Bian L, Bai Z, Shi H, Zeng Y, Liu C, He J, Zhou J, Li X, Li Z, Ni Y, Ma C, Cui L, Zhang R, Chan JKY, Ng LG, Lan Y, Ginhoux F, Liu B (2020) Deciphering human macrophage development at single-cell resolution. Nature 582 (7813):571–576. https://doi.org/10.1038/s41586-020-2316-7

7. Ginhoux F, Greter M, Leboeuf M, Nandi S, See P, Gokhan S, Mehler MF, Conway SJ, Ng LG, Stanley ER, Samokhvalov IM, Merad M (2010) Fate mapping analysis reveals that adult microglia derive from primitive macrophages. Science 330(6005):841–845. https://doi.org/10.1126/science.1194637

8. Utz SG, See P, Mildenberger W, Thion MS, Silvin A, Lutz M, Ingelfinger F, Rayan NA, Lelios I, Buttgereit A, Asano K, Prabhakar S, Garel S, Becher B, Ginhoux F, Greter M (2020) Early fate defines microglia and non-parenchymal brain macrophage development. Cell 181(3):557–573.e18. https://doi.org/10.1016/j.cell.2020.03.021

9. Van Hove H, Martens L, Scheyltjens I, De Vlaminck K, Pombo Antunes AR, De Prijck S, Vandamme N, De Schepper S, Van Isterdael G, Scott CL, Aerts J, Berx G, Boeckxstaens GE, Vandenbroucke RE, Vereecke L, Moechars D, Guilliams M, Van Ginderachter JA, Saeys Y, Movahedi K (2019) A single-cell atlas of mouse brain macrophages reveals unique transcriptional identities shaped by ontogeny and tissue environment. Nat Neurosci 22 (6):1021–1035. https://doi.org/10.1038/s41593-019-0393-4

10. Mundt S, Mrdjen D, Utz SG, Greter M, Schreiner B, Becher B (2019) Conventional DCs sample and present myelin antigens in the healthy CNS and allow parenchymal T cell entry to initiate neuroinflammation. Sci Immunol 4(31). https://doi.org/10.1126/sciimmunol.aau8380

11. Mundt S, Greter M, Flugel A, Becher B (2019) The CNS immune landscape from the viewpoint of a T cell. Trends Neurosci 42 (10):667–679. https://doi.org/10.1016/j.tins.2019.07.008

12. Ransohoff RM, Kivisakk P, Kidd G (2003) Three or more routes for leukocyte migration into the central nervous system. Nat Rev Immunol 3(7):569–581. https://doi.org/10.1038/nri1130

13. Herisson F, Frodermann V, Courties G, Rohde D, Sun Y, Vandoorne K, Wojtkiewicz GR, Masson GS, Vinegoni C, Kim J, Kim DE, Weissleder R, Swirski FK, Moskowitz MA, Nahrendorf M (2018) Direct vascular channels connect skull bone marrow and the brain surface enabling myeloid cell migration. Nat Neurosci 21(9):1209–1217. https://doi.org/10.1038/s41593-018-0213-2

14. Zeisel A, Hochgerner H, Lonnerberg P, Johnsson A, Memic F, van der Zwan J, Haring M, Braun E, Borm LE, La Manno G, Codeluppi S, Furlan A, Lee K, Skene N, Harris KD, Hjerling-Leffler J, Arenas E, Ernfors P, Marklund U, Linnarsson S (2018) Molecular architecture of the mouse nervous system. Cell 174(4):999–1014.e22. https://doi.org/10.1016/j.cell.2018.06.021

15. Zeisel A, Munoz-Manchado AB, Codeluppi S, Lonnerberg P, La Manno G, Jureus A, Marques S, Munguba H, He L, Betsholtz C, Rolny C, Castelo-Branco G, Hjerling-Leffler J, Linnarsson S (2015) Brain structure. Cell types in the mouse cortex and hippocampus revealed by single-cell RNA-seq. Science 347(6226):1138–1142. https://doi.org/10.1126/science.aaa1934

16. Darmanis S, Sloan SA, Zhang Y, Enge M, Caneda C, Shuer LM, Hayden Gephart MG, Barres BA, Quake SR (2015) A survey of

human brain transcriptome diversity at the single cell level. Proc Natl Acad Sci U S A 112(23):7285–7290. https://doi.org/10.1073/pnas.1507125112

17. Hickman S, Izzy S, Sen P, Morsett L, El Khoury J (2018) Microglia in neurodegeneration. Nat Neurosci 21(10):1359–1369. https://doi.org/10.1038/s41593-018-0242-x

18. Guerrero BL, Sicotte NL (2020) Microglia in multiple sclerosis: friend or foe? Front Immunol 11:374. https://doi.org/10.3389/fimmu.2020.00374

19. Davis BM, Salinas-Navarro M, Cordeiro MF, Moons L, De Groef L (2017) Characterizing microglia activation: a spatial statistics approach to maximize information extraction. Sci Rep 7(1):1576. https://doi.org/10.1038/s41598-017-01747-8

20. Friedman BA, Srinivasan K, Ayalon G, Meilandt WJ, Lin H, Huntley MA, Cao Y, Lee SH, Haddick PCG, Ngu H, Modrusan Z, Larson JL, Kaminker JS, van der Brug MP, Hansen DV (2018) Diverse brain myeloid expression profiles reveal distinct microglial activation states and aspects of Alzheimer's disease not evident in mouse models. Cell Rep 22(3):832–847. https://doi.org/10.1016/j.celrep.2017.12.066

21. Nimmerjahn A, Kirchhoff F, Helmchen F (2005) Resting microglial cells are highly dynamic surveillants of brain parenchyma in vivo. Science 308(5726):1314–1318. https://doi.org/10.1126/science.1110647

22. Colton CA, Chernyshev ON, Gilbert DL, Vitek MP (2000) Microglial contribution to oxidative stress in Alzheimer's disease. Ann N Y Acad Sci 899:292–307. https://doi.org/10.1111/j.1749-6632.2000.tb06195.x

23. Karperien A, Ahammer H, Jelinek HF (2013) Quantitating the subtleties of microglial morphology with fractal analysis. Front Cell Neurosci 7:3. https://doi.org/10.3389/fncel.2013.00003

24. Yamasaki R, Lu H, Butovsky O, Ohno N, Rietsch AM, Cialic R, Wu PM, Doykan CE, Lin J, Cotleur AC, Kidd G, Zorlu MM, Sun N, Hu W, Liu L, Lee JC, Taylor SE, Uehlein L, Dixon D, Gu J, Floruta CM, Zhu M, Charo IF, Weiner HL, Ransohoff RM (2014) Differential roles of microglia and monocytes in the inflamed central nervous system. J Exp Med 211(8):1533–1549. https://doi.org/10.1084/jem.20132477

25. Malm TM, Koistinaho M, Parepalo M, Vatanen T, Ooka A, Karlsson S, Koistinaho J (2005) Bone-marrow-derived cells contribute to the recruitment of microglial cells in

response to beta-amyloid deposition in APP/PS1 double transgenic Alzheimer mice. Neurobiol Dis 18(1):134–142. https://doi.org/10.1016/j.nbd.2004.09.009

26. Yuan H, Gaber MW, McColgan T, Naimark MD, Kiani MF, Merchant TE (2003) Radiation-induced permeability and leukocyte adhesion in the rat blood-brain barrier: modulation with anti-ICAM-1 antibodies. Brain Res 969(1–2):59–69. https://doi.org/10.1016/s0006-8993(03)02278-9

27. Muller A, Brandenburg S, Turkowski K, Muller S, Vajkoczy P (2015) Resident microglia, and not peripheral macrophages, are the main source of brain tumor mononuclear cells. Int J Cancer 137(2):278–288. https://doi.org/10.1002/ijc.29379

28. Kierdorf K, Katzmarski N, Haas CA, Prinz M (2013) Bone marrow cell recruitment to the brain in the absence of irradiation or parabiosis bias. PLoS One 8(3):e58544. https://doi.org/10.1371/journal.pone.0058544

29. Ajami B, Bennett JL, Krieger C, McNagny KM, Rossi FM (2011) Infiltrating monocytes trigger EAE progression, but do not contribute to the resident microglia pool. Nat Neurosci 14(9):1142–1149. https://doi.org/10.1038/nn.2887

30. Ding X, Yan Y, Li X, Li K, Ciric B, Yang J, Zhang Y, Wu S, Xu H, Chen W, Lovett-Racke AE, Zhang GX, Rostami A (2015) Silencing IFN-gamma binding/signaling in astrocytes versus microglia leads to opposite effects on central nervous system autoimmunity. J Immunol 194(9):4251–4264. https://doi.org/10.4049/jimmunol.1303321

31. Akerblom M, Sachdeva R, Quintino L, Wettergren EE, Chapman KZ, Manfre G, Lindvall O, Lundberg C, Jakobsson J (2013) Visualization and genetic modification of resident brain microglia using lentiviral vectors regulated by microRNA-9. Nat Commun 4:1770. https://doi.org/10.1038/ncomms2801

32. Yona S, Kim KW, Wolf Y, Mildner A, Varol D, Breker M, Strauss-Ayali D, Viukov S, Guilliams M, Misharin A, Hume DA, Perlman H, Malissen B, Zelzer E, Jung S (2013) Fate mapping reveals origins and dynamics of monocytes and tissue macrophages under homeostasis. Immunity 38(1):79–91. https://doi.org/10.1016/j.immuni.2012.12.001

33. Parkhurst CN, Yang G, Ninan I, Savas JN, Yates JR 3rd, Lafaille JJ, Hempstead BL, Littman DR, Gan WB (2013) Microglia promote learning-dependent synapse formation through brain-derived neurotrophic factor.

Cell 155(7):1596–1609. https://doi.org/10.1016/j.cell.2013.11.030

34. Goldmann T, Wieghofer P, Jordao MJ, Prutek F, Hagemeyer N, Frenzel K, Amann L, Staszewski O, Kierdorf K, Krueger M, Locatelli G, Hochgerner H, Zeiser R, Epelman S, Geissmann F, Priller J, Rossi FM, Bechmann I, Kerschensteiner M, Linnarsson S, Jung S, Prinz M (2016) Origin, fate and dynamics of macrophages at central nervous system interfaces. Nat Immunol 17 (7):797–805. https://doi.org/10.1038/ni.3423

35. Zhang B, Zou J, Han L, Beeler B, Friedman JL, Griffin E, Piao YS, Rensing NR, Wong M (2018) The specificity and role of microglia in epileptogenesis in mouse models of tuberous sclerosis complex. Epilepsia 59 (9):1796–1806. https://doi.org/10.1111/epi.14526

36. Snippert HJ, van der Flier LG, Sato T, van Es JH, van den Born M, Kroon-Veenboer C, Barker N, Klein AM, van Rheenen J, Simons BD, Clevers H (2010) Intestinal crypt homeostasis results from neutral competition between symmetrically dividing Lgr5 stem cells. Cell 143(1):134–144. https://doi.org/10.1016/j.cell.2010.09.016

37. Tay TL, Mai D, Dautzenberg J, Fernandez-Klett F, Lin G, Sagar DM, Drougard A, Stempfl T, Ardura-Fabregat A, Staszewski O, Margineanu A, Sporbert A, Steinmetz LM, Pospisilik JA, Jung S, Priller J, Grun D, Ronneberger O, Prinz M (2017) A new fate mapping system reveals context-dependent random or clonal expansion of microglia. Nat Neurosci 20(6):793–803. https://doi.org/10.1038/nn.4547

38. Mizutani M, Pino PA, Saederup N, Charo IF, Ransohoff RM, Cardona AE (2012) The fractalkine receptor but not CCR2 is present on microglia from embryonic development throughout adulthood. J Immunol 188 (1):29–36. https://doi.org/10.4049/jimmunol.1100421

39. Chen Z, Feng X, Herting CJ, Garcia VA, Nie K, Pong WW, Rasmussen R, Dwivedi B, Seby S, Wolf SA, Gutmann DH, Hambardzumyan D (2017) Cellular and molecular identity of tumor-associated macrophages in glioblastoma. Cancer Res 77(9):2266–2278. https://doi.org/10.1158/0008-5472.CAN-16-2310

40. Dal-Secco D, Wang J, Zeng Z, Kolaczkowska E, Wong CH, Petri B, Ransohoff RM, Charo IF, Jenne CN, Kubes P (2015) A dynamic spectrum of monocytes arising from the in situ reprogramming of CCR2+ monocytes at a site of sterile injury. J Exp Med 212(4):447–456. https://doi.org/10.1084/jem.20141539

41. Benz C, Martins VC, Radtke F, Bleul CC (2008) The stream of precursors that colonizes the thymus proceeds selectively through the early T lineage precursor stage of T cell development. J Exp Med 205(5):1187–1199. https://doi.org/10.1084/jem.20072168

42. Bowman RL, Klemm F, Akkari L, Pyonteck SM, Sevenich L, Quail DF, Dhara S, Simpson K, Gardner EE, Iacobuzio-Donahue CA, Brennan CW, Tabar V, Gutin PH, Joyce JA (2016) Macrophage ontogeny underlies differences in tumor-specific education in brain malignancies. Cell Rep 17 (9):2445–2459. https://doi.org/10.1016/j.celrep.2016.10.052

43. McKenna HJ, Stocking KL, Miller RE, Brasel K, De Smedt T, Maraskovsky E, Maliszewski CR, Lynch DH, Smith J, Pulendran B, Roux ER, Teepe M, Lyman SD, Peschon JJ (2000) Mice lacking flt3 ligand have deficient hematopoiesis affecting hematopoietic progenitor cells, dendritic cells, and natural killer cells. Blood 95 (11):3489–3497. https://doi.org/10.1182/blood.V95.11.3489

44. Werner Y, Mass E, Ashok Kumar P, Ulas T, Handler K, Horne A, Klee K, Lupp A, Schutz D, Saaber F, Redecker C, Schultze JL, Geissmann F, Stumm R (2020) Cxcr4 distinguishes HSC-derived monocytes from microglia and reveals monocyte immune responses to experimental stroke. Nat Neurosci 23(3):351–362. https://doi.org/10.1038/s41593-020-0585-y

45. McKinsey GL, Lizama CO, Keown-Lang AE, Niu A, Santander N, Larpthaveesarp A, Chee E, Gonzalez FF, Arnold TD (2020) A new genetic strategy for targeting microglia in development and disease. elife 9. https://doi.org/10.7554/eLife.54590

46. Tiedt R, Schomber T, Hao-Shen H, Skoda RC (2007) Pf4-Cre transgenic mice allow the generation of lineage-restricted gene knockouts for studying megakaryocyte and platelet function in vivo. Blood 109 (4):1503–1506. https://doi.org/10.1182/blood-2006-04-020362

47. Pham TH, Baluk P, Xu Y, Grigorova I, Bankovich AJ, Pappu R, Coughlin SR, McDonald DM, Schwab SR, Cyster JG (2010) Lymphatic endothelial cell sphingosine kinase activity is required for lymphocyte egress and lymphatic patterning. J Exp Med 207 (1):17–27. https://doi.org/10.1084/jem.20091619

48. Nawaz A, Aminuddin A, Kado T, Takikawa A, Yamamoto S, Tsuneyama K, Igarashi Y, Ikutani M, Nishida Y, Nagai Y, Takatsu K, Imura J, Sasahara M, Okazaki Y, Ueki K, Okamura T, Tokuyama K, Ando A, Matsumoto M, Mori H, Nakagawa T, Kobayashi N, Saeki K, Usui I, Fujisaka S, Tobe K (2017) CD206(+) M2-like macrophages regulate systemic glucose metabolism by inhibiting proliferation of adipocyte progenitors. Nat Commun 8(1):286. https://doi.org/10.1038/s41467-017-00231-1

49. Buttgereit A, Lelios I, Yu X, Vrohlings M, Krakoski NR, Gautier EL, Nishinakamura R, Becher B, Greter M (2016) Sall1 is a transcriptional regulator defining microglia identity and function. Nat Immunol 17 (12):1397–1406. https://doi.org/10.1038/ni.3585

50. Inoue S, Inoue M, Fujimura S, Nishinakamura R (2010) A mouse line expressing Sall1-driven inducible Cre recombinase in the kidney mesenchyme. Genesis 48 (3):207–212. https://doi.org/10.1002/dvg.20603

51. Takasato M, Osafune K, Matsumoto Y, Kataoka Y, Yoshida N, Meguro H, Aburatani H, Asashima M, Nishinakamura R (2004) Identification of kidney mesenchymal genes by a combination of microarray analysis and Sall1-GFP knockin mice. Mech Dev 121 (6):547–557. https://doi.org/10.1016/j.mod.2004.04.007

52. Krasemann S, Madore C, Cialic R, Baufeld C, Calcagno N, El Fatimy R, Beckers L, O'Loughlin E, Xu Y, Fanek Z, Greco DJ, Smith ST, Tweet G, Humulock Z, Zrzavy T, Conde-Sanroman P, Gacias M, Weng Z, Chen H, Tjon E, Mazaheri F, Hartmann K, Madi A, Ulrich JD, Glatzel M, Worthmann A, Heeren J, Budnik B, Lemere C, Ikezu T, Heppner FL, Litvak V, Holtzman DM, Lassmann H, Weiner HL, Ochando J, Haass C, Butovsky O (2017) The TREM2-APOE pathway drives the transcriptional phenotype of dysfunctional microglia in neurodegenerative diseases. Immunity 47 (3):566–581.e9. https://doi.org/10.1016/j.immuni.2017.08.008

53. Chappell-Maor L, Kolesnikov M, Kim JS, Shemer A, Haimon Z, Grozovski J, Boura-Halfon S, Masuda T, Prinz M, Jung S (2020) Comparative analysis of CreER transgenic mice for the study of brain macrophages: a case study. Eur J Immunol 50 (3):353–362. https://doi.org/10.1002/eji.201948342

54. Kaiser T, Feng G (2019) Tmem119-EGFP and Tmem119-CreERT2 transgenic mice for labeling and manipulating microglia. eNeuro 6(4). https://doi.org/10.1523/ENEURO.0448-18.2019

55. Ruan C, Sun L, Kroshilina A, Beckers L, De Jager P, Bradshaw EM, Hasson SA, Yang G, Elyaman W (2020) A novel Tmem119-tdTomato reporter mouse model for studying microglia in the central nervous system. Brain Behav Immun 83:180–191. https://doi.org/10.1016/j.bbi.2019.10.009

56. Masuda T, Amann L, Sankowski R, Staszewski O, de Lenz MP, Snaidero N, Costa Jordao MJ, Bottcher C, Kierdorf K, Jung S, Priller J, Misgeld T, Vlachos A, Luehmann MM, Knobeloch KP, Prinz M (2020) Novel Hexb-based tools for studying microglia in the CNS. Nat Immunol 21 (7):802–815. https://doi.org/10.1038/s41590-020-0707-4

57. Butovsky O, Jedrychowski MP, Moore CS, Cialic R, Lanser AJ, Gabriely G, Koeglsperger T, Dake B, Wu PM, Doykan CE, Fanek Z, Liu L, Chen Z, Rothstein JD, Ransohoff RM, Gygi SP, Antel JP, Weiner HL (2014) Identification of a unique TGF-beta-dependent molecular and functional signature in microglia. Nat Neurosci 17(1):131–143. https://doi.org/10.1038/nn.3599

58. Keren-Shaul H, Spinrad A, Weiner A, Matcovitch-Natan O, Dvir-Szternfeld R, Ulland TK, David E, Baruch K, Lara-Astaiso D, Toth B, Itzkovitz S, Colonna M, Schwartz M, Amit I (2017) A unique microglia type associated with restricting development of Alzheimer's disease. Cell 169 (7):1276–1290.e17. https://doi.org/10.1016/j.cell.2017.05.018

59. Hickman SE, Allison EK, Coleman U, Kingery-Gallagher ND, El Khoury J (2019) Heterozygous CX3CR1 deficiency in microglia restores neuronal beta-amyloid clearance pathways and slows progression of Alzheimer's like-disease in PS1-APP mice. Front Immunol 10:2780. https://doi.org/10.3389/fimmu.2019.02780

60. Sanchez-Mejias E, Navarro V, Jimenez S, Sanchez-Mico M, Sanchez-Varo R, Nunez-Diaz C, Trujillo-Estrada L, Davila JC, Vizuete M, Gutierrez A, Vitorica J (2016) Soluble phospho-tau from Alzheimer's disease hippocampus drives microglial degeneration. Acta Neuropathol 132(6):897–916. https://doi.org/10.1007/s00401-016-1630-5

61. Lund H, Pieber M, Parsa R, Han J, Grommisch D, Ewing E, Kular L, Needhamsen M, Espinosa A, Nilsson E, Overby AK, Butovsky O, Jagodic M, Zhang XM, Harris RA (2018) Competitive

repopulation of an empty microglial niche yields functionally distinct subsets of microglia-like cells. Nat Commun 9(1):4845. https://doi.org/10.1038/s41467-018-07295-7

62. Ajami B, Samusik N, Wieghofer P, Ho PP, Crotti A, Bjornson Z, Prinz M, Fantl WJ, Nolan GP, Steinman L (2018) Single-cell mass cytometry reveals distinct populations of brain myeloid cells in mouse neuroinflammation and neurodegeneration models. Nat Neurosci 21(4):541–551. https://doi.org/10.1038/s41593-018-0100-x

63. Jordao MJC, Sankowski R, Brendecke SM, Sagar LG, Tai YH, Tay TL, Schramm E, Armbruster S, Hagemeyer N, Gross O, Mai D, Cicek O, Falk T, Kerschensteiner M, Grun D, Prinz M (2019) Single-cell profiling identifies myeloid cell subsets with distinct fates during neuroinflammation. Science 363(6425). https://doi.org/10.1126/science.aat7554

64. Bennett ML, Bennett FC, Liddelow SA, Ajami B, Zamanian JL, Fernhoff NB, Mulinyawe SB, Bohlen CJ, Adil A, Tucker A, Weissman IL, Chang EF, Li G, Grant GA, Hayden Gephart MG, Barres BA (2016) New tools for studying microglia in the mouse and human CNS. Proc Natl Acad Sci U S A 113(12):E1738–E1746. https://doi.org/10.1073/pnas.1525528113

65. Mildner A, Schonheit J, Giladi A, David E, Lara-Astiaso D, Lorenzo-Vivas E, Paul F, Chappell-Maor L, Priller J, Leutz A, Amit I, Jung S (2017) Genomic characterization of murine monocytes reveals c/ebpbeta transcription factor dependence of Ly6C(−) cells. Immunity 46(5):849–862.e7. https://doi.org/10.1016/j.immuni.2017.04.018

66. Mathys H, Adaikkan C, Gao F, Young JZ, Manet E, Hemberg M, De Jager PL, Ransohoff RM, Regev A, Tsai LH (2017) Temporal tracking of microglia activation in neurodegeneration at single-cell resolution. Cell Rep 21(2):366–380. https://doi.org/10.1016/j.celrep.2017.09.039

67. Konishi H, Kobayashi M, Kunisawa T, Imai K, Sayo A, Malissen B, Crocker PR, Sato K, Kiyama H (2017) Siglec-H is a microglia-specific marker that discriminates microglia from CNS-associated macrophages and CNS-infiltrating monocytes. Glia 65(12):1927–1943. https://doi.org/10.1002/glia.23204

68. Meyer Zu Horste G, Gross CC, Klotz L, Schwab N, Wiendl H (2020) Next-generation neuroimmunology: new technologies to understand central nervous system

autoimmunity. Trends Immunol 41(4):341–354. https://doi.org/10.1016/j.it.2020.02.005

69. Lawson LJ, Perry VH, Dri P, Gordon S (1990) Heterogeneity in the distribution and morphology of microglia in the normal adult mouse brain. Neuroscience 39(1):151–170. https://doi.org/10.1016/0306-4522(90)90229-w

70. Grabert K, Michoel T, Karavolos MH, Clohisey S, Baillie JK, Stevens MP, Freeman TC, Summers KM, McColl BW (2016) Microglial brain region-dependent diversity and selective regional sensitivities to aging. Nat Neurosci 19(3):504–516. https://doi.org/10.1038/nn.4222

71. Hammond TR, Dufort C, Dissing-Olesen L, Giera S, Young A, Wysoker A, Walker AJ, Gergits F, Segel M, Nemesh J, Marsh SE, Saunders A, Macosko E, Ginhoux F, Chen J, Franklin RJM, Piao X, McCarroll SA, Stevens B (2019) Single-cell rna sequencing of microglia throughout the mouse lifespan and in the injured brain reveals complex cell-state changes. Immunity 50(1):253–271.e6. https://doi.org/10.1016/j.immuni.2018.11.004

72. Masuda T, Sankowski R, Staszewski O, Bottcher C, Amann L, Sagar SC, Nessler S, Kunz P, van Loo G, Coenen VA, Reinacher PC, Michel A, Sure U, Gold R, Grun D, Priller J, Stadelmann C, Prinz M (2019) Spatial and temporal heterogeneity of mouse and human microglia at single-cell resolution. Nature 566(7744):388–392. https://doi.org/10.1038/s41586-019-0924-x

73. Hickman SE, Kingery ND, Ohsumi TK, Borowsky ML, Wang LC, Means TK, El Khoury J (2013) The microglial sensome revealed by direct RNA sequencing. Nat Neurosci 16(12):1896–1905. https://doi.org/10.1038/nn.3554

74. Geirsdottir L, David E, Keren-Shaul H, Weiner A, Bohlen SC, Neuber J, Balic A, Giladi A, Sheban F, Dutertre CA, Pfeifle C, Peri F, Raffo-Romero A, Vizioli J, Matiasek K, Scheiwe C, Meckel S, Matz-Rensing K, van der Meer F, Thormodsson FR, Stadelmann C, Zilkha N, Kimchi T, Ginhoux F, Ulitsky I, Erny D, Amit I, Prinz M (2019) Cross-species single-cell analysis reveals divergence of the primate microglia program. Cell 179(7):1609–1622.e16. https://doi.org/10.1016/j.cell.2019.11.010

75. van der Poel M, Ulas T, Mizee MR, Hsiao CC, Miedema SSM, Adelia SKG, Helder B, Tas SW, Schultze JL, Hamann J, Huitinga I (2019) Transcriptional profiling of human

microglia reveals grey-white matter heterogeneity and multiple sclerosis-associated changes. Nat Commun 10(1):1139. https://doi.org/10.1038/s41467-019-08976-7

76. Böttcher C, Schlickeiser S, Sneeboer MAM, Kunkel D, Knop A, Paza E, Fidzinski P, Kraus L, Snijders GJL, Kahn RS, Schulz AR, Mei HE, Psy NBB, Hol EM, Siegmund B, Glauben R, Spruth EJ, de Witte LD, Priller J (2019) Human microglia regional heterogeneity and phenotypes determined by multiplexed single-cell mass cytometry. Nat Neurosci 22(1):78–90. https://doi.org/10.1038/s41593-018-0290-2

77. Prinz M, Jung S, Priller J (2019) Microglia biology: one century of evolving concepts. Cell 179(2):292–311. https://doi.org/10.1016/j.cell.2019.08.053

78. Fletcher JM, Lalor SJ, Sweeney CM, Tubridy N, Mills KH (2010) T cells in multiple sclerosis and experimental autoimmune encephalomyelitis. Clin Exp Immunol 162(1):1–11. https://doi.org/10.1111/j.1365-2249.2010.04143.x

79. Bartholomaus I, Kawakami N, Odoardi F, Schlager C, Miljkovic D, Ellwart JW, Klinkert WE, Flugel-Koch C, Issekutz TB, Wekerle H, Flugel A (2009) Effector T cell interactions with meningeal vascular structures in nascent autoimmune CNS lesions. Nature 462(7269):94–98. https://doi.org/10.1038/nature08478

80. Pesic M, Bartholomaus I, Kyratsous NI, Heissmeyer V, Wekerle H, Kawakami N (2013) 2-photon imaging of phagocyte-mediated T cell activation in the CNS. J Clin Invest 123(3):1192–1201. https://doi.org/10.1172/JCI67233

81. Tran EH, Hoekstra K, van Rooijen N, Dijkstra CD, Owens T (1998) Immune invasion of the central nervous system parenchyma and experimental allergic encephalomyelitis, but not leukocyte extravasation from blood, are prevented in macrophage-depleted mice. J Immunol 161(7):3767–3775

82. Schlager C, Korner H, Krueger M, Vidoli S, Haberl M, Mielke D, Brylla E, Issekutz T, Cabanas C, Nelson PJ, Ziemssen T, Rohde V, Bechmann I, Lodygin D, Odoardi F, Flugel A (2016) Effector T-cell trafficking between the leptomeninges and the cerebrospinal fluid. Nature 530(7590):349–353. https://doi.org/10.1038/nature16939

83. Brendecke SM, Prinz M (2015) Do not judge a cell by its cover—diversity of CNS resident, adjoining and infiltrating myeloid cells in inflammation. Semin Immunopathol 37(6):591–605. https://doi.org/10.1007/s00281-015-0520-6

84. Prinz M, Priller J (2017) The role of peripheral immune cells in the CNS in steady state and disease. Nat Neurosci 20(2):136–144. https://doi.org/10.1038/nn.4475

85. Miron VE, Boyd A, Zhao JW, Yuen TJ, Ruckh JM, Shadrach JL, van Wijngaarden P, Wagers AJ, Williams A, Franklin RJM, Ffrench-Constant C (2013) M2 microglia and macrophages drive oligodendrocyte differentiation during CNS remyelination. Nat Neurosci 16(9):1211–1218. https://doi.org/10.1038/nn.3469

86. Butovsky O, Ziv Y, Schwartz A, Landa G, Talpalar AE, Pluchino S, Martino G, Schwartz M (2006) Microglia activated by IL-4 or IFN-gamma differentially induce neurogenesis and oligodendrogenesis from adult stem/progenitor cells. Mol Cell Neurosci 31(1):149–160. https://doi.org/10.1016/j.mcn.2005.10.006

87. Piccio L, Buonsanti C, Mariani M, Cella M, Gilfillan S, Cross AH, Colonna M, Panina-Bordignon P (2007) Blockade of TREM-2 exacerbates experimental autoimmune encephalomyelitis. Eur J Immunol 37(5):1290–1301. https://doi.org/10.1002/eji.200636837

88. Asai H, Ikezu S, Tsunoda S, Medalla M, Luebke J, Haydar T, Wolozin B, Butovsky O, Kugler S, Ikezu T (2015) Depletion of microglia and inhibition of exosome synthesis halt tau propagation. Nat Neurosci 18(11):1584–1593. https://doi.org/10.1038/nn.4132

89. Liddelow SA, Guttenplan KA, Clarke LE, Bennett FC, Bohlen CJ, Schirmer L, Bennett ML, Munch AE, Chung WS, Peterson TC, Wilton DK, Frouin A, Napier BA, Panicker N, Kumar M, Buckwalter MS, Rowitch DH, Dawson VL, Dawson TM, Stevens B, Barres BA (2017) Neurotoxic reactive astrocytes are induced by activated microglia. Nature 541(7638):481–487. https://doi.org/10.1038/nature21029

90. Shi Y, Manis M, Long J, Wang K, Sullivan PM, Remolina Serrano J, Hoyle R, Holtzman DM (2019) Microglia drive APOE-dependent neurodegeneration in a tauopathy mouse model. J Exp Med 216(11):2546–2561. https://doi.org/10.1084/jem.20190980

91. Oakley H, Cole SL, Logan S, Maus E, Shao P, Craft J, Guillozet-Bongaarts A, Ohno M, Disterhoft J, Van Eldik L, Berry R, Vassar R (2006) Intraneuronal beta-amyloid aggregates, neurodegeneration, and neuron loss in

transgenic mice with five familial Alzheimer's disease mutations: potential factors in amyloid plaque formation. J Neurosci 26 (40):10129–10140. https://doi.org/10.1523/JNEUROSCI.1202-06.2006

92. Cruz JC, Tseng HC, Goldman JA, Shih H, Tsai LH (2003) Aberrant Cdk5 activation by p25 triggers pathological events leading to neurodegeneration and neurofibrillary tangles. Neuron 40(3):471–483. https://doi.org/10.1016/s0896-6273(03)00627-5

93. Giladi A, Cohen M, Medaglia C, Baran Y, Li B, Zada M, Bost P, Blecher-Gonen R, Salame TM, Mayer JU, David E, Ronchese F, Tanay A, Amit I (2020) Dissecting cellular crosstalk by sequencing physically interacting cells. Nat Biotechnol 38(5):629–637. https://doi.org/10.1038/s41587-020-0442-2

94. Plemel JR, Stratton JA, Michaels NJ, Rawji KS, Zhang E, Sinha S, Baaklini CS, Dong Y, Ho M, Thorburn K, Friedman TN, Jawad S, Silva C, Caprariello AV, Hoghooghi V, Yue J, Jaffer A, Lee K, Kerr BJ, Midha R, Stys PK, Biernaskie J, Yong VW (2020) Microglia response following acute demyelination is heterogeneous and limits infiltrating macrophage dispersion. Sci Adv 6(3):eaay6324. https://doi.org/10.1126/sciadv.aay6324

95. Mendiola AS, Ryu JK, Bardehle S, Meyer-Franke A, Ang KK, Wilson C, Baeten KM, Hanspers K, Merlini M, Thomas S, Petersen MA, Williams A, Thomas R, Rafalski VA, Meza-Acevedo R, Tognatta R, Yan Z, Pfaff SJ, Machado MR, Bedard C, Rios Coronado PE, Jiang X, Wang J, Pleiss MA, Green AJ, Zamvil SS, Pico AR, Bruneau BG, Arkin MR, Akassoglou K (2020) Transcriptional profiling and therapeutic targeting of oxidative stress in neuroinflammation. Nat Immunol 21(5):513–524. https://doi.org/10.1038/s41590-020-0654-0

96. Schirmer L, Velmeshev D, Holmqvist S, Kaufmann M, Werneburg S, Jung D, Vistnes S, Stockley JH, Young A, Steindel M, Tung B, Goyal N, Bhaduri A, Mayer S, Engler JB, Bayraktar OA, Franklin RJM, Haeussler M, Reynolds R, Schafer DP, Friese MA, Shiow LR, Kriegstein AR, Rowitch DH (2019) Neuronal vulnerability and multilineage diversity in multiple sclerosis. Nature 573(7772):75–82. https://doi.org/10.1038/s41586-019-1404-z

97. Deczkowska A, Amit I, Schwartz M (2018) Microglial immune checkpoint mechanisms. Nat Neurosci 21(6):779–786. https://doi.org/10.1038/s41593-018-0145-x

98. Deczkowska A, Weiner A, Amit I (2020) The physiology, pathology, and potential therapeutic applications of the TREM2 signaling pathway. Cell 181(6):1207–1217. https://doi.org/10.1016/j.cell.2020.05.003

99. Zhou Y, Song WM, Andhey PS, Swain A, Levy T, Miller KR, Poliani PL, Cominelli M, Grover S, Gilfillan S, Cella M, Ulland TK, Zaitsev K, Miyashita A, Ikeuchi T, Sainouchi M, Kakita A, Bennett DA, Schneider JA, Nichols MR, Beausoleil SA, Ulrich JD, Holtzman DM, Artyomov MN, Colonna M (2020) Human and mouse single-nucleus transcriptomics reveal TREM2-dependent and TREM2-independent cellular responses in Alzheimer's disease. Nat Med 26 (1):131–142. https://doi.org/10.1038/s41591-019-0695-9

100. Wang S, Mustafa M, Yuede CM, Salazar SV, Kong P, Long H, Ward M, Siddiqui O, Paul R, Gilfillan S, Ibrahim A, Rhinn H, Tassi I, Rosenthal A, Schwabe T, Colonna M (2020) Anti-human TREM2 induces microglia proliferation and reduces pathology in an Alzheimer's disease model. J Exp Med 217 (9). https://doi.org/10.1084/jem.20200785

101. Spangenberg EE, Lee RJ, Najafi AR, Rice RA, Elmore MR, Blurton-Jones M, West BL, Green KN (2016) Eliminating microglia in Alzheimer's mice prevents neuronal loss without modulating amyloid-beta pathology. Brain 139(Pt 4):1265–1281. https://doi.org/10.1093/brain/aww016

102. Sosna J, Philipp S, Albay R 3rd, Reyes-Ruiz JM, Baglietto-Vargas D, LaFerla FM, Glabe CG (2018) Early long-term administration of the CSF1R inhibitor PLX3397 ablates microglia and reduces accumulation of intraneuronal amyloid, neuritic plaque deposition and pre-fibrillar oligomers in 5XFAD mouse model of Alzheimer's disease. Mol Neurodegener 13(1):11. https://doi.org/10.1186/s13024-018-0244-x

103. Spangenberg E, Severson PL, Hohsfield LA, Crapser J, Zhang J, Burton EA, Zhang Y, Spevak W, Lin J, Phan NY, Habets G, Rymar A, Tsang G, Walters J, Nespi M, Singh P, Broome S, Ibrahim P, Zhang C, Bollag G, West BL, Green KN (2019)

Sustained microglial depletion with CSF1R inhibitor impairs parenchymal plaque development in an Alzheimer's disease model. Nat Commun 10(1):3758. https://doi.org/10.1038/s41467-019-11674-z

104. Huang Y, Xu Z, Xiong S, Sun F, Qin G, Hu G, Wang J, Zhao L, Liang YX, Wu T, Lu Z, Humayun MS, So KF, Pan Y, Li N, Yuan TF, Rao Y, Peng B (2018) Repopulated microglia are solely derived from the proliferation of residual microglia after acute depletion. Nat Neurosci 21(4):530–540. https://doi.org/10.1038/s41593-018-0090-8

105. Lloyd AF, Davies CL, Holloway RK, Labrak Y, Ireland G, Carradori D, Dillenburg A, Borger E, Soong D, Richardson JC, Kuhlmann T, Williams A, Pollard JW, des Rieux A, Priller J, Miron VE (2019) Central nervous system regeneration is driven by microglia necroptosis and repopulation. Nat Neurosci 22(7):1046–1052. https://doi.org/10.1038/s41593-019-0418-z

106. Bruttger J, Karram K, Wortge S, Regen T, Marini F, Hoppmann N, Klein M, Blank T, Yona S, Wolf Y, Mack M, Pinteaux E, Muller W, Zipp F, Binder H, Bopp T, Prinz M, Jung S, Waisman A (2015) Genetic cell ablation reveals clusters of local self-renewing microglia in the mammalian central nervous system. Immunity 43(1):92–106. https://doi.org/10.1016/j.immuni.2015.06.012

107. Henry RJ, Ritzel RM, Barrett JP, Doran SJ, Jiao Y, Leach JB, Szeto GL, Wu J, Stoica BA, Faden AI, Loane DJ (2020) Microglial depletion with CSF1R inhibitor during chronic phase of experimental traumatic brain injury reduces neurodegeneration and neurological deficits. J Neurosci 40(14):2960–2974. https://doi.org/10.1523/JNEUROSCI.2402-19.2020

108. Nissen JC, Thompson KK, West BL, Tsirka SE (2018) Csf1R inhibition attenuates experimental autoimmune encephalomyelitis and promotes recovery. Exp Neurol 307:24–36. https://doi.org/10.1016/j.expneurol.2018.05.021

109. Beckmann N, Giorgetti E, Neuhaus A, Zurbruegg S, Accart N, Smith P, Perdoux J, Perrot L, Nash M, Desrayaud S, Wipfli P, Frieauff W, Shimshek DR (2018) Brain region-specific enhancement of remyelination and prevention of demyelination by the CSF1R kinase inhibitor BLZ945. Acta Neuropathol Commun 6(1):9. https://doi.org/10.1186/s40478-018-0510-8

110. Yang X, Zhang JD, Duan L, Xiong HG, Jiang YP, Liang HC (2018) Microglia activation mediated by toll-like receptor-4 impairs brain white matter tracts in rats. J Biomed Res 32(2):136–144. https://doi.org/10.7555/JBR.32.20170033

111. Butowski N, Colman H, De Groot JF, Omuro AM, Nayak L, Wen PY, Cloughesy TF, Marimuthu A, Haidar S, Perry A, Huse J, Phillips J, West BL, Nolop KB, Hsu HH, Ligon KL, Molinaro AM, Prados M (2016) Orally administered colony stimulating factor 1 receptor inhibitor PLX3397 in recurrent glioblastoma: an Ivy Foundation Early Phase Clinical Trials Consortium phase II study. Neuro-Oncology 18(4):557–564. https://doi.org/10.1093/neuonc/nov245

112. Shin S, Walz KA, Archambault AS, Sim J, Bollman BP, Koenigsknecht-Talboo J, Cross AH, Holtzman DM, Wu GF (2014) Apolipoprotein E mediation of neuro-inflammation in a murine model of multiple sclerosis. J Neuroimmunol 271(1–2):8–17. https://doi.org/10.1016/j.jneuroim.2014.03.010

113. Wang Y, Cella M, Mallinson K, Ulrich JD, Young KL, Robinette ML, Gilfillan S, Krishnan GM, Sudhakar S, Zinselmeyer BH, Holtzman DM, Cirrito JR, Colonna M (2015) TREM2 lipid sensing sustains the microglial response in an Alzheimer's disease model. Cell 160(6):1061–1071. https://doi.org/10.1016/j.cell.2015.01.049

114. Zhong L, Xu Y, Zhuo R, Wang T, Wang K, Huang R, Wang D, Gao Y, Zhu Y, Sheng X, Chen K, Wang N, Zhu L, Can D, Marten Y, Shinohara M, Liu CC, Du D, Sun H, Wen L, Xu H, Bu G, Chen XF (2019) Soluble TREM2 ameliorates pathological phenotypes by modulating microglial functions in an Alzheimer's disease model. Nat Commun 10(1):1365. https://doi.org/10.1038/s41467-019-09118-9

115. Varvel NH, Neher JJ, Bosch A, Wang W, Ransohoff RM, Miller RJ, Dingledine R (2016) Infiltrating monocytes promote brain inflammation and exacerbate neuronal damage after status epilepticus. Proc Natl Acad Sci U S A 113(38):E5665–E5674. https://doi.org/10.1073/pnas.1604263113

116. Getts DR, Terry RL, Getts MT, Deffrasnes C, Muller M, van Vreden C, Ashhurst TM, Chami B, McCarthy D, Wu H, Ma J, Martin A, Shae LD, Witting P, Kansas GS, Kuhn J, Hafezi W, Campbell IL, Reilly D, Say J, Brown L, White MY, Cordwell SJ, Chadban SJ, Thorp EB, Bao S, Miller SD, King NJ (2014) Therapeutic inflammatory monocyte modulation using immune-modifying microparticles. Sci Transl Med 6 (219):219ra7. https://doi.org/10.1126/scitranslmed.3007563

117. Mildner A, Mack M, Schmidt H, Bruck W, Djukic M, Zabel MD, Hille A, Priller J, Prinz M (2009) CCR2+Ly-6Chi monocytes are crucial for the effector phase of autoimmunity in the central nervous system. Brain 132 (Pt 9):2487–2500. https://doi.org/10.1093/brain/awp144

Chapter 3

Activation of Astrocytes in Neurodegenerative Diseases

Jiatong Li and Song Qin

Abstract

Astrocytes are the most abundant cells in the central nervous system (CNS). Under neurodegenerative conditions, astrocytes can go through various morphological and functional changes and then transform into reactive forms. Therefore, compared with normal physiological conditions, the expression and distribution of many cellular molecules in reactive astrocytes show significant changes. It will be of great benefit for research and clinical practice if these molecular alterations of astrocytes can be used as biomarkers for the study of neurodegenerative diseases. In this chapter, we will comprehensively introduce some potential biomarkers involving various aspects of activated astrocytes, including structural and functional characteristics, and their participation in neuroinflammatory responses, such as GFAP, glutamate transporters, and S100β. We will analyze the advantages and limitations of traditional biomarkers of reactive astrocytes, such as GFAP, and provide some insights into potentially novel biomarkers. Then we will give a brief introduction of potential biomarkers in some typical neurodegenerative diseases, including Alzheimer's disease (AD), Parkinson's disease (PD), Huntington's disease (HD), and amyotrophic lateral sclerosis (ALS). Finally, we provide some methods applied in biomarker studies for reference. These methods range from astrocyte isolating, culture, and further biomarker analysis to clinical techniques such as molecular imaging. In the future, advances in the detection of biomarkers in astrocytes under neurodegenerative conditions will not only shed light on early diagnosis but also be illuminating in the treatment of neurodegenerative diseases, which is still in a rather difficult stage, by serving as potential targets for drugs. More analyses of astrocytes based on these biomarkers and a deeper understanding of the molecular pathogenesis of neurodegenerative diseases will undoubtedly bring hope to individual patients, their families, and the whole society as well.

Key words Reactive astrocytes, Neurodegenerative diseases, Biomarkers, Methodology

1 Introduction

Neurodegenerative diseases refer to a group of diseases that could lead to various clinical manifestations associated with pathological changes throughout the central nervous system (CNS) [1]. These diseases include degeneration of the brain such as that occurs in Alzheimer's disease (AD), Parkinson's disease (PD), and Huntington's disease (HD), motor neuron diseases such as amyotrophic lateral sclerosis (ALS), and some infectious diseases such as prion

Philip V. Peplow, Bridget Martinez and Thomas A. Gennarelli (eds.), *Neurodegenerative Diseases Biomarkers: Towards Translating Research to Clinical Practice*, Neuromethods, vol. 173, https://doi.org/10.1007/978-1-0716-1712-0_3,
© Springer Science+Business Media, LLC, part of Springer Nature 2022

disease [2]. One of the characteristics of neurodegenerative diseases is that they are closely related to brain aging and therefore have a higher prevalence in older adults, which has been gaining more and more attention in the increasingly aging world.

Despite the increasing cost of neurodegenerative diseases in our world, their causes and pathogenesis have not yet been clearly elucidated [3], so early diagnosis and effective treatment of such diseases remains a huge challenge. In recent years, many basic researches and clinical trials have been carried out, devoted to unravel the "mystery" of neurodegenerative diseases and to achieve progress in clinical treatment [4]. Among these studies, it has been revealed that on the one hand, neurodegeneration can lead to structural and functional alterations of cells in the CNS [5], including the activation of microglia and astrocytes [6–9]; and on the other hand, sustained inflammatory responses of these cells may contribute to the development and progression of neurodegenerative diseases [10–13].

Astrocytes are the most abundant cells in the CNS [14]. Apart from contributing to the constitution of the blood–brain barrier (BBB), astrocytes also play a crucial role in CNS homeostasis including providing structural and metabolic support for neurons [15, 16], guiding axonal growth [17], assisting synapse formation [18] and regulating ion-channel distribution [19]. Several studies have suggested that under neurodegenerative conditions, astrocytes can be activated into reactive forms and go through proliferation under the influence of certain growth factors [12, 20, 21]. This activation leads to changes in the morphology, metabolism and functions of astrocytes [22–25], which are the results of alterations of related genes [26, 27]. Studies have also shown a close relationship between neuroinflammation and neurodegenerative diseases, with many cytokines found to be expressed at higher levels during neurodegeneration [28, 29], in which astrocytes may also play an important role in leading to neurodegenerative conditions by initiating a proinflammatory and cytotoxic cascade [30].

With astrocytes undergoing the pathological processes mentioned above, the expression and contents of many cellular molecules show significant changes compared to normal physiological conditions [26, 27]. Therefore, it will be of great benefit for research and clinical practice if these molecular alterations of astrocytes can be used as biomarkers for the study of neurodegenerative diseases, which may not only help the establishment of disease models on basic research but also shed light on the early diagnosis and treatment of neurodegeneration clinically.

In this chapter, we will focus on astrocyte biomarkers associated with their structural and functional changes during neurodegeneration and list some representative neurodegenerative diseases and their potential biomarkers. Besides, we will also provide some methodological reviews for the detection of these biomarkers in basic research and clinical practice.

2 Astrocyte Activation and Altered Molecular Expression in Neurodegenerative Diseases

In the study of neurodegenerative diseases, neurons have long been regarded as the central element [31]. However, this view is being challenged with the notice that astrocytes and other glial cells play an essential role and may be centric in the pathogenesis and pathology of the diseases [32–35].

As the largest cell population in the CNS, astrocytes play an important role in maintaining the homeostasis of the CNS and are indispensable in supporting normal neuronal structure and functions [17]. So, before we discuss the activation of astrocytes in neurodegenerative diseases, let us take an initial look at the structural and functional characteristics of astroglia under physiological conditions.

Structurally, astrocytes contain a large number of interlaced fibrils in their cytoplasm, consisting of glial filaments in diameter between the microtubule and the microfilament [36]. The glial filaments are composed of glial fibrillary acid protein (GFAP), which together with vimentin, participates in constituting the intermediate filaments of the astroglial cytoskeleton [14, 37]. At least eight alternative splicing variants of GFAP have been identified in different species until now [38], including GFAP α which is the most abundant form [39], GFAP β [40], GFAP γ [41], GFAP ε [42, 43], GFAP κ [44], GFAP Δ135, GFAP Δ164, and GFAP Δexon6 [45]. These splice variants share some similar characteristics, but also show many differences in expression and distribution in the CNS. For example, GFAP β is found only in rats [40] and GFAP δ is highly expressed in proliferating astrocytes located in the subventricular zone and the subpial layer [42, 46], suggesting that these astrocytes are closely related to the migration of neurons during embryonic development and may be neural stem cells in adult brains [47–49].

Astrocytes are indispensable in maintaining the homeostasis of the CNS, which is in accordance with their structural and metabolic characteristics. The endfeet of astrocytes, together with pericytes, the basement membrane and the tight junctions of endothelial cells, constitute the blood–brain barrier which protects the CNS from harmful molecules in the peripheral circulation [50, 51]. Astrocytes also play a key role in regulating extracellular neurotransmitters. For example, extracellular glutamate released by neurons lends itself to reuptake by astrocytes through glutamate transporters EAAT1/GLAST and EAAT2/GLT-1, where it is further transformed into glutamine under the catalysis of glutamine synthetase [38, 52, 53]. This process is essential in eliminating the potential excitotoxicity of glutamate accumulation. Glutamine in astrocytes can then be transported outside and reused by neurons to synthesize glutamate again, thus completing the "glutamate–glutamine cycle" between the two cells [52, 53]. Astroglial processes are also rich in membrane transporting channels, such as aquaporin 4 (AQP4) and K^+ channels, which are critical in maintaining CNS fluid homeostasis and regulating the balance of neuronal excitability [34, 38, 54, 55]. Besides, astrocytes are coupled with each other through gap junctions made up of connexins, which may participate in both physiological and pathological conditions [34, 56].

Astrocytes also play a role in immune responses of the nervous system in some neurological disorders. Astrocytes may function as phagocytes similar to microglia, since some conserved phagocytic pathways including the Draper/Megf10 and MerTK/alpha (v)beta5 integrin pathways have been found highly enriched in astrocytes [57]. They also present as antigen-presenting cells in CNS immune responses with the expression of major histocompatibility complex molecule-II (MHC-II) [3]. Moreover, the secretion of many cytokines is also found to be upregulated in astrocytes during neuropathological processes [58, 59]. When the homeostasis of the CNS is disrupted, under the circumstances of neurodegenerative diseases for instance, the normal structure, functions and metabolism of the above-mentioned astrocytes also change, resulting in different "biomarker" expression characteristics as will be discussed below.

2.1 Morphological and Structural Changes of Activated Astrocytes

The activation of astrocytes is not an "all-or-none" phenomenon, and its morphological and structural changes present as a continuum of progression that occur with the severity of CNS insults [14, 38]. According to Sofroniew et al., astrocyte activation, namely, reactive astrogliosis, can be divided into three categories based on the severity of lesions [14]. (1) *Mild-to-moderate reactive astrogliosis.* This level is marked by variable hypertrophy of cell body and processes [60], as well as variable upregulation of GFAP expression [61]. Astrocytes do not lose their individual domain or overlap each other, nor do they undergo any proliferation [14, 60, 61]. (2) *Severe diffuse reactive astrogliosis.* Under more severe conditions, astrocyte proliferation can be found with a more pronounced hypertrophy of the cell and upregulation of GFAP expression. As a result, individual astrocytes extend over their previous domain and overlap with adjacent part, which can lead to persistent tissue reorganization [38]. (3) *Severe reactive astrogliosis with compact glial scar formation.* In addition to the milder form changes, this degree of lesion is also characterized by a distinct formation of dense, narrow and compact glial scars, which are deposited with a dense collagen extracellular matrix that may inhibit axonal and cellular migration [62]. Reactive astrocytes in these scars can interact with other types of cells, especially fibromeningeal and other glial cells [63, 64]. Activated astrocytes associated with chronic neurodegeneration often appear in the second and third forms of pathological changes, with the latter usually occurring in neurodegeneration induced by other primary insults [14].

Although pathological appearances vary, it is noteworthy that GFAP expression is usually markedly increased during neurodegenerative diseases, thus making GFAP the most widely recognized biomarker for astrocyte activation. Some studies have shown that not all astrocytes express GFAP in healthy tissues under physiological circumstances, however, GFAP expression can be found in most, if not all, reactive astrocytes after CNS insults [64–66]. In infantile neuronal ceroid lipofuscinosis (INCL), a hereditary progressive neurodegenerative disease [67–69], the expression of GFAP in astrocytes is a highly prominent first pathological sign [70, 71]. In another neurodegenerative disorder, Alexander disease, GFAP gain-of-function mutation is believed to be the primary cause of the onset of the disease [72, 73]. As described above, there are at least eight variants of GFAP in different subpopulations of astrocytes [38]. A study based on tissues of Alzheimer's disease (AD) donors revealed that the transcript levels of most GFAP isoforms increased in the hippocampus of AD patients, and that the transcript levels of GFAP α, GFAP δ, GFAP κ, and GFAP $\Delta135$ positively correlated with the Braak staging [74], which is a classical classification of AD progression [75].

The use of GFAP as a biomarker of astrocyte activation in neurodegenerative diseases has been widely accepted because of its expression closely related with their pathological progression [61]. However, there are also some limitations of this classical marker [38]: (1) GFAP is not exclusively expressed by astrocytes in the CNS. Other cells include retinal Miller glial cells, Bergmann glia in the cerebellum, tanycytes at the base of the third ventricle, pituicytes in the posterior pituitary, and neural stem cells (NSCs) in embryonic and adult brains can also express GFAP [76], thus weakening its specificity as an astroglial biomarker. (2) Regulated by a large number of intracellular and intercellular signaling molecules, GFAP expression is variable especially in the condition of mild-to-moderate reactive astrogliosis [61]. (3) Instead of presenting throughout the whole cytoplasm, GFAP expression can only be detected in main processes by immunohistochemical studies, which may lead to some errors in the estimation of astrocyte size and degree of ramification [38]. Therefore, when studying the activation of astrocytes in neurodegeneration, other markers should be applied as a reference. For example, the expression of other structural proteins, such as vimentin [3], nestin [59], synemin α and synemin β [74] also show a significant increase in reactive astrocytes, which may serve as complementary markers for GFAP.

In addition to intracellular structural changes of astrocytes in neurodegenerative diseases, studies have shown that altered communication and gap junctions between neighboring astrocytes, which is marked by the loss of connexins 43 and 30, can lead to demyelination and also show negative effects related to the progression of neurodegeneration [77].

2.2 Altered Metabolism and Functionality of Activated Astrocytes

Earlier in this chapter, we have mentioned that one of the crucial roles of astrocytes is to maintain the "glutamate–glutamine cycle," which depends on the normal function of glutamate transporters and glutamine synthetase (GS) [52, 53]. It has been confirmed that the loss of astrocytic glutamate transporters (i.e., EAAT1/GLAST and EAAT2 /GLT-1) is present in many types of neurodegenerative diseases [78–90], resulting in impediment of extracellular glutamate recycling and reuse. These functional alterations may be caused by oxidative damage, splice variants, and altered solubility of the transporters in AD [91–94]. Therefore, glutamate transporters can serve as a biomarker to detect astrocyte activation in neurodegenerative diseases. In contrast, although GS is commonly used as a marker for astrocytes [95, 96], it does not appear to be a good marker of reactive astrogliosis since studies have shown that the expression of GS does not positively correlate with GFAP expression [97]. Our recent study shows that GS expression is significantly reduced in reactive astrocytes of APP/PS1 mice and AD patients, indicating the turnover of glutamate–glutamine is impaired in AD [98] (Fig. 1). Of note, in different types of neurodegenerative diseases, functional changes in astrocytes can exhibit different characteristics, resulting in different alterations of some biomarkers, which we will discuss later in the next part of this chapter.

S100β is another widely accepted marker of astrocytes [99], and its expression has proved to be upregulated during neurodegenerative diseases [3]. S100β is a calcium-binding protein belonging to the S100 family, which is predominantly expressed and secreted by astrocytes in vertebrate brains [100, 101]. Under physiological conditions, this protein plays various roles including interacting with many putative intracellular targets related to proteins of the cytoskeleton (e.g., GFAP and CapZ), modulators of the cell cycle (e.g., p53 and Ndr kinase), protein kinase C and phosphatase 2B (calcineurin), as well as showing extracellular effects depending on its concentration [101, 102]. Therefore, the increase level of S100β in neurodegeneration may be related to the expression alterations of its targeted proteins, such as GFAP.

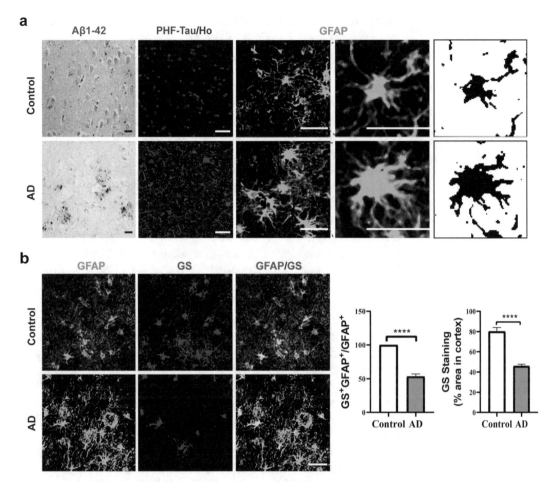

Fig. 1 Changes in astrocytic marker expression in cortical astrocytes of AD patients relative to age-matched controls. (**a**) Immunostaining of sections from cortical tissues of AD patients and age-matched controls with antibodies against Aβ1–42 (brown), PHF-Tau (red), and GFAP(green); nuclei were counterstained with hematoxylin or Hoechst 33342 (blue). Individual GFAP+ astrocytes as shown from the boxed regions were selected randomly and skeletonized for morphological analysis. Black scale bars, 10 μm. White scale bars, 40 μm. (**b**) Left, representative immunofluorescence images of cortical astrocytes from AD patients and control tissues stained for GFAP (green) and GS (red) as indicated. Right, quantification of GS immunoreactivity and the ratio of GFAP+GS+ cells to total GFAP + astrocytes in AD patients as compared with age-matched controls. Scale bars, 40 μm. *Aβ* β-amyloid, *AD* Alzheimer's disease, *GFAP* glial fibrillary acidic protein, *GS* glutamine synthetase. Pictures are from *Li et al. DOI:* https:/doi.org/10.1002/glia.23845

Apart from the expression alterations of these commonly used astrocyte markers, changes of other proteins also exist in activated astrocytes. For example, myoinositol is another marker of astrocyte activation, and is increased in normal aging brains in humans and in animal models [103, 104]. Several studies have proved that a mitochondrial membrane protein, monoamine oxidase B (MAO-B), is also increased in activated astrocytes, which can indicate ongoing astrogliosis in AD patients [105, 106].

2.3 Neuroinflammation Related to Astrocyte Activation

Neuroinflammation is believed to be closely associated with neurodegenerative diseases. One of the prominent and early features of AD is neuroinflammation, which plays an important role in AD progression via a variety of inflammatory mediators and neurotoxic molecules [107]. It has long been speculated that neuroinflammation is secondary to neurodegeneration, however, several studies have suggested that neuroinflammation might in turn help or even promote the progression of neurodegenerative diseases [108]. Microglia have been widely accepted to play a key role in neuroinflammatory response during neurodegeneration [109], while the participation of astrocytes has been given much attention to as well [110–112]. Take AD as an example, microglia produce many proinflammatory cytokines and mediators in response to β-amyloid (Aβ), which may in turn activate astrocytes. Activated astrocytes, along with microglia, secrete cytokines including interleukin-1 and tumor necrosis factor alpha (TNF-α), through which they start to participate in the process of neuroinflammation [111, 113].

According to a study by Liddelow et al. in 2017, reactive astrocytes can be divided into two categories, namely "A1" and "A2," which are formed in response of neuroinflammation and ischemia respectively [23]. Compared to reactive astrocytes induced by ischemia ("A2" type), which might have "helpful" or reparative functions [55, 114, 115], "A1" type astrocytes which are activated by neuroinflammation can upregulate many genes that have been previously shown to have "harmful" functions [108]. Under neuroinflammatory circumstances, reactive astrocytes are marked by increased secretion of proinflammatory cytokines including TNF-α and interleukin (IL)-1β as well as generation of reactive oxygen species (ROS) [116]. Of note, C3, one of the key factors in the complement system, has also been found to be specifically upregulated in A1 reactive astrocytes, but not in resting astrocytes or A2 type astrocytes [108]. In addition, transcriptome profiling has revealed that activated astrocytes can express genes related to phagocytosis, such as *Draper/Megf10* and *MerTK/integrin αVβ5*, which may participate in β-amyloid plaque degradation in AD mouse model [8, 117]. Other studies also have demonstrated that activated astrocytes can degrade β-amyloid plaques by releasing proteases, such as matrix metalloproteinase-9 (MMP-9) [118, 119]. Therefore, these proinflammatory cytokines, complement factors, and enzymes may serve as another large category of biomarkers for the detection of astrocyte activation in neurodegenerative diseases.

2.4 Mechanisms Involved in the Pathogenesis of Astrocyte Activation

As concluded in a previous review [61], the molecular triggers and modulators of reactive astrogliosis include (1) cytokines and growth factors, such as IL-6, leukemia inhibitory factor (LIF), ciliary neurotrophic factor (CNTF), IL-1, IL-10, transforming growth factor (TGF)-β, tumor necrosis factor (TNF)-α, and

Interferon (IFN)-γ [120, 121]; (2) mediators of innate immunity, such as lipopolysaccharide (LPS) and other Toll-like receptor ligands [122]; (3) neurotransmitters and modulators including glutamate and noradrenaline [123]; (4) small molecules released by cell injury such as ATP [124]; (5) molecules of oxidative stress, such as nitric oxide (NO) and reactive oxygen species (ROS) [125]; (6) hypoxia and glucose deprivation associated with ischemia [125]; (7) neurodegeneration-associated molecules such as Aβ [78], and (8) systematic metabolic toxicity such as NH_4^+ [126].

Many of the factors listed above can contribute to the activation and proliferation of astrocytes in neurodegenerative diseases. Microglia may play an important role in astrocyte activation. During neurodegeneration (e.g., AD), microglia produce a variety of proinflammatory cytokines and mediators in response to Aβ, which can trigger the activation of astrocytes. Activated astrocytes, together with microglia, secrete cytokines including IL-1 and TNFα, which in turn promote this activation effect, thus starting the development of neuroinflammatory process [111, 113]. Activated astrocytes further proliferate and increase the inflammatory response in the CNS; the molecular triggers that lead to proliferation of reactive astrocytes in vivo are incompletely characterized but include EGF, FGF, endothelin 1, and ATP [20, 21, 56].

As has been discussed previously, reactive astrocytes can be divided into two categories according to the insults they go through and their possible effects on recovery, and it has been demonstrated that whether astrocytes are transformed into "A1" or "A2" type under different circumstances might be related to the activation of different signaling pathways. Studies have found that nuclear factor kappa-B (NF-κB) is widely activated during neurodegenerative diseases, and that NF-κB-activated astrocytes might represent harmful astrocytes that promote neurodegeneration [127, 128], which is in accordance with the discovery that A1 reactive astrocytes can exhibit NF-κB activation [9]. In comparison, the Janus kinase-signal transducer and activator of transcription (JAK-STAT3) signaling pathway might mediate the activation of A2 reactive astrocytes through the induction of ischemia [108]. This pathway plays an important role in regulating cell proliferation, differentiation, and growth, as well as some inflammatory functions [129]. STAT3 is not activated in the normal brain under physiological conditions, but many studies have confirmed that, after acute injury, JAK-STAT3 is activated in scar-forming reactive astrocytes [64, 129, 130], possibly because STAT3 targets genes that promote the cell cycle and inhibit apoptosis [108]. In addition, inositol 1,4,5-trisphosphate (IP_3)-dependent Ca^{2+} signaling and the downstream functions of N-cadherin have also been demonstrated to be required for normal reactive astrogliosis and neuroprotection after brain injury [131]. In contrast to the pathways introduced above which are associated with astrocyte activation, signaling mediated by β1-integrin has the opposite effect, which is necessary for astrocytes to maintain a mature, nonreactive state [132].

3 Biomarkers of Different Types of Neurodegenerative Diseases

Neurodegenerative diseases refer to a group of diseases that have common features of neurodegeneration, including degenerative diseases of the brain such as Alzheimer's disease (AD), Parkinson's disease (PD), and Huntington's disease (HD), motor neuron diseases such as amyotrophic lateral sclerosis (ALS) and some infectious diseases such as prion disease [2]. In addition to their similarities, these diseases also show individual characteristics. In the previous part of this chapter, we introduce common points of astrocyte activation and altered molecular expression in neurodegenerative diseases, and in this part, the peculiarities of some typical neurodegenerative diseases and their potential biomarkers for research and clinical use will be further discussed.

3.1 Alzheimer's Disease (AD)

AD is one of the most common neurodegenerative diseases in older adults, characterized by progressive cognitive dysfunction and behavioral disorders [133]. AD is pathologically marked by extracellular formation of amyloid plaques (APs) and intracellular formation of neurofibrillary tangles (NFTs), consisting of insoluble aggregated β-amyloid (Aβ) peptide and abnormally phosphorylated tau protein respectively [112, 134]. At present, it has been widely accepted that the pathogenesis and development of AD is closely related to both the neurons and glial cells including astrocytes and microglia. Here, we will focus on the changes and roles of astrocytes in AD.

One of the hallmarks of astrocyte activation in AD is the upregulation of intermediate filament, that is, an increase in the level of GFAP expression [135, 136], which serves as a typical biomarker of reactive astrocytes during neurodegeneration as discussed above. AD also causes metabolic dysfunction of astrocytes, such as the disruption of glutamate homeostasis, which in turn participates in the pathogenesis and development of AD by contributing to the earliest neuronal deficits [137, 138]. In transgenic models of AD, altered mRNA and/or protein expression of glutamate transporters and altered glutamate homeostasis have been reported [139, 140]. Studies have also found inverse relationship between the expression of GFAP, excitatory amino acid transporter 2 (EAAT2) and AD progression [78, 79]. Indirect data have revealed functional impairment of glutamate transporters due to oxidative damage, splice variants and altered solubility of EAAT2 [92–94], suggesting that the expression level of glutamate transporter is another possible biomarker for reactive astrocytes in AD.

The reduced glutamine synthesis further leads to a decrease in neuronal gamma-aminobutyric acid (GABA)-mediated inhibition [141], which may lead to the reduction of spike probability of granule cells in hippocampus of AD mice and memory impairment of these mice [142].

Besides typical markers of activated astrocytes such as GFAP and glutamate transporters, there are several other molecular expression changes that may serve as biomarkers for AD. As has been discussed above, one of the pathological characteristics of AD is the aggregation of insoluble Aβ that forms amyloid plaques. The increased level of Aβ is linked to the function of β-site APP cleaving enzyme 1 (BACE1) which can cleave β-amyloid precursor protein to form the insoluble form. Strong expression of BACE1 has also been confirmed in reactive astrocytes of AD patients [143]. Although the role of reactive astrocytes in AD has not been fully understood, it has already been discovered that reactive astrocytes are intimately associated with amyloid plaques or diffuse deposits of Aβ by surrounding them with dense layers of processes as if forming miniature scars around the plaques [14]. Interestingly, some studies have shown that reactive astrocytes in AD contain substantial amounts of Aβ [135, 144], including Aβ42 and its truncated forms, suggesting these reactive astrocytes may take up and degrade extracellular Aβ deposits and may play a protective role in the progression of AD [8, 145]. One of the receptors identified as participating in uptake and clearance of Aβ by astrocytes is the low-density lipoprotein receptor-related protein 1 (LRP1) [146, 147], which also contributes to the uptake of apoE and complexes of apoE-Aβ [17]. The increased expression of presenilin (PS) is believed to be closely related to the pathogenesis of AD and is also found in reactive astrocytes in sporadic AD cases [45, 148]. Therefore, detection of these molecules may help to identify astrocyte activation in AD.

Other well-identified molecular expression changes of reactive astrocytes in AD include: downregulation of AQP4 which can interfere with normal glymphatic flow [13, 149], alterations of calpain with AD progression in the long processes of interlaminar astrocytes as well as in astrocytes in white matter which is related to the dysregulation of calcium homeostasis [150–152].

3.2 Parkinson's Disease (PD)

PD, also known as paralysis agitans, is another kind of neurodegenerative disease commonly affecting the middle-aged and the elderly, characterized clinically by static tremor, bradykinesia, myotonia, and postural balance disorder [153]. PD is marked by loss and degeneration of dopaminergic neurons in the substantia nigra compacta (SNc) area. The pathogenesis of this disease is not completely understood, and mutations of some genes, including *PARK2, ATP13A2, PTEN, PINK1* and *DJ-1* can lead to PD-like symptoms [153, 154]. The activation of astrocytes is detected primarily in the SNc of PD patients [113]. Similar to AD, typical molecular alterations of astrocyte activation have also been found in PD, including the upregulation of GFAP [155], reduced expression of glutamate transporter-1 (GLT-1) and glutamate aspartate transporter (GLAST), altered expression of aquaporin-4 (AQP4) [80], and activation of JAK/STAT3 signaling pathway [156].

Apart from the above, activated astrocytes under PD conditions also have their distinct molecular characteristics. One of the pathological hallmarks of PD is the formation of intracellular eosinophilic inclusions, named Lewy bodies, consisting of α-synuclein, ubiquitin, and heat shock proteins. Recent studies have shown that subpopulations of astrocytes can express disease-related proteins including α-synuclein, parkin, and phosphorylated tau [157], which may participate in the pathogenesis and progression of PD. It has been speculated that during PD initiation, the accumulation of α-synuclein in activated astrocytes might induce the recruitment of activated microglia, which further attack neurons in certain brain regions, leading to onset of PD symptoms clinically [158]. In accordance with this, studies have found nonfibrillized α-synuclein in activated astrocytes which distributes more broadly than Lewy bodies at early stages of PD brain [159, 160], and that the level of α-synuclein aggregates correlates with the expansion of reactive astrogliosis [80]. Therefore, α-synuclein may serve as a useful biomarker for astrocyte activation of PD.

In the previous part, we discussed the different functions of NF-κB and JAK-STAT3 signaling pathways for the activation of astrocytes during neurodegeneration. In addition, studies have shown that in PD, Wnt1/Fzd-1/β-catenin signaling pathway also plays a key role in neuron-astrocyte interactions, thus being crucial for maintaining the health of PD neurons [3], which has been demonstrated by a study showing that inhibiting this signaling pathway in SN reactive astrocytes can lead to the impairment of mesencephalic neuronal survival and this effect can be rescued by pharmacological activation of β-catenin within the SN [161].

3.3 Huntington's Disease (HD)

Unlike AD and PD, the onset of HD has a much stronger relationship with heredity. HD is associated with the mutation of the gene *interesting transcript 15* (*IT15*), which is characterized by an abnormal increase of CAG repeating copies in the 5′-end coding area, resulting in abnormal function of the coded protein Huntingtin [162, 163]. As a result, HD is considered a common chromosomal dominant disease that often occurs in middle age, but it also affects children and the elderly. Pathologically, the hallmark of HD is the degeneration of the basal ganglia and cerebral cortex, which is clinically present as chronic progressive chorea, mental disorders, and dementia.

Reactive astrogliosis is found mainly in the striatum of HD patients; however, it also exists in other brain regions such as the frontal cortex in HD [164, 165]. The role of activated astrocytes in HD has also been confirmed by many studies, showing that activated astrocytes may contribute to pathogenesis of HD in many aspects. Like other neurodegenerative diseases, the metabolism of astrocytes changes significantly during disease progression, especially the "glutamate–glutamine cycle." On the one hand, a substantial decrease of EAAT2/GLT-1 in astrocytes has been found in an HD mouse model and postmortem human tissue [164, 166–168], which can result in impairment for glutamate transporting from extracellular space into astrocytic cytoplasm; on the other hand, astrocytes also show an increased expression level of pyruvate carboxylase [31], which is a critical enzyme for de novo synthesis of glutamate, leading to augmented glutamate production and high exocytotic release of this neurotransmitter from astrocytes [169]. In addition, GS expression in reactive astrocytes is reduced, further aggravating the dysregulation of this metabolic cycle [168]. Deficient Kir4.1 potassium ion channel in astrocytes has also been observed in different HD mouse models, which may be involved in neuronal damage and further contributes to the pathogenesis of HD [170, 171]. Detecting the expression of these markers may help identify the activation of astrocytes in HD.

Like other neurodegenerative diseases, astrocytes also participate in inflammatory responses in HD and this process has been demonstrated to be mediated through the NF-κB signaling pathway [172]. Moreover, reactive astrogliosis in HD may lead to reduction in pericyte coverage of cerebral blood vessels through an IκB kinase-dependent pathway, which can result in impairment of vascular reactivity and further contribute to disease progression [173].

HD is marked by the repeating mutation of *IT1* gene and abnormality of the related protein Huntingtin as has been discussed above. Several studies have revealed abnormal Huntingtin accumulating in the nuclei of astrocytes in animal models of AD [164, 174, 175], also suggesting the crucial role of reactive astrocytes in HD pathogenesis. Further findings have revealed that the toxicity of astrocytes depends on the size of polyglutamine repeats of Huntingtin, which is a result of the abnormal enlargement of CAG repeats of *IT1* gene [174–176], suggesting that the role of astrocytes in HD progression is closely related to the severity of gene mutation.

3.4 Amyotrophic Lateral Sclerosis (ALS)

ALS belongs to the category of motor neuron diseases, characterized by the impairment of both upper and lower motor neurons. Clinically, ALS patients present progressive and irreversible muscle weakness and dystrophy, which can eventually lead to respiratory failure and death [177]. There are many hypotheses about the etiology and pathogenesis of ALS, including genetic mechanisms, oxidative stress, neuronal excitotoxicity, autoimmune mechanisms, viral infections, and environmental factors, where astrocytes may be a central element and an integral player [31].

Like other neurodegenerative diseases, ALS has long been considered a neuron-autonomous disease. However, increasing evidences have suggested the glial-centric view [12]. Recent data revealed that at least in some types of ALS, the impairment of neurons is mediated by glial cells, including astrocytes [121, 178–180], microglia [181, 182] and oligodendrocytes [183]. It has been found that a small percentage of familial ALS cases are linked to mutations of *SOD1* gene [184], which can be detected in both neurons and astrocytes [185], suggesting the possible role for astrocytes in ALS pathogenesis. In a mouse model of ALS, motor neuron degeneration is induced by transplanting the precursors of mutant SOD1 astrocytes into the spinal cords, while transplantation of wild-type astrocyte precursors into the spinal cord of ALS models results in less motor neuron deaths [186]. This is probably due to the dysfunction of glutamate uptake by astrocytes, which leads to further neuronal excitotoxicity, as another study has revealed [187].

Reactive astrocytes are observed in vulnerable regions of ALS, and the activation level may be related to the neurodegeneration degree [3]. One of the major changes of molecular expression of activated astrocytes in ALS is the decreased level of glutamate transporters, namely, EAAT2/GLT-1. This is probably due to the production of truncated EAAT2/GLT-1 protein resulting from aberrant EAAT2 mRNA processing including intron retention and exon skipping [81, 86, 89, 188–191]. The decrease in normal EAAT2/GLT-1 level further results in increased neuronal excitotoxicity, which is considered to be a possible mechanism of ALS pathogenesis [184]. Based on this biomarker, the ceftriaxone drug approved by the FDA was found in a drug screen to stimulate the expression of glutamate transporters in astrocytes, and this drug has been confirmed to successfully slow down ALS progression in animal models [192].

Astrocytes also participate in inflammatory responses in ALS, as it has been found that several inflammatory mediators including IL-6 and TGF-β may play a role in astrocyte–neuron communication [52, 193]. And the transcriptions of a variety of inflammatory molecules show an obvious increase in astrocytes derived from both familial and sporadic ALS [178]. Besides, many signaling pathways are activated during ALS. For example, IFN-γ-induced reactive astrocytes may be toxic to neurons [194], possibly through STAT3-dependent signaling pathway, as it has been shown in vitro that STAT3 activation may lead to the recruitment of reactive astrocytes to motor neurons and activate the microglia response [195], and this toxicity can be attenuated by the use of STAT3 inhibitors [196].

Alterations of other molecular expression and signaling pathways are also found in ALS, which may contribute to the pathogenesis and progression of ALS, including impaired lactate transport [185] and persistent Ca^{2+} release and apoptosis through mGluR5-mediated glutamate signaling [197].

4 Methodology

In the previous sections, we discussed molecular expression alterations of activated astrocytes from several aspects, including their structural and functional changes, participation in neuroinflammatory process, mechanism of activation and their possible role in the pathogenesis in different types of neurodegenerative diseases, which may serve as potential biomarkers for the detection of astrocyte activation in neurodegeneration for research and clinical use. In this section, we will further provide some methods applied in biomarker studies for reference.

4.1 Immunopanning (IP) Method to Isolate Astrocytes

This procedure for studying astrocyte properties in vitro is to isolate astrocytes from other cell types of the CNS, as many key genes and proteins are present or expressed by different cell types; for example, astrocytes and microglia are often activated in concert and participate in neuroinflammation simultaneously [108]. The first attempt to isolate astrocytes in vitro was in the early 1980s [198], in which the isolated astrocytes were referred to as MD-astrocytes. However, this culture system has many shortcomings, which limit the efficiency and purity of astrocyte isolation to a large extent [108]. First, the isolation involves many procedures that extend at least one week. Second, these cultures are maintained in serum-containing media, which may irreversibly alter the properties of the cells [199]. For example, MD-astrocytes show a strong dividing capacity (divide every 1.4 days), whereas adult astrocytes in vivo exhibit limited division [200, 201]. Third, the acquisition of MD-astrocytes can only be from neonatal brain, so that these cells are much more similar to radial glial and astrocyte progenitor cells instead of mature adult cells, as has been demonstrated by gene profiling studies [117, 199]. Nevertheless, almost all studies on the function of astrocytes have been based on this method until a better culture preparation system was established [202–204].

In order to overcome these limitations, a new isolation and culture method based on immunopanning was established in 2011 by Foo et al. [199]. In this isolation and culture system for astrocytes, known as IP-astrocyte isolation, cortices from rats were dissected and the meninges were removed at first, after which the tissue was enzymatically dissociated with papain enzyme then mechanically dissociated to produce single cells. Then these single cells pass over successive negative panning plates to rid the cell suspension of microglia, endothelial cells, oligodendrocyte precursor cells before passing the suspension over a positive selection plate with ITGB5 antibody for astrocytes (for more details, please refer to Foo et al., 2011) [199]. These astrocytes undergo apoptosis in culture; however, this study has shown that adding an appropriate concentration of heparin-binding epidermal growth factor-like growth factor (HBEGF) is effective in promoting the survival of astrocytes in vitro [199].

Compared to the MD-astrocyte method, immunopanning isolation has many advantages. Firstly, prospective purification ensures the representativeness of the selected astrocytes of the whole population, avoiding the selection of a minor subset. Secondly, the whole selection procedures can be completed in a day, which greatly improve the efficiency. Thirdly, the establishment of a defined, serum-free medium and the use of HBEGF enable the long-term survival of purified astrocytes in vitro. Last but not least, IP-astrocytes are much more akin to astrocytes in vivo compared with MD-astrocytes [199]. Further study has revealed that the three-dimensional polymer matrix is more appropriate for astrocyte growing than two-dimensional monolayer [205], leading to the conclusion that the immunopanning method of three-dimensional serum-free culture might be of great importance for in vitro astrocyte analysis [199].

Based on this method, subtypes of astrocytes and their activated forms can be further purified and induced, and a recent study showed that by growing purified astrocytes for 6 days, then treated for 24 h with proper concentration of Il-1α, TNF-α and C1q, A1 reactive astrocytes can be generated [23].

4.2 Three-Dimensional Culture of Astrocytes

After acquisition of purified astrocytes, the next essential step is to create an appropriate environment for their survival. As has been discussed above, three-dimensional polymer matrix (Bioactive3D) proves to be a better culture system than the standard two-dimensional monolayer [205].

Compared with traditional 2D culture system, the highlight of Bioactive 3D is the use of nanofibers. The whole process can be divided into three steps. The first step is nanofiber preparation. The solutions for electrospinning are prepared by mixing 11 wt% biocompatible polyether-based polyurethane (PU) resin in a 60:40 mixture of tetrahydrofuran (THF) and N,N-dimethylformamide (DMF). The solution is mixed for 24 h and transferred to a syringe with a metal 21G cannula for electrospinning. The second procedure is nanofiber coating. Nanofibers are treated with 70% ethanol, washed in distilled water (dH$_2$O). Bioactive 3D scaffolds are incubated with poly-L-ornithine (PDL, 10 μg/mL in dH$_2$O with 285 μL/cm^2 surface) for 2 h followed by three wash steps in dH$_2$O and subsequently with laminin (5 μg/mL in Dulbecco's phosphate buffered saline with 285 μL/cm^2 surface) overnight.

All incubation steps are conducted in a humid atmosphere at 37 °C containing 5% CO_2. Finally, astrocytes isolated and purified from tissues are cultured on this Bioactive 3D medium. For more information about the whole process of 3D culture system, please refer to [205], [206], and [207].

When astrocytes are transferred from in vivo environment to a 2D culture medium, they lose their morphological and biochemical features, which are obstacles to the study of astrocytes in vitro. However, in Bioactive 3D culture system, these features of in vivo astrocytes are well preserved [207]: for instance, the level of GFAP expression is much less upregulated in these astrocytes than 2D cultured astrocytes, and they are more morphologically complex, exhibiting many branching processes [205]. Therefore, the Bioactive 3D culture system may be another fundamental research method of analyzing astrocytes in vitro.

4.3 Immunohisto-chemistry Staining

Immunohistochemistry staining, especially immunofluorescent staining, has been widely accepted and applied as an efficient method to detect specific molecular targets in certain tissues both in research studies and clinical practice. Immunohistochemistry is based on the principle of antigen and antibody-specific binding, which gives the results of high specificity, sensitivity and positioning accuracy. In addition, this method is of great advantages for its convenient procedures and short time consuming, therefore it is particularly welcomed for clinical laboratory tests.

In order to detect activated astrocytes by immunohistochemistry staining, the targeted biomarker must possess some characteristics as follows: (1) The biomarker should be highly specific to astrocytes, so that astrocytes can be easily distinguished from other CNS cell types; (2) The expression level of the marked molecule should be significantly altered before and after activation in order to effectively detect the "active form" of astrocytes. (3) The biomarker should have enough expression in astrocytes, so that its expression alterations can be successfully visualized by microscopic technique. Considering these three prerequisites, GFAP appears to be an ideal marker for activated astrocytes, since it is relatively specifically expressed by astrocytes and even differentially expressed in various subtypes of astrocytes [117], and its expression shows a detectable increase after astrocyte activation [14, 38, 61, 74, 78, 79, 116].

For GFAP immunofluorescence staining, animals were firstly perfused with phosphate buffer saline (PBS) followed by 4% paraformaldehyde (PFA). The brains were then carefully dissected and kept in 4% PFA overnight followed by 30% sucrose solution. After this, the brains were sectioned into slices then subjected to immunofluorescence staining of GFAP (appropriate dilution ratio) and fluorescent-tagged secondary antibody. Nuclei were counterstained with 4′,6-diamidino-2-phenylindole (DAPI). Finally, immunostaining results were analyzed with a fluorescence microscope interfaced with a digital charge-coupled device camera and an image analysis system [208].

However, the use of GFAP as a biomarker for astrocyte activation also has some limitations, involving its cell specificity and expression variability (please *see* Subheading 2.1) [61, 76]. Therefore, other markers, including vimentin and EAAT2/GLT-1, may play a complementary role in the immunostaining for GFAP.

4.4 Single-Cell Isolation and Whole-Genome Analyses

As we have discussed in previous parts, activation of astrocytes in neurodegenerative diseases is a complicated process involving expression changes of many molecules and activation of various signaling pathways. In addition, even seemingly homogeneous cell populations either in vivo or in vitro may show considerable heterogeneity in expression patterns [209]. Therefore, it is necessary to find a comprehensive and quantifiable analysis method of changes in the expression of these genes, especially for research studies. Over the past few years, the spectacular progress in single-cell isolation and whole-genome analyses, namely "single-cell biology" [210], have made it possible to map alterations in gene expressions of many types of cells in parallel [13].

The single-cell biology relies on reverse transcription (RT) of RNA to complementary DNA (cDNA) followed by amplification by PCR or in vitro transcription, then deep sequencing will be completed based on the amplified cDNA [210]. This method involves a series of successive procedures. Firstly, single-cell samples were prepared, and single cells were obtained by using a microcapillary pipette or via fluorescence activating cell sorter (FACS). Each single cell was placed into a PCR tube containing cell-lysis buffer, oligo-dT primer, and deoxynucleotide (dNTP) mix with a ratio of 2:1:1, which then underwent some procedures to hybridize oligo-dT primer to the poly(A) tail of all the mRNA molecules. After that, reverse transcription was performed to generate cDNA, followed by PCR preamplification and purification, then the quality of the cDNA was checked according to the cDNA library. Tagmentation reaction was then carried out by using the Illumina Nextera XT DNA sample preparation kit and Tn5 transposase was stripped out the tagmented DNA. After another enrichment PCR, the sample was purified and quality-checked in the final cDNA library. Finally, deep DNA sequencing and data analysis could be executed. All the experiments mentioned above must be performed under a UV-sterilized hood with laminar flow, and all the surfaces must be free from RNase to prevent degradation of RNA and from DNA to prevent cross-contamination from previous samples. For more details about the procedures of "single-cell biology," please refer to the protocol by Simone et al. [211].

The advantages of this newly developed technique are considerable. By paying a fraction of the cost of currently available commercial kits, analyses of hundreds of cells can be completed

[211]. Furthermore, it allows for high multiplexing, pooling and sequencing of up to 96 samples on a single lane of an Illumina sequencer [211]. It perfectly fits the agnosticism with respect to cell type and can evaluate information on the variation in responses over a subpopulation of cells [13] so that meticulous isolation and comprehensive analysis of activated astrocytes in neurodegenerative diseases can be achieved. Base on this technique, hundreds of single-cell genomic expression data from mouse hippocampus and human cortex have been successfully resolved [212, 213]. One of the limitations of this method is that it is selective for polyadenylated (polyA) RNA, precluding analyses of RNA without polyA [211]. In addition, it only records changes in the transcriptional level of molecules [13].

4.5 Molecular Imaging Techniques

In clinical practice, the molecular imaging technique is a powerful tool for early diagnosis and treatment of diseases. Molecular imaging, which is a combination of molecular biology and imageology, refers to the visualization, characterization, and measurement of biological processes at the molecular and cellular levels of humans and other living systems [214]. One of the most widely used methodologies of molecular imaging is nuclear medical imaging, by which distribution and kinetics of a radiolabeled molecular probe can be measured using positron emission tomography (PET) and single-photon emission computed tomography (SPECT) [214]. Nuclear medical imaging, while cooperating with traditional imaging techniques such as computed tomography (CT), can connect anatomical structures with metabolic functions in high level of sensitivity and accuracy. Therefore, it is of great significance to visualize tissues of complicated structures and functions, especially the brain. For example, the definite diagnosis of AD previously could only be made through postmortem pathology, however, recent advances in molecular imaging of Aβ plaques and tau proteins have made it possible to diagnose and track AD progression in a timely manner [215].

Apart from its clinical use, molecular imaging has also proved to be of great potential help to basic research studies. Combined with specific biomarkers, it is useful in detecting and tracing molecular alterations in different cell types in neurodegenerative diseases. For example, molecular alterations of PD and atypical parkinsonisms have been tracked by the use of varieties of PD markers including dopamine transporter (DAT) and vesicular monoamine transporter type 2 (VMAT2) [216]. Back to our topic of astrocyte activation in neurodegenerative diseases, molecular imaging can provide new insights into analyzing molecular changes in activated astrocytes and determining their distribution in the CNS. Using radiolabeled $[^{11}C]$-L-deprenyl, a monoamine oxidase B (MAO-B) inhibitor present in astrocytes, one study revealed that the expression level of MAO-B is significantly upregulated in brain tissues of

AD patients [106]. This provides a good example for studying activated astrocytes and their molecular changes by specific biomarkers with the help of molecular imaging. Although the application of this newly developed technique in the study of CNS is still in its infancy, we believe that with the rapid development of this method, it will play an increasingly important role in the research of neurodegenerative diseases.

5 Future Perspectives

With the understanding of neurodegenerative diseases going deeper, the long-lasting neuron-centric view of neurodegeneration has been strongly challenged, and more and more attention has been paid to the role of glial cells. Considering many molecular changes during astrocyte activation in neurodegeneration, it becomes attractive to identify suitable biomarkers of activated astrocytes in both basic research study and clinical practice. These biomarkers involve various aspects of activated astrocytes, including structural and functional characteristics, and their participation in neuroinflammatory responses that can be detected and analyzed by different techniques as has been introduced in detail earlier in this chapter.

In basic research, it has become common to identify and analyze activated astrocytes according to certain biomarkers to describe their alterations and study their role in the pathogenesis of neurodegenerative diseases. Immunohistochemical staining is the most used method due to its sensitivity and convenience. However, as we have discussed earlier, typical biomarkers for immunostaining of activated astrocytes such as GFAP still have many shortcomings that may influence the specificity and accuracy of experiments. Therefore, finding better biomarkers or combining different kinds of biomarkers to achieve complementary effects is significant to achieve a better understanding of astrocyte activation in future studies. This requires more diversified and refined techniques, including a better method of isolating and culturing astrocytes in vitro. Single-cell whole-genome analysis has shown great potential for comprehensive detection of alterations in cellular and molecular levels. Molecular imaging technique makes it possible to study combinations of structure and function and allows for more precise and accurate research on activated astrocytes. The discovery of better biomarkers and the improvement in methodology will certainly lead to a deeper understanding of the functions of astrocytes in neurodegenerative diseases in the future.

As for clinical practice, the identification of appropriate biomarkers for activated astrocytes can also help to make early diagnosis and detect new therapeutic targets for neurodegenerative diseases. Nuclear medical imaging such as PET and SPECT, which is based on mechanisms of molecular biology, has already been widely applied to the early diagnosis of neurodegenerative diseases. Currently, the use of clinical targets of radiolabeled molecules in the neurodegenerative brain, such as Aβ plaques and tau protein in AD, remains limited. The accuracy, timeliness and sensitivity of disease diagnosis will be greatly improved if it becomes possible to introduce more molecules related to certain cell types of neurodegenerative diseases, and if the relationship between functional impairments and pathological changes can be better understood. More intriguingly, biomarkers of these activated astrocytes may shed light on treatment of neurodegenerative diseases, which is still in a rather difficult stage, by serving as potential targets for drugs. As evidence has begun to accumulate that astrocytes participate actively in the pathogenesis of neurodegeneration, astrocytes may become a hot candidate for new drug research and development in the future. One good example is ceftriaxone, a FDA-approved drug that can stimulate the expression of glutamate transporters by astrocytes, was screened out and has been confirmed to successfully slow the progression of ALS in an animal model [192]. More analyses of astrocytes based on these biomarkers, and deeper understanding of the molecular pathogenesis of neurodegenerative diseases will bring hope to individual patients and their families, and the whole society as well.

References

1. Whalley K (2014) Neurodegenerative disease: propagating pathology. Nat Rev Neurosci 15 (9):565. https://doi.org/10.1038/nrn3802

2. Montie HL, Durcan TM (2013) The cell and molecular biology of neurodegenerative diseases: an overview. Front Neurol 4:194. https://doi.org/10.3389/fneur.2013.00194

3. Li K, Li J, Zheng J, Qin S (2019) Reactive astrocytes in neurodegenerative diseases. Aging Dis 10(3):664–675. https://doi.org/10.14336/AD.2018.0720

4. Wyss-Coray T (2016) Ageing, neurodegeneration and brain rejuvenation. Nature 539 (7628):180–186. https://doi.org/10.1038/nature20411

5. Fakhoury M (2018) Microglia and astrocytes in Alzheimer's disease: implications for therapy. Curr Neuropharmacol 16(5):508–518. https://doi.org/10.2174/1570159X15666170720095240

6. Jung CK, Keppler K, Steinbach S, Blazquez-Llorca L, Herms J (2015) Fibrillar amyloid plaque formation precedes microglial activation. PLoS One 10(3):e0119768. https://doi.org/10.1371/journal.pone.0119768

7. Ries M, Sastre M (2016) Mechanisms of abeta clearance and degradation by glial cells. Front Aging Neurosci 8:160. https://doi.org/10.3389/fnagi.2016.00160

8. Wyss-Coray T, Loike JD, Brionne TC, Lu E, Anankov R, Yan F et al (2003) Adult mouse astrocytes degrade amyloid-beta in vitro and in situ. Nat Med 9(4):453–457. https://doi.org/10.1038/nm838

9. Lian H, Yang L, Cole A, Sun L, Chiang AC, Fowler SW et al (2015) NFkappaB-activated astroglial release of complement C3 compromises neuronal morphology and function associated with Alzheimer's disease. Neuron 85(1):101–115. https://doi.org/10.1016/j.neuron.2014.11.018

10. Glass CK, Saijo K, Winner B, Marchetto MC, Gage FH (2010) Mechanisms underlying inflammation in neurodegeneration. Cell 140(6):918–934. https://doi.org/10.1016/j.cell.2010.02.016

11. Perry VH, Nicoll JA, Holmes C (2010) Microglia in neurodegenerative disease. Nat Rev Neurol 6(4):193–201. https://doi.org/10.1038/nrneurol.2010.17

12. Pekny M, Pekna M (2014) Astrocyte reactivity and reactive astrogliosis: costs and benefits. Physiol Rev 94(4):1077–1098. https://doi.org/10.1152/physrev.00041.2013

13. De Strooper B, Karran E (2016) The cellular phase of Alzheimer's disease. Cell 164 (4):603–615. https://doi.org/10.1016/j.cell.2015.12.056

14. Sofroniew MV, Vinters HV (2010) Astrocytes: biology and pathology. Acta Neuropathol 119(1):7–35. https://doi.org/10.1007/s00401-009-0619-8

15. Muller HW, Matthiessen HP, Schmalenbach C, Schroeder WO (1991) Glial support of CNS neuronal survival, neurite growth and regeneration. Restor Neurol Neurosci 2(4):229–232. https://doi.org/10.3233/RNN-1991-245610

16. Yang D, Peng C, Li X, Fan X, Li L, Ming M et al (2008) Pitx3-transfected astrocytes secrete brain-derived neurotrophic factor and glial cell line-derived neurotrophic factor and protect dopamine neurons in mesencephalon cultures. J Neurosci Res 86(15):3393–3400. https://doi.org/10.1002/jnr.21774

17. Gengatharan A, Bammann RR, Saghatelyan A (2016) The role of astrocytes in the generation, migration, and integration of new neurons in the adult olfactory bulb. Front Neurosci 10:149. https://doi.org/10.3389/fnins.2016.00149

18. Theodosis DT, Piet R, Poulain DA, Oliet SH (2004) Neuronal, glial and synaptic remodeling in the adult hypothalamus: functional consequences and role of cell surface and extracellular matrix adhesion molecules. Neurochem Int 45(4):491–501. https://doi.org/10.1016/j.neuint.2003.11.003

19. Inyushin M, Kucheryavykh LY, Kucheryavykh YV, Nichols CG, Buono RJ, Ferraro TN et al (2010) Potassium channel activity and glutamate uptake are impaired in astrocytes of seizure-susceptible DBA/2 mice. Epilepsia 51(9):1707–1713. https://doi.org/10.1111/j.1528-1167.2010.02592.x

20. Gadea A, Schinelli S, Gallo V (2008) Endothelin-1 regulates astrocyte proliferation and reactive gliosis via a JNK/c-Jun signaling pathway. J Neurosci 28(10):2394–2408. https://doi.org/10.1523/JNEUROSCI.5652-07.2008

21. Levison SW, Jiang FJ, Stoltzfus OK, Ducceschi MH (2000) IL-6-type cytokines enhance epidermal growth factor-stimulated astrocyte proliferation. Glia 32(3):328–337. https://doi.org/10.1002/1098-1136(200012)32:3<328::aid-glia110>3.0.co;2-7

22. Rodriguez JJ, Yeh CY, Terzieva S, Olabarria M, Kulijewicz-Nawrot M, Verkhratsky A (2014) Complex and region-specific changes in astroglial markers in the aging brain. Neurobiol Aging 35(1):15–23. https://doi.org/10.1016/j.neurobiolaging.2013.07.002

23. Liddelow SA, Guttenplan KA, Clarke LE, Bennett FC, Bohlen CJ, Schirmer L et al (2017) Neurotoxic reactive astrocytes are induced by activated microglia. Nature 541 (7638):481–487. https://doi.org/10.1038/nature21029

24. Forster S, Grimmer T, Miederer I, Henriksen G, Yousefi BH, Graner P et al (2012) Regional expansion of hypometabolism in Alzheimer's disease follows amyloid deposition with temporal delay. Biol Psychiatry 71(9):792–797. https://doi.org/10.1016/j.biopsych.2011.04.023

25. Yao J, Rettberg JR, Klosinski LP, Cadenas E, Brinton RD (2011) Shift in brain metabolism in late onset Alzheimer's disease: implications for biomarkers and therapeutic interventions. Mol Asp Med 32(4–6):247–257. https://doi.org/10.1016/j.mam.2011.10.005

26. Matarin M, Salih DA, Yasvoina M, Cummings DM, Guelfi S, Liu W et al (2015) A genome-wide gene-expression analysis and database in transgenic mice during development of amyloid or tau pathology. Cell Rep 10 (4):633–644. https://doi.org/10.1016/j.celrep.2014.12.041

27. Karch CM, Cruchaga C, Goate AM (2014) Alzheimer's disease genetics: from the bench to the clinic. Neuron 83(1):11–26. https://doi.org/10.1016/j.neuron.2014.05.041

28. Garwood CJ, Pooler AM, Atherton J, Hanger DP, Noble W (2011) Astrocytes are important mediators of Abeta-induced neurotoxicity and tau phosphorylation in primary culture. Cell Death Dis 2:e167. https://doi.org/10.1038/cddis.2011.50

29. Jana A, Pahan K (2010) Fibrillar amyloid-beta-activated human astroglia kill primary human neurons via neutral sphingomyelinase: implications for Alzheimer's disease. J Neurosci 30(38):12676–12689. https://doi.org/10.1523/JNEUROSCI.1243-10.2010

30. Akiyama H, Barger S, Barnum S, Bradt B, Bauer J, Cole GM et al (2000) Inflammation and Alzheimer's disease. Neurobiol Aging 21 (3):383–421. https://doi.org/10.1016/s0197-4580(00)00124-x

31. Pekny M, Pekna M, Messing A, Steinhauser C, Lee JM, Parpura V et al (2016) Astrocytes: a central element in neurological diseases. Acta Neuropathol 131

(3):323–345. https://doi.org/10.1007/s00401-015-1513-1

32. Burda JE, Sofroniew MV (2014) Reactive gliosis and the multicellular response to CNS damage and disease. Neuron 81(2):229–248. https://doi.org/10.1016/j.neuron.2013.12.034

33. Pekny M, Wilhelmsson U, Pekna M (2014) The dual role of astrocyte activation and reactive gliosis. Neurosci Lett 565:30–38. https://doi.org/10.1016/j.neulet.2013.12.071

34. Seifert G, Schilling K, Steinhauser C (2006) Astrocyte dysfunction in neurological disorders: a molecular perspective. Nat Rev Neurosci 7(3):194–206. https://doi.org/10.1038/nrn1870

35. Verkhratsky A, Sofroniew MV, Messing A, de Lanerolle NC, Rempe D, Rodriguez JJ et al (2012) Neurological diseases as primary gliopathies: a reassessment of neurocentrism. ASN Neuro 4(3). https://doi.org/10.1042/AN20120010

36. Vasile F, Dossi E, Rouach N (2017) Human astrocytes: structure and functions in the healthy brain. Brain Struct Funct 222(5):2017–2029

37. Taft JR, Vertes RP, Perry GW (2005) Distribution of GFAP+ astrocytes in adult and neonatal rat brain. Int J Neurosci 115(9):1333–1343. https://doi.org/10.1080/00207450590934570

38. Guillamon-Vivancos T, Gomez-Pinedo U, Matias-Guiu J (2015) Astrocytes in neurodegenerative diseases (I): function and molecular description. Neurologia 30(2):119–129. https://doi.org/10.1016/j.nrl.2012.12.007

39. Reeves SA, Helman LJ, Allison A, Israel MA (1989) Molecular cloning and primary structure of human glial fibrillary acidic protein. Proc Natl Acad Sci U S A 86(13):5178–5182. https://doi.org/10.1073/pnas.86.13.5178

40. Condorelli DF, Nicoletti VG, Barresi V, Conticello SG, Caruso A, Tendi EA et al (1999) Structural features of the rat GFAP gene and identification of a novel alternative transcript. J Neurosci Res 56(3):219–228. https://doi.org/10.1002/(SICI)1097-4547(19990501)56:3<219::AID-JNR1>3.0.CO;2-2

41. Zelenika D, Grima B, Brenner M, Pessac B (1995) A novel glial fibrillary acidic protein mRNA lacking exon 1. Brain Res Mol Brain Res 30(2):251–258. https://doi.org/10.1016/0169-328x(95)00010-p

42. Roelofs RF, Fischer DF, Houtman SH, Sluijs JA, Van Haren W, Van Leeuwen FW et al

(2005) Adult human subventricular, subgranular, and subpial zones contain astrocytes with a specialized intermediate filament cytoskeleton. Glia 52(4):289–300. https://doi.org/10.1002/glia.20243

43. Nielsen AL, Holm IE, Johansen M, Bonven B, Jorgensen P, Jorgensen AL (2002) A new splice variant of glial fibrillary acidic protein, GFAP epsilon, interacts with the presenilin proteins. J Biol Chem 277(33):29983–29991. https://doi.org/10.1074/jbc.M112121200

44. Blechingberg J, Holm IE, Nielsen KB, Jensen TH, Jorgensen AL, Nielsen AL (2007) Identification and characterization of GFAPkappa, a novel glial fibrillary acidic protein isoform. Glia 55(5):497–507. https://doi.org/10.1002/glia.20475

45. Hol EM, Roelofs RF, Moraal E, Sonnemans MA, Sluijs JA, Proper EA et al (2003) Neuronal expression of GFAP in patients with Alzheimer pathology and identification of novel GFAP splice forms. Mol Psychiatry 8(9):786–796. https://doi.org/10.1038/sj.mp.4001379

46. van den Berge SA, Middeldorp J, Zhang CE, Curtis MA, Leonard BW, Mastroeni D et al (2010) Longterm quiescent cells in the aged human subventricular neurogenic system specifically express GFAP-delta. Aging Cell 9(3):313–326. https://doi.org/10.1111/j.1474-9726.2010.00556.x

47. Doetsch F, Caille I, Lim DA, Garcia-Verdugo JM, Alvarez-Buylla A (1999) Subventricular zone astrocytes are neural stem cells in the adult mammalian brain. Cell 97(6):703–716. https://doi.org/10.1016/s0092-8674(00)80783-7

48. Sanai N, Tramontin AD, Quinones-Hinojosa A, Barbaro NM, Gupta N, Kunwar S et al (2004) Unique astrocyte ribbon in adult human brain contains neural stem cells but lacks chain migration. Nature 427(6976):740–744. https://doi.org/10.1038/nature02301

49. Quinones-Hinojosa A, Sanai N, Soriano-Navarro M, Gonzalez-Perez O, Mirzadeh Z, Gil-Perotin S et al (2006) Cellular composition and cytoarchitecture of the adult human subventricular zone: a niche of neural stem cells. J Comp Neurol 494(3):415–434. https://doi.org/10.1002/cne.20798

50. Abbott NJ, Ronnback L, Hansson E (2006) Astrocyte-endothelial interactions at the blood-brain barrier. Nat Rev Neurosci 7(1):41–53

51. Ballabh P, Braun A, Nedergaard M (2004) The blood-brain barrier: an overview:

structure, regulation, and clinical implications. Neurobiol Dis 16(1):1–13. https://doi.org/10.1016/j.nbd.2003.12.016

52. Allaman I, Belanger M, Magistretti PJ (2011) Astrocyte-neuron metabolic relationships: for better and for worse. Trends Neurosci 34 (2):76–87. https://doi.org/10.1016/j.tins.2010.12.001

53. Schousboe A, Scafidi S, Bak LK, Waagepetersen HS, McKenna MC (2014) Glutamate metabolism in the brain focusing on astrocytes. Adv Neurobiol 11:13–30. https://doi.org/10.1007/978-3-319-08894-5_2

54. Simard M, Nedergaard M (2004) The neurobiology of glia in the context of water and ion homeostasis. Neuroscience 129(4):877–896. https://doi.org/10.1016/j.neuroscience.2004.09.053

55. Zador Z, Stiver S, Wang V, Manley GT (2009) Role of aquaporin-4 in cerebral edema and stroke. Handb Exp Pharmacol 190:159–170. https://doi.org/10.1007/978-3-540-79885-9_7

56. Nedergaard M, Ransom B, Goldman SA (2003) New roles for astrocytes: redefining the functional architecture of the brain. Trends Neurosci 26(10):523–530. https://doi.org/10.1016/j.tins.2003.08.008

57. Barres BA (2008) The mystery and magic of glia: a perspective on their roles in health and disease. Neuron 60(3):430–440

58. Borjabad A, Volsky DJ (2012) Common transcriptional signatures in brain tissue from patients with HIV-associated neurocognitive disorders, Alzheimer's disease, and multiple sclerosis. J Neuroimmune Pharmacol 7 (4):914–926. https://doi.org/10.1007/s11481-012-9409-5

59. Cotto B, Natarajaseenivasan K, Langford D (2019) Astrocyte activation and altered metabolism in normal aging, age-related CNS diseases, and HAND. J Neurovirol 25 (5):722–733. https://doi.org/10.1007/s13365-019-00721-6

60. Wilhelmsson U, Bushong EA, Price DL, Smarr B, Phung V, Terada M et al (2006) Redefining the concept of reactive astrocytes as cells that remain within their unique domains upon reaction to injury. Proc Natl Acad Sci U S A 103(46):17513–17518

61. Sofroniew MV (2009) Molecular dissection of reactive astrogliosis and glial scar formation. Trends Neurosci 32(12):638–647. https://doi.org/10.1016/j.tins.2009.08.002

62. Silver J, Miller JH (2004) Regeneration beyond the glial scar. Nat Rev Neurosci 5

(2):146–156. https://doi.org/10.1038/nrn1326

63. Bundesen LQ, Scheel TA, Bregman BS, Kromer LF (2003) Ephrin-B2 and EphB2 regulation of astrocyte-meningeal fibroblast interactions in response to spinal cord lesions in adult rats. J Neurosci 23(21):7789–7800

64. Herrmann JE, Imura T, Song B, Qi J, Ao Y, Nguyen TK et al (2008) STAT3 is a critical regulator of astrogliosis and scar formation after spinal cord injury. J Neurosci 28 (28):7231–7243

65. Bush TG, Puvanachandra N, Horner CH, Polito A, Ostenfeld T, Svendsen CN et al (1999) Leukocyte infiltration, neuronal degeneration, and neurite outgrowth after ablation of scar-forming, reactive astrocytes in adult transgenic mice. Neuron 23 (2):297–308. https://doi.org/10.1016/s0896-6273(00)80781-3

66. Voskuhl RR, Peterson RS, Song B, Ao Y, Morales LB, Tiwari-Woodruff S et al (2009) Reactive astrocytes form scar-like perivascular barriers to leukocytes during adaptive immune inflammation of the CNS. J Neurosci 29(37):11511–11522. https://doi.org/10.1523/JNEUROSCI.1514-09.2009

67. Hofmann SL, Das AK, Lu J, Wisniewski KE, Gupta P (2001) Infantile neuronal ceroid lipofuscinosis:no longer just a 'Finnish' disease. Eur J Paediatr Neurol 5:47–51

68. Hofmann SL, Das AK, Yi W, Lu JY, Wisniewski KE (1999) Genotype–phenotype correlations in neuronal ceroid lipofuscinosis due to palmitoyl-protein thioesterase deficiency. Mol Genet Metab 66(4):234–239

69. Vesa J, Hellsten E, Verkruyse LA, Camp LA, Rapola J, Santavuori P et al (1995) Mutations in the palmitoyl protein thioesterase gene causing infantile neuronal ceroid lipofuscinosis. Nature 376(6541):584–587. https://doi.org/10.1038/376584a0

70. Kielar C, Maddox L, Bible E, Pontikis CC, Macauley SL, Griffey MA et al (2007) Successive neuron loss in the thalamus and cortex in a mouse model of infantile neuronal ceroid lipofuscinosis. Neurobiol Dis 25 (1):150–162. https://doi.org/10.1016/j.nbd.2006.09.001

71. Macauley SL, Wozniak DF, Kielar C, Tan Y, Cooper JD, Sands MS (2009) Cerebellar pathology and motor deficits in the palmitoyl protein thioesterase 1-deficient mouse. Exp Neurol 217(1):124–135. https://doi.org/10.1016/j.expneurol.2009.01.022

72. Brenner M, Johnson AB, Boespflug-Tanguy-O, Rodriguez D, Goldman JE, Messing A

(2001) Mutations in GFAP, encoding glial fibrillary acidic protein, are associated with Alexander disease. Nat Genet 27 (1):117–120. https://doi.org/10.1038/83679

73. Messing A, LaPash Daniels CM, Hagemann TL (2010) Strategies for treatment in Alexander disease. Neurotherapeutics 7 (4):507–515. https://doi.org/10.1016/j.nurt.2010.05.013

74. Kamphuis W, Middeldorp J, Kooijman L, Sluijs JA, Kooi EJ, Moeton M et al (2014) Glial fibrillary acidic protein isoform expression in plaque related astrogliosis in Alzheimer's disease. Neurobiol Aging 35 (3):492–510. https://doi.org/10.1016/j.neurobiolaging.2013.09.035

75. Braak H, Braak E (1991) Neuropathological stageing of Alzheimer-related changes. Acta Neuropathol 82(4):239–259. https://doi.org/10.1007/BF00308809

76. Kriegstein A, Alvarez-Buylla A (2009) The glial nature of embryonic and adult neural stem cells. Annu Rev Neurosci 32:149–184. https://doi.org/10.1146/annurev.neuro.051508.135600

77. Lutz SE, Zhao Y, Gulinello M, Lee SC, Raine CS, Brosnan CF (2009) Deletion of astrocyte connexins 43 and 30 leads to a dysmyelinating phenotype and hippocampal CA1 vacuolation. J Neurosci 29(24):7743–7752. https://doi.org/10.1523/JNEUROSCI.0341-09.2009

78. Simpson JE, Ince PG, Lace G, Forster G, Shaw PJ, Matthews FE et al (2010) Astrocyte phenotype in relation to Alzheimer-type pathology in the ageing brain. Neurobiol Aging 31(4):578–590

79. Simpson JE, Ince PG, Shaw PJ, Heath PR, Raman R, Garwood CJ et al (2011) Microarray analysis of the astrocyte transcriptome in the aging brain: relationship to Alzheimer's pathology and APOE genotype. Neurobiol Aging 32(10):1795–1807. https://doi.org/10.1016/j.neurobiolaging.2011.04.013

80. Gu XL, Long CX, Sun L, Xie C, Lin X, Cai H (2010) Astrocytic expression of Parkinson's disease-related A53T alpha-synuclein causes neurodegeneration in mice. Mol Brain 3:12. https://doi.org/10.1186/1756-6606-3-12

81. Lin CL, Bristol LA, Jin L, Dykes-Hoberg M, Crawford T, Clawson L et al (1998) Aberrant RNA processing in a neurodegenerative disease: the cause for absent EAAT2, a glutamate transporter, in amyotrophic lateral sclerosis. Neuron 20(3):589–602. https://doi.org/10.1016/s0896-6273(00)80997-6

82. Maragakis NJ, Dykes-Hoberg M, Rothstein JD (2004) Altered expression of the glutamate transporter EAAT2b in neurological disease. Ann Neurol 55(4):469–477. https://doi.org/10.1002/ana.20003

83. Alexander GM, Deitch JS, Seeburger JL, Del Valle L, Heiman-Patterson TD (2000) Elevated cortical extracellular fluid glutamate in transgenic mice expressing human mutant (G93A) Cu/Zn superoxide dismutase. J Neurochem 74(4):1666–1673. https://doi.org/10.1046/j.1471-4159.2000.0741666.x

84. Yang Y, Gozen O, Vidensky S, Robinson MB, Rothstein JD (2010) Epigenetic regulation of neuron-dependent induction of astroglial synaptic protein GLT1. Glia 58(3):277–286. https://doi.org/10.1002/glia.20922

85. Li K, Hala TJ, Seetharam S, Poulsen DJ, Wright MC, Lepore AC (2015) GLT1 overexpression in SOD1(G93A) mouse cervical spinal cord does not preserve diaphragm function or extend disease. Neurobiol Dis 78:12–23. https://doi.org/10.1016/j.nbd.2015.03.010

86. Bristol LA, Rothstein JD (1996) Glutamate transporter gene expression in amyotrophic lateral sclerosis motor cortex. Ann Neurol 39 (5):676–679. https://doi.org/10.1002/ana.410390519

87. Flowers JM, Powell J, Leigh PN, Andersen PM, Shaw C (2001) Intron 7 retention and exon 9 skipping EAAT2 mRNA variants are not associated with amyotrophic lateral sclerosis. Ann Neurol 49(5):643–649

88. Jiang LL, Zhu B, Zhao Y, Li X, Liu T, Pina-Crespo J et al (2019) Membralin deficiency dysregulates astrocytic glutamate homeostasis leading to ALS-like impairment. J Clin Invest 129(8):3103–3120. https://doi.org/10.1172/JCI127695

89. Fray AE, Ince PG, Banner SJ, Milton ID, Usher PA, Cookson MR et al (1998) The expression of the glial glutamate transporter protein EAAT2 in motor neuron disease: an immunohistochemical study. Eur J Neurosci 10(8):2481–2489. https://doi.org/10.1046/j.1460-9568.1998.00273.x

90. Estradasanchez AM, Rebec GV (2012) Corticostriatal dysfunction and glutamate transporter 1 (GLT1) in Huntington's disease: interactions between neurons and astrocytes. Basal Ganglia 2(2):57–66

91. Duerson K, Woltjer RL, Mookherjee P, Leverenz JB, Montine TJ, Bird TD et al (2009) Detergent-insoluble EAAC1/EAAT3 aberrantly accumulates in hippocampal neurons of Alzheimer's disease patients. Brain Pathol

19(2):267–278. https://doi.org/10.1111/j.
1750-3639.2008.00186.x

92. Lauderback CM, Hackett JM, Huang FF, Keller JN, Szweda LI, Markesbery WR et al (2001) The glial glutamate transporter, GLT-1, is oxidatively modified by 4-hydroxy-2-nonenal in the Alzheimer's disease brain: the role of Aβ1–42. J Neurochem 78(2):413–416. https://doi.org/10.1046/j.1471-4159.2001.00451.x

93. Scott HA, Gebhardt FM, Mitrovic AD, Vandenberg RJ, Dodd PR (2011) Glutamate transporter variants reduce glutamate uptake in Alzheimer's disease. Neurobiol Aging 32(3):553.e1–553.11. https://doi.org/10.1016/j.neurobiolaging.2010.03.008

94. Woltjer RL, Duerson K, Fullmer JM, Mookherjee P, Ryan AM, Montine TJ et al (2010) Aberrant detergent-insoluble excitatory amino acid transporter 2 accumulates in Alzheimer disease. J Neuropathol Exp Neurol 69(7):667–676. https://doi.org/10.1097/NEN.0b013e3181e24adb

95. Norenberg MD (1979) Distribution of glutamine synthetase in the rat central nervous system. J Histochem Cytochem 27(3):756–762. https://doi.org/10.1177/27.3.39099

96. Patel AJ, Weir MD, Hunt A, Tahourdin CS, Thomas DG (1985) Distribution of glutamine synthetase and glial fibrillary acidic protein and correlation of glutamine synthetase with glutamate decarboxylase in different regions of the rat central nervous system. Brain Res 331(1):1–9. https://doi.org/10.1016/0006-8993(85)90708-5

97. Rose CF, Verkhratsky A, Parpura V (2013) Astrocyte glutamine synthetase: pivotal in health and disease. Biochem Soc Trans 41(6):1518–1524. https://doi.org/10.1042/BST20130237

98. Li K-Y, Gong P-F, Li J-T, Xu N-J, Qin S (2020) Morphological and molecular alterations of reactive astrocytes without proliferation in cerebral cortex of an APP/PS1 transgenic mouse model and Alzheimer's patients. Glia 68(11):2361–2376. https://doi.org/10.1002/glia.23845

99. Goncalves C, Leite MC, Nardin P (2008) Biological and methodological features of the measurement of S100B, a putative marker of brain injury. Clin Biochem 41(10):755–763

100. Marenholz I, Heizmann CW, Fritz G (2004) S100 proteins in mouse and man: from evolution to function and pathology (including an update of the nomenclature). Biochem Biophys Res Commun 322(4):1111–1122.

https://doi.org/10.1016/j.bbrc.2004.07.096

101. Donato R (2001) S100: a multigenic family of calcium-modulated proteins of the EF-hand type with intracellular and extracellular functional roles. Int J Biochem Cell Biol 33(7):637–668. https://doi.org/10.1016/s1357-2725(01)00046-2

102. Donato R (2003) Intracellular and extracellular roles of S100 proteins. Microsc Res Tech 60(6):540–551. https://doi.org/10.1002/jemt.10296

103. Harris JL, Yeh HW, Swerdlow RH, Choi IY, Lee P, Brooks WM (2014) High-field proton magnetic resonance spectroscopy reveals metabolic effects of normal brain aging. Neurobiol Aging 35(7):1686–1694. https://doi.org/10.1016/j.neurobiolaging.2014.01.018

104. Harris JL, Choi IY, Brooks WM (2015) Probing astrocyte metabolism in vivo: proton magnetic resonance spectroscopy in the injured and aging brain. Front Aging Neurosci 7:202. https://doi.org/10.3389/fnagi.2015.00202

105. Saura J, Luque JM, Cesura AM, Da Prada M, Chan-Palay V, Huber G et al (1994) Increased monoamine oxidase B activity in plaque-associated astrocytes of Alzheimer brains revealed by quantitative enzyme radioautography. Neuroscience 62(1):15–30. https://doi.org/10.1016/0306-4522(94)90311-5

106. Gulyas B, Pavlova E, Kasa P, Gulya K, Bakota L, Varszegi S et al (2011) Activated MAO-B in the brain of Alzheimer patients, demonstrated by [11C]-l-deprenyl using whole hemisphere autoradiography. Neurochem Int 58(1):60–68

107. Garwood CJ, Ratcliffe LE, Simpson JE, Heath PR, Ince PG, Wharton SB (2017) Review: Astrocytes in Alzheimer's disease and other age-associated dementias: a supporting player with a central role. Neuropathol Appl Neurobiol 43(4):281–298. https://doi.org/10.1111/nan.12338

108. Liddelow SA, Barres BA (2017) Reactive astrocytes: production, function, and therapeutic potential. Immunity 46(6):957–967. https://doi.org/10.1016/j.immuni.2017.06.006

109. Heppner FL, Ransohoff RM, Becher B (2015) Immune attack: the role of inflammation in Alzheimer disease. Nat Rev Neurosci 16(6):358–372. https://doi.org/10.1038/nrn3880

110. Rubio-Perez JM, Morillas-Ruiz JM (2012) A review: inflammatory process in Alzheimer's disease, role of cytokines. ScientificWorld-Journal 2012:756357. https://doi.org/10.1100/2012/756357

111. Heneka MT, O'Banion MK, Terwel D, Kummer MP (2010) Neuroinflammatory processes in Alzheimer's disease. J Neural Transm (Vienna) 117(8):919–947. https://doi.org/10.1007/s00702-010-0438-z

112. Phillips EC, Croft CL, Kurbatskaya K, Oneill MJ, Hutton M, Hanger DP et al (2014) Astrocytes and neuroinflammation in Alzheimer's disease. Biochem Soc Trans 42 (5):1321–1325

113. Heneka MT, Carson MJ, El Khoury J, Landreth GE, Brosseron F, Feinstein DL et al (2015) Neuroinflammation in Alzheimer's disease. Lancet Neurol 14(4):388–405. https://doi.org/10.1016/S1474-4422(15)70016-5

114. Gao Q, Li Y, Chopp M (2005) Bone marrow stromal cells increase astrocyte survival via upregulation of phosphoinositide 3-kinase/threonine protein kinase and mitogen-activated protein kinase kinase/extracellular signal-regulated kinase pathways and stimulate astrocyte trophic factor gene expression after anaerobic insult. Neuroscience 136 (1):123–134. https://doi.org/10.1016/j.neuroscience.2005.06.091

115. Hayakawa K, Pham LD, Arai K, Lo EH (2014) Reactive astrocytes promote adhesive interactions between brain endothelium and endothelial progenitor cells via HMGB1 and beta-2 integrin signaling. Stem Cell Res 12 (2):531–538

116. Borjabad A, Brooks AI, Volsky DJ (2010) Gene expression profiles of HIV-1-infected glia and brain: toward better understanding of the role of astrocytes in HIV-1-associated neurocognitive disorders. J Neuroimmune Pharmacol 5(1):44–62. https://doi.org/10.1007/s11481-009-9167-1

117. Cahoy JD, Emery B, Kaushal A, Foo LC, Zamanian JL, Christopherson KS et al (2008) A transcriptome database for astrocytes, neurons, and oligodendrocytes: a new resource for understanding brain development and function. J Neurosci 28 (1):264–278. https://doi.org/10.1523/JNEUROSCI.4178-07.2008

118. Yan P, Hu X, Song H, Yin K, Bateman RJ, Cirrito JR et al (2006) Matrix metalloproteinase-9 degrades amyloid-beta fibrils in vitro and compact plaques in situ. J Biol Chem 281(34):24566–24574. https://doi.org/10.1074/jbc.M602440200

119. Yin KJ, Cirrito JR, Yan P, Hu X, Xiao Q, Pan X et al (2006) Matrix metalloproteinases expressed by astrocytes mediate extracellular amyloid-beta peptide catabolism. J Neurosci 26(43):10939–10948. https://doi.org/10.1523/JNEUROSCI.2085-06.2006

120. John GR, Lee SC, Brosnan CF (2003) Cytokines: powerful regulators of glial cell activation. Neuroscientist 9(1):10–22. https://doi.org/10.1177/1073858402239587

121. Di Giorgio FP, Carrasco MA, Siao MC, Maniatis T, Eggan K (2007) Non-cell autonomous effect of glia on motor neurons in an embryonic stem cell-based ALS model. Nat Neurosci 10(5):608–614. https://doi.org/10.1038/nn1885

122. Farina C, Aloisi F, Meinl E (2007) Astrocytes are active players in cerebral innate immunity. Trends Immunol 28(3):138–145. https://doi.org/10.1016/j.it.2007.01.005

123. Bekar LK, He W, Nedergaard M (2008) Locus coeruleus alpha-adrenergic-mediated activation of cortical astrocytes in vivo. Cereb Cortex 18(12):2789–2795. https://doi.org/10.1093/cercor/bhn040

124. Neary JT, Kang Y, Willoughby KA, Ellis EF (2003) Activation of extracellular signal-regulated kinase by stretch-induced injury in astrocytes involves extracellular ATP and P2 purinergic receptors. J Neurosci 23 (6):2348–2356

125. Swanson RA, Ying W, Kauppinen TM (2004) Astrocyte influences on ischemic neuronal death. Curr Mol Med 4(2):193–205. https://doi.org/10.2174/1566524043479185

126. Norenberg MD, Rao KVR, Jayakumar AR (2009) Signaling factors in the mechanism of ammonia neurotoxicity. Metab Brain Dis 24(1):103–117

127. Migheli A, Piva R, Atzori C, Troost D, Schiffer D (1997) c-Jun, JNK/SAPK kinases and transcription factor NF-kappa B are selectively activated in astrocytes, but not motor neurons, in amyotrophic lateral sclerosis. J Neuropathol Exp Neurol 56(12):1314–1322

128. Gilmore TD (2006) Introduction to NF-kappaB: players, pathways, perspectives. Oncogene 25(51):6680–6684. https://doi.org/10.1038/sj.onc.1209954

129. Ceyzeriat K, Abjean L, Sauvage MC, Haim LB, Escartin C (2016) The complex STATes of astrocyte reactivity: how are they controlled by the JAK–STAT3 pathway? Neuroscience 330:205–218

130. Anderson MA, Burda JE, Ren Y, Ao Y, O'Shea TM, Kawaguchi R et al (2016)

Astrocyte scar formation aids central nervous system axon regeneration. Nature 532 (7598):195–200. https://doi.org/10.1038/nature17623

131. Kanemaru K, Kubota J, Sekiya H, Hirose K, Okubo Y, Iino M (2013) Calcium-dependent N-cadherin up-regulation mediates reactive astrogliosis and neuroprotection after brain injury. Proc Natl Acad Sci U S A 110 (28):11612–11617. https://doi.org/10.1073/pnas.1300378110

132. Robel S, Mori T, Zoubaa S, Schlegel J, Sirko S, Faissner A et al (2009) Conditional deletion of beta1-integrin in astroglia causes partial reactive gliosis. Glia 57 (15):1630–1647. https://doi.org/10.1002/glia.20876

133. Vakalopoulos C (2017) Alzheimer's disease: the alternative serotonergic hypothesis of cognitive decline. J Alzheimers Dis 60 (3):859–866. https://doi.org/10.3233/JAD-170364

134. McGeer PL, McGeer EG (2002) Local neuroinflammation and the progression of Alzheimer's disease. J Neurovirol 8 (6):529–538. https://doi.org/10.1080/13550280290100969

135. Nagele RG, Wegiel J, Venkataraman V, Imaki H, Wang KC, Wegiel J (2004) Contribution of glial cells to the development of amyloid plaques in Alzheimer's disease. Neurobiol Aging 25(5):663–674. https://doi.org/10.1016/j.neurobiolaging.2004.01.007

136. Li C, Zhao R, Gao K, Wei Z, Yin MY, Lau LT et al (2011) Astrocytes: implications for neuroinflammatory pathogenesis of Alzheimer's disease. Curr Alzheimer Res 8(1):67–80. https://doi.org/10.2174/156720511794604543

137. Acosta C, Anderson HD, Anderson CM (2017) Astrocyte dysfunction in Alzheimer disease. J Neurosci Res 95(12):2430–2447. https://doi.org/10.1002/jnr.24075

138. Vincent AJ, Gasperini R, Foa L, Small DH (2010) Astrocytes in Alzheimer's disease: emerging roles in calcium dysregulation and synaptic plasticity. J Alzheimers Dis 22 (3):699–714. https://doi.org/10.3233/JAD-2010-101089

139. Cassano T, Serviddio G, Gaetani S, Romano A, Dipasquale P, Cianci S et al (2012) Glutamatergic alterations and mitochondrial impairment in a murine model of Alzheimer disease. Neurobiol Aging 33 (6):1121.e1–1121.12. https://doi.org/10.1016/j.neurobiolaging.2011.09.021

140. Masliah E, Alford M, Mallory M, Rockenstein E, Moechars D, Van Leuven F (2000) Abnormal glutamate transport function in mutant amyloid precursor protein transgenic mice. Exp Neurol 163 (2):381–387. https://doi.org/10.1006/exnr.2000.7386

141. Ortinski PI, Dong J, Mungenast A, Yue C, Takano H, Watson DJ et al (2010) Selective induction of astrocytic gliosis generates deficits in neuronal inhibition. Nat Neurosci 13 (5):584–591. https://doi.org/10.1038/nn.2535

142. Jo S, Yarishkin O, Hwang YJ, Chun YE, Park M, Woo DH et al (2014) GABA from reactive astrocytes impairs memory in mouse models of Alzheimer's disease. Nat Med 20 (8):886–896

143. Cole SL, Vassar R (2007) The Alzheimer's disease beta-secretase enzyme, BACE1. Mol Neurodegener 2:22. https://doi.org/10.1186/1750-1326-2-22

144. Thal DR, Schultz C, Dehghani F, Yamaguchi H, Braak H, Braak E (2000) Amyloid beta-protein (Abeta)-containing astrocytes are located preferentially near N-terminal-truncated Abeta deposits in the human entorhinal cortex. Acta Neuropathol 100(6):608–617. https://doi.org/10.1007/s004010000242

145. Koistinaho M, Lin S, Wu X, Esterman M, Koger D, Hanson J et al (2004) Apolipoprotein E promotes astrocyte colocalization and degradation of deposited amyloid-beta peptides. Nat Med 10(7):719–726. https://doi.org/10.1038/nm1058

146. Basak JM, Verghese PB, Yoon H, Kim J, Holtzman DM (2012) Low-density lipoprotein receptor represents an apolipoprotein E-independent pathway of Abeta uptake and degradation by astrocytes. J Biol Chem 287 (17):13959–13971. https://doi.org/10.1074/jbc.M111.288746

147. Kim J, Castellano JM, Jiang H, Basak JM, Parsadanian M, Pham V et al (2009) Over-expression of low-density lipoprotein receptor in the brain markedly inhibits amyloid deposition and increases extracellular A beta clearance. Neuron 64(5):632–644. https://doi.org/10.1016/j.neuron.2009.11.013

148. Weggen S, Diehlmann A, Buslei R, Beyreuther K, Bayer TA (1998) Prominent expression of presenilin-1 in senile plaques and reactive astrocytes in Alzheimer's disease brain. Neuroreport 9(14):3279–3283

149. Iliff JJ, Wang M, Liao Y, Plogg BA, Peng W, Gundersen GA et al (2012) A paravascular pathway facilitates CSF flow through the

brain parenchyma and the clearance of interstitial solutes, including amyloid β. Sci Transl Med 4(147)

150. Garwood C, Faizullabhoy A, Wharton SB, Ince PG, Heath PR, Shaw PJ et al (2013) Calcium dysregulation in relation to Alzheimer-type pathology in the ageing brain. Neuropathol Appl Neurobiol 39 (7):788–799. https://doi.org/10.1111/nan.12033

151. Kobayashi K, Hayashi M, Nakano H, Fukutani Y, Sasaki K, Shimazaki M et al (2002) Apoptosis of astrocytes with enhanced lysosomal activity and oligodendrocytes in white matter lesions in Alzheimer's disease. Neuropathol Appl Neurobiol 28 (3):238–251. https://doi.org/10.1046/j.1365-2990.2002.00390.x

152. Sjobeck M, Englund E (2003) Glial levels determine severity of white matter disease in Alzheimer's disease: a neuropathological study of glial changes. Neuropathol Appl Neurobiol 29(2):159–169. https://doi.org/10.1046/j.1365-2990.2003.00456.x

153. Panmontojo F, Anichtchik O, Dening Y, Knels L, Pursche S, Jung R et al 2010 Progression of Parkinson's disease pathology is reproduced by intragastric administration of rotenone in mice. PLoS One 5(1)

154. Wang HL, Chou AH, Wu AS, Chen SY, Weng YH, Kao YC et al (2011) PARK6 PINK1 mutants are defective in maintaining mitochondrial membrane potential and inhibiting ROS formation of substantia nigra dopaminergic neurons. Biochim Biophys Acta 1812 (6):674–684. https://doi.org/10.1016/j.bbadis.2011.03.007

155. Ciesielska A, Joniec I, Kurkowska-Jastrzebska I, Cudna A, Przybylkowski A, Czlonkowska A et al (2009) The impact of age and gender on the striatal astrocytes activation in murine model of Parkinson's disease. Inflamm Res 58(11):747–753. https://doi.org/10.1007/s00011-009-0026-6

156. Sriram K, Benkovic SA, Hebert MA, Miller DB, O'Callaghan JP (2004) Induction of gp130-related cytokines and activation of JAK2/STAT3 pathway in astrocytes precedes up-regulation of glial fibrillary acidic protein in the 1-methyl-4-phenyl-1,2,3,6-tetrahydropyridine model of neurodegeneration: key signaling pathway for astrogliosis in vivo? J Biol Chem 279(19):19936–19947. https://doi.org/10.1074/jbc.M309304200

157. Song YJ, Halliday GM, Holton JL, Lashley T, O'Sullivan SS, McCann H et al (2009) Degeneration in different parkinsonian syndromes relates to astrocyte type and astrocyte protein expression. J Neuropathol Exp Neurol 68(10):1073–1083. https://doi.org/10.1097/NEN.0b013e3181b66f1b

158. Halliday GM, Stevens CH (2011) Glia: initiators and progressors of pathology in Parkinson's disease. Mov Disord 26(1):6–17. https://doi.org/10.1002/mds.23455

159. Lee HJ, Suk JE, Patrick C, Bae EJ, Cho JH, Rho S et al (2010) Direct transfer of alpha-synuclein from neuron to astroglia causes inflammatory responses in synucleinopathies. J Biol Chem 285(12):9262–9272. https://doi.org/10.1074/jbc.M109.081125

160. Barcia C, Ros CM, Annese V, Gomez A, Ros-Bernal F, Aguado-Llera D et al (2012) IFN-gamma signaling, with the synergistic contribution of TNF-alpha, mediates cell specific microglial and astroglial activation in experimental models of Parkinson's disease. Cell Death Dis 3:e379. https://doi.org/10.1038/cddis.2012.123

161. L'Episcopo F, Serapide MF, Tirolo C, Testa N, Caniglia S, Morale MC et al (2011) A Wnt1 regulated Frizzled-1/β-catenin-signaling pathway as a candidate regulatory circuit controlling mesencephalic dopaminergic neuron-astrocyte crosstalk: therapeutical relevance for neuron survival and neuroprotection. Mol Neurodegener 6(1):49. https://doi.org/10.1186/1750-1326-6-49

162. Bates GP, Dorsey R, Gusella JF, Hayden MR, Kay C, Leavitt BR et al (2015) Huntington disease. Nat Rev Dis Primers 1:15005. https://doi.org/10.1038/nrdp.2015.5

163. Walker FO (2007) Huntington's disease. Lancet 369(9557):218–228. https://doi.org/10.1016/S0140-6736(07)60111-1

164. Faideau M, Kim J, Cormier K, Gilmore R, Welch M, Auregan G et al (2010) In vivo expression of polyglutamine-expanded huntingtin by mouse striatal astrocytes impairs glutamate transport: a correlation with Huntington's disease subjects. Hum Mol Genet 19(15):3053–3067. https://doi.org/10.1093/hmg/ddq212

165. Vonsattel JP, Myers RH, Stevens TJ, Ferrante RJ, Bird ED, Richardson EP Jr (1985) Neuropathological classification of Huntington's disease. J Neuropathol Exp Neurol 44 (6):559–577. https://doi.org/10.1097/00005072-198511000-00003

166. Behrens PF, Franz P, Woodman B, Lindenberg KS, Landwehrmeyer GB (2002) Impaired glutamate transport and glutamate-glutamine cycling: downstream effects of the Huntington mutation. Brain 125(Pt 8):1908–1922. https://doi.org/10.1093/brain/awf180

167. Hassel B, Tessler S, Faull RLM, Emson PC (2008) Glutamate uptake is reduced in prefrontal cortex in Huntington's disease. Neurochem Res 33(2):232–237

168. Lievens JC, Woodman B, Mahal A, Spasic-Boscovic O, Samuel D, Kerkerian-Le Goff L et al (2001) Impaired glutamate uptake in the R6 Huntington's disease transgenic mice. Neurobiol Dis 8(5):807–821. https://doi.org/10.1006/nbdi.2001.0430

169. Lee W, Reyes RC, Gottipati MK, Lewis K, Lesort M, Parpura V et al (2013) Enhanced Ca(2+)-dependent glutamate release from astrocytes of the BACHD Huntington's disease mouse model. Neurobiol Dis 58:192–199. https://doi.org/10.1016/j.nbd.2013.06.002

170. Khakh BS, Sofroniew MV (2014) Astrocytes and Huntington's disease. ACS Chem Neurosci 5(7):494–496

171. Tong X, Ao Y, Faas GC, Nwaobi SE, Xu J, Haustein MD et al (2014) Astrocyte Kir4.1 ion channel deficits contribute to neuronal dysfunction in Huntington's disease model mice. Nat Neurosci 17(5):694–703. https://doi.org/10.1038/nn.3691

172. Hsiao HY, Chen YC, Chen HM, Tu PH, Chern Y (2013) A critical role of astrocyte-mediated nuclear factor-kappaB-dependent inflammation in Huntington's disease. Hum Mol Genet 22(9):1826–1842. https://doi.org/10.1093/hmg/ddt036

173. Hsiao HY, Chen YC, Huang CH, Chen CC, Hsu YH, Chen HM et al (2015) Aberrant astrocytes impair vascular reactivity in Huntington disease. Ann Neurol 78(2):178–192. https://doi.org/10.1002/ana.24428

174. Bradford J, Shin JY, Roberts M, Wang CE, Sheng G, Li S et al (2010) Mutant huntingtin in glial cells exacerbates neurological symptoms of Huntington disease mice. J Biol Chem 285(14):10653–10661. https://doi.org/10.1074/jbc.M109.083287

175. Bradford J, Shin JY, Roberts M, Wang CE, Li XJ, Li S (2009) Expression of mutant huntingtin in mouse brain astrocytes causes age-dependent neurological symptoms. Proc Natl Acad Sci U S A 106(52):22480–22485. https://doi.org/10.1073/pnas.0911503106

176. Juopperi TA, Kim WR, Chiang CH, Yu H, Margolis RL, Ross CA et al (2012) Astrocytes generated from patient induced pluripotent stem cells recapitulate features of Huntington's disease patient cells. Mol Brain 5:17. https://doi.org/10.1186/1756-6606-5-17

177. Kiernan MC, Vucic S, Cheah BC, Turner MR, Eisen A, Hardiman O et al (2011) Amyotrophic lateral sclerosis. Lancet 377 (9769):942–955. https://doi.org/10.1016/S0140-6736(10)61156-7

178. Haidet-Phillips AM, Hester ME, Miranda CJ, Meyer K, Braun L, Frakes A et al (2011) Astrocytes from familial and sporadic ALS patients are toxic to motor neurons. Nat Biotechnol 29(9):824–828. https://doi.org/10.1038/nbt.1957

179. Nagai M, Re DB, Nagata T, Chalazonitis A, Jessell TM, Wichterle H et al (2007) Astrocytes expressing ALS-linked mutated SOD1 release factors selectively toxic to motor neurons. Nat Neurosci 10(5):615–622. https://doi.org/10.1038/nn1876

180. Yamanaka K, Chun SJ, Boillee S, Fujimoritonou N, Yamashita H, Gutmann DH et al (2008) Astrocytes as determinants of disease progression in inherited amyotrophic lateral sclerosis. Nat Neurosci 11 (3):251–253

181. Boillee S, Yamanaka K, Lobsiger CS, Copeland NG, Jenkins NA, Kassiotis G et al (2006) Onset and progression in inherited ALS determined by motor neurons and microglia. Science 312(5778):1389–1392. https://doi.org/10.1126/science.1123511

182. Wang L, Sharma K, Grisotti G, Roos RP (2009) The effect of mutant SOD1 dismutase activity on non-cell autonomous degeneration in familial amyotrophic lateral sclerosis. Neurobiol Dis 35(2):234–240. https://doi.org/10.1016/j.nbd.2009.05.002

183. Kang SH, Li Y, Fukaya M, Lorenzini I, Cleveland DW, Ostrow LW et al (2013) Degeneration and impaired regeneration of gray matter oligodendrocytes in amyotrophic lateral sclerosis. Nat Neurosci 16(5):571–579. https://doi.org/10.1038/nn.3357

184. Ferrer I (2017) Diversity of astroglial responses across human neurodegenerative disorders and brain aging. Brain Pathol 27 (5):645–674. https://doi.org/10.1111/bpa.12538

185. Ferraiuolo L, Higginbottom A, Heath PR, Barber S, Greenald D, Kirby J et al (2011) Dysregulation of astrocyte-motoneuron cross-talk in mutant superoxide dismutase 1-related amyotrophic lateral sclerosis. Brain 134(Pt 9):2627–2641. https://doi.org/10.1093/brain/awr193

186. Papadeas ST, Kraig SE, O'Banion C, Lepore AC, Maragakis NJ (2011) Astrocytes carrying the superoxide dismutase 1 (SOD1G93A) mutation induce wild-type motor neuron degeneration in vivo. Proc Natl Acad Sci U S

A 108(43):17803–17808. https://doi.org/10.1073/pnas.1103141108

187. Pardo AC, Wong V, Benson LM, Dykes M, Tanaka K, Rothstein JD et al (2006) Loss of the astrocyte glutamate transporter GLT1 modifies disease in SOD1(G93A) mice. Exp Neurol 201(1):120–130. https://doi.org/10.1016/j.expneurol.2006.03.028

188. Maragakis NJ, Rothstein JD (2006) Mechanisms of disease: astrocytes in neurodegenerative disease. Nat Clin Pract Neurol 2 (12):679–689. https://doi.org/10.1038/ncpneuro0355

189. Meyer K, Ferraiuolo L, Miranda CJ, Likhite S, McElroy S, Renusch S et al (2014) Direct conversion of patient fibroblasts demonstrates non-cell autonomous toxicity of astrocytes to motor neurons in familial and sporadic ALS. Proc Natl Acad Sci U S A 111(2):829–832. https://doi.org/10.1073/pnas.1314085111

190. Rothstein JD, Dykes-Hoberg M, Pardo CA, Bristol LA, Jin L, Kuncl RW et al (1996) Knockout of glutamate transporters reveals a major role for astroglial transport in excitotoxicity and clearance of glutamate. Neuron 16(3):675–686. https://doi.org/10.1016/s0896-6273(00)80086-0

191. Rothstein JD, Van Kammen M, Levey AI, Martin LJ, Kuncl RW (1995) Selective loss of glial glutamate transporter GLT-1 in amyotrophic lateral sclerosis. Ann Neurol 38 (1):73–84. https://doi.org/10.1002/ana.410380114

192. Rothstein JD, Patel S, Regan MR, Haenggeli C, Huang YH, Bergles DE et al (2005) Beta-lactam antibiotics offer neuroprotection by increasing glutamate transporter expression. Nature 433(7021):73–77. https://doi.org/10.1038/nature03180

193. Phatnani H, Guarnieri P, Friedman BA, Carrasco MA, Muratet M, Okeeffe S et al (2013) Intricate interplay between astrocytes and motor neurons in ALS. Proc Natl Acad Sci U S A 110(8):201222361

194. Hashioka S, Klegeris A, Schwab C, McGeer PL (2009) Interferon-gamma-dependent cytotoxic activation of human astrocytes and astrocytoma cells. Neurobiol Aging 30 (12):1924–1935. https://doi.org/10.1016/j.neurobiolaging.2008.02.019

195. Hashioka S, Klegeris A, Qing H, McGeer PL (2011) STAT3 inhibitors attenuate interferon-gamma-induced neurotoxicity and inflammatory molecule production by human astrocytes. Neurobiol Dis 41 (2):299–307. https://doi.org/10.1016/j.nbd.2010.09.018

196. Shibata N, Yamamoto T, Hiroi A, Omi Y, Kato Y, Kobayashi M (2010) Activation of STAT3 and inhibitory effects of pioglitazone on STAT3 activity in a mouse model of SOD1-mutated amyotrophic lateral sclerosis. Neuropathology 30(4):353–360. https://doi.org/10.1111/j.1440-1789.2009.01078.x

197. Martorana F, Brambilla L, Valori CF, Bergamaschi C, Roncoroni C, Aronica E et al (2012) The BH4 domain of Bcl-X L rescues astrocyte degeneration in amyotrophic lateral sclerosis by modulating intracellular calcium signals. Hum Mol Genet 21 (4):826–840

198. McCarthy KD, de Vellis J (1980) Preparation of separate astroglial and oligodendroglial cell cultures from rat cerebral tissue. J Cell Biol 85 (3):890–902. https://doi.org/10.1083/jcb.85.3.890

199. Foo LC, Allen NJ, Bushong EA, Ventura PB, Chung WS, Zhou L et al (2011) Development of a method for the purification and culture of rodent astrocytes. Neuron 71 (5):799–811. https://doi.org/10.1016/j.neuron.2011.07.022

200. Haas R, Werner J, Fliedner TM (1970) Cytokinetics of neonatal brain cell development in rats as studied by the 'complete 3H-thymidine labelling' method. J Anat 107:421–437

201. Skoff RP, Knapp PE (1991) Division of astroblasts and oligodendroblasts in postnatal rodent brain: evidence for separate astrocyte and oligodendrocyte lineages. Glia 4 (2):165–174. https://doi.org/10.1002/glia.440040208

202. Christopherson KS, Ullian EM, Stokes CC, Mullowney CE, Hell JW, Agah A et al (2005) Thrombospondins are astrocyte-secreted proteins that promote CNS synaptogenesis. Cell 120(3):421–433

203. Eroglu Ç, Allen NJ, Susman MW, O'Rourke NA, Park CY, Özkan E et al (2009) Gabapentin receptor α2δ-1 is a neuronal thrombospondin receptor responsible for excitatory CNS synaptogenesis. Cell 139(2):380–392. https://doi.org/10.1016/j.cell.2009.09.025

204. Allen NJ, Bennett ML, Foo LC, Wang GX, Chakraborty C, Smith SJ et al (2012) Astrocyte glycans 4 and 6 promote formation of excitatory synapses via GluA1 AMPA receptors. Nature 486(7403):410–414. https://doi.org/10.1038/nature11059

205. Puschmann TB, Zanden C, Lebkuechner I, Philippot C, de Pablo Y, Liu J et al (2014) HB-EGF affects astrocyte morphology,

proliferation, differentiation, and the expression of intermediate filament proteins. J Neurochem 128(6):878–889. https://doi.org/10.1111/jnc.12519

206. Puschmann TB, de Pablo Y, Zanden C, Liu J, Pekny M (2014) A novel method for three-dimensional culture of central nervous system neurons. Tissue Eng Part C Methods 20 (6):485–492. https://doi.org/10.1089/ten.TEC.2013.0445

207. Puschmann TB, Zanden C, De Pablo Y, Kirchhoff F, Pekna M, Liu J et al (2013) Bioactive 3D cell culture system minimizes cellular stress and maintains the in vivo-like morphological complexity of astroglial cells. Glia 61(3):432–440. https://doi.org/10.1002/glia.22446

208. Chiu C, Yao N, Guo JH, Shen C, Lee H, Chiu Y et al (2017) Inhibition of astrocytic activity alleviates sequela in acute stages of intracerebral hemorrhage. Oncotarget 8 (55):94850–94861

209. Wilkinson DJ (2009) Stochastic modelling for quantitative description of heterogeneous biological systems. Nat Rev Genet 10 (2):122–133. https://doi.org/10.1038/nrg2509

210. Sandberg R (2014) Entering the era of single-cell transcriptomics in biology and medicine. Nat Methods 11(1):22–24. https://doi.org/10.1038/nmeth.2764

211. Picelli S, Faridani OR, Bjorklund AK, Winberg G, Sagasser S, Sandberg R (2014) Full-length RNA-seq from single cells using Smart-seq2. Nat Protoc 9(1):171–181. https://doi.org/10.1038/nprot.2014.006

212. Zeisel A, Munoz-Manchado AB, Codeluppi S, Lonnerberg P, La Manno G, Jureus A et al (2015) Brain structure. Cell types in the mouse cortex and hippocampus revealed by single-cell RNA-seq. Science 347 (6226):1138–1142. https://doi.org/10.1126/science.aaa1934

213. Darmanis S, Sloan SA, Zhang Y, Enge M, Caneda C, Shuer LM et al (2015) A survey of human brain transcriptome diversity at the single cell level. Proc Natl Acad Sci U S A 112 (23):7285–7290. https://doi.org/10.1073/pnas.1507125112

214. Saji H (2017) In vivo molecular imaging. Biol Pharm Bull 40(10):1605–1615. https://doi.org/10.1248/bpb.b17-00505

215. Anderson CJ, Lewis JS (2017) Current status and future challenges for molecular imaging. Philos Trans A Math Phys Eng Sci 375:2107. https://doi.org/10.1098/rsta.2017.0023

216. Strafella AP, Bohnen NI, Perlmutter JS, Eidelberg D, Pavese N, Van Eimeren T et al (2017) Molecular imaging to track Parkinson's disease and atypical parkinsonisms: new imaging frontiers. Mov Disord 32 (2):181–192. https://doi.org/10.1002/mds.26907

Chapter 4

Tau Pathology in Neurodegenerative Diseases

Thomas Vogels and Tomáš Hromádka

Abstract

Aggregation and cellular accumulation of tau protein is a defining feature of tauopathies, a class of histopathologically and clinically heterogeneous neurodegenerative diseases. Tauopathies include diseases as diverse as Alzheimer's disease (AD), Pick's disease, progressive supranuclear palsy, corticobasal degeneration, and chronic traumatic encephalopathy. Tau pathology affects different cell types in various tauopathies and strongly correlates with clinical symptoms. The complexity of tau pathology and the structural diversity of tau aggregates in different tauopathies represent an active area of research in neurodegeneration. The initiation, spreading, and cellular clearance of tau pathology play important roles in the disease process of tauopathies. Spreading of tau pathology throughout the brain may underlie the progressive nature of these diseases. Understanding of mechanisms underlying the spreading of tau pathology between neurons and glial cells is essential for the development and use of emerging therapeutics and tau biomarkers. Recent biomarkers have enabled identification of the earliest stages of AD pathology and tracking the anatomical progression of tau pathology over time. An increasing number of tau-targeting therapeutics are entering clinical trials and might lead to the development of a treatment for these devastating neurodegenerative disorders.

Key words Alzheimer's disease, Tauopathies, Tau protein, Tau propagation, Neurodegeneration, Cerebrospinal fluid, Positron emission tomography, Therapy

---

1 Introduction

Aggregation and subsequent intracellular accumulation of tau protein is a common denominator of a heterogeneous group of neurodegenerative disorders termed tauopathies [1]. Alzheimer's disease (AD) is the most common tauopathy and currently affects approximately 50 million people worldwide [2]. AD is additionally associated with parenchymal plaques composed of aggregated amyloid-β (Aβ) [3]. The other tauopathies are primarily characterized by tau pathology and are therefore referred to as primary tauopathies. This group includes Pick's disease (PiD), corticobasal degen-

Philip V. Peplow, Bridget Martinez and Thomas A. Gennarelli (eds.), *Neurodegenerative Diseases Biomarkers: Towards Translating Research to Clinical Practice*, Neuromethods, vol. 173, https://doi.org/10.1007/978-1-0716-1712-0_4,
© Springer Science+Business Media, LLC, part of Springer Nature 2022

Table 1
Tau pathology and cellular inclusions in different tauopathies

Tauopathy	Tau Isoforms; fibril core	Predominately affected cells	Neuronal pathology	Astrocytic pathology	Oligodendrocytic pathology
AD	3R/4R; aa306-378	Neurons	NFTs, NTs, neuritic plaques	–	–
CTE	3R/4R; aa305-379	Neurons, glia	NFTs, NTs	Granular-fuzzy astrocytes	Coiled bodies
PiD	3R; aa254-378	Neurons	Pick bodies, Pick cells	Ramified astrocytes	–
CBD	4R; aa274-380	Neurons, glia	NFTs, NTs, balloon neurons	Astrocytic plaques	Coiled bodies
PSP	4R; unknown	Neurons, glia	NFTs, NTs	Tufted astrocytes	Coiled bodies
AGD	4R; unknown	Neurons, glia	Grains	Granular-fuzzy astrocytes	–
GGT	4R; unknown	Glia	–	Globular inclusions	Globular inclusions
ARTAG	4R; unknown	Glia	–	Granular-fuzzy and thorn-shaped astrocytes	–

Abbreviations: *AD* Alzheimer's disease, *CTE* chronic traumatic encephalopathy, *PiD* pick's disease, *CBD* corticobasal degeneration, *PSP* progressive supranuclear palsy, *AGD* argyrophilic grain disease, *GGT* globular glial tauopathy, *ARTAG* aging-related tau astrogliopathy, *NFTs* neurofibrillary tangles, *NTs* neuropil threads

eration (CBD), progressive supranuclear palsy (PSP), chronic traumatic encephalopathy (CTE), argyrophilic grain disease (AGD), globular glial tauopathy (GGT), aging-related tau astrogliopathy (ARTAG), and many others [4]. These diseases vary in their clinical presentations, affected brain regions, affected cell types, and structural features of tau aggregates (Table 1).

Given its high prevalence compared to the primary tauopathies, most research on the role of tau in neurodegeneration has focused on AD. Tau pathology in AD is strongly linked to neuroinflammation and synapse loss [5], and it closely mirrors neurodegeneration and cognitive decline [6, 7]. As the defining feature, tau pathology also plays a central role in primary tauopathies [1, 4] In this chapter we will focus on the basic pathobiology of tau pathology and its progressive propagation throughout the brain. In addition, we will briefly describe the associated recent developments in imaging tau pathology in vivo, tau biomarkers in peripheral fluids, and tau-based therapeutics.

2 Tau Protein Physiology

Tau protein is highly abundant in the axonal compartment of neurons and detectable at lower levels in the neuronal nucleus, neuronal dendrites, and the extracellular space [8, 9]. Its expression in glial cells is controversial; tau is likely present in low amounts in oligodendrocytes, extremely low amounts in astrocytes, but not in microglia [10, 11]. Tau was originally described as a protein that plays a central role in the assembly, bundling, and spacing of the microtubules [12]. Other physiological roles for tau include regulating axonal transport, protection of nuclear DNA and RNA, protection of ribosomal DNA, regulation of myelination, synaptic plasticity, and insulin signaling [12, 13]. New protein interaction partners and functions of tau are still being discovered [12].

Human tau protein is encoded by microtubule-associated protein tau (MAPT) gene, which contains 16 exons and is located on chromosome 17q21 [8]. Alternative splicing of exons 2, 3, and 10 leads to the generation of six tau isoforms in the central nervous system (CNS) [14]. Each isoform contains the total of 0, 1, or 2N-terminal inserts encoded by exons 2 and 3, with 2N isoforms being less abundant in the human brain. Exclusion or inclusion of exon 10 leads to either 3 or 4-repeat (R) domains, respectively, with 3R and 4R isoforms being expressed in equal ratio in the human brain (Fig. 1). Human tau isoforms range from 352–441 amino acids (aa) and molecular weights of 36.7–45.9 kilodalton (kDa). The relative abundance of these isoforms varies between different subcellular compartments and anatomical regions in the brain. Tau protein is divided into four domains: a highly disordered

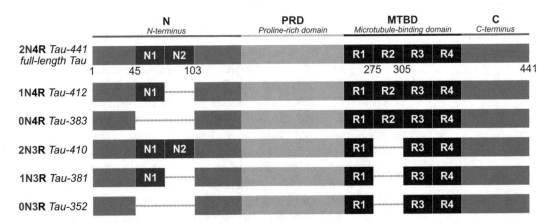

Fig. 1 The six human isoforms of tau protein family are the product of alternative splicing of the MAPT gene. Each isoform contains up to two inserts (N1, N2) in the N-terminal part of the protein, which are the product of exons 2 and 3. Alternative slicing of exon 10 leads to the absence or presence of the second repeat (R2) in the MTBD. Depending on the number of inserts, human tau protein isoforms range from 352 amino acids to the full-length tau protein that contains 441 amino acids

N-terminal, a proline-rich mid-domain with seven PxxP motifs to which signaling proteins can bind, microtubule-binding domain (MTBD) containing several microtubule-binding repeats (MTBR) which bind the microtubules, and a C-terminal domain [8].

Tau is classified as an intrinsically disordered or natively unfolded protein, because its molecular conformation under physiological conditions is highly dynamic and lacks clearly defined secondary or tertiary structures [8]. Tau does, however, seem to have a global physiological structure in which the C-terminal folds over the MTBR, and both ends of the protein approach each other [15]. Additionally, tau displays local conformations upon interaction with other proteins, such as when it is associated with the microtubules.

Tau protein is subject to a wide range of post-translational modifications (PTMs), including phosphorylation, acetylation, glycosylation, glycation, deamidation, isomerization, nitration, methylation, ubiquitylation, sumoylation, and truncation [8, 16]. These PTMs influence the subcellular localization, conformation, and ultimately also the function of tau [9]. For example, phosphorylation of residues in the MTBR or the flanking regions changes the affinity of tau for tubulin, thereby regulating the on-off kinetics of tau association with the microtubules [8]. However, tau protein has at least 86 potential phosphorylation sites and PTMs occur in a wide variety of combinations [16, 17]. It is therefore challenging to link individual PTMs to physiological tau functions.

3   Tau Aggregation

3.1   Structural Characteristics of Tau Aggregates in Different Tauopathies

The MTBR of tau contains two hexapeptides that can form intermolecular beta sheet-rich structures: aa275–280 (VCIINK) in R2 and aa306–311 (VQIVYK) in R3 [18, 19]. Under pathological conditions tau monomers can misfold to expose these motifs [20], which then leads to the self-aggregation and formation of higher order aggregates that incorporate physiological tau monomers. This process is referred to as templated misfolding or seeded-nucleation. After the initial aggregation phase, seeding of new physiological tau molecules leads to rapid elongation and subsequent fragmentation of tau aggregates [21, 22]. The seeding process leads to the formation of soluble oligomers and larger insoluble fibrils that ultimately appear as neurofibrillary tangles (NFTs) or other forms of intracellular tau aggregates [21, 22]. Not only the amount of seeds but also the conformation of tau aggregates determines the seeding potency [23]. This could be of high importance, since the seed-competence of the tau species isolated from AD patient brains correlates with the aggressiveness of their clinical symptoms [24].

Tau fibrils viewed under an electron microscope (EM) have a characteristic appearance. Their core is mostly composed of the repeat domains and is clearly visible with silver staining [25]. In addition, there is a difficult-to-visualize "fuzzy coat" polyelectrolyte brush which mainly consists of the disordered N-terminal of tau and the distal part of the C-terminal [26]. Tau fibrils in AD come in two forms: paired helical filaments (PHF) and the less abundant straight filaments (SF). The filaments from different tauopathies show distinct structural features when examined using electron microscopy [27]. Fibrils from different tauopathies also incorporate distinct tau isoforms: AD and CTE-derived fibrils contain both 3R and 4R tau, PiD fibrils contain 3R tau, and CBD and PSP fibrils contain 4R tau (Table 1).

A groundbreaking study showed that tau filaments derived from the brains of AD patients contain a core that is located between amino acid residues 306 and 378 of tau, which includes repeats 3 and 4 (R3, R4), and a part of the C-terminus [28]. The absence of the second repeat in the core also explains why AD-tau can incorporate both 3R and 4R tau isoforms. The structure of AD filament core was shown to be the same among different sporadic and familial AD cases [29]. The core of CTE fibrils also includes R3 and R4, but with a distinct amino acid residue range and fold compared to AD fibrils. Nevertheless, these findings explain why both AD and CTE can incorporate all 6 isoforms of tau. Unlike AD fibrils, however, tau filaments from CTE also display an additional density that is not part of tau, suggesting that the aggregation of tau in CTE may be caused by a yet unidentified cellular nucleation-promoting co-factor [30].

Tau filaments derived from PiD adopt a different fold with the core located between residues 254 and 378 of 3R tau, spanning R1, R3, R4, and the 9 amino acids after R4. This configuration does not include the second repeat, which explains why PiD is a 3R tauopathy. PiD fibrils also expose surface residues and phosphoepitopes that are distinct from AD-derived fibrils [31]. The core of tau fibrils from CBD (a 4R tauopathy) is located between residues 274 and 380, spanning the last amino acid of R1 repeat, R2, R3, and R4 repeats, and the 12 amino acids of the C-terminus after R4-explaining why 3R isoforms cannot be incorporated into CBD tau filaments [32, 33]. The cryo-EM structures of tau filaments of other primary tauopathies and commonly used mouse models of tauopathy are eagerly awaited. Additionally, polyanion heparin-induced recombinant tau filaments were shown to come in a wide variety of forms, but none of them having a core that would resemble AD, CTE, PiD, or CBD fibrils [34]. Interestingly, monomers derived from the brains of patients with different tauopathies display distinct conformations [20, 35]. which may underlie the heterogeneity of tau fibrils found in these tauopathies (Fig. 2).

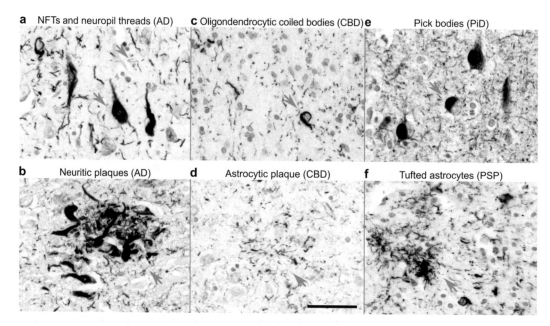

Fig. 2 Heterogeneity of tau aggregates in several tauopathies. (**a, b**) Neuronal tau aggregates are the dominant intracellular pathology present in the brains of patients with Alzheimer's disease (AD). Neurofibrillary tangles (NFTs), neuropil threads (**a**, arrows), and neuritic plaques (**b**, arrow) in the hippocampus of an AD patient. (**c, d**) Tau pathology in glial cells is characteristic for corticobasal degeneration (CBD). Oligodendrocytic coiled bodies (**c**, arrow) and astrocytic plaques (**d**, arrow) in the caudate nucleus of a CBD patient. (**e**) Pick bodies (arrow) in the granular cells of the hippocampus of a patient with Pick's disease (PiD). (**f**) Tufted astrocytes in the caudate nucleus of a patient with progressive supranuclear palsy (PSP). All panels show AT8 (p202/205) staining on paraffin sections. Scale bar shows 50 μm

3.2 Tau Aggregation in Familial Tauopathies

Tau is an unusually soluble protein with a very low potential for aggregation [8]. The reasons behind its increased aggregation under pathological conditions are still puzzling. Several clues were obtained from patients with genetic predisposition toward developing tauopathy. Europeans with the H1 haplotype of MAPT have higher risk of developing tauopathy compared to subjects with the H2 haplotype [36]. The H1 has an inversion of the MAPT gene and leads to a slight overexpression of tau [37]. The MAPT gene can also be duplicated, which likewise leads to an overexpression of tau [38–40]. Furthermore, abnormal methylation of the MAPT gene has also been linked to development of PSP and was associated with overexpression of tau mRNA [41]. These studies indicate that higher baseline levels of tau over decades can increase the likelihood of the aggregation event. Indeed, transgenic mice overexpressing all six human brain isoforms of tau develop NFTs at advanced age [42].

Furthermore, at least 50 mutations in the MAPT gene have been found that all uniquely influence the behavior of tau protein and show increased risk of developing familial tauopathies [43]. For example, certain MAPT mutations decrease association of tau with the microtubules [44], which indicates that a relative increase of free tau compared to microtubule-bound tau might increase the risk for aggregation [45]. Other mutations change splicing of tau, leading to increased abundance of the 4R isoform that is believed to be more aggregation-prone [46]. Finally, specific mutations change the conformation of tau such that it is more likely to aggregate [19]. These genetic studies indicate that subtle but prolonged abnormalities in the tau proteome can increase the risk for aggregation and developing tauopathy [47]. Mouse models that over-express human tau with pathogenic mutations show early onset NFTs and tau-induced neurodegeneration [48]. The complexity of tau pathology is also illustrated by the fact that the same pathogenic tau mutation can lead to pathological tau aggregates with a wide structural and phenotypical diversity in mice or patients with primary tauopathies [49].

3.3 Tau Aggregation in Sporadic Tauopathies

What causes the initial tau aggregation in sporadic tauopathies? Phosphorylation and acetylation at certain sites can mimic the effects of pathogenic mutations by decreasing the affinity of tau for the microtubules [43]. Additionally, mislocalization of tau from axonal to the somatodendritic compartment is one of the earliest pathological events and several PTMs have been shown to increase this event [9]. Finally, different patterns of PTMs may also change tau conformation and likely increase its aggregation propensity [50]. Prolonged dysregulation of enzymes that regulate these PTMs (kinases, phosphatases, acetylases) may thus increase the chance of aggregation. In addition, similar to RNA-binding proteins, phosphorylated or mutated tau were also recently shown to be able to undergo liquid-liquid phase transition and this was shown to initiate aggregation [51, 52]. The presence of co-factors such as RNA is also important for the initiation of aggregation and maintenance of the aggregates in vitro [53].

Proteomic analyses of pathological tau from the brains of large cohorts of AD patients and other tauopathies have revealed a dramatic heterogeneity in PTM profiles—both between different tauopathies and between individual AD patients [16, 24, 54]. The PTM heterogeneity between different AD patients is particularly striking in the sarkosyl soluble, oligomeric tau fraction [16, 24]. Given that the AD tau fibril core seems to be the same in different AD patients [28, 29], it is possible that tau oligomers are more structurally diverse and do not yet have the structured core of more mature detergent-insoluble tau filaments. Alternatively, PTMs could impose structural diversity and differential seed competence on tau aggregates that share the same filament core. Quantitative mapping of PTMs of non–seed-competent tau, soluble seed-competent tau, and tau fibrils from the brains of a large sample of AD patients, and tau PTMs from control donor brains, suggested that the process of tau aggregation starts with phosphorylation in the proline-rich domain and subsequent cleavage of the C-terminus. Tau fibrillization is then potentially promoted further by additional phosphorylation in the proline-rich domain, and acetylation and ubiquitination in the microtubule binding domain [16].

Tau is also subject to extensive cleavage. Most tau aggregates in AD contain tau truncated at the N-terminus [55–57] and/or C-terminus [16, 58, 59]. Tau fragments containing the repeat domains have higher propensity for aggregation, and human truncated tau leads to rapid formation of neurofibrillary tangles when overexpressed in adult wild-type mice [60, 61]. In vivo imaging in combination with histological analysis revealed that caspase-3 mediated cleavage of aspartate 421 on the C-terminus of tau leads to rapid formation of neurofibrillary tangles [62]. Why is truncated tau more prone to aggregation? Physiological tau molecules have a global structure in which the N- and C-terminals approach each other to cover the MTBR—thereby preventing aggregation [15]. This protective structure is lost on seed-competent tau monomers, which expose the VQIINK and VQIVYK motifs in the MTBR [20]. C- or N-terminally truncated tau molecules may therefore derive their toxicity from decreasing the barrier to aggregation [63].

4   Cellular Pathways of Tau Degradation

It is possible that tau aggregation happens continuously but is effectively resolved by the cellular protein degradation system. Subtle functional decreases in this system associated with cellular stress or aging may shift the clearance balance and lead to accumulation of pathological tau aggregates [64–66]. Furthermore, accumulation of aggregated proteins can start a vicious cycle by negatively affecting the proteostasis system [67]. The proteostasis machinery can be found in all cell types of the brain and consists of multiple integrated systems: molecular chaperones in the endoplasmic reticulum (ER), the ubiquitin-proteasome system (UPS), and the autophagy-lysosomal system.

Molecular chaperones in the ER play a crucial role in maintaining the structure and function of tau. Chaperones can regulate the refolding or degradation of misfolded tau. Improper function of chaperones may lead to ER stress and tau aggregation [68, 69]. Indeed, an autosomal dominant mutation in valosin-containing protein, a chaperone involved in dis-aggregation of tau, has been associated with a novel human tauopathy with AD-like tau pathology [70].

When misfolding or aggregation of tau does occur, pathological tau can be tagged by ubiquitin and subsequently degraded by the UPS. Indeed, the UPS is often activated in neurons with hyperphosphorylated tau inclusions [71]. Improper function of the UPS leads to tau aggregation [66, 72], and aggregated tau in turn impairs and perhaps overwhelms UPS function [73]. Ubiquitin is often found in tau aggregates in the brains of tauopathy patients [16, 74]. Aggregated tau is found both pre- and post-synaptically and associated with ubiquitin and other proteasomal components [75]. This indicates that tau aggregates are marked for degradation, but unsuccessfully degraded by the UPS. Indeed, endogenously formed tau deposits are ubiquitin-positive, but are not recruited to macroautophagosomes, and thereby escape clearance [76].

The autophagy-lysosome system is specialized in degrading larger structures such as damaged mitochondria or protein aggregates. This system regulates neurotransmission, synapse morphology, and is recruited to dendritic spines in response to synaptic activity [77–79]. Dysfunction of the autophagy-lysosomal system is involved in the activity-dependent accumulation of tau oligomers in synapses [80, 81] and potentially spreading of tau pathology [82]. Compromised autophagy can lead to granulovacuolar degeneration, and granulovacuolar degeneration bodies (GVB) are induced by tau pathology in both cellular and animal-based tau seeding models [83]. Indeed, some neurons with neurofibrillary tangles contain GVBs; although this is subregion-dependent and GVBs also occur in neighboring neurons [83–86]. It is currently unclear if tau-induced GVBs are protective or damaging [85], but alterations in mitophagy and induction of necroptosis have been associated with tau pathology and granulovacuolar degeneration [87, 88].

5 Tau Propagation

5.1 Evidence for Tau Propagation in Human AD Brains

Cross-sectional histopathological staging of tau pathology in AD describes a stereotypical progression of tau pathology, which starts in the entorhinal cortex and then seemingly follows neuronal connectivity [89]. Tau pathology begins in Braak stage I/II, then progresses to the limbic regions (e.g., hippocampus) in Braak stage III/IV, and further to the neocortex in Braak stage V/VI [90, 91]. Tau pathology staging schemes have also been devised for primary tauopathies such as PSP [92], ARTAG [93], and CTE [94], however, the evidence for tau spreading in primary tauopathies is only emerging (see for example [95]).

Tau PET scans can recapitulate Braak staging in human AD patients [96, 97]. Computational models or correlations with functional MRI data predict that most of the progression of tau pathology detected with tau PET could be explained by neuronal connectivity [98, 99]. In fact, individualized prediction of tau PET signal progression based on functional connectivity patterns outperforms the predictive power of Braak staging [100].

Tau histopathology has been compared with tau seeding using sensitive cellular biosensor assays in a series of studies [101–103]. Tau seeding was detected in the entorhinal cortex and synaptically connected areas before hyperphosphorylated tau inclusions or NFTs could be detected [102, 103]. Tau seeds were also isolated from white matter tracts and synaptosomes from regions without overt tau pathology, providing further evidence that tau pathology can propagate along neuronal connections [103].Cumulative results from human AD brains provide tentative evidence that tau seeds spread along fibers and synapses and induce tau pathology in connected regions.

5.2 Evidence for Tau Propagation Along Neuronal Connections in AD Model Systems

Individual cells readily take up tau aggregates and spread the pathology to neighboring cells in vitro, which could also be blocked by anti-tau antibodies [104–107]. These observations were corroborated in vivo as brain extracts from transgenic mice with abundant tau pathology injected into the brains of mice overexpressing human tau rapidly led to tau pathology when tau was not immunodepleted from the extracts [108]. Intracerebral injections of tau oligomers immunoprecipitated from AD brains, or synthetic recombinant tau fibrils, induced tau pathology in mice [109, 110], providing further evidence that the delivered tau was responsible for the pathology. Tau propagation to other brain regions initiated by truncated tau fibrils was blocked by monoclonal anti-tau antibodies that did not detect domains on the injected tau, indicating that the seeded endogenous tau spread to other neurons [111].

Transgenic mouse models that overexpress human tau selectively in the entorhinal cortex show predominant propagation of human tau to synaptically connected brain regions, although local spreading to neighboring neurons and astrocytes was also observed [112–114]. Application of adeno-associated viral (AAV) vectors that induced overexpression of human tau in the entorhinal cortex led to rapid propagation of human tau to synaptically connected nontransduced neurons [61, 115, 116]. Analogous spreading has been observed in seeding-based models [117] (*see* also [118] for comprehensive review). Interestingly, both AAV- and seeding-based animal models indicate that tau propagation accelerates in aged mice [116, 119].

Thus, results obtained from post-mortem human AD brains and cellular and animal model systems strongly suggest that pathological tau can spread along neuronal connections and lead to templated misfolding in recipient cells [1].

5.3 Mechanisms of Tau Secretion

Accumulating evidence from cell cultures suggests that tau spreading via synaptic connections could be blocked by tau antibodies [120–127]. In addition, pathological tau is frequently symmetrically detected at pre- and postsynaptic sites in AD patients [128], and seed-competent tau was detected in AD synaptosomes before overt neurofibrillary pathology [103]. Neurons with tau pathology are stressed but still viable and can efficiently propagate tau pathology to other cells [129]. Both physiological and pathological tau release is increased in response to neuronal activity [124, 130–132]. Tau can be released via a number of unconventional mechanisms, such as direct translation across the membrane [133–135] and autophagy or lysosomal-based secretion (reviewed in [82]). In addition, exosomes from human AD patients, tau transgenic animals, and animals with traumatic brain injury contain pathological tau and seed tau pathology in vitro and in vivo (reviewed in [136]). Tau spreading via tunneling nanotubes has also been described in vitro, both between neurons and from neurons to astrocytes [76, 137, 138]. Although the existence of these structures has been controversial and assumed to be an artefact of in vitro culturing conditions, they have recently been shown to regulate neurovascular coupling in retinal pericytes in vivo [139]. Finally, pathological tau can appear in the extracellular space when neurons degenerate. The relative importance of each these mechanisms is an active topic of research.

5.4 Mechanisms of Tau Uptake

Heparan sulfate proteoglycans (HSPG) in the neuronal plasma membrane likely play a key role in macropinocytosis-mediated uptake of tau aggregates [140–149]. HSPGs act in conjunction with more specific receptors such as low-density lipoprotein receptor-related protein 1 (LRP1) to promote endocytosis of extracellular molecules [150]. Indeed, LRP1 mediates uptake of monomeric and oligomeric tau in human induced pluripotent stem cell-derived neurons, as well as uptake of AD-derived tau in other cells [151–153]. Moreover, knockdown of LRP1 reduced tau spreading in an AAV-based tau propagation animal model [151]. Other receptors like APP, muscarinic receptors, and the cellular prion protein have been shown to bind extracellular tau aggregates and may therefore also be involved in subsequent endocytosis [154–157]. Perhaps unsurprisingly, knocking down genes at various levels downstream of endocytosis influences uptake of tau aggregates [152, 158]. In addition to macropinocytosis, other pathways such as bulk endocytosis [122] and clathrin-mediated endocytosis [159, 160] have also been proposed. Once endocytosed, tau aggregates rupture the endocytic plasma membrane and escape to seed physiological tau [153, 160–164].

5.5 The Role of Microglia, Astrocytes, and the Glymphatic System in Tau Propagation

The tau seeding process depends on the transfer of tau seeds from neurons with tau pathology to naïve neurons and the induction of templated misfolding. However, several in vivo microdialysis studies in mouse models of tauopathy have detected seed-competent tau in the interstitial fluid [121, 165, 166], indicating that tau propagation is potentially not confined just to narrow synaptic clefts or extracellular vesicles. It is possible that glial phagocytosis of extracellular tau or other extracellular clearance mechanisms may critically influence the tau propagation process as well.

Microglial depletion has been shown to reduce AAV-mediated tau propagation from the entorhinal cortex to the dentate gyrus [167]. Indeed, human primary microglia efficiently phagocytose tau [168], and, when cultured from AD and other tauopathy patients, contain tau seeds and secrete seed-competent tau into culture medium [169]. Microglia were also shown to secrete seed-competent tau in exosomes, which may be more efficiently taken up by recipient neurons than 'naked' tau seeds, thereby potentially facilitating the propagation of extracellular tau seeds [167, 170, 171] and increasing the efficiency of tau propagation.

Astrocytes take up human tau after it is overexpressed in neurons in mice [112, 151]. Different forms of tau are also efficiently phagocytosed after application to cultured astrocytes [83, 172, 173]. AAV-mediated overexpression of Transcription Factor EB (TFEB) in astrocytes reduced tau propagation in an in vivo seeding mouse model [174]. Since TFEB stimulates lysosomal biogenesis and autophagy, this study indicates that promoting degradation of extracellular tau in astrocytes reduces tau propagation [174]. Interestingly, propagation of astrocytic tau pathology to other astrocytes was reduced after selective knockout of the *MAPT* gene in neurons—suggesting that propagation of astrocytic tau pathology is dependent on seeding in neurons [10]. In contrast, oligodendrocytes were shown to propagate tau pathology to other oligodendrocytes in the absence of neuronal tau along white matter tracts [10]. More research on the role of astrocytes and oligodendrocytes in tau propagation between neurons and/or glia is, therefore, clearly needed.

Astrocytic end feet create perivascular spaces through which cerebrospinal fluid (CSF) enters the brain interstitium, mixes with the fluid in the extracellular space, and ultimately exits the central nervous system via the meningeal and cervical lymphatic system [175]. This glymphatic system plays a critical role in clearing waste from the interstitial fluid [175]. Indeed, tau clearance was significantly impaired in mice without functional lymphatic vessels in the brain [176] or after pharmacological reduction of glymphatic function [177]. Reducing the function of the glymphatic system also promoted traumatic brain injury–induced tau pathology [178]. Importantly, glymphatic function is also impaired in an aggressive mouse model of tauopathy which raises the possibility that tau pathology creates a vicious cycle by indirectly impairing its own clearance [177].

5.6 Amyloid Beta and Tau Propagation

Given their co-occurrence in AD, amyloid beta and tau pathologies can potentially interact at many levels (e.g., neuroinflammation, neuronal networks). The propagation of tau pathology to the dentate gyrus has been sped up in mice with selective expression of mutated human tau in the entorhinal cortex which were crossed with an amyloid mouse model [179]. Interestingly, injecting pathological tau derived from an AD brain into a mouse model of amyloidosis led to increased formation of tau-positive dystrophic neurites around the plaques, and also accelerated tau propagation [180, 181]. AD brain extract–induced tau pathology and propagation was increased even further in an amyloid model with a humanized *MAPT* sequence [182]. These results suggest that the presence of amyloid pathology can accelerate the propagation of tau pathology.

6   Clinical Tau-Based Diagnostics

Given the progressive nature of tau pathology and its close association with clinical cognitive or motor symptoms in AD and other tauopathies, there has been considerable research effort to identify patients with tau pathology as early as possible and to track their disease progression. Tau-based diagnostics can broadly be divided into two categories: (1) imaging-based biomarkers based on positron emission tomography (PET), and (2) detection of tau in body fluids, such as CSF and blood [183]. PET-based detection of tau pathology provides spatial information on the anatomical distribution of the tau pathology and its progression but can be limited by its high cost. The first successful tau PET tracer (flortaucipir) has been approved by the US Food and Drug Administration and plays a critical role in assessing tau pathology in clinical and in research settings. Detection of tau in body fluids has been initially restricted to detection in CSF, which is generally considered too invasive to be used as a screening tool in primary care settings. However, recent advances in sensitive biochemical detection methods have made it possible to detect tau as a correlate of early pathological changes in the blood of AD patients and opened the door for a potential tau-based blood test for AD in primary care settings. Here, we do not provide a historical overview of developments in these fields, but rather focus on the current state of the art in the tau-based diagnosis field (for more details see also [183–185]).

6.1   PET-Based Detection of Tau Pathology

The recent FDA approval of flortaucipir marks a significant improvement in understanding the stage of tau pathology of AD patients in clinical settings [186]. Despite this success, this tracer has several limitations: (1) its sensitivity to detect tau in early Braak stages is relatively limited, (2) it has significant off-target binding in the choroid plexus and brain stem nuclei, and (3) it only binds to tau pathology in AD patients, but not in other tauopathies [186]. Second generation tau PET ligands are thus being developed with the aim to improve sensitivity and limit off-target binding. Although these ligands are generally more sensitive than flortaucipir for early stages of AD tau pathology, their off-target binding still seems to be an issue or remains uncharacterized for most of these ligands [185]. Two of the next-generation tau PET ligands (PI-2620, PM-PBB3) are notable for their apparent ability to bind to tau pathology in non-AD tauopathies [187, 188], but further characterization of their sensitivity and off-target binding profiles is required. The ultimate aim is to have a toolkit that allows for sensitive anatomical detection and tracking of tau pathology in all tauopathies, with minimal off-target binding. The validation of existing ligands and identification of new ligands are therefore still ongoing and represent an exciting area of research in tauopathies.

6.2 Body Fluid-Based Biomarkers of Tau Pathology

Initial research into detection of tau in body fluids was mostly focused on the CSF, as it provided a more direct connection to tau levels in the brain, and the levels of brain-derived tau in peripheral blood were under the detection limits of existing methods. However, the development of ultra-sensitive detection techniques has made it possible to distinguish AD patients from controls by detecting brain-derived tau species in blood. The initial successful biomarker was tau phosphorylated at threonine-181 (p-tau181) [189–192]. This was rapidly followed by the detection of tau phosphorylated at threonine-217 (p-tau217) in plasma, which accurately distinguished AD patients from controls and was detectable even before a positive tau PET scan [193–195]. Current efforts are made to further validate these blood-based biomarkers and prepare them for clinical implementation [196]. These developments could soon make a sensitive and affordable screening tool for AD in primary care a reality, which could result in more efficient identification of patients for clinical trials and, importantly, correctly diagnosing patients before the occurrence of cognitive symptoms [197]. CSF and plasma p-tau181 and p-tau217 recognize N-terminal fragments that are already increased in amyloid-positive individuals in the absence of a positive tau PET scan. Current efforts are therefore also underway to measure the microtubule binding region in body fluids, which may more strongly correlate with tau pathology and therefore has the potential to stage Alzheimer's disease and provide a readout for tau-directed therapeutics [198].

The current tau-based biomarkers seem to be mostly useful for the detection of AD. More work therefore needs to be done to also identify blood-based biomarkers for non-AD neurodegenerative disorders, such as the primary tauopathies [199]. The neurodegeneration-related plasma biomarker neurofilament light chain (NfL) does seem to be increased across all neurodegenerative disorders [199]. Importantly, in vivo plasma NfL levels correlate with the presence of neurofibrillary tangles and neurodegeneration confirmed post-mortem, indicating that NfL could be used as a readout of the degree of neurodegeneration in tauopathies [200].

7   Tau-Based Therapeutics

The high-profile failures of amyloid beta targeting drugs in late stage clinical trials, along with the close correlation of tau pathology with clinical symptoms have contributed to a pivot toward the targeting of tau pathology [201–203]. Tau targeting drugs currently in clinical trials can be broadly classified as (1) immunotherapeutics, (2) antisense oligonucleotides, and (3) small molecules [204]. So far the only tau-targeting therapeutic that has been tested in a phase 3 clinical trial is the small-molecule aggregation inhibitor LMTM, which was found to be inactive in patients with mild to moderate AD [205].

7.1   Tau Immunotherapy

The active immunotherapy approach aims to stimulate the patient's immune system and make antibodies against pathological variants of tau, akin to vaccines against infectious diseases. Currently one vaccine is in phase 1 clinical trials (ACI-35), whereas another has recently successfully completed phase 2 clinical trials (AADvac1) [204, 206, 207]. Passive immunotherapy, a related approach, is based on application of monoclonal antibodies, which bind to specific domains on tau. Currently, several monoclonal antibodies are in clinical trials against the N-terminal of tau (Gosuranemab, Semorinemab, Tilavonemab), mid-domain (BIIB076, Bepranemab), phosphorylated tau epitopes (JNJ-63733657, Lu AF87908, PNT001), and conformational epitopes of tau (Zagotenemab) [208]. The main strength of the immunotherapy approach is the ability of antibodies to target specific tau epitopes. The first generation of therapeutic tau antibodies targeted the N-terminus of tau. These antibodies performed poorly compared to mid-domain antibodies in cellular and mouse seeding models using AD-derived tau aggregates [56, 111]. Given that some N-terminus targeting tau antibodies have failed clinical trials in primary tauopathies and AD, future clinical studies will have to determine whether targeting other epitopes on tau would be more efficacious [209].

The brain uptake of tau-specific antibodies via intact blood–brain barrier is limited, only around 0.1% of plasma level of IgG is detectable in the CSF, and even lower levels can be expected in the brain [210]. Furthermore, no or low uptake of antibodies into neurons suggests that tau antibodies mostly target extracellular tau [208]. The next generation of passive tau immunotherapeutics currently tested in preclinical studies consists of monoclonal tau antibodies coupled to brain shuttles [211], and the use of viral vectors to express antibodies (or their fragments) directly in the brain tissue [212, 213]. In addition, multispecific antibodies engineered to bind to tau and other neurodegeneration-related targets may represent another promising approach [211, 214].

7.2 Antisense Oligonucleotides

Tau-targeting antisense oligonucleotides aim to lower tau expression, which might potentially reduce tau pathology, presumably by limiting the fresh pool of monomers that would be incorporated into progressing tau pathology [215]. BIIB080 is currently the sole representative of the class being tested in a clinical trial. The strength of this approach is that it targets tau in the intracellular compartment. On the other hand, removing physiological tau could lead to potential adverse effects, although preclinical studies in nonhuman primates suggest that this is likely not problematic [215]. The oligonucleotide must be applied directly into the CSF, which is considered relatively invasive and might limit efficient distribution throughout the brain compared to drugs delivered via the bloodstream. The delivery could be potentially improved by coupling the drug to a brain shuttle and administering it intravenously [216], similar to a described approach for monoclonal antibodies [217].

7.3 Small Molecules

Despite the high-profile failure of LMTM, small molecules remain an attractive class of therapeutics as they can potentially be designed to have low cost, good brain uptake, and efficient cellular uptake. Two tau-targeting small-molecule therapeutics are currently being studied in clinical trials. ACI-3024 is an aggregation inhibitor which is thought to prevent the aggregation of tau and even disaggregate preexisting tau pathology [218]. LY3372689 is an inhibitor of the O-GlcNAcase enzyme, promoting tau glycosylation, and thereby preventing aggregation by stabilizing tau in its nonpathogenic form [219]. More small molecules are on the horizon, most notably those based on the proteolysis targeting chimera (PROTAC) platform, including one candidate that binds pathological tau and targets it for intracellular degradation [220]. Whether these drugs can truly lower tau pathology in patients remains to be determined. However, the advent of tau PET imaging has made it possible to at least test this proof of principle in a small clinical study [204].

8 Conclusion

These are exciting times to be a "Tauist." Recent years have brought a dramatic increase in research focus on tau pathology in AD and other tauopathies. New intriguing insights into the process of tau aggregation, its cellular degradation, and tau spreading throughout the brain appear regularly. Although beyond the scope of this chapter, there is also a strong research focus on how tau pathology affects the brain immune system, neuronal networks, and neurodegeneration [5, 221]. Tau research has contributed to the dramatic increase of our ability to visualize and track tau pathology in the brains of living patients. Furthermore, tau-based blood tests are currently making their way into primary care, potentially allowing for early identification of people at risk for the development of AD. Similar efforts are also ongoing to identify accurate blood tests for Frontotemporal Dementia (FTD). Several promising avenues for tau-targeting therapies are currently being pursued, which will hopefully lead to an effective management of these devastating neurodegenerative conditions.

References

1. Vogels T, Leuzy A, Cicognola C et al (2020) Propagation of tau pathology: integrating insights from postmortem and in vivo studies. Biol Psychiatry 87(9):808–818. https://doi.org/10.1016/j.biopsych.2019.09.019

2. Prince M (2017) Progress on dementia-leaving no one behind. Lancet 390 (10113):51–53. https://doi.org/10.1016/S0140-6736(17)31757-9

3. Scheltens P, Blennow K, Breteler MMB et al (2017) Alzheimer's disease. Lancet 388 (10043):505–517. https://doi.org/10.1016/S0140-6736(15)01124-1

4. Hoglinger GU, Respondek G, Kovacs GG (2018) New classification of tauopathies. Rev Neurol 174(9):664–668. https://doi.org/10.1016/j.neurol.2018.07.001

5. Vogels T, Murgoci A-N, Hromadka T (2019) Intersection of pathological tau and microglia at the synapse. Acta Neuropathol Comm 7 (1):109. https://doi.org/10.1186/s40478-019-0754-y

6. Gomez-Isla T, Hollister R, West H et al (1997) Neuronal loss correlates with but exceeds neurofibrillary tangles in Alzheimer's disease. Ann Neurol 41(1):17–24. https://doi.org/10.1002/ana.410410106

7. Arriagada PV, Growdon JH, Hedley-Whyte ET et al (1992) Neurofibrillary tangles but not senile plaques parallel duration and severity of Alzheimer's disease. Neurology 42 (3):631–639

8. Wang Y, Mandelkow E (2016) Tau in physiology and pathology. Nat Rev Neurosci 17 (1):5–21. https://doi.org/10.1038/nrn.2015.1

9. Zempel H, Mandelkow E (2014) Lost after translation: missorting of tau protein and consequences for Alzheimer disease. Trends Neurosci 37(12):721–732. https://doi.org/10.1016/j.tins.2014.08.004

10. Narasimhan S, Changolkar L, Riddle DM et al (2020) Human tau pathology transmits glial tau aggregates in the absence of neuronal tau.

J Exp Med 217(2):e20190783. https://doi.org/10.1084/jem.20190783

11. Zhang Y, Chen K, Sloan SA et al (2014) An RNA-sequencing transcriptome and splicing database of glia, neurons, and vascular cells of the cerebral cortex. J Neurosci 34 (36):11929–11947. https://doi.org/10.1523/JNEUROSCI.1860-14.2014

12. Sotiropoulos I, Galas MC, Silva JM et al (2017) Atypical, non-standard functions of the microtubule associated Tau protein. Acta Neuropathol Comm 5(1):91. https://doi.org/10.1186/s40478-017-0489-6

13. Kent SA, Spires-Jones TL, Durrant CS (2020) The physiological roles of tau and Aβ: implications for Alzheimer's disease pathology and therapeutics. Acta Neuropathol 140 (4):417–447. https://doi.org/10.1007/s00401-020-02196-w

14. Gibbons GS, Lee VMY, Trojanowski JQ (2018) Mechanisms of cell-to-cell transmission of pathological tau: A Review. JAMA Neurol 76(1):101–108. https://doi.org/10.1001/jamaneurol.2018.2505

15. Jeganathan S, von Bergen M, Brutlach H et al (2006) Global hairpin folding of tau in solution. Biochemistry 45(7):2283–2293. https://doi.org/10.1021/bi0521543

16. Wesseling H, Mair W, Kumar M et al (2020) Tau PTM profiles identify patient heterogeneity and stages of Alzheimer's disease. Cell 183 (6):1699–1713. https://doi.org/10.1016/j.cell.2020.10.029

17. Guo T, Noble W, Hanger DP (2017) Roles of tau protein in health and disease. Acta Neuropathol 133(5):665–704. https://doi.org/10.1007/s00401-017-1707-9

18. von Bergen M, Friedhoff P, Biernat J et al (2000) Assembly of tau protein into Alzheimer paired helical filaments depends on a local sequence motif ((306)VQIVYK(311)) forming beta structure. PNAS 97(10):5129–5134

19. von Bergen M, Barghorn S, Li L et al (2001) Mutations of tau protein in frontotemporal dementia promote aggregation of paired

helical filaments by enhancing local beta-structure. J Biol Chem 276 (51):48165–48174. https://doi.org/10.1074/jbc.M105196200

20. Mirbaha H, Chen D, Morazova OA et al (2018) Inert and seed-competent tau monomers suggest structural origins of aggregation. eLife 7:e36584. https://doi.org/10.7554/eLife.36584

21. Kundel F, Hong L, Falcon B et al (2018) Measurement of tau filament fragmentation provides insights into prion-like spreading. ACS Chem Neurosci 9(6):1276–1282. https://doi.org/10.1021/acschemneuro.8b00094

22. Marreiro A, Van Kolen K, Sousa C et al (2020) Comparison of size distribution and (Pro249-Ser258) epitope exposure in in vitro and in vivo derived tau fibrils. BMC Mol and Cell Biol 21(1):81. https://doi.org/10.1186/s12860-020-00320-y

23. Falcon B, Cavallini A, Angers R et al (2015) Conformation determines the seeding potencies of native and recombinant Tau aggregates. J Biol Chem 290(2):1049–1065. https://doi.org/10.1074/jbc.M114.589309

24. Dujardin S, Commins C, Lathuiliere A et al (2020) Tau molecular diversity contributes to clinical heterogeneity in Alzheimer's disease. Nat Med 26(8):1256–1263. https://doi.org/10.1038/s41591-020-0938-9

25. Crowther T, Goedert M, Wischik CM (1989) The repeat region of microtubule-associated protein tau forms part of the core of the paired helical filament of Alzheimer's disease. Ann Med 21(2):127–132

26. Wischik CM, Novak M, Edwards PC et al (1988) Structural characterization of the core of the paired helical filament of Alzheimer disease. PNAS 85(13):4884–4888

27. Yagishita S, Itoh Y, Nan W et al (1981) Reappraisal of the fine structure of Alzheimer's neurofibrillary tangles. Acta Neuropathol 54 (3):239–246

28. Fitzpatrick AWP, Falcon B, He S et al (2017) Cryo-EM structures of tau filaments from Alzheimer's disease. Nature 547 (7662):185–190. https://doi.org/10.1038/nature23002

29. Falcon B, Zhang W, Schweighauser M et al (2018) Tau filaments from multiple cases of sporadic and inherited Alzheimer's disease adopt a common fold. Acta Neuropathol 136(5):699–708. https://doi.org/10.1007/s00401-018-1914-z

30. Falcon B, Zivanov J, Zhang W et al (2019) Novel tau filament fold in chronic traumatic encephalopathy encloses hydrophobic molecules. Nature 568(7752):420–423. https://doi.org/10.1038/s41586-019-1026-5

31. Falcon B, Zhang W, Murzin AG et al (2018) Structures of filaments from Pick's disease reveal a novel tau protein fold. Nature 561 (7721):137–140. https://doi.org/10.1038/s41586-018-0454-y

32. Zhang W, Tarutani A, Newell KL et al (2020) Novel tau filament fold in corticobasal degeneration. Nature 580(7802):283–287. https://doi.org/10.1038/s41586-020-2043-0

33. Arakhamia T, Lee CE, Carlomagno Y et al (2020) Posttranslational modifications mediate the structural diversity of tauopathy strains. Cell 180(4):633–644.e12. https://doi.org/10.1016/j.cell.2020.01.027

34. Zhang W, Falcon B, Murzin AG et al (2018) Heparin-induced tau filaments are polymorphic and differ from those in Alzheimer's and Pick's disease. BioRxiv:468892. https://doi.org/10.1101/468892

35. Sharma AM, Thomas TL, Woodard DR et al (2018) Tau monomer encodes strains. eLife 7:e37813. https://doi.org/10.7554/eLife.37813

36. Heckman MG, Brennan RR, Labbe C et al (2019) Association of MAPT subhaplotypes with risk of progressive supranuclear palsy and severity of tau pathology. JAMA Neurol 76 (6):710–717. https://doi.org/10.1001/jamaneurol.2019.0250

37. Myer AJ, Pittman AM, Zhao AS et al (2007) The MAPT H1c risk haplotype is associated with increased expression of tau and especially of 4 repeat containing transcripts. Neurobiol Dis 25(3):561–570. https://doi.org/10.1016/j.nbd.2006.10.018

38. Rovelet-Lecrux A, Hannequin D, Guillin O et al (2010) Frontotemporal dementia phenotype associated with MAPT gene duplication. J Alzheimers Dis 21(3):897–902. https://doi.org/10.3233/JAD-2010-100441

39. Le Guennec K, Quenez O, Nicolas G et al (2017) 17q21.31 duplication causes prominent tau-related dementia with increased MAPT expression. Mol Psychiatry 22 (8):1119–1125. https://doi.org/10.1038/mp.2016.226

40. Chen Z, Chen JA, Shatunov A et al (2019) Genome-wide survey of copy number variants finds MAPT duplications in progressive supranuclear palsy. Movement Disord 34

(7):1049–1059. https://doi.org/10.1002/mds.27702

41. Huin V, Deramecourt V, Caparros-Lefebvre D et al (2016) The MAPT gene is differentially methylated in the progressive supranuclear palsy brain. Movement Disord 31 (12):1883–1890. https://doi.org/10.1002/mds.26820

42. Andorfer C, Kress Y, Espinoza M et al (2003) Hyperphosphorylation and aggregation of tau in mice expressing normal human tau isoforms. J Neurochem 86(3):582–590

43. Strang KH, Golde TE, Giasson BI (2019) MAPT mutations, tauopathy, and mechanisms of neurodegeneration. Lab Investig 99 (7):912–928. https://doi.org/10.1038/s41374-019-0197-x

44. Rizzu P, Van Swieten JC, Joosse M et al (1999) High prevalence of mutations in the microtubule-associated protein tau in a population study of frontotemporal dementia in the Netherlands. Am J Hum Genet 64 (2):414–421. https://doi.org/10.1086/302256

45. Di Primio C, Quercioli V, Siano G et al (2017) The distance between N and C termini of tau and of FTDP-17 mutants Is modulated by microtubule interactions in living cells. Front Mol Neurosci 10:210. https://doi.org/10.3389/fnmol.2017.00210

46. Hutton M, Lendon CL, Rizzu P et al (1998) Association of missense and 5′-splice-site mutations in tau with the inherited dementia FTDP-17. Nature 393(6686):702–705. https://doi.org/10.1038/31508

47. Hong M, Zhukareva V, Vogelsberg-Ragaglia V et al (1998) Mutation-specific functional impairments in distinct tau isoforms of hereditary FTDP-17. Science 282 (5395):1914–1917

48. Gotz J, Bodea LG, Goedert M (2018) Rodent models for Alzheimer disease. Nat Rev Neurosci 19(10):583–598. https://doi.org/10.1038/s41583-018-0054-8

49. Daude N, Kim C, Kang SG et al (2020) Diverse, evolving conformer populations drive distinct phenotypes in frontotemporal lobar degeneration caused by the same MAPT-P301L mutation. Acta Neuropathol 139(6):1045–1070. https://doi.org/10.1007/s00401-020-02148-4

50. Drombosky KW, Chen D, Woodard D et al (2018) Native tau structure is disrupted by disease-associated mutations that promote aggregation. BioRxiv:330266. https://doi.org/10.1101/330266

51. Ambadipudi S, Biernat J, Riedel D et al (2017) Liquid-liquid phase separation of the microtubule-binding repeats of the Alzheimer-related protein tau. Nat Commun 8(1):275. https://doi.org/10.1038/s41467-017-00480-0

52. Wegmann S, Eftekharzadeh B, Tepper K et al (2018) Tau protein liquid-liquid phase separation can initiate tau aggregation. EMBO J 37(7):e98049. https://doi.org/10.15252/embj.201798049

53. Fichou Y, Lin Y, Rauch JN et al (2018) Cofactors are essential constituents of stable and seeding-active tau fibrils. PNAS 115 (52):13234–13239. https://doi.org/10.1073/pnas.1810058115

54. Kametani F, Yoshida M, Matsubara T et al (2020) Comparison of common and disease-specific post-translational modifications of pathological tau associated with a wide range of tauopathies. Front Neurosci 14:1110. https://doi.org/10.3389/fnins.2020.581936

55. Li L, Jiang Y, Hu W, Tung YC et al (2019) Pathological alterations of tau in Alzheimer's disease and 3xTg-AD mouse brains. Mol Neurobiol 56(9):6168–6183. https://doi.org/10.1007/s12035-019-1507-4

56. Courade JP, Angers R, Mairet-Coello G et al (2018) Epitope determines efficacy of therapeutic anti-tau antibodies in a functional assay with human Alzheimer tau. Acta Neuropathol 136(5):729–745. https://doi.org/10.1007/s00401-018-1911-2

57. Derisbourg M, Leghay C, Chiappetta G et al (2015) Role of the Tau N-terminal region in microtubule stabilization revealed by new endogenous truncated forms. Sci Rep 5:9659. https://doi.org/10.1038/srep09659

58. Novak M, Kabat J, Wischik CM (1993) Molecular characterization of the minimal protease resistant tau unit of the Alzheimer's disease paired helical filament. EMBO J 12 (1):365–370

59. Nicholls SB, DeVos SL, Commins C et al (2017) Characterization of TauC3 antibody and demonstration of its potential to block tau propagation. PLoS One 12(5): e0177914. https://doi.org/10.1371/journal.pone.0177914

60. Zilka N, Kovacech B, Barath P et al (2012) The self-perpetuating tau truncation circle. Biochem Soc T 40(4):681–686. https://doi.org/10.1042/BST20120015

61. Vogels T, Vargová G, Brezováková V et al (2020) Viral delivery of non-mutated human

truncated tau to neurons recapitulates key features of human tauopathy in wild-type mice. J Alzheimers Dis 77(2):551–568. https://doi.org/10.3233/JAD-200047

62. de Calignon A, Fox LM, Pitstick R et al (2010) Caspase activation precedes and leads to tangles. Nature 464(7292):1201–1204. https://doi.org/10.1038/nature08890

63. Wang YP, Biernat J, Pickhardt M et al (2007) Stepwise proteolysis liberates tau fragments that nucleate the Alzheimer-like aggregation of full-length tau in a neuronal cell model. PNAS 104(24):10252–10257. https://doi.org/10.1073/pnas.0703676104

64. Bodea LG, Evans HT, Van der Jeugd A et al (2017) Accelerated aging exacerbates a pre-existing pathology in a tau transgenic mouse model. Aging Cell 16(2):377–386. https://doi.org/10.1111/acel.12565

65. Kundra R, Ciryam P, Morimoto RI et al (2017) Protein homeostasis of a metastable subproteome associated with Alzheimer's disease. PNAS 114(28):5703–5711. https://doi.org/10.1073/pnas.1618417114

66. Wang P, Joberty G, Buist A et al (2017) Tau interactome mapping based identification of Otub1 as Tau deubiquitinase involved in accumulation of pathological Tau forms in vitro and in vivo. Acta Neuropathol 133 (5):731–749. https://doi.org/10.1007/s00401-016-1663-9

67. Zhang Y, Chen X, Zhao Y et al (2017) The role of ubiquitin proteasomal system and autophagy-lysosome pathway in Alzheimer's disease. Rev Neurosci 28(8):861–868. https://doi.org/10.1515/revneuro-2017-0013

68. Kim E, Sakata K, Liao FF (2017) Bidirectional interplay of HSF1 degradation and UPR activation promotes tau hyperphosphorylation. PLoS Genet 13(7):e1006849. https://doi.org/10.1371/journal.pgen.1006849

69. Shelton LB, Baker JD, Zheng D et al (2017) Hsp90 activator Aha1 drives production of pathological tau aggregates. PNAS 114 (36):9707–9712. https://doi.org/10.1073/pnas.1707039114

70. Darwich NF, Phan JM, Kim B et al (2020) Autosomal dominant VCP hypomorph mutation impairs disaggregation of PHF-tau. Science 370(6519):eaay8826. https://doi.org/10.1126/science.aay8826

71. Hoozemans JJM, van Haastert ES, Nijholt DAT et al (2009) The unfolded protein response is activated in pretangle neurons in Alzheimer's disease hippocampus. Am J Pathol 174(4):1241–1251. https://doi.org/10.2353/ajpath.2009.080814

72. Liu YH, Wei W, Yin J et al (2009) Proteasome inhibition increases tau accumulation independent of phosphorylation. Neurobiol Aging 30(12):1949–1961. https://doi.org/10.1016/j.neurobiolaging.2008.02.012

73. Keck S, Nitsch R, Grune T et al (2003) Proteasome inhibition by paired helical filament-tau in brains of patients with Alzheimer's disease. J Neurochem 85(1):115–122

74. Mori H, Kondo J, Ihara Y (1987) Ubiquitin is a component of paired helical filaments in Alzheimer's disease. Science 235 (4796):1641–1644

75. Tai HC, Serrano-Pozo A, Hashimoto T et al (2012) The synaptic accumulation of hyperphosphorylated tau oligomers in Alzheimer disease is associated with dysfunction of the ubiquitin-proteasome system. Am J Pathol 181(4):1426–1435. https://doi.org/10.1016/j.ajpath.2012.06.033

76. Chastagner P, Loria F, Vargas JY et al (2020) Fate and propagation of endogenously formed Tau aggregates in neuronal cells. EMBO Mol Med 12:e12025. https://doi.org/10.15252/emmm.202012025

77. Goo MS, Sancho L, Slepak N et al (2017) Activity-dependent trafficking of lysosomes in dendrites and dendritic spines. J Cell Biol 216(8):2499–2513. https://doi.org/10.1083/jcb.201704068

78. Inoue K, Rispoli J, Yang L et al (2013) Coordinate regulation of mature dopaminergic axon morphology by macroautophagy and the PTEN signaling pathway. PLoS Genet 9 (10):e1003845. https://doi.org/10.1371/journal.pgen.1003845

79. Hernandez D, Torres CA, Setlik W et al (2012) Regulation of presynaptic neurotransmission by macroautophagy. Neuron 74 (2):277–284. https://doi.org/10.1016/j.neuron.2012.02.020

80. Akwa Y, Gondard E, Mann A et al (2017) Synaptic activity protects against AD and FTD-like pathology via autophagic-lysosomal degradation. Mol Psychiatry 23 (6):1530–1540. https://doi.org/10.1038/mp.2017.142

81. Kimura T, Suzuki M, Akagi T (2017) Age-dependent changes in synaptic plasticity enhance tau oligomerization in the mouse hippocampus. Acta Neuropathol Comm 5 (1):67. https://doi.org/10.1186/s40478-017-0469-x

82. Jiang S, Bhaskar (2020) Degradation and transmission of tau by autophagic-

endolysosomal networks and potential therapeutic targets for tauopathy. Front Mol Neurosci 13:586731. https://doi.org/10.3389/fnmol.2020.586731

83. Wiersma VI, van Ziel AM, Vazquez-Sanchez S et al (2019) Granulovacuolar degeneration bodies are neuron-selective lysosomal structures induced by intracellular tau pathology. Acta Neuropathol 138(6):943–970. https://doi.org/10.1007/s00401-019-02046-4

84. Puladi B, Dinekov M, Arzberger T et al (2020) The relation between tau pathology and granulovacuolar degeneration of neurons. Neurobiol Dis 147:105138. https://doi.org/10.1016/j.nbd.2020.105138

85. Wiersma VI, Hoozemans JJM, Scheper W (2020) Untangling the origin and function of granulovacuolar degeneration bodies in neurodegenerative proteinopathies. Acta Neuropathol Comm 8(1):153. https://doi.org/10.1186/s40478-020-00996-5

86. Thal DR, Del Tredici K, Ludolph AC et al (2011) Stages of granulovacuolar degeneration: their relation to Alzheimer's disease and chronic stress response. Acta Neuropathol 122(5):577–589. https://doi.org/10.1007/s00401-011-0871-6

87. Koper MJ, Van Schoor E, Ospitalieri S et al (2020) Necrosome complex detected in granulovacuolar degeneration is associated with neuronal loss in Alzheimer's disease. Acta Neuropathol 139(3):463–484. https://doi.org/10.1007/s00401-019-02103-y

88. Hou X, Watzlawik JO, Cook C et al (2020) Mitophagy alterations in Alzheimer's disease are associated with granulovacuolar degeneration and early tau pathology. Alzheimers Dement. https://doi.org/10.1002/alz.12198

89. Braak H, Braak E (1991) Neuropathological staging of Alzheimer-related changes. Acta Neuropathol 82(4):239–259

90. Nelson PT, Alafuzoff I, Bigio EH et al (2012) Correlation of Alzheimer disease neuropathologic changes with cognitive status: a review of the literature. J Neuropath Exp Neur 71(5):362–381. https://doi.org/10.1097/NEN.0b013e31825018f7

91. Scholl M, Lockhart SN, Schonhaut DR et al (2016) PET imaging of tau deposition in the aging human brain. Neuron 89(5):971–982. https://doi.org/10.1016/j.neuron.2016.01.028

92. Kovacs GG, Lukic MJ, Irwin DJ et al (2020) Distribution patterns of tau pathology in progressive supranuclear palsy. Acta Neuropathol 140(2):99–119. https://doi.org/10.1007/s00401-020-02158-2

93. Kovacs GG, Xie SX, Robinson JL et al (2018) Sequential stages and distribution patterns of aging-related tau astrogliopathy (ARTAG) in the human brain. Acta Neuropathol Comm 6:50. https://doi.org/10.1186/s40478-018-0552-y

94. Alosco ML, Cherry JD, Huber BR et al (2020) Characterizing tau deposition in chronic traumatic encephalopathy (CTE): utility of the McKee CTE staging scheme. Acta Neuropathol 140(4):495–512. https://doi.org/10.1007/s00401-020-02197-9

95. Kim EJ, Hwang JHL, Gaus SE et al (2020) Evidence of corticofugal tau spreading in patients with frontotemporal dementia. Acta Neuropathol 139(1):27–43. https://doi.org/10.1007/s00401-019-02075-z

96. Schwarz AJ, Yu P, Miller BB et al (2016) Regional profiles of the candidate tau PET ligand 18F-AV-1451 recapitulate key features of Braak histopathological stages. Brain 139(Pt 5):1539–1550. https://doi.org/10.1093/brain/aww023

97. Lowe VJ, Wiste HJ, Senjem ML et al (2018) Widespread brain tau and its association with ageing, Braak stage and Alzheimer's dementia. Brain 141(1):271–287. https://doi.org/10.1093/brain/awx320

98. Vogel JW, Iturria-Medina Y, Strandberg OT et al (2020) Spread of pathological tau proteins through communicating neurons in human Alzheimer's disease. Nat Commun 11(1):2612. https://doi.org/10.1038/s41467-020-15701-2

99. Franzmeier N, Neitzel J, Rubinski A et al (2020) Functional brain architecture is associated with the rate of tau accumulation in Alzheimer's disease. Nat Commun 11(1):347. https://doi.org/10.1038/s41467-019-14159-1

100. Franzmeier N, Dewenter A, Frontzkowski L et al (2020) Patient-centered connectivity-based prediction of tau pathology spread in Alzheimer's disease. Sci Adv 6(48):eabd1327. https://doi.org/10.1126/sciadv.abd1327

101. Kaufman SK, Thomas TL, Del Tredici K et al (2017) Characterization of tau prion seeding activity and strains from formaldehyde-fixed tissue. Acta Neuropathol Comm 5(1):41. https://doi.org/10.1186/s40478-017-0442-8

102. Kaufman SK, Del Tredici K, Thomas TL et al (2018) Tau seeding activity begins in the transentorhinal/entorhinal regions and anticipates phospho-tau pathology in

Alzheimer's disease and PART. Acta Neuropathol 136(1):57–67. https://doi.org/10.1007/s00401-018-1855-6

103. DeVos SL, Corjuc BT, Oakley DH et al (2018) Synaptic tau seeding precedes tau pathology in human Alzheimer's disease Brain. Front Neurosci 12:267. https://doi.org/10.3389/fnins.2018.00267

104. Frost B, Jacks RL, Diamond MI (2009) Propagation of Tau misfolding from the outside to the inside of a cell. J Biol Chem 284 (19):12845–12852. https://doi.org/10.1074/jbc.M808759200

105. Guo JL, Lee VMY (2011) Seeding of normal tau by pathological tau conformers drives pathogenesis of Alzheimer-like tangles. J Biol Chem 286(17):15317–15331. https://doi.org/10.1074/jbc.M110.209296

106. Kfoury N, Holmes BB, Jiang H et al (2012) Trans-cellular propagation of Tau aggregation by fibrillar species. J Biol Chem 287 (23):19440–19451. https://doi.org/10.1074/jbc.M112.346072

107. Santa-Maria I, Varghese M, Ksiezak-Reding H et al (2012) Paired helical filaments from Alzheimer disease brain induce intracellular accumulation of Tau protein in aggresomes. J Biol Chem 287(24):20522–20533. https://doi.org/10.1074/jbc.M111.323279

108. Clavaguera F, Bolmont T, Crowther RA et al (2009) Transmission and spreading of tauopathy in transgenic mouse brain. Nat Cell Biol 11(7):909–913. https://doi.org/10.1038/ncb1901

109. Lasagna-Reeves CA, Castillo-Carranza DL, Sengupta U et al (2012) Alzheimer brain-derived tau oligomers propagate pathology from endogenous tau. Sci Rep 2:700. https://doi.org/10.1038/srep00700

110. Iba M, Guo JL, McBride JD et al (2013) Synthetic tau fibrils mediate transmission of neurofibrillary tangles in a transgenic mouse model of Alzheimer's-like tauopathy. J Neurosci 33(3):1024–1037. https://doi.org/10.1523/JNEUROSCI.2642-12.2013

111. Albert M, Mairet-Coello G, Danis C et al (2019) Prevention of tau seeding and propagation by immunotherapy with a central tau epitope antibody. Brain 142(6):1736–1750. https://doi.org/10.1093/brain/awz100

112. de Calignon A, Polydoro M, Suarez-Calvet M et al (2012) Propagation of tau pathology in a model of early Alzheimer's disease. Neuron 73(4):685–697. https://doi.org/10.1016/j.neuron.2011.11.033

113. Harris JA, Koyama A, Maeda S et al (2012) Human P301L-mutant tau expression in

mouse entorhinal-hippocampal network causes tau aggregation and presynaptic pathology but no cognitive deficits. PLoS One 7(9):e45881. https://doi.org/10.1371/journal.pone.0045881

114. Liu L, Drouet V, Wu JW et al (2012) Trans-synaptic spread of tau pathology in vivo. PLoS One 7(2):e31302. https://doi.org/10.1371/journal.pone.0031302

115. Wegmann S, Maury EA, Kirk MJ et al (2015) Removing endogenous tau does not prevent tau propagation yet reduces its neurotoxicity. EMBO J 34(24):3028–3041. https://doi.org/10.15252/embj.201592748

116. Wegmann S, Bennett RE, Delorme L et al (2019) Experimental evidence for the age dependence of tau protein spread in the brain. Sci Adv 5(6):eaaw6404. https://doi.org/10.1126/sciadv.aaw6404

117. Ahmed Z, Cooper J, Murray TK et al (2014) A novel in vivo model of tau propagation with rapid and progressive neurofibrillary tangle pathology: the pattern of spread is determined by connectivity, not proximity. Acta Neuropathol 127(5):667–683. https://doi.org/10.1007/s00401-014-1254-6

118. McAllister BB, Lacoursiere SG, Sutherland RJ et al (2020) Intracerebral seeding of amyloid-β and tau pathology in mice: factors underlying prion-like spreading and comparisons with α-synuclein. Neurosci Biobehav Rev 112:1–27. https://doi.org/10.1016/j.neubiorev.2020.01.026

119. Guo JL, Narasimhan S, Changolkar L et al (2016) Unique pathological tau conformers from Alzheimer's brains transmit tau pathology in nontransgenic mice. J Exp Med 213 (12):2635–2654. https://doi.org/10.1084/jem.20160833

120. Calafate S, Buist A, Miskiewicz K et al (2015) Synaptic contacts enhance cell-to-cell Tau pathology propagation. Cell Rep 11 (8):1176–1183. https://doi.org/10.1016/j.celrep.2015.04.043

121. Takeda S, Wegmann S, Cho H et al (2015) Neuronal uptake and propagation of a rare phosphorylated high-molecular-weight tau derived from Alzheimer's disease brain. Nat Commun 6:8490. https://doi.org/10.1038/ncomms9490

122. Wu JW, Herman M, Liu L et al (2013) Small misfolded tau species are internalized via bulk endocytosis and anterogradely and retrogradely transported in neurons. J Biol Chem 288(3):1856–1870. https://doi.org/10.1074/jbc.M112.394528

123. Dujardin S, Lécolle K, Caillierez R et al (2014) Neuron-to-neuron wild-type Tau protein transfer through a trans-synaptic mechanism: relevance to sporadic tauopathies. Acta Neuropathol Comm 2:14. https://doi.org/10.1186/2051-5960-2-14

124. Wu JW, Hussaini SA, Bastille IM et al (2016) Neuronal activity enhances tau propagation and tau pathology in vivo. Nat Neurosci 19 (8):1085–1092. https://doi.org/10.1038/nn.4328

125. Wang Y, Balaji V, Kaniyappan S et al (2017) The release and trans-synaptic transmission of Tau via exosomes. Mol Neurodegener 12 (1):5. https://doi.org/10.1186/s13024-016-0143-y

126. Nobuhara CK, DeVos SL, Commins C et al (2017) Tau antibody-targeting pathological species block neuronal uptake and interneuron propagation of tau in vitro. Am J Pathol 187(6):1399–1412. https://doi.org/10.1016/j.ajpath.2017.01.022

127. Katsikoudi A, Ficulle E, Cavallini A et al (2020) Quantitative propagation of assembled human tau from Alzheimer's disease brain in microfluidic neuronal cultures. J Biol Chem 295(37):13079–13093. https://doi.org/10.1074/jbc.RA120.013325

128. Tai HC, Wang BY, Serrano-Pozo A et al (2014) Frequent and symmetric deposition of misfolded tau oligomers within presynaptic and postsynaptic terminals in Alzheimer's disease. Acta Neuropathol Comm 2:146. https://doi.org/10.1186/s40478-014-0146-2

129. Hallinan GI, Vargas-Caballero M, West J et al (2019) Tau misfolding efficiently propagates between individual intact hippocampal neurons. J Neurosci 39(48):9623–9632. https://doi.org/10.1523/JNEUROSCI.1590-19.2019

130. Pooler AM, Phillips EC, Lau DHW et al (2013) Physiological release of endogenous tau is stimulated by neuronal activity. EMBO Rep 14(4):389–394. https://doi.org/10.1038/embor.2013.15

131. Yamada K, Holth JK, Liao F et al (2014) Neuronal activity regulates extracellular tau in vivo. J Exp Med 211(3):387–393. https://doi.org/10.1084/jem.20131685

132. Sokolow S, Henkins KM, Bilousova T et al (2015) Pre-synaptic C-terminal truncated tau is released from cortical synapses in Alzheimer's disease. J Neurochem 133 (3):368–379. https://doi.org/10.1111/jnc.12991

133. Chai X, Dage JL, Citron M (2012) Constitutive secretion of tau protein by an unconventional mechanism. Neurobiol Dis 48 (3):356–366. https://doi.org/10.1016/j.nbd.2012.05.021

134. Merezhko M, Brunello CA, Yan X et al (2018) Secretion of tau via an unconventional non-vesicular mechanism. Cell Rep 25 (8):2027–2035.e4. https://doi.org/10.1016/j.celrep.2018.10.078

135. Katsinelos T, Zeitler M, Dimou E et al (2018) Unconventional secretion mediates the transcellular spreading of tau. Cell Rep 23 (7):2039–2055. https://doi.org/10.1016/j.celrep.2018.04.056

136. Pérez M, Avila J, Hernández F (2019) Propagation of tau via extracellular vesicles. Front Neurosci 13:698. https://doi.org/10.3389/fnins.2019.00698

137. Abounit S, Wu JW, Duff K et al (2016) Tunneling nanotubes: a possible highway in the spreading of tau and other prion-like proteins in neurodegenerative diseases. Prion 10 (5):344–351. https://doi.org/10.1080/19336896.2016.1223003

138. Tardivel M, Begard S, Bousset L et al (2016) Tunneling nanotube (TNT)-mediated neuron-to neuron transfer of pathological tau protein assemblies. Acta Neuropathol Comm 4(1):117. https://doi.org/10.1186/s40478-016-0386-4

139. Alarcon-Martinez L, Villafranca-Baughman D, Quintero H et al (2020) Interpericyte tunnelling nanotubes regulate neurovascular coupling. Nature 585(7823):91–95. https://doi.org/10.1038/s41586-020-2589-x

140. Holmes BB, DeVos SL, Kfoury N et al (2013) Heparan sulfate proteoglycans mediate internalization and propagation of specific proteopathic seeds. PNAS 110(33):E3138–E3147. https://doi.org/10.1073/pnas.1301440110

141. Mirbaha H, Holmes BB, Sanders DW et al (2015) Tau trimers are the minimal propagation unit spontaneously internalized to seed intracellular aggregation. J Biol Chem 290 (24):14893–14903. https://doi.org/10.1074/jbc.M115.652693

142. Funk KE, Mirbaha H, Jiang H et al (2015) Distinct therapeutic mechanisms of tau antibodies: promoting microglial clearance versus blocking neuronal uptake. J Biol Chem 290 (35):21652–21662. https://doi.org/10.1074/jbc.M115.657924

143. Rauch JN, Chen JJ, Sorum AW et al (2018) Tau internalization is regulated by 6-O sulfation on heparan sulfate proteoglycans

(HSPGs). Sci Rep 8(1):6382. https://doi.org/10.1038/s41598-018-24904-z

144. Stopschinski BE, Holmes BB, Miller GM et al (2018) Specific glycosaminoglycan chain length and sulfation patterns are required for cell uptake of tau vs. alpha-synuclein and beta-amyloid aggregates. J Biol Chem 293 (27):10826–10840. https://doi.org/10.1074/jbc.RA117.000378

145. Weisová P, Cehlár O, Škrabana R et al (2019) Therapeutic antibody targeting microtubule-binding domain prevents neuronal internalization of extracellular tau via masking neuron surface proteoglycans. Acta Neuropathol Comm 7(1):129. https://doi.org/10.1186/s40478-019-0770-y

146. Hudák A, Kusz E, Domonkos I et al (2019) Contribution of syndecans to cellular uptake and fibrillation of α-synuclein and tau. Sci Rep 9(1):16543. https://doi.org/10.1038/s41598-019-53038-z

147. Zhao J, Zhu Y, Song X et al (2020) 3-O-sulfation of heparan sulfate enhances tau interaction and cellular uptake. Angew Chem Int Edit 59(5):1818–1827. https://doi.org/10.1002/anie.201913029

148. Puangmalai N, Bhatt N, Montalbano M et al (2020) Internalization mechanisms of brain-derived tau oligomers from patients with Alzheimer's disease, progressive supranuclear palsy and dementia with Lewy bodies. Cell Death Dis 11(5):314. https://doi.org/10.1038/s41419-020-2503-3

149. Stopschinski BE, Thomas TL, Nadji S et al (2020) A synthetic heparinoid blocks Tau aggregate cell uptake and amplification. J Biol Chem 295(10):2974–2983. https://doi.org/10.1074/jbc.RA119.010353

150. Kanekiyo T, Zhang J, Liu Q et al (2011) Heparan sulphate proteoglycan and the low-density lipoprotein receptor-related protein 1 constitute major pathways for neuronal amyloid-beta uptake. J Neurosci 31 (5):1644–1651. https://doi.org/10.1523/JNEUROSCI.5491-10.2011

151. Rauch JN, Luna G, Guzman E et al (2020) LRP1 is a master regulator of tau uptake and spread. Nature 580(7803):381–385. https://doi.org/10.1038/s41586-020-2156-5

152. Evans LD, Strano A, Campbell A et al (2020) Whole genome CRISPR screens identify LRRK2-regulated endocytosis as a major mechanism for extracellular tau uptake by human neurons. BioRxiv 2020.08.11.246363. https://doi.org/10.1101/2020.08.11.246363

153. Cooper JM, Lathuiliere A, Migliorini M et al (2020) LRP1 and SORL1 regulate tau internalization and degradation and enhance tau seeding. BioRxiv 2020.11.17.386581. https://doi.org/10.1101/2020.11.17.386581

154. De Cecco E, Celauro L, Vanni S et al (2020) The uptake of tau amyloid fibrils is facilitated by the cellular prion protein and hampers prion propagation in cultured cells. J Neurochem 155(5):577–591. https://doi.org/10.1111/jnc.15040

155. Takahashi M, Miyata H, Kametani F et al (2015) Extracellular association of APP and tau fibrils induces intracellular aggregate formation of tau. Acta Neuropathol 129 (6):895–907. https://doi.org/10.1007/s00401-015-1415-2

156. Morozova V, Cohen LS, Makki AEH et al (2019) Normal and pathological tau uptake mediated by M1/M3 muscarinic receptors promotes opposite neuronal changes. Front Cell Neurosci 13:403. https://doi.org/10.3389/fncel.2019.00403

157. Corbett GT, Wang Z, Hong W et al (2020) PrP is a central player in toxicity mediated by soluble aggregates of neurodegeneration-causing proteins. Acta Neuropathol 139 (3):503–526. https://doi.org/10.1007/s00401-019-02114-9

158. Zhong Z, Grasso L, Sibilla C et al (2018) Prion-like protein aggregates exploit the RHO GTPase to cofilin-1 signaling pathway to enter cells. EMBO J 37(6). https://doi.org/10.15252/embj.201797822

159. Evans LD, Wassmer T, Fraser G et al (2018) Extracellular monomeric and aggregated Tau efficiently enter human neurons through overlapping but distinct pathways. Cell Rep 22(13):3612–3624. https://doi.org/10.1016/j.celrep.2018.03.021

160. Calafate S, Flavin W, Verstreken P et al (2016) Loss of Bin1 promotes the propagation of tau pathology. Cell Rep 17(4):931–940. https://doi.org/10.1016/j.celrep.2016.09.063

161. Flavin WP, Bousset L, Green ZC et al (2017) Endocytic vesicle rupture is a conserved mechanism of cellular invasion by amyloid proteins. Acta Neuropathol 134 (4):629–653. https://doi.org/10.1007/s00401-017-1722-x

162. Chen JJ, Nathaniel DL, Raghavan P et al (2019) Compromised function of the ESCRT pathway promotes endolysosomal escape of tau seeds and propagation of tau aggregation. J Biol Chem 294 (50):18952–18966. https://doi.org/10.1074/jbc.RA119.009432

163. Ugbode C, Fort-Aznar L, Sweeney ST (2019) Leaky endosomes push tau over the seed limit. J Biol Chem 294 (50):18967–18968. https://doi.org/10.1074/jbc.H119.011687

164. Falcon B, Noad J, McMahon H et al (2017) Galectin-8-mediated selective autophagy protects against seeded tau aggregation. J Biol Chem 293(7):2438–2451. https://doi.org/10.1074/jbc.M117.809293

165. Takeda S, Commins C, DeVos SL et al (2016) Seed-competent high-molecular-weight tau species accumulates in the cerebrospinal fluid of Alzheimer's disease mouse model and human patients. Ann Neurol 80 (3):355–367. https://doi.org/10.1002/ana.24716

166. Barini E, Plotzky G, Mordashova Y et al (2020) Tau in the brain interstitial fluid is fragmented and seeding-competent. BioRxiv 2020.07.15.205724. https://doi.org/10.1101/2020.07.15.205724

167. Asai H, Ikezu S, Tsunoda S et al (2015) Depletion of microglia and inhibition of exosome synthesis halt tau propagation. Nat Neurosci 18(11):1584–1593. https://doi.org/10.1038/nn.4132

168. Zilkova M, Nolle A, Kovacech B et al (2020) Humanized tau antibodies promote tau uptake by human microglia without any increase of inflammation. Acta Neuropathol Comm 8(1):74. https://doi.org/10.1186/s40478-020-00948-z

169. Hopp SC, Lin Y, Oakley D et al (2018) The role of microglia in processing and spreading of bioactive tau seeds in Alzheimer's disease. J Neuroinflamm 15(1):269. https://doi.org/10.1186/s12974-018-1309-z

170. Ruan Z, Delpech JC, Venkatesan Kalavai S et al (2020) P2RX7 inhibitor suppresses exosome secretion and disease phenotype in P301S tau transgenic mice. Mol Neurodegener 15(1):47. https://doi.org/10.1186/s13024-020-00396-2

171. Crotti A, Sait HR, McAvoy KM et al (2019) BIN1 favors the spreading of Tau via extracellular vesicles. Sci Rep 9(1):9477. https://doi.org/10.1038/s41598-019-45676-0

172. Perea JR, Lopez E, Diez-Ballesteros JC et al (2019) Extracellular monomeric tau is internalized by astrocytes. Front Neurosci 13:442. https://doi.org/10.3389/fnins.2019.00442

173. Piacentini R, Li Puma DD, Mainardi M et al (2017) Reduced gliotransmitter release from astrocytes mediates tau-induced synaptic dysfunction in cultured hippocampal neurons. Glia 65(8):1302–1316. https://doi.org/10.1002/glia.23163

174. Martini-Stoica H, Cole AL, Swartzlander DB et al (2018) TFEB enhances astroglial uptake of extracellular tau species and reduces tau spreading. J Exp Med 215(9):2355–2377. https://doi.org/10.1084/jem.20172158

175. Nedergaard M, Goldman SA (2020) Glymphatic failure as a final common pathway to dementia. Science 370(6512):50–56. https://doi.org/10.1126/science.abb8739

176. Patel TK, Habimana-Griffin L, Gao X et al (2019) Dural lymphatics regulate clearance of extracellular tau from the CNS. Mol Neurodegener 14(1):11. https://doi.org/10.1186/s13024-019-0312-x

177. Harrison IF, Ismail O, Machhada A et al (2020) Impaired glymphatic function and clearance of tau in an Alzheimer's disease model. Brain 143(8):2576–2593. https://doi.org/10.1093/brain/awaa179

178. Iliff JJ, Chen MJ, Plog BA et al (2014) Impairment of glymphatic pathway function promotes tau pathology after traumatic brain injury. J Neurosci 34(49):16180–16193. https://doi.org/10.1523/JNEUROSCI.3020-14.2014

179. Pooler AM, Polydoro M, Maury EA et al (2015) Amyloid accelerates tau propagation and toxicity in a model of early Alzheimer's disease. Acta Neuropathol Comm 3:14. https://doi.org/10.1186/s40478-015-0199-x

180. He Z, Guo JL, McBride JD et al (2018) Amyloid-β plaques enhance Alzheimer's brain tau-seeded pathologies by facilitating neuritic plaque tau aggregation. Nat Med 24 (1):29–38. https://doi.org/10.1038/nm.4443

181. Vergara C, Houben S, Suain V et al (2019) Amyloid-beta pathology enhances pathological fibrillary tau seeding induced by Alzheimer PHF in vivo. Acta Neuropathol 137 (3):397–412. https://doi.org/10.1007/s00401-018-1953-5

182. Saito T, Mihira N, Matsuba Y et al (2019) Humanization of the entire murine Mapt gene provides a murine model of pathological human tau propagation. J Biol Chem 294 (34):12754–12765. https://doi.org/10.1074/jbc.RA119.009487

183. Ehrenberg AJ, Khatun A, Coomans E et al (2020) Relevance of biomarkers across different neurodegenerative diseases. Alzheimers Res Ther 12(1):56. https://doi.org/10.1186/s13195-020-00601-w

184. Karikari TK, Benedet AL, Ashton NJ et al (2020) Diagnostic performance and prediction of clinical progression of plasma phospho-tau181 in the Alzheimer's disease neuroimaging initiative. Mol Psychiatry. https://doi.org/10.1038/s41380-020-00923-z

185. Young PNE, Estarellas M, Coomans E et al (2020) Imaging biomarkers in neurodegeneration: current and future practices. Alzheimers Res Ther 12(1):49. https://doi.org/10.1186/s13195-020-00612-7

186. Mattay VS, Fotenos AF, Ganley CJ et al (2020) Brain tau imaging: Food and Drug Administration approval of (18)F-Flortaucipir injection. J Nucl Med 61:1411–1412. https://doi.org/10.2967/jnumed.120.252254

187. Brendel M, Barthel H, van Eimeren T et al (2020) Assessment of 18F-PI-2620 as a biomarker in progressive supranuclear palsy. JAMA Neurol 77(11):1408–1419. https://doi.org/10.1001/jamaneurol.2020.2526

188. Tagai K, Ono M, Kubota M et al (2020) High-contrast in vivo imaging of tau pathologies in Alzheimer's and non-Alzheimer's disease tauopathies. Neuron 16:S0896-6273(20)30766-2. https://doi.org/10.1016/j.neuron.2020.09.042

189. Karikari TK, Pascoal TA, Ashton NJ et al (2020) Blood phosphorylated tau 181 as a biomarker for Alzheimer's disease: a diagnostic performance and prediction modelling study using data from four prospective cohorts. Lancet Neurol 19(5):422–433. https://doi.org/10.1016/S1474-4422(20)30071-5

190. Thijssen EH, La Joie R, Wolf A et al (2020) Diagnostic value of plasma phosphorylated tau181 in Alzheimer's disease and frontotemporal lobar degeneration. Nat Med 26(3):387–397. https://doi.org/10.1038/s41591-020-0762-2

191. Janelidze S, Mattsson N, Palmqvist S et al (2020) Plasma P-tau181 in Alzheimer's disease: relationship to other biomarkers, differential diagnosis, neuropathology and longitudinal progression to Alzheimer's dementia. Nat Med 26(3):379–386. https://doi.org/10.1038/s41591-020-0755-1

192. Mielke MM, Hagen CE, Xu J et al (2018) Plasma phospho-tau181 increases with Alzheimer's disease clinical severity and is associated with tau- and amyloid-positron emission tomography. Alzheimers Dement 14(8):989–997. https://doi.org/10.1016/j.jalz.2018.02.013

193. Mattsson-Carlgren N, Janelidze S, Palmqvist S et al (2020) Longitudinal plasma p-tau217 is increased in early stages of Alzheimer's disease. Brain 143(11):3234–3241. https://doi.org/10.1093/brain/awaa286

194. Palmqvist S, Janelidze S, Quiroz YT et al (2020) Discriminative accuracy of plasma phospho-tau217 for Alzheimer disease vs other neurodegenerative disorders. JAMA 324(8):772–781. https://doi.org/10.1001/jama.2020.12134

195. Janelidze S, Berron D, Smith R et al (2020) Associations of plasma phospho-Tau217 levels with tau positron emission tomography in early Alzheimer disease. JAMA Neurol. https://doi.org/10.1001/jamaneurol.2020.4201

196. Hampel H, O'Bryant SE, Molinuevo JL et al (2018) Blood-based biomarkers for Alzheimer disease: mapping the road to the clinic. Nat Rev Neurol 14(11):639–652. https://doi.org/10.1038/s41582-018-0079-7

197. Gauthier S, Therriault J, Pascoal T et al (2020) Impact of p-tau181 and p-tau217 levels on enrolment for randomized clinical trials and future use of anti-amyloid and anti-tau drugs. Expert Rev Neurother 20(12):1211–1213. https://doi.org/10.1080/14737175.2020.1841637

198. Horie K, Barthélemy NR, Sato C et al (2020) CSF tau microtubule binding region identifies tau tangle and clinical stages of Alzheimer's disease. Brain awaa373. https://doi.org/10.1093/brain/awaa373

199. Ashton NJ, Hye A, Rajkumar AP et al (2020) An update on blood-based biomarkers for non-Alzheimer neurodegenerative disorders. Nat Rev Neurol 16(5):265–284. https://doi.org/10.1038/s41582-020-0348-0

200. Ashton NJ, Leuzy A, Lim YM et al (2019) Increased plasma neurofilament light chain concentration correlates with severity of post-mortem neurofibrillary tangle pathology and neurodegeneration. Acta Neuropathol Comm 7(1):5. https://doi.org/10.1186/s40478-018-0649-3

201. Congdon EE, Sigurdsson EM (2018) Tau-targeting therapies for Alzheimer disease. Nat Rev Neurol 14(7):399–415. https://doi.org/10.1038/s41582-018-0013-z

202. Panza F, Lozupone M, Seripa D et al (2020) Development of disease-modifying drugs for frontotemporal dementia spectrum disorders. Nat Rev Neurol 16(4):213–228. https://doi.org/10.1038/s41582-020-0330-x

203. Li C, Gotz J (2017) Tau-based therapies in neurodegeneration: opportunities and

challenges. Nat Rev Drug Discov 16 (12):863–883. https://doi.org/10.1038/nrd.2017.155

204. Jadhav S, Avila J, Scholl M et al (2019) A walk through tau therapeutic strategies. Acta Neuropathol Comm 7(1):22. https://doi.org/10.1186/s40478-019-0664-z

205. Gauthier S, Feldman HH, Schneider LS et al (2016) Efficacy and safety of tau-aggregation inhibitor therapy in patients with mild or moderate Alzheimer's disease: a randomised, controlled, double-blind, parallel-arm, phase 3 trial. Lancet 388(10062):2873–2884. https://doi.org/10.1016/S0140-6736(16)31275-2

206. Novak P, Schmidt R, Kontsekova E et al (2016) Safety and immunogenicity of the tau vaccine AADvac1 in patients with Alzheimer's disease: a randomised, double-blind, placebo-controlled, phase 1 trial. Lancet Neurol 16(2):123–134. https://doi.org/10.1016/S1474-4422(16)30331-3

207. Novak P, Schmidt R, Kontsekova E et al (2018) FUNDAMANT: an interventional 72-week phase 1 follow-up study of AADvac1, an active immunotherapy against tau protein pathology in Alzheimer's disease. Alzheimers Res Ther 10(1):108. https://doi.org/10.1186/s13195-018-0436-1

208. Colin M, Dujardin S, Schraen-Maschke S et al (2020) From the prion-like propagation hypothesis to therapeutic strategies of anti-tau immunotherapy. Acta Neuropathol 139 (1):3–25. https://doi.org/10.1007/s00401-019-02087-9

209. Mullard A (2020) Failure of first anti-tau antibody in Alzheimer disease highlights risks of history repeating. Nat Rev Drug Discov doi. https://doi.org/10.1038/d41573-020-00217-7

210. Pardridge WM (2020) Blood-brain barrier and delivery of protein and gene therapeutics to brain. Front Aging Neurosci 11:373

211. Kariolis MS, Wells RC, Getz JA et al (2020) Brain delivery of therapeutic proteins using an Fc fragment blood-brain barrier transport vehicle in mice and monkeys. Sci Transl Med 12(545):eaay1359. https://doi.org/10.1126/scitranslmed.aay1359

212. Ising C, Gallardo G, Leyns CEG et al (2017) AAV-mediated expression of anti-tau scFvs decreases tau accumulation in a mouse model of tauopathy. J Exp Med 214 (5):1227–1238. https://doi.org/10.1084/jem.20162125

213. Liu W, Zhao L, Blackman B et al (2016) Vectored intracerebral immunization with the anti-tau monoclonal antibody PHF1 markedly reduces tau pathology in mutant tau transgenic mice. J Neurosci 36 (49):12425–12435. https://doi.org/10.1523/JNEUROSCI.2016-16.2016

214. Quint WH, Matečko-Burmann I, Schilcher I et al (2020) Bispecific tau antibodies with additional binding to C1q or alpha-synuclein. BioRxiv 2020.11.10.376301. https://doi.org/10.1101/2020.11.10.376301

215. DeVos SL, Miller RL, Schoch KM et al (2017) Tau reduction prevents neuronal loss and reverses pathological tau deposition and seeding in mice with tauopathy. Sci Transl Med 9(374):eaag0481. https://doi.org/10.1126/scitranslmed.aag0481

216. Lee HJ, Boado RJ, Braasch DA et al (2002) Imaging gene expression in the brain in vivo in a transgenic mouse model of Huntington's disease with an antisense radiopharmaceutical and drug-targeting technology. J Nucl Med 43(7):948–956

217. Pardridge WM (2016) Re-engineering therapeutic antibodies for Alzheimer's disease as blood-brain barrier penetrating bi-specific antibodies. Exp Opin Biol Ther 16 (12):1455–1468. https://doi.org/10.1080/14712598.2016.1230195

218. ACI-3024. Alzforum (nd) https://www.alzforum.org/therapeutics/aci-3024. Accessed 14 Dec 2020

219. LY3372689. Alzforum (nd) https://www.alzforum.org/therapeutics/ly3372689. Accessed 14 Dec 2020

220. Konstantinidou M, Li J, Zhang B et al (2019) PROTACs- a game-changing technology. Exp Opin Drug Disc 14(12):1255–1268. https://doi.org/10.1080/17460441.2019.1659242

221. Vargova G, Vogels T, Kostecka Z et al (2018) Inhibitory interneurons in Alzheimer's disease. Bratisl Med J 119(4):205–209

Chapter 5

Role of SIRT3 and Mitochondrial Dysfunction in Neurodegeneration

Jin-Hui Hor, Munirah Mohamad Santosa, and Shi-Yan Ng

Abstract

Loss of function of Sirtuin-3 (SIRT3) has been associated with multiple neurodegenerative diseases, including Alzheimer's disease, Parkinson's disease, Huntington's disease, and amyotrophic lateral sclerosis. Given the involvement of SIRT3 in maintaining healthy mitochondrial respiration and redox homeostasis, it is widely speculated that reduction of this mitochondrial enzyme contributes to the pathogenesis of these neurodegenerative disorders. More recently, SIRT3 activation has also been shown in model organisms and human cells to be an effective strategy to slow the progression of neurodegeneration. In this chapter, we describe the roles of SIRT3 in regulating cellular metabolism, and review recent research on the roles of SIRT3 in neurodegeneration. Finally, we provide a list of biochemical assays for the investigation of SIRT3 levels and activities in cultured cells.

Keywords Sirtuin, SIRT3, Mitochondrial respiration, Alzheimer's disease, Parkinson's disease, Huntington's disease, Amyotrophic lateral sclerosis, Multiple sclerosis

1 Introduction

SIRT3, one of the seven mammalian sirtuin homologs of the yeast Sir2 gene, is a mitochondrial NAD^+-dependent deacetylase. It acts as a cellular metabolic sensor as its activity directly correlates with $NAD^+/NADH$ nutrient levels in the cells [1, 2]. Compared to other mitochondrial sirtuins (SIRT4 and SIRT5), SIRT3 regulates the majority of mitochondrial lysine deacetylation [3]. SIRT3 is highly expressed in metabolically active tissues like heart, liver, brain, brown adipose tissue, skeletal muscle, and kidney [4].

SIRT3, being one of the most abundant proteins in the brain, is involved in regulating mitochondrial oxidative phosphorylation, amino acid and fatty acid metabolism and reactive oxygen species (ROS) homeostasis [3, 5–7]. For instance, SIRT3-mediated

---

Jin-Hui Hor and Munirah Mohamad Santosa contributed equally with all other contributors.

Philip V. Peplow, Bridget Martinez and Thomas A. Gennarelli (eds.), *Neurodegenerative Diseases Biomarkers: Towards Translating Research to Clinical Practice*, Neuromethods, vol. 173, https://doi.org/10.1007/978-1-0716-1712-0_5,
© Springer Science+Business Media, LLC, part of Springer Nature 2022

deacetylation of the transcription factor forkhead box O3 (FOXO3), is shown to have an effect in survival, metabolism, ROS defense, apoptosis, and differentiation, suggesting the multiple roles of SIRT3 in cellular homeostasis [8, 9].

Studies have shown a direct link between mitochondrial dysfunction and neurodegeneration with SIRT3 emerging as a protein of interest due to its mitochondrial localization and its involvement as a longevity protein [10]. It was demonstrated that under stress conditions such as oxidative, genotoxic, or metabolic stress, SIRT3 knockout mice exhibited an accelerated pace of age-related diseases like cancer, neurodegenerative diseases, cardiovascular diseases, and metabolic syndromes [11]. A global acetylation of mitochondrial proteins was observed in SIRT3 knockout animals [12–14]. These acetylated proteins included anti-oxidants, genes involved in the mitochondrial membrane potential [15, 16], fatty acid oxidation [5, 17], and the urea cycle [5]. Moreover, behavioral observations of SIRT3-knockout mice include poor remote memory, observed by deviated long-term potentiation and contributed by the loss of anterior cingulate cortex neurons [18]. It became evident that SIRT3 has important functions in neurological disorders led by mitochondrial dysfunction. Hence, in this chapter, we review the well-documented function of SIRT3 in the mitochondria and evidence linking the involvement of SIRT3 in neurodegenerative diseases.

2 Mitochondrial Functions of SIRT3

Mitochondria are important organelles in the nervous system that have a myriad of functions in maintaining lipid and energy metabolism, calcium ion homeostasis, combating ROS stress responses and preserving mitochondrial quality [19]. SIRT3 regulates the acetylation status of mitochondrial proteins involved in antioxidant defenses, energy metabolism, and mitochondrial dynamics (Fig. 1).

2.1 SIRT3 in Mitochondrial Fission and Fusion

Mitochondria are profoundly dynamic cellular organelles that continually fuse, divide and regenerate to maintain energy homeostasis to meet the demands of the cells. Hence, any disturbance in mitochondrial dynamics could lead to mitochondrial malfunction. It is reported that the structure of mitochondria is regulated by two counteracting processes, mitochondrial fusion and fission, with dynamin-related protein 1 (DRP1), fission 1 (Fis1), mitochondrial fission factor (MFF), mitofusin 1 (MFN1), mitofusin 2 (MFN2), and optic atrophy 1 (OPA1) as the key proteins involved in mitochondrial dynamics [20].

Several SIRT3 targets are found to be involved in mitochondrial dynamics. A study demonstrated that SIRT3-mediated deacetylation of FOXO3 leads to up-regulation of genes associated with mitochondrial fusion (MFN2) and fission (DRP1 and Fis1),

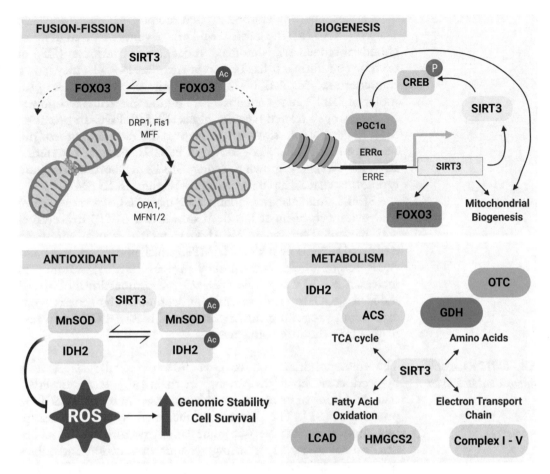

Fig. 1 The roles of SIRT3 in the mitochondria. SIRT3 deacetylates the proteins in the mitochondria involved in fusion–fission, biogenesis, antioxidant defenses, and metabolism. SIRT3 deacetylates FOXO3, that upregulates mitochondrial fusion (MFN2) and fission (DRP1 and FIS1) genes to promote this activity. In mitochondrial biogenesis, PGC-1α promotes SIRT3 expression by mediating ERRα binding to this sequence motif. Additionally, SIRT3 also upregulates PGC-1α via a positive feedback mechanism by deacetylating LKB1, which activates AMPK to phosphorylate CREB, thus increasing the expression of PGC-1α. SIRT3 also deacetylates antioxidants like MnSOD and IDH2 to combat ROS and activating the glutathione antioxidant protection system, thereby increasing genomic stability and survival. Finally, under caloric restriction, SIRT3 regulates a wide range of proteins in the TCA cycle (IDH2, ACS), amino acid metabolism (GDH, OTC), fatty acid oxidation (LCAD, HMGCS2) and electron transport chain (Complex I–V), suggesting that SIRT3 is important in cellular metabolic homeostasis

mitophagy to remove damaged mitochondria (Bnip, nix, and LC3-II/LC3-I), and upregulation of both TFAM and PGC-1α genes to promote mitochondrial biogenesis [21]. Consecutively, this preserves mitochondrial quality in the cells. Moreover, SIRT3 have been found to facilitate mitochondrial fusion via deacetylating OPA1, maintaining mitochondrial networking which prevented cell death in cardiomyocytes under stress conditions [22].

Peroxisome proliferator-activated receptor gamma coactivator 1-alpha (PGC-1α), the master regulator of genes involved in mitochondrial biogenesis, is induced under caloric restriction (CR) or exercise conditions. It has been reported that PGC-1α may have a transcriptional regulation on SIRT3 gene expression. The promoter of SIRT3 gene contains a recognition site that is recognized by the estrogen related receptor-alpha (ERRα). PGC-1α promotes SIRT3 gene expression by mediating ERRα binding to this sequence motif [23]. Knockdown of PGC-1α via small interfering RNA (siRNA) have shown to reduce SIRT3 gene expression, indicating the relationship between PGC-1α and SIRT3 [24].

SIRT3 could also potentially upregulate PGC-1α via a positive feedback mechanism. It has been shown that SIRT3 deacetylates and activates liver kinase B1 (LKB1) which in turn stimulates AMP-activated protein kinase (AMPK) which phosphorylates cyclic AMP response element binding protein (CREB), leading to increased PGC-1α expression [25–29]. The knockdown of SIRT3 reduced PGC-1α and several mitochondrial biogenesis gene expression, suggesting the importance of SIRT3-PGC-1α in regulating mitochondrial biogenesis.

2.2 SIRT3 Regulates Antioxidant Defenses

The mitochondria produce over 90% of cellular ROS, as a by-product of the electron transport chain [30]. It was observed that the brains of aged rats exhibit increased mitochondrial ROS production [31]. The presence of increased ROS could disrupt mitochondrial oxidative phosphorylation, perturbing mitochondrial axonal transport, inducing synaptic dysfunction and subsequently neurodegeneration [32, 33]. SIRT3 has been found to reduce ROS levels under caloric restriction (CR) by directly regulating with manganese superoxide dismutase (MnSOD), which is an antioxidant responsible for combating ROS [16, 34, 35], isocitrate dehydrogenase 2 (IDH2) where it mediates glutathione antioxidant protection system [36], and acyl coenzyme A (acyl CoA) dehydrogenase, which regulates fatty-acid oxidation [37]. The SIRT3-dependent deacetylation of FOXO3 could also defend against ROS by upregulating MnSOD, which in turn alleviates cardiac hypertrophy in mice [7]. SIRT3 also deacetylates Cyclophilin D (Cyp D), where it protects the cells from oxidative stress by preventing cytochrome c and glutathione (GSH) release, hence blocking mitochondrial permeability transition pore formation, and thus limiting cell apoptosis [38].

2.3 SIRT3 Roles in Energy Metabolism

SIRT3 is dependent on the availability of NAD^+ to mediate its activity and its activity governs the energy status of the cells. Under caloric restriction (CR), SIRT3 activates a wide range of mitochondrial proteins that are associated with the tricarboxylic acid (TCA) cycle, amino acid metabolism, fatty acid oxidation and the electron transport chain [39].

In skeletal muscles, SIRT3 also activates pyruvate dehydrogenase (PDH) via deacetylating the E1α subunit of PDH, promoting carbohydrate oxidation [40]. To promote TCA cycle, SIRT3 deacetylates and activates important TCA cycle enzymes such as acetyl-CoA synthetase and isocitrate dehydrogenase 2 [41, 42]. SIRT3 was also found to interact with the NADH dehydrogenase ubiquinone 1 α sub complex 9 (NDUFA9), a subunit of the mitochondrial Complex I. SIRT3 KO mice exhibited reduced complex I activity and ATP production that was ameliorated by the addition of exogenous SIRT3 [15]. The knockdown of PGC-1α was shown to indirectly affect Complex IV activity in skeletal muscle cells via a SIRT3-mediated manner [24].

During low caloric intake, amino acids become important as energy fuel. SIRT3 plays a role in amino acid metabolism by deacetylating glutamate dehydrogenase, thereby activating its functions in promoting the conversion of glutamate to α-ketoglutarate, which can enter the TCA cycle [43–45]. SIRT3 also activates the urea cycle by deacetylating ornithine transcarbamoylase (OTC), which could aid the amino acid catabolism under fasting conditions [5]. Interestingly, SIRT3 KO mouse embryonic fibroblast (MEFs) exhibited an increased accumulation of many amino acids, suggesting a defective amino acid catabolism [14].

Under fasting conditions, large amounts of acetate are released to the blood and utilized by acetyl-CoA synthetase (ACS), an enzyme that aids the formation of acetyl-CoA from acetate and CoA for energy production [46, 47]. SIRT3 promotes this reaction by directly deacetylating ACS2, thereby activating its function in the mitochondria [48, 49]. SIRT3 was also found to be involved in ketone body formation during fasting by deacetylating 3-hydroxy-3-methylglutaryl CoA synthase 2 (HMGCS2) which enhances β-hydroxybutyrate production. The usage of triacylglycerols was also enhanced by SIRT3-mediated deacetylation of long-chain acyl coenzyme A dehydrogenase (LCAD). Interestingly, SIRT3 deficient mice exhibited hyperacetylated LCAD, which contributed to reduced ATP levels and deregulation of fatty acid oxidation [37].

3   SIRT3 and Neurodegenerative Diseases

Since mitochondrial dysfunction occurs in most of the age-related neurodegenerative diseases, SIRT3, which is fundamentally active and important to maintain mitochondrial homeostasis, would be an important target for therapeutic strategies. Here, we explore evidences surrounding the involvement of SIRT3 in neurodegenerative diseases such as Alzheimer's disease, Parkinson's disease, Huntington's disease, Amyotrophic Lateral Sclerosis and Multiple Sclerosis.

3.1 Alzheimer's Disease

Alzheimer's disease (AD) is a progressive neurodegenerative disease which is characterized by memory and cognitive judgement impairment. AD is the most common form of dementia that accounts for 60–80% of all dementia cases [50]. Neurofibrillary tangles, consisting of phosphorylated tau paired helical filaments, and amyloid plagues composed of abnormally folded amyloid β peptide (Aβ), which are by-products of Amyloid Precursor Protein (APP) metabolism, are common pathologies found in AD [51]. Accumulation of Aβ could in turn disrupt the mitochondrial membrane potential, increase ROS generation and reduce ATP production [52, 53]. Aβ has also been shown to impair mitochondrial anterograde transport in Aβ-treated neurons, leading to synaptic degeneration [54]. Other AD pathologies include synaptic dysfunction and metabolic deficiencies. APP plays a critical role in synaptic plasticity. Hence, its variation in the synaptic network could lead to cognitive decline [55]. In line with this, mitochondrial metabolism generating ATP and combating ROS damage is vital to support these synaptic activities in neurons. For instance, pituitary adenylate cyclase activating polypeptide (PACAP), a neurotrophin, which induces Sirt3 levels, was found to be downregulated in 3xTg-AD mice brains and post-mortem brain tissues of AD patient brains which had high Aβ and tau protein levels [56]. The neuroprotective effect of PACAP to reduce Aβ toxicity, was further abolished in SIRT3 knockout mice, suggesting the importance of synaptic activities and AD disease pathogenesis [56].

In post-mortem brain slices, reduced SIRT3 expression was observed in AD patients compared to healthy control in multiple studies [57–59]. SIRT3 levels were also found to be downregulated in the cortex of the Amyloid Precursor Protein/Presenilin 1 (APP/PS1) double transgenic AD mouse model [60]. In cortical neurons of APP mice, SIRT3 levels were reduced [58]. Interestingly, reduced nicotinamide adenine dinucleotide (NAD$^+$) levels and ATP production were also observed [58], suggesting that SIRT3 expression and activity impacts mitochondrial function. Furthermore, learning and memory performance was also affected in these APP mice. Importantly, mitochondrial function and ATP production was restored in SIRT3 overexpressing mouse hippocampal neurons, reverting the damaged caused by Aβ [58].

Increased tau acetylation levels were observed with reduced SIRT3 expression. Further genetic reduction of SIRT3 revealed that increasing acetylated tau levels was mediated via Aβ in cortical neurons from mice that carry human tau protein and hippocampal HT22 cells [58, 59]. On the other hand, tau acetylation decreased with SIRT3 overexpression in hippocampal HT22 cells, indicating the regulatory role of SIRT3 in AD pathogenesis [58]. SIRT3 overexpression was also found to be neuroprotective in primary hippocampal cultures as it was able to alleviate oxidative stress [61]. Other than tau and Aβ, apolipoprotein E4 (APOE4) is one of the sporadic genes that is associated with AD. It was identified that APOE4 carriers had a downregulation of SIRT3 in the frontal cortex as compared to noncarriers, suggesting that reduced SIRT3 expression may be responsible for sporadic AD [62, 63].

Some recent studies reported the therapeutic potential of enhancing SIRT3 activity and expression in alleviating AD. Two studies demonstrated that honokiol, [2-(4-hydroxy-3-prop-2-enyl-phenyl)-4-prop-2-enyl-phenol], could significantly lower Aβ production plaque deposition. Honokiol was found to be able to activate SIRT3 and increase the expression of $5'$ adenosine monophosphate-activated protein kinase (AMPK), PGC1-a and cAMP response element-binding protein (CREB), which then reduces Aβ levels by blocking β secretase activity [64]. Furthermore, honokiol treated cells carrying the APP and Presenilin PS1 mutation demonstrated enhanced mitochondrial metabolism and anti-oxidant activity, suggesting that SIRT3 activation enhances mitochondrial function in AD [64]. Moreover, honokiol was able to elevate SIRT3 expression, enhancing ATP production and reducing ROS production, restoring mitochondrial Aβ-mediated dysfunction in neurons of $PS1_{V97L}$ AD transgenic mice [65]. The treatment with honokiol also improved memory deficits, suggesting that SIRT3 activation with honokiol could delay the onset and pathogenesis of AD [65].

One study identified that NAD^+ supplementation, using nico-tinamide riboside (NR) in their $3xTgAD/Polβ^{+/-}$ AD mice model ameliorates major AD features [66]. They demonstrated that in these NR-treated AD mice, increased SIRT3 expression, alleviated pTau pathology, neuroinflammation, reduced DNA damage and reduced cell death was observed in the hippocampal neurons [37]. Although it did not improve Aβ build-up, they observed that these mice had improved hippocampal synaptic plasticity, sug-gesting NAD^+ supplementation for SIRT3 activity could be a potential therapy for AD [37].

3.2 Parkinson's Disease

Parkinson's disease (PD) is the second most common age-related degenerative neurological disorder after AD. PD affects 1–2% of the adult population over the age of 60 and a small percentage of people (4% of all PD cases) are diagnosed before the age of 50. Some of its pathophysiologic motor symptoms are rigidity, tremor, postural imbalance and bradykinesia which are the result of dopa-minergic neuronal loss in the substantia nigra pars compacta (SNc) and the accumulation of cytoplasmic α-synuclein Lewy body aggre-gates [67, 68].

The susceptibility of dopaminergic neurons to damage could be contributed by a combination of a loss in mitochondrial homeo-stasis and bioenergetics, overproduction of ROS, and protein aggregation. Naturally, some of these affected mitochondrial pro-teins are targets or indirect targets of SIRT3-mediated deacetylation. MPTP (1-methyl-4-phenyl-1,2,3,6-tetrahydropyri-dine)-induced Parkinsonism mice, which metabolizes to the MPP^+ neurotoxin that specifically destroys dopaminergic cells, have been used as a model for PD [69]. A reduced SIRT3 expression was observed in the midbrain, which further exacerbated dopaminergic neuronal loss in the SNc of MPTP-induced PD mice [70]. Overexpression of SIRT3 was shown to reduce dopaminergic neuronal loss in these MPTP-induced PD mice [71]. These MPTP-induced PD mice also suffered from the loss of ROS defense machinery, such as of MnSOD [70] and glutathione peroxidase [70]. High levels of acetylation also affects ATP synthase β function, resulting in reduced ATP production [71].

In a similar study utilizing the rotenone-induced PD cell model, the genetic deletion of SIRT3 reduces antioxidant SOD and glutathione (GSH) levels, hence reducing SNc dopaminergic neurons membrane potential and aggravated cell death [72]. Conversely, SIRT3 overexpression increased SOD and GSH levels, thereby decreasing ROS generation and prevented α-synuclein accumulation, accentuating the neuroprotective functions of SIRT3 [72]. MPP-treated neuroblastoma cells, SH-SY5Y, also exhibited reduced SIRT3 expression which affected the deacetylation of a key mitochondrial enzyme, citrate synthase, thus affecting mitochondrial enzyme activities [73].

The aggregation of α-synuclein is one of the central pathological characteristics of PD. Although α-synuclein mostly resides in the cytosol, studies have identified α-synuclein to associated with the mitochondria [74]. Research has shown that with the enhancement of SIRT3 in mutant rat models of parkinsonism, it was able to improve mitochondrial bioenergetics and decrease oxidative stress, thus conferring its neuroprotective effects on dopaminergic neurons [75]. Another study suggested that SIRT3 is involved in α-synuclein-mediated mitochondrial dysfunction as SIRT3 downregulation in the presence of α-synuclein oligomerization demonstrated mitochondrial respiration impairment and disruption of mitochondrial dynamics by decreasing AMPK and CREB phosphorylation and increasing dynamin-related protein 1 (DRP1) phosphorylation [76]. SIRT3 restoration on the other hand decreases α-synuclein accumulation and boosts mitochondrial function, supporting the neuroprotective role of SIRT3 in counteracting α-synuclein accumulation [76].

Dysregulation in mitochondrial fission and fusion dynamics is widely reported and is another predominant aspect of PD pathogenesis [77]. SIRT3 was found to modulate some genes that are involved in mitochondrial dynamism via its deacetylation activity. Notably, SIRT3 deacetylates OPA1, a protein important in mitochondrial fusion, to regulate mitochondrial dynamics during stress [22]. Furthermore, several PD models have detected OPA1 mutations and its association with PD [78, 79], indicating mitochondrial fusion dysregulation involved in PD that could be further aggravated by reduced SIRT3 levels. Mutations in PINK1, a mitochondrial Ser/Thr kinase, has important implications in familial PD. Mutant PINK1 can increase mitochondrial fission and promote mitochondrial damage by stimulating the DRP1 mitochondrial translocation, reducing its degradation together with mitochondrial fission 1 protein (FIS1) by Parkin [80].

SIRT3 has been shown to indirectly affect PINK1-Parkin pathway through Lon protease 1. Lon degrades PINK1 to the matrix of the mitochondria which may disrupt mitophagy and mitochondria quality control [81]. This process is often uncoordinated in PD models [82]. SIRT3 deacetylates FOXO3, which promotes the expression of mitochondrial dynamics genes, such as FIS1, DRP1 and MFN2 [21]. Furthermore, SIRT3-mediated deacetylation of FOXO3 promotes expression of other genes like Nix, LC3-II/ LC3-I and Bnip3, that are required to clear mitochondrial damage [21]. Hence, SIRT3 plays a neuroprotective role by regulating the mitochondrial quality control via mitophagy [83].

3.3 Huntington's Disease

Huntington's disease (HD) is a lethally inherited, autosomal dominant and progressive neurodegenerative disorder characterized by impaired cognitive, behavior and motor functions. In HD, the gene encoding huntingtin (HTT), located at chromosome 4, is mutated in the exon-1 region, resulting in an expansion of CAG repeats in this exon ($\geq$40 CAGs) [84, 85]. The expansion of CAG repeats encodes for a stretch of glutamine residues, resulting in a change of conformation in HTT protein and aggregations in neurons [84]. HTT is expressed throughout the body and has roles in gene transcription, protein trafficking and mitochondrial function [84, 86]. On the other hand, mutant HTT (mHTT) was shown to affect the brain striatum and cortex by increasing glutamatergic signaling, in addition to mitochondrial function in corticostriatal neurons, leading to neurodegeneration [87]. Other studies show that single allele deletion of HTT in mice demonstrated significant motor and cognitive abnormalities, further emphasizing the neuroprotective role of HTT [88].

Notably, a decline in SIRT3 levels and activity has been observed in cells expressing mHTT [89]. Treatment with ε-viniferin, a stilbene resveratrol dimer, increased SIRT3 protein levels and activated AMPK in a SIRT3-dependent pathway [89]. Other than AMPK, activation of SIRT3 also leads to LKB1 activation, enhancing mitochondrial biogenesis and activating ROS detoxification mechanisms in mHTT cells [89]. SIRT3 acts as a ROS regulator by directly interacting with MnSOD, regulating its deacetylation status, thereby enhancing MnSOD antioxidant activity. The neuroprotective role of SIRT3 mediated by viniferin was further confirmed by SIRT3 knockdown in mHTT expressing striatal cell culture [89, 90].

Moreover, mitochondrial electron transport chain (ETC) complex II and III activity was found to be decreased in HD patients [91]. This phenomenon was also observed in the striatal cells of HD mice where mitochondrial respiration and ATP production was significantly reduced, resulting in massive ROS production [91]. In addition, mHTT have shown to affect mitochondrial dynamics by promoting mitochondrial fission via Drp1 and Fis1 activation and reduced mitochondrial fusion by decreasing MFN1/2 and OPA1 expression [92–94]. Direct deacetylation of OPA1 by SIRT3 was shown to promote mitochondrial fusion [22]. Moreover, deacetylation of FOXO3 by SIRT3 promotes MFN2 expression which delays disordered mitochondrial dynamics, maintains mitochondrial function, and potentially slows down the progression of striatal lesions [21, 22].

3.4 Amyotrophic Lateral Sclerosis

Amyotrophic lateral sclerosis (ALS) is an adult-onset neurodegenerative disease characterized by progressive loss of upper and lower motor neurons (LMNs) located at the spinal and bulbar level. There are two forms of ALS, familial and sporadic. The most common form is sporadic ALS (SALS) which accounts for 90–95% of all ALS cases where there is no known etiology. The other 10% of patients have familial ALS where mutations are commonly found in SOD1, C9ORF72, and TDP43 [95]. Despite being intrinsically heterogeneous, the disease manifestation between sporadic and familial ALS is clinically indistinguishable, suggesting a possible converging pathogenic mechanism.

There have been a few investigations on the neuroprotective role of SIRT3 in ALS pathogenesis, suggesting the influence of SIRT3 in ALS disease progression. Inducing glutamate in neurons lacking SIRT3 have shown that these neurons are significantly vulnerable to excitotoxity [96], a phenomenon often seen in ALS motor neurons. It has been suggested that mutant SOD1 promotes apoptosis by binding to antiapoptotic B-cell lymphoma 2 (Bcl-2) and inhibits its function [97]. SIRT3 overexpression was demonstrated to prevent activation of apoptotic cascades in cultured motor neurons from SOD1G93A mutant mice and reduce mitochondrial fragmentation, hence preserving mitochondrial function [98]. Another instance where the role of SIRT3 was reported in ALS was when SIRT3 overexpression prevented motor neuron death induced by primary astrocytes from SOD1^{G93A} mutant in a coculture system, suggesting that elevating SIRT3 activity may provide a protective role in ALS pathogenesis [99].

More recently, reduced mitochondrial bioenergetics as a result of reduced SIRT3 activity, was demonstrated in both induced pluripotent stem cells (iPSC)-derived familial and sporadic ALS motor neurons [100]. Motor neurons rely on mitochondrial respiration to fuel their metabolic needs. Hence, any disruption in mitochondrial function will lead to neurological disorders. Recent studies have found a causative relationship between defective energy metabolism and the survival rate of patients with ALS [101, 102]. Other studies show that the loss of SIRT3 function results in an elevated global mitochondrial acetylation, affecting mitochondrial respiration and bioenergetics [96, 100, 103]. Moreover, increased MnSOD-K68ac signals [100] and dysregulated mitochondrial acetylproteome which could potentially result in reduced ETC activity was observed in post-mortem spinal cord tissue of sporadic ALS patients [104, 105]. Activation of SIRT3 by 7-hydroxy-3-(4′-methoxyphenyl) coumarin (C12) or a NAD$^+$ precursor improved mitochondrial bioenergetics and neuronal survival in familial and sporadic ALS motor neurons, suggesting that SIRT3 is a potential target for ALS therapeutics [100].

3.5 Multiple Sclerosis

Multiple sclerosis (MS) is a chronic inflammatory disorder in which the body's immune system targets the central nervous system, leading to demyelination of neurons. Inflammation-induced neuronal damage to the axon and the lack of myelin regeneration in denuded axons cause these neurons to be much more vulnerable to stress. Demyelination and axonal injury in MS relates with stroke, suggesting energy metabolism and hypoxia could have a pathogenetic role in MS pathology. Impairment in NADH dehydrogenase activity and an increased complex IV activity of the electron transport chain have been observed in MS [106]. Furthermore, mitochondrial DNA (mtDNA) dysfunction and increased ROS generation have been shown to be implicated in MS [107, 108].

Several studies have explored the roles of SIRT3 in MS pathogenesis. Post-mortem brain tissues have demonstrated reduced SIRT3 expression in the grey matter of a MS-affected brain [109]. mtDNA alterations in the grey matter have also been attributed to neurodegeneration in MS [110]. Under oxidative stress conditions, mtDNA variants have been shown to affect SIRT3 expression, via an unknown mechanism, resulting in reduced ATP levels [111]. Furthermore, single nucleotide polymorphisms (SNPs) in the SIRT3 gene have been found in MS affected brains, suggesting that sequence variability of the SIRT3 gene may contribute to MS disease progression [112].

SIRT3 was further demonstrated recently as a therapeutic target for MS in cuprizone-induced demyelinated mice [113]. These mice have reduced mitochondrial dehydrogenases activity, inhibited respiratory chain-complex activities and increased ROS. Reduced levels of SIRT3 expression in the skeletal muscle of cuprizone-induced MS mice was also observed. Supplementation of ellagic acid, a polyphenolic natural compound found in fruits and nuts, demonstrated decreased neuroinflammation and reduced population of activated macrophage in cuprizone-induced MS mice. Ellagic acid supplementation increased SIRT3 expression and improved mitochondrial function, recovering muscular dysfunction and motor incoordination in a dose-dependent manner [113].

4   Research Methods

Different research methods have been employed frequently to evaluate SIRT3 expression and activity to study mitochondrial dysfunction in neurodegenerative diseases. In this section, we will briefly describe common research methods to study SIRT3 expression and activity in vitro.

4.1   Transcriptional Activity of SIRT3 Promoter

4.1.1   Luciferase-Based Reporter System

Genomic DNA was extracted from a single 10 cm plate or T75 cultured flask using DNA extraction kit. SIRT3 promoter fragment was amplified from 50 ng of genomic DNA using specific primers sequences via polymerase chain reaction (PCR) [114]. Primers can also be designed to introduce site-specific mutations into the promoter construct to study specific sequence elements within the promoter. The PCR product was purified, cloned into luciferase-based reporter vector (ampicillin resistant) and transformed into electrocompetent bacteria. Colonies were grown in LB- ampicillin cultures and sequences of DNA construct were validated via DNA sequencing before large scale harvesting of plasmid DNA. The construct was transfected into cells and luciferase activity was assayed 24–48 h after the initial transfection. The luminescence signal was measured with a fluorescence spectrophotometer at a wavelength of 560 nm.

4.2   SIRT3 Gene and Protein Expression

4.2.1   Gene Expression Analysis

For gene expression studies, total RNA can be isolated from control and diseased brain tissues or cell cultures. Then, the total RNA (1 μg) is converted to cDNA and used for real time-PCR (RT-PCR) with primers that are specific to SIRT3 gene, together with two housekeeping genes, glyceraldehyde-3-phosphate dehydrogenase (GAPDH) and beta-actin (ACTINB) [57].

SIRT3 gene expression is analyzed using the threshold cycle (CT) relative quantification method. Briefly, CT values of SIRT3 gene of both control and diseased samples are first normalized to each housekeeping CT values [$\Delta CT = CT$ (SIRT3 gene) $-$ CT (Housekeeping)]. Then, diseased ΔCT was compared to the control ΔCT using the $\Delta\Delta CT$ formula [$\Delta\Delta CT = \Delta CT$ (diseased) $-$ ΔCT (control)]. To calculate the fold change difference, the $2^{-\Delta\Delta CT}$ formula was utilized.

4.2.2   Immunoblotting

Protein lysates can be harvested from control and diseased samples in RIPA buffer. Homogenates were resolved in 12% SDS-PAGE gels or 4–20% precast gels in Tris-glycine- SDS buffer. Proteins were then transferred to a nitrocellulose membrane and blocked with 5% milk in TBST buffer. SIRT3 antibody (1:500) and a loading control antibody were diluted in 5% milk and incubated with the membranes overnight at 4 °C Membranes were washed thrice in TBST buffer [100]. The corresponding horseradish peroxidase secondary antibodies were then diluted 1:5000 in 5% milk and incubated at room temperature for 90 min. Blots were washed thrice before exposing to enhanced chemiluminescence (ECL) reagent for imaging. Densitometric analysis can be done by measuring protein bands with image analysis software [115].

4.3 SIRT3 Enzymatic Activity

4.3.1 Fluorometric Assay

The deacetylase activity of SIRT3 can be determined by a fluorometric assay using a fluorometric kit according to the manufacturer's protocol [116]. Briefly, the assay is done by detecting a fluorescent signal upon the deacetylation of an acetylated substrate peptide when treated with the developer provided by the fluorometric kit. The intensity of the fluorescence signal can be measured on a microplate reader at an excitation of 355 nm and emission detection of 460 nm, according to the manufacturer's protocol. Suramin sodium, an inhibitor of SIRT3, can be used as experimental control.

4.3.2 MnSOD Acetylation Levels as a Proxy for SIRT3 Activity

Protein lysates can be harvested from control and diseased samples in RIPA buffer. Homogenates were resolved in 12% SDS-PAGE gels or 4–20% precast gels in Tris-glycine-SDS buffer. Proteins were then transferred to a nitrocellulose membrane and blocked with 5% milk in TBST buffer. Total MnSOD antibody (1:500), MnSOD-K68ac antibody (1:250) and a loading control antibody were diluted in 5% milk and incubated with the membranes overnight at 4 °C Membranes were washed thrice in TBST buffer [100]. The corresponding horseradish peroxidase secondary antibodies were then diluted 1:5000 in 5% milk and incubated at room temperature for 90 min. Blots were washed thrice before exposing to ECL for imaging. Densitometric analysis can be done by measuring protein bands with image analysis software [115]. The higher the MnSOD-K68ac signals after normalized to total MnSOD illustrates a reduced activity of SIRT3.

5 Conclusions

In this chapter, we summarize the important neuroprotective roles of SIRT3 in several neurodegenerative diseases. SIRT3 could directly or indirectly affect mitochondrial dynamics and quality, antioxidant defenses, and mitochondrial metabolism and homeostasis (Fig. 1). It is interesting that although these neurodegenerative diseases have different pathogenesis and affect different neurons, they all converge to progressive mitochondrial dysfunction and dysregulation (Fig. 2). This usually comes with an increased age, where reduced SIRT3 was observed [117]. For

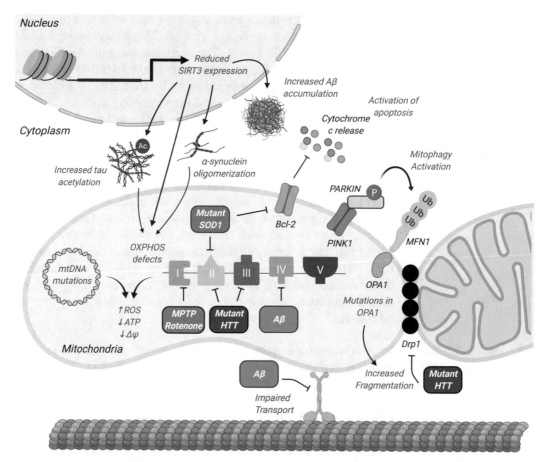

Fig. 2 The roles of SIRT3 and mitochondrial dysfunction in neurodegenerative diseases. A reduction in SIRT3 expression increases Aβ accumulation and tau acetylation in AD. Aβ accumulation also results in oxidative phosphorylation defects, increased cellular ROS, and mitochondrial transport defects. In PD, reduced SIRT3 levels results in increased α-synuclein oligomerization, which also affects the PINK1-Parkin pathway, by disrupting mitophagy and mitochondrial quality control. In HD, mutant HTT affects the mitochondrial dynamics and mitochondrial bioenergetics, which has been attributed to decreased SIRT3 levels. Reduced SIRT3 activity was also demonstrated in ALS, resulting in reduced mitochondrial bioenergetics and leading to motor neuron death. Reduced SIRT3 expression is also responsible for mitochondrial DNA mutations in MS, resulting in reduced mitochondrial metabolism and respiration

instance, reduced SIRT3 expression in AD could lead to an increased Aβ accumulation and tau acetylation, resulting in mitochondrial oxidative phosphorylation (OXPHOS), antioxidant and transport defects. In PD, reduced SIRT3 levels could lead to further α-synuclein oligomerization, affecting mitochondrial respiration and mitochondrial dynamics. It could also affect the PINK1-Parkin pathway to disrupt mitophagy and mitochondrial quality control. A decline in SIRT3 levels and activity was observed in mHTT HD cells, which corresponded with defects in mitochondrial dynamics, respiration, and ATP production. Reduced SIRT3 activity in iPSC-derived familial and sporadic ALS motor neurons was responsible for reduced mitochondrial bioenergetics, ETC activity and eventually contributed to motor neuron death. In MS models, reduced SIRT3 expression result in mitochondrial DNA alterations, accompanied by increased ROS, and reduced mitochondrial metabolism and respiration. Hence, it is becoming clear that SIRT3 is involved in maintaining mitochondrial homeostasis in neurodegenerative diseases.

Although we reviewed some reports about the effects and benefits of SIRT3 activity, more focus needs to be put on delineating the roles of SIRT3 in neurodegeneration as well as therapies and supplements, so as to design or repurpose drugs to treat or mitigate these common neurodegenerative diseases. It is vital that potential neurodegenerative drugs should include those that target or alleviate the problems faced in these areas of mitochondrial dysfunction that we have reviewed and perhaps modulate SIRT3 expression and activity.

As such, some of these reports we summarized have looked at some potential therapeutic strategies to enhance SIRT3 expression or activity as a neuroprotectant.. All these compounds (honokiol, ε-viniferin, C12, NAD$^+$ precursors, ellagic acid), increased either SIRT3 expression and/or activity, have similar effects of enhancing mitochondrial bioenergetics, reducing ROS levels, increasing mitochondrial biogenesis, reducing DNA damage and reducing neuroinflammation. These neuroprotective effects from increasing SIRT3 activity/expression decrease cell death and alleviate disease pathology, suggesting SIRT3 could be a potential target for neurodegenerative disease therapeutics. Moreover, the reduction of SIRT3 below a threshold could possibly serve as a biomarker in the propensity for an aged individual to acquire neurodegenerative diseases. Hence, preparedness and precautions could be taken since there is currently no cure for these neurodegenerative diseases.

Taken together, SIRT3 modulation in neurodegenerative disease could enhance neuronal survival by improving mitochondrial and cellular health in neurons, and hence slow down disease progression.

References

1. Berger F, Lau C, Dahlmann M, Ziegler M (2005) Subcellular compartmentation and differential catalytic properties of the three human nicotinamide mononucleotide adenylyltransferase isoforms. J Biol Chem 280 (43):36334–36341. https://doi.org/10.1074/jbc.M508660200

2. Haigis MC, Sinclair DA (2010) Mammalian sirtuins: biological insights and disease relevance. Annu Rev Pathol 5:253–295. https://doi.org/10.1146/annurev.pathol.4.110807.092250

3. Lombard DB, Alt FW, Cheng HL, Bunkenborg J, Streeper RS, Mostoslavsky R, Kim J, Yancopoulos G, Valenzuela D, Murphy A, Yang Y, Chen Y, Hirschey MD, Bronson RT, Haigis M, Guarente LP, Farese RV Jr, Weissman S, Verdin E, Schwer B (2007) Mammalian Sir2 homolog SIRT3 regulates global mitochondrial lysine acetylation. Mol Cell Biol 27(24):8807–8814. https://doi.org/10.1128/MCB.01636-07

4. Onyango P, Celic I, Mccaffery JM, Boeke JD, Feinberg AP (2002) SIRT3, a human SIR2

homologue, is an NAD-dependent deacety-lase localized to mitochondria. Proc Natl Acad Sci U S A 99(21):13653–13658. https://doi.org/10.1073/pnas.222538099

5. Hallows WC, Yu W, Smith BC, Devries MK, Ellinger JJ, Someya S, Shortreed MR, Prolla T, Markley JL, Smith LM, Zhao S, Guan KL, Denu JM (2011) Sirt3 promotes the urea cycle and fatty acid oxidation during dietary restriction. Mol Cell 41(2):139–149. https://doi.org/10.1016/j.molcel.2011.01. 002

6. Hiromasa Y, Fujisawa T, Aso Y, Roche TE (2004) Organization of the cores of the mammalian pyruvate dehydrogenase complex formed by E2 and E2 plus the E3-binding protein and their capacities to bind the E1 and E3 components. J Biol Chem 279 (8):6921–6933. https://doi.org/10.1074/ jbc.M308172200

7. Sundaresan NR, Gupta M, Kim G, Rajamo-han SB, Isbatan A, Gupta MP (2009) Sirt3 blocks the cardiac hypertrophic response by augmenting Foxo3a-dependent antioxidant defense mechanisms in mice. J Clin Invest 119(9):2758–2771. https://doi.org/10. 1172/JCI39162

8. Calnan DR, Brunet A (2008) The FoxO code. Oncogene 27(16):2276–2288. https://doi.org/10.1038/onc.2008.21

9. Van Der Horst A, Burgering BM (2007) Stressing the role of FoxO proteins in lifespan and disease. Nat Rev Mol Cell Biol 8 (6):440–450. https://doi.org/10.1038/ nrm2190

10. Hurst LD, Williams EJ, Pal C (2002) Natural selection promotes the conservation of link-age of co-expressed genes. Trends Genet 18 (12):604–606. https://doi.org/10.1016/ s0168-9525(02)02813-5

11. Mcdonnell E, Peterson BS, Bomze HM, Hirschey MD (2015) SIRT3 regulates pro-gression and development of diseases of aging. Trends Endocrinol Metab 26 (9):486–492. https://doi.org/10.1016/j. tem.2015.06.001

12. Brown K, Xie S, Qiu X, Mohrin M, Shin J, Liu Y, Zhang D, Scadden DT, Chen D (2013) SIRT3 reverses aging-associated degenera-tion. Cell Rep 3(2):319–327. https://doi. org/10.1016/j.celrep.2013.01.005

13. Fritz KS, Galligan JJ, Hirschey MD, Verdin E, Petersen DR (2012) Mitochondrial acety-lome analysis in a mouse model of alcohol-induced liver injury utilizing SIRT3 knockout mice. J Proteome Res 11(3):1633–1643. https://doi.org/10.1021/pr2008384

14. Hebert AS, Dittenhafer-Reed KE, Yu W, Bai-ley DJ, Selen ES, Boersma MD, Carson JJ, Tonelli M, Balloon AJ, Higbee AJ, Westphall MS, Pagliarini DJ, Prolla TA, Assadi-Porter F, Roy S, Denu JM, Coon JJ (2013) Calorie restriction and SIRT3 trigger global repro-gramming of the mitochondrial protein acet-ylome. Mol Cell 49(1):186–199. https://doi. org/10.1016/j.molcel.2012.10.024

15. Ahn BH, Kim HS, Song S, Lee IH, Liu J, Vassilopoulos A, Deng CX, Finkel T (2008) A role for the mitochondrial deacetylase Sirt3 in regulating energy homeostasis. Proc Natl Acad Sci U S A 105(38):14447–14452. https://doi.org/10.1073/pnas. 0803790105

16. Qiu X, Brown K, Hirschey MD, Verdin E, Chen D (2010) Calorie restriction reduces oxidative stress by SIRT3-mediated SOD2 activation. Cell Metab 12(6):662–667. https://doi.org/10.1016/j.cmet.2010.11. 015

17. Bharathi SS, Zhang Y, Mohsen AW, Uppala R, Balasubramani M, Schreiber E, Uechi G, Beck ME, Rardin MJ, Vockley J, Verdin E, Gibson BW, Hirschey MD, Goetzman ES (2013) Sir-tuin 3 (SIRT3) protein regulates long-chain acyl-CoA dehydrogenase by deacetylating conserved lysines near the active site. J Biol Chem 288(47):33837–33847. https://doi. org/10.1074/jbc.M113.510354

18. Kim H, Kim S, Choi JE, Han D, Koh SM, Kim HS, Kaang BK (2019) Decreased neuron number and synaptic plasticity in SIRT3-knockout mice with poor remote memory. Neurochem Res 44(3):676–682. https:// doi.org/10.1007/s11064-017-2417-3

19. Kausar S, Wang F, Cui H (2018) The role of mitochondria in reactive oxygen species gen-eration and its implications for neurodegener-ative diseases. Cells 7(12). https://doi.org/ 10.3390/cells7120274

20. Eisner V, Picard M, Hajnoczky G (2018) Mitochondrial dynamics in adaptive and mal-adaptive cellular stress responses. Nat Cell Biol 20(7):755–765. https://doi.org/10. 1038/s41556-018-0133-0

21. Tseng AH, Shieh SS, Wang DL (2013) SIRT3 deacetylates FOXO3 to protect mitochondria against oxidative damage. Free Radic Biol Med 63:222–234. https://doi.org/10. 1016/j.freeradbiomed.2013.05.002

22. Samant SA, Zhang HJ, Hong Z, Pillai VB, Sundaresan NR, Wolfgeher D, Archer SL, Chan DC, Gupta MP (2014) SIRT3 deacety-lates and activates OPA1 to regulate mito-chondrial dynamics during stress. Mol Cell

Biol 34(5):807–819. https://doi.org/10.1128/MCB.01483-13

23. Ranhotra HS (2009) Up-regulation of orphan nuclear estrogen-related receptor alpha expression during long-term caloric restriction in mice. Mol Cell Biochem 332 (1–2):59–65. https://doi.org/10.1007/s11010-009-0174-6

24. Kong X, Wang R, Xue Y, Liu X, Zhang H, Chen Y, Fang F, Chang Y (2010) Sirtuin 3, a new target of PGC-1alpha, plays an important role in the suppression of ROS and mitochondrial biogenesis. PLoS One 5(7):e11707. https://doi.org/10.1371/journal.pone.0011707

25. Bergeron R, Ren JM, Cadman KS, Moore IK, Perret P, Pypaert M, Young LH, Semenkovich CF, Shulman GI (2001) Chronic activation of AMP kinase results in NRF-1 activation and mitochondrial biogenesis. Am J Physiol Endocrinol Metab 281(6):E1340–E1346. https://doi.org/10.1152/ajpendo.2001.281.6.E1340

26. Pillai VB, Sundaresan NR, Kim G, Gupta M, Rajamohan SB, Pillai JB, Samant S, Ravindra PV, Isbatan A, Gupta MP (2010) Exogenous NAD blocks cardiac hypertrophic response via activation of the SIRT3-LKB1-AMP-activated kinase pathway. J Biol Chem 285 (5):3133–3144. https://doi.org/10.1074/jbc.M109.077271

27. Thomson DM, Herway ST, Fillmore N, Kim H, Brown JD, Barrow JR (1985) Winder WW (2008) AMP-activated protein kinase phosphorylates transcription factors of the CREB family. J Appl Physiol 104 (2):429–438. https://doi.org/10.1152/japplphysiol.00900.2007

28. Woods A, Johnstone SR, Dickerson K, Leiper FC, Fryer LG, Neumann D, Schlattner U, Wallimann T, Carlson M, Carling D (2003) LKB1 is the upstream kinase in the AMP-activated protein kinase cascade. Curr Biol 13(22):2004–2008. https://doi.org/10.1016/j.cub.2003.10.031

29. Zong H, Ren JM, Young LH, Pypaert M, Mu J, Birnbaum MJ, Shulman GI (2002) AMP kinase is required for mitochondrial biogenesis in skeletal muscle in response to chronic energy deprivation. Proc Natl Acad Sci U S A 99(25):15983–15987. https://doi.org/10.1073/pnas.252625599

30. Finkel T, Holbrook NJ (2000) Oxidants, oxidative stress and the biology of ageing. Nature 408(6809):239–247. https://doi.org/10.1038/35041687

31. Sawada M, Carlson JC (1987) Changes in superoxide radical and lipid peroxide formation in the brain, heart and liver during the lifetime of the rat. Mech Ageing Dev 41 (1–2):125–137. https://doi.org/10.1016/0047-6374(87)90057-1

32. Baloh RH, Schmidt RE, Pestronk A, Milbrandt J (2007) Altered axonal mitochondrial transport in the pathogenesis of Charcot-Marie-Tooth disease from mitofusin 2 mutations. J Neurosci 27(2):422–430. https://doi.org/10.1523/JNEUROSCI.4798-06.2007

33. Reynolds IJ, Malaiyandi LM, Coash M, Rintoul GL (2004) Mitochondrial trafficking in neurons: a key variable in neurodegeneration? J Bioenerg Biomembr 36(4):283–286. https://doi.org/10.1023/B:JOBB.0000041754.78313.c2

34. Tao R, Coleman MC, Pennington JD, Ozden O, Park SH, Jiang H, Kim HS, Flynn CR, Hill S, Hayes Mcdonald W, Olivier AK, Spitz DR, Gius D (2010) Sirt3-mediated deacetylation of evolutionarily conserved lysine 122 regulates MnSOD activity in response to stress. Mol Cell 40(6):893–904. https://doi.org/10.1016/j.molcel.2010.12.013

35. Tao R, Vassilopoulos A, Parisiadou L, Yan Y, Gius D (2014) Regulation of MnSOD enzymatic activity by Sirt3 connects the mitochondrial acetylome signaling networks to aging and carcinogenesis. Antioxid Redox Signal 20(10):1646–1654. https://doi.org/10.1089/ars.2013.5482

36. Someya S, Yu W, Hallows WC, Xu J, Vann JM, Leeuwenburgh C, Tanokura M, Denu JM, Prolla TA (2010) Sirt3 mediates reduction of oxidative damage and prevention of age-related hearing loss under caloric restriction. Cell 143(5):802–812. https://doi.org/10.1016/j.cell.2010.10.002

37. Hirschey MD, Shimazu T, Goetzman E, Jing E, Schwer B, Lombard DB, Grueter CA, Harris C, Biddinger S, Ilkayeva OR, Stevens RD, Li Y, Saha AK, Ruderman NB, Bain JR, Newgard CB, Farese RV Jr, Alt FW, Kahn CR, Verdin E (2010) SIRT3 regulates mitochondrial fatty-acid oxidation by reversible enzyme deacetylation. Nature 464 (7285):121–125. https://doi.org/10.1038/nature08778

38. Bause AS, Haigis MC (2013) SIRT3 regulation of mitochondrial oxidative stress. Exp Gerontol 48(7):634–639. https://doi.org/10.1016/j.exger.2012.08.007

39. Yang W, Nagasawa K, Munch C, Xu Y, Satterstrom K, Jeong S, Hayes SD, Jedrychowski MP, Vyas FS, Zaganjor E, Guarani V, Ringel AE, Gygi SP, Harper JW, Haigis MC (2016) Mitochondrial sirtuin

network reveals dynamic SIRT3-dependent deacetylation in response to membrane depolarization. Cell 167(4):985–1000.e1021. https://doi.org/10.1016/j.cell.2016.10.016

40. Jing E, O'neill BT, Rardin MJ, Kleinridders A, Ilkeyeva OR, Ussar S, Bain JR, Lee KY, Verdin EM, Newgard CB (2013) Sirt3 regulates metabolic flexibility of skeletal muscle through reversible enzymatic deacetylation. Diabetes 62(10):3404–3417. https://doi.org/10.2337/db12-1650

41. Kincaid B, Bossy-Wetzel E (2013) Forever young: SIRT3 a shield against mitochondrial meltdown, aging, and neurodegeneration. Front Aging Neurosci 5:48. https://doi.org/10.3389/fnagi.2013.00048

42. Schwer B, Verdin E (2008) Conserved metabolic regulatory functions of sirtuins. Cell Metab 7(2):104–112. https://doi.org/10.1016/j.cmet.2007.11.006

43. Dong K, Pelle E, Yarosh DB, Pernodet N (2012) Sirtuin 4 identification in normal human epidermal keratinocytes and its relation to sirtuin 3 and energy metabolism under normal conditions and UVB-induced stress. Exp Dermatol 21(3):231–233. https://doi.org/10.1111/j.1600-0625.2011.01439.x

44. Schlicker C, Gertz M, Papatheodorou P, Kachholz B, Becker CF, Steegborn C (2008) Substrates and regulation mechanisms for the human mitochondrial sirtuins Sirt3 and Sirt5. J Mol Biol 382(3):790–801. https://doi.org/10.1016/j.jmb.2008.07.048

45. Verdin E, Hirschey MD, Finley LW, Haigis MC (2010) Sirtuin regulation of mitochondria: energy production, apoptosis, and signaling. Trends Biochem Sci 35(12):669–675. https://doi.org/10.1016/j.tibs.2010.07.003

46. Fujino T, Kondo J, Ishikawa M, Morikawa K, Yamamoto TT (2001) Acetyl-CoA synthetase 2, a mitochondrial matrix enzyme involved in the oxidation of acetate. J Biol Chem 276(14):11420–11426. https://doi.org/10.1074/jbc.M008782200

47. Shimazu T, Hirschey MD, Hua L, Dittenhafer-Reed KE, Schwer B, Lombard DB, Li Y, Bunkenborg J, Alt FW, Denu JM, Jacobson MP, Verdin E (2010) SIRT3 deacetylates mitochondrial 3-hydroxy-3-methylglutaryl CoA synthase 2 and regulates ketone body production. Cell Metab 12(6):654–661. https://doi.org/10.1016/j.cmet.2010.11.003

48. Hallows WC, Lee S, Denu JM (2006) Sirtuins deacetylate and activate mammalian acetyl-CoA synthetases. Proc Natl Acad Sci U S A 103(27):10230–10235. https://doi.org/10.1073/pnas.0604392103

49. Schwer B, Bunkenborg J, Verdin RO, Andersen JS, Verdin E (2006) Reversible lysine acetylation controls the activity of the mitochondrial enzyme acetyl-CoA synthetase 2. Proc Natl Acad Sci U S A 103(27):10224–10229. https://doi.org/10.1073/pnas.0603968103

50. Alzheimer's A (2015) 2015 Alzheimer's disease facts and figures. Alzheimers Dement 11(3):332–384. https://doi.org/10.1016/j.jalz.2015.02.003

51. Hyman BT, Phelps CH, Beach TG, Bigio EH, Cairns NJ, Carrillo MC, Dickson DW, Duyckaerts C, Frosch MP, Masliah E, Mirra SS, Nelson PT, Schneider JA, Thal DR, Thies B, Trojanowski JQ, Vinters HV, Montine TJ (2012) National Institute on Aging-Alzheimer's Association guidelines for the neuropathologic assessment of Alzheimer's disease. Alzheimers Dement 8(1):1–13. https://doi.org/10.1016/j.jalz.2011.10.007

52. Manczak M, Anekonda TS, Henson E, Park BS, Quinn J, Reddy PH (2006) Mitochondria are a direct site of Aβ accumulation in Alzheimer's disease neurons: implications for free radical generation and oxidative damage in disease progression. Hum Mol Genet 15(9):1437–1449. https://doi.org/10.1093/hmg/ddl066

53. Mungarro-Menchaca X, Ferrera P, Moran J, Arias C (2002) beta-Amyloid peptide induces ultrastructural changes in synaptosomes and potentiates mitochondrial dysfunction in the presence of ryanodine. J Neurosci Res 68(1):89–96. https://doi.org/10.1002/jnr.10193

54. Calkins MJ, Reddy PH (2011) Amyloid beta impairs mitochondrial anterograde transport and degenerates synapses in Alzheimer's disease neurons. Biochim Biophys Acta 1812(4):507–513. https://doi.org/10.1016/j.bbadis.2011.01.007

55. Ludewig S, Korte M (2017) Novel insights into the physiological function of the APP (gene) family and its proteolytic fragments in synaptic plasticity. Front Mol Neurosci 9:161. https://doi.org/10.3389/fnmol.2016.00161

56. Han P, Tang Z, Yin J, Maalouf M, Beach TG, Reiman EM, Shi J (2014) Pituitary adenylate cyclase-activating polypeptide protects against beta-amyloid toxicity. Neurobiol Aging 35(9):2064–2071. https://doi.org/10.1016/j.neurobiolaging.2014.03.022

57. Lee J, Kim Y, Liu T, Hwang YJ, Hyeon SJ, Im H, Lee K, Alvarez VE, Mckee AC, Um SJ, Hur M, Mook-Jung I, Kowall NW, Ryu H (2018) SIRT3 deregulation is linked to mitochondrial dysfunction in Alzheimer's disease. Aging Cell 17(1):e12679. https://doi.org/10.1111/acel.12679

58. Li S, Yin J, Nielsen M, Beach TG, Guo L, Shi J (2019) Sirtuin 3 mediates tau deacetylation. J Alzheimers Dis 69(2):355–362. https://doi.org/10.3233/JAD-190014

59. Yin J, Han P, Song M, Nielsen M, Beach TG, Serrano GE, Liang WS, Caselli RJ, Shi J (2018) Amyloid-beta increases tau by mediating sirtuin 3 in Alzheimer's disease. Mol Neurobiol 55(11):8592–8601. https://doi.org/10.1007/s12035-018-0977-0

60. Yang W, Zou Y, Zhang M, Zhao N, Tian Q, Gu M, Liu W, Shi R, Lü Y, Yu W (2015) Mitochondrial Sirt3 expression is decreased in APP/PS1 double transgenic mouse model of Alzheimer's disease. Neurochem Res 40(8):1576–1582. https://doi.org/10.1007/s11064-015-1630-1

61. Weir HJ, Murray TK, Kehoe PG, Love S, Verdin EM, O'neill MJ, Lane JD, Balthasar N (2012) CNS SIRT3 expression is altered by reactive oxygen species and in Alzheimer's disease. PLoS One 7(11):e48225. https://doi.org/10.1371/journal.pone.0048225

62. Ansari A, Rahman MS, Saha SK, Saikot FK, Deep A, Kim KH (2017) Function of the SIRT 3 mitochondrial deacetylase in cellular physiology, cancer, and neurodegenerative disease. Aging Cell 16(1):4–16. https://doi.org/10.1111/acel.12538

63. Yin J, Nielsen M, Carcione T, Li S, Shi J (2019) Apolipoprotein E regulates mitochondrial function through the PGC-1a-sirtuin 3 pathway. Aging (Albany NY) 11(23):11148-11156. https://doi.org/10.18632/aging.102516

64. Ramesh S, Govindarajulu M, Lynd T, Briggs G, Adamek D, Jones E, Heiner J, Majrashi M, Moore T, Amin R, Suppiramaniam V, Dhanasekaran M (2018) SIRT3 activator Honokiol attenuates beta-Amyloid by modulating amyloidogenic pathway. PLoS One 13(1):e0190350. https://doi.org/10.1371/journal.pone.0190350

65. Li H, Jia J, Wang W, Hou T, Tian Y, Wu Q, Xu L, Wei Y, Wang X (2018) Honokiol alleviates cognitive deficits of Alzheimer's disease (PS1 V97L) transgenic mice by activating mitochondrial SIRT3. J Alzheimers Dis 64(1):291–302. https://doi.org/10.3233/jad-180126

66. Hou Y, Lautrup S, Cordonnier S, Wang Y, Croteau DL, Zavala E, Zhang Y, Moritoh K, O'Connell JF, Baptiste BA, Stevnsner TV, Mattson MP, Bohr VA (2018) NAD(+) supplementation normalizes key Alzheimer's features and DNA damage responses in a new AD mouse model with introduced DNA repair deficiency. Proc Natl Acad Sci U S A 115(8):E1876–E1885. https://doi.org/10.1073/pnas.1718819115

67. Blesa J, Phani S, Jackson-Lewis V, Przedborski S (2012) Classic and new animal models of Parkinson's disease. J Biomed Biotechnol 2012:845618. https://doi.org/10.1155/2012/845618

68. Savitt JM, Dawson VL, Dawson TM (2006) Diagnosis and treatment of Parkinson disease: molecules to medicine. J Clin Invest 116(7):1744–1754. https://doi.org/10.1172/JCI29178

69. Dauer W, Przedborski S (2003) Parkinson's disease: mechanisms and models. Neuron 39(6):889–909. https://doi.org/10.1016/s0896-6273(03)00568-3

70. Liu L, Peritore C, Ginsberg J, Kayhan M, Donmez G (2015) SIRT3 attenuates MPTP-induced nigrostriatal degeneration via enhancing mitochondrial antioxidant capacity. Neurochem Res 40(3):600–608. https://doi.org/10.1007/s11064-014-1507-8

71. Zhang X, Ren X, Zhang Q, Li Z, Ma S, Bao J, Li Z, Bai X, Zheng L, Zhang Z, Shang S, Zhang C, Wang C, Cao L, Wang Q, Ji J (2016) PGC-1alpha/ERRalpha-Sirt3 pathway regulates DAergic neuronal death by directly deacetylating SOD2 and ATP synthase beta. Antioxid Redox Signal 24(6):312–328. https://doi.org/10.1089/ars.2015.6403

72. Zhang JY, Deng YN, Zhang M, Su H, Qu QM (2016) SIRT3 acts as a neuroprotective agent in rotenone-induced Parkinson cell model. Neurochem Res 41(7):1761–1773. https://doi.org/10.1007/s11064-016-1892-2

73. Cui XX, Li X, Dong SY, Guo YJ, Liu T, Wu YC (2017) SIRT3 deacetylated and increased citrate synthase activity in PD model. Biochem Biophys Res Commun 484(4):767–773. https://doi.org/10.1016/j.bbrc.2017.01.163

74. Nakamura K, Nemani VM, Wallender EK, Kaehlcke K, Ott M, Edwards RH (2008) Optical reporters for the conformation of alpha-synuclein reveal a specific interaction with mitochondria. J Neurosci 28(47):12305–12317. https://doi.org/10.1523/JNEUROSCI.3088-08.2008

75. Gleave JA, Arathoon LR, Trinh D, Lizal KE, Giguere N, Barber JH, Najarali Z, Khan MH, Thiele SL, Semmen MS (2017) Sirtuin 3 rescues neurons through the stabilisation of mitochondrial biogenetics in the virally-expressing mutant α-synuclein rat model of parkinsonism. Neurobiol Dis 106:133–146. https://doi.org/10.1016/j.nbd.2017.06.009

76. Park J-H, Burgess JD, Faroqi AH, Demeo NN, Fiesel FC, Springer W, Delenclos M, Mclean PJ (2020) Alpha-synuclein-induced mitochondrial dysfunction is mediated via a sirtuin 3-dependent pathway. Mol Neurodegener 15(1):1–19

77. West A, Brummel BE, Braun AR, Rhoades E, Sachs JN (2016) Membrane remodeling and mechanics: experiments and simulations of alpha-Synuclein. Biochim Biophys Acta 1858 (7 Pt B):1594–1609. https://doi.org/10.1016/j.bbamem.2016.03.012

78. Patterson VL, Zullo AJ, Koenig C, Stoessel S, Jo H, Liu X, Han J, Choi M, Dewan AT, Thomas JL, Kuan CY, Hoh J (2014) Neural-specific deletion of Htra2 causes cerebellar neurodegeneration and defective processing of mitochondrial OPA1. PLoS One 9(12): e115789. https://doi.org/10.1371/journal.pone.0115789

79. Stafa K, Tsika E, Moser R, Musso A, Glauser L, Jones A, Biskup S, Xiong Y, Bandopadhyay R, Dawson VL, Dawson TM, Moore DJ (2014) Functional interaction of Parkinson's disease-associated LRRK2 with members of the dynamin GTPase superfamily. Hum Mol Genet 23(8):2055–2077. https://doi.org/10.1093/hmg/ddt600

80. Deng H, Dodson MW, Huang H, Guo M (2008) The Parkinson's disease genes pink1 and parkin promote mitochondrial fission and/or inhibit fusion in Drosophila. Proc Natl Acad Sci U S A 105(38):14503–14508. https://doi.org/10.1073/pnas.0803998105

81. Thomas RE, Andrews LA, Burman JL, Lin W-Y, Pallanck LJ (2014) PINK1-Parkin pathway activity is regulated by degradation of PINK1 in the mitochondrial matrix. PLoS Genet 10(5):e1004279. https://doi.org/10.1371/journal.pgen.1004279

82. Pickrell AM, Youle RJ (2015) The roles of PINK1, parkin, and mitochondrial fidelity in Parkinson's disease. Neuron 85(2):257–273. https://doi.org/10.1016/j.neuron.2014.12.007

83. Huang W, Huang Y, Huang RQ, Huang CG, Wang WH, Gu JM, Dong Y (2016) SIRT3 expression decreases with reactive oxygen species generation in rat cortical neurons during early brain injury induced by experimental subarachnoid hemorrhage. Biomed Res Int 2016:8263926. https://doi.org/10.1155/2016/8263926

84. Arrasate M, Finkbeiner S (2012) Protein aggregates in Huntington's disease. Exp Neurol 238(1):1–11. https://doi.org/10.1016/j.expneurol.2011.12.013

85. Myers RH (2004) Huntington's disease genetics. NeuroRx 1(2):255–262. https://doi.org/10.1602/neurorx.1.2.255

86. Valor LM (2015) Transcription, epigenetics and ameliorative strategies in Huntington's disease: a genome-wide perspective. Mol Neurobiol 51(1):406–423. https://doi.org/10.1007/s12035-014-8715-8

87. Raymond LA, Andre VM, Cepeda C, Gladding CM, Milnerwood AJ, Levine MS (2011) Pathophysiology of Huntington's disease: time-dependent alterations in synaptic and receptor function. Neuroscience 198:252–273. https://doi.org/10.1016/j.neuroscience.2011.08.052

88. Nasir J, Floresco SB, O'kusky JR, Diewert VM, Richman JM, Zeisler J, Borowski A, Marth JD, Phillips AG, Hayden MR (1995) Targeted disruption of the Huntington's disease gene results in embryonic lethality and behavioral and morphological changes in heterozygotes. Cell 81(5):811–823. https://doi.org/10.1016/0092-8674(95)90542-1

89. Fu J, Jin J, Cichewicz RH, Hageman SA, Ellis TK, Xiang L, Peng Q, Jiang M, Arbez N, Hotaling K, Ross CA, Duan W (2012) Trans-(−)-epsilon-Viniferin increases mitochondrial sirtuin 3 (SIRT3), activates AMP-activated protein kinase (AMPK), and protects cells in models of Huntington disease. J Biol Chem 287(29):24460–24472. https://doi.org/10.1074/jbc.M112.382226

90. Herskovits AZ, Guarente L (2013) Sirtuin deacetylases in neurodegenerative diseases of aging. Cell Res 23(6):746–758. https://doi.org/10.1038/cr.2013.70

91. Browne SE, Beal MF (2004) The energetics of Huntington's disease. Neurochem Res 29 (3):531–546. https://doi.org/10.1023/b:nere.0000014824.04728.dd

92. Jodeiri Farshbaf M, Ghaedi K (2017) Huntington's disease and mitochondria. Neurotox Res 32(3):518–529. https://doi.org/10.1007/s12640-017-9766-1

93. Kim J, Moody JP, Edgerly CK, Bordiuk OL, Cormier K, Smith K, Beal MF, Ferrante RJ (2010) Mitochondrial loss, dysfunction and

altered dynamics in Huntington's disease. Hum Mol Genet 19(20):3919–3935. https://doi.org/10.1093/hmg/ddq306

94. Song W, Chen J, Petrilli A, Liot G, Klinglmayr E, Zhou Y, Poquiz P, Tjong J, Pouladi MA, Hayden MR, Masliah E, Ellisman M, Rouiller I, Schwarzenbacher R, Bossy B, Perkins G, Bossy-Wetzel E (2011) Mutant huntingtin binds the mitochondrial fission GTPase dynamin-related protein-1 and increases its enzymatic activity. Nat Med 17(3):377–382. https://doi.org/10.1038/nm.2313

95. Valdmanis PN, Rouleau GA (2008) Genetics of familial amyotrophic lateral sclerosis. Neurology 70(2):144–152. https://doi.org/10.1212/01.wnl.0000296811.19811.db

96. Cheng A, Yang Y, Zhou Y, Maharana C, Lu D, Peng W, Liu Y, Wan R, Marosi K, Misiak M, Bohr VA, Mattson MP (2016) Mitochondrial SIRT3 mediates adaptive responses of neurons to exercise and metabolic and excitatory challenges. Cell Metab 23(1):128–142. https://doi.org/10.1016/j.cmet.2015.10.013

97. Pasinelli P, Belford ME, Lennon N, Bacskai BJ, Hyman BT, Trotti D, Brown RH Jr (2004) Amyotrophic lateral sclerosis-associated SOD1 mutant proteins bind and aggregate with Bcl-2 in spinal cord mitochondria. Neuron 43(1):19–30. https://doi.org/10.1016/j.neuron.2004.06.021

98. Song W, Song Y, Kincaid B, Bossy B, Bossy-Wetzel E (2013) Mutant SOD1G93A triggers mitochondrial fragmentation in spinal cord motor neurons: neuroprotection by SIRT3 and PGC-1alpha. Neurobiol Dis 51:72–81. https://doi.org/10.1016/j.nbd.2012.07.004

99. Harlan BA, Pehar M, Sharma DR, Beeson G, Beeson CC, Vargas MR (2016) Enhancing NAD+ salvage pathway reverts the toxicity of primary astrocytes expressing amyotrophic lateral sclerosis-linked mutant superoxide dismutase 1 (SOD1). J Biol Chem 291(20):10836–10846. https://doi.org/10.1074/jbc.M115.698779

100. Hor J-H, Santosa MM, Lim VJW, Xuan Ho B, Taylor A, Khong ZJ, Ravits J, Fan Y, Liou Y-C, Soh B-S, Ng S-Y (2021) ALS motor neurons exhibit hallmark metabolic defects that are rescued by SIRT3 activation. Cell Death Differ 28(4): 1379–1397. https://doi.org/10.1038/s41418-020-00664-0.

101. Jawaid A, Murthy SB, Wilson AM, Qureshi SU, Amro MJ, Wheaton M, Simpson E, Harati Y, Strutt AM, York MK, Schulz PE (2010) A decrease in body mass index is associated with faster progression of motor symptoms and shorter survival in ALS. Amyotroph Lateral Scler 11(6):542–548. https://doi.org/10.3109/17482968.2010.482592

102. Peter RS, Rosenbohm A, Dupuis L, Brehme T, Kassubek J, Rothenbacher D, Nagel G, Ludolph AC (2017) Life course body mass index and risk and prognosis of amyotrophic lateral sclerosis: results from the ALS registry Swabia. Eur J Epidemiol 32(10):901–908. https://doi.org/10.1007/s10654-017-0318-z

103. Parodi-Rullan RM, Chapa-Dubocq XR, Javadov S (2018) Acetylation of mitochondrial proteins in the heart: the role of SIRT3. Front Physiol 9:1094. https://doi.org/10.3389/fphys.2018.01094

104. Borthwick GM, Johnson MA, Ince PG, Shaw PJ, Turnbull DM (1999) Mitochondrial enzyme activity in amyotrophic lateral sclerosis: implications for the role of mitochondria in neuronal cell death. Ann Neurol 46(5):787–790. https://doi.org/10.1002/1531-8249(199911)46:5<787::aid-ana17>3.0.co;2-8

105. Wiedemann FR, Manfredi G, Mawrin C, Beal MF, Schon EA (2002) Mitochondrial DNA and respiratory chain function in spinal cords of ALS patients. J Neurochem 80(4):616–625. https://doi.org/10.1046/j.0022-3042.2001.00731.x

106. Lassmann H, Van Horssen J, Mahad D (2012) Progressive multiple sclerosis: pathology and pathogenesis. Nat Rev Neurol 8(11):647–656. https://doi.org/10.1038/nrneurol.2012.168

107. Friese MA, Schattling B, Fugger L (2014) Mechanisms of neurodegeneration and axonal dysfunction in multiple sclerosis. Nat Rev Neurol 10(4):225–238. https://doi.org/10.1038/nrneurol.2014.37

108. Zundorf G, Reiser G (2011) Calcium dysregulation and homeostasis of neural calcium in the molecular mechanisms of neurodegenerative diseases provide multiple targets for neuroprotection. Antioxid Redox Signal 14(7):1275–1288. https://doi.org/10.1089/ars.2010.3359

109. Rice CM, Sun M, Kemp K, Gray E, Wilkins A, Scolding NJ (2012) Mitochondrial sirtuins—a new therapeutic target for repair and protection in multiple sclerosis. Eur J Neurosci 35(12):1887–1893. https://doi.org/10.1111/j.1460-9568.2012.08150.x

110. Campbell GR, Ziabreva I, Reeve AK, Krishnan KJ, Reynolds R, Howell O, Lassmann H, Turnbull DM, Mahad DJ (2011)

Mitochondrial DNA deletions and neurode-generation in multiple sclerosis. Ann Neurol 69(3):481–492. https://doi.org/10.1002/ana.22109

111. D'aquila P, Rose G, Panno ML, Passarino G, Bellizzi D (2012) SIRT3 gene expression: a link between inherited mitochondrial DNA variants and oxidative stress. Gene 497 (2):323–329. https://doi.org/10.1016/j.gene.2012.01.042

112. Inkster B, Strijbis EM, Vounou M, Kappos L, Radue EW, Matthews PM, Uitdehaag BM, Barkhof F, Polman CH, Montana G, Geurts JJ (2013) Histone deacetylase gene variants predict brain volume changes in multiple sclerosis. Neurobiol Aging 34(1):238–247. https://doi.org/10.1016/j.neurobiolaging.2012.07.007

113. Khodaei F, Rashedinia M, Heidari R, Rezaei M, Khoshnoud MJ (2019) Ellagic acid improves muscle dysfunction in cuprizone-induced demyelinated mice via mitochondrial Sirt3 regulation. Life Sci 237:116954. https://doi.org/10.1016/j.lfs.2019.116954

114. Satterstrom FK, Haigis MC (2014) Luciferase-based reporter to monitor the transcriptional activity of the SIRT3 promoter. Methods Enzymol 543:141–163. https://doi.org/10.1016/B978-0-12-801329-8.00007-6

115. Schneider CA, Rasband WS, Eliceiri KW (2012) NIH image to ImageJ: 25 years of image analysis. Nat Methods 9(7):671–675. https://doi.org/10.1038/nmeth.2089

116. Guan X, Lin P, Knoll E, Chakrabarti R (2014) Mechanism of inhibition of the human sirtuin enzyme SIRT3 by nicotinamide: computational and experimental studies. PLoS One 9 (9):e107729. https://doi.org/10.1371/journal.pone.0107729

117. Camacho-Pereira J, Tarrago MG, Chini CCS, Nin V, Escande C, Warner GM, Puranik AS, Schoon RA, Reid JM, Galina A, Chini EN (2016) CD38 dictates age-related NAD decline and mitochondrial dysfunction through an SIRT3-dependent mechanism. Cell Metab 23(6):1127–1139. https://doi.org/10.1016/j.cmet.2016.05.006

Chapter 6

Oxidative Stress and Cellular Dysfunction in Neurodegenerative Disease

Anju Singh, Ritushree Kukreti, and Shrikant Kukreti

Abstract

Neurodegenerative diseases are conditions characterized by irreversible loss of neurons and loss of cognitive and motor function, mostly occurring in elderly people. The increasing age has been the most consistent risk factor for developing a neurodegenerative disorder. Interlinking of oxidative stress and neurodegeneration is well established, indicating that the overproduction of reactive oxygen/nitrogen/sulfur species (ROS/RNS/RSS) correlates with progressive neurodegeneration. It is well defined that neurons are prone to oxidative damage due to the enrichment of neuronal membranes with polyunsaturated fatty acids (PUFA) along with highly dynamic and regiospecific oxygen consumption by brain cells and a weak antioxidant defense. Oxidative stress (OS) refers to the condition where imbalance between oxidants and antioxidants leads to the elevation in ROS/RNS/RSS levels in biological systems. Increasing evidence has demonstrated the dysfunction of redox signaling and its regulation leads to oxidative stress in the cellular environment. Thus, disruption as well as dysregulation of redox signaling is found to be linked with various neurodegenerative disorders such as Alzheimer's disease (AD), Parkinson's disease (PD), Huntington's disease (HD), and amyotrophic lateral sclerosis (ALS). The present chapter gives insight about how redox mediated OS can lead to neurodegenerative diseases.

Keywords Redox signaling, Alzheimer's disease (AD), Parkinson's disease (PD), Huntington's disease (HD), Amyotrophic lateral sclerosis (ALS), Mitochondrial dysfunction, ROS/RNS/RSS

1 Introduction

Physiological aging is accompanied by alterations in the functions of central cholinergic neurons, multiplying progressively over time leading to a morphological and functional deterioration of the brain. This time-dependent, sequential deterioration of brain physiology is linked specifically to numerous neurodegenerative diseases such as Alzheimer's disease (AD), Parkinson's disease (PD), Huntington's disease (HD), and amyotrophic lateral sclerosis (ALS). The prevalence of these neurological disorders increases with age. These diseases are mainly characterized by loss of neurons and motor skills, leading to the impairment of cognitive behavior and

Philip V. Peplow, Bridget Martinez and Thomas A. Gennarelli (eds.), *Neurodegenerative Diseases Biomarkers: Towards Translating Research to Clinical Practice*, Neuromethods, vol. 173, https://doi.org/10.1007/978-1-0716-1712-0_6,
© Springer Science+Business Media, LLC, part of Springer Nature 2022

thus, becoming a primary health concern for the majority of an ageing population that face the consequences of these diseases [1, 2]. The exact etiology of neurodegenerative disease is still unclear; however, oxidative stress (OS) is found to be one of the correlates [3, 4]. Reports indicate that PD is the second deadliest neurodegenerative disease in the United States with 1–2% of the population affected whereas AD is the sixth most prevalent and leading cause of death in older people [5–9]. The malicious role played by reactive chemical species such as ROS/RNS/RSS at the onset of neurodegenerative diseases has recently been reviewed [10]. Reactive oxygen/nitrogen species are usually produced spontaneously in biological systems during the course of various chemical reactions. These reactive species are involved in mediating many cellular activities like cell survival, inflammation, stress/response, and many diseases like muscular dystrophy, cardiovascular diseases, and cancers [11, 12].

Cells require an optimum level of reactive oxygen/nitrogen species for proper functioning of cellular machinery. Elevation of reactive species concentration in a biological system leads to major critical complications and health problems. ROS/RNS/RSS are highly reactive and when present at higher concentration cause an imbalance in pro-oxidant and antioxidants, resulting in oxidative stress. Mounting scientific evidence indicates a decisive role played by these reactive species, as an elevated level of oxidative stress was observed in the brains of patients suffering from neurodegeneration with deterioration and dysfunction of neurons [11, 13]. Though free radicals do not directly trigger the neurodegeneration, they aggravate the disease indirectly by causing oxidative stress and mitochondrial dysfunction. Neurons are more vulnerable to oxidative damage owing to their enrichment in cell membranes of polyunsaturated fatty acids (PUFA) and high consumption of oxygen along with a weaker antioxidant defense system [14].

Spontaneous generation of free radicals in various cellular metabolic processes is indispensable for various signaling processes. Reactive oxygen species ($O_2 \bullet^-$, $OH\bullet$, H_2O_2, etc.), reactive nitrogen species ($\bullet NO$, $OONO^-$, $\bullet NO_2$, etc.), and reactive sulfur species ($RS\bullet$, $RSNO$, etc.) are considered the key elements of signal transduction pathways as well as redox signaling [15, 16]. An optimum level of these reactive species/free radicals is required in cells, while elevated levels can cause deleterious effects on biomolecules. They modify amino acid residues in proteins, cause lesions on nucleic acids via oxidation, and alter their structure and function [16, 17]. Elevation of free radicals causes an imbalance in redox signaling and debilitates antioxidant systems. Redox imbalance leads to many dysfunctions of vital organs such as skin, liver, eyes, lungs, liver, and brain. As the brain is composed of various cell types, the antioxidant defense system of the brain is inefficient to

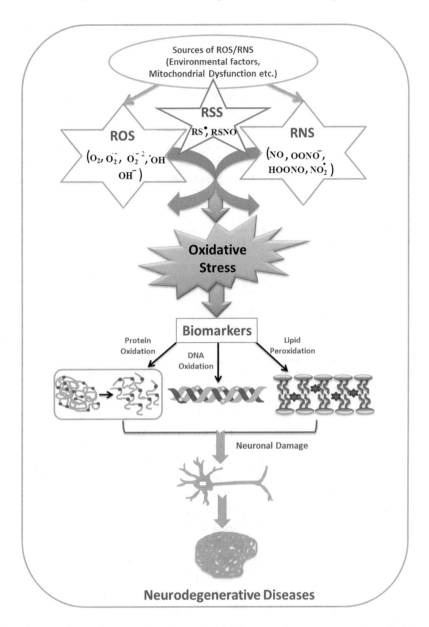

Fig. 1 Diagrammatic representation of various ROS/RNS/RSS species and their role in oxidative stress and neurodegenerative diseases

cope up with redox imbalance. Figure 1 displayed a schematic representation of the source of ROS/RNS/RSS which leads to OS biomarkers of OS results in neuronal damage and neurodegenerative diseases.

The structure, function and interaction of various brain cells with other cellular components together with elevated consumption of oxygen make neuronal brain cells susceptible to neurodegeneration. The brain uses 20% of inhaled oxygen in spite of the fact that it represents only 2% of total body weight. Metabolically, being the most active organ in the body, it contributes to generating an excessive number of free radicals [18–21]. Polyunsaturated fatty acids (PUFAs), eicosapentanoic acid (C20:5) and docosahexanoic acid (C22:6) abundantly present in neuronal membranes, are prone to be attacked by hydroxyl radicals (•OH), resulting in lipid peroxidation via a chain reaction. This is one of the reasons for neuronal deterioration, ultimately leading to neurodegeneration [22]. This chapter focuses on the basic principle of oxidative stress and redox biology, and their pivotal role in developing neurodegenerative diseases.

2 Reactive Free Radicals and Their Production

Although oxygen is important in biological systems, an elevation in the oxygen level becomes potentially harmful and can cause malicious effects. Aerobic respiration requires oxygen for the ultimate generation of adenosine triphosphate (ATP) via oxidative phosphorylation in mitochondria [23]. Numerous chemical entities, with one or more unpaired electrons in their molecular orbitals, termed free radicals, are capable of existing independently in biological systems [24–26].

Oxygen-centered free radicals include hydroxyl radical (•OH), superoxide ($O_2^{•-}$), singlet oxygen ($^1O_2$), and hydrogen peroxide (H_2O_2). Nitrogen-centered free radicals called reactive nitrogen species (RNS), are also present in biological systems, these include nitric oxide (•NO), peroxynitrite ($ONOO^-$), and nitrogen dioxide (•NO_2). Reactive sulfur species (RSS) including sulfenic acid, thiosulfinate, and thiyl radical. Are also found in biological systems along with ROS and RNS (Fig. 2). The type of reactions which involve transfer of electrons between two species, causing simultaneous oxidation and reduction, are known as redox reactions. The following section includes the description of free radical species, their importance as well as their effect on cellular components in biological systems.

2.1 Reactive Oxygen Species (ROS)

Numerous reactions as well as enzymatic processes occurring in vivo contribute to the generation of these species. ROS entities are involved in various types of signaling processes. They play a pivotal role in intracellular killing of bacteria by neutrophilic granulocytes, detoxification by the liver, as well as cell signaling [27–31]. In living organisms, being a natural by-product, $O_2^•$ is the most common reactive oxygen species, produced abundantly in the mitochondria during oxidative phosphorylation. As a biradical, molecular oxygen contains two unpaired electrons with parallel spin configuration. On complete reduction of molecular oxygen,

Free Radical Species	Examples of Reactive Species
ROS (Reactive Oxygen Species)	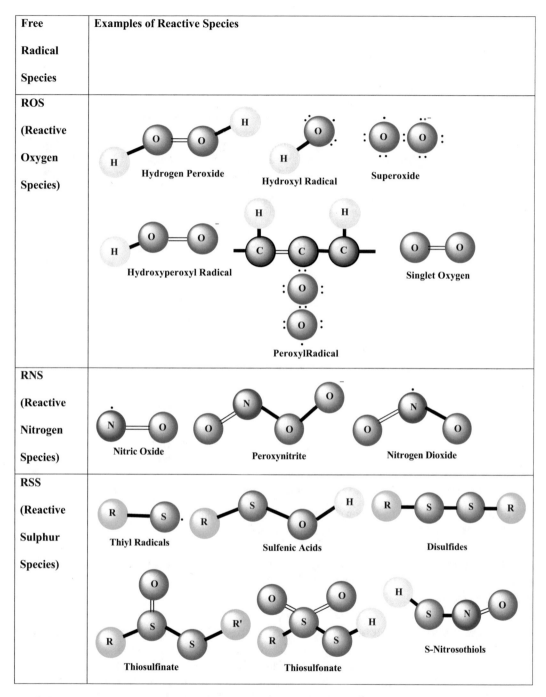
RNS (Reactive Nitrogen Species)	
RSS (Reactive Sulphur Species)	

Fig. 2 Examples of free radical species having Oxygen, Nitrogen and Sulphur as central atom (ROS, RNS and RSS)

water is formed with the generation of highly reactive intermediates, that is, H_2O_2 and •OH [32, 33]. The schematic representation of reactions is depicted in Fig. 3. The steps involved in the complete reduction of oxygen are as follows.

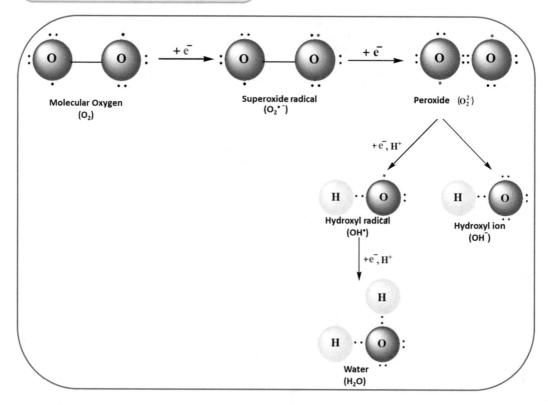

$$O_2 + e^{\cdot} \longrightarrow O_2^{\cdot-} \quad \text{Superoxide radical}$$
$$O_2^{\cdot-} + H_2O \longrightarrow HO_2^{\cdot} + OH^- \quad \text{Hydroperoxyl radical}$$
$$HO_2^{\cdot} + e^- + H \longrightarrow H_2O_2 \quad \text{Hydroperoxide}$$
$$H_2O_2 + e^- \longrightarrow OH + OH^- \quad \text{Hydroxyl radical}$$

Fig. 3 Reduction of molecular oxygen into its free radical components i.e. superoxide radical, peroxide, hydroxyl radical, hydroxyl ion and at the end water molecule

2.2 ROS and Their Implication in Biological System

A brief description of various oxygen centered reactive species is given here (*see* Fig. 3):

2.2.1 Superoxide Anion ($O_2^{\cdot-}$)

Undoubtedly, oxygen carries a very important role in biological systems and thus its free radical state or ROS have also a very significant role to play in living beings. Macrophages and neutrophils, best known for engulfing harmful foreign entities in cells, are prominent sources of $O_2^{\cdot-}$. During the invasion of pathogens, phagocytic cells become active and produce $O_2^{\cdot-}$ as a part of a defense mechanism to combat the pathogens and cause inflammation. Superoxide anions are also waste products in purine metabolism produced together with uric acid from xanthine. This reaction is catalyzed by xanthine oxidase [34]. Several reactions taking place in an intracellular milieu, mediated by cytochrome P450, peroxisomal oxidases and NAD-(P)-H oxidases, are responsible for the production of ROS [35–37].

The importance of the superoxide anion in the body is due to its tendency to produce other free radicals which have deleterious effect on biomolecules and cells [30, 31]. Superoxides significantly affect enzyme function [38]. It plays a central role in extracting iron and sulfur from the iron–sulfur protein clusters via inactivating the iron regulatory protein-1 (IRP-1), leading to release of iron from the protein cluster [35]. Superoxide anion can also interact with the cell-signaling molecule nitric oxide (•NO) to form the other free radical product peroxynitrile (ONOO•) [39]. There are many other sources of ROS production such as autoxidation reactions of reduced flavins, metalloproteins, metal ions, quinones, and ionizing radiation or photochemical irradiation, all actively involved in the generation of superoxide anion.

2.2.2 Hydrogen Peroxide
(H$_2$O$_2$)

Although hydrogen peroxide does not come under the category of free radicals, it is a very important biological oxidant owing to its active participation in the production of the hydroxyl radical. The presence of H_2O_2 in higher concentration in the cellular environment results in an adverse effect on macromolecules. It is also involved in various signaling pathways as well as inducing detrimental damage in pathophysiological conditions. It actively participates in the production of other free radicals such as hydroxyl radical (•OH), either by Fenton reaction or the Haber-Weiss reaction [39, 40]. Due to its high diffusion constant, hydrogen peroxide diffuses from the hydrophobic membranes. It also escapes from mitochondria and enters cell membranes. The breakdown of H_2O_2 into oxygen and water molecules is catalyzed by an antioxidant enzyme catalase [41, 42]. Another enzyme, glutathione peroxidase, can also catalyze breaking down of hydrogen peroxide along with other peroxides that are generated on lipids within the biological system to produce water with oxidation of glutathione [42].

2.2.3 Hydroxyl Radical
(•OH)

Hydroxyl radicals are produced from hydrogen peroxide and are considered highly reactive species having a very short half-life of the order of 10^{-9} s. They have a detrimental effect on various cellular components such as DNA, protein, lipids and carbohydrates. They can result in the formation of DNA-protein cross links, single- and double-strand DNA breaks, and damage in nucleotide bases as well as protein fragmentation [43, 44]. Enrichment of brain and cell membrane in lipids makes them more prone to peroxidation on exposure to •OH radical. Hydroxyl radical production is enhanced in the presence of transition metals such as manganese, zinc, iron, cobalt, and molybdenum, having a partially filled d subshell. These elements can participate in bond formation via electrons in the d subshell and exist in various oxidation states and thus act as a redox active element. Transition metals like iron and copper can transfer an electron and exist in two oxidation states (for iron $3^+/2^+$ and for copper $2^+/1^+$). Though these transition metal ions play a substantial role in biological redox reactions, they can also play a critical role in the generation of free radicals [45]. Thus, the reaction involving as substrates ferrous ions and hydrogen peroxide results in the products hydroxyl ion and hydroperoxyl radical is known as the Fenton reaction. These free radicals enter into the oxidation of biomolecules (DNA, proteins, and lipids).

$$Fe^{2+} + H_2O_2 \longrightarrow Fe^{3+} + {\bullet}OH + OH^-$$

$$Fe^{3+} + H_2O_2 \longrightarrow Fe^{2+} + {\bullet}OOH + H^+$$

2.2.4 Peroxyl Radical (ROO•)

The major pathway of peroxyl radical production in cells is autoxidation of biomolecules, such as lipid peroxidation. Free radicals attack lipids or the organs rich in PUFAs containing Carbon/Carbon double bonds and cause peroxidation [46–48]. Various enzymes like lipoxygenase, cyclooxygenase, and cytochrome P450 are also involved in catalyzing lipid peroxidation. PUFAs are attacked by peroxyl radicals creating a chain reaction and generating more free radicals. The free radicals thus produced cause damage of important classes of biomolecules (DNA, proteins, and lipids). This deleterious effect of peroxyl radicals can be overcome by an antioxidant, that is, vitamin E or tocopherol which mediate in a free radical chain breaking reaction.

2.2.5 Hydroperoxyl Radical (•HO₂)

The other very important ROS, known as hydroperoxyl radical, is just the protonated form of $O_2{\bullet}^-$. This reactive species is also termed as perhydroxyl radical. It is documented that around 0.3% of $O_2{\bullet}^-$ exists as protonated form in cellular environment [49]. Lipid peroxidation can be induced by hydroperoxyl radical and may promote tumor formation.

2.2.6 Singlet Oxygen (¹O₂)

Singlet oxygen is an electronically highly reactive state of molecular oxygen and at the same time a highly toxic reactive species. Produced from molecular oxygen, on activation it first gets excited to the first state and then next to the singlet state, where each orbital is occupied by one electron, with opposite spins. In vivo production of singlet oxygen also takes place via activation of neutrophils as well eosinophils [50–52]. Various enzymatic reactions catalyzed by lipoxygenases, dioxygenases, and lactoperoxidase also give rise to singlet oxygen species [53–55]. Being a very reactive species, it has detrimental effects on biomolecules usually leading to DNA damage as well as tissue damage [56].

2.3 RNS and Their Implication in Biological System

Evidences now prove that other free radicals such as reactive nitrogen species (RNS) also contribute to redox imbalance in biological processes. The following is a brief description of some RNS:

2.3.1   *Nitric Oxide (•NO)*

This is the most important RNS that plays a pivotal role in cell signaling processes and is produced in tissues by different nitric oxide synthases (NOS) [57, 58]. NOS is able to convert L-arginine into L-citrulline in mammals. It involves a reaction in which one of the terminal guanido nitrogen atoms undergoes an oxidation process and results in the production of •NO [59]. All of the three isoforms of NOS, that is, neuronal NOS (nNOS), endothelial NOS (eNOS), and inducible NOS (iNOS) participate in generation of NO radicals.

$$ \text{L-Arginine} + O_2 + \text{NADPH} \xrightarrow{\text{NOS}} \text{L-Citrulline} + \overset{\bullet}{N}O + \text{NADP}^+ $$

Owing to its solubility nitric oxide is able to permeate the cytoplasm and plasma membrane, that is, it is soluble in aqueous as well as in lipid medium. This reactive species also becomes a part of intracellular second messenger and induces stimulation of guanylate cyclase and protein kinase which results in relaxation of smooth muscle in blood vessels. It is also involved in various critical vascular functions such as muscle contraction/dilatation to inflammation and neuroplasticity [60]. Its role in cellular redox regulation and regulation of enzymatic activity via nitrosylating proteins is well established [61, 62]. Nitric oxide is involved in many crucial processes such as blood pressure, neurotransmission, immune regulation, and defensive mechanism [63].

2.3.2   *Peroxynitrite*
(ONOO$^-$)

The peroxynitrite radical is a highly reactive nitrogen species generated by the reaction of •NO and $O_2\bullet^-$ and mediates many deleterious reactions in the cellular environment [64]. It is short lived and exerts a pronounced degenerating effect on biomolecules. It reacts with CO_2 and forms the highly reactive and toxic nitrosoperoxocarboxylate ($ONOOCO_2^-$) or peroxynitrous acid (ONOOH) [65]. OH• and NO_2 are produced on hemolysis of peroxynitrous acid or it can also rearrange to form NO_3. Peroxynitrite can oxidize lipids and oxidize methionine and tyrosine residues in proteins. It is also involved in oxidation of guanine in nitroguanine on any DNA strand and thus inhibits various biological processes. The presence of nitrotyrosine residues is an indication of peroxynitrite-induced cellular damage [66]. Production and formation of the peroxynitrite radical suppresses the bioavailability of •NO and thus leads to the obstruction of crucial •NO-mediated cellular processes.

2.3.3   *Nitrogen Dioxide*
(•NO$_2$)

Nitrogen dioxide is predominantly considered as an environmental pollutant, however its endogenous generation via enzyme myeloperoxidases makes it one of the significant reactive nitrogen species. A hydrophobic environment facilitates the reaction of •NO and O_2 to produce toxic nitrogen dioxide, which leads to the deterioration of lipids, finally resulting in apoptosis [67, 68]. It is well documented that generation of 3-nitrotyrosine in vivo depends on the concentration of $\cdot NO_2$ radical available in the cells. Moreover, $\cdot NO_2$ adversely affects antioxidants, the significant scavengers of this radical such as ascorbic acid, alpha-tocopherol, and bilirubin [69].

2.4 RSS and Their Implication in Biological Systems

For years, the focus on redox biology centered around reactive species containing oxygen and nitrogen free radicals. Sulfur was thought primarily a part of building blocks of proteins and iron–sulfur proteins etc. Various sulfur containing compounds such as sulfur dioxide (SO_2) and hydrogen sulfide (H_2S) found extensively in the environment have an adverse effect on human health due to their toxicity. Recent studies including some in vivo reports indicated active participation and a substantial role of RSS in redox biology as well as in cell signaling [70, 71]. Thiol oxidation products of proteins such as the disulfides, sulfenic acids, sulfinic acids, and sulfonic acids are well known. These thiols react with many oxidizing agents like H_2O_2, $^1O_2$, $ONOO^-$, and $O_2 \cdot^-$ leading to RSS formation. The involvement of many sulfur compounds in mediating redox buffering and causing distortion of macromolecules, which lead to oxidative stress, is well established. The extensively studied glutathionylation, resulting in cysteine modification, reflects the crucial role of sulfur played in biological processes [72, 73].

The existence of numerous reactive sulfur species is accredited to its multivalent oxidation states varying from -2 (in H_2S) to $+6$ states (SO_4^{2-}). The term RSS first emerged in 2001 with the introduction of RSS species, namely, thiyl radical (RS•), sulfenic acids (RSOH), disulfides (RSSR), thiosulfinate [RS (O) SR], thiosulfonate [RS (O)$_2$ SR], and S-nitrosothiols (Fig. 2). These species act as oxidizing agents and are considered to be formed in an oxidative stress environment. Sulfide oxidation, coupled to oxidative phosphorylation, occurs in mitochondria; thus, this cellular organelle plays a crucial role in the generation of RSS. In recent times, besides thiosulfate and sulfate, other RSS such as glutathione persulfide (GSSH) originating in mitochondria, have also drawn much attention [74, 75]. Sulfide oxidation catalyzed by sulfide quinone oxidoreductase, occurs via a two electron oxidation of hydrogen sulfide, and resulting in a persulfide intermediate [76]. The GSSH thus formed is utilized in a sulfur-transferase reaction and transfers the sulfane sulfur to sulfite, yielding thiosulfate in the reaction. Enzymes such as eosinophil peroxidase, prostaglandin H synthase, or myeloperoxidase catalyze an oxidation reaction of sulfite, resulting in the formation of the sulfite anion radical, peroxymonosulfate radical, and sulfate radical, ultimately causing destruction of proteins [77].

Recent evidence has shown that the gasotransmitter H_2S to be the utmost important candidate for the signaling process. H_2S, originated from cysteine, is involved in various processes from inflammation to cardiovascular functions. It has been observed that H_2S plays a crucial role in physiological functions of brain such as involvement in regulation of the N-methyl-D-aspartate (NMDA) receptor (NMDAR) leading to neuromodulation. It has a cytoprotective effect and also plays a role in rescuing neurons in many processes. Many other processes such as post-translational modification of reactive cysteine residues also envisage its role in signaling.

3   Markers and Fatalities of Free Radicals and Reactive Species

Reactive species are abundantly generated in biological systems; however, their concentration and optimum level is under the continuous surveillance of the body's cellular machinery. Antioxidants combat continuously with elevated level of reactive species and scavenge them for the benefit of human health. Yet, sometimes situations arise when the antioxidants defense gets hampered as levels of free radicals become enhanced abruptly, creating an imbalance between free radicals and antioxidants, leading to oxidative stress and nitrosative stress. These elevated levels of free radicals are highly reactive and have deleterious effect on all the three biomolecules, that is, nucleic acids, proteins, and lipids [78].

3.1   Morbid Effect on Deoxyribonucleic Acid (DNA)

DNA is the most important and unique member of the biomolecules responsible for the transfer of hereditary characters and thus called the blueprint of life. Free radicals may have a damaging effect on DNA via oxidation of its nitrogenous bases specifically guanine (G). In addition to the nucleus of the cell, the cell organelle mitochondria also contain DNA. The mitochondrial DNA is more vulnerable than nuclear DNA because the former resides in the vicinity of a reactive species source which is mitochondria. In particular, OH• radicals significantly react with all DNA components, that is, purine, pyrimidine bases, and sugar–phosphate backbone. It leads to alterations in the DNA strand including double stranded breaks [79]. A number of reactions including purine and pyrimidine with OH• radical occur via extracting a proton, thus yielding modified base products, as well as DNA–protein crosslinks. The nucleobase pyrimidine is usually attacked by OH• radical and produces pyrimidine adducts such as thymine glycol, uracil glycol, 5-hydroxydeoxy uridine, 5-hydroxydeoxycytidine, and hydantoin [80]. Purine adducts like 8-hydroxydeoxy guanosine, 8-hydroxydeoxy adenosine, and 2,6-diamino-4-hydroxy-5-formamidopyrimidine are also formed on reaction with OH• radicals. Further, the sugar backbone is also attacked by OH• radicals and results in adducts like glycolic acid, 2-deoxytetrodialdose, and erythrose implementation. The 8-hydroxy deoxyguanosine, known as a significant oxidative DNA marker, has been shown to play a pivotal role in mutagenesis, carcinogenesis, and aging [81]. Surprisingly, when compared with the nuclear DNA, the elevated level of guanine modification as 8-hydroxydeoxyguanosine is only reflected in mitochondrial DNA [81].

The reactive nitrogen species peroxynitrite, which is not a free radical by chemical nature but a powerful oxidant, may attack the nucleobases causing DNA single strand breakage, nitration and oxidation of guanine. The Peroxynitrite induced-DNA modification commonly includes the formation of 8-nitroguanine and 8-oxodeoxyguanosine. 8-nitoguanine, when formed is not stable and is removed spontaneously creating apurinic sites [82]. During DNA synthesis adenine can be paired with 8-nitroguanine and cause G-T transversions. 8-nitroguanine is also linked with mutagenesis and carcinogenesis [83]. The DNA damage induced by peroxynitrite also initiates the activation of a DNA repair system.

3.2 Morbid Effect on Protein

The amino acid side chains cause the proteins to be vulnerable to oxidation by free radical species. The free radical species such as $O_2^{\cdot-}$, OH$\cdot$, peroxyl, hydroperoxyl as well as nonradical species like H_2O_2, O_3, HOCl, and OONO$^-$ can cause oxidation in a protein moiety by interacting with amino acid side chains [84]. The oxidation of proteins via various free radical reactive species causes aberrations such as protein–protein cross-links, denaturation of proteins, loss of enzyme activity, receptors, transport proteins, along with loss of protein functioning involving various significant biological processes [85]. As mentioned in the previous section, sulfur-containing amino acid residues such as cysteine and methionine are more prone to oxidation by ROS, leading to formation of disulfides and methionine sulfoxide, respectively [86, 87]; however, in the presence of enzymes disulfide reductase and methionine sulfoxide reductase, these oxidized products are converted back to their native forms [88–91].

ROS can also mediate and induce oxidative damage to various amino acids such as lysine, proline, threonine and arginine producing their carbonyl derivatives. These carbonyl derivatives are significant biomarkers for ROS-mediated lesions caused in proteins. O-Tyrosine is a significant marker for hydroxyl radical–mediated oxidation, while 3-nitrotyrosine is a known marker for RNS-mediated reaction. The elevated level of protein carbonyls is observed in the patients of Alzheimer's disease, Parkinson's disease, muscular dystrophy, rheumatoid arthritis, Werner's syndrome, and aging [3].

3.3 Morbid Effect on Lipids

As lipids are the integral part of all cell membranes, the component PUFA residue of phospholipids is more susceptible to lipid peroxidation via reactive free radical species. The lipid peroxidation plays a vital role in vivo owing to its indulgence in various pathological conditions. Peroxidation of lipids causes reduced membrane fluidity, loss of membrane functioning, and reduced activity of membrane-bound enzymes and receptors [92, 93]. Formation of

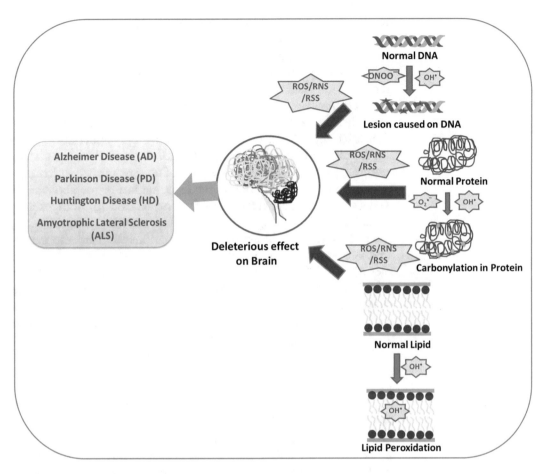

Fig. 4 Deleterious effect on brain: free radical-mediated lesions caused in biomolecules

a carbon-centered lipid radical takes place when any free radical approaches the lipid and causes proton removal from the methylene groups of fatty acids. The generated lipid radical then reacts with molecular oxygen and a lipid peroxyl radical is originated, which later undergoes rearrangement via cyclization reaction yielding endoperoxides. The resultant endoperoxides finally produce the highly toxic substances malondialdehyde (MDA) and 4-hydroxylnonenal (4-HNA), which cause lesions in DNA and proteins [94]. The lipid peroxyl radical is not eliminated and propagates further the peroxidation process via removing hydrogen atoms from other lipid molecules, finally deteriorating the biomolecules. Figure 4 depicts the role of ROS on biomolecules and neurodegenerative diseases.

4   ROS Mediated Mitochondrial Dysfunction

The literature is rich in reports on the central role played by mitochondrial dysfunction in the pathophysiology and etiology of neurodegenerative diseases. Maintenance of mitochondrial integrity is very crucial and a key challenge in order to protect neuronal damage. As electron transfer and electron transport chain reactions take place in mitochondria, it is considered the hub of free radicals generation, hence it is involved in OS. More than 90% of ROS production takes place in mitochondria. The mitochondrion is undoubtedly a cell organelle of the utmost importance and mitochondrial impairment can hinder oxidative phosphorylation leading to cell death. A cell is totally dependent on ATPs for its energy requirement. Apart from production of energy, mitochondria perform myriad other functions such as oxidation of metabolites via Krebs's cycle and β-oxidation of fatty acids, calcium homeostasis, production and scavenging of free radicals, cellular differentiation, and apoptosis [95]. Along with ATPs, various ROS are also generated as by-products of oxidative phosphorylation in mitochondria. Elevation in ROS may mediate oxidative damage to nuclear DNA (nDNA) as well as mitochondrial DNA (mt DNA), which might be linked with various age-related neurodegenerative diseases [96]. Mt. DNA is more prone to oxidative damage in comparison to nDNA owing to the absence of histone proteins bound to nDNA, and also because mtDNA resides lie near the site of ROS generating respiratory chain. Mitochondria play a crucial role in apoptotic pathways because ROS also affect significantly the vital biomolecules such as lipids, DNA and proteins via the processes of lipid peroxidation, DNA oxidation, and protein carbonylation. Apart from ROS, other reasons, such as replication errors as well as repair mechanism, might also involve mtDNA mutations [97]. mtDNA replication is not dependent on cellular division unlike nDNA, therefore mtDNA replication rate is higher than nDNA [98].

Mitochondrial dynamics can be controlled by fusion and fission proteins. Dysfunctional mitochondria are first recognized and exclusively removed by autophagic process known as mitophagy. Dysfunctional mitochondria have been chosen for mitochondrial fragmentation via the fission process then undergo mitophagy and are engulfed/degraded in lysosomes [99]. Fission process plays a substantial role in maintaining standard, quality and integrity of mitochondria. Ageing affects the fusion (combining) and fission (splitting) process of mitochondria, resulting in the hampering of the dysfunctional mitochondria elimination process. Excessive mitochondrial fragmentation results in increased ROS production. Further, dysfunctional mitochondria are involved in genomic instability, apoptosis, inflammation and neurodegenerative diseases [100, 101].

Mitochondrial dysfunction has a deleterious effect on ATP generation by hampering the activity of enzymes involved in electron transport chain, overproduction of ROS, reduction of mt DNA, and release of caspase 3. Accumulation of mutant protein and mitochondrial dysfunction is also aggravated by ageing which leads to the functional and structural changes in neuronal activity resulting in cell death. However, a plethora of evidences support the fact that aging-related neurodegeneration results from mitochondrial dysfunction. Figure 5 summarizes the role of mitochondrial dysfunction in neurodegenerative disease.

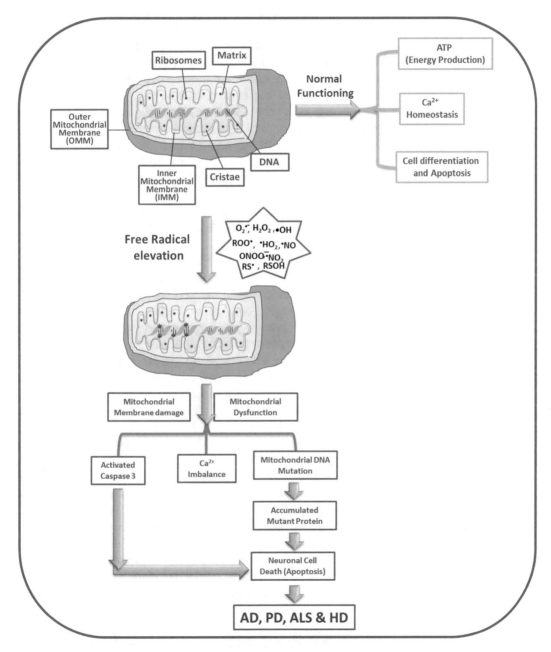

Fig. 5 Substantial role of mitochondrial dysfunction in neurodegenerative diseases (AD, PD, ALS and HD) via elevation in ROS/RNS/RSS, mitochondrial DNA mutation, mutant protein

Mitochondria are also involved in cell differentiation and cell death. Various extrinsic and intrinsic factors play a crucial role in mitochondria-mediated cell death. Stress processes and cell injury aggravate the intrinsic factors leading to apoptosis [102]. Cell injury and stress causes deterioration in mitochondrial membrane, which causes pro-apoptotic factors in the cytoplasm. A special class of proteins known as Bcl-2 control and maintain apoptotic mechanism [103]. Bcl-2 family is classified further into three groups, namely, antiapoptotic (Bcl-2, Bcl-xL, Bcl-w, A1, Mcl-1, and Bcl-b), proapoptotic BH3 proteins (Bid, Bik, Bmf, Bad, Bim, Puma, BNip3, Hrk, and Noxa), and proapoptotic Bak/Bax proteins [102]. Many experimental models of neurodegenerative diseases have witnessed the overexpression of these proteins. These evidences suggested that Bcl-2 proteins are actively involved in controlling mitochondrial dysfunction as well as apoptotic pathways which significantly play critical roles in neurological disorder. Therefore, mitochondrial integrity is crucial for proper functioning of all the biological processes as well as neuronal well-being.

5   Oxidative Stress–Mediated Ferroptosis and Neurodegenerative Diseases

Cell death or apoptosis is an undoubtedly unavoidable critical process to maintain every aspect of mammalian growth, development, and homeostatic regulation to sustain life under physiological as well as pathological conditions. Earlier, cell death was divided into apoptosis and necrosis but in addition to these two modes, in recent studies other programmed cell deaths such as autophagy and necrotic apoptosis are also established with unique characteristics. The term ferroptosis was first proposed in 2012 by Dixon et al., which is different from apoptosis, pyroptosis, necroptosis, and other programmed cell death [104]. Ferroptosis is identified as a nonapoptotic, iron-dependent, oxidative cell death mechanism and a type of regulated cell death [105]. Ferroptosis is considered to be distinct from other modes of cell death in context of morphology, biochemistry and heredity. It is found to be caused by oxidative alteration in cellular microenvironment critically maintained by glutathione peroxidase 4 (GPX4), which can be obstructed by iron chelators and lipophilic antioxidants [106]. Recent studies have suggested that ferroptosis plays a pivotal role in the pathology of brain and neurological disorders [107]. PUFAs-enriched brain cells make them susceptible to lipid peroxidation. Along with lipid peroxidation, accumulation of iron in different brain regions is also found to be involved in neurodegeneration leading to cell death [108].

Iron is an abundantly found transition metal in brain which can participate in crucial functions of the central nervous system, such as   mitochondrial   energy   transduction,   enzyme   catalysis, myelination, synaptic plasticity, and neurotransmitter synthesis [109, 110]. Iron gets ingested via receptor-mediated endocytosis of transferrin in brain capillary endothelial cells which leads to transportation of iron into the cerebral cytoplasm through astrocytes or via divalent cation binding protein (DMT1), thereby maintaining a steady saturated level of iron in brain. This pathway maintains normal physiological function of the nervous system [111]. Brain iron homeostasis is controlled and regulated by iron movement between blood and brain tissue, intracellular and extracellular and via many iron pools. Protein-bound iron is safe in a cellular environment whereas accumulation of excessive free iron produce molecules via Fenton's reaction which can have deleterious effect on protein and nucleotides [112]. Thus, iron accumulation is crucially involved in the onset and progression of neurodegenerative diseases, involving oxidative stress induction, mitochondrial dysfunction, elevation in ROS production along with morbid effect on nDNA and mtDNA. It is established in recent reports that accumulation of iron in nerve cells might play a key role in the ferroptosis-associated neurodegenerative diseases.

6 Free Radical–Mediated Neurodegenerative Diseases

Free radicals usually cause severe damage and lesions in biomolecules which ultimately lead to an adverse effect on the physiology and vital organs of the body. Brain is one of the most metabolically active organs, which consumes most of the oxygen available in the body. Enrichment of PUFA in the brain makes it more susceptible to oxidative damage. The brain is the main reservoir of the polyunsaturated ω-3 fatty acids, specifically DHA [95]. Apart from DHA, the myelin sheath synthesis of neurons also requires lipid molecules such as cholesterol, of which 20% of whole-body cholesterol is in the brain [113].

Various transition metals such as iron present in the brain can also mediate reactions for the generation of free radicals like •OH. Specific regions of the brain such as the hippocampus, substantia nigra, and striatum, are more vulnerable to the attack of free radical reactive species [114–116]. Figure 6 depicted the region of brain get affected in various neurodegenerative diseases.

Brain susceptibility toward oxidative damage, owing to the degeneration or deterioration of neurons by reactive free radicals, leads to the main cause of aging and neurodegenerative diseases, namely, Alzheimer's disease, Parkinson's disease, Huntington's disease, and amyotrophic lateral sclerosis [117–120].

6.1 Alzheimer's Disease (AD)

Alzheimer's disease (AD) is considered one of the most important neurodegenerative diseases, characterized by cognitive impairment and progressive degeneration of motor neurons and behavioral deficits. Neuronal degeneration leads to difficulties in performing

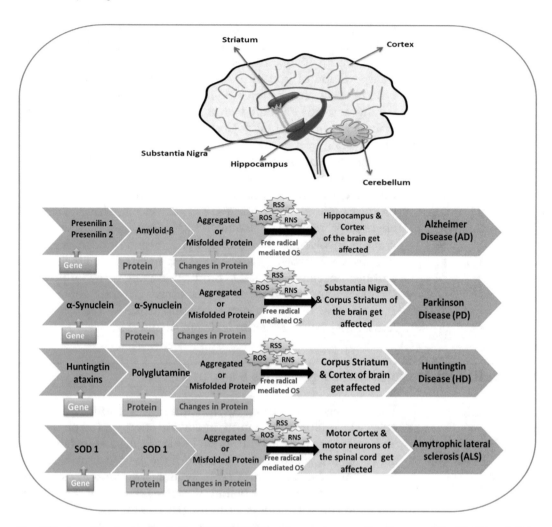

Fig. 6 Region of the brain affected by ROS/RNS/RSS and neurodegenerative diseases (AD, PD, HD and ALS)

daily life activities of AD patients [121]. AD affects approximately 16 million of the population worldwide and causes a high mortality rate owing to this disorder. One promising biomarker of AD is beta amyloid deposition in the brain. Formation of amyloid plaques via chelation of amyloid-β-peptide (Aβ) with transition metal ions such as Cu^{2+}, Zn^{2+}, or Fe^{3+} is the characteristic feature of this deadly disorder. The key morbid role played by Aβ is due to the presence of a histidine amino acid residue at position 6, 13 and 14, respectively, facilitating a site for transition metals binding. Along with Aβ plaques, accumulation of intracellular tau neurofibrillary tangles (NFT) is also recognized in the brain of AD patients [122, 123]. The formation of plaques reduces the calcium (Ca^{2+}) storage in the endoplasmic reticulum, which leads to accumulation of Ca^{2+} in the cytosol. Ca^{2+} overload in the cytosol leads to the reduced secretion of GSH (glutathione), resulting in an elevated level of ROS inside the cellular environment. Glutathione is a pervasive tripeptide thiol (γ-L-glutamyl-L-cysteinyl-glycine) significantly present as an intracellular as well as extracellular defending antioxidant. It is synthesized by the condensation reaction between glutamic acid, cysteine and glycine. Free GSH is exclusively present in its reduced form in normal cellular conditions. Total GSH exists as free and bound to proteins in cells. The enzyme glutathione reductase converts oxidized glutathione into its reduced form GSSH. The ratio of GSH:GSSH is used as a marker specifically for cellular toxicity [124]. In normal cellular condition GSH:GSSH ratio exists in range of 1:10 mM whereas in OS conditions the ratio is found to be changed to 10:1 and even 1:1 [125, 126]. GSH, as an antioxidant, plays a substantial role in the treatment of liver diseases, tumors, poisoning, and aging diseases. It was demonstrated recently that ROS-mediated OS plays a prominent role in the pathophysiology of AD. Mitochondrial dysregulation also plays a crucial role in the overproduction and accumulation of ROS, as reduced generation of ATP alters the Ca^{2+} homeostasis as well as inducing excitotoxicity [127, 128].

Accumulating evidence suggests that in severe cases of AD, NMDARs (N-methyl-D-aspartate-type glutamate receptors) are found to be overactivated, leading to excessive influx of Ca^{2+} via inducing cell permeability and assisting in the production of a neurotoxic concentration of ROS/RNS [129]. These reactive species play a key role in stress-activated protein kinase pathways, directly involved in hyperphosphorylation of tau protein and A-β-induced apoptosis. Aβ-induced overproduction of ROS/RNS also involves cellular signaling pathways and inhibits the functioning of significant cellular processes. It is known that Aβ can induce calcineurin activity, thereby triggering the Bcl-2 associated promoter death, which in turn leads to the release of cytochrome c from mitochondria. Aβ can also directly be associated with caspases and subsequent involvement in neuronal apoptosis [130]. Besides these aspects, a number of environmental factors such as pollutants, chemicals and radiations are also associated with elevation of reactive free radicals. An increasing deposition of iron in the body can induce ROS formation. In contrast, Cu^{2+} and Zn^{2+} supplements can be given to reduce Aβ-related ROS production and metal-catalyzed Aβ-deposition [131, 132]. The earlier belief that A-β-plaque deposition is the main cause of AD, is now being revisited based on recent reports that this plaque is a physiological antioxidant but this function is deteriorating because of deposition of high level of Aβ-plaque. In future, this key property will be treated as a potent therapy to combat AD [10].

6.2 Parkinson's Disease (PD)

Parkinson's disease (PD), fundamentally a movement disorder, is the second most prevalent chronic progressive neurodegenerative disorder affecting the nervous system. Neuropathology indicates that PD involves degeneration of dopaminergic neurons in the substantia nigra pars compacta (SNpc) and is characterized by the accumulation of inclusion bodies (also known as Lewy bodies) of an intracellular protein α-synuclein in SNpc. These inclusion bodies are exclusively expressed in the brain and mutation in this protein is reported in familial forms of AD. Over 200 hundred years ago, PD was first recognized as shaking palsy and was held responsible for motor nerve impairment in patients suffering from this disease [133]. Loss of dopaminergic neurons in turn leads to the depletion of the neurotransmitter dopamine. A patient having this disease does not only face problems in motor skills but also experiences impaired cognitive behavior, insomnia, and abnormal sense of smell. An elevated concentration of PUFA along with reduced level of lipid peroxidation markers, that is, malondialdehyde and 4-hydroxynonenal was observed in the PD brain. Further, the presence of protein carbonyls, resulting from oxidatively damaged protein in the PD brain, revealed the role of reactive species in nitration and nitrosylation of some proteins [134, 135]. An increased level of 8-hydroxydeoxyguanosine is also observed in the PD brain associated with a specific deletion in mitochondrial DNA in the remaining dopaminergic neurons in the substantia nigra. As mitochondrial DNA is more prone to oxidative damage via ROS, it was hypothesized that this deletion is caused by oxidative stress [136, 137].

The elevation and accumulation of ROS in the PD brain has been observed to be triggered when individuals are exposed to environmental factors like pesticides, neurotoxins, and dopamine. Many other studies indicated that the presence of neuromelanin can also be interlinked with loss of dopaminergic neurons, as highly pigmented neurons are more vulnerable to oxidative damage. Neuromelanin produced on autoxidation of dopamine, is thought to be induced by ROS overproduction [138]. In some reports mitochondrial complex I defect is observed to be related with PD, the said defect is linked with the mutation of phosphatase and tensin homolog (PTEN)-induced putative kinase I, also known as PINK1 [139]. PINK1 is one of the most significant proteins universally expressed in human tissues and critically involved in combating OS and maintenance of mitochondrial potential for normal cellular function. PINK1 mutation is directly linked with PD occurrence. It has been implicated in the causation of an AR (autosomal recessive) form of Parkinsonism. Various other mutations are also shown to be implicated in the progression of PD such as DJ-1, parkin, α-synnuclein, and leucine-rich repeat kinase 2 (LRRK2) [140–142]. These gene mutations can lead to mitochondrial dysfunction by aggravating ROS production and mediating oxidative stress [142].

Though lacking effective measures to check the progression of PD, the exact molecular mechanism and role of ROS in aggravating PD can provide insight to designing strategies for the control of

PD. Fruits and many antioxidants are recognized to decrease the harmful effects of free radicals [11, 143]. Many neuroprotective strategies are designed to help in eliminating the detrimental effect of mitochondrial-mediated OS in dopaminergic neurons. Lipoic acid (LA) a coenzyme, also plays a key role in the recycling of crucial antioxidants like vitamins C and E and GSH. LA protects neurons by repairing OS-induced mitochondrial dysregulation via producing GSH as well as by inhibiting the lipid peroxidation [144]. A number of studies have demonstrated that substances with an antioxidant property such as coenzyme Q10, vitamin C, tocopherol, DHA and polyphenols failed in treating patients with PD. These studies open new avenues for combination therapies to attenuate mitochondrial dysfunction along with inhibiting ROS overproduction in PD patients [145].

6.3 Huntington's Disease (HD)

Huntington's disease (HD) is an autosomal dominant genetic neurodegenerative disorder associated with mutant protein deposits that generally affect people between the ages 35–55 years [146–149]. In the early stages of HD, neuronal degeneration takes place in the striatum, followed by deterioration of the cerebral cortex and thalamus in advanced stages. The characteristic symptoms of HD are emotional as well as psychiatric disturbances, mood swings, bradykinesia, rigidity, dementia, and impaired motor skills, along with loss of cognitive function [150]. It is known that OS plays a substantial role in the pathophysiology of Huntington's disease, a disease manifested by a polyglutamine expansion in the Huntingtin (Htt) protein [151]. The genetic aberration identified in HD includes an expansion of unstable CAG trinucleotide repeats in exon 1 of the huntingtin gene, on chromosome 4. In the Western population HD has a prevalence of 10.6–13.7 individuals per 100,000. Htt is involved in various abnormalities of protein function such as protein aggregation, transcriptional dysregulation, excitotoxicity and inflammation [152–154]. The HD phenotype is observed when the extension of consecutive CAG trinucleotide repeats in HTT gene is more than 36 times. It is well established that mitochondrial dysregulation-mediated OS plays a crucial role in the later stages of HD progression. Any malfunction in the mitochondrial electron transport chain (ETC) is a major cause of elevated level of ROS/RNS generation leading to the etiopathogenesis of HD. HD is also involved in modulating energy levels in the brain, a phenomenon now being exploited by researchers to gain more insights on the pathophysiology of the disease [155].

Several studies have shown that when compared to healthy people, an elevation in lipid peroxidation in blood plasma was seen in patients having symptoms of HD [156, 157]. Expression of various genes like AHCY1, ACO-2, and OXCT-1, which are involved in encoding S-adenosyl-L-homocysteine hydrolase 1, aconitase and enzymes linked in expression of 3-oxoacid CoA transferase 1 proteins, were found to be associated with oxidative stress. This led to the downregulation of mitochondrial energy metabolism in the peripheral leucocytes of HD genetic patients as well as carriers [158]. In other studies involving HD patients, elevations in various OS markers such as 8-OHDG, plasma MDA levels, advanced oxidation protein product (AOPP) were observed along with lower levels of antioxidants in erythrocytes such as Cu/Zn- superoxide dismutase (SOD) and glutathione peroxidase 1 (GPX1). AOPP was shown to crucially involve with motor sensitivity in HD patients [159]. Previous studies found that an optimum level of ROS in cells is essential for proper signaling in biological processes; however, the elevated level of ROS interferes crucially with various biological processes, specifically transcriptional process. Many transcription factors are linked actively in monitoring the redox states of NAD, NADP, or glutathione. Reports demonstrated that in order to compensate the illicit effect of oxidative stress, glutathione depletion can enhance acetylation of Sp1 and Sp3 transcription factors. Class I histone deacetylase (HDACs) modulation can enhance Sp1/Sp3 acetylation, resulting in subsequent processes such as increased Sp1/Sp3 DNA binding, and resistance in glutathione depletion. 3-Nitropropionic acid causing striatal damage leads to transgenic overexpression of mutant huntingtin (mHtt) [160]. mHtt is also indicated to be involved in suppression of transcription factors Sp1/Sp3, thus leading to depletion of the genes involved in alleviating OS. Furthermore, it was established that OS dramatically induced DNA binding activity of Sp1/Sp3 in cortical neurons, along with enhanced levels of these proteins [160].

It has been reported extensively that mitochondria are not only responsible for the production of ROS, but rather they are also critically involved in significant scavenging of ROS [161–163]. The literature is rich in reports that OS is involved in the physiopathology of neurodegenerative disease including HD, though the exact pathways and molecular mechanisms are still unresolved. Detailed studies are required to gain a clear picture of the exact mechanisms and pathways of HD and the role of ROS/RNS in the progression of HD.

6.4 Amyotrophic Lateral Sclerosis (ALS)

Amyotrophic lateral sclerosis (ALS) is a group of rare neurodegenerative diseases manifested by selective degeneration of motor neurons controlling voluntary muscle movements, resulting in severe muscle weakness [164, 165]. ALS is the most common type of motor neuron disease. ALS can be categorized into sporadic (90%) as well as familial (10%) cases depending on the clear indication of involvement of a genetic element. Though the commonest form of ALS is sporadic ALS (sALS), its exact etiology is still not clear. The familial ALS (fALS) arises due to the mutation in the genes CuZn superoxide dismutase 1 (SOD1), Tar DNA binding protein

(TARDP), fused in sarcoma (FUS), ubiquilin 2 (UBQLN2), and optineurin (OPTN) [166–171]. A major genetic mutation shown as an expansion in GGGGCC hexanucleotide repeat harboring in noncoding region of the C9Orf72 gene present on chromosome 9p21 is found to play a crucial role in the etiology of ALS [172]. Although the exact cause of sALS is still illusive, it was found to be associated with many risk factors like smoking, level of physical fitness, age, environmental toxins such as exposure to pesticides, other chemicals, or radiation. According to a hypothesis, some overlapping pathological mechanism is involved in sALS and fALS owing to the presence of common clinical symptoms shared by sALS and fALS patients [173]. Involvement of glutamate excitotoxicity, defective axonal transport, protein aggregation and misfolding, obstructed RNA metabolism, dysregulated protein trafficking, ROS-mediated OS, and mitochondrial dysfunction were found to play a substantial role in ALS [174].

It is well recognized that SOD1 is crucially involved in diverse activities such as scavenging excessive superoxide radicals, modulation of cellular respiration, as well as post-translational modifications. The evidences indicating the role of SOD1 mutation in 20% of fALS cases, signifies that the mutation in SOD1 leads to disruption of its antioxidant property resulting in ROS-mediated morbidity to biomolecules (DNA, protein and lipids) [175, 176]. A study by Bastow et al. reported that amino acid biosynthesis of cells in a yeast model is disrupted by mutant SOD1, and manifested in cellular mortality responsible for the neural degeneration observed in ALS [177]. Under normal cellular conditions, there is a balance between ROS and antioxidants levels; however in diseased conditions, rapid elevation in ROS/RNS along with disruption in antioxidant defense systems leads to the increased DNA damage, lipid peroxidation, and cell death. Conversion of superoxide to hydrogen peroxide (H_2O_2) and molecular oxygen is done by normal SOD1. The mutant SOD1 is responsible for elevations in ROS production, directly affecting the neuronal mortality in ALS. Furthermore, SOD1 is also thought to be involved in maintenance and well-being of motor neurons via activation of ERK/AKT signaling. It has been demonstrated that in OS condition there is over secretion of SOD1 in neurons which aggravates the neurodegeneration [178].

The USA FDA (Food and Drug Administration) has approved two drugs Riluzole and Edaravone for the treatment of ALS. Riluzole is found to be an effectively potent drug which acts as antagonist to glutamate [179]. Edaravone is found to play a crucial role in scavenging free radical species such as lipid peroxyl radical (LOO•), and peroxynitrite ($ONOO^-$) via exploiting its electron donating properties, leading ultimately to the protection of neurons [180]. Edaravone is also effective in elimination of nitrosative stress in cerebrospinal fluid of ALS patients. Many other potential antioxidants such as acetylcysteine and creatine have not been much effective against ALS progression [181].

Several laboratories worldwide are working to investigate the exact pathways of the disease, etiology, pathophysiology, and treatment for ALS; however, the complete mechanism of the disease is still unknown [182, 183]. Therefore, more information about the disease is needed to work on strategies and development of drugs to halt the disease progression.

7   Concluding Remarks and Future Perspective

Recognizing the factors that contribute to neurodegenerative progressions in the brain is one of the major goals of modern medicine. In this chapter we have revisited recent advances made in our understanding of the intracellular pathways involved in the generation of reactive species (free radicals) and their effective role in selective neuronal degeneration in various neurological disorders. As more is learned about neurodegenerative disorders, OS appears to be a main contributor to disease pathogenesis. A growing body of scientific literature on the impact of ROS/RNS/RSS in the development and pathophysiology of prevalent AD, PD, HD, and ALS supports the idea that OS leads to the development of neurodegenerative diseases [184–186].

It should be kept in mind that low or moderate concentrations of ROS are necessary for the maturation processes of cells and can act as armaments for the host defense system, supporting cell proliferation and survival pathways. Thus, a fine balance between an optimum level of ROS and antioxidants is essential for normal functioning of the cells. There have been many studies on the therapeutic effects of antioxidants on neurodegenerative diseases/disorders and shown to have inconsistent results. In normal conditions upregulation of antioxidant defense systems effectively scavenge the ROS by neutralization, obstruct the initiation of chain reactions; ultimately inhibiting OS-mediated health disorders, aging, and so on. Elevation in ROS causing mitochondrial dysfunction is used in studying the deleterious effect on mitochondrial DNA [187]. ROS/RNS/RSS also play critical roles in signaling processes in cell organelles; for example, in lysosomes, autophagy is linked to redox regulation [188, 189]. In several neurodegenerative diseases, impaired autophagy is found to be involved in the progression of misfolding along with aggregation of proteins in cell organelles, compromising neuroprotection. The common feature of these diseases is extensive evidence of oxidative stress. Undoubtedly, the link of mitochondrial dysfunction with OS-mediated neurodegeneration is undeniable because they are predominantly linked with ROS/RNS generation. Furthermore, most of the neurodegenerative diseases are characterized by extensive oxidative damage to biomolecules, leading to cell death by a variety of mechanisms. Unsaturated lipids which are particularly susceptible to oxidative modification make lipid peroxidation a sensitive marker of oxidative stress.

Although an extensive literature is available on neurodegenerative diseases, understanding the varied susceptibility of specific parts of the brain toward distinct neurodegenerative diseases/disorders is still ongoing. Involvement of environmental factors, radiation, chemicals, pesticides, mutations, and genetic and epigenetic factors in initiation as well as progression of neurodegenerative diseases has revealed that combinations of these factors are responsible for neuron mortality. More focus is needed on the investigations of specific molecular mechanisms and etiology of neurodegenerative diseases. As far as the pharmacological strategies to slow down or halt neurodegeneration, are concerned, antioxidant supplementation is shown to have limited therapeutic benefit. Myriad reasons are advanced for the failure of antioxidants in curing the disease or halting the progression of neurodegeneration, or to prolong the lives of patients. The most favorable stage and time, and dose of antioxidants administration, is still not known. In some instances, cysteine and GSH supplementation has been found to be beneficial in curing age-related diseases; however, scarcity/paucity of the antioxidant or molecule remains a matter of concern. Therefore, treatment is completely dependent on the bioavailability of these compounds. Various reports are apprehensive about the studies on mouse models of disease, as it cannot be applied/translated to human patients owing to the differences in antioxidant networks in biological systems. For instance, vitamin E is effective in treating familial ALS in a mouse model but ineffective in treating ALS patients. Refining the understanding of ALS pathogenesis is critical in developing earlier diagnostic methods, as well as suggesting new effective treatments. Antioxidant efficacy in removing all types of free radical species is still debatable.

However, in keeping all pros and cons in mind it is suggested that there is a dire need of designing molecules with low cytotoxicity, good cell permeability, ability to pass the blood–brain barrier, low cost, commercially availability, and so on. Advent of new therapeutics and high-throughput methodologies provide platforms to investigate and design drugs to treat neurodegenerative diseases. Targeting various transcription factors which play a key role in modulation of redox biology also pave the way for the possible treatment of neurodegeneration. Elucidation of the pathways important in the production of and defense from free radicals may be important in devising new therapeutic strategies to slow down neurodegenerative processes and improve our understanding of the relevance of OS in the pathobiology of these disease conditions.

The scientific community believes that ongoing research worldwide will open new avenues for the development of novel strategies to encounter the redox mediated OS and neurodegenerative diseases/disorders. Owing to this research aspiration and by improving and accelerating therapeutic development, scientists would turn the loss into hope.

References

1. Albers DS, Beal MF (2000) Mitochondrial dysfunction and oxidative stress in aging and neurodegenerative disease. J Neural Transm Suppl 59:133–154. https://doi.org/10.1007/978-3-7091-6781-6_16

2. Hamer M, Chida Y (2009) Physical activity and risk of neurodegenerative disease: a systematic review of prospective evidence. Psychol Med 39:3–11. https://doi.org/10.1017/S0033291708003681

3. Liu Z, Zhou T, Ziegler AC et al (2017) Oxidative stress in neurodegenerative diseases: from molecular mechanisms to clinical applications. Oxid Med Cell Longev 2017:2525967. 11 pages. https://doi.org/10.1155/2017/2525967

4. Singh A, Kukreti R, Saso L et al (2019) Oxidative stress: a key modulator in neurodegenerative diseases. Molecules 24(8):1583, 1-20. https://doi.org/10.3390/molecules24081583

5. Silvade HR, Khan NL, Wood NW (2000) The genetics of Parkinson's disease. Curr Opin Genet Dev 10:292–298. https://doi.org/10.1016/s0959-437x(00)00082-4

6. Rijkde MC, Launer LJ, Berger K et al (2000) Prevalence of Parkinson's disease in Europe: a collaborative study of population-based cohorts. Neurology 54:S21–SS3

7. Bekris LM, Mata IF, Zabetian CP (2010) The genetics of Parkinson disease. J Geriatr Psychiatry Neurol 23:228–242. https://doi.org/10.1177/0891988710383572

8. Farrer MJ (2006) Genetics of Parkinson disease: paradigm shifts and future prospects. Nat Rev Genet 7:306–318. https://doi.org/10.1038/nrg1831

9. Alzheimer's Association (2011) 2011 Alzheimer's disease facts and figures. Alzheimers Dement 7:208–244. https://doi.org/10.1016/j.jalz.2011.02.004

10. Sbodio JI, Snyder SH, Paul BD (2019) Redox mechanisms in neurodegeneration: from disease outcomes to therapeutic opportunities. Antioxid Redox Signal 30(11):1450–1499. https://doi.org/10.1089/ars.2017.7321

11. Zuo L, Zhou T, Pannell BK et al (2015) Biological and physiological role of reactive oxygen species—the good, the bad and the ugly. Acta Physiol (Oxf) 214:329–348. https://doi.org/10.1111/apha.12515

12. He F, Zuo L (2015) Redox roles of reactive oxygen species in cardiovascular diseases. Int J Mol Sci 16:27770–27780. https://doi.org/10.3390/ijms161126059

13. Dias V, Junn E, Mouradian MM (2013) The role of oxidative stress in Parkinson's disease. J Parkinsons Dis 3(4):461–491. https://doi.org/10.3233/JPD-130230

14. Rego AC, Oliveira CR (2003) Mitochondrial dysfunction and reactive oxygen species in excitotoxicity and apoptosis: implications for the pathogenesis of neurodegenerative diseases. Neurochem Res 28:1563–1574. https://doi.org/10.1023/a:1025682611389

15. Fukai T, Ushio-Fukai M (2011) Superoxide dismutases: role in redox signaling, vascular function, and diseases. Antioxid Redox Signal 15:1583–1606. https://doi.org/10.1089/ars.2011.3999

16. Gadoth N, Goebel HH (2011) Oxidative stress and free radical damage in neurology. Humana Press, New York, viii, 323 p

17. Floyd RA, Carney JM (1992) Free radical damage to protein and DNA: mechanisms involved and relevant observations on brain undergoing oxidative stress. Ann Neurol 32(Suppl):S22–S27. https://doi.org/10.1002/ana.410320706

18. Salim S (2017) Oxidative stress and the central nervous system. J Pharmacol Exp Ther 360:201–205. https://doi.org/10.1124/jpet.116.237503

19. Nathan J, Maria C, Fiorello L et al (2018) 13 reasons why the brain is susceptible to oxidative stress. Redox Biol 15:490–503. https://doi.org/10.1016/j.redox.2018.01.008

20. Gemma C, Vila J, Bachstetter A et al (2007) Chapter 15-oxidative stress and the aging brain: from theory to prevention. In: Riddle DR (ed) Brain aging: models, methods, and mechanisms. CRC Press, Boca Raton, FL

21. Chiurchiù V, Orlacchio A, Maccarrone M (2016) Is modulation of oxidative stress an answer? The state of the art of redox therapeutic actions in neurodegenerative diseases.

Oxidative Med Cell Longev 2016:1–11. https://doi.org/10.1155/2016/7909380

22. Saxena S, Caroni P (2011) Selective neuronal vulnerability in neurodegenerative diseases: from stressor thresholds to degeneration. Neuron 71:35–48. https://doi.org/10.1016/j.neuron.2011.06.031

23. Harvey L, Arnold B, Lawrence Z et al (1999) Molecular cell biology, 4th edn. W.H. Freeman, pp 197–433. 4Rev Ed edition

24. Commoner B, Townsend J, Pake GE (1954) Free radicals in biological materials. Nature 174:689–691. https://doi.org/10.1038/174689a0

25. McCord JM (2000) The evolution of free radicals and oxidative stress. Am J Med 108:652–659. https://doi.org/10.1016/s0002-9343(00)00412-5

26. Pham-Huy LA, He H, Pham-Huy C (2008) Free radicals, antioxidants in disease and health. Int J Biomed Sci 4:89–96

27. Anas AA, Wiersinga WJ, de Vos AF et al (2010) Recent insights into the pathogenesis of bacterial sepsis. Neth J Med 68 (4):147–152

28. Victor VM, Rocha M, De la Fuente M (2004) Immune cells: free radicals and antioxidants in sepsis. Int Immunopharmacol 4:327–347. https://doi.org/10.1016/j.intimp.2004.01.020

29. Webster NR, Nunn JF (1988) Molecular structure of free radicals and their importance in biological reactions. Br J Anaesth 60 (1):98–108

30. Cadenas E (2004) Mitochondrial free radical production and cell signaling. Mol Aspects Med 25(1- 2):17–26. https://doi.org/10.1016/j.mam.2004.02.005

31. Pacher P, Beckman JS, Liaudet L (2007) Nitric oxide and peroxynitrite in health and disease. Physiol Rev 87(1):315–424. https://doi.org/10.1152/physrev.00029.2006

32. Aruoma OI, Halliwell B, Gajewski E et al (1991) Copper-ion-dependent damage to the bases in DNA in the presence of hydrogen peroxide. Biochem J 273(Pt 3):601–604. https://doi.org/10.1042/bj2730601

33. Lenaz G (2001) The mitochondrial production of reactive oxygen species: mechanisms and implications in human pathology. IUBMB Life 52(3–5):159–164. 30. https://doi.org/10.1080/15216540152845957

34. Matesanz N, Lafuente N, Azcutia V et al (2007) Xanthine oxidase derived extracellular superoxide anions stimulate activator protein 1 activity and hypertrophy in human vascular smooth muscle via c-Jun N-terminal kinase and p38 mitogen-activated protein kinases. J Hypertens 25(3):609–618. https://doi.org/10.1097/HJH.0b013e328013e7c4

35. Zangar RC, Davydov DR, Verma S (2004) Mechanisms that regulate production of reactive oxygen species by cytochrome P450. Toxicol Appl Pharmacol 199(3):316–331. https://doi.org/10.1016/j.taap.2004.01.018

36. Schrader M, Fahimi HD (2004) Mammalian peroxisomes and reactive oxygen species. Histochem Cell Biol 122(4):383–393. https://doi.org/10.1007/s00418-004-0673-1

37. Li WG, Miller FJ Jr, Zhang HJ et al (2001) H (2)O(2)-induced O(2) production by a non-phagocytic NAD(P)H oxidase causes oxidant injury. J Biol Chem 276 (31):29251–29256

38. D'Autreaux B, Toledano MB (2007) ROS as signalling molecules: mechanisms that generate specificity in ROS homeostasis. Nat Rev Mol Cell Biol 8(10):813–824. https://doi.org/10.1038/nrm2256

39. Imlay JA, Chin SM, Linn S (1988) Toxic DNA damage by hydrogen peroxide through the Fenton reaction in vivo and in vitro. Science 240:640–642. https://doi.org/10.1126/science.2834821

40. Yamazaki I, Piette LH (1990) ESR spin-trapping studies on the reaction of Fe^{2+} ions with H2O2-reactive species in oxygen toxicity in biology. J Biol Chem 265:13589–13594

41. Yu BP (1994) Cellular defenses against damage from reactive oxygen species. Physiol Rev 74(1):139–162. https://doi.org/10.1152/physrev.1994.74.1.139

42. Mates JM, Sanchez-Jimenez F (1999) Antioxidant enzymes and their implications in pathophysiologic processes. Front Biosci 4: D339–D345. https://doi.org/10.2741/mates

43. Lloyd RV, Hanna PM, Mason RP (1997) The origin of the hydroxyl radical oxygen in the Fenton reaction. Free Rad Biol Med 22:885–888. https://doi.org/10.1016/s0891-5849(96)00432-7

44. Stohs SJ, Bagchi D (1995) Oxidative mechanisms in the toxicity of metal ions. Free Rad Biol Med 18:321–336. https://doi.org/10.1016/0891-5849(94)00159-h

45. Clayton PT (2017) Inherited disorders of transition metal metabolism: an update. J Inherit Metab Dis 40:519–529. https://doi.org/10.1007/s10545-017-0030-x

46. Reed GA (1987) Co-oxidation of xenobiotics: lipid peroxyl derivatives as mediators of metabolism. Chem Phys Lipids 44:127–148.

https://doi.org/10.1016/0009-3084(87)
90047-8

47. Spiteller G, Afzal M (2014) The action of peroxyl radicals, powerful deleterious reagents, explains why neither cholesterol nor saturated fatty acids cause atherogenesis and age-related diseases. Chemistry 20:14928–14945. https://doi.org/10.1002/chem.201404383

48. Ayala A, Munoz MF, Arguelles S (2014) Lipid peroxidation: production, metabolism, and signaling mechanisms of malondialdehyde and 4-hydroxy-2-nonenal. Oxidative Med Cell Longev 2014:360438. https://doi.org/10.1155/2014/360438

49. De Grey AD (2002) HO_2^*: the forgotten radical. DNA Cell Biol 21:251–257. https://doi.org/10.1089/104454902753759672

50. Agnez-Lima LF, Melo JT, Silva AE et al (2012) Review DNA damage by singlet oxygen and cellular protective mechanisms. Mutat Res Rev Mutat Res 751(1):1–14. https://doi.org/10.1016/j.mrrev.2011.12.005

51. Hampton MB, Kettle AJ, Winterbourn CC (1998) Inside the neutrophil phagosome: oxidants, myeloperoxidase, and bacterial killing. Blood 92(9):3007–3017

52. Kanovasky JR (1989) Singlet oxygen production by biological systems. Chem Biol Interact 70(1–2):1–28. https://doi.org/10.1016/0009-2797(89)90059-8

53. Chan HWS (1971) Singlet oxygen analogs in biological systems: coupled oxygenation of 1,3-dienes by soybean lipoxidase. J Am Chem Soc 93(9):2357–2358. https://doi.org/10.1021/ja00738a064

54. Hayaishi O, Nozaki M (1969) Nature and mechanisms of oxygenases. Science 164:389–396. https://doi.org/10.1126/science.164.3878.389

55. Kanofsky JR (1983) Singlet oxygen production by lactoperoxidase. J Biol Chem 258 (10):5991–5993. https://doi.org/10.1016/S0021-9258(18)32358-5

56. Sies H, Menck CF (1992) Singlet oxygen induced DNA damage. Mutat Res 275:367–375. https://doi.org/10.1016/0921-8734(92)90039-r

57. Bian K, Murad F (2003) Nitric oxide (NO)—biogeneration, regulation, and relevance to human diseases. Front Biosci 8:d264–d278. https://doi.org/10.2741/997

58. Ignarro LJ (1990) Biosynthesis and metabolism of endothelium-derived nitric oxide. Annu Rev Pharmacol Toxicol 30:535–560. https://doi.org/10.1146/annurev.pa.30.040190.002535

59. Andrew PJ, Mayer B (1999) Enzymatic function of nitric oxide synthases. Cardiovasc Res 43(3):521–531. https://doi.org/10.1016/s0008-6363(99)00115-7

60. Ignarro LJ, Buga GM, Wood KS et al (1987) Endothelium-derived relaxing factor produced and released from artery and vein is nitric oxide. Proc Natl Acad Sci U S A 84:9265–9269. https://doi.org/10.1073/pnas.84.24.9265

61. Wink DA, Mitchell JB (1998) Chemical biology of nitric oxide: insights into regulatory, cytotoxic, and cytoprotective mechanisms of nitric oxide. Free Radic Biol Med 25 (4–5):434–456. https://doi.org/10.1016/s0891-5849(98)00092-6

62. Stamler JS (1994) Redox signaling: nitrosylation and related target interactions of nitric oxide. Cell 78(6):931–936. https://doi.org/10.1016/0092-8674(94)90269-0

63. Koshland DE Jr (1992) The molecule of the year. Science 258(5090):1861. https://doi.org/10.1126/science.1470903

64. Radi R (2013) Peroxynitrite, a stealthy biological oxidant. J Biol Chem 288:26464–26472. https://doi.org/10.1074/jbc.R113.472936

65. Beckman JS, Koppenol WH (1996) Nitric oxide, superoxide, and peroxynitrite: the good, the bad, and ugly. Am J Phys 271: C1424–C1437. https://doi.org/10.1152/ajpcell.1996.271.5.C1424

66. Ischiropoulos H, Al-Mehdi AB (1995) Peroxynitrite mediated oxidative protein modifications. FEBS Lett 364(3):279–282. https://doi.org/10.1016/0014-5793(95)00307-u

67. Luc R, Vergely C (2008) Forgotten radicals in biology. Int J Biomed Sci 4:255–259

68. Kirsch M, Korth HG, Sustmann R et al (2002) The pathobiochemistry of nitrogen dioxide. Biol Chem 383:389–399. https://doi.org/10.1515/BC.2002.043

69. Halliwell B, Hu ML, Louie S et al (1992) Interaction of nitrogen dioxide with human plasma. Antioxidant depletion and oxidative damage. FEBS Lett 313:62–66. https://doi.org/10.1016/0014-5793(92)81185-O

70. Giles GI, Nasim MJ, Ali W et al (2017) The reactive sulfur species concept: 15 years on. Antioxidants 6:E38. https://doi.org/10.3390/antiox6020038

71. Gruhlke MC, Slusarenko AJ (2012) The biology of reactive sulfur species (RSS). Plant Physiol Biochem 59:98–107. https://doi.org/10.1016/j.plaphy.2012.03.016

72. Ghezzi P, Bonetto V, Fratelli M (2005) Thiol-disulfide balance: from the concept of oxidative stress to that of redox regulation.

Antioxid Redox Signal 7:964–972. https://doi.org/10.1089/ars.2005.7.964

73. Ghezzi P, Chan P (2017) Redox proteomics applied to the thiol secretome. Antioxid Redox Signal 26:299–312. https://doi.org/10.1089/ars.2016.6732

74. Libiad M, Yadav PK, Vitvitsky V et al (2014) Organization of the human mitochondrial hydrogen sulfide oxidation pathway. J Biol Chem 289:30901–30910. https://doi.org/10.1074/jbc.M114.602664

75. Mishanina TV, Libiad M, Banerjee R (2015) Biogenesis of reactive sulfur species for signaling by hydrogen sulfide oxidation pathways. Nat Chem Biol 11:457–464. https://doi.org/10.1038/nchembio.1834

76. Jackson MR, Melideo SL, Jorns MS (2012) Human sulfide: quinone oxidoreductase catalyzes the first step in hydrogen sulfide metabolism and produces a sulfane sulfur metabolite. Biochemistry 51:6804–6815. https://doi.org/10.1021/bi300778t

77. Ranguelova K, Rice AB, Lardinois OM et al (2013) Sulfite-mediated oxidation of myeloperoxidase to a free radical: immuno-spin trapping detection in human neutrophils. Free Radic Biol Med 60:98–106. https://doi.org/10.1016/j.freeradbiomed.2013.01.022

78. Droge W (2002) Review Free radicals in the physiological control of cell function. Physiol Rev 82(1):47–95. https://doi.org/10.1152/physrev.00018.2001

79. Halliwell B, Gutteridge JM (1999) Free radicals in biology and medicine, 3rd edn. Oxford University Press, Midsomer Norton

80. Dizdaroglu M, Jaruga P, Birincioglu M et al (2002) Free radical-induced damage to DNA: mechanisms and measurement. Free Radic Biol Med 32(11):1102–1115. https://doi.org/10.1016/s0891-5849(02)00826-2

81. Barja G (2000) The flux of free radical attack through mitochondrial DNA is related to aging rate. Aging 12(5):342–355. https://doi.org/10.1007/BF03339859

82. Yermilov V, Rubio J, Ohshima H (1995) Formation of 8-nitroguaninein DNA treated with peroxynitrite in vitro and its rapid removal from DNA by depurination. FEBS Lett 376(3):207–210. https://doi.org/10.1016/0014-5793(95)01281-6

83. Loeb LA, Preston BD (1986) Mutagenesis by apurinic/apyrimidinic sites. Annu Rev Genet 20:201–230. https://doi.org/10.1146/annurev.ge.20.120186.001221

84. Dean RT, Fu S, Stocker R et al (1997) Biochemistry and pathology of radical-mediated protein oxidation. Biochem J 324:1–18. https://doi.org/10.1042/bj3240001

85. Butterfield DA, Koppal T, Howard B et al (1998) Structural and functional changes in proteins induced by free radical-mediated oxidative stress and protective action of the antioxidants N-tert-butyl-alpha-phenylnitrone and vitamin E. Ann N Y Acad Sci 854:448–462. https://doi.org/10.1111/j.1749-6632.1998.tb09924.x

86. Brodie E, Reed DJ (1990) Cellular recovery of glyceraldehyde-3- phosphate dehydrogenase activity and thiol status after exposure to hydroperoxide. Arch Biochem Biophys 276(1):210–212. https://doi.org/10.1016/0003-9861(90)90028-w

87. Pryor WA, Jin X, Squadrito GL (1994) One- and two-electron oxidations of methionine by peroxynitrite. Proc Natl Acad Sci U S A 91(23):11173–11177. https://doi.org/10.1073/pnas.91.23.11173

88. Berlett BS, Stadtman E (1997) Protein oxidation in aging, disease, and oxidative stress. J Bio Chem 272(33):20313–20316. https://doi.org/10.1074/jbc.272.33.20313

89. Kikugawa K, Kato T, Okamoto Y (1994) Damage of amino acids and proteins induced by nitrogen dioxide, a free radical toxin, in air. Free Rad Biol Med 16(3):373–382. https://doi.org/10.1016/0891-5849(94)90039-6

90. Uchida K, Kawakishi S (1993) 2-oxohistidine as a novel biological marker for oxidatively modified proteins. FEBS Lett 332(3):208–210. https://doi.org/10.1016/0014-5793(93)80632-5

91. Garrison WM (1987) Reaction mechanisms in radiolysis of peptides, polypeptides, and proteins. Chem Rev 8792:381–398. https://doi.org/10.1021/cr00078a006.

92. Spickett CM (2013) The lipid peroxidation product 4-hydroxy-2-nonenal: advances in chemistry and analysis. Redox Biol 1:145–152. https://doi.org/10.1016/j.redox.2013.01.007

93. Yin H, Xu L, Porter NA (2011) Free radical lipid peroxidation: mechanisms and analysis. Chem Rev 111:5944–5972. https://doi.org/10.1021/cr200084z

94. Marnett LJ (1999) Lipid peroxidation—DNA damage by malondialdehyde. Mutat Res 424(1–2):83–95

95. Halliwell B, Gutteridge JMC (2015) Free radicals in biology & medicine. Oxford University Press, 5th edition

96. Hazra TK, Das A, Das S et al (2007) Oxidative DNA damage repair in mammalian cells: a new perspective. DNA Repair (Amst) 6

(4):470–480.   https://doi.org/10.1016/j.
dnarep.2006.10.011

97. Szczepanowska K, Trifunovic A (2015) Different faces of mitochondrial DNA mutators. Biochim Biophys Acta 1847(11):1362–1372. https://doi.org/10.1016/j.bbabio.2015.05. 016

98. Pinto M, Moraes CT (2015) Mechanisms linking mtDNA damage and aging. Free Radic Biol Med 85:250–258. https://doi. org/10.1016/j.freeradbiomed.2015.05.005

99. Wu Y, Chena M, Jiang J (2019) Mitochondrial dysfunction in neurodegenerative diseases and drug targetsvia apoptotic signaling. Mitochondrion 49:35–45. https://doi.org/ 10.1016/j.mito.2019.07.003

100. Scheibye-Knudsen M, Fang EF, Croteau DL et al (2015) Protecting the mitochondrial powerhouse. Trends Cell Biol 25:158–170. https://doi.org/10.1016/j.tcb.2014.11. 002

101. Wallace DC (2013) A mitochondrial bioenergetic etiology of disease. J Clin Invest 123:1405–1412. https://doi.org/10.1172/ JCI61398

102. Farshbaf MJ, Ghaedi K (2017) Huntington's disease and mitochondria. Neurotox Res 32:518–529.   https://doi.org/10.1007/ s12640-017-9766-1

103. Carmo C, Naia L, Lopes C et al (2018) Mitochondrial dysfunction in Huntington's disease, polyglutamine disorders. Adv Exp Med Biol 1049:59–83. https://doi.org/10.1007/ 978-3-319-71779-1_3

104. Dixon SJ, Lemberg KM, Lamprecht MR et al (2012) Ferroptosis: an iron-dependent form of   nonapoptotic   cell   death.   Cell 149:1060–1072. https://doi.org/10.1016/ j.cell.2012.03.042

105. Dixon SJ (2017) Ferroptosis: bug or feature? Immunol Rev 277:150–157. https://doi. org/10.1111/imr.12533

106. Galluzzi L, Vitale I, Aaronson SA et al (2018) Molecular mechanisms of cell death: recommendations of the nomenclature committee on cell death 2018. Cell Death Differ 25:486–541.   https://doi.org/10.1038/ s41418-017-0012-4

107. Kenny EM, Fidan E, Yang Q et al (2019) Ferroptosis contributes to neuronal death and functional outcome after traumatic brain injury. Crit Care Med 47:410–418. https:// doi.org/10.1097/ccm.0000000000003555

108. Guiney SJ, Adlard PA, Bush AI et al (2017) Ferroptosis and cell death mechanisms in Parkinson's disease. Neurochem Int 104:34–48.

https://doi.org/10.1016/j.neuint.2017.01. 004

109. Devos D, Moreau C, Devedjian JC et al (2014) Targeting chelatable iron as a therapeutic modality in Parkinson's disease. Antioxid Redox Signal 21:195–210. https://doi. org/10.1089/ars.2013.5593

110. Lane DJR, Ayton S, Bush AI (2018) Iron and Alzheimer's disease: an update on emerging mechanisms.   J   Alzheimers   Dis   64: S379–S395.   https://doi.org/10.3233/jad- 179944

111. Belaidi AA, Bush AI (2016) Iron neurochemistry in Alzheimer's disease and Parkinson's disease: targets for therapeutics. J Neurochem 139(Suppl. 1):179–197. https://doi.org/10. 1111/jnc.13425

112. Wu Y, Song J, Wang Y (2019) The potential role of ferroptosis in neonatal brain injury. Front Neurosci 13:115. https://doi.org/10. 3389/fnins.2019.00115

113. Schönfeld P, Reiser G (2013) Why does brain metabolism not favor burning of fatty acids to provide energy?—reflections on disadvantages of the use of free fatty acids as fuel for brain.   J   Cereb   Blood   Flow   Metab 33:1493–1499.   https://doi.org/10.1038/ jcbfm.2013.128

114. Pollack M, Leeuwenburgh C (1999) Molecular mechanisms of oxidative stress in aging: free radicals, aging, antioxidants and disease. In: Handbook of oxidants and antioxidants in exercise. Elsevier Science B.V., Part X, Chapter 30, pp. 881–923

115. Rivas-Arancibia   S,   Guevara-Guzma'n   R, Lo'pez-Vidal Y et al (2010) Oxidative stress caused by ozone exposure induces loss of brain repair in the hippocampus of adult rats. Toxicol Sci 113(1):187–197. https://doi. org/10.1093/toxsci/kfp252

116. Santiago-Lo'pez JA, Bautista-Martı'nez CI, Reyes-Hernandez M et al (2010) Oxidative stress, progressive damage in the substantia nigra and plasma dopamine oxidation, in rats chronically exposed to ozone. Toxicol Lett 197(3):193–200. https://doi.org/10.1016/ j.toxlet.2010.05.020

117. Pan XD, Zhu YG, Lin N et al (2011) Microglial   phagocytosis   induced   by   fibrillar b-amyloid   is   attenuated   by   oligomeric b-amyloid: implications for Alzheimer's disease.   Mol   Neurodegener   6(45):1–17. https://doi.org/10.1186/1750-1326-6-45

118. Cioffia F, Adam RHI, Broersen K (2019) Molecular mechanisms and genetics of oxidative stress in Alzheimer's disease. J Alzheimers

Dis 72:981–1017. https://doi.org/10.3233/JAD-190863

119. Sevcsik E, Trexler AJ, Dunn JM et al (2011) Allostery in a disordered protein: oxidative modifications to a-synuclein act distally to regulate membrane binding. J Am Chem Soc 133(18):7152–7158. https://doi.org/10.1021/ja2009554

120. Zhao W, Varghese M, Yemul S et al (2011) Peroxisome proliferator activator receptor gamma coactivator- 1alpha (PGC-1a) improves motor performance and survival in a mouse model of amyotrophic lateral sclerosis. Mol Neurodegener 6(1):1–8. https://doi.org/10.1186/1750-1326-6-51

121. Zuo L, Hemmelgarn BT, Chuang CC et al (2015) The role of oxidative stress-induced epigenetic alterations in amyloid-beta production in Alzheimer's disease. Oxid Med Cell Longev 2015:604658, 13 pages. https://doi.org/10.1155/2015/604658

122. Querfurth HW, LaFerla FM (2010) Alzheimer's disease. N Engl J Med 362:329–344. https://doi.org/10.1056/NEJMra0909142

123. Butterfield DA (2014) The 2013 SFRBM discovery award: selected discoveries from the butterfield laboratory of oxidative stress and its sequela in brain in cognitive disorders exemplified by Alzheimer disease and chemotherapy induced cognitive impairment. Free Rad Biol Med 74:157–174. https://doi.org/10.1016/j.freeradbiomed.2014.06.006

124. Townsend DM, Tew KD, Tapiero H (2003) The importance of glutathione in human disease. Biomed Pharmacother 57:145–155. https://doi.org/10.1016/s0753-3322(03)00043-x

125. Chai YC, Ashraf SS, Rokutan K, Johnston RB Jr, Thomas JA (1994) S-thiolation of individual human neutrophil proteins including actin by stimulation of the respiratory burst: evidence against a role for glutathione disulfide. Arch Biochem Biophys 310:273–281. https://doi.org/10.1006/abbi.1994.1167

126. Zitka O, Skalickova S, Gumulec J et al (2012) Redox status expressed as GSH:GSSG ratio as a marker for oxidative stress in paediatric tumour patients. Oncol Lett 4:1247–1253. https://doi.org/10.3892/ol.2012.931

127. Huang WJ, Zhang X, Chen WW (2016) Role of oxidative stress in Alzheimer's disease. Biomed Rep 4:519–522. https://doi.org/10.3892/br.2016.630

128. Wang W, Zhao F, Ma X et al (2020) Mitochondria dysfunction in the pathogenesis of Alzheimer's disease: recent advances. Mol Neurodegener 15:30. https://doi.org/10.1186/s13024-020-00376-6

129. Nakamura T, Lipton SA (2011) Redox modulation by S-nitrosylation contributes to protein misfolding, mitochondrial dynamics, and neuronal synaptic damage in neurodegenerative diseases. Cell Death Differ 18:1478–1486. https://doi.org/10.1038/cdd.2011.65

130. Awasthi A, Matsunaga Y, Yamada T (2005) Amyloid-beta causes apoptosis of neuronal cells via caspase cascade, which can be prevented by amyloid-beta-derived short peptides. Exp Neurol 196:282–289. https://doi.org/10.1016/j.expneurol.2005.08.001

131. Nizzari M, Thellung S, Corsaro A et al (2012) Neurodegeneration in Alzheimer disease: role of amyloid precursor protein and presenilin 1 intracellular signaling. J Toxicol 2012:187297, 13 pages. https://doi.org/10.1155/2012/187297

132. Curtain CC, Ali F, Volitakis I et al (2001) Alzheimer's disease amyloid-beta binds copper and zinc to generate an allosterically ordered membrane-penetrating structure containing superoxide dismutase-like subunits. J Biol Chem 276:20466–20473. https://doi.org/10.1074/jbc.M100175200

133. Poewe W, Seppi K, Tanner CM et al (2017) Parkinson disease. Nat Rev Dis Prim 3:17013. https://doi.org/10.1038/nrdp.2017.13

134. Brown GC, Borutaite V (2004) Inhibition of mitochondrial respiratory complex I by nitric oxide, peroxynitrite and S nitrosothiols. Biochim Biophys Acta 1658(1–2):44–49. https://doi.org/10.1016/j.bbabio.2004.03.016

135. Seet RCS, Lee CYJ, Lim ECH et al (2010) Oxidative damage in Parkinson disease: measurement using accurate biomarkers. Free Radic Biol Med 48(4):560–566. https://doi.org/10.1016/j.freeradbiomed.2009.11.026

136. Bender A, Krishnan KJ, Morris CM et al (2006) High levels of mitochondrial DNA deletions in substantia nigra neurons in aging and Parkinson disease. Nat Genet 38(5):515–517. https://doi.org/10.1038/ng1769

137. Zuo L, Motherwell MS (2013) The impact of reactive oxygen species and genetic mitochondrial mutations in Parkinson's disease. Gene 532:18–23. https://doi.org/10.1016/j.gene.2013.07.085

138. Perfeito R, Cunha-Oliveira T, Rego AC (2012) Revisiting oxidative stress and mitochondrial dysfunction in the pathogenesis of Parkinson disease–resemblance to the effect of amphetamine drugs of abuse. Free Radic Biol Med 53:1791–1806. https://doi.org/10.1016/j.freeradbiomed.2012.08.569

139. Valente EM, Abou-Sleiman PM, Caputo V et al (2004) Hereditary early-onset Parkinson's disease caused by mutations in PINK1. Science 304:1158–1160. https://doi.org/10.1126/science.1096284

140. Jiang H, Ren Y, Zhao J, Feng J (2004) Parkin protects human dopaminergic neuroblastoma cells against dopamine induced apoptosis. Hum Mol Genet 13:1745–1754. https://doi.org/10.1093/hmg/ddh180

141. Shapira AH (2008) Mitochondria in the aetiology and pathogenesis of Parkinson's disease. Lancet Neurol 7:97–109. https://doi.org/10.1016/S1474-4422(07)70327-7

142. Ganguly G, Chakrabarti S, Chatterjee U et al (2017) Proteinopathy, oxidative stress and mitochondrial dysfunction: cross talk in Alzheimer's disease and Parkinson's disease. Drug Des Devel Ther 11:797–810. https://doi.org/10.2147/DDDT.S130514

143. Mazo NA, Echeverria V, Cabezas R et al (2017) Medicinal plants as protective strategies against Parkinson's disease. Curr Pharma Design 23(28):4180–4188. https://doi.org/10.2174/1381612823666170316142803

144. Moreira PI, Zhu X, Wang X et al (2010) Mitochondria: a therapeutic target in neurodegeneration. Biochim Biophys Acta 1802:212–220. https://doi.org/10.1016/j.bbadis.2009.10.007

145. Yan MH, Wang X, Zhu X (2013) Mitochondrial defects and oxidative stress in Alzheimer disease and Parkinson disease. Free Radic Biol Med 62:90–101. https://doi.org/10.1016/j.freeradbiomed.2012.11.014

146. Vonsattel JP, DiFiglia M (1998) Huntington disease. J Neuropathol Exp Neurol 57:369–384. https://doi.org/10.1097/00005072-199805000-00001

147. Nekrasov ED, Vigont VA, Klyushnikov SA et al (2016) Manifestation of Huntington's disease pathology in human induced pluripotent stem cell-derived neurons. Mol Neurodegener 11:27. https://doi.org/10.1186/s13024-016-0092-5

148. Gonzalez-Alegre P, Afifi AK (2006) Clinical characteristics of childhood-onset (juvenile) Huntington disease: report of 12 patients and review of the literature. J Child Neurol 21:223–229. https://doi.org/10.2310/7010.2006.00055

149. Walke F (2007) Huntington's disease. Lancet 369:218–228. https://doi.org/10.1016/S0140-6736(07)60111-1

150. Ross CA, Tabrizi SJ (2011) Huntington's disease: from molecular pathogenesis to clinical treatment. Lancet Neurol 10(1):83–98. https://doi.org/10.1016/S1474-4422(10)70245-3

151. Gipson TA, Neueder A, Wexler NS et al (2013) Aberrantly spliced HTT, a new player in Huntington's disease pathogenesis. RNA Biol 10(11):1647–1652. https://doi.org/10.4161/rna.26706

152. Hatters DM (2012) Putting huntingtin "aggregation" in view with windows into the cellular milieu. Curr Top Med Chem 12 (22):2611–2622

153. Kumar A, Vaish M, Ratan RR (2014) Transcriptional dys-regulation in Huntington's disease: a failure of adaptive transcriptional homeostasis. Drug Discov Today 19 (7):956–962. https://doi.org/10.1016/j.drudis.2014.03.016

154. Sepers MD, Raymond LA (2014) Mechanisms of synaptic dysfunction and excitotoxicity in Huntington's disease. Drug Discov Today 19(7):990–996. https://doi.org/10.1016/j.drudis.2014.02.006

155. Molero AE, Arteaga-Bracho EE, Chen CH et al (2016) Selective expression of mutant huntingtin during development recapitulates characteristic features of Huntington's disease. Proc Natl Acad Sci U S A 113 (20):5736–5741. https://doi.org/10.1073/pnas.1603871113

156. Tunez I, Sanchez-Lopez F, Aguera E et al (2011) Important role of oxidative stress biomarkers in Huntington's disease. J Med Chem 54:5602–5606. https://doi.org/10.1021/jm200605a

157. Stoy N, Mackay GM, Forrest CM et al (2005) Tryptophan metabolism and oxidative stress in patients with Huntington's disease. J Neurochem 93(3):611–623. https://doi.org/10.1111/j.1471-4159.2005.03070.x

158. Christofides J, Bridel M, Egerton M et al (2006) Blood 5-hydroxytryptamine, 5-hydroxyindoleacetic acid and melatonin levels in patients with either Huntington's disease or chronic brain injury. J Neurochem 97(4):1078–1088. https://doi.org/10.1111/j.1471-4159.2006.03807.x

159. Chang KH, Chen YC, Wu YR et al (2012) Downregulation of genes involved in metabolism and oxidative stress in the peripheral leukocytes of Hunt-ington's disease patients. PLoS One 7(9):e46492. https://doi.org/10.1371/journal.pone.0046492

160. Pena-Sanchez M, Riveron-Forment G, Zaldivar-Vaillant T et al (2015) Association of status redox with demographic, clinical and imaging parameters in patients with Huntington's disease. Clin Biochem 48

(18):1258–1263. https://doi.org/10.3233/jhd-160205

161. Paul BD, Sbodio JI, Xu R et al (2014) Cystathionine gammalyase deficiency mediates neurodegeneration in Huntington's disease. Nature 509(7498):96–100. https://doi.org/10.1038/nature13136

162. Starkov AA, Andreyev AY, Zhang SF et al (2014) Scavenging of H2O2 by mouse brain mitochondria. J Bioenerg Biomembr 46 (6):471–477. https://doi.org/10.1007/s10863-014-9581-9

163. Kumar A, Ratan RR (2016) Oxidative stress and Huntington's disease: the good, the bad, and the ugly. J Huntingtons Dis 5 (3):217–237. https://doi.org/10.3233/JHD-160205

164. Kiernan MC, Vucic S, Cheah BC et al (2011) Amyotrophic lateral sclerosis. Lancet 377:942–955. https://doi.org/10.1016/S0140-6736(10)61156-7

165. Taylor JP, Brown RH Jr, Cleveland DW (2016) Decoding ALS: from genes to mechanism. Nature 539:197–206. https://doi.org/10.1038/nature20413

166. Deng HX, Hentati A, Tainer JA et al (1993) Amyotrophic lateral sclerosis and structural defects in Cu, Zn superoxide dismutase. Science 261(5124):1047–1051. https://doi.org/10.1126/science.8351519

167. Kaur SJ, McKeown SR, Rashid S (2016) Mutant SOD1 mediated pathogenesis of amyotrophic lateral sclerosis. Gene 577 (2):109–118. https://doi.org/10.1016/j.gene.2015.11.049

168. Rosen DR, Siddique T, Patterson D et al (1993) Mutations in Cu/Zn superoxide dismutase gene are associated with familial amyotrophic lateral sclerosis. Nature 362 (6415):59–62. https://doi.org/10.1016/j.gene.2015.11.049

169. Synofzik M, Ronchi D, Keskin I et al (2012) Mutant superoxide dismutase-1 indistinguishable from wild-type causes ALS. Hum Mol Genet 21(16):3568–3574. https://doi.org/10.1093/hmg/dds188

170. Sreedharan J, Blair IP, Tripathi VB et al (2008) TDP-43 mutations in familial and sporadic amyotrophic lateral sclerosis. Science 319(5870):1668–1672. https://doi.org/10.1126/science

171. Kwiatkowski TJ, Bosco DA, LeClerc AL et al (2009) Mutations in the FUS/TLS gene on chromosome 16 cause familial amyotrophic lateral sclerosis. Science 323 (5918):1205–1208. https://doi.org/10.1126/science.1166066

172. Mendez EF, Sattler R (2015) Biomarker development for C9orf72 repeat expansion in ALS. Brain Res 1607:26–35. https://doi.org/10.1016/j.brainres.2014.09.041

173. Oskarsson B, Gendron TF, Saff NP (2018) Amyotrophic lateral sclerosis: an update for 2018. Mayo Clin Proc 93(11):1617–1628. https://doi.org/10.1016/j.mayocp.2018.04.007

174. Pizzino G, Irrera N, Cucinotta M et al (2017) Oxidative stress: harms and benefits for human health. Oxid Med Cell Longev 2017:8416763, 13 pages. https://doi.org/10.1155/2017/8416763

175. Gamez J, Corbera-Bellalta M, Nogales G et al (2006) Mutational analysis of the cu/Zn superoxide dismutase gene in a Catalan ALS population: should all sporadic ALS cases also be screened for SOD1? J Neurol Sci 247:21–28. https://doi.org/10.1016/j.jns.2006.03.006

176. Saccon RA, Bunton-Stasyshyn RK, Fisher EM et al (2013) Is SOD1 loss of function involved in amyotrophic lateral sclerosis? Brain 136:2342–2358. https://doi.org/10.1093/brain/awt097

177. Bastow EL, Peswani AR, Tarrant DSJ et al (2016) New links between SOD1 and metabolic dysfunction from a yeast model of amyotrophic lateral sclerosis. J Cell Sci 129:4118–4129. https://doi.org/10.1242/jcs.190298

178. Damiano S, Petrozziello T, Ucci V et al (2013) Cu-Zn superoxide dismutase activates muscarinic acetylcholine M1 receptor pathway in neuroblastoma cells. Mol Cell Neurosci 52:31–37. https://doi.org/10.1016/j.mcn.2012.11.001

179. Yoshino H, Kimura A (2006) Investigation of the therapeutic effects of edaravone, a free radical scavenger, on amyotrophic lateral sclerosis (phase II study). Amyotrop Later Scler 7:241–245. https://doi.org/10.1080/17482960600881870

180. Louwerse ES, Weverling GJ, Bossuyt PMM et al (1995) Randomized, double-blind, controlled trial of acetylcysteine in amyotrophic lateral sclerosis. Arch Neurol 52:559–564. https://doi.org/10.1001/archneur.1995.00540300031009

181. Guo C, Sun L, Chen X et al (2013) Oxidative stress, mitochondrial damage and neurodegenerative diseases. Neural Regen Res 8:2003–2014. https://doi.org/10.3969/j.issn.1673-5374.2013.21.009

182. Wang H, Guo W, Mitra J et al (2018) Mutant FUS causes DNA ligation defects to inhibit

oxidative damage repair in Amyotrophic Lateral Sclerosis. Nat Commun 9:3683. https://doi.org/10.1038/s41467-018-06111-6

183. Kovacic P, Weston W (2018) Unifying mechanism for multiple sclerosis and amyotrophic lateral sclerosis: reactive oxygen species, oxidative stress, and antioxidants. J Biopharm Ther Chal 2:1–8

184. Wojsiat J, Zoltowska KM, Laskowska-Kaszub K et al (2018) Oxidant/antioxidant imbalance in alzheimer's disease: therapeutic and diagnostic prospects. Oxid Med Cell Longev 2018:6435861, 16 pages. https://doi.org/10.1155/2018/6435861

185. Tönnies E, Trushina E (2017) Oxidative stress, synaptic dysfunction, and Alzheimer's disease. J Alzheimer's Dis 57:1105–1121. https://doi.org/10.3233/JAD-161088

186. Puspita L, Chung SY, Shim J-W (2017) Oxidative stress and cellular pathologies in Parkinson's disease. Mol Brain 10(53):1–12. https://doi.org/10.1186/s13041-017-0340-9

187. Lee J, Giordano S, Zhang J (2012) Autophagy, mitochondria and oxidative stress: cross-talk and redox signalling. Biochem J 441:523–540. https://doi.org/10.1042/BJ20111451

188. Zhang X, Yu L, Xu H (2016) Lysosome calcium in ROS regulation of autophagy. Autophagy 12:1954–1955. https://doi.org/10.1080/15548627.2016.1212787

189. Hrelia P, Sita G, Ziche M et al (2020) Common protective strategies in neurodegenerative disease: focusing on risk factors to target the cellular redox system. Oxid Med Cell Logev 2020:8363245, 1–18. https://doi.org/10.1155/2020/8363245

Chapter 7

Biomarkers in Parkinson's Disease

Andrei Surguchov

Abstract

Parkinson's disease (PD) is the second most common neurodegenerative disorder, with a documented significant increase in its prevalence in the past three decades. Both environmental and genetic factors contribute to the pathophysiology of this disorder. The diagnosis relies mainly on clinical findings, and currently, there is no reliable and trustworthy method of early identification of PD except for genetic testing limited to rare cases of monogenic forms of the disease. Current scientific research is focused on the identification of new diagnostic criteria and new disease biomarkers, allowing correct and timely diagnosis. Various types of potential PD biomarkers have been tested, including clinical, imaging, pathological, biochemical, and genetic, but none of them can be considered excellent or even satisfactory. However, several recent developments raise hope that early diagnosis of PD will be possible in the near future. The finding of reliable PD biomarkers may happen due to the successes in the field of neuroimaging, including positron emission tomography (PET), single-photon emission CT (SPECT), novel MRI techniques, as well as a result of the progress of new biochemical methods. Examples of these methods include aggregated α-synuclein measurements by real-time quaking-induced conversion (RT-QuIC), protein misfolding cyclic amplification (PMCA), and development of epigenetic-based biomarkers. Besides, a combination of biomarkers looks very promising to maximize their utility and ensure an accurate diagnosis of premotor or early-stage PD. Our goal here is to provide the reader with new developments in biomarkers suitable for early PD identification and to highlight the advantages and limitations of an existing individual biomarker and their combination.

Keywords Parkinson's disease, Neurodegeneration, α-synuclein, Gut microbiota, Antioxidants, Dopamine, Protein aggregation, Epigenetics

1 Introduction: Parkinson's Disease Pathology

Parkinson's disease (PD) is a common neurodegenerative disease affecting more than six million people around the world, associated with a massive socioeconomic burden and enormous human suffering. The calculated number of patients will double in the next few decades. PD is a heterogeneous multisystem neurological disorder of movement. It is characterized by the presence of α-synuclein containing Lewy bodies (LB) and the loss of dopaminergic neurons in substantia nigra (SN) pars compacta [1, 2]. The

Philip V. Peplow, Bridget Martinez and Thomas A. Gennarelli (eds.), *Neurodegenerative Diseases Biomarkers: Towards Translating Research to Clinical Practice*, Neuromethods, vol. 173, https://doi.org/10.1007/978-1-0716-1712-0_7,
© Springer Science+Business Media, LLC, part of Springer Nature 2022

identification of α-synuclein as a major component of LB, a pathological hallmark of PD leads to the classification of PD as a protein misfolding disease [3]. The dopamine deficiency developed as a result of dopaminergic neuron loss leads to classic parkinsonian motor symptoms. Currently, the literature has not yet identified reliable biomarkers, which can stratify an individual at risk for developing PD before most of the dopaminergic neurons have been lost. What are the signs of molecular and cellular neuropathology of PD?

- Pathological changes in the brain of PD patients have been defined in detail. Braak and coauthors [4, 5] described neuropathological stages of PD based on extension and topographical distribution of the LB lesions across midbrain and brain. The pathology involves specific brain areas with ascending processes from the brain stem to the cortex. This classification is valuable and convenient, but it is based only on postmortem examination.

- The beginning of molecular and cellular neuropathology of PD occurs without clinical manifestations of PD decades before the onset of the motor symptoms typical for the disease. Patients exhibiting these symptoms have already undergone a significant dopaminergic neuronal loss.

- The long latent period of neurodegeneration suggests that early identification of emerging PD is necessary. It would permit the initiation of neuroprotective or disease-modifying therapies at a stage when this intervention is the most effective.

Earlier treatment will allow slowing or prevention of the degeneration of dopaminergic neurons. Therefore, accurate early diagnosis of PD and other parkinsonian syndromes is essential in controlling clinical management, evaluating prognosis, and predicting the patients' response to treatment.

Therefore, the identification of biomarkers on the early presymptomatic stages of PD is of paramount importance to reduce the burden of this severe pathology and to establish a more accurate diagnosis at the very onset of the disease.

2   Definition and Classification of Biomarkers. Tools and Procedures in Biomarker Research

The NIH Biomarkers Working Group [6] defined biomarker as "a characteristic that is objectively measured and evaluated as an indicator of normal biological processes, pathogenic processes or pharmacologic response to a therapeutic intervention." Thus, biomarkers are molecules, imaging features, or functional measures that represent an easily identified and quantified pathological

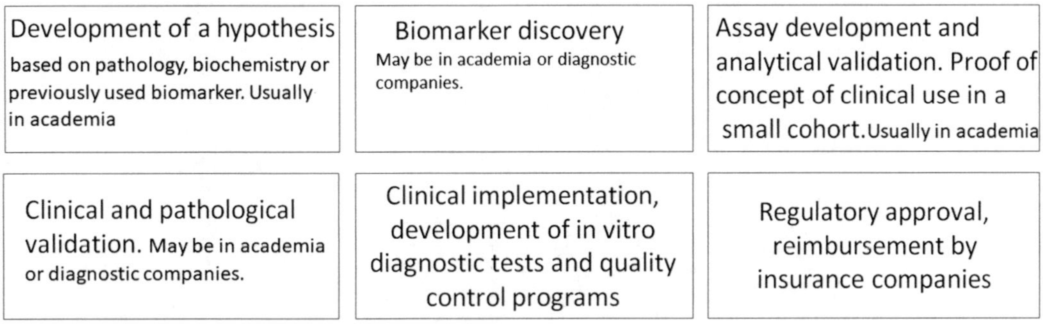

Development of a hypothesis based on pathology, biochemistry or previously used biomarker. Usually in academia	Biomarker discovery May be in academia or diagnostic companies.	Assay development and analytical validation. Proof of concept of clinical use in a small cohort.Usually in academia
Clinical and pathological validation. May be in academia or diagnostic companies.	Clinical implementation, development of in vitro diagnostic tests and quality control programs	Regulatory approval, reimbursement by insurance companies

Fig. 1 Steps of biomarker development

signature specific for a particular disease (Fig. 1). Biomarkers may help to establish PD diagnosis, to understand disease progression or disease-related features, and to monitor therapeutic effects. Due to the overlap of clinical symptoms at an early stage of the disease, it is difficult to differentiate PD from other neurological disorders. PD may be confused with progressive supranuclear palsy (PSP), multiple system atrophy (MSA), dementia with Lewy bodies (DLB), Parkinson's disease dementia (PDD), and other atypical parkinsonian disorders. According to published data, between 20 and 40% of PD and MSA patients may be misdiagnosed during the lifetime [7]. These results further confirm the urgent necessity of finding new biomarkers and the improvement of existing candidates and methodology.

Biomarkers can be categorized into several types (Fig. 2).

- A biomarker can be based on clinical data, for example, using a quantitatively measured physical examination finding.
- It can be grounded on the genetic examination of a genotype of a patient or be imaging-based.
- An important group of biomarkers relies on biochemical analysis. In the last case, several methods can be applied to identify specific features of biochemical alterations, such as the following.
  - Proteomics, that is, the study of the structure and function of proteins,
  - Metabolomics based on the study of small-molecule metabolites or analysis of gene expression profiling [8, 9].

Several programs currently serve as a portal for biosamples and data requests. For example, the PD Biomarkers Program, (PDBP, pdbp.ninds.nih.gov) was initiated in 2012 by the National Institute for Neurological Disorders and Stroke (NINDS). The PDBP possesses a biorepository (BioSEND) of DNA, RNA, and biofluid samples from several US centers. Imaging data exists on a

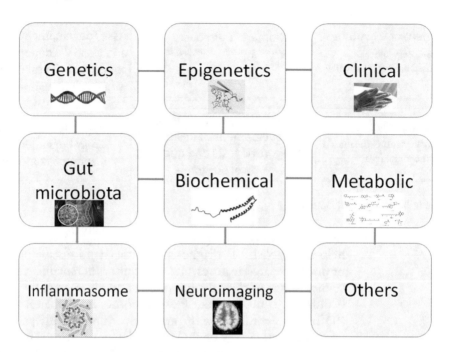

Fig. 2 Biomarkers of PD. PD is a complex disease with diverse molecular mechanisms of pathogenesis. A method based on the analysis of several biomarkers may be necessary for identification of early steps of the disease

subsample of this cohort. Another program—the Parkinson's Progression Marker Initiative (PPMI, www.ppmi-info.org)—is a multicenter cohort with an associated biorepository. The characteristics of other similar programs are described in reference [10].

3   Clinical Markers

Despite considerable efforts to find reliable and efficient biomarkers for presymptomatic early stages of PD the most important diagnostic marker remains clinical signs. The most significant clinical markers are the motor features of bradykinesia, rigidity, and resting tremor. However, very often, the typical motor symptoms may be preceded by nonmotor features, for example, REM sleep behavioral disorders, olfactory dysfunction, depression, reduced interest in new experiences, bowel dysfunction, and others. The screening of such nonmotor function changes is easy and inexpensive. Another advantage of examining these nonmotor functions is that they are noninvasive. These tests may have some additional diagnostic value, especially if conducted in a combination of several screenings. However, the specificity and sensitivity of these tests are limited, and unfortunately, none of these manifestations is specific for PD [11].

4   Biologic Specimens Used for Biomarker Analysis

Several biologic specimens are suitable for identifying biochemical biomarkers in PD, including: cerebrospinal fluid (CSF), blood components (plasma, serum), urine, saliva, tears, and skin [12–17].

The candidate biomarkers analyzed in these specimens can be divided into different categories, including:

- Neuroinflammatory biomarkers.

- Axonal damage proteins.

- Various forms of α-synuclein.

- Epigenetic markers, etc. [8, 9]. Despite considerable efforts and funds that were spent to identify potential clinical, biochemical, genetic, and imaging biomarkers for PD [14–17] no validated biomarkers exist for early diagnosis ("trait" biomarkers) or its neuropathological progression ("state" biomarkers) [18].

4.1   Neuroimaging Biomarkers

Contemporary technology allows the detection of brain defects using imaging techniques. Transcranial B-mode sonography (TCS), susceptibility-weighted imaging (SWI), diffusion-weighted imaging (DWI) [19], positron emission tomography (PET) scan, MR spectroscopy, and single-photon emission computed tomography (SPECT) scan are successfully used for PD [20–37](Table 1).

4.1.1   Transcranial B-Mode Sonography

(TCS) monitors the blood flow velocity of the brain's vessels by measuring the frequency of ultrasonic waves and their echoes. This inexpensive and reliable method shows the higher echogenicity of the SN in PD brains compared to a normal group that possibly occurs due to increased iron and gliosis levels in SN of PD patients

Table 1
Neuroimaging biomarkers

Method	Mechanism, substances used	References
Transcranial B-mode sonography (TCS)	Measures the blood flow velocity in the brain vessels	[19, 20]
	Measurements of the frequency of ultrasounds waves and their echoes	
Magnetic resonance Imaging (MRI)	Measures the rate of water diffusion through the tissue to determine the structural details of that tissue	[18, 21, 22]
MR spectroscopy	Measurements of N-acetylaspartate, creatine, and myoinositol	[24]
SPECT	Use of radiotracers and computer techniques to generate 3D images	[23–32]
PET scan	Use of radiolabeled analogs, such as 18F-dopa, 18F-FE-PE2I, and 18F-β-CFT, 11C-methylphenidate, VMAT2, DTBZ, 11C-MP4A	[37–40]

[20]. The increased iron level in PD can be due to either alteration or malfunction of the blood-brain barrier (BBB). In PD, the increase in the number of iron transferring receptors, including both transferrin receptors of BBB and iron-binding receptors of neurons, can lead to the accumulation of iron in SN [21].

4.1.2 Magnetic Resonance Imaging (MRI)

DWI is a form of MRI that measures the rate of water diffusion through the tissue to determine structural details of that tissue. The higher measured diffusivity means the greater mobility of water molecules that can be due to changes of cell shape or the reduction of the region of cellular volume. This technique can differentiate PD from MSA at the early stage, while the clinical symptoms of these disorders are very similar. In particular, the higher diffusivity of water in middle cerebellar peduncles in MSA patients in comparison to PD patients has been reported based on the DWI examination [19]. DWI can also differentiate patients with PSP from PD patients by detecting abnormalities in basal ganglia [22]. More traditional MRI methods, such as high-resolution 3-Tesla T1-weighted MRI can also detect the reduced volume of caudate and putamen in PD patients compared to controls [23].

4.1.3 Single-Photon Emission Computed Tomography (SPECT)

Both PET and SPECT scans can detect the early onset of PD and the loss of dopaminergic neurons using radiotracers 3D imaging. The majority of radiotracers are noninvasive radiopharmaceuticals with a short half-life so that they usually decay soon after the imaging is complete. Moreover, the 3D images of PET and SPECT scans reveal some functions of the brain, whereas MRI can only monitor the anatomy and structure [25]. The advantages of SPECT are the following:

- Radiotracers used for SPECT scan have a longer half-life compared to PET radiotracers. They mostly are 123iodine (123I) and 99mtechnetium (99mTc) to emit gamma rays. The DAT-SPECT imaging can detect the degeneration of presynaptic terminals in dopaminergic neurons by visualizing DAT (dopamine transporter) quantity. This method is a good way to diagnose the reduction of DAT in the brain, but it cannot distinguish PD from other Parkinsonian Syndromes. The DAT gamma-emitting ligands such as 123I-iometopane (123I-β-CIT), 123I-ioflupane (123I-FP-CIT), and 123I-altropane (123I-IPT) are the most common DAT-density SPECT tracers. These ligands are derivatives of tropane and dopamine reuptake inhibitors that target DAT [26, 27].

- SPECT also can employ dopamine D2 receptor radioligands that are dopamine antagonists. They include 123I-iodobenzamide (123I-IBZM) [28], 123I-IBF [29], and 123I-epidepride [30]. 123I-2'-iodospiperone (2'-ISP) has been also used in some studies to monitor D2 dopamine receptors. This radiotracer can distinguish between PD and other forms of Parkinsonism due to the pattern of its uptake in basal ganglia. The disadvantage is a high imaging background that has insufficient signal-to-noise ratio. The method should be modified to improve its performance [31].

- Vesicular acetylcholine transporter (VAChT) can be monitored by SPECT radiotracers and used as an approach for PD early diagnosis. Acetylcholine (ACh) is a neurotransmitter that is in balance with dopamine in healthy people. The death of dopaminergic neurons and the reduction of dopamine levels in PD are associated with increased ACh. 123I-iodobenzovesamicol (123I-IBVM) binds to VAChT and relates to the density of acetylcholine containing vesicles. Reduction of VAChT in parietal and occipital lobes in PD patients without dementia and reduced VAChT in all lobes of the cerebral cortex in PD patients with dementia has been established by this SPECT radiotracer [32].

- 123I-metaiodo-benzylguanidine (123I-MIBG) is another radiotracer that can distinguish between PD and MSA [33]. The heart-to-mediastinum ratio of 123I-MIBG uptake is impaired in idiopathic PD patients, but not in patients with MSA. PET/CT scanning of these PD patients showed decreased FP-CIT striatal uptake [34, 35]. MIBG scintigraphy distinguishes not only PD from MSA but also PD from DLB [33, 36].

4.1.4 Positron Emission Tomography (PET) Scan

PET scan radiotracers emit electrons and positively charged antiparticles (positrons) with a similar mass as electrons. The presence of presynaptic dopamine transporter (DAT) in dopaminergic neurons of striatum and SN can be assessed with 18F and/or 11C radiolabeled dopamine analogs [35].

- These DAT radioligands include 18F-dopamine (18F-dopa), 18F-FE-PE2I [34], 18F-β-CFT [37], and 18F-LBT999 and 11C-methylphenidate [34–38].

- The vesicular monoamine transporter 2 (VMAT2) quantification is also possible using either 11C or 18F radiolabeled dihydrotetrabenazine (DTBZ) [39, 40]. Because of dopaminergic cell loss and subsequent loss of VMAT2, the PET signal of radiolabeled DTBZ is lower in PD patients than in controls. Both DAT and VMAT2 radioligands can detect the early signs of dopaminergic damage, although PD may not be differentiated from atypical Parkinsonism with dopaminergic dysfunction. 11C-MP4A is another PET radiotracer that detects the level of acetylcholinesterase (AChE) activity [41]. AChE deactivates ACh and terminates the signal. Impairment of the cholinergic system and reduction of cortical AChE has been assessed by 11C-MP4A-PET scan. AChE activity reduces cortical AChE more in PDD than in PD, indicating that cholinergic dysfunction is correlated with dementia in PD [41].

4.1.5 Proton 1H Magnetic Resonance (MR) Spectroscopy

Proton 1H magnetic resonance (MR) spectroscopy may be an important tool for PD diagnosis. Importantly, this method also helps to evaluate the efficacy of treatment in PD. N-acetylaspartate, total creatine, and myoinositol are significantly lower in the putamen of patients with parkinsonian syndromes compared to healthy volunteers [24]. Importantly, treatment of patients with L-DOPA restored N-acetylaspartate and total creatine. On the other hand, L-DOPA does not change myoinositol level. These results demonstrate that N-acetylaspartate, total creatine, and myoinositol in the putamen may be considered as putative PD biomarkers. The authors assume that changes in N-acetylaspartate and total creatine levels in the putamen of PD patients could be explained by the impairment in mitochondrial energy [24].

4.2 Biochemical Biomarkers

4.2.1 α-Synuclein as a PD Biomarker

α-Synuclein is a small structurally unfolded protein, which forms a significant part of LB [42]. Since the discovery of α-synuclein as a protein linked to familial forms of PD and its identification as the major protein component of the neuropathological hallmark of idiopathic PD, considerable attention has focused on various forms of misfolded and post-translationally modified α-synuclein as PD biomarkers. The pathogenesis of PD involves the accumulation of aggregated and misfolded forms of α-synuclein, and therefore PD is considered protein misfolding disease. No significant difference has been reported in the levels of monomeric α-synuclein between PD patients and controls [43]. Therefore, attention shifted to modified forms of the protein. The interest is now directed toward oligomeric and phosphorylated α-synuclein at serine 129 (pS129 α-synuclein) [44–46] (Table 2). Several recent results demonstrate that S129-α-synuclein plasma level may give important information about PD progression. In one study, plasma levels of total and pS129-α-synuclein were significantly higher in PD patients than controls [47]. These data demonstrate that pS129-α-synuclein levels correlate with motor severity and progression, but not cognitive decline. This analysis may be especially valuable when combined with the results of other biomarkers.

In another study, the examination of pS129-α-synuclein together with another member of the synuclein family –γsynuclein oxidized on methionine-39 (oxi-γ-syn) showed their colocalization in LB [48] (Fig. 3). In these inclusions, α-synuclein was localized almost exclusively in peripheral ring-like structures (Fig. 3b). At the same time, oxi-γ-syn immunoreactivity was also found in the internal parts forming a dot-like pattern of staining (Fig. 3a). Further

Table 2
Analysis of α-synuclein (α-syn) as a biomarker in PD and related diseases

Type of α-syn	Source	Alterations	Comments	References
pS129 α-syn	Plasma	Higher in PD patients	Correlate with motor severity and progression	[47]
Oligomeric α-syn	Saliva	Higher in PD patients based on comparison of 20 analysis		[49]
Monomeric α-syn	CSF	Similar in PD and controls	Total α-syn is not reliable biomarker	[39]
Oligomeric α-syn	CSF	Higher in PDD, DLB and AD	Higher in PDD compared to AD	[41]
pS129 α-syn	CSF	Higher in PD patients	Association depends on disease stage	[42]
Total-α-syn and pS129α-syn	Peripheral tissues	pS129 α-syn higher in PD	Reviewed from several studies	[46]
		Total α-syn is lower	Significant variation across different studies	
Strain-specific	CSF	RT-QuiC analysis is a reliable the method reveals 95% sensitivity and 100%	Aggregated α-syn tool for PD diagnostic Specificity for PD-specific α-syn strains	[52–58]
Strain-specific	CSF	PMCA can differentiate PD and the method has high reliability in recognition		[59, 60]
Aggregated α-*syn*	MSA with sensitivity of 95.4% of PD-specific α-syn strain			

studies might reveal what combination of biomarkers possesses high predictive and prognostic power.

Recent studies point to a good chance that analysis of α-synuclein in saliva may become a reliable diagnostic marker. The oligomeric form of salivary α-synuclein is higher in PD patients

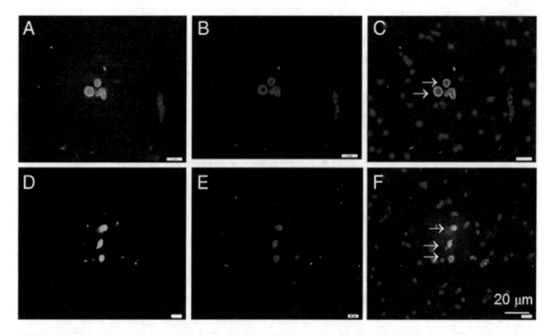

Fig. 3 Paraffin sections of brain specimens stained with α- and γ-synuclein specific antibodies. Reproduced from Surgucheva et al. [48]. The article was published in Acta Neuropathol Commun. Surgucheva I, Newell KL, Burns J, Surguchov A (2014) New α- and γ-synuclein immunopathological lesions in human brain. Acta Neuropathol Commun. 2:132. doi: 10.1186/s40478-014-0132-8

[49]. Although the analysis of various modified forms of α-synuclein may improve the sensitivity and specificity for diagnosing PD, significant variabilities exist in the levels of α-synuclein across different studies [50]. Importantly, accumulation of aggregated α-synuclein occurs not only in PD but also in Alzheimer's disease, DLB, MSA, and other neurodegenerative diseases collectively called synucleinopathies [51, 52]. An important breakthrough in diagnostic methods is the development of methods that allow the differentiation of various forms (strains) of misfolded proteins [54–62].

4.2.2 Development of New Methods that Can Improve α-Synuclein-Based PD Diagnostics

Advancement of analysis may significantly improve the reliability of α-synuclein as a PD biomarker [54–56]. Recently developed new techniques raise hope that they will ensure more consistent results for the diagnosis of early stages of PD.

- For example, Real-time Quaking-induced conversion (RT-QuIC) is a highly sensitive assay based on thioflavin T visible spectrum fluorescence detection initially developed for prion analysis. It allows for the identification of misfolded proteins in CSF and other biofluids. The method initially developed for prion detection [57] can now be used for diagnosis of PD and other protein misfolding diseases, based on the assay of α-synuclein in CSF [58, 59]. The principle of the method is the measurement of the ability of the aggregated α-synuclein to induce further aggregation of non-aggregated α-synuclein in a cyclic manner. The technique can detect abnormal α-synuclein in CSF of patients with PD or DLB with sensitivities of 95% and 92%, respectively, and with an overall specificity of 100% when compared to Alzheimer's and control CSF [58, 59]. Importantly, patients with tauopathies, that is, PSP and corticobasal degeneration, gave negative results, confirming the high specificity of the method. These results suggest that RT-QuIC analysis of CSF has a good perspective for the early clinical assessment of patients with PD and other α-synucleinopathies [58, 59].

- Another new method, in addition to RT-QuIC that may significantly improve the diagnostic of PD and other neurodegenerative diseases, is PMCA (protein misfolding cyclic amplification) [60]. This technique permits the identification of distinct α-synuclein strains that are specific for various protein misfolding disorders. The method showed its high specificity and reliability in the diagnosis of patients with PD and MSA [62]. A brief comparison of RT-QuIC and PMCA and their application for diagnosis of protein folding diseases can be found in recent publications [58–61].

4.2.3 microRNAs (miRNAs)

MicroRNAs (miRNAs) are conserved short RNAs playing the role of regulators of gene expression on the post-transcriptional level. miRNAs binding to complementary sequences in the 3′UTR of target RNA transcripts inhibits protein expression. More than 2500 human miRNAs have been detected, and more than 500 of them are identified in the blood [63].

- Deregulation of miRNAs often occurs in human diseases, and therefore they can potentially be used as biomarkers of diagnosis, prognosis, and response to treatment. Recent publications contain evidence that drug response and efficacy also can be modulated by miRNA-mediated mechanisms [64]. One of the important features of miRNAs for translational use is their ability to regulate the expression of genes, which are targets for pharmacological interventions.

- Another potential application with miRNAs is the restoration of miRNA composition altered in disease. It may be done by increasing or decreasing the level of specific mRNA(s), which was altered due to the disease. The presence of miRNAs in serum, CSF, saliva, urine, and other body fluids makes them a convenient tool for analysis of the onset of a disease.

- Schwienbacher et al. [65] described changes in miRNAs profile in PD patients and the effect of treatment on the composition of miRNAs. The authors found the upregulation of miR-30b-5p in plasma of drug-naïve patients and upregulation of miR-30a-5p in l-dopa-treated PD patients [65]. In another study an over-expression of miR-29a-3p, miR-30b-5p, and miR-103a-3p in l-dopa-treated PD patients was described [66].

In several recent publications, attempts to use cell-free miRNAs as biomarkers of PD have been described [67].

- Some of these dysregulated miRNAs have been associated with main PD-related or neuron-specific processes. For example, miR-124-3p is an abundant neuron-specific miRNA implicated in neuronal plasticity, differentiation, and fate determination.

- Another miRNA, miR-132-3p regulates dopamine neuron differentiation. Further analysis demonstrated that the level of approximately 60% of deregulated miRNAs was reduced in the plasma, serum, and CSF of PD patients. These findings presume that mRNA targets specific for these miRNAs will be increased. Interestingly, PD deregulated miRNA targets often include transcription factors. Therefore, after the first regulatory step directed on transcription factors, the second step will include alterations in the transcription specificity, and finally, these modulations will change proteome composition. Importantly, many of the deregulated miRNAs are clustered together on chromosome 14 near the site 14q32. This chromosomal region is intrinsically unstable because it contains TGG nucleotide repeats often involved in various human abnormalities [65].

- Several miRNAs are dysregulated in PD brain. This causes changes in expression of such PD-related genes as α-synuclein, LRRK2, PARK2, GBA, and other genes [68]. Dysregulation of α-synuclein affects expression of downstream targets, since α-synuclein is a modulator of specific gene expression [69].

- More about miRNAs can be found in Section 4.5.2 entitled "Epigenetic markers in body fluids".

4.2.4 Inflammasomes   Potential biomarkers are also located downstream of the α-synuclein signaling pathway.

- α-Synuclein triggers the activation of NOD-like receptor-3 (NLRP3) inflammasome. These cytosolic multiprotein oligomers of the innate immune system are responsible for the activation of inflammatory responses, and inflammasome-induced inflammation is an essential process in PD pathogenesis [70–76].

- NLRP3 inflammasome activation in microglia plays a central role in dopaminergic neurodegeneration. NLRP3 inflammasomes are activated in the peripheral blood mononuclear cells from PD patients [71] and therefore, the analysis of patient's blood may be an indicator of the level of inflammation and the state of the innate immune system.

- This analysis could be used as a noninvasive biomarker to identify the early steps of PD and to monitor the severity and progression of this disease if inflammasomes can be quantified in blood [72]. Further investigation will be then necessary to establish the relation of clinical features of PD and the inflammasome-mediated inflammatory response.

- Other prospective biomarkers of PD are cytokines—a family of secreted proteins involved in immunoregulatory and inflammatory processes. For example, concentrations of chemokine CCL28 in CSF are elevated in CSF of PD patients; these differences allow the differentiation of PD from MSA [77].

4.2.5 Ubiquitin C-Terminal Hydrolase L1

The presence of α-synuclein and ubiquitin aggregates in LB and Lewy neurites looks nonaccidental. Their co-occurrence is a result of a deficiency of the ubiquitin-proteasome system (UPS) or the autophagic pathway. Ubiquitin C-terminal hydrolase L1 (UCHL1)—an essential component of UPS—modulates ubiquitin level and ensures ubiquitin stability within cells. Plasma ubiquitin C-terminal hydrolase L1 (UCH-L1) in PD patients was found significantly higher than in healthy controls [78]. Across all PD patients, UCHL1 correlated significantly with Unified Parkinson's Disease Rating Scale (UPDRS), Part III motor scores ($\beta = 3.87$, 95% CI $= 0.43$–7.31, $p = 0.028$), but not with global cognition. Thus, UCHL1 correlates with motor function in PD, with higher levels seen in later disease stages. However, in CSF UCH-L1 levels are significantly decreased in PD patients [79].

4.2.6 Neurofilament Light Chain (NFL) as a PD Biomarker

Neurofilament proteins, including NFL and the *p*hosphorylated *n*eurofilament *h*eavy chain (pNF-H) are present in CSF and peripheral blood. Most promising as a biomarker is the neurofilament light chain (NFL) in CSF [13, 80, 81]. NFL is an abundant cytoskeletal 68 kDa protein encoded by the NFL gene that may distinguish atypical parkinsonian disorders from PD. An assay of NFL in combination with tau may differentiate MSA from PD [80].

Further investigations showed that CSF NFL, together with a panel of other biomarkers, predicted a diagnosis of PD versus atypical parkinsonian syndromes. Additional studies confirmed that increased levels of NFL in CSF offered clinically relevant, high accuracy discrimination between PD and MSA [13, 81, 82]. Recent results also demonstrated the efficacy of CSF NFL as a biomarker for disease severity and progression in PD [83].

4.2.7 DJ-1 (Parkinson Disease Protein 7)

Parkinson disease protein 7 is a dimeric, highly conserved, and ubiquitously expressed protein encoded by the PARK7 gene linked to early-onset, familial forms of PD [84–89]. Several groups have analyzed DJ-1 in CSF as a potential biomarker for PD and received uncertain and contradictory results [86, 89]. However, analysis of DJ-1 in exosomes of neural origin demonstrated its increase in PD patients compared to controls. Moreover, there is a positive correlation between levels of DJ-1 and α-synuclein in plasma neuronal-derived exosomes. This finding points to exosomal DJ-1 as a potential PD marker [89].

4.3 Nonprotein Markers (Neurochemical Biomarkers)

4.3.1 Orexin

Orexin (also called hypocretin) is a neuropeptide hormone expressed in the dorsolateral hypothalamus. Orexin regulates various physiological functions, including the sleep-wake cycle, cardiovascular responses, heart rate, and blood pressure [90]. According to some measurements, orexin-A level is decreased in PD patients proportionally to the severity of the disease [91] such that the more severe the disease, the lower orexin levels were present in the CSF of a patient. In the late stages of PD, decreased orexin levels may be responsible for daytime sleepiness [92].

4.3.2 8-Hydroxy-2'-Deoxyguanosine (8-OHdG)

8-OHdG is an oxidized form of 8-hydroxyguanine (8-OHG). Serum 8-OHdG is a prospective biomarker for PD since its levels are higher in PD patients compared to healthy individuals [93]. Measurement of 8-OHdG may help to differentiate between patients with different forms of PD, for example, between individuals carrying PD-associated mutations in leucine-rich repeat kinase 2 (LRRK2 CTL) and healthy people [94].

4.3.3 Biogenic Amine 3-Methoxy-4-Hydroxyphenylglycol (MHPG)

Recent studies have pointed to 3-methoxy-4-hydroxyphenylglycol (MHPG) as a valuable biomarker to distinguish several forms of neurodegenerative diseases. MHPG can pass through the blood-brain barrier (BBB), and analysis of its level in serum and CSF may help to determine cognitive staging in PD and distinguish PD from non-PD controls [95, 96].

4.3.4 Dopamine Metabolites

As found by several investigators, decreased concentrations of 3,4-dihydroxyphenylacetic acid (DOPAC) and DOPA in CSF may identify a preclinical form of PD [97, 98]. According to a recent theory, PD might develop when 3, 4-dihydroxyphenyl-acetaldehyde (DOPAL) oligomerizes and aggregates α-synuclein. These data provide a link between synucleinopathy and catecholamine neuron loss in Lewy body disease [96].

4.3.5 Microbiota Based Biomarkers

One of the first changes detected in PD patients is gastrointestinal abnormalities, including constipation and hyperpermeability. In many studies, alterations of microbiota in PD patients have also been described [99–101]. The altered microbiota may cause metabolic changes in PD patients, which play an essential role in the onset and progression of the disease. Importantly, these alterations may occur at the early stages of PD long before motor and gait symptoms appear. Finding early changes in microbiota composition opens the possibility of using gut microbial imbalances as specific microbial signatures and predictive biomarkers for early PD diagnosis [102].

- For example, a significant increase in genera Akkermansia, a slight increase in Bifidobacterium, and the reduction of Prevotella have been described in PD patients [103].

- In other studies, a decrease in species from the Lachnospiraceae family, Faecalibacterium sp., and Bacteroides sp. in PD patients has been found [104]. The changes in gut microbiota occur in the mouse model of PD. Interestingly, these studies have demonstrated that the gut microbiota is involved in α-synuclein aggregation pathology via their production of short-chain fatty acids [105].

4.4 Genetic Biomarkers of PD

The genetically inherited forms of PD are relatively rare, with disease-causing mutations present in approximately 2% of patients [106]. However, the investigation of biochemistry, molecular biology, and immunopathology of these forms provides significant insights into the pathogenesis of PD. The most common PD-causing genes with Mendelian inheritance are α-synuclein, leucine-rich repeat kinase 2 (LRRK2), PARKIN, PTEN-induced putative kinase 1 (PINK1), DJ-1 (Daisuke-Junko-1) and ATP13A2. PD susceptibility genes include, among others, glucocerebrosidase beta acid (GBA) and microtubule-associated protein tau (MAPT) [106].

- In α-synuclein gene point mutations, gene duplication and triplication events have been identified in families with autosomal dominant early-onset PD [107–109].

- Investigation of glucocerebrosidase (GBA) gene show an essential role in the understanding of PD genetics. It encodes a lysosomal enzyme involved in sphingolipid degradation. Mutations in the GBA gene are the single largest risk factor for PD [110, 111].

- New genetic studies point to the existence of additional untested variants in the primary Mendelian and genome-wide association study (GWAS) genes that contribute to the genetic etiology of sporadic PD [106]. Many cases of PD are the result of complex gene-environmental interactions affecting the development of the disease.

4.5 Epigenetic-Based Biomarkers

4.5.1 Definition of Epigenetics

What Is Epigenetics?

There is an increasing body of evidence pointing to a growing role of epigenetics in the development and progression of PD. The shortest definition of epigenetics refers to alterations in gene expression or function without changes in the DNA sequence. Epigenetics can also be defined as the study of changes in gene function that are heritable, and that do not entail a change in the DNA sequence. There are other definitions of epigenetics; for example, "… a change in the state of expression of a gene that does not involve a mutation, but that is nevertheless inherited in the absence of the signal (or event) that initiated the change" [112].

Generally speaking, epigenetics includes alterations that affect gene activity and expression; however, the term may be used to define any heritable phenotypic change. For example, epigenetics comprises histone methylation and acetylation, which often occur at actively transcribed genes. DNA methylation at particular CpG regions (cytosine and guanine separated by phosphate group) and noncoding RNAs-mediated changes regulate gene expression (Fig. 4). Due to these processes, a direct connection between specific DNA sequences and particular physical features, including predisposition to a disease, can be masked. Epigenetic modifications may contribute to PD-related pathogenesis and provide a molecular explanation to bridge the gap between the genome and environmental signals. For example, abnormal DNA methylation of specific genes or histone acetylation may serve as a possible biomarker suitable for the early diagnosis of PD.

Epigenetic modifications in PD act as mediators between environmental exposure and genes, contributing to PD-related neurodegeneration. Abnormal DNA methylation of specific genes or histone acetylation may be a potential clinical biomarker useful in the identification and diagnosis of the disease. A comprehensive understanding of aberrant epigenetic modifications involved in PD not only provides better insights into the pathogenesis of PD but also may lead to the development of new biomarkers and novel therapeutic strategies [113]. Epigenetic changes are already being incorporated as valuable candidates in the biomarker field. Importantly, these biomarkers have a reversible nature and therefore offer an opportunity to affect disease symptoms by using epigenetic-based therapy.

4.5.2 Epigenetic Markers in Body Fluids

Several diagnostic markers based on epigenomics are successfully used for body fluids analysis.

- For example, hypomethylation of genes encoding α-synuclein and leucine-rich repeat kinase 2 (LRRK2 or dardarin) has been used as a noninvasive biomarker for early PD diagnosis [114].

- In another study, the level of several microRNAs was found to be significantly different in blood samples from PD patients and

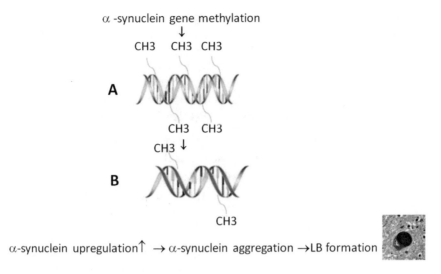

α -synuclein gene methylation
↓
CH3   CH3   CH3

A

CH3   CH3
CH3 ↓

B

CH3

α-synuclein upregulation↑ → α-synuclein aggregation →LB formation

Fig. 4 DNA methylation at CpG regions and noncoding RNA-mediated changes regulate gene expression

controls. For example, levels of miR-1, miR-22, and miR-29 were lower in PD patients than in healthy controls [115]. Contrary to this, the levels of miR-16-2, miR-26a2, and miR-30a were increased in the group treated by levodopa/carbidopa compared to the nontreated PD patients [115].

Epigenetics is a potentially relevant field for the development of new PD biomarkers. Currently, this area is at the preliminary stage of development in search of the most reliable epigenetics-based biomarkers. There are several approaches where there is a good chance of finding reliable biomarkers. For example, hypomethylation of the α-synuclein gene promoter region has been described in the SN of PD patients [116]. Importantly, this promotor hypomethylation has been shown to upregulate α-synuclein expression in cell culture, conceivably contributing to the pathology of PD by this mechanism. Interestingly, the treatment by L-dopa has been associated with hypermethylation of the α-synuclein promotor region, pointing to a possibility that PD therapy may affect methylation [117]. In a longitudinal genome-wide methylation study examining approximately 850,000 CpG sites in the blood of 189 PD patients and corresponding control individuals different patterns of methylation in PD patients compared to controls were detected [118]. The data demonstrate that DNA methylation is dynamic in PD, and specific changes occur over time during disease progression and in response to dopaminergic treatment. These results indicate that the methylome of PD patients in both brain and blood tissues is changed and that these alterations may be used as a basis for the development of biomarkers for early PD diagnosis.

**4.6 Advancement
of Biomarkers into
Clinical Practice**

Recently, significant progress has been achieved in the development of biomarkers for neurological diseases and their implementation in clinical practice. Their application for diagnostic purposes is successfully used in several diseases. The number of instances where biomarkers are beneficial for the diagnosis of neurological diseases is quickly growing. Below are several examples of the translational application of biomarkers developed for neurological diseases.

- The presence of oligoclonal IgG bands in CSF has become a basis of multiple sclerosis diagnosis [119].

- The analysis of antibodies against aquaporin 4 has entirely altered the diagnostic approach toward neuromyelitis optica; this finding highlighted new pathological and biochemical features of this disease [120].

- The analysis of aggregated prion proteins and 14-3-3 proteins and their utility in Creutzfeldt–Jacob's disease diagnosis [121]. And lastly, the analysis of phospho-Tau181, total-Tau, Abeta42, and Abeta40 in CSF improved the diagnostic capabilities of Alzheimer's disease [122].

There is hope that reliable PD biomarkers will also be translated into clinical practice as a result of the following advancements.

1. Modification of biochemical methods for existing biomarkers. *See*, for example, Chapter 7.2 "Real-time Quaking-induced conversion (RT-QuIC)" [58, 59] and PMCA (protein misfolding cyclic amplification) [60–62].

2. Discovery of new candidates in groups of biomarkers relatively recently identified, such as inflammasomes (Chapter 7.6.4), microbiota-based biomarkers (Chapter 7.7.5), and epigenetic-based biomarkers (Chapter 7.9.2).

3. Finding an optimal combination of existing biomarkers that will enhance their diagnostic value.

4. Discovery of brand-new biomarkers. An illustration that new candidates can be discovered is a recently published analysis of blood-based PD biomarkers [123]. The authors measured levels of 1129 proteins using an aptamer-based platform. The blood was taken from 527 people with PD, ALS, or no neurological disease to discover new diagnostic and prognostic biomarkers. The following proteins were robustly associated with PD: aminoacylase-1 (ACY1), bone sialoprotein (BSP), growth hormone receptor (GHR), and osteomodulin (OMD). These peripheral blood samples may be developed for both disease characterization and prediction of future disease progression in PD.

5   Conclusion. Four Big Questions Waiting for Answers

The etiology of PD is multifaceted and is based on the interaction of genes, environmental factors, and aging. Identification of new PD biomarkers and their validation is essential for successful prediction of clinical outcomes, patient monitoring, and drug efficiency assessment. The critical challenge is how to assimilate preclinical data to find a reliable biomarker that can be used with acceptable costs in clinical practice. A straightforward methodology would be to identify the relevant triggers of the disease and to target the physiopathological mechanisms causing the death of dopaminergic neurons. The recent development of new methods and modifications of existing methods of biomarkers identification and validation raise hope that new reliable biomarkers will appear in the near future. The latest insights point toward a more significant role for epigenetics as an important field where new biomarkers will be developed. There is hope that the study of histone modifiers and human dopamine neurons will suggest approaches for early PD diagnosis. Furthermore, these new developments will allow the finding of novel therapeutic strategies for PD treatment.

About 4 years ago, *Nature* published a short review about PD entitled "Parkinson's disease: 4 big questions" [124]. These questions were:

1. How does Parkinson's disease begin?
2. What is the role of the α-synuclein protein?
3. What is the role of the gut in Parkinson's disease?
4. What is the best way to divide people with the disease into subtypes?

Now 4 years later, we can confess that there are no considerable breakthroughs in any of these questions. However, the results of recent investigations accumulate quickly, and we do believe that in 4 next years, we will be able to find answers to at least some of these questions. Furthermore, the important first step helping to unwrap mysteries of this disorder and to find answers to these questions will be developing biomarkers of early steps of PD.

Acknowledgments

This research was supported by the Veterans Affairs grants I01BX000361 and the Glaucoma Foundation grant QB42308.

Funding: *Andrei Surguchov received research grants from VA Merit Review grant 1I01BX000361 and the Glaucoma Foundation grant QB42308.*

References

1. Kalia LV, Lang AE (2015) Parkinson's disease. Lancet 386(9996):896–912. https://doi.org/10.1016/S0140-6736(14)61393-3

2. Espay AJ, Lang AE (2018) Parkinson diseases in the 2020s and beyond: replacing clinicopathologic convergence with systems biology divergence. J Parkinsons Dis 8(s1):S59–S64. https://doi.org/10.3233/JPD-181465

3. Tan JM, Wong ES, Lim KL (2009) Protein misfolding and aggregation in Parkinson's disease. Antioxid Redox Signal 11 (9):2119–2134. https://doi.org/10.1089/ARS.2009.2490

4. Braak H, Del Tredici K, Rüb U et al (2003) Staging of brain pathology related to sporadic Parkinson's disease. Neurobiol Aging 24 (2):197–211. https://doi.org/10.1016/s0197-4580(02)00065-9

5. Braak H, Rüb U, Gai WP et al (2003) Idiopathic Parkinson's disease: possible routes by which vulnerable neuronal types may be subject to neuroinvasion by an unknown pathogen. J Neural Transm 110(5):517–536. https://doi.org/10.1007/s00702-002-0808-2

6. GBD 2016 Parkinson's disease Collaborators (2018) Global, regional, and national burden of Parkinson's disease, 1990–2016: a systematic analysis for the Global Burden of Disease Study. Lancet Neurol 17(11):939–953

7. Koga S, Aoki N, Uitti RJ et al (2015) When DLB, PD, and PSP masquerade as MSA: an autopsy study of 134 patients. Neurology 85 (5):404–412. https://doi.org/10.1212/WNL.0000000000001807

8. Chahine LM, Stern MB, Chen-Plotkin A (2014) Blood-based biomarkers for Parkinson's disease. Parkinsonism Relat Disord 20 (Suppl 1):S99–S103. https://doi.org/10.1016/S1353-8020(13)70025-7

9. Chahine LM, Stern MB (2017) Parkinson's disease biomarkers: where are we and where do we go next? Mov Disord Clin Pract 4 (6):796–805. https://doi.org/10.1002/mdc3.12545

10. Chen-Plotkin AS, Zetterberg H (2018) Updating our definitions of Parkinson's disease for a molecular age. J Parkinsons Dis 8 (s1):S53–S57. https://doi.org/10.3233/JPD-181487

11. Delenclos M, Jones DR, PJ ML, Uitti RJ (2016) Biomarkers in Parkinson's disease: advances and strategies. Parkinsonism Relat Disord 22(Suppl 1):S106–S110. https://doi.org/10.1016/j.parkreldis.2015.09.048

12. LeWitt PAS, Huber BR, Zhang J (2013) An update on CSF biomarkers of Parkinson's disease. In: Mandel S (ed) Neurodegenerative diseases: integrative PPPM approach as the medicine of the future. Springer, Dordrecht, pp 161–184

13. Magdalinou N, Lees AJ, Zetterberg H (2014) Cerebrospinal fluid biomarkers in parkinsonian conditions: an update and future directions. J Neurol Neurosurg Psychiatry 85 (10):1065–1075. https://doi.org/10.1136/jnnp-2013-307539

14. Gramotnev G, Gramotnev DK, Gramotnev A (2019) Parkinson's disease prognostic scores for progression of cognitive decline. Sci Rep 9 (1):17485. https://doi.org/10.1038/s41598-019-54029-w

15. Mollenhauer B, Caspell-Garcia CJ, Coffey CS et al (2017) Longitudinal CSF biomarkers in patients with early Parkinson's disease and healthy controls. Neurology 89 (19):1959–1969. https://doi.org/10.1212/WNL.0000000000004609

16. Parnetti L, Gaetani L, Eusebi P et al (2019) CSF and blood biomarkers for Parkinson's disease. Lancet Neurol 18(6):573–586. https://doi.org/10.1016/S1474-4422(19)30024-9

17. Ge F, Ding J, Liu Y, Lin H, Chang T (2018) Cerebrospinal fluid NFL in the differential diagnosis of parkinsonian disorders: a meta-analysis. Neurosci Lett 685:35–41. https://doi.org/10.1016/j.neulet.2018.07.030

18. Gasser T (2009) Genomic and proteomic biomarkers for Parkinson disease. Neurology 17 (7 Suppl):S27–S31. https://doi.org/10.1212/WNL.0b013e318198e054

19. Chung EJ, Kim EG, Bae JS et al (2009) Usefulness of diffusion-weighted MRI for differentiation between Parkinson's disease and Parkinson variant of multiple system atrophy. J Mov Disord 2(2):64–68. https://doi.org/10.14802/jmd.09017

20. Skoloudík D, Jelínková M, Blahuta J et al (2014) Transcranial sonography of the substantia nigra: digital image analysis. AJNR Am J Neuroradiol 35(12):2273–2278. https://doi.org/10.3174/ajnr.A4049

21. Hare D, Ayton S, Bush A, Lei P (2013) A delicate balance: iron metabolism and diseases of the brain. Front Aging Neurosci 5:34. https://doi.org/10.3389/fnagi.2013.00034

22. Seppi K, Schocke MF, Esterhammer R et al (2003) Diffusion-weighted imaging discriminates progressive supranuclear palsy from PD,

but not from the parkinson variant of multiple system atrophy. Neurology 60(6):922–927. https://doi.org/10.1212/01.WNL.0000049911.91657.9D

23. Saeed U, Compagnone J, Aviv RI et al (2017) Imaging biomarkers in Parkinson's disease and Parkinsonian syndromes: current and emerging concepts. Transl Neurodegener 6:8. https://doi.org/10.1186/s40035-017-0076-6

24. Mazuel L, Chassain C, Jean B et al (2016) Proton MR spectroscopy for diagnosis and evaluation of treatment efficacy in Parkinson disease. Radiology 278(2):505–513. https://doi.org/10.1148/radiol.2015142764

25. Histed SN, Lindenberg ML, Mena E et al (2012) Review of functional/anatomic imaging in oncology. Nucl Med Commun 33(4):349–361. https://doi.org/10.1097/MNM.0b013e32834ec8a5

26. Wang L, Zhang Q, Li H, Zhang H (2012) SPECT molecular imaging in Parkinson's disease. J Biomed Biotechnol 2012:412486. https://doi.org/10.1155/2012/412486

27. Brooks DJ (2016) Molecular imaging of dopamine transporters. Ageing Res Rev 30:114–121. https://doi.org/10.1016/j.arr.2015.12.009

28. Reiche W, Grundmann M, Huber G (1995) Dopamine (D2) receptor SPECT with 123I-iodobenzamide (IBZM) in diagnosis of Parkinson syndrome. Radiologie 35(11):838–843

29. Sasaki T, Amano T, Hashimoto J, Itoh Y, Muramatsu K, Kubo A et al (2003) SPECT imaging using [123I] beta-CIT and [123I] IBF in extrapyramidal diseases. No To Shinkei 55(1):57–64

30. Pirker W, Asenbaum S, Wenger SL et al (1997) Iodine-123-epidepride-SPECT: studies in Parkinson's disease, multiple system atrophy and Huntington's disease. J Nucl Med 38(11):1711–1717

31. Yonekura Y, Saji H, Iwasaki Y, Tsuchida T, Fukuyama H, Shimatsu A et al (1995) Initial clinical experiences with dopamine D2 receptor imaging by means of 2′-iodospiperone and single-photon emission computed tomography. Ann Nucl Med 9(3):131–136. https://doi.org/10.1007/BF03165039

32. Niethammer M, Feigin A, Eidelberg D (2012) Functional neuroimaging in Parkinson's disease. Cold Spring Harb Perspect Med 2(5):a009274. https://doi.org/10.1101/cshperspect.a009274

33. Goldstein DS (2001) Cardiac sympathetic neuroimaging to distinguish multiple system atrophy from Parkinson disease. Clin Auton Res 11(6):341–342. https://doi.org/10.1007/BF02292764

34. Fazio P, Svenningsson P, Forsberg A, Jönsson EG, Amini N, Nakao R, Nag S, Halldin C, Farde L, Varrone A (2015) Quantitative analysis of ^{18}F-(E)-N-(3-Iodoprop-2-Enyl)-2-β-carbofluoroethoxy-3β-(4′-methyl-phenyl) nortropane binding to the dopamine transporter in Parkinson disease. J Nucl Med 56(5):714–720. https://doi.org/10.2967/jnumed.114.152421

35. Oh JK, Choi EK, Song IU, Kim JS, Chung YA (2015) Comparison of I-123 MIBG planar imaging and SPECT for the detection of decreased heart uptake in Parkinson disease. J Neural Transm (Vienna) 122(10):1421–1427. https://doi.org/10.1007/s00702-015-1409-1

36. Goldstein DS (2013) Sympathetic neuroimaging. Handb Clin Neurol 117:365–370. https://doi.org/10.1016/B978-0-444-53491-0.00029-8

37. Rinne J, Ruottinen H, Bergman J et al (1999) Usefulness of a dopamine transporter PET ligand [(18)F] β-CFT in assessing disability in Parkinson's disease. J Neurol Neurosurg Psychiatry 67(6):737–741. https://doi.org/10.1136/jnnp.67.6.737

38. Arlicot N, Vercouillie J, Malherbe C, Bidault R, Gissot V, Maia S et al (2019) PET imaging of dopamine transporter with [18F] LBT-999: initial evaluation in healthy volunteers. Q J Nucl Med Mol Imaging. https://doi.org/10.23736/S1824-4785.19.03175-3

39. Lin K, Weng YH, Hsieh CJ et al (2013) Brain imaging of vesicular monoamine transporter type 2 in healthy aging subjects by 18F-FP-(+)-DTBZ PET. PLoS One 8(9):e75952. https://doi.org/10.1371/journal.pone.0075952

40. Tong J, Wilson A, Boileau I, Houle S, Kish SJ (2008) Dopamine modulating drugs influence striatal (+)-[11C] DTBZ binding in rats: VMAT2 binding is sensitive to changes in vesicular dopamine concentration. Synapse 62(11):873–876. https://doi.org/10.1002/syn.20573

41. Bohnen NI, Albin RL, Koeppe RA et al (2006) Positron emission tomography of monoaminergic vesicular binding in aging and Parkinson disease. J Cereb Blood Flow Metab 26(9):1198–1212. https://doi.org/10.1038/sj.jcbfm.9600276

42. Surguchov A (2015) Intracellular dynamics of synucleins: here, there and everywhere. Int

Rev Cell Mol Biol 320:103–169. https://doi.org/10.1016/bs.ircmb.2015.07.007

43. Borghi R, Marchese R, Negro A et al (2000) Full length alpha-synuclein is present in cerebrospinal fluid from Parkinson's disease and normal subjects. Neurosci Lett 287 (1):65–67. https://doi.org/10.1016/s0304-3940(00)01153-8

44. Pronin AN, Morris AJ, Surguchov A, Benovic JL (2000) Synucleins are a novel class of substrates for G protein-coupled receptor kinases. J Biol Chem 275(34):26515–26522. https://doi.org/10.1074/jbc.M003542200

45. Hansson O, Hall S, Ohrfelt A et al (2014) Levels of cerebrospinal fluid α-synuclein oligomers are increased in Parkinson's disease with dementia and dementia with Lewy bodies compared to Alzheimer's disease. Alzheimers Res Ther 63(25). https://doi.org/10.1186/alzrt255

46. Stewart T, Sossi V, Aasly JO et al (2015) Phosphorylated α-synuclein in Parkinson's disease: correlation depends on disease severity. Acta Neuropathol Commun 3(7). https://doi.org/10.1186/s40478-015-0185-3

47. Lin CH, Liu HC, Yang SY et al (2019) Plasma pS129-α-synuclein is a surrogate biofluid marker of motor severity and progression in Parkinson's disease. J Clin Med 8(10). https://doi.org/10.3390/jcm8101601

48. Surgucheva I, Newell KL, Burns J, Surguchov A (2014) New α- and γ-synuclein immunopathological lesions in human brain. Acta Neuropathol Commun 2(132). https://doi.org/10.1186/s40478-014-0132-8

49. Figura M, Friedman A (2020) In search of Parkinson's disease biomarkers—is the answer in our mouths? A systematic review of the literature on salivary biomarkers of Parkinson's disease. Neurol Neurochir Pol 54 (1):14–20. https://doi.org/10.5603/PJNNS.a2020.0011

50. Fayyad M, Salim S, Majbour N et al (2019) Parkinson's disease biomarkers based on α-synuclein. J Neurochem 150(5):626–636. https://doi.org/10.1111/jnc.14809

51. Visanji NP, Lang AE, Kovacs GG (2019) Beyond the synucleinopathies: alpha synuclein as a driving force in neurodegenerative comorbidities. Transl Neurodegener 8:28. https://doi.org/10.1186/s40035-019-0172-x

52. Twohig D, Nielsen HM (2019) α-Synuclein in the pathophysiology of Alzheimer's disease. Mol Neurodegener 14(1):23. https://doi.org/10.1186/s13024-019-0320-x

53. Bongianni M, Ladogana A, Capaldi S et al (2019) α-Synuclein RT-QuIC assay in cerebrospinal fluid of patients with dementia with Lewy bodies. Ann Clin Transl Neurol 6 (10):2120–2126. https://doi.org/10.1002/acn3.50897

54. De Luca CMG, Elia AE, Portaleone SM et al (2019) Efficient RT-QuIC seeding activity for α-synuclein in olfactory mucosa samples of patients with Parkinson's disease and multiple system atrophy. Transl Neurodegener 8:24. https://doi.org/10.1186/s40035-019-0164-x

55. Garrido A, Fairfoul G, Tolosa ES, Martí MJ, Green A Barcelona LRRK2 Study Group (2019) α-synuclein RT-QuIC in cerebrospinal fluid of LRRK2-linked Parkinson's disease. Ann Clin Transl Neurol 6(6):1024–1032. https://doi.org/10.1002/acn3.772

56. van Rumund A, Green AJE, Fairfoul G et al (2019) α-Synuclein real-time quaking-induced conversion in the cerebrospinal fluid of uncertain cases of Parkinsonism. Ann Neurol 85(5):777–781. https://doi.org/10.1002/ana.25447

57. Atarashi R, Sano K, Satoh K, Nishida N (2011) Real-time quaking-induced conversion. A highly sensitive assay for prion detection. Prion 5(3):150–153. https://doi.org/10.4161/pri.5.3.16893

58. Schmitz M, Cramm M, Llorens F et al (2016) The real-time quaking-induced conversion assay for detection of human prion disease and study of other protein misfolding diseases. Nat Protoc 11(11):2233–2242. https://doi.org/10.1038/nprot.2016.120

59. Fairfoul G, McGuire LI, Pal S et al (2016) Alpha-synuclein RT-QuIC in the CSF of patients with alpha-synucleinopathies. Ann Clin Transl Neurol 3(10):812–818. https://doi.org/10.1002/acn3.338

60. Herva ME, Zibaee S, Fraser G et al (2014) Anti-amyloid compounds inhibit α-synuclein aggregation induced by protein misfolding cyclic amplification (PMCA). J Biol Chem 289(17):11897–11905. https://doi.org/10.1074/jbc.M113.542340

61. Surguchov A (2020) Analysis of protein conformational strains—a key for new diagnostic methods of human diseases. Int J Mol Sci 21 (8):E2801. https://doi.org/10.3390/ijms21082801

62. Shahnawaz M, Mukherjee A, Pritzkow S et al (2020) Discriminating α-synuclein strains in Parkinson's disease and multiple system atrophy. Nature 578(7794):273–277. https://doi.org/10.1038/s41586-020-1984-7

63. Burgos K, Malenica I, Metpally R et al (2014) Profiles of extracellular miRNA in cerebrospinal fluid and serum from patients with Alzheimer's and Parkinson's diseases correlate with disease status and features of pathology. PLoS One 9(5):e94839. https://doi.org/10.1371/journal.pone.0094839

64. Nuzziello N, Ciaccia L, Liguori M (2019) Precision medicine in neurodegenerative diseases: some promising tips coming from the microRNAs' world. Cell 9(1):75. https://doi.org/10.3390/cells9010075

65. Schwienbacher C, Foco L, Picard A et al (2017) Plasma and white blood cells show different miRNA expression profiles in Parkinson's disease. J Mol Neurosci 62 (2):244–254. https://doi.org/10.1007/s12031-017-0926-9

66. Serafin A, Foco L, Zanigni S et al (2015) Overexpression of blood microRNAs 103a, 30b, and 29a in L-dopa-treated patients with PD. Neurology 84(7):645–653. https://doi.org/10.1212/WNL.0000000000125860

67. Doxakis E (2020) Cell-free microRNAs in Parkinson's disease: potential biomarkers that provide new insights into disease pathogenesis. Ageing Res Rev 58:101023. https://doi.org/10.1016/j.arr.2020.101023

68. Martinez B, Peplow PV (2017) MicroRNAs in Parkinson's disease and emerging therapeutic targets. Neural Regen Res 12 (12):1945–1959. https://doi.org/10.4103/1673-5374.221147

69. Surguchev AA, Surguchov A (2017) Synucleins and gene expression: ramblers in a crowd or cops regulating traffic? Front Mol Neurosci 10:224. https://doi.org/10.3389/fnmol.2017.00224

70. Fan Z, Pan YT, Zhang ZY et al (2020) Systemic activation of NLRP3 inflammasome and plasma α-synuclein levels are correlated with motor severity and progression in Parkinson's disease. J Neuroinflammation 17 (1):11. https://doi.org/10.1186/s12974-019-1670-6

71. Lee E, Hwang I, Park S et al (2019) MPTP-driven NLRP3 inflammasome activation in microglia plays a central role in dopaminergic neurodegeneration. Cell Death Differ 26 (2):213–228. https://doi.org/10.1038/s41418-018-0124-5

72. Martinon F, Burns K, Tschopp J (2002) The inflammasome: a molecular platform triggering activation of inflammatory caspases and processing of proIL-beta. Mol Cell 10

(2):417–426. https://doi.org/10.1016/s1097-2765(02)00599-3

73. Gordon R, Albornoz EA, Christie DC et al (2018) Inflammasome inhibition prevents alpha-synuclein pathology and dopaminergic neurodegeneration in mice. Sci Transl Med 10(456). https://doi.org/10.1126/scitranslmed.aah4066

74. Alcocer-Gómez E, Castejón-Vega B, López-Sánchez M, Cordero MD (2018) Inflammasomes in clinical practice: a brief introduction. Exp Suppl 108:1–8. https://doi.org/10.1007/978-3-319-89390-7_1

75. Zhou K, Shi L, Wang Y, Chen S, Zhang J (2016) Recent advances of the NLRP3 inflammasome in central nervous system disorders. J Immunol Res 2016:9238290. https://doi.org/10.1155/2016/9238290

76. Campolo M, Paterniti I, Siracusa R et al (2019) TLR4 absence reduces neuroinflammation and inflammasome activation in Parkinson's diseases in vivo model. Brain Behav Immun 76:236–247. https://doi.org/10.1016/j.bbi.2018.12.003

77. Santaella A, Kuiperij HB, van Rumund A et al (2020) Inflammation biomarker discovery in Parkinson's disease and atypical parkinsonisms. BMC Neurol 20(1):26. https://doi.org/10.1186/s12883-020-1608-8

78. Ng ASL, Tan YJ, Lu Z et al (2020) Plasma ubiquitin C-terminal hydrolase L1 levels reflect disease stage and motor severity in Parkinson's disease. Aging 12(2):1488–1495. https://doi.org/10.18632/aging.102695

79. Mondello S, Constantinescu R, Zetterberg H et al (2014) CSF α-synuclein and UCH-L1 levels in Parkinson's disease and atypical parkinsonian disorders. Parkinsonism Relat Disord 20(4):382–387. https://doi.org/10.1016/j.parkreldis.2014.01.011

80. Abdo WF, De Jong D, Hendriks JC et al (2004) Cerebrospinal fluid analysis differentiates multiple system atrophy from Parkinson's disease. Mov Disord 19(5):571–579. https://doi.org/10.1002/mds.10714

81. Sako W, Murakami N, Izumi Y, Kaji R (2015) Neurofilament light chain level in cerebrospinal fluid can differentiate Parkinson's disease from atypical Parkinsonism: evidence from a meta-analysis. J Neurol Sci 352(1–2):84–87. https://doi.org/10.1016/j.jns.2015.03.041

82. Herbert MK, Aerts MB, Beenes M et al (2015) CSF neurofilament light chain but not FLT3 ligand discriminates Parkinsonian

disorders. Front Neurol 5(6):91. https://doi.org/10.3389/fneur.2015.00091

83. Lin CH, Li CH, Yang KC et al (2019) Blood NfL: a biomarker for disease severity and progression in Parkinson disease. Neurology 93 (11):e1104–e1111. https://doi.org/10.1212/WNL.0000000000008088

84. Drechsel J, Mandl FA, Sieber SA (2018) Chemical probe to monitor the Parkinsonism-associated protein DJ-1 in live cells. ACS Chem Biol 13(8):2016–2019. https://doi.org/10.1021/acschembio.8b00633

85. Yamagishi Y, Saigoh K, Saito Y et al (2018) Diagnosis of Parkinson's disease and the level of oxidized DJ-1 protein. Neurosci Res 128:58–62. https://doi.org/10.1016/j.neures.2017.06.008

86. Farotti L, Paciotti S, Tasegian A, Eusebi P, Parnetti L (2017) Discovery, validation and optimization of cerebrospinal fluid biomarkers for use in Parkinson's disease. Expert Rev Mol Diagn 17(8):771–780. https://doi.org/10.1080/14737159.2017.1341312

87. DosSantos MCT, Scheller D, Schulte C et al (2018) Evaluation of cerebrospinal fluid proteins as potential biomarkers for early stage Parkinson's disease diagnosis. PLoS One 13 (11):e0206536. https://doi.org/10.1371/journal.pone.0206536

88. Shi M, Furay AR, Sossi V, Aasly JO et al (2012) DJ-1 and αSYN in LRRK2 CSF do not correlate with striatal dopaminergic function. Neurobiol Aging 33(4):836.e5–836.e7. https://doi.org/10.1016/j.neurobiolaging.2011.09.01

89. Zhao ZH, Chen ZT, Zhou RL et al (2019) Increased DJ-1 and α-synuclein in plasma neural-derived exosomes as potential markers for Parkinson's disease. Front Aging Neurosci 10:438. https://doi.org/10.3389/fnagi.2018.00438

90. Imperatore R, Palomba L, Cristino L (2017) Role of Orexin-a in hypertension and obesity. Curr Hypertens Rep 19(4):34. https://doi.org/10.1007/s11906-017-0729-y

91. Fronczek R, Overeem S, Lee SY et al (2007) Hypocretin (orexin) loss in Parkinson's disease. Brain 130(6):1577–1585. https://doi.org/10.1093/brain/awm090

92. Wienecke M, Werth E, Poryazova R et al (2012) Progressive dopamine and hypocretin deficiencies in Parkinson's disease: is there an impact on sleep and wakefulness? J Sleep Res 21(6):710–717. https://doi.org/10.1111/j.1365-2869.2012.01027

93. Kikuchi Y, Yasuhara T, Agari T et al (2011) Urinary 8-OHdG elevations in a partial lesion rat model of Parkinson's disease correlate with behavioral symptoms and nigrostriatal dopaminergic depletion. J Cell Physiol 226 (5):1390–1398. https://doi.org/10.1002/jcp.22467

94. Loeffler DA, Aasly JO, LeWitt PA, Coffey MP (2019) What have we learned from cerebrospinal fluid studies about biomarkers for detecting LRRK2 Parkinson's disease patients and healthy subjects with Parkinson's-associated LRRK2 mutations? J Parkinsons Dis 9 (3):467–488. https://doi.org/10.3233/JPD-191630

95. van der Zee S, Vermeiren Y, Fransen E et al (2018) Monoaminergic markers across the cognitive spectrum of Lewy body disease. J Parkinsons Dis 8(1):71–84. https://doi.org/10.3233/JPD-171228

96. Vermeiren Y, De Deyn PP (2017) Targeting the norepinephrinergic system in Parkinson's disease and related disorders: the locus coeruleus story. Neurochem Int 102:22–32. https://doi.org/10.1016/j.neuint.2016.11.009

97. Goldstein DS, Kopin IJ, Sharabi Y (2014) Catecholamine autotoxicity. Implications for pharmacology and therapeutics of Parkinson disease and related disorders. Pharmacol Ther 144(3):268–282. https://doi.org/10.1016/j.pharmthera.2014.06.006

98. Goldstein DS, Holmes C, Lopez GJ, Wu T, Sharabi Y (2018) Cerebrospinal fluid biomarkers of central dopamine deficiency predict Parkinson's disease. Parkinsonism Relat Disord 50:108–112. https://doi.org/10.1016/j.parkreldis.2018.02.023

99. Cersosimo MG, Benarroch EE (2012) Autonomic involvement in Parkinson's disease: pathology, pathophysiology, clinical features and possible peripheral biomarkers. J Neurol Sci 313(1–2):57–63. https://doi.org/10.1016/j.jns.2011.09.030

100. Scheperjans F, Aho V, Pereira PA et al (2015) Gut microbiota are related to Parkinson's disease and clinical phenotype. Mov Disord 30 (3):350–358. https://doi.org/10.1002/mds.26069

101. Mulak A, Bonaz B (2015) Brain-gut-microbiota axis in Parkinson's disease. World J Gastroenterol 21(37):10609–10620. https://doi.org/10.3748/wjg.v21.i37

102. Scheperjans F, Derkinderen P, Borghammer P (2018) The gut and Parkinson's disease: hype or hope? J Parkinsons Dis 8(s1):S31–S39. https://doi.org/10.3233/JPD-181477

103. Vidal-Martinez G, Chin B, Camarillo C et al (2019) A pilot microbiota study in Parkinson's disease patients versus control subjects, and effects of FTY720 and FTY720-mitoxy therapies in parkinsonian and multiple system atrophy mouse models. J Parkinsons Dis 10 (1):185–192. https://doi.org/10.3233/JPD-191693

104. van Kessel SP, El Aidy S (2019) Bacterial metabolites mirror altered gut microbiota composition in patients with Parkinson's disease. J Parkinsons Dis 9(s2):S359–S370. https://doi.org/10.3233/JPD-191780

105. Sampson TR, Debelius JW, Thron T et al (2016) Gut microbiota regulate motor deficits and neuron inflammation in a model of Parkinson's disease. Cell 167(6):1469–1480. https://doi.org/10.1016/j.cell.2016.11.018

106. Benitez BA, Davis AA, Jin SC et al (2016) Resequencing analysis of five Mendelian genes and the top genes from genome-wide association studies in Parkinson's disease. Mol Neurodegener 11:29. https://doi.org/10.1186/s13024-016-0097-0

107. Chartier-Harlin MC, Kachergus J, Roumier C (2004) Alpha-synuclein locus duplication as a cause of familial Parkinson's disease. Lancet 364(9):1167–1169. https://doi.org/10.1016/S0140-6736(04)17103-1

108. Polymeropoulos MH, Lavedan C, Leroy E et al (1997) Mutation in the alpha-synuclein gene identified in families with Parkinson's disease. Science 276(5321):2045–2047. https://doi.org/10.1126/science.276.5321.2045

109. Singleton AB, Farrer M, Johnson J et al (2003) Alpha-synuclein locus triplication causes Parkinson's disease. Science 302 (5646):841. https://doi.org/10.1126/science.1090278

110. Bozi M, Papadimitriou D, Antonellou R et al (2014) Genetic assessment of familial and early-onset Parkinson's disease in a Greek population. Eur J Neurol 21 (7):963–968.12. https://doi.org/10.1111/ene.12315

111. Alcalay RN, Caccappolo E, Mejia-Santana H et al (2012) Cognitive performance of GBA mutation carriers with early-onset PD: the CORE-PD study. Neurology 78 (18):1434–1440. https://doi.org/10.1212/WNL.0b013e318253d54b

112. Ptashne M (2007) On the use of the word 'epigenetic'. Curr Biol 17(7):R233–R236. https://doi.org/10.1016/j.cub.2007.02.030

113. Feng Y, Jankovic J, Wu YC (2015) Epigenetic mechanisms in Parkinson's disease. J Neurol Sci 349(1–2):3–9. https://doi.org/10.1016/j.jns.2014.12.017

114. Tan YY, Wu L, Zhao ZB et al (2014) Methylation of alpha-synuclein and leucine-rich repeat kinase 2 in leukocyte DNA of Parkinson's disease patients. Parkinsonism Relat Disord 20(3):308–313. https://doi.org/10.1016/j.parkreldis.2013.12.002

115. Margis R, Margis R, Rieder CR (2011) Identification of blood microRNAs associated to Parkinson's disease. J Biotechnol 152 (3):96–101. https://doi.org/10.1016/j.jbiotec.2011.01.023

116. Jowaed A, Schmitt I, Kaut O, Wüllner U (2010) Methylation regulates alpha-synuclein expression and is decreased in Parkinson's disease patients' brains. J Neurosci 30 (18):6355–6359. https://doi.org/10.1523/JNEUROSCI.6119-09.2010

117. Schmitt I, Kaut O, Khazneh H et al (2015) L-dopa increases alpha-synuclein DNA methylation in Parkinson's disease patients in vivo and in vitro. Mov Disord 30(13):1794–1801. https://doi.org/10.1002/mds.26319

118. Henderson-Smith A, Fisch KM, Hua J et al (2019) DNA methylation changes associated with Parkinson's disease progression: outcomes from the first longitudinal genome-wide methylation analysis. Epigenetics 14 (4):365–382. https://doi.org/10.1080/15592294.2019.1588682

119. Thompson AJ, Banwell BL, Barkhof F et al (2017) Diagnosis of multiple sclerosis: 2017 revisions of the McDonald criteria. Lancet Neurol 17(2):162–173. https://doi.org/10.1016/S1474-4422(17)30470-2

120. Wingerchuk DM, Banwell B, Bennett JL et al (2015) International consensus diagnostic criteria for neuromyelitis optica spectrum disorders. Neurology 85(2):177–189. https://doi.org/10.1212/WNL.0000000000001729

121. Zerr I, Zafar S, Schmitz M, Llorens F (2018) Cerebrospinal fluid in Creutzfeldt-Jakob disease. Handb Clin Neurol 146:115–124. https://doi.org/10.1016/B978-0-12-804279-3.00008-3

122. Bousiges O, Cretin B, Lavaux T et al (2016) Diagnostic value of cerebrospinal fluid biomarkers (phospho-Tau181, total-Tau, Abeta42, and Abeta40) in prodromal stage

of Alzheimer's disease and dementia with Lewy bodies. J Alzheimers Dis 51 (4):1069–1083. https://doi.org/10.3233/JAD-150731

123. Posavi M, Diaz-Ortiz M, Liu B et al (2019) Characterization of Parkinson's disease using blood-based biomarkers: a multicohort proteomic analysis. PLoS Med 16(10): e1002931. https://doi.org/10.1371/journal.pmed.1002931

124. Deweerdt S (2016) Parkinson's disease: 4 big questions. Nature 538(7626):S17. https://doi.org/10.1038/538S17a

Chapter 8

Metabolomic Biomarkers in Parkinson's Disease

Yaping Shao, Xiaojiao Xu, Nanxing Wang, Guowang Xu, and Weidong Le

Abstract

Metabolomics analysis has developed rapidly in recent decades and has become a powerful tool to understand comprehensive metabolic changes in biological systems. In Parkinson's disease (PD) research, great efforts have been made toward the discovery of novel biomarkers and biochemical pathways to improve diagnosis, prognosis, and therapy. With the recent achievements in metabolomics, the identification of multiple novel biomarkers for PD has been greatly improved. In this chapter, we outline a detailed metabolomics workflow including sample pretreatment, instrumental analysis, and data interpretation and summarize recent advances in technology and bioinformatics implemented in metabolomics studies. Based on the current metabolomic findings in PD research, this chapter covers the most promising metabolic biomarkers and pathway disturbances in PD and discusses the progress made toward translating these findings to clinical practice, emphasizing the potential importance of endogenous small molecular metabolites in disease research.

Key words Parkinson's disease, Metabolomics, Metabolic biomarker, Metabolic disturbance, Mass spectrometry

1 Introduction

Parkinson's disease (PD) is the second most common age-related progressive neurodegenerative disorder worldwide, mainly affecting people over 65 years [1]. With the expanding of an aging population, the incidence of PD is increasing, which imposes a great burden on the public healthcare system. Inadequate knowledge of the pathophysiological processes that governs PD entails a lack of biomarkers and reliable diagnostic tools as well as the absence of effective preventive therapies or treatments to reverse the consequences of this devastating disease. The pathological hallmarks of PD are the progressive loss of dopaminergic neurons in the substantia nigra pars compacta coupled with intracytoplasmic accumulation of abnormal α-synuclein fibrils known as Lewy bodies that the postmortem detection has been considered as the definitive diagnosis criteria for PD [2, 3]. At present, the clinical diagnosis for

Philip V. Peplow, Bridget Martinez and Thomas A. Gennarelli (eds.), *Neurodegenerative Diseases Biomarkers: Towards Translating Research to Clinical Practice*, Neuromethods, vol. 173, https://doi.org/10.1007/978-1-0716-1712-0_8,
© Springer Science+Business Media, LLC, part of Springer Nature 2022

PD is still highly dependent on clinical symptoms. Due to the late onset of motor symptoms and the heterogeneity of clinical presentations, PD is difficult to diagnose accurately, especially in its early phase [2–5]. Therefore, the discovery of biomarkers capable of monitoring its evolution and progression especially in the preclinical phases is challenging and represents a high research priority.

Biomarkers for PD could be used for different purposes, such as risk assessment, disease screening, diagnosis, monitoring disease severity, and evaluation of therapeutic efficacy [6]. A number of studies have been dedicated to the identification of potential biomarkers for PD, which derive from clinical, biochemical, imaging, and genetic observations [7, 8]. So far, the most promising biomarker for PD diagnosis may be α-synuclein. Reduced level of α-synuclein has been observed in cerebrospinal fluid (CSF) of PD patients [7, 9, 10]. It has been reported that the phosphorylated form of α-synuclein is significantly increased in the plasma of PD [10, 11]. Additionally, DJ-1, urate, and apolipoprotein A1 are also interesting candidates for PD biomarkers; however, the results are not clear and several relevant studies are still underway [6, 8, 11, 12].

Metabolomics is an emerging high-throughput technology, which can be used to elucidate disease pathogenesis and the search for potential biomarkers by comprehensively analyzing endogenous and exogenous low-molecular weight metabolites in biofluids or tissues [13]. Due to its abilities to detect slight changes in metabolic pathways and alterations in homeostasis, metabolomics is a powerful tool to characterize biological phenotype during disease development and progression [14–16]. In recent years, metabolomics has been widely applied to study neurological and neurodegenerative disorders [16]. Indeed, previous studies have demonstrated the applicability of metabolomics in the identification of biomarkers for PD diagnosis, disease onset, progression, and prognosis, as well as the discovery of novel mechanisms underlying PD [14, 17–21]. Using metabolomics and enrichment analysis tools, the PD related pathway disturbances in purine metabolism, oxidative stress, fatty acid (FFA) metabolism, amino acid metabolism, caffeine metabolism, and intestinal dyshomeostasis, have been revealed in PD.

In this chapter, an overview of the recent advancements and applications of metabolomics to investigate PD is presented. We have reviewed a detailed metabolomics workflow and methodologies including sample pretreatment, instrumental analysis, and data interpretation, and summarized recent advances in technology and bioinformatics implemented in metabolomics studies. Based on the current metabolomic findings in PD, we also summarized the most promising metabolic biomarkers in PD and discussed their importance for improving diagnosis, monitoring progression, and identifying novel targets for drug development of the disease.

2   Workflow and Technology Advancements of Metabolomics Analysis

A simplified workflow for metabolomics analysis is presented in Fig. 1 [22]. The essential procedures include scientific question determination, metabolomics strategy selection, sample pretreatment, high-throughput instrumental analysis, and data processing.

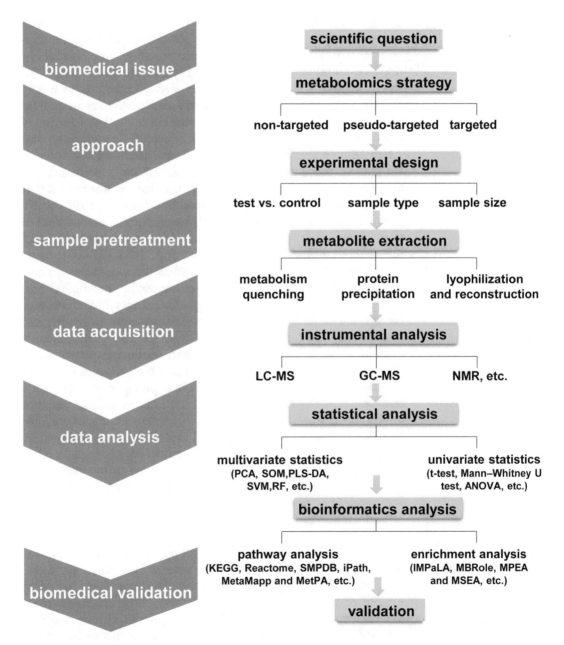

Fig. 1 Workflow for metabolomics analysis. Typically, a metabolomics analysis consists of experimental design, sample pretreatment, data acquisition, data analysis, and biomedical validation

A highly repeatable and robust analytical method for the detection of a broad coverage of metabolome is crucial to the acquisition of a meaningful metabolic data. Due to the highly diverse physicochemical properties of metabolites and complex matrices of analytical samples, sample pretreatment including extraction solvents, extraction strategies, and instrumental analysis conditions should be optimized to maximize the coverage, analytical stability, and sensitivity of the detection.

2.1 Sample Pretreatment

The choice of sample pretreatment strategy is crucial to the success of a given experiment, because these procedures dramatically influence data quality and ultimately interpretation and conclusions drawn from metabolomics analysis [23]. For the global metabolomics analysis of biofluids, simple unselective methods including "dilution-injection" and solvent precipitation are commonly used strategies, which enables high metabolite coverage. The selection of a specific approach depends on the inherent characteristics and complexity of the sample. The commonly used organic solvents for protein precipitation are acetonitrile, ethanol, methanol, and acetone or their combination [24–26]. It has been reported that precipitation with acetonitrile or acetone achieved better protein removal efficiency, whereas precipitation with methanol, ethanol or the mixture of methanol and ethanol performed better in terms of metabolite coverage and method reproducibility [27–29]. Apart from one phase solvent system, liquid–liquid extraction systems are also used to separate the polar and nonpolar metabolites into two phases, which enables achieving a broad metabolite coverage using small amounts of samples. Chloroform–methanol–water and methyl tert-butyl ether–methanol–water are two predominantly used systems [25, 30]. Different from biofluids, quenching in liquid nitrogen or homogenization at low temperatures is needed to minimize metabolism before protein precipitation for tissue samples [31]. For the sample preparation of intracellular metabolomics, the conventional procedures include quenching of metabolism and chemical cellular lysis using organic solvents, followed by sonication and centrifugation steps to precipitate proteins or cellular debris. Recently, a microfluidic device–based automated sample preparation method has been reported by combining organic solvent and bulk electroporation, which allows high-throughput, reproducible measurements [32].

2.2 Mass Spectrometry–Based Instrumental Analysis

Considering the distinct chemical and physical properties and the high dynamics of the metabolites, several instrument platforms have been developed for metabolomics data acquisition including nuclear magnetic resonance (NMR) [33], chromatography [34], and mass spectrometry (MS) [35]. Among them, chromatography coupled to MS is the most widely adopted platform in current metabolomics studies [36]. The prior separation by chromatography helps to reduce the complexity of analytical samples and improves the response and detection capability of the MS detector. Moreover, the retention time of metabolites during chromatographic separation is also an important index along with the precursor mass and fragment fingerprinting for the annotation of the metabolites [37]. Gas chromatography-MS (GC-MS) and liquid chromatography-MS (LC-MS) are two predominant techniques.

GC is a useful method to analyze volatile and thermally stable metabolites (naturally or made by derivatization) such as FFAs [38, 39], sterols [40], amino acids [41], and organic acids [42], among others. Electron ionization (EI) is a typically used ionization mode in GC-MS analysis [43]. During EI process, metabolites are dissociated into specific fragments, which can be searched by spectral library such as NIST/EPA/NIH Mass Spectral Library. Thus, it can greatly facilitate the identification of metabolites. The introduction of GC×GC strategy greatly increased the throughput and coverage of metabolites, which is constituted by two GC columns with orthogonal retention properties [44]. A GC×GC-based metabolomic platform optimized by Yu et al. was demonstrated to be able to detect 600 molecular features, 165 of which were annotated including amino acids, FFAs, lipids, carbohydrates and nucleotides [45]. However, the information of the precursor ions is usually lost during the EI process, which makes it difficult to obtain the molecular weight of the compound. Moreover, the time-consuming and laborious derivation processes hamper the high throughput and high reproducibility of the analysis.

The implementation of LC-MS could alleviate the above problem, and have been widely used in recent years [46]. According to the polarity of metabolites, reverse phase liquid chromatography (RPLC) and hydrophilic interaction liquid chromatography (HILIC) are implemented to analyze medium to high polarity metabolites, respectively [47]. The combination of RPLC and HILIC as 2D-LC could provide higher peak capacity and better resolution [48]. Recently, a novel online 3D-LC system developed by Wang et al. via coupling of preseparation and comprehensive 2D-LC using a stop-flow interface greatly improved the coverage of metabolites and facilitated large-scale metabolomics studies [49].

Untargeted and targeted methods are two typically used strategies in metabolomics studies. Untargeted methods can theoretically detect all compounds in biological samples, which provide nonbiased, high coverage of the metabolome. The typically used MS detectors are high-resolution MS (HRMS) including time of flight-MS (TOF-MS), quadrupole-TOF-MS (Q-TOF-MS), and Orbitrap-MS, among others. Targeted methods are usually applied to analyze metabolites of interest with high sensitivity, high specificity, and excellent quantification ability. Single Q-MS, QQQ-MS, QTrap-MS are typically equipped MS detectors [17]. With the development of MS technology, the detection of a large number of metabolites within a targeted LC-HRMS run is possible. Recently, new strategies such as pseudotargeted metabolomics and larger-scale targeted metabolomics quantification have been reported, which can detect hundreds to thousands of metabolites with reliable quantitative arrays similar to targeted methods [50–52]. Pseudotargeted methods provide a new alternative to the untargeted methods for large-scale metabolomics analysis.

2.3 Data Analysis of the Metabolomics Data

In order to extract valuable information from metabolomics analysis, it is necessary to use informatics techniques to analyze the acquired data. Machine learning, feature selection and functional and biological analysis are prominently incorporated techniques in this area.

Machine learning can be divided into two categories, which includes unsupervised and supervised learning techniques. Unsupervised learning method processes data without labels and identifies pattern based on data naturally, which allows to discover new classes [24]. The commonly used unsupervised learning techniques for metabolomics studies include principal component analysis (PCA), clustering and self-organizing map (SOM) [24]. Among them, PCA was the most widely used technique, which can be used to visualize similarities and or differences among known classes of the studied samples. Supervised learning is based on a set of given labeled data, this kind of classification algorithm also can be used to select features/metabolites providing the best distinction between two groups. Partial least squares (PLS), support vector machine (SVM) and random forest (RF) are the most commonly utilized supervised classification algorithms. In the area of data mining, there are three main methods for feature selection, including filter, wrapper, and embedded methods [53]. Both wrapper and embedded methods are machine learning algorithm dependent. In contrast, filter methods are independent from the classification algorithm and based on quantified scores. Therefore, filter methods such as Student's *t*-test or Mann–Whitney *U*-test nonparametric testing are widely used in metabolomics. Functional interpretation analysis is key to gaining a biological understanding of the metabolomics data. Pathway mapping and visualization and enrichment analysis are two main approaches implemented toward this purpose [54]. Pathway analysis helps to mapping candidate metabolites onto the specific metabolic pathways. A number of databases are publicly available for pathway mapping and visualization, including KEGG [55], Reactome [56], SMPDB [57], iPath [58], MetaMapp [59], and MetPA [60]. On the other hand, enrichment analysis provides information of enrichment of a group of metabolites in a predefined set of annotated functionally related metabolites or pathways [24]. The available tools for metabolite enrichment analysis are IMPaLA [61], MBRole [62], MPEA [63], and MSEA [64], among others [24].

3 Metabolomic Findings in PD

Based on clinical population and different types of PD models, a number of potential metabolic biomarkers have been identified to predict diseases risk, improve diagnosis, monitor disease progression, and provide novel perspectives for the pathogenesis and drug development of PD. In this section, we include the reported promising metabolic biomarkers in PD, mainly involved in metabolism of catecholamine, caffeine, FFA, bile acid, and tryptophan, and oxidative stress processes as well as co–interaction between host and intestinal bacteria, among others (Fig. 2).

3.1 Catecholamine Metabolism Biomarkers

The loss of dopamine (DA) neurons in the substantia nigra striatum and the production of toxic catecholamine metabolites are one of the accepted hypotheses for the occurrence and development of PD [65]. Given the marked central dopaminergic lesion, measurements

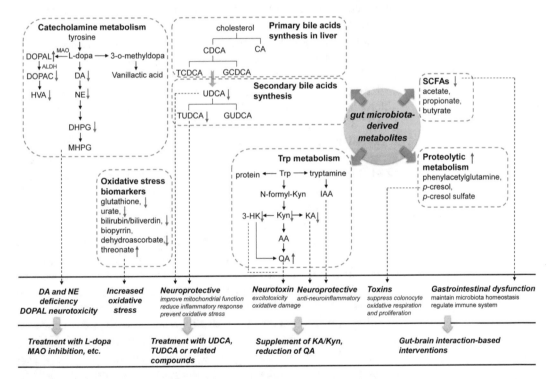

Fig. 2 Potential metabolic biomarkers of PD and their associations with the pathogenesis and therapy of the disease. Presently, the reported metabolic biomarkers are mainly catecholamines, bile acids, tryptophan catabolites, SCFAs, toxic amino acids and metabolites involved in oxidative stress. *AA* anthranilic acid, *ALDH* aldehyde dehydrogenase, *CA* cholic acid, *CDCA* chenodeoxycholic acid, *DA* dopamine, *DHPG* dihydroxyphenylglycol, *DOPAC* dihydroxyphenylacetate, *DOPAL* 3,4-dihydroxyphenylacetaldehyde, *GCDCA* glycochenodeoxycholic acid, *GUDCA* glycoursodeoxycholic acid, *3-HK* 3-hydroxykynurenine, *HVA* homovanillic acid, *IAA* indoleacetic acid, *KA* kynurenic acid, *Kyn* kynurenine, *MAO* monoamine oxidase, *MHPG* 3-methoxy-4-hydroxyphenylglycol, *NE* norepinephrine, *QA* quinolinic acid, *SCFA* short-chain fatty acid, *TCDCA* taurochenodeoxycholic acid, *Trp* tryptophan, *TUDCA* tauroursodeoxycholic acid, *UDCA* ursodeoxycholic acid

of CSF levels of DA and its metabolites have become a research priority in the investigation of biomarkers for PD. In line with the degeneration of catecholaminergic neurons, reduced levels of DA and its metabolites dihydroxyphenylacetate (DOPAC) and homovanillic acid (HVA) have been observed in the CSF of PD patients compared with controls [66, 67]. Besides, patients with PD also showed neurochemical evidence for central norepinephrine (NE) deficiency, which is mainly concentrated in locus coeruleus. Consistently, lower levels of NE and dihydroxyphenylglycol (DHPG) were reported in the CSF of PD patients compared with healthy controls [67, 68]. Autopsy also revealed that the intracranial NE and DHPG of PD patients were reduced, although the decline was not as significant as the DA metabolites [69].

In contrast to the decline of the above mentioned DA metabolites, elevated levels of 3,4-dihydroxyphenylacetaldehyde (DOPAL) have been found in the substantia nigra, putamen and caudate nucleus of PD patients compared with controls [69, 70]. DOPAL is indeed neurotoxic and can directly damage dopamine neurons in the substantia nigra by several routes, including triggering α-synuclein oligomerization, affecting lysine function and causing abnormal protein aggregation, promoting oxidative stress and mitochondrial dysfunction [71, 72]. It was reported that decreased vesicular uptake of DA and decreased DOPAL detoxification by aldehyde dehydrogenase might explain its increased level [72, 73]. Recently, elevated levels of DA and NE have been reported in the tear fluid of untreated PD patients, which were especially pronounced on the side with motor symptoms [74]. Notably, increased levels of NE in the tear fluid were also found in the preclinical and clinical stages of mouse models of PD, which suggested that NE in tears may be a promising biomarker for early diagnosis of PD [74].

The decline in CSF DA, DOPAC, HVA, and DHPG may be efficient in detecting central DA deficiency in PD, however, these metabolites might not be perfect biomarkers for the diagnosis of PD. LeWitt et al. indicated that CSF HVA might not a reliable biomarker for PD. They reported a ratio of xanthine to HVA as a possible biomarker to distinguish PD from controls, which showed significantly negative correlation with clinical scores of Unified Parkinson's Disease Rating Scale [75]. Besides, both PD and other synucleinopathies are characterized by central DA and NE deficiency, and the detections of CSF catecholamines cannot distinguish PD from other synucleinopathies such as multiple system atrophy (MSA) and pure autonomic failure (PAF) [67]. It has been revealed that the decrease of CSF DOPAC is more pronounced in PD than in PAF, whereas the decrease of CSF DHPG is more prominent in PAF than that of PD [65]. Therefore, differential patterns of DA and NE deficiency may contribute to different clinical manifestations of these diseases.

3.2 Caffeine Metabolism Biomarkers

Convergent epidemiological and preclinical data indicate that caffeine consumption is associated with a reduced risk of PD [1, 76, 77]. Several metabolomics studies noted that caffeine and its downstream metabolites were significantly lower in the serum/plasma of patients with PD than in healthy controls and that the deceased levels of these metabolites in early PD were correlated with disease progression [76, 78, 79]. Likewise, significantly decreased caffeine concentrations were observed in PD patients with motor complications compared with those without motor complications [80]. It was revealed that the reductions of caffeine and downstream metabolites had no significant correlation with habitual caffeine intake and lower serum/plasma levels of these metabolites might due to caffeine malabsorption, hypermetabolism, or increased clearance of caffeine in PD patients [76, 78, 80].

Accumulating evidence supported the neuroprotective effects of caffeine in PD, which made caffeine a promising therapeutic agent for this disease. It has been reported that both caffeine and caffeine metabolites can protect against nigral neurodegeneration, attenuate 1-Methyl-4-phenyl-1,2,3,6-tetrahydropyridine (MPTP) induced neurotoxicity and influence the onset and progression of PD in rodent PD models [80–82]. Moreover, the present experimental and preclinical data also suggested that caffeine can improve both motor deficits and nonmotor symptoms of patients with PD [83, 84]. Detailed description of the antiparkinsonian effects of caffeine and relevant metabolites have been reviewed elsewhere [82, 84]. The main explanation for the psychostimulant effects of caffeine is that it acts as an antagonist against adenosine A2A receptor that regulates the release of glutamine and dopamine [85]. Although the exact molecular mechanisms remined to be elucidated, the neuroprotection exerted by caffeine and selective A2A receptor antagonists have been extensively demonstrated. However, the potential neuroprotective effects of caffeine can be negated or even reversed in the presence of endogenous or exogenous estrogen, evidenced by both epidemiological and experimental studies [86, 87]. Considering this, it is necessary to stratify the clinical and disease modifying effects according to sex and hormonal status in the investigation of the use of caffeine and relevant agents to treat PD.

However, caffeine does not appear to be an ideal diagnostic marker for PD. Precious studies have indicated that plasma caffeine level was also decreased in patients with amyotrophic lateral sclerosis [78, 88]. A recent study reported that the reductions of caffeine and downstream metabolites were also observed in patients with progressive supranuclear palsy (PSP) and MSA, which limits the use of these metabolites as biomarkers for the differential diagnosis of these diseases [89].

3.3 Fatty Acid Metabolism Biomarkers

FFAs, as components of lipids, are essential for membrane formation, energy generation, and cell signaling transduction [90]. Increasing evidence indicated that aberrant FFA metabolism played an important role in the pathophysiology of PD. Changes in FFA metabolism also have been observed in both PD patients and different PD models such as MPTP and 6-hydroxydopamine (6-OHDA)-treated rodent models [91, 92], MPTP-induced goldfish [93], and zebrafish larvae presenting with PD-like symptoms [94]. FFA can be divided into short-chain fatty acids (SCFAs), medium-chain fatty acids (MCFAs), long-chain fatty acids (LCFAs), and very-long chain fatty acids according to the length of the carbon chains.

SCFAs are produced by gut microbiota through the fermentation of dietary fiber, which will be discussed in the following section. MCFAs and LCFAs can be obtained from diet or synthesized from other FFAs endogenously. MCFAs are crucial in maintaining the integrity of neurons and oligodendrocytes in neurological disorders. FFA with different saturation levels could exhibit different impacts on neuroinflammation and oxidative damage [90]. Using LC-MS and GC-MS approaches, Peter and colleagues found that the levels of FFAs were strongly correlated to PD progression [79]. Besides, higher levels of MCFAs (caprylate, caprate, laurate, and 5-dodecenoate) and polyunsaturated fatty acids (PUFAs) were observed in serum of PD patients carrying PARK2 mutation compared with control subjects [95]. Willkommen et al. reported increased levels of PUFAs in the CSF of PD patient compared with healthy controls [96]. However, findings from different laboratories were not always consistent. Significantly reduced levels of LCFAs, especially PUFAs with 16–22 carbon have been reported in the blood of PD patients [97, 98].

PUFAs are essential in maintaining neuronal membrane fluidity and permeability [99]. It was reported that PUFAs, including α-linolenic acid, eicosapentaenoic acid (EPA), docosapentaenoic acid, docosahexaenoic acid (DHA), and arachidonic acid, are neuroprotective and can modulate dopaminergic activity in the basal ganglia [100]. Emerging evidence showed that altered PUFAs level is associated with PD pathogenesis [18, 96, 97, 101–104]. Fan et al. identified a total of 225 differentially expressed genes (DEGs) in the blood by comparing the expression profiles of PD patients with slow and rapid disease progression rates. These DEGs were significantly enriched in pathways of FFA metabolism [102]. Higher levels of PUFAs in the brain could induce increasing oxidative stress and FFA peroxidation [90, 96]. Recent reports suggested that FFAs, particularly PUFAs, can bind to monomeric α-synuclein and accelerate the formation of α-synuclein assemblies [105, 106]. Due to the anti-inflammatory, anti-oxidative, and neuroprotective properties of n-3 PUFAs, dietary supplementation with n-3 PUFAs such as DHA and EPA has been widely used in various PD models and brought about neuroprotection [100, 107, 108]. It was revealed that a lower intake of n-3 PUFAs was associated with a higher risk of PD onset [109, 110]. Considering this, targeting FFA metabolism may represent an attractive therapeutic strategy for PD.

3.4 Tryptophan Metabolism Biomarkers

The imbalance of tryptophan (Trp) metabolism plays an important role in the pathogenesis of PD by contributing to mitochondrial disturbances, metabolic disturbances, and impairment of brain energy metabolism. Kynurenine (Kyn) pathway is the main catabolic rote of Trp, which accounts for more than 90% of Trp metabolism [111, 112]. Metabolites of Kyn pathway play crucial roles in maintaining normal brain function. In the brain, Kyn can be metabolized into the neurotoxic 3-hydroxykynurenine (3-HK) or quinolinic acid (QA) and the neuroprotective kynurenic acid (KA) [113].

Several metabolomics studies have reported a dysregulation of peripheral Kyn pathway in PD [19, 114]. Lower levels of KA/Kyn ratio and elevations of Trp, Kyn, QA and QA/KA ratio have been observed in the plasma of PD patients, and further stratification of PD patients suggested that these metabolic changes were more pronounced in patients at advanced stages [19]. It was revealed that QA/KA was a potential biomarker for detecting PD from both healthy control and patients with Huntington's disease [19]. Jesper and colleagues reported a higher 3-HK/KA ratio in plasma as a potential biomarker of L-DOPA-induced dyskinesia (LID) [115]. 3-HK is a precursor of the excitotoxin QA and it can cause oxidative damage through producing hydroxyl radicals [5]. The elevations of these ratios in PD indicated a biased Kyn pathway toward producing excess oxidative stress and excitotoxicity [19]. Likewise, accumulated levels of Kyn, xanthurenic acid and hydroxytryptophan involved in Trp metabolism were also found in the urine of idiopathic PD patients [116]. In addition, increased 3-HK concentration was also reported in the brain tissues, plasma and CSF of PD patients, which further supported the involvement of excitotoxicity in the pathogenesis of PD [76, 117, 118].

A number of studies have been dedicated to developing therapeutic intervention for PD by restoring the altered Kyn metabolism. One strategy is to increase the level of KA. Several in vitro and in vivo studies demonstrated that the supplement of KA could attenuate 1-Methyl-4-phenylpyridinium (MPP^+)-induced neuronal cell death and toxic effects [111, 119]. Furthermore, synthetic derivatives of both Kyn and KA have also been designed to improve pharmacological properties, and have been reviewed in detail previously [119–121]. Another potential pharmacological approach could be regulation of intrinsic activities of enzymes involved in Trp metabolism to increase the level of neuroprotective intermediates and decrease those of potential toxic ones. It was reported that nicotinylalanine-induced elevated level of Kyn by inhibiting the activities of kynureninase and kynurenine 3-monooxygenase, could protect against QA toxicity [112, 122]. Phytochemicals, notably phenolic compounds that contain a hydroxyl functional group, are also promising new treatments for PD by reducing striatal lesion size, reducing inflammation and preventing lipid peroxidation caused by QA [123].

3.5 Bile Acid Biomarkers

Bile acids are amphipathic steroid acids derived from cholesterol. They are widely known for their role as detergents to expedite the digestion and absorption of dietary lipids and lipid-soluble nutrients [124, 125]. Bile acids also function as signaling molecules, participating as ligands in both nuclear and cell-surface receptors and ion channel and regulating glucose and lipid metabolism [125, 126]. To date, a few studies have implicated bile acids in neurodegenerative disorders and suggested a possible neuroprotective role in neurons and the central nervous system [126, 127].

Generally, most of the studies have focused on the neuroprotection and the underlying molecular mechanism of ursodeoxycholic acid (UDCA) and its derivatives in PD. It has been reported that UDCA could reverse the LRRK2^{G2019S}-induced mitochondrial abnormalities in both LRRK2^{G2019S} fly model and fibroblasts from patients with the LRRK2^{G2019S} mutation, which might be dependent on the activation of glucocorticoid receptor and increased phosphorylation of Akt [124, 128, 129]. Likewise, it was found that UDCA treatment could prominently improve motor performance, prevent the decline of striatal dopamine content, improve mitochondrial function and reduce the inflammatory response in a rotenone-induced PD rat model [127]. These salutary effects open new horizons to its therapeutic intervention in PD. Tauroursodeoxycholic acid (TUDCA), a taurine-conjugated product of UDCA, also has been found to possess neuroprotective properties. It has been reported that TUDCA treatment could prevent the decline of dopaminergic fibers and ATP levels, ameliorate mitochondrial dysfunction and neuroinflammation in mouse models of PD [126, 130–132]. Moreover, TUDCA administration, either before or after MPTP, significantly improved motor symptoms in the MPTP-treated mouse model [126]. It was revealed that TUDCA can prevent both MPP$^+$ and α-synuclein-induced oxidative stress through modulation of the Nrf2 signaling pathway in both in vivo and in vitro studies [133]. In the perspective of future therapies, UDCA and TUDCA may represent promising therapeutic agents that might prevent further neurodegeneration, by halting disease progress after the initial diagnosis of PD. It is worth mentioning that significantly reduced levels of ω-muricholic acid (MCAo), TUDCA and UDCA were documented in a prodromal PD model, which demonstrated the potential of bile acids for the prediction of patients at greatest risk of developing PD, particularly in the prodromal phase [134].

Bile acids also have been reported to be implicated in the pathogenesis of PD through pathway-based genome-wide association studies (GWAS). In a meta-analysis of GWAS data based on the genotypes of PD patients, three pathways including primary bile acid biosynthesis were identified to be significantly associated with PD, providing further genetic evidence for the important role of bile acids in the disease [135]. A recently published study indicated that the intestinal dysbiosis-induced bile acid abnormalities might explain the lipid dysregulation in PD patients [136]. By integrating microbiome analysis and longitudinal metabolome data in PD populations, Johannes and colleagues highlighted that taurine-conjugated bile acids are stable markers of variability in the severity of motor symptoms and low levels of sulfated taurolithocholate were associated with PD incidence in the general population [137]. A newly published study reported that the fecal bile acids were significantly increased in the SN with an adeno-associated virus (AAV) α-synuclein overexpressing rat model of PD, which highlighted the crosslink between brain and gut microbiota as an important mechanism in PD pathology [138].

In spite of this, bile acids are a relatively large group of structurally similar molecules, and knowledge of the potential efficacy of other bile acids and the specific role of circulating or endogenous bile acids derived from CNS are still limited. Determining the specific role and its precise molecular mechanism of each bile acid in PD will provide novel insights into future therapeutic strategies.

3.6 Oxidative Stress Metabolic Biomarkers

Several molecular mechanisms of PD pathogenesis have highlighted the important role of oxidative stress in PD. Impaired antioxidative capacities might make cells more vulnerable to free radical damage. Glutathione is one of the major antioxidative peptides, which is considered to be a marker of oxidative damage [139]. The level of glutathione was found to be decreased in both plasma and CSF of PD patients, which indicated a diminished protection against oxidative stress in PD [118, 140]. Mutations in the DJ-1 gene PARK7 are associated with a familial autosomal recessive form of PD [5]. Meiser and colleagues demonstrated that loss of DJ-1 resulted in decreased glutamine influx and serine biosynthesis, which affected the de novo synthesis of glutathione and provoked oxidative damage in DJ-1 deficient neuronal cells [141]. Additionally, reduced glutathione concentrations were also reported previously in the substantia nigra of PD patients and MPTP-treated rats and monkey models [142–144]. These findings provided further evidences that the utilization of antioxidant glutathione for free radical detoxification was activated in the brain of PD.

Urate, another important endogenous antioxidant, was also found at a reduced level in the plasma and CSF of PD patients [17, 145, 146]. It has been revealed that a higher level of urate might be associated with a lower risk and a slower progression of PD [147]. Generally, the bases in DNA/RNA can be hydroxylated and oxidized under the attack of reactive oxidative species, the resulting products 8-hydroxy-2-deoxyguanosine (8-OHdG) and 8-hydroxyguanosine (8-OHG) are regarded as indicators of oxidative DNA/RNA damage [5, 17]. Significant elevations of 8-OHdG and 8-OHG were documented in the CSF of PD patients [148, 149]. Moreover, an elevated level of 8-OHdG was also reported in the serum and urine of PD patients as well as the SN of the brain tissues [17, 150, 151]. It was indicated that urinary 8-OHdG might be a promising biomarker for PD discrimination and progression [152, 153].

Heme oxygenases (HOs) are considered as dynamic sensors of cellular stress, which are responsible for the degradation of heme to bilirubin/biliverdin, free iron, and carbon monoxide [78, 154]. In PD, HO-1 was markedly overexpressed in astrocytes of the substantia nigra and was identified as a component of Lewy bodies [78, 154]. A decrease of bilirubin–biliverdin ratio was observed in the serum of PD patients, which implied an increased systemic oxidative stress [78]. Urinary biopyrrin, an oxidative product of bilirubin, was identified as a novel biomarker for the prediction of PD [155].

A lower level of dehydroascorbic acid and a higher level of threonic acid (an ascorbic acid catabolite) were also reported, which further demonstrated an increased oxidative stress response in PD [156, 157]. In addition, given that membrane phospholipid, comprising of esterified PUFAs, was a major cellular target for free radical damage, decreased levels of PUFAs and cholesterols in the brain of MPTP-treated PD models might be directly ascribed to the increased oxidative stress [93].

3.7 Gut Microbiota-Derived Metabolic Biomarkers

The critical role of the enteric microbiota in bidirectional gut–brain interactions in PD has been increasingly recognized. It has been postulated that α-synuclein pathology spreads from gut to the brain although the exact etiology of PD is still unclear [158]. Recently, a number of studies have reported an altered gut microbiome in PD compared with healthy controls. The microbiota-derived metabolites including bile acids, Trp catabolites, proteolytic metabolism products, SCFAs. and neurotransmitters may act as crucial signaling molecules for the microbial effects on gut–brain interactions [159]. An overview of the gut microbiota–derived metabolic alterations in PD patients is given in Table 1.

Significantly reduced fecal SCFA concentrations including acetate, propionate, and butyrate were reported in patients with PD using GC-based quantitative analysis [160]. The reduction of SCFAs might contribute to the gastrointestinal dysmotility in PD. Using ^{1}H nuclear magnetic resonance spectroscopy, Shiek and collogues found a reduced level of acetate in the blood of PD

Table 1
Gut microbiota-derived metabolic alterations in PD patients compared with healthy controls

Analytical platform	Subjects	Differential metabolites/ metabolic pathways	Sample type	References
GC	PD ($n = 34$); control ($n = 34$)	Decreased acetate, propionate, and butyrate	Feces	Unger et al. [160]
1H-NMR	Drug-naïve PD ($n = 43$); healthy control ($n = 37$)	Decreased acetate	Plasma	Ahmed et al. [161]
GC-MS	PD ($n = 38$); normal control ($n = 33$)	Increased acetate	Plasma	Shin et al. [162]
NMR	PD ($n = 76$); healthy control ($n = 37$)	Increased acetate, propionate	Saliva	Kumari et al. [163]
RPLC-MS HILIC-MS	PD ($n = 75$); control ($n = 50$)	Increased p-cresol, p-cresol sulfate and phenylacetylglutamine	Serum	Cirstea et al. [166]
LC-MS GC-MS	Idiopathic PD ($n = 35$); healthy control ($n = 15$)	Decreased indoleacetic acid	Serum	Hatano et al. [78]
LC-MS GC-MS	Idiopathic PD ($n = 92$); normal control ($n = 65$)	Increased indoleacetic acid	urine	Luan et al. [116]
LC-MS	PD ($n = 106$); normal control ($n = 104$)	Increased indoleacetic acid and tryptamine	Urine	Luan et al. [172]
LC-MS	PARK2 PD ($n = 15$); healthy control ($n = 19$)	Decreased benzoate metabolites (hippurate, 3-hydroxyhippurate, catechol sulfate, guaiacol sulfate, 3-methyl catechol sulfate, etc.)	Serum	Okuzumi et al. [168]
Targeted approach	drug-naïve PD ($n = 30$); control ($n = 30$)	Transsulfuration pathway (increased homoserine, methionine, serine, and α-aminobutyrate), bile acid metabolism	Plasma	Hertel et al. [137]

patients [161]. However, a case–control study using GC-MS approach observed a higher plasma level of acetate in PD compared with controls after correction with covariates. They also indicated that the levels of SCFAs were associated with age, disease severity, and antiparkinsonian medications [162]. It was assumed that the contrary results in feces and plasma might result from leakage of intestinal SCFAs caused by altered microbiota-induced subclinical inflammation. Elevated levels of salivary acetate, propionate were also reported in PD, and concentrations of propionate and butyrate were correlated with disease duration and Hoehn and Yahr Disability Scale (H&Y) stage respectively [163]. SCFAs can be absorbed in the gut and capable of maintaining microbiota homeostasis and regulating the immune system [164]. Significant decrease of SCFAs producers such as *Lachnospiraceae* family, *Faecalibacterium* species, and *Bacteroides* species were also observed in PD [158, 165].

Recently, Mihai and collogues integrated microbiota sequencing and serum metabolomics to investigate the associations between microbiota composition, gastrointestinal (GI) features and systemic microbial metabolites in PD. They found a pathway shift from carbohydrate fermentation toward proteolytic metabolism in the gut of PD patients, which might be mechanistically related to GI dysfunction [166]. Phenylacetylglutamine, *p*-cresol, and *p*-cresol sulfate, by-products of protein degradation by intestinal bacteria, were significantly increased in PD patients [166]. These bioactive metabolites are considered to be detrimental, especially *p*-cresol, which might suppress colonocyte oxidative respiration and proliferation [167]. In contrast, the level of catechol sulfate was decreased in PD patients, which may be due to the higher abundance of bacteria relevant to the benzoate degradation pathway [168–170]. It has been indicated that higher bacterial metabolism of catechol might be associated with inflammation [171]. Therefore, gut–brain interaction–based interventions might provide benefit for GI symptom management and perhaps even disease treatment and prevention.

Apart from Kyn metabolites, which have been detailed in the previous section, other Trp catabolites including indoleacetic acid (IAA) and tryptamine were also dysregulated in PD. Decreased serum level and increased urine level in IAA were observed in PD patients [78, 116, 168, 172]. The CSF level of IAA was reported to be correlated to the disease progression of PD [79]. Congruously, increased urinary tryptamine was also documented in PD patients [172]. IAA can be generated via the deamination of tryptamine, both of which are strongly dependent on *Ruminococcus gnavus* and *Clostridium sporogenes* [158, 173]. It has been demonstrated that IAA could modulate liver inflammatory responses through attenuating the release of pro-inflammatory cytokines and the cytokine-mediated upregulation of lipogenesis in macrophages and hepatocytes [174]. Moreover, a study reported that indole alkaloid derivatives including IAA showed anti-neuroinflammatory effects in lipopolysaccharide-stimulated BV2 cells [175]. Therefore, the dysregulated IAA might be relevant to inflammation in PD.

By integrating longitudinal metabolome and metagenomics data of gut microbiota in a PD cohort, a recent study indicated that the host–microbial transsulfuration cycle driven by the continuous removal of taurine by *B. wadsworthia* and sulfur metabolites generated by *A. muciniphila* and *B. wadsworthia* may contribute to PD severity [137]. Accordingly, the metabolites including homoserine, methionine, serine, and α-aminobutyrate involved in transsulfuration pathway were increased in the plasma of PD patients. Taken together, investigating the cross-link between intestinal bacteria and gut function might open new horizons in the etiology, pathophysiology, and therapeutic strategies of the disease.

3.8 Others

In addition to the metabolites mentioned in the above sections, several other metabolites have been reported as biomarkers for PD, mainly involved in glucose metabolism, energy metabolism, lipid metabolism, inflammatory process, and so on.

A metabolic shift from mitochondrial oxidative phosphorylation to glycolysis was observed in both clinical and animal models of PD [161, 176]. Deprivations of citrate, succinate and malate as well as accumulation of pyruvate have been noticed in the plasma of PD patients, which indicated an insufficient energy production and impaired mitochondrial function in PD [161]. It has been accepted that PD is a multifactorial disease with a complex pathogenesis including aging, genetic factors, and environmental exposures [177]. Exposure to environmental toxins such as pesticides has been considered an important risk factor for PD [178]. It has been revealed that the overexpression of α-synuclein and paraquat treatment worked synergistically in promoting dopaminergic cell death via stimulating glucose uptake/transportation and AMPK activity, and directing the glucose metabolism toward the pentose phosphate pathway to supply paraquat redox cycling [177, 179]. The inhibition of glucose metabolism/transport and the pentose phosphate pathway could alleviate the induced toxic synergism [178]. A recent study indicated that the decreased activity of the glycolytic enzyme glyceraldehyde-3-phosphate dehydrogenase (GAPDH) by its binding to the monomeric and oligomeric forms of α-synuclein might be one of the causes of impaired glucose metabolism in PD [180]. Furthermore, deficiency in the enzyme phosphoglycerate kinase 1 (PGK1), the first ATP generating enzyme in the glycolysis pathway, was associated with PD [181]. A number of studies have demonstrated that terazosin, which can bind to and enhance the activity of PGK1, could attenuate neurodegeneration and improve clinical symptoms in both animal models and PD patients [181–183].

Significantly increased levels of acylcarnitines were documented in the serum of idiopathic early-stage PD patients [169]. Receiver operator curve assessment confirmed that the combination of nine acylcarnitines could achieve good predictive accuracy for early-stage PD. Besides, significant associations were identified between acylcarnitines and mild cognitive impairment as well as the expression of serological brain-derived neurotrophic factor levels. Acylcarnitines, especially medium and long chain acylcarnitines, are important components for the transportation of FFA into mitochondria for β-oxidation [172]. Consistently, significantly increased levels of acylcarnitines were found in the urine of PD patients [116, 172]. However, decreased levels of seven long-chain acylcarnitines were also reported in the plasma of PD [184]. Despite the inconsistent results, all these studies highlighted the important role of FFA β-oxidation in the PD pathophysiology.

Additionally, dysregulated levels of corticosterone, N-acetylaspartate [144, 185], lipids [14, 92, 186, 187], inositol metabolites [93, 103, 161] and purines [188] were also reported in PD. Marked reduction of corticosterone was found in the CSF of PD patients. It was revealed that corticosteroids in the brain might help to regulate chronic inflammatory responses [118]. The level of N-acetylaspartate was decreased in the substantia nigra and striatum of rodent PD models and PD patients [144, 189]. N-acetylaspartate is a recognized marker of neuronal integrity, and the reduction of its concentration in PD may be attributed to the loss of dopaminergic neurons [144, 190]. An abnormally high myoinositol level has been reported in the plasma of PD patients and the brain of MPTP-induced PD goldfish model [93, 161]. The accumulated myo-inositol might impair the osmoregulation process and membrane metabolism of the cell and directly affect neuronal development, survival, and function. A number of studies suggested that altered lipid homeostasis is part of PD pathogenesis underlying the α-synuclein aggregation process. In general, reduced levels of lipids were documented in the brain of PD models, while the levels of monooxygenated cardiolipins (CLs) and several lysophosphatidylcholines were significantly increased [187]. The altered PUFA-CLs and oxidized CLs suggested possible mitophagy and apoptosis processes in the development of PD [17, 187]. The important role of lipids in modulating α-synuclein toxicity and neurodegeneration in PD has been reviewed recently [191–193]. Therefore, we will not elaborate in detail here. Generally, the interaction of α-synuclein as multimers with synaptic vesicle membranes composed of phospholipids and other lipids is crucial to the formulation of α-synuclein aggregates and the resultant pathological α-synuclein conversion [192]. Recently, Sergio et al. indicated that disturbances in lipid metabolism might directly affect the autophagy pathway and thereby have an impact on the neuronal and synaptic function, causing the aggregation of dysfunctional proteins and further promoting neurodegeneration [191]. Therefore, the concept of controlling the formation of α-synuclein aggregates by regulating lipid metabolism might be a promising therapeutic strategy for PD.

4   Conclusions and Future Perspectives

Presently no biomarkers have been used in clinical practice for PD diagnosis or prognosis, given the many factors that confuse interpretation of most biomarker levels. Nevertheless, very promising findings have been achieved using metabolomics analysis, which opens a new perspective for biomarkers research. The overview of the promising biomarkers for PD diagnosis and progression has been summarized in Table 2. These metabolic biomarkers have been reported in terms of a single metabolite or a metabolite panel. In general, metabolites of catecholamine and caffeine metabolism seem not specific to PD, because such alterations have been also noted in other neurodegenerative disorders such as MSA, PSP, and others. It is likely that 8-OHdG is a stable marker of oxidative damage in PD and has the potential to monitor the disease progression. Furthermore, the current findings suggest that Kyn metabolites or their ratios might be promising biomarkers for the diagnosis of PD. In addition, the detection of Kyn metabolites may also provide a clue to monitor treated PD patients for increasing risk of LID. Moreover, it is worth noting that the ratio of KA/QA may have a potential to discriminate PD from healthy controls, and other neurodegenerative diseases. Besides, detection of bile acids and its metabolites may help for the early diagnosis of PD. It seems that multiple metabolic pathways are usually involved in PD pathogenesis, therefore, the combination of several metabolites derived from different metabolic pathways using mathematical models such as binary logistic regression, partial least squares discriminant analysis (PLS-DA), and random forest might prove to be effective strategies to identify novel biomarkers.

Despite this, there is much work needed before metabolomics can be implemented in the clinic. A majority of studies were based on relatively small sample size, some of which have reported very controversial results. In future studies, it will be important to confirm the candidates of biomarkers using targeted approaches in larger-scale clinical cohorts. From the perspective of analytical methods, it is necessary to establish standardized workflow from sample collection and pretreatment, instrumental analysis to data processing techniques, which will maximize the reduction of systematic deviations across different laboratories and different analytical batches. Due to the high sensitivity of metabolic levels to various external factors such as diets, genetic background, disease history, and lifestyle, the exact metabolic changes that result from the disease itself might be obscured. Therefore, the possible influences from these factors should be taken into account at the beginning of the study. Also, due to the very similar symptoms, it is difficult to distinguish PD from atypical Parkinsonian syndromes such as PSP and MSA even by experienced neurologists [11, 194–196]. Future studies to investigate the metabolic differences among these diseases are highly recommended, which might provide novel insights into the differential diagnosis of these diseases.

Table 2
Overview of potential metabolic biomarkers in PD diagnosis and progression

Analytical platform	Sample type	Metabolic biomarker	Statistics	PD diagnosis/differential diagnosis	References
ELISA	Urine	8-OHdG (↑)	Student t-test	Monitor PD progression Not suitable for early-stage PD	Sato et al. [153]
Untargeted metabolomics ELISA	Urine	Biopyrrin (↑)	AUC = 0.950–0.980	Distinguish idiopathic PD from HC	Luan et al. [155]
LC-MS GC-MS	Urine	A metabolite panel consisting of 18 metabolites	AUC = 0.870 AUC = 0.990 AUC = 1.000	Distinguish early-stage PD from HC Distinguish mid-stage PD from HC Distinguish late-stage PD from HC	Luan et al. [116]
LC-MS	Serum	N8-acetyl spermidine (↑)	OPLS-DA	Distinguish PD from HC and increased with PD progression	Roede et al. [149]
LC-MS	Serum	Hexanoylglutamine, decanoylcarnitine, myristoleoylcarnitine, octanoylcarnitine, oleoylcarnitine, palmitoleoylcarnitine, suberoylcarnitine, octadecanedioate, and 3-hydroxysebacate	AUC = 0.857	Distinguish idiopathic early-stage PD from HC	Burte et al. [169]
LC-MS	Serum	Caffeine and its main metabolites	AUC = 0.870	Distinguish PD from HC	Fujimaki et al. [80]
LC-MS	Serum	TLCA, GCDCA, TUDCA	AUC = 0.906	Distinguish PD in prodromal phase	Graham et al. [134]
LC-MS	Serum	A 5-metabolite panel An 8-metabolite panel	AUC = 0.955 AUC = 0.862	Distinguish PD from HC Distinguish PD with no dementia from PD with incipient dementia	Han et al. [76]
LC-MS	Plasma	QA/KA (↑)	AUC = 0.856 AUC = 0.837 AUC = 0.784	Discriminate PD from HC Discriminate PD from Huntington's disease Discriminate advanced PD from early PD	Chang et al. [19]

Method	Sample	Metabolites	Statistics	Purpose	Reference
LC-MS	Plasma	3-HK/KA (↑)		Monitor L-DOPA-treated PD patients for increasing risk of LID	Havelund et al. [115]
LCECA	Plasma	8-OHdG (↑), glutathione (↑), urate (↓)	PLS-DA	Discriminate PD from HC	Bogdanov et al. [140]
LC-MS and CE-MS	Plasma	Long-chain acylcarnitines (↓)	AUC = 0.846–0.895	Discriminate PD from HC	Saiki et al. [184]
LCECA	CSF	DOPAC (↓)	Sensitivity = 89% Specificity = 80%	Distinguish PD from HC, but cannot separate PD from MSA	Goldstein et al. [67]
LCECA	CSF	Xanthine/HVA (↑)	t-test	Discriminate PD from HC	LeWitt et al. [75]
GC-MS	CSF	Threonate (↑), Mannose (↑), Fructose (↑)	AUC = 0.800–0.833	Discriminate early-stage PD from HC	Trezzi et al. [157]
LC-MS	CSF	A 19-metabolite panel	AUC = 0.790–0.930	Discriminate PD from HC	LeWitt et al. [118]
NMR	Saliva	Histidine, propionate, tyrosine, isoleucine, acetoin, N-acetylglutamate, acetoacetate, and valine	AUC = 0.670–0.720	Discriminate PD from HC	Kumari et al. [163]

AUC area under the curve, CE-MS capillary electrophoresis- mass spectrometry, CSF cerebrospinal fluid, DOPAC dihydroxyphenylacetate, ELISA enzyme-linked immunosorbent assay, GCDCA glycochenodeoxycholic acid, GC-MS gas chromatography-mass spectrometry, HC healthy control, 3-HK 3-hydroxykynurenine, HVA homovanillic acid, KA kynurenic acid, LCECA liquid chromatography coupled with electrochemical coulometric array detection, LC-MS liquid chromatography-mass spectrometry, LID levodopa-induced dyskinesia, MSA multiple system atrophy, NMR nuclear magnetic resonance, 8-OHdG 8-hydroxy-2-deoxyguanosine, OPLS-DA orthogonal partial least squares discriminant analysis, PD Parkinson's disease, PLS-DA partial least squares discriminant analysis, QA quinolinic acid, TLCA taurolithocholic acid, TUDCA tauroursodeoxycholic acid

Acknowledgments

This work was supported by National Natural Science Foundation of China (NSFC 81771521), Key Research & Development Plan of Liaoning Science and Technology Department (2018225051) and Doctoral Scientific Research Foundation of Liaoning Science and Technology Department (2020-BS-200).

References

1. Ascherio A, Schwarzschild MA (2016) The epidemiology of Parkinson's disease: risk factors and prevention. Lancet Neurol 15 (12):1257–1272. https://doi.org/10.1016/S1474-4422(16)30230-7

2. Kalia LV, Lang AE (2015) Parkinson's disease. Lancet 386(9996):896–912. https://doi.org/10.1016/s0140-6736(14)61393-3

3. Hayes MT (2019) Parkinson's disease and parkinsonism. Am J Med 132(7):802–807. https://doi.org/10.1016/j.amjmed.2019.03.001

4. Gershanik OS (2017) Past, present, and future of Parkinson's disease. Mov Disord 32 (9):1263. https://doi.org/10.1002/mds.27113

5. Andersen AD, Binzer M, Stenager E, Gramsbergen JB (2017) Cerebrospinal fluid biomarkers for Parkinson's disease—a systematic review. Acta Neurol Scand 135(1):34–56. https://doi.org/10.1111/ane.12590

6. Cova I, Priori A (2018) Diagnostic biomarkers for Parkinson's disease at a glance: where are we? J Neural Transm 125 (10):1417–1432. https://doi.org/10.1007/s00702-018-1910-4

7. Emamzadeh FN, Surguchov A (2018) Parkinson's disease: biomarkers, treatment, and risk factors. Front Neurosci 12:612. https://doi.org/10.3389/fnins.2018.00612

8. Delenclos M, Jones DR, McLean PJ, Uitti RJ (2016) Biomarkers in Parkinson's disease: advances and strategies. Parkinsonism Relat Disord 22(Suppl 1):S106–S110. https://doi.org/10.1016/j.parkreldis.2015.09.048

9. Goldman JG, Andrews H, Amara A, Naito A, Alcalay RN, Shaw LM, Taylor P, Xie T, Tuite P, Henchcliffe C, Hogarth P, Frank S, Saint-Hilaire MH, Frasier M, Arnedo V, Reimer AN, Sutherland M, Swanson-Fischer C, Gwinn K, Fox Investigation of New Biomarker D, Kang UJ (2018) Cerebrospinal fluid, plasma, and saliva in the BioFIND study: relationships among biomarkers and Parkinson's disease features. Mov Disord 33 (2):282–288 https://doi.org/10.1002/mds.27232

10. Htike TT, Mishra S, Kumar S, Padmanabhan P, Gulyas B (2019) Peripheral biomarkers for early detection of Alzheimer's and Parkinson's diseases. Mol Neurobiol 56 (3):2256–2277. https://doi.org/10.1007/s12035-018-1151-4

11. Ren R, Sun Y, Zhao X, Pu X (2015) Recent advances in biomarkers for Parkinson's disease focusing on biochemicals, omics and neuroimaging. Clin Chem Lab Med 53 (10):1495–1506. https://doi.org/10.1515/cclm-2014-0783

12. Chen-Plotkin AS, Albin R, Alcalay R, Babcock D, Bajaj V, Bowman D, Buko A, Cedarbaum J, Chelsky D, Cookson MR, Dawson TM, Dewey R, Foroud T, Frasier M, German D, Gwinn K, Huang X, Kopil C, Kremer T, Lasch S, Marek K, Marto JA, Merchant K, Mollenhauer B, Naito A, Potashkin J, Reimer A, Rosenthal LS, Saunders-Pullman R, Scherzer CR, Sherer T, Singleton A, Sutherland M, Thiele I, van der Brug M, Van Keuren-Jensen K, Vaillancourt D, Walt D, West A, Zhang J (2018) Finding useful biomarkers for Parkinson's disease. Sci Transl Med 10(454). https://doi.org/10.1126/scitranslmed.aam6003

13. Nicholson JK, Lindon JC (2008) Systems biology: metabonomics. Nature 455 (7216):1054–1056. https://doi.org/10.1038/4551054a

14. Stoessel D, Schulte C, Teixeira Dos Santos MC, Scheller D, Rebollo-Mesa I, Deuschle C, Walther D, Schauer N, Berg D, Nogueira da Costa A, Maetzler W (2018) Promising metabolite profiles in the plasma and CSF of early clinical Parkinson's disease. Front Aging Neurosci 10:51. https://doi.org/10.3389/fnagi.2018.00051

15. Peng B, Li H, Peng XX (2015) Functional metabolomics: from biomarker discovery to metabolome reprogramming. Protein Cell 6 (9):628–637. https://doi.org/10.1007/s13238-015-0185-x

16. Ibanez C, Cifuentes A, Simo C (2015) Recent advances and applications of metabolomics to investigate neurodegenerative diseases. Int Rev Neurobiol 122:95–132. https://doi.org/10.1016/bs.irn.2015.05.015

17. Shao Y, Le W (2019) Recent advances and perspectives of metabolomics-based investigations in Parkinson's disease. Mol Neurodegener 14(1):3. https://doi.org/10.1186/s13024-018-0304-2

18. Havelund JF, Heegaard NHH, Faergeman NJK, Gramsbergen JB (2017) Biomarker research in Parkinson's disease using metabolite profiling. Metabolites 7(3). https://doi.org/10.3390/metabo7030042

19. Chang KH, Cheng ML, Tang HY, Huang CY, Wu YR, Chen CM (2018) Alternations of metabolic profile and kynurenine metabolism in the plasma of Parkinson's disease. Mol Neurobiol 55(8):6319–6328. https://doi.org/10.1007/s12035-017-0845-3

20. Mesa-Herrera F, Taoro-Gonzalez L, Valdes-Baizabal C, Diaz M, Marin R (2019) Lipid and lipid raft alteration in aging and neurodegenerative diseases: a window for the development of new biomarkers. Int J Mol Sci 20 (15). https://doi.org/10.3390/ijms20153810

21. Kim A, Nigmatullina R, Zalyalova Z, Soshnikova N, Krasnov A, Vorobyeva N, Georgieva S, Kudrin V, Narkevich V, Ugrumov M (2019) Upgraded methodology for the development of early diagnosis of Parkinson's disease based on searching blood markers in patients and experimental models. Mol Neurobiol 56(5):3437–3450. https://doi.org/10.1007/s12035-018-1315-2

22. Sussulini A (2017) Metabolomics: from fundamentals to clinical applications. Springer

23. Zheng X, Yu J, Cairns TC, Zhang L, Zhang Z, Zhang Q, Zheng P, Sun J, Ma Y (2019) Comprehensive improvement of sample preparation methodologies facilitates dynamic metabolomics of Aspergillus niger. Biotechnol J 14(3):1800315. https://doi.org/10.1002/biot.201800315

24. Kusonmano K, Vongsangnak W, Chumnanpuen P (2016) Informatics for metabolomics. Adv Exp Med Biol 939:91–115. https://doi.org/10.1007/978-981-10-1503-8_5

25. Ren S, Shao Y, Zhao X, Hong CS, Wang F, Lu X, Li J, Ye G, Yan M, Zhuang Z, Xu C, Xu G, Sun Y (2016) Integration of metabolomics and transcriptomics reveals major metabolic pathways and potential biomarker involved in prostate cancer. Mol Cell Proteomics 15(1):154–163. https://doi.org/10.1074/mcp.M115.052381

26. Shao Y, Ye G, Ren S, Piao HL, Zhao X, Lu X, Wang F, Ma W, Li J, Yin P, Xia T, Xu C, Yu JJ, Sun Y, Xu G (2018) Metabolomics and transcriptomics profiles reveal the dysregulation of the tricarboxylic acid cycle and related mechanisms in prostate cancer. Int J Cancer 143(2):396–407. https://doi.org/10.1002/ijc.31313

27. Bruce SJ, Tavazzi I, Parisod VR, Rezzi S, Kochhar S, Guy PA (2009) Investigation of human blood plasma sample preparation for performing metabolomics using ultrahigh performance liquid chromatography/mass spectrometry. Anal Chem 81(9):3285–3296

28. Polson C, Sarkar P, Incledon B, Raguvaran V, Grant R (2003) Optimization of protein precipitation based upon effectiveness of protein removal and ionization effect in liquid chromatography–tandem mass spectrometry. J Chromatograph B Anal Technol Biomed Life Sci 785(2):263–275

29. Want EJ, O'Maille G, Smith CA et al (2006) Solvent-dependent metabolite distribution, clustering, and protein extraction for serum profiling with mass spectrometry. Anal Chem 78(3):743–752

30. Chen S, Hoene M, Li J, Li Y, Zhao X, Haring HU, Schleicher ED, Weigert C, Xu G, Lehmann R (2013) Simultaneous extraction of metabolome and lipidome with methyl tert-butyl ether from a single small tissue sample for ultra-high performance liquid chromatography/mass spectrometry. J Chromatogr A 1298:9–16. https://doi.org/10.1016/j.chroma.2013.05.019

31. Rammouz RE, Létisse F, Durand S, Portais JC, Moussa ZW, Fernandez X (2010) Analysis of skeletal muscle metabolome: evaluation of extraction methods for targeted metabolite quantification using liquid chromatography tandem mass spectrometry. Anal Biochem 398(2):169–177

32. Filla LA, Sanders KL, Filla RT, Edwards JL (2016) Automated sample preparation in a microfluidic culture device for cellular metabolomics. The Analyst 141(12):3858-3865. https://doi.org/10.1039/c6an00237d

33. Silva RA, Pereira TCS, Souza AR, Ribeiro PR (2020) H-1 NMR-based metabolite profiling for biomarker identification. Clinica Chimica Acta 502:269–279. https://doi.org/10.1016/j.cca.2019.11.015

34. Witting M, Böcker S (2020) Current status of retention time prediction in metabolite identification. J Sep Sci 43(9–10):1746–1754. https://doi.org/10.1002/jssc.202000060

35. Liu X, Zhou L, Shi X, Xu G (2019) New advances in analytical methods for mass spectrometry-based large-scale metabolomics study. TrAC Trends Anal Chem 121:115665. https://doi.org/10.1016/j.trac.2019.115665

36. Sun QS, Fan TWM, Lane AN, Higashi RM (2020) Applications of chromatography-ultra high-resolution MS for stable isotope-resolved metabolomics (SIRM) reconstruction of metabolic networks. Trac-Trends Anal Chem 123:9. https://doi.org/10.1016/j.trac.2019.115676

37. Kind T, Wohlgemuth G, Lee DY, Lu Y, Palazoglu M, Shahbaz S, Fiehn O (2009) FiehnLib: mass spectral and retention index libraries for metabolomics based on quadrupole and time-of-flight gas chromatography/mass spectrometry. Anal Chem 81(24):10038–10048

38. Lin S, Liu N, Yang Z, Song W, Wang P, Chen H, Lucio M, Schmitt-Kopplin P, Chen G, Cai Z (2010) GC/MS-based metabolomics reveals fatty acid biosynthesis and cholesterol metabolism in cell lines infected with influenza A virus. Talanta 83(1):262–268

39. Li MH, Liu YM, Li QL, Yang M, Pi YZ, Yang N, Zheng Y, Yue XQ (2020) Comparative exploration of free fatty acids in donkey colostrum and mature milk based on a metabolomics approach. J Dairy Sci 103(7):6022–6031. https://doi.org/10.3168/jds.2019-17720

40. Shackleton CH (2012) Role of a disordered steroid metabolome in the elucidation of sterol and steroid biosynthesis. Lipids 47(1):1–12

41. Koek MM, Muilwijk B, van der Werf MJ, Hankemeier T (2006) Microbial metabolomics with gas chromatography/mass spectrometry. Anal Chem 78(4):1272–1281

42. Zhou Y, Song R, Zhang Z, Lu X, Zeng Z, Hu C, Liu X, Li Y, Hou J, Sun Y (2016) The development of plasma pseudotargeted GC-MS metabolic profiling and its application in bladder cancer. Anal Bioanal Chem 408(24):6741–6749

43. Koek MM, Jellema RH, van der Greef J, Tas AC, Hankemeier T (2011) Quantitative metabolomics based on gas chromatography mass spectrometry: status and perspectives. Metabolomics 7(3):307–328. https://doi.org/10.1007/s11306-010-0254-3

44. Li X, Xu Z, Lu X, Yang X, Yin P, Kong H, Yu Y, Xu G (2009) Comprehensive two-dimensional gas chromatography/time-of-flight mass spectrometry for metabonomics: biomarker discovery for diabetes mellitus. Anal Chim Acta 633(2):257–262

45. Yu Z, Huang H, Reim A, Charles PD, Northage A, Jackson D, Parry I, Kessler BM (2017) Optimizing 2D gas chromatography mass spectrometry for robust tissue, serum and urine metabolite profiling. Talanta 165:685–691. https://doi.org/10.1016/j.talanta.2017.01.003

46. Cui L, Lu H, Lee YH (2018) Challenges and emergent solutions for LC-MS/MS based untargeted metabolomics in diseases. Mass Spectrom Rev 37(6):772–792. https://doi.org/10.1002/mas.21562

47. Tang D-Q, Zou L, Yin X-X, Ong CN (2016) HILIC-MS for metabolomics: an attractive and complementary approach to RPLC-MS. Mass Spectrom Rev 35(5):574–600. https://doi.org/10.1002/mas.21445

48. Wang S, Li J, Shi X, Qiao L, Lu X, Xu G (2013) A novel stop-flow two-dimensional liquid chromatography–mass spectrometry method for lipid analysis. J Chromatograp A 1321:65–72

49. Wang S, Shi X, Xu G (2017) Online three dimensional liquid chromatography/mass spectrometry method for the separation of complex samples. Anal Chem 89(3):1433–1438

50. Shao Y, Zhu B, Zheng R, Zhao X, Yin P, Lu X, Jiao B, Xu G, Yao Z (2015) Development of urinary pseudotargeted LC-MS-based metabolomics method and its application in hepatocellular carcinoma biomarker discovery. J Proteome Res 14(2):906–916. https://doi.org/10.1021/pr500973d

51. Chen SL, Kong HW, Lu X, Li Y, Yin PY, Zeng ZD, Xu GW (2013) Pseudotargeted metabolomics method and its application in serum biomarker discovery for hepatocellular carcinoma based on ultra high-performance liquid chromatography/triple quadrupole mass spectrometry. Anal Chem 85(17):8326–8333. https://doi.org/10.1021/ac4016787

52. Zheng F, Zhao X, Zeng Z, Wang L, Lv W, Wang Q, Xu G (2020) Development of a plasma pseudotargeted metabolomics

method based on ultra-high-performance liquid chromatography–mass spectrometry. Nature Protocols 15(8):2519–2537. https://doi.org/10.1038/s41596-020-0341-5

53. Saeys Y, Inza I, Larrañaga P (2007) A review of feature selection techniques in bioinformatics. Bioinformatics 23(19):2507–2517. https://doi.org/10.1093/bioinformatics/btm344

54. Monica C, Florencio P (2013) Tools for the functional interpretation of metabolomic experiments. Brief Bioinformatics 6:737–744

55. Kanehisa M, Goto S (2000) KEGG: kyoto encyclopedia of genes and genomes. Nucleic Acids Res 28(1):27-30. https://doi.org/10.1093/nar/28.1.27

56. Croft D, O'Kelly G, Wu G, Haw R, Stein L (2010) Reactome: a database of reactions, pathways and biological processes. Nucleic Acids Res 39(Database issue):D691–D697

57. Jewison T, Su Y, Disfany FM, Liang Y, Knox C, Maciejewski A, Poelzer J, Huynh J, Zhou Y, Arndt D, Djoumbou Y, Liu Y, Deng L, Guo AC, Han B, Pon A, Wilson M, Rafatnia S, Liu P, Wishart DS (2014) SMPDB 2.0: big improvements to the Small Molecule Pathway Database. Nucleic Acids Res 42 (Database issue):D478–484. https://doi.org/10.1093/nar/gkt1067

58. Takuji Y, Ivica L, Shujiro O, Minoru K, Peer B (2011) iPath2.0: interactive pathway explorer. Nucleic Acids Res 39(Web Server issue): W412–W415

59. Barupal DK, Haldiya PK, Wohlgemuth G, Kind T, Kothari SL, Pinkerton KE, Fiehn O (2012) MetaMapp: mapping and visualizing metabolomic data by integrating information from biochemical pathways and chemical and mass spectral similarity. BMC Bioinformatics 13:99–99. https://doi.org/10.1186/1471-2105-13-99

60. Xia J, Wishart DS (2010) MetPA: a web-based metabolomics tool for pathway analysis and visualization. Bioinformatics 26 (18):2342–2344. https://doi.org/10.1093/bioinformatics/btq418

61. Keun HC (2011) Integrated pathway-level analysis of transcriptomics and metabolomics data with IMPaLA. Bioinformatics 27 (20):2917–2918

62. Pazos F (2011) MBRole: enrichment analysis of metabolomic data. Bioinformatics 27 (5):730–731

63. Kankainen M, Gopalacharyulu P, Holm L, Oresic M (2011) MPEA--metabolite pathway enrichment analysis. Bioinformatics 27spi2; (13):1878–1879. https://doi.org/10.1093/bioinformatics/btr278

64. Persicke M, Rückert C, Plassmeier J, Stutz LJ, Kessler N, Kalinowski J, Goesmann A, Neuweger H (2012) MSEA: metabolite set enrichment analysis in the MeltDB metabolomics software platform: metabolic profiling of Corynebacterium glutamicum as an example. Metabolomics 8(2):310–322. https://doi.org/10.1007/s11306-011-0311-6

65. Goldstein DS (2013) Biomarkers, mechanisms, and potential prevention of catecholamine neuron loss in Parkinson disease. Adv Pharmacol 68:235–272. https://doi.org/10.1016/B978-0-12-411512-5.00012-9

66. Goldstein DS, Holmes C, Lopez GJ, Wu T, Sharabi Y (2018) Cerebrospinal fluid biomarkers of central dopamine deficiency predict Parkinson's disease. Parkinsonism Relat Disord 50:108–112. https://doi.org/10.1016/j.parkreldis.2018.02.023

67. Goldstein DS, Holmes C, Sharabi Y (2012) Cerebrospinal fluid biomarkers of central catecholamine deficiency in Parkinson's disease and other synucleinopathies. Brain 135 (Pt 6):1900–1913. https://doi.org/10.1093/brain/aws055

68. Cerroni R, Liguori C, Stefani A, Conti M, Garasto E, Pierantozzi M, Mercuri N, Bernardini S, Fucci G, Massoud R (2020) Increased noradrenaline as an additional cerebrospinal fluid biomarker in PSP-like parkinsonism. Front Aging Neurosci 12:126. https://doi.org/10.3389/fnagi.2020.00126

69. Goldstein DS, Sullivan P, Holmes C, Kopin IJ, Basile MJ, Mash DC (2011) Catechols in post-mortem brain of patients with Parkinson disease. Eur J Neurol 18(5):703–710. https://doi.org/10.1111/j.1468-1331.2010.03246.x

70. Mattammal MB, Chung HD, Strong R, Hsu FF (1993) Confirmation of a dopamine metabolite in parkinsonian brain tissue by gas chromatography-mass spectrometry. J Chromatogr 614(2):205–212. https://doi.org/10.1016/0378-4347(93)80310-z

71. Burke W, Li S, Williams E, Nonneman R, Zahm D (2003) 3,4-Dihydroxyphenylacetaldehyde is the toxic dopamine metabolite in vivo: implications for Parkinson's disease pathogenesis.

Brain Res 989(2):205–213. https://doi.org/10.1016/s0006-8993(03)03354-7

72. Masato A, Plotegher N, Boassa D, Bubacco L (2019) Impaired dopamine metabolism in Parkinson's disease pathogenesis. Mol Neurodegenerat 14:35. https://doi.org/10.1186/s13024-019-0332-6

73. Ohmichi T, Kasai T, Kosaka T, Shikata K, Tatebe H, Ishii R, Shinomoto M, Mizuno T, Tokuda T (2018) Biomarker repurposing: therapeutic drug monitoring of serum theophylline offers a potential diagnostic biomarker of Parkinson's disease. PLoS One 13 (7):e0201260. https://doi.org/10.1371/journal.pone.0201260

74. Kim AR, Nodel MR, Pavlenko TA, Chesnokova NB, Yakhno NN, Ugrumov MV (2019) Tear fluid catecholamines as biomarkers of the Parkinson's disease: a clinical and experimental study. Acta Nat 11(4):99–103. https://doi.org/10.32607/20758251-2019-11-4-99-103

75. Le Witt P, Schultz L, Auinger P, Lu M, Parkinson Study Group DI (2011) CSF xanthine, homovanillic acid, and their ratio as biomarkers of Parkinson's disease. Brain Res 1408:88–97. https://doi.org/10.1016/j.brainres.2011.06.057

76. Han W, Sapkota S, Camicioli R, Dixon RA, Li L (2017) Profiling novel metabolic biomarkers for Parkinson's disease using in-depth metabolomic analysis. Mov Disord 32 (12):1720–1728. https://doi.org/10.1002/mds.27173

77. Abbas MM, Xu Z, Tan LCS (2018) Epidemiology of Parkinson's disease-east versus west. Mov Disord Clin Pract 5(1):14–28. https://doi.org/10.1002/mdc3.12568

78. Hatano T, Saiki S, Okuzumi A, Mohney RP, Hattori N (2016) Identification of novel biomarkers for Parkinson's disease by metabolomic technologies. J Neurol Neurosurg Psyc 87(3):295–301. https://doi.org/10.1136/jnnp-2014-309676

79. LeWitt PA, Li J, Lu M, Guo L, Auinger P (2017) Metabolomic biomarkers as strong correlates of Parkinson disease progression. Neurology 88(9):862

80. Fujimaki M, Saiki S, Li Y, Kaga N, Taka H, Hatano T, Ishikawa KI, Oji Y, Mori A, Okuzumi A, Koinuma T, Ueno SI, Imamichi Y, Ueno T, Miura Y, Funayama M, Hattori N (2018) Serum caffeine and metabolites are reliable biomarkers of early Parkinson disease. Neurology 90(5):e404–e411. https://doi.org/10.1212/WNL.0000000000004888

81. Xu K, Xu YH, Chen JF, Schwarzschild MA (2010) Neuroprotection by caffeine: time course and role of its metabolites in the MPTP model of Parkinson's disease. Neuroscience 167(2):475–481. https://doi.org/10.1016/j.neuroscience.2010.02.020

82. Prediger RD (2010) Effects of caffeine in Parkinson's disease: from neuroprotection to the management of motor and non-motor symptoms. J Alzheimers Dis 20(Suppl 1): S205–S220. https://doi.org/10.3233/JAD-2010-091459

83. Tadaiesky MT, Dombrowski PA, Figueiredo CP, Cargnin-Ferreira E, Da Cunha C, Takahashi RN (2008) Emotional, cognitive and neurochemical alterations in a premotor stage model of Parkinson's disease. Neuroscience 156(4):830–840. https://doi.org/10.1016/j.neuroscience.2008.08.035

84. Obeso JA, Rodriguez-Oroz M, Marin C, Alonso F, Zamarbide I, Lanciego JL, Rodriguez-Diaz M (2004) The origin of motor fluctuations in Parkinson's disease. Neurology 62(1 suppl 1):S17. https://doi.org/10.1212/WNL.62.1_suppl_1.S17

85. Ferre S, Diaz-Rios M, Salamone JD, Prediger RD (2018) New developments on the adenosine mechanisms of the central effects of caffeine and their implications for neuropsychiatric disorders. J Caffeine Adenosine Res 8(4):121–131. https://doi.org/10.1089/caff.2018.0017

86. Xu K, Xu Y, Brown-Jermyn D, Chen J-F, Ascherio A, Dluzen DE, Schwarzschild MA (2006) Estrogen prevents neuroprotection by caffeine in the mouse 1-methyl-4-phenyl-1,2,3,6-tetrahydropyridine model of Parkinson's disease. J Neurosci 26(2):535–541. https://doi.org/10.1523/JNEUROSCI.3008-05.2006

87. Palacios N, Gao X, McCullough ML, Schwarzschild MA, Shah R, Gapstur S, Ascherio A (2012) Caffeine and risk of Parkinson's disease in a large cohort of men and women. Mov Disord 27(10):1276–1282. https://doi.org/10.1002/mds.25076

88. Lawton KA, Cudkowicz ME, Brown MV, Alexander D, Caffrey R, Wulff JE, Bowser R, Lawson R, Jaffa M, Milburn MV (2012) Biochemical alterations associated with ALS. Amyotrophic Lateral Sclerosis 13 (1):110–118

89. Takeshige-Amano H, Saiki S, Fujimaki M, Ueno SI, Li Y, Hatano T, Ishikawa KI, Oji Y, Mori A, Okuzumi A, Tsunemi T, Daida K, Ishiguro Y, Imamichi Y, Nanmo H, Nojiri S, Funayama M, Hattori N (2020) Shared metabolic profile of caffeine in

Parkinsonian disorders. Mov Disord 35 (8):1438–1447. https://doi.org/10.1002/mds.28068

90. Bogie JFJ, Haidar M, Kooij G, Hendriks JJA (2020) Fatty acid metabolism in the progression and resolution of CNS disorders. Adv Drug Deliv Rev 159:198–213. https://doi.org/10.1016/j.addr.2020.01.004

91. Shah A, Han P, Wong MY, Chang RC, Legido-Quigley C (2019) Palmitate and stearate are increased in the plasma in a 6-OHDA model of Parkinson's disease. Metabolites 9 (2). https://doi.org/10.3390/metabo9020031

92. Li XZ, Zhang SN, Lu F, Liu CF, Wang Y, Bai Y, Wang N, Liu SM (2013) Cerebral metabonomics study on Parkinson's disease mice treated with extract of Acanthopanax senticosus harms. Phytomedicine 20 (13):1219–1229. https://doi.org/10.1016/j.phymed.2013.06.002

93. Lu Z, Wang J, Li M, Liu Q, Wei D, Yang M, Kong L (2014) (1)H NMR-based metabolomics study on a goldfish model of Parkinson's disease induced by 1-methyl-4-phenyl-1,2,3,6-tetrahydropyridine (MPTP). Chem Biol Interact 223:18–26. https://doi.org/10.1016/j.cbi.2014.09.006

94. Ren C, Hu X, Li X, Zhou Q (2016) Ultratrace graphene oxide in a water environment triggers Parkinson's disease-like symptoms and metabolic disturbance in zebrafish larvae. Biomaterials 93:83–94. https://doi.org/10.1016/j.biomaterials.2016.03.036

95. Okuzumi A, Hatano T, Ueno SI, Ogawa T, Saiki S, Mori A, Koinuma T, Oji Y, Ishikawa KI, Fujimaki M, Sato S, Ramamoorthy S, Mohney RP, Hattori N (2019) Metabolomics-based identification of metabolic alterations in PARK2. Ann Clin Transl Neurol 6(3):525–536. https://doi.org/10.1002/acn3.724

96. Willkommen D, Lucio M, Moritz F, Forcisi S, Kanawati B, Smirnov KS, Schroeter M, Sigaroudi A, Schmitt-Kopplin P, Michalke B (2018) Metabolomic investigations in cerebrospinal fluid of Parkinson's disease. PLoS One 13(12):e0208752. https://doi.org/10.1371/journal.pone.0208752

97. Trupp M, Jonsson P, Ohrfelt A, Zetterberg H, Obudulu O, Malm L, Wuolikainen A, Linder J, Moritz T, Blennow K, Antti H, Forsgren L (2014) Metabolite and peptide levels in plasma and CSF differentiating healthy controls from patients with newly diagnosed Parkinson's disease. J Parkinsons Dis 4(3):549–560. https://doi.org/10.3233/JPD-140389

98. Schulte EC, Elisabeth A, Berger HS, Trinh DK, Gabi K, Simone W, Jerzy A, Annette P, Jan K, Karsten S (2016) Alterations in lipid and inositol metabolisms in two dopaminergic disorders. PLoS One 11(1):e0147129

99. Alecu I, Bennett SAL (2019) Dysregulated lipid metabolism and its role in alpha-synucleinopathy in Parkinson's disease. Front Neurosci 13:328. https://doi.org/10.3389/fnins.2019.00328

100. Li P, Song C (2020) Potential treatment of Parkinson's disease with omega-3 polyunsaturated fatty acids. Nutritional Neuroscience (18):1–12

101. Saiki S, Hatano T, Fujimaki M, Ishikawa KI, Mori A, Oji Y, Okuzumi A, Fukuhara T, Koinuma T, Imamichi Y, Nagumo M, Furuya N, Nojiri S, Amo T, Yamashiro K, Hattori N (2017) Decreased long-chain acylcarnitines from insufficient β-oxidation as potential early diagnostic markers for Parkinson's disease. Sci Rep 7(1):7328. https://doi.org/10.1038/s41598-017-06767-y

102. Fan Y, Xiao S (2018) Progression rate associated peripheral blood biomarkers of Parkinson's disease. J Molecular Neurosci 65 (3):312–318. https://doi.org/10.1007/s12031-018-1102-6

103. Schulte EC, Altmaier E, Berger HS, Do KT, Kastenmuller G, Wahl S, Adamski J, Peters A, Krumsiek J, Suhre K, Haslinger B, Ceballos-Baumann A, Gieger C, Winkelmann J (2016) Alterations in lipid and inositol metabolisms in two dopaminergic disorders. PLoS One 11 (1):e0147129. https://doi.org/10.1371/journal.pone.0147129

104. Lee PH, Lee G, Paik MJ (2008) Polyunsaturated fatty acid levels in the cerebrospinal fluid of patients with Parkinson's disease and multiple system atrophy. Mov Disord 23 (2):309–310. https://doi.org/10.1002/mds.21846

105. Kawahata I, Bousset L, Melki R, Fukunaga K (2019) Fatty acid-binding protein 3 is critical for alpha-synuclein uptake and MPP(+)-induced mitochondrial dysfunction in cultured dopaminergic neurons. Int J Mol Sci 20(21). https://doi.org/10.3390/ijms20215358

106. Sharon R, Bar-Joseph I, Frosch MP, Walsh DM, Hamilton JA, Selkoe DJ (2003) The formation of highly soluble oligomers of α-synuclein is regulated by fatty acids and enhanced in Parkinson's disease. Neuron 37 (4):583–595. https://doi.org/10.1016/S0896-6273(03)00024-2

107. Mori MA, Delattre AM, Carabelli B, Pudell C, Bortolanza M, Staziaki PV,

Visentainer JV, Montanher PF, Del Bel EA, Ferraz AC (2018) Neuroprotective effect of omega-3 polyunsaturated fatty acids in the 6-OHDA model of Parkinson's disease is mediated by a reduction of inducible nitric oxide synthase. Nutr Neurosci 21 (5):341–351. https://doi.org/10.1080/1028415x.2017.1290928

108. Chitre NM, Wood BJ, Ray A, Moniri NH, Murnane KS (2020) Docosahexaenoic acid protects motor function and increases dopamine synthesis in a rat model of Parkinson's disease via mechanisms associated with increased protein kinase activity in the striatum. Neuropharmacology 167:107976. https://doi.org/10.1016/j.neuropharm.2020.107976

109. Ådén E, Carlsson M, Poortvliet E, Stenlund H, Linder J, Edström M, Forsgren L, Håglin L (2011) Dietary intake and olfactory function in patients with newly diagnosed Parkinson's disease: a case-control study. Nutr Neurosci 14(1):25–31. https://doi.org/10.1179/174313211x12966635733312

110. de Lau LM, Bornebroek M, Witteman JC, Hofman A, Koudstaal PJ, Breteler MM (2005) Dietary fatty acids and the risk of Parkinson disease: the Rotterdam study. Neurology 64(12):2040–2045. https://doi.org/10.1212/01.wnl.0000166038.67153.9f

111. Bohar Z, Toldi J, Fulop F, Vecsei L (2015) Changing the face of kynurenines and neurotoxicity: therapeutic considerations. Int J Mol Sci 16(5):9772–9793. https://doi.org/10.3390/ijms16059772

112. Szabo N, Kincses ZT, Toldi J, Vecsei L (2011) Altered tryptophan metabolism in Parkinson's disease: a possible novel therapeutic approach. J Neurol Sci 310(1-2):256–260. https://doi.org/10.1016/j.jns.2011.07.021

113. Németh H, Toldi J, Vécsei L (2006) Kynurenines, Parkinson's disease and other neurodegenerative disorders: preclinical and clinical studies. In: Riederer P, Reichmann H, Youdim MBH, Gerlach M (eds). Parkinson's disease and related disorders, Springer, pp 285–304

114. Wang Q, Liu D, Song P, Zou MH (2015) Tryptophan-kynurenine pathway is dysregulated in inflammation, and immune activation. Front Biosci 20:1116–1143

115. Havelund JF, Andersen AD, Binzer M, Blaabjerg M, Heegaard NHH, Stenager E, Faergeman NJ, Gramsbergen JB (2017) Changes in kynurenine pathway metabolism in Parkinson patients with L-DOPA-induced dyskinesia. J Neurochem 142(5):756–766. https://doi.org/10.1111/jnc.14104

116. Luan H, Liu L-F, Tang Z, Zhang M, Chua K-K, Song J-X, Mok VCT, Li M, Cai Z (2015) Comprehensive urinary metabolomic profiling and identification of potential non-invasive marker for idiopathic Parkinson's disease. Sci Report 5(1):13888. https://doi.org/10.1038/srep13888

117. Ogawa T, Matson WR, Beal MF, Myers RH, Bird ED, Milbury P, Saso S (1992) Kynurenine pathway abnormalities in Parkinson's disease. Neurology 42(9):1702–1706. https://doi.org/10.1212/wnl.42.9.1702

118. Lewitt PA, Li J, Lu M, Beach TG, Adler CH, Guo L, Arizona Parkinson's Disease C (2013) 3-hydroxykynurenine and other Parkinson's disease biomarkers discovered by metabolomic analysis. Mov Disord 28 (12):1653–1660. https://doi.org/10.1002/mds.25555

119. Schwarcz R (2004) The kynurenine pathway of tryptophan degradation as a drug target. Curr Opin Pharmacol 4(1):12–17

120. Zadori D, Klivenyi P, Toldi J, Fulop F, Vecsei L (2012) Kynurenines in Parkinson's disease: therapeutic perspectives. J Neural Transm 119(2):275–283. https://doi.org/10.1007/s00702-011-0697-3

121. Fulop F, Szatmári I, Vamos E, Zadori D, Toldi J, Vecsei L (2009) Syntheses, transformations and pharmaceutical applications of kynurenic acid derivatives. Curr Med Chem 16(36):4828–4842

122. Miranda AF, Boegman RJ, Beninger RJ, Jhamandas K (1997) Protection against quinolinic acid-mediated excitotoxicity in nigrostriatal dopaminergic neurons by endogenous kynurenic acid. Neuroscience 78 (4):967

123. Parasram K (2018) Phytochemical treatments target kynurenine pathway induced oxidative stress. Redox Rep 23(1):25–28. https://doi.org/10.1080/13510002.2017.1343223

124. Ackerman HD, Gerhard GS (2016) Bile acids in neurodegenerative disorders. Front Aging Neurosci 8:263. https://doi.org/10.3389/fnagi.2016.00263

125. Kiriyama Y, Nochi H (2019) The biosynthesis, signaling, and neurological functions of bile acids. Biomolecules 9(6):232. https://doi.org/10.3390/biom9060232

126. Rosa AI, Duarte-Silva S, Silva-Fernandes A, Nunes MJ, Carvalho AN, Rodrigues E, Gama MJ, Rodrigues CMP, Maciel P, Castro-Caldas M (2018) Tauroursodeoxycholic acid improves motor symptoms in a mouse

model of Parkinson's disease. Mol Neurobiol 55(12):9139–9155. https://doi.org/10.1007/s12035-018-1062-4

127. Abdelkader NF, Safar MM, Salem HA (2016) Ursodeoxycholic acid ameliorates apoptotic cascade in the rotenone model of Parkinson's disease: modulation of mitochondrial perturbations. Mol Neurobiol 53(2):810–817. https://doi.org/10.1007/s12035-014-9043-8

128. Mortiboys H, Furmston R, Bronstad G, Aasly J, Elliott C, Bandmann O (2015) UDCA exerts beneficial effect on mitochondrial dysfunction in LRRK2(G2019S) carriers and in vivo. Neurology 85(10):846–852. https://doi.org/10.1212/wnl.0000000000001905

129. Mortiboys H, Aasly J, Bandmann O (2013) Ursocholanic acid rescues mitochondrial function in common forms of familial Parkinson's disease. Brain 136(Pt 10):3038–3050. https://doi.org/10.1093/brain/awt224

130. Castro-Caldas M, Carvalho AN, Rodrigues E, Henderson CJ, Wolf CR, Rodrigues CM, Gama MJ (2012) Tauroursodeoxycholic acid prevents MPTP-induced dopaminergic cell death in a mouse model of Parkinson's disease. Mol Neurobiol 46(2):475–486. https://doi.org/10.1007/s12035-012-8295-4

131. Mendes MO, Rosa AI, Carvalho AN, Nunes MJ, Dionisio P, Rodrigues E, Costa D, Duarte-Silva S, Maciel P, Rodrigues CMP, Gama MJ, Castro-Caldas M (2019) Neurotoxic effects of MPTP on mouse cerebral cortex: modulation of neuroinflammation as a neuroprotective strategy. Mol Cell Neurosci 96:1–9. https://doi.org/10.1016/j.mcn.2019.01.003

132. Rosa AI, Fonseca I, Nunes MJ, Moreira S, Rodrigues E, Carvalho AN, Rodrigues CMP, Gama MJ, Castro-Caldas M (2017) Novel insights into the antioxidant role of tauroursodeoxycholic acid in experimental models of Parkinson's disease. Biochim Biophys Acta Mol Basis Dis 1863(9):2171–2181. https://doi.org/10.1016/j.bbadis.2017.06.004

133. Moreira S, Fonseca I, Nunes MJ, Rosa A, Lemos L, Rodrigues E, Carvalho AN, Outeiro TF, Rodrigues CMP, Gama MJ, Castro-Caldas M (2017) Nrf2 activation by tauroursodeoxycholic acid in experimental models of Parkinson's disease. Exp Neurol 295:77–87. https://doi.org/10.1016/j.expneurol.2017.05.009

134. Graham SF, Rey NL, Ugur Z, Yilmaz A, Sherman E, Maddens M, Bahado-Singh RO,

Becker K, Schulz E, Meyerdirk LK, Steiner JA, Ma J, Brundin P (2018) Metabolomic profiling of bile acids in an experimental model of prodromal Parkinson's disease. Metabolites 8(4). https://doi.org/10.3390/metabo8040071

135. Huang A, Martin ER, Vance JM, Cai X (2014) Detecting genetic interactions in pathway-based genome-wide association studies. Genet Epidemiol 38(4):300–309. https://doi.org/10.1002/gepi.21803

136. Hasuike Y, Endo T, Koroyasu M, Matsui M, Mori C, Yamadera M, Fujimura H, Sakoda S (2019) Bile acid abnormality induced by intestinal dysbiosis might explain lipid metabolism in Parkinson's disease. Med Hypotheses 134:109436. https://doi.org/10.1016/j.mehy.2019.109436

137. Hertel J, Harms AC, Heinken A, Baldini F, Thinnes CC, Glaab E, Vasco DA, Pietzner M, Stewart ID, Wareham NJ, Langenberg C, Trenkwalder C, Kruger R, Hankemeier T, Fleming RMT, Mollenhauer B, Thiele I (2019) Integrated analyses of microbiome and longitudinal metabolome data reveal microbial-host interactions on sulfur metabolism in Parkinson's disease. Cell Rep 29(7):1767–1777.e1768. https://doi.org/10.1016/j.celrep.2019.10.035

138. O'Donovan SM, Crowley EK, Brown JR, O'Sullivan O, O'Leary OF, Timmons S, Nolan YM, Clarke DJ, Hyland NP, Joyce SA, Sullivan AM, O'Neill C (2020) Nigral overexpression of alpha-synuclein in a rat Parkinson's disease model indicates alterations in the enteric nervous system and the gut microbiome. Neurogastroenterol Motil 32(1):e13726. https://doi.org/10.1111/nmo.13726

139. Quinones MP, Kaddurah-Daouk R (2009) Metabolomics tools for identifying biomarkers for neuropsychiatric diseases. Neurobiol Dis 35(2):165–176. https://doi.org/10.1016/j.nbd.2009.02.019

140. Bogdanov M, Matson WR, Wang L, Matson T, Saunders-Pullman R, Bressman SS, Flint Beal M (2008) Metabolomic profiling to develop blood biomarkers for Parkinson's disease. Brain 131(Pt 2):389–396. https://doi.org/10.1093/brain/awm304

141. Meiser J, Delcambre S, Wegner A, Jager C, Ghelfi J, d'Herouel AF, Dong X, Weindl D, Stautner C, Nonnenmacher Y, Michelucci A, Popp O, Giesert F, Schildknecht S, Kramer L, Schneider JG, Woitalla D, Wurst W, Skupin A, Weisenhorn DM, Kruger R, Leist M, Hiller K (2016) Loss of DJ-1 impairs antioxidant

response by altered glutamine and serine metabolism. Neurobiol Dis 89:112–125. https://doi.org/10.1016/j.nbd.2016.01.019

142. Sofic E, Lange KW, Jellinger K, Riederer P (1992) Reduced and oxidized glutathione in the substantia nigra of patients with Parkinson's disease. Neurosci Lett 142(2):128–130

143. Kumar P, Kaundal RK, More S, Sharma SS (2009) Beneficial effects of pioglitazone on cognitive impairment in MPTP model of Parkinson's disease. Behav Brain Res 197 (2):398–403

144. Heo H, Ahn JB, Lee HH, Kwon E, Yun JW, Kim H, Kang BC (2017) Neurometabolic profiles of the substantia nigra and striatum of MPTP-intoxicated common marmosets: an in vivo proton MRS study at 9.4 T. NMR Biomed 30(2). https://doi.org/10.1002/nbm.3686

145. Ascherio A, Lewitt PA, Xu K, Eberly S, Watts A, Matson WR, Marras C, Kieburtz K, Rudolph A, Bogdanov MB, Schwid SR, Tennis M, Tanner CM, Beal MF, Lang AE, Oakes D, Fahn S, Shoulson I, Schwarzschild MA, Parkinson Study Group DI (2009) Urate as a predictor of the rate of clinical decline in Parkinson disease. Arch Neurol 66 (12):1460–1468. https://doi.org/10.1001/archneurol.2009.247

146. Cipriani S, Chen X, Schwarzschild MA (2010) Urate: a novel biomarker of Parkinson's disease risk, diagnosis and prognosis. Biomark Med 4(5):701–712. https://doi.org/10.2217/bmm.10.94

147. Scheperjans F, Pekkonen E, Kaakkola S, Auvinen P (2015) Linking smoking, coffee, urate, and Parkinson's disease—a Role for gut microbiota? J Parkinsons Dis 5(2):255–262. https://doi.org/10.3233/JPD-150557

148. Stoessel D, Stellmann JP, Willing A, Behrens B, Rosenkranz SC, Hodecker SC, Sturner KH, Reinhardt S, Fleischer S, Deuschle C, Maetzler W, Berg D, Heesen C, Walther D, Schauer N, Friese MA, Pless O (2018) Metabolomic profiles for primary progressive multiple sclerosis stratification and disease course monitoring. Front Hum Neurosci 12:226. https://doi.org/10.3389/fnhum.2018.00226

149. Roede JR, Uppal K, Park Y, Lee K, Tran V, Walker D, Strobel FH, Rhodes SL, Ritz B, Jones DP (2013) Serum metabolomics of slow vs. rapid motor progression Parkinson's disease: a pilot study. PLoS One 8(10): e77629. https://doi.org/10.1371/journal.pone.0077629

150. Alam ZI, Jenner A, Daniel SE, Lees AJ, Cairns N, Marsden CD, Jenner P, Halliwell B (2002) Oxidative DNA damage in the parkinsonian brain: an apparent selective increase in 8-hydroxyguanine levels in Substantia Nigra. J Neurochem 69(3):1196–1203. https://doi.org/10.1046/j.1471-4159.1997.69031196.x

151. Kikuchi A, Takeda A, Onodera H, Kimpara T, Hisanaga K, Sato N, Nunomura A, Castellani RJ, Perry G, Smith MA, Itoyama Y (2002) Systemic increase of oxidative nucleic acid damage in Parkinson's disease and multiple system atrophy. Neurobiol Dis 9 (2):244–248. https://doi.org/10.1006/nbdi.2002.0466

152. Bolner A, Pilleri M, De Riva V, Nordera GP (2011) Plasma and urinary HPLC-ED determination of the ratio of 8-OHdG/2-dG in Parkinson's disease. Clin Lab 57 (11-12):859–866

153. Sato S, Mizuno Y, Hattori N (2005) Urinary 8-hydroxydeoxyguanosine levels as a biomarker for progression of Parkinson disease. Neurology 64(6):1081

154. Schipper HM, Song W, Zukor H, Hascalovici JR, Zeligman D (2009) Heme oxygenase-1 and neurodegeneration: expanding frontiers of engagement. J Neurochem 110 (2):469–485. https://doi.org/10.1111/j.1471-4159.2009.06160.x

155. Luan H, Liu LF, Tang Z, Mok VC, Li M, Cai Z (2015) Elevated excretion of biopyrrin as a new marker for idiopathic Parkinson's disease. Parkinsonism Relat Disord 21 (11):1371–1372. https://doi.org/10.1016/j.parkreldis.2015.09.009

156. Glaab E, Trezzi JP, Greuel A, Jager C, Hodak Z, Drzezga A, Timmermann L, Tittgemeyer M, Diederich NJ, Eggers C (2019) Integrative analysis of blood metabolomics and PET brain neuroimaging data for Parkinson's disease. Neurobiol Dis 124:555–562. https://doi.org/10.1016/j.nbd.2019.01.003

157. Trezzi JP, Galozzi S, Jaeger C, Barkovits K, Brockmann K, Maetzler W, Berg D, Marcus K, Betsou F, Hiller K, Mollenhauer B (2017) Distinct metabolomic signature in cerebrospinal fluid in early Parkinson's disease. Mov Disord 32(10):1401–1408. https://doi.org/10.1002/mds.27132

158. van Kessel SP, El Aidy S (2019) Bacterial metabolites mirror altered gut microbiota composition in patients with Parkinson's disease. J Parkinsons Dis 9(s2):S359–S370. https://doi.org/10.3233/JPD-191780

159. Luan H, Wang X, Cai Z (2019) Mass spectrometry-based metabolomics: Targeting the crosstalk between gut microbiota and brain in neurodegenerative disorders. Mass Spectrom Rev 38(1):22–33. https://doi.org/10.1002/mas.21553

160. Unger MM, Spiegel J, Dillmann KU, Grundmann D, Philippeit H, Burmann J, Fassbender K, Schwiertz A, Schafer KH (2016) Short chain fatty acids and gut microbiota differ between patients with Parkinson's disease and age-matched controls. Parkinsonism Relat Disord 32:66–72. https://doi.org/10.1016/j.parkreldis.2016.08.019

161. Ahmed SS, Santosh W, Kumar S, Christlet HT (2009) Metabolic profiling of Parkinson's disease: evidence of biomarker from gene expression analysis and rapid neural network detection. J Biomed Sci 16:63. https://doi.org/10.1186/1423-0127-16-63

162. Shin C, Lim Y, Lim H, Ahn T-B (2020) Plasma short-chain fatty acids in patients with Parkinson's disease. Mov Disord 35 (6):1021–1027. https://doi.org/10.1002/mds.28016

163. Kumari S, Goyal V, Kumaran SS, Dwivedi SN, Srivastava A, Jagannathan NR (2020) Quantitative metabolomics of saliva using proton NMR spectroscopy in patients with Parkinson's disease and healthy controls. Neurol Sci 41(5):1201–1210. https://doi.org/10.1007/s10072-019-04143-4

164. Erny D, Hrabě de Angelis AL, Jaitin D, Wieghofer P, Staszewski O, David E, Keren-Shaul H, Mahlakoiv T, Jakobshagen K, Buch T, Schwierzeck V, Utermöhlen O, Chun E, Garrett WS, McCoy KD, Diefenbach A, Staeheli P, Stecher B, Amit I, Prinz M (2015) Host microbiota constantly control maturation and function of microglia in the CNS. Nat Neurosci 18(7):965–977. https://doi.org/10.1038/nn.4030

165. Flint HJ, Duncan SH, Scott KP, Louis P (2015) Links between diet, gut microbiota composition and gut metabolism. Proc Nutr Soc 74(1):13–22. https://doi.org/10.1017/s0029665114001463

166. Cirstea MS, Yu AC, Golz E, Sundvick K, Kliger D, Radisavljevic N, Foulger LH, Mackenzie M, Huan T, Finlay BB, Appel-Cresswell S (2020) Microbiota Composition and Metabolism Are Associated With Gut Function in Parkinson's Disease. Mov Disord 35(7):1208–1217. https://doi.org/10.1002/mds.28052

167. Diether N, Willing B (2019) Microbial fermentation of dietary protein: an important factor in diet–microbe–host interaction. Microorganisms 7(1):19

168. Okuzumi A, Hatano T, Ueno S-I, Ogawa T, Saiki S, Mori A, Koinuma T, Oji Y, Ishikawa K-I, Fujimaki M (2019) Metabolomics-based identification of metabolic alterations in PARK2. Ann Clin Transl Neurol 6 (3):525–536

169. Burte F, Houghton D, Lowes H, Pyle A, Nesbitt S, Yarnall A, Yu-Wai-Man P, Burn DJ, Santibanez-Koref M, Hudson G (2017) Metabolic profiling of Parkinson's disease and mild cognitive impairment. Mov Disord 32 (6):927–932. https://doi.org/10.1002/mds.26992

170. Rooks MG, Veiga P, Wardwell-Scott LH, Tickle T, Segata N, Michaud M, Gallini CA, Beal C, Van Hylckama-Vlieg JE, Ballal SA (2014) Gut microbiome composition and function in experimental colitis during active disease and treatment-induced remission. ISME J 8(7):1403–1417

171. Keshavarzian A, Green SJ, Engen PA, Voigt RM, Naqib A, Forsyth CB, Mutlu E, Shannon KM (2015) Colonic bacterial composition in Parkinson's disease. Mov Disord 30 (10):1351–1360

172. Luan H, Liu LF, Meng N, Tang Z, Chua KK, Chen LL, Song JX, Mok VC, Xie LX, Li M, Cai Z (2015) LC-MS-based urinary metabolite signatures in idiopathic Parkinson's disease. J Proteome Res 14(1):467–478. https://doi.org/10.1021/pr500807t

173. Williams BB, Van Benschoten AH, Cimermancic P, Donia MS, Zimmermann M, Taketani M, Ishihara A, Kashyap PC, Fraser JS, Fischbach MA (2014) Discovery and characterization of gut microbiota decarboxylases that can produce the neurotransmitter tryptamine. Cell Host Microbe 16(4):495–503. https://doi.org/10.1016/j.chom.2014.09.001

174. Krishnan S, Ding Y, Saedi N, Choi M, Sridharan GV, Sherr DH, Yarmush ML, Alaniz RC, Jayaraman A, Lee K (2018) Gut microbiota-derived tryptophan metabolites modulate inflammatory response in hepatocytes and macrophages. Cell Rep 23 (4):1099–1111. https://doi.org/10.1016/j.celrep.2018.03.109

175. Kim DC, Quang TH, Yoon CS, Ngan NTT, Lim SI, Lee SY, Kim YC, Oh H (2016) Anti-neuroinflammatory activities of indole alkaloids from kanjang (Korean fermented soy source) in lipopolysaccharide-induced BV2 microglial cells. Food Chem 213:69–75. https://doi.org/10.1016/j.foodchem.2016.06.068

176. Amo T, Oji Y, Saiki S, Hattori N (2019) Metabolomic analysis revealed mitochondrial dysfunction and aberrant choline metabolism in MPP(+)-exposed SH-SY5Y cells. Biochem Biophys Res Commun 519(3):540–546. https://doi.org/10.1016/j.bbrc.2019.09.031

177. Lei S, Zavala-Flores L, Garcia-Garcia A, Nandakumar R, Huang Y, Madayiputhiya N, Stanton RC, Dodds ED, Powers R, Franco R (2014) Alterations in energy/redox metabolism induced by mitochondrial and environmental toxins: a specific role for glucose-6-phosphate-dehydrogenase and the pentose phosphate pathway in paraquat toxicity. ACS Chem Biol 9(9):2032–2048. https://doi.org/10.1021/cb400894a

178. Powers R, Lei S, Anandhan A, Marshall DD, Worley B, Cerny RL, Dodds ED, Huang Y, Panayiotidis MI, Pappa A, Franco R (2017) Metabolic investigations of the molecular mechanisms associated with Parkinson's disease. Metabolites 7(2). https://doi.org/10.3390/metabo7020022

179. Anandhan A, Lei S, Levytskyy R, Pappa A, Panayiotidis MI, Cerny RL, Khalimonchuk O, Powers R, Franco R (2017) Glucose metabolism and AMPK signaling regulate dopaminergic cell death induced by gene (alpha-synuclein)-environment (paraquat) interactions. Mol Neurobiol 54(5):3825–3842. https://doi.org/10.1007/s12035-016-9906-2

180. Melnikova A, Pozdyshev D, Barinova K, Kudryavtseva S, Muronetz VI (2020) α-Synuclein overexpression in SH-SY5Y human neuroblastoma cells leads to the accumulation of thioflavin S-positive aggregates and impairment of glycolysis. Biochemistry 85(5):604–613. https://doi.org/10.1134/s0006297920050090

181. Tang BL (2020) Glucose, glycolysis, and neurodegenerative diseases. J Cell Physiol 235(11):7653–7662. https://doi.org/10.1002/jcp.29682

182. Cai R, Zhang Y, Simmering JE, Schultz JL, Liu L (2019) Enhancing glycolysis attenuates Parkinson's disease progression in models and clinical databases. J Clin Investig 129(Suppl 3):4539–4549

183. Chen X, Zhao C, Li X et al (2015) Terazosin activates Pgk1 and Hsp90 to promote stress resistance. Nat Chem Biol 11(1):19–25

184. Saiki S, Hatano T, Fujimaki M, Ishikawa KI, Mori A, Oji Y, Okuzumi A, Fukuhara T, Koinuma T, Imamichi Y, Nagumo M, Furuya N, Nojiri S, Amo T, Yamashiro K, Hattori N (2017) Decreased long-chain acyl-carnitines from insufficient beta-oxidation as potential early diagnostic markers for Parkinson's disease. Sci Rep 7(1):7328. https://doi.org/10.1038/s41598-017-06767-y

185. Zheng H, Zhao L, Xia H, Xu C, Wang D, Liu K, Lin L, Li X, Yan Z, Gao H (2016) NMR-based metabolomics reveal a recovery from metabolic changes in the striatum of 6-OHDA-induced rats treated with basic fibroblast growth factor. Mol Neurobiol 53(10):6690–6697. https://doi.org/10.1007/s12035-015-9579-2

186. Rappley I, Myers DS, Milne SB, Ivanova PT, Lavoie MJ, Brown HA, Selkoe DJ (2009) Lipidomic profiling in mouse brain reveals differences between ages and genders, with smaller changes associated with alpha-synuclein genotype. J Neurochem 111(1):15–25. https://doi.org/10.1111/j.1471-4159.2009.06290.x

187. Tyurina YY, Polimova AM, Maciel E, Tyurin VA, Kapralova VI, Winnica DE, Vikulina AS, Domingues MR, McCoy J, Sanders LH, Bayir H, Greenamyre JT, Kagan VE (2015) LC/MS analysis of cardiolipins in substantia nigra and plasma of rotenone-treated rats: implication for mitochondrial dysfunction in Parkinson's disease. Free Radic Res 49(5):681–691. https://doi.org/10.3109/10715762.2015.1005085

188. Johansen KK, Wang L, Aasly JO, White LR, Matson WR, Henchcliffe C, Beal MF, Bogdanov M (2009) Metabolomic profiling in LRRK2-related Parkinson's disease. PLoS One 4(10):e7551. https://doi.org/10.1371/journal.pone.0007551

189. Seraji-Bozorgzad N, Bao F, George E, Krstevska S, Gorden V, Chorostecki J, Santiago C, Zak I, Caon C, Khan O (2015) Longitudinal study of the substantia nigra in Parkinson disease: a high-field 1H-MR spectroscopy imaging study. Mov Disord 30(10):1400–1404

190. Boska MD, Lewis TB, Destache CJ et al (2005) Quantitative 1H magnetic resonance spectroscopic imaging determines therapeutic immunization efficacy in an animal model of Parkinson's disease. J Neuroence 25(7):1691–1700

191. Hernandez-Diaz S, Soukup SF (2020) The role of lipids in autophagy and its implication in neurodegeneration. Cell Stress 4(7):167–186. https://doi.org/10.15698/cst2020.07.225

192. Mori A, Imai Y, Hattori N (2020) Lipids: key players that modulate alpha-synuclein toxicity and neurodegeneration in Parkinson's disease. Int J Mol Sci 21(9). https://doi.org/10.3390/ijms21093301

193. Fanning S, Selkoe D, Dettmer U (2021) Vesicle trafficking and lipid metabolism in synucleinopathy. Acta Neuropathol 141 (4):491–510. https://doi.org/10.1007/s00401-020-02177-z

194. Rizzo G, Copetti M, Arcuti S, Martino D, Fontana A, Logroscino G (2016) Accuracy of clinical diagnosis of Parkinson disease: a systematic review and meta-analysis. Neurology 86(6):566–576. https://doi.org/10.1212/wnl.0000000000002350

195. Fayyad M, Salim S, Majbour N, Erskine D, Stoops E, Mollenhauer B, El-Agnaf OMA (2019) Parkinson's disease biomarkers based on alpha-synuclein. J Neurochem 150 (5):626–636. https://doi.org/10.1111/jnc.14809

196. Atik A, Stewart T, Zhang J (2016) Alpha-synuclein as a biomarker for Parkinson's disease. Brain Pathol 26(3):410–418. https://doi.org/10.1111/bpa.12370

Chapter 9

Neuroinflammation in Huntington's Disease

John D. Lee, Martin W. Lo, Jenny N. T. Fung, and Trent M. Woodruff

Abstract

Huntington's disease (HD) is a debilitating inherited neurodegenerative disorder characterized by motor, cognitive and psychiatric deficits. Microglial and astrocyte activation, part of the process termed neuroinflammation, is one hallmark of HD, and modulation of neuroinflammation has been suggested as a potential target for therapeutic intervention. Although the relationship between neuroinflammation markers and the disease pathology is not completely understood, there is now compelling evidence to suggest that microglial and astrocyte activation signatures, identified as soluble factors in the cerebrospinal fluid or blood, or identified using PET imaging, could be used as potential complementary biomarkers to monitor and evaluate disease progression in HD patients. Identification of neuroinflammation markers prior to clinical symptoms opens up the possibility of evaluating disease-modifying treatments in the premanifest phase. Hence, neuroinflammatory biofluid and imaging biomarkers could provide an objective measurement for assessing HD severity and would also be valuable in the clinical care of existing HD patients. Neuroinflammatory biomarkers would also be useful in a clinical trial context, potentially serving as surrogate endpoints. This chapter will explore the evidence of roles for activated microglia, astrocytes, and peripheral immune cells in HD, and explore possible biomarkers of neuroinflammation.

Key words Microglia, Astrocytes, Monocytes, Macrophages, Cytokines, Biomarkers, Huntingtin, Huntington's disease, Neuroinflammation

1 Introduction

Huntington's disease (HD) is a debilitating neurodegenerative condition characterized by motor, cognitive, and psychiatric abnormalities [1]. The disease stems from an autosomal dominant polyglutamine (polyQ) mutation in the Huntingtin (*HTT*) gene located on chromosome 4p16.3 [1]. PolyQ tracts longer than 39 repeats are pathognomonic of the disease and tract length is associated with disease severity, both in terms of onset and progression [2–4]. The *HTT* gene encodes a ubiquitously expressed adapter protein involved in many signaling pathways [5–7] and in its mutant form may obtain loss-of-function [8], gain-of-function [9], and direct toxic effects including proteome dyshomeostasis

Philip V. Peplow, Bridget Martinez and Thomas A. Gennarelli (eds.), *Neurodegenerative Diseases Biomarkers: Towards Translating Research to Clinical Practice*, Neuromethods, vol. 173, https://doi.org/10.1007/978-1-0716-1712-0_9,
© Springer Science+Business Media, LLC, part of Springer Nature 2022

and prionism [10]. These effects result in widespread cortical and subcortical atrophy typified by striatal degeneration [11–14], and extracranial disease including cardiac failure, osteoporosis, sarcopenia, and gastrointestinal abnormalities [15, 16]. Overlaying this degenerative process is a hyperactive and dysregulated inflammatory response [17, 18], which in animal models has been shown to contribute to disease progression and be amenable to therapeutic modification [19, 20]. This novel aspect of HD is incompletely understood, and further research is needed to pursue drug development in a condition currently without any disease-modifying treatments.

In HD, hyperactivity and dysregulation of the immune system are widely seen. In the central nervous system (CNS), microglia and astrocytes become activated in response to inflammatory stimulus such as damage-associated molecular patterns (DAMPs), and from mutant HTT protein released from dying neurons and their own cell autonomous production of mutant HTT [17, 21]. As a result, microglia and astrocytes elicit an exaggerated immune response characterized by cytokine release [22], reactive oxygen species production [23], and complement activation [24] that eventuates in a self-perpetuating cycle of cytotoxicity and inflammation, which has been correlated with neuronal loss and disease severity [25–31]. Similarly, monocytes and macrophages from HD patients express mutant HTT and exhibit hyperactive cytokine production [32], proinflammatory transcriptome alterations [32–34], impaired chemotaxis [35], and upregulated phagocytosis [36], although the role of peripheral leukocytes in HD has yet to be completely elucidated [17]. Nevertheless, HD is associated with a hyperactive and dysfunctional immune response that exacerbates the underlying pathology. Therefore, it is possible that this aberrant immune activity may be amenable to pharmacological modification or utilized as a diagnostic or prognostic marker for HD. The present chapter will review the current evidence on the role of microglia, astrocytes and the peripheral immune system in driving neuroinflammation and its contribution to HD pathogenesis. It will also address the potential use of the immune system as the objective biomarker to monitor disease progression and treatment response in HD.

2   Microglia Activation in HD

Microglia have long been considered the resident macrophage-like cells in the CNS that are capable of orchestrating a potent inflammatory response following an inflammatory stimulus. Under physiological conditions, the number and function of microglia are tightly controlled and are characterized by a small cell body and ramified processes that patrol the entire brain parenchyma every few hours, during which they actively contribute to brain homeostasis

[37]. The patrolling microglia are actively involved in synaptic organization, phagocytosis of apoptotic cells and debris, secretion of homeostatic neuronal trophic factors such as transforming growth factor β, scavenging activity, and synaptic pruning [38, 39]. In response to infection or tissue damage, microglia become activated, where they rapidly modify their morphology to an amoeboid appearance (i.e., larger cell body and shortened thick processes), increase proliferation, phagocytic activity, antigen presentation and initiate an innate immune response by releasing various inflammatory factors such as cytokines and chemokines [40–44]. Chronic activation of microglia, such as that observed in HD, can lead to persistent expression of inflammatory mediators, driving inflammation and contributing to further tissue damage and disease progression. Microglial activation is an important feature of neuroinflammation that is prominent in almost all neurodegenerative diseases, including HD.

Atrophy of the striatum, which includes the regions caudate nucleus, putamen, and globus pallidus, is the neuropathological hallmark of HD [45]. Immunohistochemical analyses of postmortem brain tissues from HD patients demonstrates a marked increase in reactive microglia in the cortex, striatum and globus pallidus, that is absent in healthy control brains [45–47]. It was further noted that the number of activated microglia in the striatum and cortex of HD patients shows a direct correlation with the degree of neuronal loss, suggesting that microglia activation could be the driving factor for ongoing neuronal degeneration [46]. Indeed, another study using in vivo positron emission tomography (PET) demonstrated that microglial activation was greater in HD patients with greater disease severity and striatal neuronal loss [29]. Interestingly, other PET studies also detected microglial activation in HD patients prior to disease manifestations [30, 48]. In addition, increased microglial activation was evident in the striatum of the R6/2 mouse model of HD at a presymptomatic stage [49]. Taken together, these studies identified microglial activation as an early change in HD and highlights microglial activation as an integral component of HD pathogenesis.

3 Mutant HTT and Microglial Activation

Many studies have now reported mutant HTT protein accumulation in neurons increases microglia activation [50]. In primary neuron cultures, expression of mutant HTT initiated microglia responses, where elevated numbers of microglia with an activated amoeboid phenotype were observed along the degenerating neurites [51]. This hypothesis was further supported by another study where lowering neuronal mutant HTT protein expression led to a drastic amelioration of concomitant microglia activation [52], suggesting that activation of microglia occurs in response to neuronal expression of mutant HTT protein and ultimately drives neuronal death through increased secretion of inflammatory mediators and phagocytotic activity. Although, neuronal expression of mutant HTT protein activates microglia, adding exogenous primary microglia to mutant HTT expressing neurons also increased their survival and this effect was proportional to the number of microglia added to the neurons. This suggests that mutant HTT-expressing microglia could have increased basal gene expression and enhanced responses to exogenous stimuli [51]. Indeed, using a genome-wide approach, a study provided evidence that the overexpression of mutant HTT solely in microglia is sufficient to confer a cell-autonomous increase in the proinflammatory gene expression and exaggerated neurotoxic effects on healthy neurons [33]. This increase in proinflammatory gene expression by mutant HTT protein was driven through increased expression and transcriptional activities of the myeloid lineage determining factors PU.1 and CCAT/enhancer-binding protein [33]. The enhanced transcriptional activities of PU.1 and CCAT/enhancer-binding protein is associated with increased expression of proinflammatory cytokines such as interleukin-6 (IL-6) and tumor necrosis factor-α (TNFα), which can perpetuate further inflammation and tissue damage. This heightened response in mutant HTT-expressing microglia could be due to a priming mechanism, as microglial priming results in an exaggerated response to a second stimulus. In the context of HD, mutant HTT induced neuronal death leading to generation of cellular debris could be the second stimulus for mutant HTT expressing microglia [53]. Contrary to multiple lines of evidence suggesting detrimental effects of mutant HTT expressing microglia, a recent study demonstrated that depletion of mutant HTT in microglia of BACHD mice showed no significant improvement in behavioral performance and neuropathological phenotype [54]. However, depleting mutant HTT in neurons and astrocytes resulted in rescue of behavioral performance and neuropathological phenotype [54]. Although this observation is from one mouse model of HD, this clearly suggests that overexpression of mutant HTT only in microglia may not be sufficient to cause a direct HD phenotype, which instead requires ubiquitous expression of mutant HTT in the brain. Whether mutant HTT expression in microglia alone is sufficient to drive the production and release of inflammatory mediators and increase phagocytotic activity (and how this contributes to HD neuropathology) will need to be tested in multiple animal models to corroborate these findings.

4   Major Signaling Pathways Underlying Microglial Activation in HD

A number of key signaling pathways including Nuclear Factor Kappa B (NF-κB) pathway and the kynurenine pathway have been implicated in the activation of microglia in HD. The NF-κB pathway is a key pathway through which mutant HTT induces activation of microglia in HD [55, 56]. Mutant HTT activates the I kappa B kinase (IKK) γ subunit of the IKK complex, which in turn upregulates the NF-κB signaling pathway and promotes the gene expression of pro-inflammatory cytokines [56–60]. This finding was supported by a separate study using proximity ligation assays to demonstrate that mutant HTT interacts with IKKγ and increases its activity to mediate transcriptional changes of NF-κB pathway, resulting in elevated levels of pro-inflammatory cytokines such as IL-1β, IL-6, IL-8, and TNFα [32]. Furthermore, lowering mutant HTT protein levels in microglia by using targeted siRNA technology alleviated excessive NF-κB pathway activation and reduced the levels of proinflammatory cytokines [32]. Microglia also express toll-like receptors to recognize extracellular stimuli like mutant HTT protein, and its signaling involves intracellular adaptor protein myeloid differentiated 88, which triggers downstream NF-κB signaling cascade and thus increases the production of pro-inflammatory cytokines such as TNFα, IL-6, and IL-10 [61].

Another pathway of interest is the kynurenine pathway, where its activation has been clearly documented in HD. The kynurenine pathway produces several metabolites with neuroactive properties via metabolism of L-tryptophan [62]. Tryptophan is metabolized to neurotoxic quinolinic acid (QUIN) and 3-hydroxykynurenine (3HK) by the enzyme kynurenine 3-monooxygenase (KMO). QUIN and 3HK have been postulated to have neurotoxic roles as QUIN is a selective agonist for N-methyl-D-aspartic acid receptors while 3HK can potentiate QUIN activity. Interestingly, genetic deletion of KMO showed reduction in mutant HTT induced neurotoxicity, suggesting mutant HTT toxicity is mediated through the production of QUIN and 3HK [63]. KMO is mainly expressed by microglia [64] and the metabolites QUIN and 3HK have been reported to be upregulated in the striatum and cortex of low-grade HD brains, suggesting that a neurotoxic contribution of the kynurenine pathway could be involved in the early stages of HD pathophysiology [65]. A successive study also demonstrated that microglia isolated from R6/2 mice had increased KMO activity and increased secreted levels of 3HK, which contributed to the elevated levels in the striatum and cortex of these mice [66]. Further, this increase in KMO activity and 3HK levels were reduced in R6/2 mice through the use of histone deacetylase inhibitors suggesting that the use of these inhibitors as potential therapeutics to treat HD by reducing microglial activation [66]. Another

220   John D. Lee et al.

rate-limiting enzyme of the kynurenine pathway is the indoleamine 2,3-dioxgenase (IDO), which is localized to the microglia [67]. A recent study using the transgenic N171-82Q mouse model showed that IDO activity was significantly upregulated in these mice [68]. Moreover, elevated IDO activity in the early stage of HD contributed to the increased levels of QUIN and 3HK. A schematic summary of microglia activation involved in HD is presented in Fig. 1.

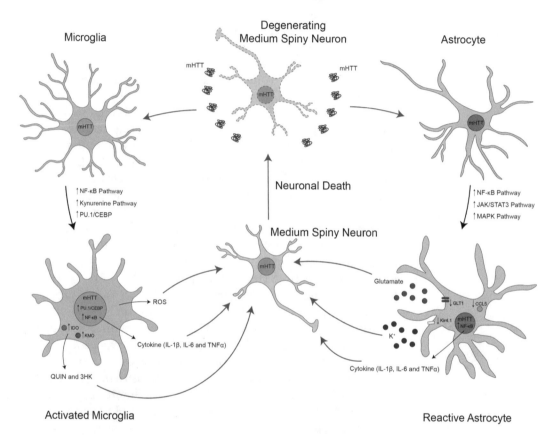

Fig. 1 Microglial and astrocyte activation and its contribution to neuronal death in Huntington's disease (HD). Expression of mutant Huntingtin (mHTT) in medium spiny neurons within the striatum triggers cell-autonomous neuronal degeneration and release mHTT. In parallel, expression of mHTT in microglia and mHTT released from neurons triggers activation of microglia via NF-κB and the kynurenine pathway and increases expression and transcriptional activities of myeloid lineage determine factors PU.1 and CCAT/enhancer-binding protein (CEBP). Microglial activation leads to the release of proinflammatory cytokines (IL-1β, IL-6 and TNFα), reactive oxygen species (ROS) and neurotoxic metabolites such as quinolinic acid (QUIN) and 3-hydroxykynurenine (3HK). These molecules released from activated microglia can induce further neuronal damage and death. Expression of mHTT in astrocytes leads to reduction in CCL5, Kir4.1, and GLT1, which leads to glutamate excitotoxicity and hyperexcitability of neurons. Expression of mHTT can also induce a neurotoxic phenotype of astrocytes via JAK/STAT3, NF-κB, and MAPK pathways, which leads to the production and release of proinflammatory cytokines (IL-1β, IL-6, and TNFα). Enhanced microglial and astrocyte activation with continued neuronal death results in a self-perpetuating cycle of cytotoxicity and inflammation propagating neurodegeneration in HD

5 Potential Microglial Biomarkers Implicated in HD

Many studies have now provided compelling evidence that microglia are able to be activated through mutant HTT expression in neurons, or microglia, via different pathways. The NF-κB and kynurenine pathways appear to be crucial, and their activation can lead to increase in production of proinflammatory biomarkers such as cytokines (IL-1β, IL-6, IL-10, and TNF-α), chemokines (IL-8, CCL2, and CCL20) and neurotoxic metabolites (QUIN and 3HK), that can contribute to the progression of HD pathogenesis. Biomarkers in biofluids such as cerebrospinal fluid (CSF) and blood have previously been quantified due to advantageous properties which include minimal invasiveness, reliable quantification and bulk processing of samples which can be performed retrospectively. Additionally, a single sample of CSF and blood can also produce results for multiple analytes of interest. Biofluid inflammatory markers that can reflect the neuropathology evident in HD have great potential to reliably track disease progression and demonstrate measurable responses to evaluate anti-inflammatory therapeutic interventions. Indeed, microglial associated proinflammatory markers have been heavily investigated with numerous studies demonstrating increased levels of proinflammatory products such as IL-6 and IL-8 in the CSF and blood of HD patients utilizing enzyme-linked immunosorbent assay (ELISA) and Mesoscale Discovery electrochemiluminescence assays [27, 69]. These studies also reported elevated levels of IL-6 in premanifest HD patients (16 years prior to predicted onset of clinical symptoms), suggesting that microglial activation occurs prior to any clinical phenotypes [69]. This is crucial as it opens up the possibility of evaluating disease modifying anti-inflammatory treatments in the premanifest phase for future clinical trials. Furthermore, other studies have also shown that IL-6 levels inversely correlate with the unified HD rating scale, functional capacity and cognitive tests [25, 27], indicating that circulating IL-6 levels could be a good prognostic marker for HD. More importantly these studies also suggested that levels of IL-6 and IL-8 in the CSF and blood show correlation, indicating a key link between central and peripheral immune activation in HD [69]. Although these products have been shown to cross the blood brain barrier directly [70] and enter the bloodstream, one counter to this assumption is that these abnormalities in the CNS and blood could be due to ubiquitous expression of mutant HTT, causing parallel changes centrally (microglia) and peripherally (monocytes/macrophages).

In addition to the biofluid inflammatory markers, many studies have also focused on PET imaging using $[^{11}C]$PK11195 tracer and second generation $[^{11}C]$PBR28 tracers to investigate microglial activation [29, 30, 48, 71]. As mentioned above, PET studies demonstrated that microglial activation was greater in HD patients and correlated with disease progression (i.e., disease severity). In support of the studies showing elevated IL-6 levels in premanifest HD patients, PET imaging also detected microglial activation prior to disease phenotypes [30, 48]. To further strengthen the claim that microglial activation is an early contributor to HD pathology, a recent study combined [^{11}C]PK11195 PET imaging and quantification of blood inflammatory biomarkers. This study demonstrated that plasma levels of IL-1β, IL-6, IL-8, and TNF-α positively correlated with increased [^{11}C]PK11195 binding in the cortex of premanifest HD patients, who were more than 10 years from their predicted symptom onset [72]. These studies further demonstrate a direct link between peripheral and central immune dysfunction in HD and support the role of neuroinflammation in pathogenesis of HD. Although neuroinflammation remains relatively underexplored in HD compared to other neurodegenerative diseases, this provides an exciting avenue of research as there is compelling evidence to suggest that microglial activation is a key driver of HD pathogenesis and microglial associated morphological and molecular changes could be a potential biomarker to monitor disease progression and treatment response in HD.

6 Astrocytes Activation in HD

Astrocytes are one of the most abundant glial cell types with various structures and functions and are ubiquitously expressed in the CNS [73, 74]. Astrocytes are critical components of neural circuits, where they provide physical support to neurons and microglia by forming a matrix-type structure. There is increasing evidence that astrocytes are also associated with various aspects of physiological function, including neurovascular coupling, stabilizing the blood brain barrier, regulation of synaptic function, ion homeostasis, neurotransmitter clearance, metabolic support of neurons, and regulating action potential waveforms and synaptic plasticity [73, 75–80]. Astrocytes are also required to remove toxic materials such as excess extracellular glutamate from the microenvironment. Astrocytes are actively maintained in a resting state. However, following an insult to the CNS, astrocytes respond by undergoing a process known as reactive astrogliosis, where they undergo significant morphological and gene expression alterations as they become hypertrophic, and present with increased expression of the key astrocyte marker glial fibrillary acidic protein [81, 82]. Reactive astrocytes also display functional changes in glutamate, ion homeostasis, and energy metabolism [50], suggesting a shift in their normal housekeeping functions during CNS injury. This reactive astrogliosis process has been shown to be both protective and detrimental as many studies have demonstrated that astrocytes can amplify the inflammatory process initiated by microglia [83], assist in cleaning up cellular debris [84] and also be involved in tissue repair [85].

Astrocyte dysfunction is one of the major contributors to HD pathology. Increased numbers of astrocytes have been demonstrated in postmortem brain tissue of HD patients [47]. In contrast to microglial activation that is detected prior to symptom onset and neuropathological changes, no reactive astrocytes are observed in premanifest (Grade 0) HD brains. Reactive astrocytes were only observed after neurodegeneration occurred, and the number of reactive astrocytes correlated with the striatal neurodegeneration (i.e., disease severity) [11, 45, 86]. This indicates that unlike microgliosis, astrogliosis may not be a contributor to the early stage of HD pathology. A prominent upregulation of astrocytes has also been detected in HD animal models at later-disease stages [73, 87], suggesting that reactive astrogliosis could be a key driver of late-stage neuroinflammation that contributes to HD progression.

7 Mutant HTT and Astrocyte Activation

Numerous studies have now indicated that mutant HTT-expressing astrocytes are defective in performing housekeeping functions that maintain neuronal health. Previous studies have documented that mutant HTT protein is expressed and aggregates in astrocytes, which contributes to neuronal excitotoxicity through reduced expression of glutamate transporter EAAT2/GLT1 [87]. This reduction in glutamate transporter has been noted as one of the earliest signs of astrocyte dysfunction in HD patients and several mouse models of HD [88–93], which results in impaired glutamine uptake and leads to alterations in synaptic transmission [94]. A recent study showed that Ceftriaxone administration (a beta-lactam antibiotic known to elevate GLT1) to R6/2 mice attenuated several HD behavioral phenotypes [95, 96]. This improvement was correlated with increased striatal and cortical GLT1 levels, and increased striatal glutamate uptake [97]. It has also been reported that mutant HTT-expressing astrocytes can contribute to HD pathogenesis through reducing the availability of chemokine ligand 5 (CCL5) to neurons [98]. The mutant HTT protein in astrocytes reduced the transcription of *CCL5* gene in an NF-κB dependent pathway and inhibited astrocyte secretion of CCL5 [98]. The lack of CCL5 in neurons can lead to detrimental effects on neuronal functions, ultimately contributing to neuronal death [98]. Astrocytes with nuclear mutant HTT inclusions also show decreased expression of the Kir4.1 channel, leading to impaired K^+ homeostasis in HD [99]. The reduction in Kir4.1 channels was further supported by data from R6/2 and Q175 mouse HD models, where decreased levels of Kir4.1 channel was observed in the striatum, mainly due to a reduction in striatal astrocytes [99]. Defective K^+ homeostasis leads to increased extracellular K^+ levels, which causes hyperexcitability-mediated death of medium spiny neurons in HD [100]. Viral delivery of Kir4.1 channel in the striatum ameliorated medium spiny neuron dysfunction, diminished motor phenotypes and extended survival in R6/2 mice [99]. Furthermore, mutant HTT protein expression in astrocytes has also been shown to reduce the release of brain derived neurotrophic factor that is essential in promoting neuronal survival [101–103]. Taken together, these studies suggest that mutant HTT-expressing astrocytes contribute to HD pathogenesis through their inability to perform their standard housekeeping function in order to promote and maintain a healthy microenvironment for neurons. In addition to their defective physiological functions, exciting recent studies have also provided compelling evidence that reactive astrocytes can contribute to HD pathology through transcriptional activation of proinflammatory genes to perpetuate a chronic inflammatory state [50, 104, 105]. The neurotoxic reactive astrocytes termed A1 astrocytes have been shown to be induced by activated microglia and consequently result in a similar expression profile of proinflammatory cytokines, which can exacerbate neuronal death in HD [106, 107]. These neurotoxic astrocytes are associated with intracellular pathways such as the Janus kinase/signal transducer and activator of transcription 3 (JAK/STAT3) [82], NF-κB [108], and mitogen-activated protein kinase (MAPK), which are all involved in the production and release of proinflammatory cytokines and chemokines [81].

Overall, the precise role of reactive astrocytes in HD progression is still not well understood, however many studies have now demonstrated that reactive astrocytes predominantly possess detrimental roles in HD pathogenesis through the loss of neuroprotective function, or gain of neurotoxic properties [83]. Dysfunctional astrocytes expressing mutant HTT lose both their neuronal support abilities as well as gaining a neurotoxic phenotype that can exacerbate neuronal death and contribute to HD pathology. Similar to microglial activation, as neurotoxic astrocytes can release toxic signals such as proinflammatory cytokines (IL-1β, IL-6, IL-10, and TNF-α), this has the potential to be used as an objective biomarker to monitor disease progression and treatment response in HD. A schematic summary of astrocyte activation involved in HD is presented in Fig. 1.

8 Peripheral Inflammation in HD

In addition to reactive gliosis in the CNS, HD is also associated with a hyperactive and dysregulated peripheral immune response. Here, mutant HTT is expressed ubiquitously and primes leukocytes for activation whilst also imparting a number of cellular defects. During the asymptomatic phase, these cell-autonomous processes drive a form of low-grade inflammation, but as the blood brain barrier breaks down, hyperactive immune cells and their neurotoxic cytokines begin to invade the CNS and rapidly accelerate disease progression. However, despite this simplistic view, a number of caveats exist in this area of research. Indeed, HD has a number of extracranial pathologies that may contribute to or confound peripheral neuroinflammation and some studies even suggest that immune stimulation may ameliorate certain defects and protect against neurodegeneration. Thus, further research into the peripheral immune response in HD is needed to capitalize on its potential as a biomarker and therapeutic target.

In HD, mutant HTT has a cell autonomous effect on peripheral leukocytes that drives their hyperactivity and dysregulation [109]. Foremostly, in monocytes mutant HTT directly interacts with the NF-κB transcription factor and upregulates the expression of a number of proinflammatory cytokines (e.g., IL-1β, IL-6, TNFα) that can be detected in the plasma of pre- and manifest individuals [32, 34, 69]. In addition, these monocytes have increased phagocytic capacity and reduced chemotactic ability [35, 36] and mutant HTT can also downregulate mast cell cytokine production and ostensibly prime natural killer cells for activation [110, 111]. Interestingly, these processes may precede immune activity in the CNS and thus offer a window of opportunity for drug development to intervene prior to overt neurodegeneration [112]. Indeed, pharmacological modification of these intrinsic pathways can improve survival in HD animal models [113]. Thus, these cell autonomous processes play a key role in priming leukocytes for proinflammatory and neurotoxic action in HD.

Overlying these intrinsic inflammatory alterations, HD is also associated with a peripheral neuroinflammatory response that is akin to that in other neurodegenerative diseases. Initially, inflammation is confined to the CNS, but as the disease progresses the blood–brain barrier breaks down and allow various immunostimulatory substances to leak into the circulatory system and activate peripheral leukocytes [114]. However, myeloid cells in HD have chemotactic deficits that limit their invasion into the CNS and force them to exert their influence at a distance via the production of various proinflammatory cytokines (e.g., IL-4, IL-6, IL-10, IL-12, TNFα) [17, 25, 27, 35, 69, 112, 115, 116]. Hence, peripheral inflammation in HD is likely to be primarily mediated by lymphoid cells [117]. Indeed, HD patients have increased numbers of Th17.1 cell in their cerebrospinal fluid [118] and a small number of studies suggest the presence of a humoral response directed against HD-specific CNS antigens and the angiotensin II type 1 receptor, which may further promote T cell activation [119, 120]. Many of these processes have been correlated with disease progression and their pharmacological inhibition has been shown to improve neurological deficits in HD animal models [19, 59, 121]. Thus, further research is needed to harness their therapeutic and biomarker potential.

However, there are a number of caveats to peripheral immunity in HD. Firstly, owing to the ubiquitous expression of the huntingtin gene, HD is associated with a number of extracranial manifestations including cardiac dysfunction, skeletal muscle atrophy, osteoporosis, weight loss, and impaired glucose tolerance [16]. These processes are themselves linked to a number of inflammatory processes including elevated levels of IL-6, IL-8, and TNFα and thus interpretation of peripheral immune findings in HD must be placed in the context of multiple disease processes [122–124]. Secondly, there is some evidence to suggest that peripheral inflammation may be protective in HD. Indeed, in animal models of HD, chronic exposure to low-dose LPS can increase survival whilst genetic ablation of interleukin-1 receptor exacerbates neurological symptoms and increases mutant HTT protein accumulation in the CNS [125, 126]. Hence, further research into the peripheral immune response in HD is needed to delineate its protective and detrimental effects. These caveats highlight the complexity of HD immunology and underscore the need for a precise immunomodulatory approach in therapeutic development.

9   Concluding Remarks

HD is a debilitating neurodegenerative condition with no currently available disease modifying treatments. However, emerging evidence suggests that a hyperactive central and peripheral immune response plays a key role in accelerating disease progression in HD patients and animal models. Although peripheral and CNS inflammation has been established as one of the main hallmarks of HD, and modulation of neuroinflammation has been suggested as a potential target for therapeutic intervention, the relationship between inflammation markers and the disease pathology is still poorly understood. Based on prior studies as summarized in this chapter, it is highly plausible that specific microglial and astrocyte activation signatures, identified as either soluble factors present in the CSF or blood, or identified using brain imaging, can offer complementary information, providing a comprehensive evaluation of disease progression to inform clinical trial design and to serve as surrogate endpoints. Although utilization of neuroinflammatory biomarkers have made a significant progress, future developments and further longitudinal validation in this field are greatly anticipated.

References

1. Gusella JF, Wexler NS, Conneally PM, Naylor SL, Anderson MA, Tanzi RE et al (1983) A polymorphic DNA marker genetically linked to Huntington's disease. Nature 306 (5940):234–238. https://doi.org/10.1038/306234a0

2. Andrew SE, Goldberg YP, Kremer B, Telenius H, Theilmann J, Adam S et al (1993) The relationship between trinucleotide (CAG) repeat length and clinical features of Huntington's disease. Nat Genet 4 (4):398–403. https://doi.org/10.1038/ng0893-398

3. Duyao M, Ambrose C, Myers R, Novelletto A, Persichetti F, Frontali M et al (1993) Trinucleotide repeat length instability and age of onset in Huntington's disease. Nat Genet 4(4):387–392. https://doi.org/10.1038/ng0893-387

4. Snell RG, MacMillan JC, Cheadle JP, Fenton I, Lazarou LP, Davies P et al (1993) Relationship between trinucleotide repeat expansion and phenotypic variation in Huntington's disease. Nat Genet 4(4):393–397. https://doi.org/10.1038/ng0893-393

5. Ross CA, Tabrizi SJ (2011) Huntington's disease: from molecular pathogenesis to clinical treatment. Lancet Neurol 10(1):83–98. https://doi.org/10.1016/S1474-4422(10)70245-3

6. Tourette C, Li B, Bell R, O'Hare S, Kaltenbach LS, Mooney SD et al (2014) A large scale Huntingtin protein interaction network implicates Rho GTPase signaling pathways in Huntington disease. J Biol Chem 289 (10):6709–6726. https://doi.org/10.1074/jbc.M113.523696

7. Yanai A, Huang K, Kang R, Singaraja RR, Arstikaitis P, Gan L et al (2006) Palmitoylation of huntingtin by HIP14 is essential for its trafficking and function. Nat Neurosci 9 (6):824–831. https://doi.org/10.1038/nn1702

8. Cattaneo E, Zuccato C, Tartari M (2005) Normal huntingtin function: an alternative approach to Huntington's disease. Nat Rev Neurosci 6(12):919–930. https://doi.org/10.1038/nrn1806

9. Bates GP, Dorsey R, Gusella JF, Hayden MR, Kay C, Leavitt BR et al (2015) Huntington disease. Nat Rev Dis Primers 1:15005. https://doi.org/10.1038/nrdp.2015.5

10. Masnata M, Sciacca G, Maxan A, Bousset L, Denis HL, Lauruol F et al (2019) Demonstration of prion-like properties of mutant huntingtin fibrils in both in vitro and in vivo paradigms. Acta Neuropathol 137 (6):981–1001. https://doi.org/10.1007/s00401-019-01973-6

11. Rub U, Vonsattel JP, Heinsen H, Korf HW (2015) The Neuropathology of Huntington's disease: classical findings, recent developments and correlation to functional neuroanatomy. Adv Anat Embryol Cell Biol 217:1–146

12. Tabrizi SJ, Langbehn DR, Leavitt BR, Roos RA, Durr A, Craufurd D et al (2009) Biological and clinical manifestations of Huntington's disease in the longitudinal TRACK-HD study: cross-sectional analysis of baseline data. Lancet Neurol 8(9):791–801. https://doi.org/10.1016/S1474-4422(09)70170-X

13. Tabrizi SJ, Reilmann R, Roos RA, Durr A, Leavitt B, Owen G et al (2012) Potential endpoints for clinical trials in premanifest and early Huntington's disease in the TRACK-HD study: analysis of 24 month observational data. Lancet Neurol 11(1):42–53. https://doi.org/10.1016/S1474-4422(11)70263-0

14. Tabrizi SJ, Scahill RI, Owen G, Durr A, Leavitt BR, Roos RA et al (2013) Predictors of phenotypic progression and disease onset in premanifest and early-stage Huntington's disease in the TRACK-HD study: analysis of 36-month observational data. Lancet Neurol 12(7):637–649. https://doi.org/10.1016/S1474-4422(13)70088-7

15. Carroll JB, Bates GP, Steffan J, Saft C, Tabrizi SJ (2015) Treating the whole body in Huntington's disease. Lancet Neurol 14 (11):1135–1142. https://doi.org/10.1016/S1474-4422(15)00177-5

16. van der Burg JM, Bjorkqvist M, Brundin P (2009) Beyond the brain: widespread pathology in Huntington's disease. Lancet Neurol 8 (8):765–774. https://doi.org/10.1016/S1474-4422(09)70178-4

17. Crotti A, Glass CK (2015) The choreography of neuroinflammation in Huntington's disease. Trends Immunol 36(6):364–373. https://doi.org/10.1016/j.it.2015.04.007

18. Ellrichmann G, Reick C, Saft C, Linker RA (2013) The role of the immune system in Huntington's disease. Clin Dev Immunol 2013:541259. https://doi.org/10.1155/2013/541259

19. Kwan W, Magnusson A, Chou A, Adame A, Carson MJ, Kohsaka S et al (2012) Bone marrow transplantation confers modest benefits in mouse models of Huntington's disease. J

Neurosci 32(1):133–142. https://doi.org/10.1523/JNEUROSCI.4846-11.2012

20. Woodruff TM, Crane JW, Proctor LM, Buller KM, Shek AB, de Vos K et al (2006) Therapeutic activity of C5a receptor antagonists in a rat model of neurodegeneration. FASEB J 20 (9):1407–1417. https://doi.org/10.1096/fj.05-5814com

21. Yang HM, Yang S, Huang SS, Tang BS, Guo JF (2017) Microglial activation in the pathogenesis of Huntington's disease. Front Aging Neurosci 9:193. https://doi.org/10.3389/fnagi.2017.00193

22. Silvestroni A, Faull RL, Strand AD, Moller T (2009) Distinct neuroinflammatory profile in post-mortem human Huntington's disease. Neuroreport 20(12):1098–1103. https://doi.org/10.1097/WNR.0b013e32832e34ee

23. Sorolla MA, Reverter-Branchat G, Tamarit J, Ferrer I, Ros J, Cabiscol E (2008) Proteomic and oxidative stress analysis in human brain samples of Huntington disease. Free Radic Biol Med 45(5):667–678. https://doi.org/10.1016/j.freeradbiomed.2008.05.014

24. Singhrao SK, Neal JW, Morgan BP, Gasque P (1999) Increased complement biosynthesis by microglia and complement activation on neurons in Huntington's disease. Exp Neurol 159 (2):362–376. https://doi.org/10.1006/exnr.1999.7170

25. Bouwens JA, van Duijn E, Cobbaert CM, Roos RA, van der Mast RC, Giltay EJ (2016) Plasma cytokine levels in relation to neuropsychiatric symptoms and cognitive dysfunction in Huntington's disease. J Huntingtons Dis 5(4):369–377. https://doi.org/10.3233/JHD-160213

26. Bouwens JA, van Duijn E, Cobbaert CM, Roos RAC, van der Mast RC, Giltay EJ (2017) Disease stage and plasma levels of cytokines in Huntington's disease: a 2-year follow-up study. Mov Disord 32 (7):1103–1104. https://doi.org/10.1002/mds.26950

27. Chang KH, Wu YR, Chen YC, Chen CM (2015) Plasma inflammatory biomarkers for Huntington's disease patients and mouse model. Brain Behav Immun 44:121–127. https://doi.org/10.1016/j.bbi.2014.09.011

28. Chen CM, Wu YR, Cheng ML, Liu JL, Lee YM, Lee PW et al (2007) Increased oxidative damage and mitochondrial abnormalities in the peripheral blood of Huntington's disease patients. Biochem Biophys Res Commun 359 (2):335–340. https://doi.org/10.1016/j.bbrc.2007.05.093

29. Pavese N, Gerhard A, Tai YF, Ho AK, Turkheimer F, Barker RA et al (2006) Microglial activation correlates with severity in Huntington disease: a clinical and PET study. Neurology 66(11):1638–1643. https://doi.org/10.1212/01.wnl.0000222734.56412.17

30. Politis M, Pavese N, Tai YF, Kiferle L, Mason SL, Brooks DJ et al (2011) Microglial activation in regions related to cognitive function predicts disease onset in Huntington's disease: a multimodal imaging study. Hum Brain Mapp 32(2):258–270. https://doi.org/10.1002/hbm.21008

31. Rodrigues FB, Byrne LM, McColgan P, Robertson N, Tabrizi SJ, Zetterberg H et al (2016) Cerebrospinal fluid inflammatory biomarkers reflect clinical severity in Huntington's disease. PLoS One 11(9):e0163479. https://doi.org/10.1371/journal.pone.0163479

32. Trager U, Andre R, Lahiri N, Magnusson-Lind A, Weiss A, Grueninger S et al (2014) HTT-lowering reverses Huntington's disease immune dysfunction caused by NFkappaB pathway dysregulation. Brain 137 (Pt 3):819–833. https://doi.org/10.1093/brain/awt355

33. Crotti A, Benner C, Kerman BE, Gosselin D, Lagier-Tourenne C, Zuccato C et al (2014) Mutant Huntingtin promotes autonomous microglia activation via myeloid lineage-determining factors. Nat Neurosci 17 (4):513–521. https://doi.org/10.1038/nn.3668

34. Miller JR, Lo KK, Andre R, Hensman Moss DJ, Trager U, Stone TC et al (2016) RNA-Seq of Huntington's disease patient myeloid cells reveals innate transcriptional dysregulation associated with proinflammatory pathway activation. Hum Mol Genet 25 (14):2893–2904. https://doi.org/10.1093/hmg/ddw142

35. Kwan W, Trager U, Davalos D, Chou A, Bouchard J, Andre R et al (2012) Mutant huntingtin impairs immune cell migration in Huntington disease. J Clin Invest 122 (12):4737–4747. https://doi.org/10.1172/JCI64484

36. Trager U, Andre R, Magnusson-Lind A, Miller JR, Connolly C, Weiss A et al (2015) Characterisation of immune cell function in fragment and full-length Huntington's disease mouse models. Neurobiol Dis 73:388–398. https://doi.org/10.1016/j.nbd.2014.10.012

37. Li Q, Barres BA (2018) Microglia and macrophages in brain homeostasis and disease. Nat

Rev Immunol 18(4):225–242. https://doi.org/10.1038/nri.2017.125

38. Ransohoff RM, Perry VH (2009) Microglial physiology: unique stimuli, specialized responses. Annu Rev Immunol 27:119–145. https://doi.org/10.1146/annurev.immunol.021908.132528

39. Weinstein JR, Koerner IP, Moller T (2010) Microglia in ischemic brain injury. Fut Neurol 5(2):227–246. https://doi.org/10.2217/fnl.10.1

40. Cuadros MA, Navascues J (1998) The origin and differentiation of microglial cells during development. Prog Neurobiol 56(2):173–189. https://doi.org/10.1016/s0301-0082(98)00035-5

41. Gomez-Nicola D, Fransen NL, Suzzi S, Perry VH (2013) Regulation of microglial proliferation during chronic neurodegeneration. J Neurosci 33(6):2481–2493. https://doi.org/10.1523/JNEUROSCI.4440-12.2013

42. Jimenez-Ferrer I, Jewett M, Tontanahal A, Romero-Ramos M, Swanberg M (2017) Allelic difference in Mhc2ta confers altered microglial activation and susceptibility to alpha-synuclein-induced dopaminergic neurodegeneration. Neurobiol Dis 106:279–290. https://doi.org/10.1016/j.nbd.2017.07.016

43. Kettenmann H, Hanisch UK, Noda M, Verkhratsky A (2011) Physiology of microglia. Physiol Rev 91(2):461–553. https://doi.org/10.1152/physrev.00011.2010

44. Sierra A, Abiega O, Shahraz A, Neumann H (2013) Janus-faced microglia: beneficial and detrimental consequences of microglial phagocytosis. Front Cell Neurosci 7:6. https://doi.org/10.3389/fncel.2013.00006

45. Vonsattel JP, Myers RH, Stevens TJ, Ferrante RJ, Bird ED, Richardson EP Jr (1985) Neuropathological classification of Huntington's disease. J Neuropathol Exp Neurol 44(6):559–577. https://doi.org/10.1097/00005072-198511000-00003

46. Sapp E, Kegel KB, Aronin N, Hashikawa T, Uchiyama Y, Tohyama K et al (2001) Early and progressive accumulation of reactive microglia in the Huntington disease brain. J Neuropathol Exp Neurol 60(2):161–172. https://doi.org/10.1093/jnen/60.2.161

47. Vonsattel JP, Keller C, Cortes Ramirez EP (2011) Huntington's disease—neuropathology. Handb Clin Neurol 100:83–100. https://doi.org/10.1016/B978-0-444-52014-2.00004-5

48. Tai YF, Pavese N, Gerhard A, Tabrizi SJ, Barker RA, Brooks DJ et al (2007) Microglial activation in presymptomatic Huntington's disease gene carriers. Brain 130(Pt 7):1759–1766. https://doi.org/10.1093/brain/awm044

49. Simmons DA, Casale M, Alcon B, Pham N, Narayan N, Lynch G (2007) Ferritin accumulation in dystrophic microglia is an early event in the development of Huntington's disease. Glia 55(10):1074–1084. https://doi.org/10.1002/glia.20526

50. Palpagama TH, Waldvogel HJ, Faull RLM, Kwakowsky A (2019) The role of microglia and astrocytes in Huntington's disease. Front Mol Neurosci 12:258. https://doi.org/10.3389/fnmol.2019.00258

51. Kraft AD, Kaltenbach LS, Lo DC, Harry GJ (2012) Activated microglia proliferate at neurites of mutant huntingtin-expressing neurons. Neurobiol Aging 33(3):621 e17–621 e33. https://doi.org/10.1016/j.neurobiolaging.2011.02.015

52. Wang N, Gray M, Lu XH, Cantle JP, Holley SM, Greiner E et al (2014) Neuronal targets for reducing mutant huntingtin expression to ameliorate disease in a mouse model of Huntington's disease. Nat Med 20(5):536–541. https://doi.org/10.1038/nm.3514

53. Li JW, Zong Y, Cao XP, Tan L, Tan L (2018) Microglial priming in Alzheimer's disease. Ann Transl Med 6(10):176. https://doi.org/10.21037/atm.2018.04.22

54. Petkau TL, Hill A, Connolly C, Lu G, Wagner P, Kosior N et al (2019) Mutant huntingtin expression in microglia is neither required nor sufficient to cause the Huntington's disease-like phenotype in BACHD mice. Hum Mol Genet 28(10):1661–1670. https://doi.org/10.1093/hmg/ddz009

55. Khoshnan A, Ko J, Watkin EE, Paige LA, Reinhart PH, Patterson PH (2004) Activation of the IkappaB kinase complex and nuclear factor-kappaB contributes to mutant huntingtin neurotoxicity. J Neurosci 24(37):7999–8008. https://doi.org/10.1523/JNEUROSCI.2675-04.2004

56. Khoshnan A, Patterson PH (2011) The role of IkappaB kinase complex in the neurobiology of Huntington's disease. Neurobiol Dis 43(2):305–311. https://doi.org/10.1016/j.nbd.2011.04.015

57. Hacker H, Karin M (2006) Regulation and function of IKK and IKK-related kinases. Sci STKE 2006(357):re13. https://doi.org/10.1126/stke.3572006re13

58. Hsiao HY, Chen YC, Chen HM, Tu PH, Chern Y (2013) A critical role of astrocyte-mediated nuclear factor-kappaB-dependent

inflammation in Huntington's disease. Hum Mol Genet 22(9):1826–1842. https://doi.org/10.1093/hmg/ddt036

59. Hsiao HY, Chiu FL, Chen CM, Wu YR, Chen HM, Chen YC et al (2014) Inhibition of soluble tumor necrosis factor is therapeutic in Huntington's disease. Hum Mol Genet 23 (16):4328–4344. https://doi.org/10.1093/hmg/ddu151

60. Siew JJ, Chen HM, Chen HY, Chen HL, Chen CM, Soong BW et al (2019) Galectin-3 is required for the microglia-mediated brain inflammation in a model of Huntington's disease. Nat Commun 10(1):3473. https://doi.org/10.1038/s41467-019-11441-0

61. Takeda K, Akira S (2004) TLR signaling pathways. Semin Immunol 16(1):3–9. https://doi.org/10.1016/j.smim.2003.10.003

62. Schwarcz R, Guidetti P, Sathyasaikumar KV, Muchowski PJ (2010) Of mice, rats and men: revisiting the quinolinic acid hypothesis of Huntington's disease. Prog Neurobiol 90 (2):230–245. https://doi.org/10.1016/j.pneurobio.2009.04.005

63. Giorgini F, Guidetti P, Nguyen Q, Bennett SC, Muchowski PJ (2005) A genomic screen in yeast implicates kynurenine 3-monooxygenase as a therapeutic target for Huntington disease. Nat Genet 37 (5):526–531. https://doi.org/10.1038/ng1542

64. Guillemin GJ, Smith DG, Smythe GA, Armati PJ, Brew BJ (2003) Expression of the kynurenine pathway enzymes in human microglia and macrophages. Adv Exp Med Biol 527:105–112. https://doi.org/10.1007/978-1-4615-0135-0_12

65. Guidetti P, Luthi-Carter RE, Augood SJ, Schwarcz R (2004) Neostriatal and cortical quinolinate levels are increased in early grade Huntington's disease. Neurobiol Dis 17 (3):455–461. https://doi.org/10.1016/j.nbd.2004.07.006

66. Giorgini F, Moller T, Kwan W, Zwilling D, Wacker JL, Hong S et al (2008) Histone deacetylase inhibition modulates kynurenine pathway activation in yeast, microglia, and mice expressing a mutant huntingtin fragment. J Biol Chem 283(12):7390–7400. https://doi.org/10.1074/jbc.M708192200

67. Heyes MP, Achim CL, Wiley CA, Major EO, Saito K, Markey SP (1996) Human microglia convert l-tryptophan into the neurotoxin quinolinic acid. Biochem J 320(Pt 2):595–597. https://doi.org/10.1042/bj3200595

68. Donley DW, Olson AR, Raisbeck MF, Fox JH, Gigley JP (2016) Huntingtons disease mice infected with Toxoplasma gondii demonstrate early Kynurenine pathway activation, altered CD8+ T-cell responses, and premature mortality. PLoS One 11(9):e0162404. https://doi.org/10.1371/journal.pone.0162404

69. Björkqvist M, Wild EJ, Thiele J, Silvestroni A, Andre R, Lahiri N et al (2008) A novel pathogenic pathway of immune activation detectable before clinical onset in Huntington's disease. J Exp Med 205(8):1869–1877. https://doi.org/10.1084/jem.20080178

70. Banks WA, Plotkin SR, Kastin AJ (1995) Permeability of the blood-brain barrier to soluble cytokine receptors. Neuroimmunomodulation 2(3):161–165. https://doi.org/10.1159/000096887

71. Lois C, Gonzalez I, Izquierdo-Garcia D, Zurcher NR, Wilkens P, Loggia ML et al (2018) Neuroinflammation in Huntington's disease: new insights with (11)C-PBR28 PET/MRI. ACS Chem Nerosci 9 (11):2563–2571. https://doi.org/10.1021/acschemneuro.8b00072

72. Politis M, Lahiri N, Niccolini F, Su P, Wu K, Giannetti P et al (2015) Increased central microglial activation associated with peripheral cytokine levels in premanifest Huntington's disease gene carriers. Neurobiol Dis 83:115–121. https://doi.org/10.1016/j.nbd.2015.08.011

73. Khakh BS, Sofroniew MV (2015) Diversity of astrocyte functions and phenotypes in neural circuits. Nat Neurosci 18(7):942–952. https://doi.org/10.1038/nn.4043

74. von Bartheld CS, Bahney J, Herculano-Houzel S (2016) The search for true numbers of neurons and glial cells in the human brain: a review of 150 years of cell counting. J Comp Neurol 524(18):3865–3895. https://doi.org/10.1002/cne.24040

75. Halassa MM, Haydon PG (2010) Integrated brain circuits: astrocytic networks modulate neuronal activity and behavior. Annu Rev Physiol 72:335–355. https://doi.org/10.1146/annurev-physiol-021909-135843

76. Allen NJ (2014) Astrocyte regulation of synaptic behavior. Annu Rev Cell Dev Biol 30:439–463. https://doi.org/10.1146/annurev-cellbio-100913-013053

77. Araque A, Carmignoto G, Haydon PG, Oliet SH, Robitaille R, Volterra A (2014) Gliotransmitters travel in time and space. Neuron 81(4):728–739. https://doi.org/10.1016/j.neuron.2014.02.007

78. Volterra A, Liaudet N, Savtchouk I (2014) Astrocyte Ca(2)(+) signalling: an unexpected

complexity. Nat Rev Neurosci 15 (5):327–335. https://doi.org/10.1038/nrn3725

79. Chung WS, Allen NJ, Eroglu C (2015) Astrocytes control synapse formation, function, and elimination. Cold Spring Harb Perspect Biol 7(9):a020370. https://doi.org/10.1101/cshperspect.a020370

80. Shigetomi E, Patel S, Khakh BS (2016) Probing the complexities of astrocyte calcium signaling. Trends Cell Biol 26(4):300–312. https://doi.org/10.1016/j.tcb.2016.01.003

81. Ben Haim L, Carrillo-de Sauvage MA, Ceyzeriat K, Escartin C (2015) Elusive roles for reactive astrocytes in neurodegenerative diseases. Front Cell Neurosci 9:278. https://doi.org/10.3389/fncel.2015.00278

82. Ben Haim L, Ceyzeriat K, Carrillo-de Sauvage MA, Aubry F, Auregan G, Guillermier M et al (2015) The JAK/STAT3 pathway is a common inducer of astrocyte reactivity in Alzheimer's and Huntington's diseases. J Neurosci 35(6):2817–2829. https://doi.org/10.1523/JNEUROSCI.3516-14.2015

83. Zamanian JL, Xu L, Foo LC, Nouri N, Zhou L, Giffard RG et al (2012) Genomic analysis of reactive astrogliosis. J Neurosci 32 (18):6391–6410. https://doi.org/10.1523/JNEUROSCI.6221-11.2012

84. Wakida NM, Cruz GMS, Ro CC, Moncada EG, Khatibzadeh N, Flanagan LA et al (2018) Phagocytic response of astrocytes to damaged neighboring cells. PLoS One 13(4):e0196153. https://doi.org/10.1371/journal.pone.0196153

85. Gallo V, Deneen B (2014) Glial development: the crossroads of regeneration and repair in the CNS. Neuron 83(2):283–308. https://doi.org/10.1016/j.neuron.2014.06.010

86. Myers RH, Vonsattel JP, Paskevich PA, Kiely DK, Stevens TJ, Cupples LA et al (1991) Decreased neuronal and increased oligodendroglial densities in Huntington's disease caudate nucleus. J Neuropathol Exp Neurol 50 (6):729–742. https://doi.org/10.1097/00005072-199111000-00005

87. Shin JY, Fang ZH, Yu ZX, Wang CE, Li SH, Li XJ (2005) Expression of mutant huntingtin in glial cells contributes to neuronal excitotoxicity. J Cell Biol 171(6):1001–1012. https://doi.org/10.1083/jcb.200508072

88. Menalled LB, Kudwa AE, Miller S, Fitzpatrick J, Watson-Johnson J, Keating N et al (2012) Comprehensive behavioral and molecular characterization of a new knock-in mouse model of Huntington's disease: zQ175. PLoS One 7(12):e49838. https://doi.org/10.1371/journal.pone.0049838

89. Lievens JC, Woodman B, Mahal A, Spasic-Boscovic O, Samuel D, Kerkerian-Le Goff L et al (2001) Impaired glutamate uptake in the R6 Huntington's disease transgenic mice. Neurobiol Dis 8(5):807–821. https://doi.org/10.1006/nbdi.2001.0430

90. Langfelder P, Cantle JP, Chatzopoulou D, Wang N, Gao F, Al-Ramahi I et al (2016) Integrated genomics and proteomics define huntingtin CAG length-dependent networks in mice. Nat Neurosci 19(4):623–633. https://doi.org/10.1038/nn.4256

91. Huang K, Kang MH, Askew C, Kang R, Sanders SS, Wan J et al (2010) Palmitoylation and function of glial glutamate transporter-1 is reduced in the YAC128 mouse model of Huntington disease. Neurobiol Dis 40 (1):207–215. https://doi.org/10.1016/j.nbd.2010.05.027

92. Grewer C, Gameiro A, Rauen T (2014) SLC1 glutamate transporters. Pflugers Arch 466 (1):3–24. https://doi.org/10.1007/s00424-013-1397-7

93. Behrens PF, Franz P, Woodman B, Lindenberg KS, Landwehrmeyer GB (2002) Impaired glutamate transport and glutamate-glutamine cycling: downstream effects of the Huntington mutation. Brain 125 (Pt 8):1908–1922. https://doi.org/10.1093/brain/awf180

94. O'Donovan SM, Sullivan CR, McCullumsmith RE (2017) The role of glutamate transporters in the pathophysiology of neuropsychiatric disorders. NPJ Schizophr 3 (1):32. https://doi.org/10.1038/s41537-017-0037-1

95. Sari Y, Prieto AL, Barton SJ, Miller BR, Rebec GV (2010) Ceftriaxone-induced up-regulation of cortical and striatal GLT1 in the R6/2 model of Huntington's disease. J Biomed Sci 17:62. https://doi.org/10.1186/1423-0127-17-62

96. Miller BR, Dorner JL, Shou M, Sari Y, Barton SJ, Sengelaub DR et al (2008) Up-regulation of GLT1 expression increases glutamate uptake and attenuates the Huntington's disease phenotype in the R6/2 mouse. Neuroscience 153(1):329–337. https://doi.org/10.1016/j.neuroscience.2008.02.004

97. Estrada-Sanchez AM, Rebec GV (2012) Corticostriatal dysfunction and glutamate transporter 1 (GLT1) in Huntington's disease: interactions between neurons and astrocytes. Basal Ganglia 2(2):57–66. https://doi.org/10.1016/j.baga.2012.04.029

98. Chou SY, Weng JY, Lai HL, Liao F, Sun SH, Tu PH et al (2008) Expanded-polyglutamine huntingtin protein suppresses the secretion and production of a chemokine (CCL5/RANTES) by astrocytes. J Neurosci 28 (13):3277–3290. https://doi.org/10.1523/JNEUROSCI.0116-08.2008

99. Tong X, Ao Y, Faas GC, Nwaobi SE, Xu J, Haustein MD et al (2014) Astrocyte Kir4.1 ion channel deficits contribute to neuronal dysfunction in Huntington's disease model mice. Nat Neurosci 17(5):694–703. https://doi.org/10.1038/nn.3691

100. Rebec GV, Barton SJ, Ennis MD (2002) Dysregulation of ascorbate release in the striatum of behaving mice expressing the Huntington's disease gene. J Neurosci 22(2):RC202

101. Boussicault L, Herard AS, Calingasan N, Petit F, Malgorn C, Merienne N et al (2014) Impaired brain energy metabolism in the BACHD mouse model of Huntington's disease: critical role of astrocyte-neuron interactions. J Cereb Blood Flow Metab 34 (9):1500–1510. https://doi.org/10.1038/jcbfm.2014.110

102. Hong Y, Zhao T, Li XJ, Li S (2016) Mutant Huntingtin impairs BDNF release from Astrocytes by disrupting conversion of Rab3a-GTP into Rab3a-GDP. J Neurosci 36 (34):8790–8801. https://doi.org/10.1523/JNEUROSCI.0168-16.2016

103. Wang L, Lin F, Wang J, Wu J, Han R, Zhu L et al (2012) Truncated N-terminal huntingtin fragment with expanded-polyglutamine (htt552-100Q) suppresses brain-derived neurotrophic factor transcription in astrocytes. Acta Biochim Biophys Sin (Shanghai) 44 (3):249–258. https://doi.org/10.1093/abbs/gmr125

104. Iglesias J, Morales L, Barreto GE (2017) Metabolic and inflammatory adaptation of reactive astrocytes: role of PPARs. Mol Neurobiol 54(4):2518–2538. https://doi.org/10.1007/s12035-016-9833-2

105. Sochocka M, Diniz BS, Leszek J (2017) Inflammatory response in the CNS: friend or foe? Mol Neurobiol 54(10):8071–8089. https://doi.org/10.1007/s12035-016-0297-1

106. Liddelow SA, Barres BA (2017) Reactive astrocytes: production, function, and therapeutic potential. Immunity 46(6):957–967. https://doi.org/10.1016/j.immuni.2017.06.006

107. Sofroniew MV (2009) Molecular dissection of reactive astrogliosis and glial scar formation. Trends Neurosci 32(12):638–647. https://doi.org/10.1016/j.tins.2009.08.002

108. Fan Y, Mao R, Yang J (2013) NF-kappaB and STAT3 signaling pathways collaboratively link inflammation to cancer. Protein Cell 4 (3):176–185. https://doi.org/10.1007/s13238-013-2084-3

109. Weiss A, Träger U, Wild EJ, Grueninger S, Farmer R, Landles C et al (2012) Mutant huntingtin fragmentation in immune cells tracks Huntington's disease progression. J Clin Invest 122(10):3731–3736. https://doi.org/10.1172/jci64565

110. Park HJ, Lee SW, Im W, Kim M, Van Kaer L, Hong S (2019) iNKT cell activation exacerbates the development of Huntington's disease in R6/2 transgenic mice. Mediators Inflamm 2019:3540974. https://doi.org/10.1155/2019/3540974

111. Pérez-Rodríguez MJ, Ibarra-Sánchez A, Román-Figueroa A, Pérez-Severiano F, González-Espinosa C (2020) Mutant Huntingtin affects toll-like receptor 4 intracellular trafficking and cytokine production in mast cells. J Neuroinflammation 17(1):95. https://doi.org/10.1186/s12974-020-01758-9

112. Pido-Lopez J, Andre R, Benjamin AC, Ali N, Farag S, Tabrizi SJ et al (2018) In vivo neutralization of the protagonist role of macrophages during the chronic inflammatory stage of Huntington's disease. Sci Rep 8(1):11447. https://doi.org/10.1038/s41598-018-29792-x

113. Gelderblom H, Wüstenberg T, McLean T, Mütze L, Fischer W, Saft C et al (2017) Bupropion for the treatment of apathy in Huntington's disease: a multicenter, randomised, double-blind, placebo-controlled, prospective crossover trial. PLoS One 12(3): e0173872. https://doi.org/10.1371/journal.pone.0173872

114. Sciacca G, Cicchetti F (2017) Mutant huntingtin protein expression and blood-spinal cord barrier dysfunction in huntington disease. Ann Neurol 82(6):981–994. https://doi.org/10.1002/ana.25107

115. Ajami B, Samusik N, Wieghofer P, Ho PP, Crotti A, Bjornson Z et al (2018) Single-cell mass cytometry reveals distinct populations of brain myeloid cells in mouse neuroinflammation and neurodegeneration models. Nat Neurosci 21(4):541–551. https://doi.org/10.1038/s41593-018-0100-x

116. Wild E, Magnusson A, Lahiri N, Krus U, Orth M, Tabrizi SJ et al (2011) Abnormal peripheral chemokine profile in Huntington's

disease. PLoS Curr 3:Rrn1231. https://doi.org/10.1371/currents.RRN1231

117. Niemelä V, Burman J, Blennow K, Zetterberg H, Larsson A, Sundblom J (2018) Cerebrospinal fluid sCD27 levels indicate active T cell-mediated inflammation in premanifest Huntington's disease. PLoS One 13(2):e0193492. https://doi.org/10.1371/journal.pone.0193492

118. von Essen MR, Hellem MNN, Vinther-Jensen T, Ammitzboll C, Hansen RH, Hjermind LE et al (2020) Early intrathecal T helper 17.1 cell activity in Huntington disease. Ann Neurol 87(2):246–255. https://doi.org/10.1002/ana.25647

119. Husby G, Li L, Davis LE, Wedege E, Kokmen E, Williams RC Jr (1977) Antibodies to human caudate nucleus neurons in Huntington's chorea. J Clin Invest 59(5):922–932. https://doi.org/10.1172/jci108714

120. Lee D-H, Heidecke H, Schröder A, Paul F, Wachter R, Hoffmann R et al (2014) Increase of angiotensin II type 1 receptor auto-antibodies in Huntington's disease. Mol Neurodegen 9(1):49. https://doi.org/10.1186/1750-1326-9-49

121. Bouchard J, Truong J, Bouchard K, Dunkelberger D, Desrayaud S, Moussaoui S et al (2012) Cannabinoid receptor 2 signaling in peripheral immune cells modulates disease onset and severity in mouse models of

Huntington's disease. J Neurosci 32(50):18259–18268. https://doi.org/10.1523/jneurosci.4008-12.2012

122. Fong Y, Moldawer LL, Marano M, Wei H, Barber A, Manogue K et al (1989) Cachectin/TNF or IL-1 alpha induces cachexia with redistribution of body proteins. Am J Physiol 256(3 Pt 2):R659–R665. https://doi.org/10.1152/ajpregu.1989.256.3.R659

123. Langen RC, Schols AM, Kelders MC, Wouters EF, Janssen-Heininger YM (2001) Inflammatory cytokines inhibit myogenic differentiation through activation of nuclear factor-kappaB. FASEB J 15(7):1169–1180

124. Spiegelman BM, Hotamisligil GS (1993) Through thick and thin: wasting, obesity, and TNF alpha. Cell 73(4):625–627

125. Lee SW, Park HJ, Im W, Kim M, Hong S (2018) Repeated immune activation with low-dose lipopolysaccharide attenuates the severity of Huntington's disease in R6/2 transgenic mice. Anim Cells Syst (Seoul) 22(4):219–226. https://doi.org/10.1080/19768354.2018.1473291

126. Wang CE, Li S, Li XJ (2010) Lack of interleukin-1 type 1 receptor enhances the accumulation of mutant huntingtin in the striatum and exacerbates the neurological phenotypes of Huntington's disease mice. Mol Brain 3:33. https://doi.org/10.1186/1756-6606-3-33

Chapter 10

Biomarkers in Huntington's Disease

Annie Killoran

Abstract

Huntington's disease (HD) is a neurodegenerative disorder caused by a trinucleotide repeat expansion mutation. The expansion length is the diagnostic biomarker for the disease. It can result in four different scenarios depending on its number of repeats: the intermediate allele, adult onset disease with either reduced or full penetrance, and juvenile HD. Along with various genetic modifiers, the mutation predicts the age of clinical onset. By convention, the clinical diagnosis of HD is reached when there is sufficient motor dysfunction, as judged by the Unified HD Rating Scale. This scale is widely used in clinical trials, but is limited by its subjectivity, categorical ratings, and insensitivity to subtle features in early disease. The need for sensitive, objective, continuous biomarkers has heralded a wave of digital biomarkers, which range from devices embedded with force transducers to smartphone applications. These have become a standard component of clinical trial assessments. However, they provide no insight into the pathological status of the disease. Certain wet biomarkers can identify the earliest indication of pathological change, when disease-modifying therapy would ideally be initiated to prevent or delay the clinical manifestations of the disease. Such treatments are being developed, but their evaluation in subjects with minimal clinical findings will depend on sensitive pathological biomarkers of progression. The most promising protein biomarkers are neurofilament light chain and mutant huntingtin, the latter having shown utility as a pharmacodynamic biomarker for a novel huntingtin-lowering treatment. Of the genetic biomarkers, certain microRNAs and telomere length have proven to be early indicators of pathological change. These are discussed, along with other biomarker candidates.

Key words Huntington's disease, Biomarkers, Genetic modifiers, Mutant huntingtin, Neurofilament light chain

1 Diagnostic Biomarker

1.1 Huntington's Disease Gene

Huntington's disease (HD) is an autosomal dominant neurodegenerative condition, that leads to an early death. It is characterized by psychiatric problems, and deterioration of both cognitive function and motor control. HD is caused by a cytosine-adenine-guanine (CAG) trinucleotide repeat expansion mutation in exon 1 of the huntingtin (HTT) gene, on chromosome 4 [1]. The HD gene test is the diagnostic biomarker for HD, and it is commercially available. The result is reported as the expansion's number of CAG repeats,

Philip V. Peplow, Bridget Martinez and Thomas A. Gennarelli (eds.), *Neurodegenerative Diseases Biomarkers: Towards Translating Research to Clinical Practice*, Neuromethods, vol. 173, https://doi.org/10.1007/978-1-0716-1712-0_10,

Table 1
Summary of the HTT gene mutation's clinical implications

Expansion length (CAGn)	Outcome	Description
27–35	Intermediate allele	Subtle features Offspring at-risk due to anticipation
36–39	Low penetrance	May develop mild disease late in life
40–59	Full penetrance	Adult onset
60 +	Juvenile HD	Presents before age 20 years Dystonic-rigid phenotype

which can have different implications, (Table 1). The expansion length inversely correlates with the timing of disease onset, such that the longer the expansion, the earlier the disease [2]. This inverse correlation is most pronounced at longer expansions. For example, greater than 60 CAG repeats invariably results in juvenile HD, starting before age 20. Whereas age of onset is less predictable for those with repeats numbering in the low 40s [2]. For most adult patients, the CAG expansion length only accounts for roughly 66% of the variability in the age of onset [3], with the remainder mostly being due to genetic modifiers, (discussed below). For this reason, the expansion length is generally not used to estimate age of disease onset on an individual basis. For reference, the mean age of HD onset is 40 years [4], and 45 is the mean number of repeats in affected individuals [3]. At 40 or more trinucleotide repeats, the mutation is fully penetrant [5]. Expansion lengths of only 36 to 39 CAG repeats are considered low penetrance alleles; they may or may not lead to disease manifestation [6]. When they do, it is typically a relatively mild phenotype with onset somewhat late in life. Expansions of only 27–35 CAG repeats are considered intermediate alleles, or "premutations," which are relatively common, occurring in 1 of every 16 people [7]. These are traditionally considered nonpathogenic, such that carriers are unaffected. However, their offspring are at risk because the HD mutation is unstable and susceptible to anticipation, especially during paternal transmission [8]. Anticipation entails further CAG repeat expansion during gametogenesis, resulting in an earlier age of disease onset in the offspring. For intermediate allele carriers, anticipation could result in children with a disease-causing, fully penetrant mutation [9]. This can appear as a de novo mutation with no known family history of HD, as the parent (with the premutation) is asymptomatic. However, recent studies from 3 independent cohorts have challenged the notion of asymptomatic intermediate alleles, as carriers were found to have subtle (but significant) HD features [10–12]. These abnormalities are presumably caused by a low level of toxicity from the mutant HTT. The mild features appear to be relatively benign in that they generally do not lead to fully manifest HD. However, there have been a handful of case reports describing patients with the intermediate allele who do develop HD [13–16]. Such cases are thought to be at least partly due to somatic mosaicism, in which different cell populations have repeat expansions of different lengths. In cases of manifest HD with an intermediate allele, (as detected in blood), presumably the affected patient's relevant brain tissue harbors the destructive fully expanded allele [17].

2 Genetic Modifiers of Onset: Prognostic Biomarkers

2.1 Loss of Interrupting Allele

The variability in timing of HD onset is partly due to genetic modifiers, which can function as prognostic biomarkers. With the advent of genome wide association studies, such variants are increasingly coming to light. One of the most consequential occurs in the sequence immediately downstream of the HTT gene's pathogenic CAG repeat expansion [18]. This recently discovered single-nucleotide polymorphism (SNP) involves an adenosine being replaced by a guanine. This simple change is referred to as the "loss of interrupting" (LOI) allele as it results in a longer series of uninterrupted CAG repeats, (without altering the length of the polyglutamine tract). It also leads to an increase in somatic mosaicism, which is related to earlier disease [19] and is thought to be the cause for the premature onset seen with this SNP [18]. In the population studied, affected subjects' phenoconversion was on average 25 years earlier than predicted. This dramatic effect on the age of onset was greatest for those with a reduced penetrance HD gene mutation (36–39 CAG repeats) in which the loss of interrupting allele occurs with high frequency. The variant was seen in 84.6% of patients with a reduced penetrance allele who had manifest HD; as opposed to only 3.3% of the tested study population. The loss of interrupting allele may explain the case reports of individuals with the intermediate allele who develop manifest HD. One study did indeed document this variant in a manifest HD patient with an expansion of only 35 CAG repeats. No other genetic variant identified to date has been found to have such a large effect size. Ideally, wide scale testing will be performed to determine an accurate frequency of this clinically meaningful genetic variant.

2.2 rs13102260 (G > A)

Another notable modifier is a SNP in the HTT promoter region, rs13102260 (G > A) [20]. This allele-specific variant causes reduced binding of the transcription factor NF-κB to the HTT promoter, resulting in less protein production. When the variant is present in the *wild-type* HTT allele, relatively less wild type HTT (wtHTT) is produced (compared to mutant HTT) and the age of onset occurs roughly 4 years *earlier* than expected. Whereas when the variant occurs in the *mutant* HTT allele, it causes relatively less mutant HTT (mHTT) to be produced (compared to wtHTT) and disease onset is a mean of 10 years *later*. This delay in onset was inversely correlated with the length of the CAG repeat expansion, such that shorter expansions had greater delays in onset. Specifically, an expansion of 41 CAG repeats was associated with a 17.3-year delay in onset, whereas an expansion length of 55 repeats saw only a 2.7-year delay in onset. The authors suggest that the imbalance between mutant and wild-type huntingtin affects the pathology and progression of HD [20].

There are several other variants that have been identified that either accelerate or delay disease onset [21–36], and presumably, there are many more to be discovered. The Genetic Modifiers of Huntington's Disease (GeM-HD) Consortium is investigating the whole genome aiming to identify these variants [36]. The group has a cohort of over 9000 subjects. Discovery of more of these influential variants will enable a comprehensive approach to genetic testing for people at-risk for HD. Though currently impractical and costly, it may become standard procedure to test at-risk individuals for the HD gene mutation along with a panel of genetic variants. Such prognostic biomarkers may improve the accuracy of the predicted age of onset, such that they could be used for prognostication on individual patients, representing a step toward personalized medicine. They could also be utilized in premanifest cohorts in clinical trials to reduce the variance in the expected age of onset, which is currently determined by a standard calculation that relies on the mutation expansion length and age alone [2].

3   Genetic Modifiers: Phenotypic Biomarkers

3.1   Val158Met Polymorphism

Some genetic modifiers affect the clinical manifestations of the disease. An example of this is the tau H2 haplotype, which is associated with an increased rate of cognitive decline in HD [37]. A more complicated example is the Val158Met polymorphism of the chromosome 22 gene for catechol-O-methyltransferase (COMT), which degrades dopamine [38]. The presence of valine (Val) instead of methionine (Met) increases COMT activity, resulting in lower dopamine levels. One study demonstrated that in the early to middle stages of HD, Met/Met homozygotes had better cognitive scores than those with the Val allele (Met/Val or Val/Val) [38]. However, this effect reversed after around 10 years of disease, at which point the Met/Met homozygotes' cognition was relatively worse. The authors proposed that the relatively higher dopamine is initially beneficial, but in the long-term has a toxic effect, leading to neuronal damage and relative worsening of cognitive performance. This COMT polymorphism may be a useful biomarker to help with cohort stratification in clinical trials investigating experimental therapeutics that affect dopaminergic activity. Alternatively, if combined with other genetic variants that affect clinical manifestations, it could be used as part of a comprehensive phenotypic biomarker.

4   Clinical Diagnosis

4.1   Unified Huntington's Disease Rating Scale

The genetic diagnosis, which relies on the presence of the HD mutation, is distinct from a clinical diagnosis, which is a subjective measure. These two diagnoses can arise several years apart and do not depend on one other. A clinical diagnosis of HD is (by convention) based on the presence of motor abnormalities detected on neurological examination of a person who is known to have the gene mutation or is at-risk for HD. This evaluation is typically completed by a certified rater using the motor section of the Unified HD Rating Scale (UHDRS), which was developed by the Huntington Study Group [39]. It attributes scores to 15 different motor aspects including oculomotor function, motor impersistence, rapid alternating movements, rigidity, dystonia, gait, and chorea (which is heavily weighted as it requires scores for the face, mouth, trunk and each of the extremities). Each aspect is rated from 0 to 4. These are combined to yield the UHDRS "total motor score" (TMS), which ranges from 0 to 124, with higher scores signifying relatively increased dysfunction. There is no suggested cutoff score or any particular examination features that are required for an HD clinical diagnosis. Instead, diagnosis depends on the clinician's "diagnostic confidence level" (DCL), an item on the UHDRS. The DCL ranges from zero (normal) to 4, which is diagnostic for HD. A DCL of 4 indicates that the examiner feels 99% confident that the patient's motor features are "unequivocal" signs of HD. Whereas a DCL of 3 indicates that the examiner feels only 90% to 98% confident in the clinical diagnosis, because the patient's motor features are only "likely" signs of HD. The DCL criterion might suggest that patients' diagnoses depend less on his or her motor abnormalities and more on the examiner's character and experience. Despite the DCL's "pseudo-precision" [40] with its percentages and arbitrary probability thresholds, it has long been considered the standard tool amongst HD clinicians and researchers in clinical trials. In the PREDICT study, the TMS at disease onset ranged from 2 to 56 [41], signifying the variability in motor severity at which examiners were confident in a diagnosis. This also highlights the subjectivity of DCL's diagnostic benchmark. There is also the question of reliability, given the variability in patients' severity of motor dysfunction. For example, chorea can dramatically worsen during periods of heightened emotions [42]. So, it is possible for an individual to attain a DCL of 4 during one evaluation and then only a DCL of 3 at the subsequent one, essentially reversing his or her diagnosis. Another problem is that the UHDRS may not be sufficiently inclusive, due to HD's variable phenotype. Some adults have a predominantly hypokinetic-rigid presentation, with minimal chorea [43, 44], similar to juvenile HD [45]. While others, notably the elderly, can present primarily with a gait disorder [46]. A cluster analysis of the PREDICT-HD cohort revealed that there are also many individuals with a predominantly nonmotor presentation [47]. It was shown that using the UHDRS's other domains (cognitive, behavioral, and functional capacity), instead of just the motor-focused DCL, would lead to over a third of cases being diagnosed sooner.

It is important to note that a DCL of 4 signifies a major milestone for a patient. This is considered "disease onset," when he or she is considered to have phenoconverted to HD. This conventional approach to HD's clinical diagnosis artificially makes it into a dichotomous time point. However, HD is slowly progressive. By the time of "motor onset," patients have been showing subtle but gradually worsening abnormalities for several years [48, 49]. Longitudinal observational studies have found that motor and cognitive deficits can be detected up to 14 and 15 years (respectively) in advance of predicted motor onset [50, 51]. During this time period, patients also have notable psychiatric manifestations [52]. These clinical findings are supported by pathological and imaging studies. As many as half of the neurons in the caudate nucleus have died by the time of "disease onset" [53], and longitudinal imaging studies have detected structural brain changes (striatal and gray matter volume loss) as far back as 20 years prior to predicted phenoconversion [54, 55]. It is for this reason that an HD-focused task force has proposed certain terminology that recognizes the early progression of HD [40]. The terms "premanifest" and "manifest" describe the periods before and after traditional motor onset (as recognized by a DCL of 4). The premanifest period is divided into "presymptomatic" HD, (in which there are no abnormalities), and the subsequent "prodromal" stage, during which subtle features are apparent in the years leading up to traditional "motor onset."

It is the premanifest period which is of interest currently due to the recent development of potential disease-modifying therapeutics. It is felt that the first indication of neurodegeneration would be the optimal time to administer a neuroprotective treatment, with the goal of trying to prevent (or at least delay) this otherwise-incurable condition. A major limitation is that the commonly used UHDRS was designed for individuals with manifest HD. Its 5-point categorical rating scheme is not sensitive enough for scoring the subtle abnormalities of the prodromal period. Thus, any clinical trials exploring the potential benefits of experimental drugs during the premanifest period would need a more sensitive way to evaluate participants, hence the need for biomarkers, notably continuous measures that can detect subtle manifestations in premanifest disease. The farthest from onset, the better.

5   Digital Biomarkers

Many of the features that we test for on clinical examination can easily be quantified with the use of digital biomarkers. This transforms subjective measures into objective ones and resolves the issue of inter- and intrarater variability. It also allows for better accuracy over the UHDRS's limited selection of categorical choices. Technology also improves the sensitivity of evaluations, notably for the subtle abnormalities of the premanifest individuals. Most clinical trial evaluations now supplement the traditional UHDRS with quantitative motor testing, commonly referred to as "Q-motor." The Q-motor battery is generally composed of assessments of finger tapping, chorea and grip force variability. Also available are tests for foot tapping and/or alternating pronation-supination of the hand, and tongue force variability.

One of the first quantitative assessments in HD clinical research was for motor impersistence. As part of the UHDRS, clinicians are challenged with trying to judge patients' ability to keep their tongue fully protruded for 10 s. However, a force transducer can easily and accurately measure variability in tongue protrusion force [56]. Quantitative scores correlate with the TMS and the disease burden score (which is calculated according to CAG repeat number and age) and can distinguish the three groups of subjects from one another: manifest, premanifest and controls. In premanifest disease, baseline tongue force variability was shown to be a strong predictor of disease progression [55]. Motor impersistence can also be measured with grip force variability. For this assessment, subjects are tasked with isometrically maintaining hold of an object between their thumb and index finger [57, 58]. Grip force variability is increased in manifest compared to premanifest HD, and in premanifest cases compared to controls [58]. Grip force variability correlates with the TMS, but is a more sensitive indicator of motor progression, including for those with premanifest disease [55].

Chorea is inherently difficult to subjectively score. Early quantitative assessments involved subjects gripping and holding in place a small device with an imbedded 3-dimentional electromagnetic sensor to detect movements in all planes [59]. Scores were found to correlate with the TMS and the disease burden score. Both the position and orientation indices were increased in the manifest HD group compared to premanifest and control subjects, whereas, in premanifest individuals, only the position index was increased compared to those of the controls.

Finger tapping is used to assess fine motor control. On the UHDRS, it is judged by its slowness, fatiguing and motor arrests. This can be more accurately assessed with digitomotography. In premanifest HD, results of these force-transducer-based tests (both speeded and evenly self-paced) correlate with predicted time to HD motor onset [60]. The variability of inter-tap tapping intervals was found to distinguish between manifest, premanifest and healthy controls [50]. Even in premanifest disease, tapping variability correlates with the TMS, caudate atrophy (on MRI), and the disease burden score [50, 61–63]. Over 36 months, the inter-tap interval changed significantly in HD gene carriers who were at least 10.8 years ahead of their predicted motor onset [55]. This indicates an early detectable clinical change. In contrast, the rate of change in the TMS for individuals who were far (or near) to their predicted phenoconversion was not distinguishable from that of healthy controls. Additionally, unlike the UHDRS, digitomotography appears to be immune to the placebo effect in clinical trials [64].

Postural stability is assessed as part of the UHDRS by pulling patients backward to throw off their balance and seeing how easily they correct their footing. Quantitatively, this can be achieved by having subjects stand on a force plate equipped with force transducers. Posturographic measures are increased in subjects with premanifest and manifest HD compared to controls. These scores have been shown to correlate with the TMS, the disease burden score and the Total Functional Capacity (TFC), which measures the ability to perform daily tasks [65, 66].

On the UHDRS, gait evaluation includes walking in tandem ("heel to toe") for 10 steps and counting the number of missteps that occur. Early quantitative HD gait analyses made use of an electronic mat with sensors embedded in its surface. Relative to controls, individuals with premanifest HD were found to have significantly reduced gait velocity and stride length, as well as increased variability in stride length and step time [67]. These features correlated with expected time to phenoconversion, whereas the UHDRS was not sufficiently sensitive to detect these gait abnormalities. When evaluated longitudinally, the largest gait differences in premanifest HD were seen in the variability of stride length and swing time coefficient of variation [68]. These changes were evident in as little as one year. And swing time variability correlated with predicted time to age of onset, at both 1 and 5 years. From baseline to the fifth year, in addition to the increase in variability of stride length and swing time, increased step length difference and decreased gait velocity were also detectable.

Three-dimensional (3D) Inertial Measurement Units (IMUs), also called 'wearable sensors', function to detect position and 3D movement. Being comprised of accelerometers and gyroscopes, these tools are now widely used in gait analyses. One of the benefits of wearable sensors is that they can be used remotely. One HD gait study had participants wearing five accelerometers to capture movements of all four limbs and chest. The gait analysis occurred both in the clinic and at home, with comparable results [69], suggesting that remote data collection from digital biomarkers is reliable, and that this type of study is feasible in HD. This supports the notion that remote monitoring could replace site visits, thereby reducing the burden of clinical trials, with the goal of making study enrollment easier.

Remote monitoring also allows for information to be collected from the patient's real-world environment, and this can provide new insights that may have otherwise been missed. For example, a study of subjects wearing inertial sensors at home revealed that the HD gene carriers spent a surprising 50% of their time lying down, compared to only 34% in the healthy controls [70]. The sensors were well tolerated, with participants reporting a high degree of willingness to wear them again.

In contrast to the snapshot assessment performed at a clinic visit, remote digital monitoring allows for data collection over a range of time to cover patients' fluctuations. The Digital-HD Study is an ongoing observational trial using a Roche-designed digital monitoring platform to evaluate cognitive and motor features seen in HD [71, 72]. Participants were provided with a smartphone for continuous passive monitoring of activity levels and for weekly assessments using an application consisting of tests of cognition (Symbol Digit Modalities Test and Stroop Word Reading test) and motor function, (Speeded tapping, Chorea, Draw-a-shape, Balance, 2-minute walk, U-turn). Preliminary data from a 5-week period showed that results corresponded well to those seen with traditional in-clinic testing. Results significantly differed between the manifest group and controls, and between the manifest and premanifest groups. However, only through an exploratory analysis did any results show a significant difference between the premanifest subjects and the control group. This was on the Speeded tapping test and the Draw-a-shape test. Using the same application and testing protocol, another study (NCT03342053) noted that adherence was generally high, with 80% of the weekly tests being performed over a 20-week period [73]. The Roche digital monitoring platform is also being incorporated into two of the companies' other clinical trials: the HD Natural History Study and Generation HD1 [74].

A group at the University of Rochester has developed their own smartphone HD application, called 'George', which is being evaluated for use in remote assessments. It assesses chorea as well as tapping speed [75]. Also targeting the HD population, researchers at the Kaunas University of Technology in Lithuania created an application designed to detect early motor abnormalities with different finger tapping tests [76]. One potential concern regarding remote monitoring is the sheer amount of data generated and how to store it. For example, one HD study captured more than 14,000 gait assessments over 1 week [77], and a 6-month study collected 33,000 h of accelerometer data from the wearables of only 10 participants [78]. Another limitation to the use of digital biomarkers in clinical trials is that they do not differentiate symptomatic benefit from underlying pathological change.

6   Protein Biomarkers of Progression

6.1   Huntingtin Protein

The huntingtin (HTT) gene has 67 exons and encodes for HTT, a large protein comprised of 3144 amino acids. HTT appears to have many different functions, though these have not been fully elucidated. As HTT knockout models are embryonic lethal, it is clear that HTT is essential for embryogenesis [79]. It also plays a role in the early development of the central nervous system [80]. HTT appears to act as a molecular scaffold, interacting with various other proteins and being involved in transcriptional regulation, cell signaling and trafficking [81]. In HD, there is loss of function of these various cellular processes due to the mutant HTT (mHTT) [82–84]. However, part of HD's pathology also arises from toxic aggregates made up of fragments of the mutant protein's polyglutamine tract-containing N-terminal [85], disrupting a wide range of cellular processes [82].

HTT's role in the pathogenesis of HD make it an obvious biomarker candidate. HTT is expressed outside of the brain in various tissue types [86] so it can easily be measured peripherally. An early study used Western blotting and an enzyme-linked immunosorbent assay (ELISA) to detect total (wild type and mutant) HTT in peripheral blood mononuclear cells [87]. Compared to healthy controls, HD gene carriers had significantly reduced total HTT. Levels were progressively lower over four stages of disease (premanifest, early, moderate, and advanced). It has also been detected in saliva, where total salivary HTT was significantly increased in manifest HD participants, (compared to the control group), but not premanifest ones [88]. Total salivary HTT also correlated with scores on the UHDRS and TFC, but not the Mini-Mental State Examination (MMSE), a standard cognitive test [89]. Levels showed no intra-individual variation, but they were affected by age. Looking specifically at peripheral mHTT, in buffy coat samples and peripheral blood mononuclear cells, a homogeneous time-resolved Förster resonance energy transfer (HTRF) assay showed that mHTT levels were higher in HD gene carriers, compared to controls [90, 91]. Using a similar method (a time-resolved Förster resonance energy transfer (TR-FRET) immunoassay), mean mHTT levels in peripheral blood leukocytes were found to be significantly different between healthy controls, premanifest HD and manifest HD [92]. In addition, levels in monocyte and T cells were associated with disease burden scores and rates of caudate atrophy.

With the aim of detecting the low extracellular levels of mHTT in human cerebrospinal fluid (CSF), researchers developed a single-molecule counting (SMC) mHTT bead-based immunoassay that involves labeled antibodies emitting fluorescent light upon binding to their target [93]. They utilized this SMC assay with the mouse monoclonal MW1 antibody (which binds to the HTT polygluta-mine tract) [94] and the 2B7 antibody (which binds to the HTT N-terminal 17 amino acids) [95]. The test was sufficiently sensitive to detect femtomolar levels of mHTT, and they used it on the CSF of two independent cohorts [93]. The assay had high specificity, detecting no mHTT in any of the control subjects, whereas mHTT was identified in the CSF of most of the HD gene carriers. In the premanifest group, levels were associated with predicted age of onset, and they were 3-times lower than in the manifest group. In manifest individuals, the CSF mHTT levels correlated with clinical measures of disease progression, including the TMS, symbol-digit modality test, Stroop color naming test, Stroop word reading test, and Stroop interference test. CSF mHTT was also shown to correlate with CSF tau and neurofilament light chain markers of neuronal injury (discussed below) [96]. A different team utilized this same ultrasensitive assay, but with a different antibody pair: MW1 (as above) and HDB410 (which binds to HTT's amino acids 1844–2131) [97]. They reported similar findings except for a lack of correlation between CSF mHTT levels and clinical measures in subjects with advanced HD. The Singulex SMC mHTT immuno-assay for quantifying CSF mHTT has since been validated as per guidelines established by the US Food and Drug Administration and the European Medicine Agency [98]. The SMC assay was able to demonstrate its accuracy, precision, specificity, and reproducibil-ity. With mHTT being the reliably measured entity of HD pathol-ogy, detectable in premanifest disease and correlating with clinical measures, it is a worthy biomarker for monitoring disease progres-sion. In addition, CSF mHTT elevation was the earliest detectable feature of HD, according to event-based modelling, which charted the progression of probabilistic changes through the premanifest period [99]. This suggests that it could be used to signal the earliest neurodegenerative change. This event-based model predicted that the next probabilistic changes were rises in both plasma and CSF neurofilament light chain (NfL) [99].

6.2 Neurofilament Light Chain

Neurofilaments are cylindrical polymers [100] that make up the cytoskeleton in myelinated neurons and play a major role in estab-lishing and maintaining the structural integrity of axons [101]. Any neuronal insult that leads to axonal damage results in the release of neurofilaments, which can be measured in both CSF and blood. These serve as nonspecific markers of neuronal dysfunction [102].

Early quantification of CSF NfL was done by inexpensive and commercially available ELISAs. However, these tests were only sensitive enough to detect NfL in CSF, and there was considerable variation between different laboratories, reportedly due to different protein standards [103]. Electrochemiluminescence (ECL)-based assays then came into use, as these were more sensitive [104]. However, since then, single-molecule array (SiMoA) technology has become the standard immunoassay for NfL. SiMoA kits are commercially available and generate reliable results, even between different laboratories. This digital technique for measuring NfL is 25- and 126-times more sensitive than ELISAs and ECL-based assays, respectively [105]. SiMoA detects NfL in the femtomolar range, at the low concentrations in the blood of young healthy people, where it is 50 times less than that of the CSF [106]. It is this ultra-sensitivity that has made possible many recent studies investigating NfL as a potential biomarker in neurodegenerative conditions. These include Alzheimer's disease, Parkinson's disease, and multiple sclerosis, in which CSF and plasma NfL are elevated and correlate with disease severity [107–109]. Also, in clinical trials, NfL declines in response to effective disease-modifying therapeutics, supporting its use as a potential pharmacodynamic biomarker in neurodegeneration [110, 111].

In the HD population specifically, CSF NfL correlates well with serum levels, which makes it is easily accessible via a minimally invasive blood draw [96, 99, 112]. Levels are very stable over several weeks, with low intraindividual variation. NfL progressively increases in premanifest and manifest HD, with concentrations correlating with predicted time to onset [96, 112] as well as with clinical and imaging measures in both cross-sectional [113–115] and longitudinal studies [99, 112, 116, 117]. These correlations were more robust than those seen with CSF mHTT [99, 118], so, it appears that NfL qualifies as a relatively more suitable biomarker for HD, at least in regard to monitoring progression. In particular, plasma (rather than CSF) Nfl was found to be the best marker for predicting brain atrophy and clinical measures [99, 118]. A recent clinical trial for a putative neuroprotective treatment (tominersen) reported transient rises in CSF NfL in some participants who had been treated with the two highest doses [119]. This might suggest that the treatment itself causes some initial neuronal damage.

Calculations determined that with use of NfL or mHTT reductions as outcome measures, clinical trials would only require 35 subjects per study arm to detect a therapeutic effect size as low as 20% [99]. This is a dramatic difference to the large numbers that are required to see treatment benefit using traditional clinical measures. Such low enrollment needs would have obvious benefits in regard to costs, time, manpower etc. Moreover, it would be less draining on the limited pool of potential study participants, so there would be more subjects available for clinical trials investigating other novel therapies.

6.3 Brain-Derived Neurotrophic Factor

Brain-derived neurotrophic factor (BDNF) is required for the survival and protection of striatal neurons (those most susceptible to damage in HD) [120, 121], but it is relatively reduced in the striatum of HD animal models and patients [84]. This is presumably, or at least partly, due to the fact that mHTT represses BDNF gene transcription [84] and disrupts BDNF's transportation from the cortex to the striatum [122]. Early studies reported that BDNF was decreased in the blood of HD patients compared to controls [123–125], but this has since been refuted [126–129]. Likewise, no differences in CSF BDNF levels were detected between manifest, premanifest and control subjects [129]. However, a recent study detected reduced levels of salivary BDNF (detected by ELISA) in subjects with premanifest and manifest HD, compared to healthy controls [128]. In the premanifest cohort, those who were sooner-to-diagnosis had significantly lower BDNF levels compared to subjects who were more than 10 years out from predicted motor onset. There were no correlations to clinical measures, including the TMS, TFC, MMSE, or Hospital Anxiety and Depression Scale, (a self-assessment tool used to measure depression and anxiety) [130]. This indicates that BDNF may not be useful as a marker of progression, but it may have use as an indicator of pathological change in early premanifest individuals. A longitudinal study in people at-risk for HD who are several years out from their motor onset would determine how many years prior to phenoconversion a detectable reduction in saliva BDNF occurs, and whether it correlates with changes in other biomarkers or with predicted onset. Saliva's easy access would make it a very convenient, noninvasive test option. Simple-to-use testing kits would make it feasible for at-risk individuals to screen for any changes at-home, to signify when they may need to undergo in-person confirmatory testing. Unfortunately, salivary BDNF changes are not specific to HD. It has circadian variations [131] and is relatively reduced in people who are sedentary [132]. So, levels would need to be checked at the same time of day and trusted only in individuals with regular wake-sleep cycles and activity levels. The potential influence of other variables would also have to be evaluated to avoid misinterpretation of results. For example, stress increases salivary BDNF in animal models [133], whereas in humans, serum (or plasma) BDNF is reduced with stress [134], as well as with infection [135], depression [136], and anxiety [137]. This BDNF association with psychiatric symptoms is not specific to HD. However, methylation at CpG sites in the BDNF promotor IV was increased in HD gene carriers' blood-derived DNA compared to that of controls. Levels inversely correlated with the Hospital Anxiety and Depression Scale scores, though not with motor or cognitive measures (including the UHDRS, the TFC and the MMSE) [128]. This would suggest a possible risk biomarker for psychiatric features specific for manifest HD, though it is unclear if there is a need for this. BDNF's utility as a convenient salivary marker of pathological change would likely be more useful, assuming the effect of other variables is negligible.

6.4 Tau Protein

Tau protein, which is involved in microtubule assembly and stabilization [138], is known as a nonspecific marker of neuronal damage and is present in numerous neurodegenerative diseases, such as frontotemporal dementia and Alzheimer's disease. Tau and tau fragments are found in HD brains in post-mortem studies and are thought to play a role in the pathology of HD [37, 139, 140]. For this reason, it has been investigated as a possible HD biomarker. CSF tau (detected by ELISA) was found to be increased in HD gene carriers, compared to controls, [96, 113, 141, 142] though levels were widely dispersed with notable overlap with those of the healthy group [114, 141, 142]. CSF tau levels correlated with mHTT [96] and NfL [113], and one study also reported correlations with the TMS, the TFC, and cognitive measures in HD [141]. However, these findings could not be replicated in other cohorts [96, 113, 142]. Yet another study found that relatively high CSF tau associated with the presence of psychiatric symptoms in HD subjects, but there was no correlation with symptom severity [116]. Thus, this may represent another possible susceptibility biomarker for psychiatric features in HD, but tau appears to have limited potential as an HD biomarker of progression, notably when other CSF degenerative markers are available that have reliably shown clinical correlations.

7 Genetic Biomarkers of Progression

7.1 MicroRNAs

In recent years, there has been a growing understanding of the role of noncoding RNA sequences. Included in this large class of transcripts are microRNAs (miRNAs), which have become a major focus in clinical research [143]. MicroRNAs are short noncoding nucleotide sequences that regulate transcription in an organ- or cell-specific manner [144]. Due to widespread dysregulation of miRNAs in HD mouse models and the human HD cortex, miRNAs are thought to be involved in the pathology of HD [145]. In post-mortem brain tissue, next generation sequencing identified 75 miRNAs that were differentially expressed in HD subjects compared to controls [146]. MicroRNA-10b-5p was the most strongly expressed and it correlated with HD striatal pathology. In addition, along with miR-196a-5p, its levels were associated with age of onset. In plasma, both miR-10b and miR-196a-5p were found to be elevated in manifest HD subjects compared to controls, but the relationship with age of onset was no longer apparent.

More recently, results from a bioinformatics study added value to these two potential biomarkers. SRNAlytics is a biotechnology company with a proprietary computational platform that uses next generation sequencing with big data analytics to detect genetic markers of disease. They found miR-10b and miR-196a-5p expressed in HD brain tissue and CSF, but not in that of controls [147]. Both miRNAs were validated by targeted RT-qPCR in different post-mortem HD brain tissue and in CSF (from the PREDICT-HD study). The levels of these miRNAs correlated with both striatal degeneration and motor dysfunction. In addition, they could be detected in CSF samples from premanifest subjects who were not expected to reach motor onset for another 12–20 years. How exactly miR-10b and miR-196a contribute to HD pathology is unclear, however both molecules target HTT. In addition, they interact with Repressor Element 1 Silencing transcription factor (REST) [147], which regulates genes that have a putative role in HD pathogenesis, including the BDNF gene [148]. The fact that these miRNAs were validated, are associated with clinical progression, and were detected in early premanifest HD suggests that they would be suitable biomarkers of progression.

7.2 Telomere Length

Human telomeres are repetitive hexa-nucleotide (TTAGGG) sequences that cap the ends of chromosomes and function to protect DNA [149]. With repeated cell division, telomeres increasingly shorten, such that their length is a measure of biological age [150]. Telomere shortening is also increased in age-related diseases, such as Alzheimer's disease and Parkinson's disease [151, 152], and it is accelerated with oxidative stress [153], which is known to play a role in the pathology of HD [154]. Telomere length in peripheral blood leukocytes was measured by real-time PCR [155]. In patients with manifest HD it was shorter compared to that of premanifest individuals, and it was nearly half that of the age-matched healthy controls. In premanifest subjects, it was only after the age of 30 years that telomere length was significantly shorter than in healthy controls. This suggests that HD leukocyte telomeres presumably only start to shorten when the HD pathology reaches a certain threshold. In the premanifest subjects, there was a positive relationship between telomere length and estimated years to diagnosis. Two other studies have also reported telomere shortening in HD compared to controls, suggesting reproducibility [156, 157]. In terms of biomarker suitability, the fact that results are obtained with widely available RT-PCR from a simple inexpensive, noninvasive blood draw is highly favorable. Verification in a longitudinal study would be needed to determine if telomere length predicts time to HD motor onset, or to at least see how it associates with other biomarkers of progression in the premanifest stage. It would also be useful to evaluate how it compares to clinical features in manifest disease.

8 Pharmacodynamic Biomarker

Mutant HTT is a natural choice for use as an outcome measure and pharmacodynamic biomarker for clinical trials investigating novel, HTT-lowering, experimental therapeutics. One the most compelling new treatments being investigated is antisense oligonucleotide (ASO) therapy, which is administered by repeat intrathecal injections. ASOs are short synthetic single strands of nucleotides that bind to and trigger degradation of complimentary mRNA [158]. In the case of HD, the ASO is designed to target the HTT mRNA, leading to a reduction of toxic mHTT, thus halting HD's neurodegeneration. The ASO was first tested in HD mouse models, where it led to a decline in mHTT levels in murine brain lysates and CSF, as measured by SMC [97]. The first-in-human study of an HTT-targeting ASO (called 'tominersen') was a dose-escalation phase 1–2a trial in 46 adults with early HD [119]. Being effectively used as a proof of concept biomarker, and showing target engagement, CSF mHTT decreased in a dose-dependent fashion. At the highest monthly ASO dose, (120 mg), CSF mHTT dropped by an average of about 40%. Based on preclinical studies, this reduction should correspond to a 55%-70% decrease of cortical mHTT [159] and be clinically beneficial [160, 161]. A post hoc analysis did suggest some benefit in clinical outcome measures, but the study had not been designed or sufficiently powered to assess for therapeutic benefit. The 120 mg dose was selected for the open-label extension study (NCT03342053), which included all original 46 trial participants. Subjects were treated either monthly or every other month. At 15 months, this resulted in CSF mHTT reductions of 70% and 44% from baseline, respectively, (per preliminary analysis) [162]. Again, the CSF mHTT levels would guide the dosing regimen for the subsequent clinical trial of the ASO, in this case a high-stakes, phase III study. Generation HD1 (NCT03761849) is a large-scale, 2-year, international trial with 800 participants, (with recruitment completed). Subjects are being randomly assigned to 120 mg of tominersen every 8 or 16 weeks. As secondary outcome measures, this pivotal trial will include CSF mHTT, NfL and digital biomarkers. It was the levels of CSF mHTT that informed the dosing protocol at each stage in the evaluation of this putative treatment, functioning as a crucial pharmacodynamic biomarker.

9   Conclusion

In conclusion, the HD gene mutation is a complex diagnostic biomarker. This expansion mutation, along with a growing array of genetic modifiers, may ultimately establish a comprehensive genetic profile that more accurately predicts age of motor onset. The limitations with the traditional HD clinical evaluation have led to the evolution of HD digital biomarkers, which are now standard use in clinical trials. Remote assessments are only just starting to be explored in HD studies, but they will presumably lead to less need for in-person evaluations, thereby reducing a major barrier to clinical trial recruitment. The most promising wet biomarkers of progression are mHTT and NfL, as these have been shown to be early indicators of neuronal dysfunction, signaling the pathological onset of disease. In addition, their levels correlate with predicted time to motor onset and other clinical measures. CSF mHTT has also proven its use as a pharmacodynamic biomarker for HTT-lowering treatment. The convenience of a salivary protein biomarker is intriguing, but the research here is only in its infancy. Finally, specific microRNAs (miR-10b and miR-196a) and telomere length are promising genetic biomarkers of progression. Interestingly, these are also potential pharmacological targets, and therefore may have future roles as pharmacodynamic biomarkers as well. Table 2 provides a summary of the biomarkers discussed.

Table 2
Summary of biomarkers discussed

Category	Biomarker	Key points
Diagnostic	HTT gene mutation	Outcome depends on expansion length For "genetic diagnosis" only Not for clinical status or disease stage
Modifiers of Onset	Loss of interrupting allele rs13102260 (G > A)	Numerous SNPs exist Have wide range of effect sizes
Phenotypic	Tau H2 Val158Met	Impact clinical features (e.g., cognitive or psychiatric)
Digital	Q-motor assessments and smartphone applications	Continuous measures of clinical abnormalities Not useful for pathological change Can be used remotely Potential for data overload
Prognostic	CSF mHTT Plasma and CSF NfL CSF miRNAs Blood telomere length	Early indicators of pathological change Prognostic of disease progression
Pathologic change	Salivary BDNF	Convenient but needs further study
Pharmacodynamic	CSF mHTT	Indicator of target engagement for mHTT-lowering drugs

References

1. The Huntington's Disease Collaborative Research Group (1993) A novel gene containing a trinucleotide repeat that is expanded and unstable on Huntington's disease chromosomes. Cell 72(6):971–983
2. Langbehn DR, Brinkman RR, Falush D, Paulsen JS, Hayden MR, International Huntington's Disease Collaborative Group (2004) A new model for prediction of the age of onset and penetrance for Huntington's disease based on CAG length. Clin Genet 65 (4):267–277
3. Lee JM, Ramos EM, Lee JH, Gillis T, Mysore JS, Hayden MR, Warby SC, Morrison P, Nance M, Ross CA, Margolis RL, Squitieri F, Orobello S, Di Donato S, Gomez-Tortosa E, Ayuso C, Suchowersky O, Trent RJ, McCusker E, Novelletto A, Frontali M, Jones R, Ashizawa T, Frank S, Saint-Hilaire MH, Hersch SM, Rosas HD, Lucente D, Harrison MB, Zanko A, Abramson RK, Marder K, Sequeiros J, Paulsen JS, Landwehrmeyer GB, Myers RH, MacDonald ME, Gusella JF (2012) CAG repeat expansion in Huntington disease determines age at onset in a fully dominant fashion. Neurology 78(10):690–695. https://doi.org/10.1212/WNL. 0b013e318249f683
4. Myers RH (2004) Huntington's disease genetics. NeuroRx 1(2):255–262. https://doi.org/10.1602/neurorx.1.2.255
5. Bates GP (2005) History of genetic disease: the molecular genetics of Huntington disease—a history. Nat Rev Genet 6 (10):766–773. https://doi.org/10.1038/nrg1686
6. Rubinsztein DC, Leggo J, Coles R, Almqvist E, Biancalana V, Cassiman JJ, Chotai K, Connarty M, Crauford D, Curtis A, Curtis D, Davidson MJ, Differ AM, Dode C, Dodge A, Frontali M, Ranen NG, Stine OC, Sherr M, Abbott MH, Franz ML, Graham CA, Harper PS, Hedreen JC, Jackson A, Kaplan JC, Losekoot M, MacMillan JC, Morrison P, Trottier Y, Novelletto A, Simpson SA, Theilmann J, Whittaker JL, Folstein SE, Ross CA, Hayden MR (1996) Phenotypic characterization of individuals with 30-40 CAG repeats in the Huntington disease (HD) gene reveals HD cases with 36 repeats and apparently normal elderly individuals with 36-39 repeats. Am J Hum Genet 59 (1):16–22
7. Kay C, Collins JA, Miedzybrodzka Z, Madore SJ, Gordon ES, Gerry N, Davidson M, Slama RA, Hayden MR (2016) Huntington disease reduced penetrance alleles occur at high frequency in the general population. Neurology 87(3):282–288. https://doi.org/10.1212/WNL.0000000000002858
8. Ranen NG, Stine OC, Abbott MH, Sherr M, Codori AM, Franz ML, Chao NI, Chung AS, Pleasant N, Callahan C, Kasch LM, Ghaffari M, Chase GA, Kazazian HH, Brandt J, Folstein SE, Ross CA (1995) Anticipation and instability of IT-15 (CAG)n repeats in parent-offspring pairs with Huntington disease. Am J Hum Genet 57 (3):593–602
9. Goldberg YP, Kremer B, Andrew SE, Theilmann J, Graham RK, Squitieri F, Telenius H, Adam S, Sajoo A, Starr E, Heiberg A, Wolff G, Hayden MR (1993) Molecular analysis of new mutations for Huntington's disease: intermediate alleles and sex of origin effects. Nat Genet 5(2):174–179. https://doi.org/10.1038/ng1093-174
10. Killoran A, Biglan KM, Jankovic J, Eberly S, Kayson E, Oakes D, Young AB, Shoulson I (2013) Characterization of the Huntington intermediate CAG repeat expansion phenotype in PHAROS. Neurology 80 (22):2022–2027. https://doi.org/10.1212/WNL.0b013e318294b304
11. Ha AD, Beck CA, Jankovic J (2012) Intermediate CAG Repeats in Huntington's disease: analysis of COHORT. Tremor Other Hyperkinet Mov (N Y) 2:tre-02-64-287-4. https://doi.org/10.7916/D8FF3R2P
12. Cubo E, Ramos-Arroyo MA, Martinez-Horta S, Martinez-Descalls A, Gil-Polo C (2016) Intermediate CAG repeats in Huntington's disease. A longitudinal analysis of the European Huntington's Disease Network REGISTRY Cohort (S25. 003) [Abstract]. Neurology 86:571–578
13. Kenney C, Powell S, Jankovic J (2007) Autopsy-proven Huntington's disease with 29 trinucleotide repeats. Mov Disord 22 (1):127–130. https://doi.org/10.1002/mds.21195
14. Ha AD, Jankovic J (2011) Exploring the correlates of intermediate CAG repeats in Huntington disease. Postgrad Med 123 (5):116–121. https://doi.org/10.3810/pgm.2011.09.2466
15. Squitieri F, Esmaeilzadeh M, Ciarmiello A, Jankovic J (2011) Caudate glucose hypometabolism in a subject carrying an unstable allele of intermediate CAG(33) repeat length in the Huntington's disease gene. Mov

Disord 26(5):925–927. https://doi.org/10.1002/mds.23623

16. Semaka A, Kay C, Belfroid RDM, Bijlsma EK, Losekoot M, van Langen IM, van Maarle MC, Oosterloo M, Hayden MR, van Belzen MJ (2015) A new mutation for Huntington disease following maternal transmission of an intermediate allele. Eur J Med Genet 58(1):28–30. https://doi.org/10.1016/j.ejmg.2014.11.005

17. Gonitel R, Moffitt H, Sathasivam K, Woodman B, Detloff PJ, Faull RL, Bates GP (2008) DNA instability in postmitotic neurons. Proc Natl Acad Sci U S A 105(9):3467–3472. https://doi.org/10.1073/pnas.0800048105

18. Wright GEB, Collins JA, Kay C, McDonald C, Dolzhenko E, Xia Q, Bečanović K, Drögemöller BI, Semaka A, Nguyen CM, Trost B, Richards F, Bijlsma EK, Squitieri F, Ross CJD, Scherer SW, Eberle MA, Yuen RKC, Hayden MR (2019) Length of uninterrupted CAG, independent of polyglutamine size, results in increased somatic instability, hastening onset of huntington disease. Am J Hum Genet 104(6):1116–1126. https://doi.org/10.1016/j.ajhg.2019.04.007

19. Swami M, Swami M, Hendricks AE, Gillis T, Massood T, Mysore J, Myers RH, Wheeler VC et al Hum Mol Genet 18(16):3039–3047. https://doi.org/10.1093/hmg/ddp242

20. Bečanović K, Nørremølle A, Neal SJ, Kay C, Collins JA, Arenillas D, Lilja T, Gaudenzi G, Manoharan S, Doty CN, Beck J, Lahiri N, Portales-Casamar E, Warby SC, Connolly C, De Souza RA, REGISTRY Investigators of the European Huntington's Disease Network, Tabrizi SJ, Hermanson O, Langbehn DR, Hayden MR, Wasserman WW, Leavitt BR (2015) A SNP in the HTT promoter alters NF-κB binding and is a bidirectional genetic modifier of Huntington disease. Nat Neurosci 18(6):807–816. https://doi.org/10.1038/nn.4014

21. Long JD, Lee JM, Aylward EH, Gillis T, Mysore JS, Abu Elneel K, Chao MJ, Paulsen JS, MacDonald ME, Gusella JF (2018) Genetic modification of Huntington disease acts early in the prediagnosis phase. Am J Hum Genet 103(3):349–357. https://doi.org/10.1016/j.ajhg.2018.07.017

22. Holbert S, Denghien I, Kiechle T, Rosenblatt A, Wellington C, Hayden MR, Margolis RL, Ross CA, Dausset J, Ferrante RJ, Néri C (2001) The Gln-Ala repeat transcriptional activator CA150 interacts with huntingtin: neuropathologic and genetic evidence for a role in Huntington's disease pathogenesis. Proc Natl Acad Sci U S A 98:1811–1816. https://doi.org/10.1073/pnas.041566798

23. Metzger S, Saukko M, Van Che H, Tong L, Puder Y, Riess O, Nguyen HP (2010) Age at onset in Huntington's disease is modified by the autophagy pathway: implication of the V471A polymorphism in Atg7. Hum Genet 128:453–459. https://doi.org/10.1007/s00439-010-0873-9

24. Soyal SM, Felder TK, Auer S, Hahne P, Oberkofler H, Witting A, Paulmichl M, Landwehrmeyer GB, Weydt P, Patsch W, European Huntington Disease Network (2012) A greatly extended PPARGC1A genomic locus encodes several new brain-specific isoforms and influences Huntington disease age of onset. Hum Mol Genet 21:3461–3473. https://doi.org/10.1093/hmg/dds177

25. Xu EH, Tang Y, Li D, Jia JP (2009) Polymorphism of HD and UCHL-1 genes in Huntington's disease. J Clin Neurosci 16:1473–1477. https://doi.org/10.1016/j.jocn.2009.03.027

26. Kloster E, Saft C, Epplen JT, Arning L (2013) CNR1 variation is associated with the age at onset in Huntington disease. Eur J Med Genet 56:416–419. https://doi.org/10.1016/j.ejmg.2013.05.007

27. Gayán J, Brocklebank D, Andresen JM, Alkorta-Aranburu G, US-Venezuela Collaborative Research Group, Zameel Cader M, Roberts SA, Cherny SS, Wexler NS, Cardon LR, Housman DE (2008) Genomewide linkage scan reveals novel loci modifying age of onset of Huntington's disease in the Venezuelan HD kindreds. Genet Epidemiol 32:445–453. https://doi.org/10.1002/gepi.20317

28. Genetic Modifiers of Huntington's Disease (GeM-HD) Consortium (2015) Identification of genetic factors that modify clinical onset of Huntington's disease. Cell 162(3):516–526. https://doi.org/10.1016/j.cell.2015.07.003

29. Moss DJH, Pardiñas AF, Langbehn D, Lo K, Leavitt BR, Roos R, Durr A, Mead S, TRACK-HD Investigators; REGISTRY Investigators, Holmans P, Jones L, Tabrizi SJ (2017) Identification of genetic variants associated with Huntington's disease progression: a genome-wide association. Lancet Neurol 16(9):701–711. https://doi.org/10.1016/S1474-4422(17)30161-8

30. Dhaenens CM, Burnouf S, Simonin C, Van Brussel E, Duhamel A, Defebvre L, Duru C,

Vuillaume I, Cazeneuve C, Charles P, Maison P, Debruxelles S, Verny C, Gervais H, Azulay JP, Tranchant C, Bachoud-Levi AC, Dürr A, Buée L, Krystkowiak P, Sablonnière B, Blum D, Huntington French Speaking Network (2009) A genetic variation in the ADORA2A gene modifies age at. onset in Huntington's disease. Neurobiol Dis 35:474–476. https://doi.org/10.1016/j.nbd.2009.06.009

31. Djoussé L, Knowlton B, Hayden MR, Almqvist EW, Brinkman RR, Ross CA, Margolis RL, Rosenblatt A, Durr A, Dode C, Morrison PJ, Novelletto A, Frontali M, Trent RJ, McCusker E, Gómez-Tortosa E, Mayo Cabrero D, Jones R, Zanko A, Nance M, Abramson RK, Suchowersky O, Paulsen JS, Harrison MB, Yang Q, Cupples LA, Mysore J, Gusella JF, MacDonald ME, Myers RH (2004) Evidence for a modifier of onset age in Huntington disease linked to the HD gene in 4p16. Neurogenetics 5 (2):109–114. https://doi.org/10.1007/s10048-004-0175-2

32. Taherzadeh-Fard E, Saft C, Andrich J, Wieczorek S, Arning L (2009) PGC-1alpha as modifier of onset age in Huntington disease. Mol Neurodegen 4:10. https://doi.org/10.1186/1750-1326-4-10

33. Metzger S, Bauer P, Tomiuk J, Laccone F, Didonato S, Gellera C, Mariotti C, Lange HW, Weirich-Schwaiger H, Wenning GK, Seppi K, Melegh B, Havasi V, Balikó L, Wieczorek S, Zaremba J, Hoffman-Zacharska D, Sulek A, Basak AN, Soydan E, Zidovska J, Kebrdlova V, Pandolfo M, Ribaï P, Kadasi L, Kvasnicova M, Weber BH, Kreuz F, Dose M, Stuhrmann M, Riess O (2006) Genetic analysis of candidate genes modifying the age-at-onset in Huntington's disease. Hum Genet 120(2):285–292. https://doi.org/10.1007/s00439-006-0221-2

34. Lee JM, Chao MJ, Harold D, Abu Elneel K, Gillis T, Holmans P, Jones L, Orth M, Myers RH, Kwak S, Wheeler VC, MacDonald ME, Gusella JF (2017) A modifier of Huntington's disease onset at the MLH1 locus. Hum Mol Genet 26(19):3859–3867. https://doi.org/10.1093/hmg/ddx286

35. Vuono R, Kouli A, Legault EM, Chagnon L, Allinson KS, La Spada A, REGISTRY Investigators of the European Huntington's Disease Network, Biunno I, Barker RA, Drouin-Ouellet J (2020) Association between Toll-Like Receptor 4 (TLR4) and triggering receptor expressed on myeloid cells 2 (TREM2) genetic variants and clinical progression of Huntington's disease. Mov Disord 35 (3):401–408. https://doi.org/10.1002/mds.27911

36. Correia K, Harold D, Kim KH, Holmans P, Jones L, Orth M, Myers RH, Kwak S, Wheeler VC, MacDonald ME, Gusella JF, Lee JM (2015) The genetic modifiers of motor onset age (GeM MOA) website: genome-wide association analysis for genetic modifiers of Huntington's disease. J Huntingtons Dis 4(3):279–284. https://doi.org/10.3233/JHD-150169

37. Vuono R, Winder-Rhodes S, de Silva R, Cisbani G, Drouin-Ouellet J, REGISTRY Investigators of the European Huntington's Disease Network, Spillantini MG, Cicchetti F, Barker RA (2015) The role of tau in the pathological process and clinical expression of Huntington's disease. Brain 138:1907–1918. https://doi.org/10.1093/brain/awv107

38. de Diego-Balaguer R, Schramm C, Rebeix I, Dupoux E, Durr A, Brice A, Charles P, Cleret de Langavant L, Youssov K, Verny C, Damotte V, Azulay JP, Goizet C, Simonin C, Tranchant C, Maison P, Rialland A, Schmitz D, Jacquemot C, Fontaine B, Bachoud-Lévi AC, French Speaking Huntington Group (2016) COMT Val158Met polymorphism modulates Huntington's disease progression. PLoS One 11(9):e0161106. https://doi.org/10.1371/journal.pone.0161106

39. Huntington Study Group (1996) Unified Huntington's disease rating scale: reliability and consistency. Mov Disord 11 (2):136–142. https://doi.org/10.1002/mds.870110204

40. Ross CA, Reilmann R, Cardoso F, McCusker EA, Testa CM, Stout JC, Leavitt BR, Pei Z, Landwehrmeyer B, Martinez A, Levey J, Srajer T, Bang J, Tabrizi SJ (2019) Movement disorder society task force viewpoint: Huntington's disease diagnostic categories. Mov Disord Clin Pract 6(7):541–546. https://doi.org/10.1002/mdc3.12808

41. Paulsen JS, Long JD, Ross CA, Harrington DL, Erwin CJ, Williams JK, Westervelt HJ, Johnson HJ, Aylward EH, Zhang Y, Bockholt HJ, Barker RA, PREDICT-HD Investigators and Coordinators of the Huntington Study Group (2014) Prediction of manifest Huntingtons disease with clinical and imaging measures: a prospective observational study. Lancet Neurol 13(12):1193–1201. https://doi.org/10.1016/S1474-4422(14)70238-8

42. Pla P, Orvoen S, Saudou F, David DJ, Humbert S (2014) Mood disorders in Huntington's disease: from behavior to cellular and

molecular mechanisms. Front Behav Neurosci 8:135. https://doi.org/10.3389/fnbeh.2014.00135

43. Jacobs M, Hart EP, van Zwet EW, Bentivoglio AR, Burgunder JM, Craufurd D, Reilmann R, Saft C, Roos RA, REGISTRY Investigators of the European Huntington's Disease Network (2016) Progression of motor subtypes in Huntington's disease: a 6-year follow-up study. J Neurol 263(10):2080–2085. https://doi.org/10.1007/s00415-016-8233-x

44. Louis ED, Anderson KE, Moskowitz C, Thorne DZ, Marder K (2000) Dystonia-predominant adult-onset Huntington disease: association between motor phenotype and age of onset in adults. Arch Neurol 57 (9):1326–1330. https://doi.org/10.1001/archneur.57.9.1326

45. Fusilli C, Migliore S, Mazza T, Consoli F, De Luca A, Barbagallo G, Ciammola A, Gatto EM, Cesarini M, Etcheverry JL, Parisi V, Al-Oraimi M, Al-Harrasi S, Al-Salmi Q, Marano M, Vonsattel JG, Sabatini U, Landwehrmeyer GB, Squitieri F (2018) Biological and clinical manifestations of juvenile Huntington's disease: a retrospective analysis. Lancet Neurol 17(11):986–993. https://doi.org/10.1016/S1474-4422(18)30294-1

46. McCusker E (2010) Commentary: Huntington disease in a nonagenarian mistakenly diagnosed as normal pressure hydrocephalus. J Clin Neurosci 17(8):1068. https://doi.org/10.1016/j.jocn.2010.01.003

47. Biglan KM, Zhang Y, Long JD, Geschwind M, Kang GA, Killoran A, Lu W, McCusker E, Mills JA, Raymond LA, Testa C, Wojcieszek J, Paulsen JS, PREDICT-HD Investigators of the Huntington Study Group (2013) Refining the diagnosis of Huntington disease: the PREDICT-HD study. Front Aging Neurosci 5:12. https://doi.org/10.3389/fnagi.2013.00012

48. Huntington Study Group PHAROS Investigators (2006) At risk for Huntington disease: the PHAROS (Prospective Huntington At Risk Observational Study) cohort enrolled. Arch Neurol 63:991–996. https://doi.org/10.1001/archneur.63.7.991

49. Paulsen JS, Hayden M, Stout JC, Langbehn DR, Aylward E, Ross CA, Guttman M, Nance M, Kieburtz K, Oakes D, Shoulson I, Kayson E, Johnson S, Penziner E, Predict-HD Investigators of the Huntington Study Group (2006) Preparing for preventive clinical trials: the Predict-HD study. Arch Neurol 63:883–890. https://doi.org/10.1001/archneur.63.6.883

50. Bechtel N, Scahill RI, Rosas HD, Acharya T, van den Bogaard SJ, Jauffret C, Say MJ, Sturrock A, Johnson H, Onorato CE, Salat DH, Durr A, Leavitt BR, Roos RA, Landwehrmeyer GB, Langbehn DR, Stout JC, Tabrizi SJ, Reilmann R (2010) Tapping linked to function and structure in premanifest and symptomatic Huntington disease. Neurology 75(24):2150–2160. https://doi.org/10.1212/WNL.0b013e3182020123

51. Stout JC, Paulsen JS, Queller S, Solomon AC, Whitlock KB, Campbell JC, Carlozzi N, Duff K, Beglinger LJ, Langbehn DR, Johnson SA, Biglan KM, Aylward EH (2011) Neurocognitive signs in prodromal Huntington disease. Neuropsychology 25(1):1–14. https://doi.org/10.1037/a0020937

52. Epping EA, Kim JI, Craufurd D, Brashers-Krug TM, Anderson KE, McCusker E, Luther J, Long JD, Paulsen JS, PREDICT-HD Investigators and Coordinators of the Huntington Study Group (2016) Longitudinal psychiatric symptoms in prodromal Huntington's disease: a decade of data. Am J Psychiatry 173(2):184–192. https://doi.org/10.1176/appi.ajp.2015.14121551

53. Vonsattel JP, Myers RH, Stevens TJ, Ferrante RJ, Bird ED, Richardson EP Jr (1985) Neuropathological classification of Huntington's disease. J Neuropathol Exp Neurol 44 (6):559–577. https://doi.org/10.1097/00005072-198511000-00003

54. Aylward EH, Sparks BF, Field KM, Yallapragada V, Shpritz BD, Rosenblatt A, Brandt J, Gourley LM, Liang K, Zhou H, Margolis RL, Ross CA (2004) Onset and rate of striatal atrophy in preclinical Huntington disease. Neurology 63(1):66–72. https://doi.org/10.1212/01.wnl.0000132965.14653.d1

55. Tabrizi SJ, Scahill RI, Owen G, Durr A, Leavitt BR, Roos RA, Borowsky B, Landwehrmeyer B, Frost C, Johnson H, Craufurd D, Reilmann R, Stout JC, Langbehn DR, Investigators TRACK-HD (2013) Predictors of phenotypic progression and disease onset in premanifest and early-stage Huntington's disease in the TRACK-HD study: analysis of 36-month observational data. Lancet Neurol 12(7):637–649. https://doi.org/10.1016/S1474-4422(13)70088-7

56. Reilmann R, Bohlen S, Klopstock T, Bender A, Weindl A, Saemann P, Auer DP, Ringelstein EB, Lange HW (2010) Tongue force analysis assesses motor phenotype in premanifest and symptomatic Huntington's

disease. Mov Disord 25(13):2195–2202. https://doi.org/10.1002/mds.23243

57. Reilmann R, Kirsten F, Quinn L, Henningsen H, Marder K, Gordon AM (2001) Objective assessment of progression in Huntington's disease: a 3-year follow-up study. Neurology 57(5):920–924. https://doi.org/10.1212/wnl.57.5.920

58. Reilmann R, Bohlen S, Klopstock T, Bender A, Weindl A, Saemann P, Auer DP, Ringelstein EB, Lange HW (2010) Grasping premanifest Huntington's disease—shaping new endpoints for new trials. Mov Disord 25 (16):2858–2862. https://doi.org/10.1002/mds.23300

59. Reilmann R, Bohlen S, Kirsten F, Ringelstein EB, Lange HW (2011) Assessment of involuntary choreatic movements in Huntington's disease—toward objective and quantitative measures. Mov Disord 26(12):2267–2273. https://doi.org/10.1002/mds.23816

60. Paulsen JS, Langbehn DR, Stout JC, Aylward E, Ross CA, Nance M, Guttman M, Johnson S, MacDonald M, Beglinger LJ, Duff K, Kayson E, Biglan K, Shoulson I, Oakes D, Hayden M, Predict-HD Investigators and Coordinators of the Huntington Study Group (2008) Detection of Huntington's disease decades before diagnosis: the Predict-HD study. J Neurol Neurosurg Psychiatry 79(8):874–880. https://doi.org/10.1136/jnnp.2007.128728

61. Penney JB Jr, Vonsattel JP, MacDonald ME, Gusella JF, Myers RH (1997) CAG repeat number governs the development rate of pathology in Huntington's disease. Ann Neurol 41:689–692. https://doi.org/10.1002/ana.410410521

62. Michell AW, Goodman AO, Silva AH, Lazic SE, Morton AJ, Barker RA (2008) Hand tapping: a simple, reproducible, objective marker of motor dysfunction in Huntington's disease. J Neurol 255(8):1145–1152. https://doi.org/10.1007/s00415-008-0859-x

63. Saft C, Andrich J, Meisel NM, Przuntek H, Müller T (2006) Assessment of simple movements reflects impairment in Huntington's disease. Mov Disord 21(8):1208–1212. https://doi.org/10.1002/mds.20939

64. Reilmann R, Rouzade-Dominguez ML, Saft C, Süssmuth SD, Priller J, Rosser A, Rickards H, Schöls L, Pezous N, Gasparini F, Johns D, Landwehrmeyer GB, Gomez-Mancilla B (2015) A randomized, placebo-controlled trial of AFQ056 for the treatment of chorea in Huntington's disease.

Mov Disord 30(3):427–431. https://doi.org/10.1002/mds.26174

65. Reilmann R, Rumpf S, Beckmann H, Koch R, Ringelstein EB, Lange HW (2012) Huntington's disease: objective assessment of posture—a link between motor and functional deficits. Mov Disord 27(4):555–559. https://doi.org/10.1002/mds.24908

66. Reyes A, Salomonczyk D, Teo WP, Medina LD, Bartlett D, Pirogovsky-Turk E, Zaenker P, Bloom JC, Simmons RW, Ziman M, Gilbert PE, Cruickshank T (2018) Computerised dynamic posturography in premanifest and manifest individuals with Huntington's disease. Sci Rep 8(1):14615. https://doi.org/10.1038/s41598-018-32924-y

67. Rao AK, Muratori L, Louis ED, Moskowitz CB, Marder KS (2008) Spectrum of gait impairments in presymptomatic and symptomatic Huntington's disease. Mov Disord 23(8):1100–1107. https://doi.org/10.1002/mds.21987

68. Rao AK, Mazzoni P, Wasserman P, Marder K (2011) Longitudinal change in gait and motor function in pre-manifest Huntington's disease. PLoS Curr 3:RRN1268. https://doi.org/10.1371/currents.RRN1268

69. Andrzejewski KL, Dowling AV, Stamler D, Felong TJ, Harris DA, Wong C, Cai H, Reilmann R, Little MA, Gwin JT, Biglan KM, Dorsey ER (2016) Wearable sensors in Huntington's disease: a pilot study. J Huntingtons Dis 5(2):199–206. https://doi.org/10.3233/JHD-160197

70. Adams JL, Dinesh K, Xiong M, Tarolli CG, Sharma S, Sheth N, Aranyosi AJ, Zhu W, Goldenthal S, Biglan KM, Dorsey ER, Sharma G (2017) Multiple wearable sensors in Parkinson and Huntington disease individuals: a pilot study in clinic and at home. Digit Biomark 1(1):52–63. https://doi.org/10.1159/000479018

71. Tortelli R, Simillion C, Lipsmeier F, Kilchenmann T, Rodrigues FB, Byrne LM, Bamdadian A, Gossens C, Schobel S, Lindemann M, Wild E (2020) The Digital-HD study: smartphone-based remote testing to assess cognitive and motor symptoms in Huntington's disease. [Abstract]. Neurology 94(15 Supplement):1816

72. Simillion C, Lipsmeier F, Bamdadian A, Smith A, Schobel SA, Tortelli R, Rodrigues FB, Byrne LM, Wild E, Lindemann M (2020) Application of a digital monitoring platform to track severity and progression in Huntington's disease. Poster. In: 10th European

Conference on Rare Diseases & Orphan Products. Virtual

73. Lipsmeier F, Simillion C, Bamdadian A, Smith A, Schobel S, Czech C, Gossens C, Weydt P, Wild E, Lindemann M (2019) Preliminary reliability and validity of a novel digital biomarker smartphone application to assess cognitive and motor symptoms in Huntington's disease (HD) [Abstract]. (P1.8-042). Neurology 92(15 Supplement):P1.8-042

74. Roberts B (2020) Digitalization and personalized health care. https://www.roche.com/dam/jcr:61545ef8-94ee-4b3b-8084-c21505d09cd9/en/irp20190916_digitalisation.pdf. Accessed 25 June

75. Waddell EM, Dinesh K, Spear KL, Tarolli CG, Elson MJ, Curtis MJ, Mitten DJ, Sharma G, Dorsey ER, Adams JL (2019) GEORGE®—The first smartphone application for Huntington disease—Pilot study. [Abstract]. Neurotherapeutics 16:1350–1390

76. Lauraitis A, Maskeliunas R, Damasevicius R, Polap D, Wozniak M (2019) A smartphone application for automated decision support in cognitive task based evaluation of central nervous system motor disorders. IEEE J Biomed Health Inform 23(5):1865–1876. https://doi.org/10.1109/JBHI.2019.2891729

77. Dorsey ER, Papapetropoulos S, Xiong M, Kieburtz K (2017) The First Frontier: digital biomarkers for neurodegenerative disorders. Digit Biomark 1(1):6–13. https://doi.org/10.1159/000477383

78. Gordon MF, Grachev ID, Mazeh I, Dolan Y, Reilmann R, Loupe PS, Fine S, Navon-Perry-L, Gross N, Papapetropoulos S, Savola JM, Hayden MR (2019) Quantification of motor function in Huntington disease patients using wearable sensor devices. Digit Biomark 3(3):103–115. https://doi.org/10.1159/000502136

79. Nasir J, Floresco SB, O'Kusky JR, Diewert VM, Richman JM, Zeisler J, Borowski A, Marth JD, Phillips AG, Hayden MR (1995) Targeted disruption of the Huntington's disease gene results in embryonic lethality and behavioral and morphological changes in heterozygotes. Cell 81(5):811–823. https://doi.org/10.1016/0092-8674(95)90542-1

80. Reiner A, Dragatsis I, Zeitlin S, Goldowitz D (2003) Wild-type huntingtin plays a role in brain development and neuronal survival. Mol Neurobiol 28(3):259–276. https://doi.org/10.1385/MN:28:3:259

81. Harjes P, Wanker EE (2003) The hunt for huntingtin function: interaction partners tell many different stories. Trends Biochem Sci 28(8):425–433. https://doi.org/10.1016/S0968-0004(03)00168-3

82. Jimenez-Sanchez M, Licitra F, Underwood BR, Rubinsztein DC (2017) Huntington's disease: mechanisms of pathogenesis and therapeutic strategies. Cold Spring Harb Perspect Med 7(7):a024240. https://doi.org/10.1101/cshperspect.a024240

83. Brandstaetter H, Kruppa AJ, Buss F (2014) Huntingtin is required for ER-to-Golgi transport and for secretory vesicle fusion at the plasma membrane. Dis Model Mech 7(12):1335–1340. https://doi.org/10.1242/dmm.017368

84. Zuccato C, Ciammola A, Rigamonti D, Leavitt BR, Goffredo D, Conti L, MacDonald ME, Friedlander RM, Silani V, Hayden MR, Timmusk T, Sipione S, Cattaneo E (2001) Loss of huntingtin-mediated BDNF gene transcription in Huntington's disease. Science 293:493–498. https://doi.org/10.1126/science.1059581

85. Wetzel R (2012) Physical chemistry of polyglutamine: intriguing tales of a monotonous sequence. J Mol Biol 421:466–449. https://doi.org/10.1016/j.jmb.2012.01.030

86. Trottier Y, Devys D, Imbert G, Saudou F, An I, Lutz Y, Weber C, Agid Y, Hirsch EC, Mandel JL (1995) Cellular localization of the Huntington's disease protein and discrimination of the normal and mutated form. Nat Genet 10(1):104–110. https://doi.org/10.1038/ng0595-104

87. Massai L, Petricca L, Magnoni L, Rovetini L, Haider S, Andre R, Tabrizi SJ, Süssmuth SD, Landwehrmeyer BG, Caricasole A, Pollio G, Bernocco S (2013) Development of an ELISA assay for the quantification of soluble huntingtin in human blood cells. BMC Biochem 14:34. https://doi.org/10.1186/1471-2091-14-34

88. Corey-Bloom J, Haque AS, Park S, Nathan AS, Baker RW, Thomas EA (2018) Salivary levels of total huntingtin are elevated in Huntington's disease patients. Sci Rep 8(1):7371. https://doi.org/10.1038/s41598-018-25095-3

89. Folstein MF, Folstein SE, McHugh PR (1975) "Mini-Mental State": a practical method for grading the cognitive state of patients for the clinician. J Psychiatr Res 12(3):189–198. https://doi.org/10.1016/0022-3956(75)90026-6

90. Moscovitch-Lopatin M, Weiss A, Rosas HD, Ritch J, Doros G, Kegel KB, Difiglia M, Kuhn R, Bilbe G, Paganetti P, Hersch S (2010) Optimization of an HTRF assay for the detection of soluble mutant huntingtin

in human buffy coats: a potential biomarker in blood for Huntington disease. PLoS Curr 2: RRN1205.   https://doi.org/10.1371/currents.RRN1205

91. Moscovitch-Lopatin M, Goodman RE, Eberly S, Ritch JJ, Rosas HD, Matson S, Matson W, Oakes D, Young AB, Shoulson I, Hersch SM, Huntington Study Group PHAROS Investigators (2013) HTRF analysis of soluble huntingtin in PHAROS PBMCs. Neurology 81(13):1134–1140. https://doi.org/10.1212/WNL.0b013e3182a55ede

92. Weiss A, Träger U, Wild EJ, Grueninger S, Farmer R, Landles C, Scahill RI, Lahiri N, Haider S, Macdonald D, Frost C, Bates GP, Bilbe G, Kuhn R, Andre R, Tabrizi SJ (2012) Mutant huntingtin fragmentation in immune cells tracks Huntington's disease progression. J Clin Invest 122(10):3731–3736. https://doi.org/10.1172/JCI64565

93. Wild EJ, Boggio R, Langbehn D, Robertson N, Haider S, Miller JR, Zetterberg H, Leavitt BR, Kuhn R, Tabrizi SJ, Macdonald D, Weiss A (2015) Quantification of mutant huntingtin protein in cerebrospinal fluid from Huntington's disease patients. J Clin Invest 125(5):1979–1986. https://doi.org/10.1172/JCI80743

94. Macdonald D, Tessari MA, Boogaard I, Smith M, Pulli K, Szynol A, Albertus F, Lamers MB, Dijkstra S, Kordt D, Reindl W, Herrmann F, McAllister G, Fischer DF, Munoz-Sanjuan I (2014) Quantification assays for total and polyglutamine-expanded huntingtin proteins. PLoS One 9(5):e96854. https://doi.org/10.1371/journal.pone.0096854

95. Weiss A, Abramowski D, Bibel M, Bodner R, Chopra V, DiFiglia M, Fox J, Kegel K, Klein C, Grueninger S, Hersch S, Housman D, Régulier E, Rosas HD, Stefani M, Zeitlin S, Bilbe G, Paganetti P (2009) Single-step detection of mutant huntingtin in animal and human tissues: a bioassay for Huntington's disease. Anal Biochem 395 (1):8–15.   https://doi.org/10.1016/j.ab.2009.08.001

96. Wild EJ, Boggio R, Langbehn D, Robertson N, Haider S, Miller JR, Zetterberg H, Leavitt BR, Kuhn R, Tabrizi SJ, Macdonald D, Weiss A (2015) Quantification of mutant huntingtin protein in cerebrospinal fluid from Huntington's disease patients. J Clin Investig 125(5):1979–1986. https://doi.org/10.1172/JCI80743

97. Southwell AL, Smith SE, Davis TR, Caron NS, Villanueva EB, Xie Y, Collins JA, Ye ML, Sturrock A, Leavitt BR, Schrum AG, Hayden MR (2015) Ultrasensitive measurement of huntingtin protein in cerebrospinal fluid demonstrates increase with Huntington disease stage and decrease following brain huntingtin suppression. Sci Rep 5:12166. https://doi.org/10.1038/srep12166

98. Fodale V, Boggio R, Daldin M, Cariulo C, Spiezia MC, Byrne LM, Leavitt BR, Wild EJ, Macdonald D, Weiss A, Bresciani A (2017) Validation of ultrasensitive mutant huntingtin detection in human cerebrospinal fluid by single molecule counting immunoassay. J Huntingtons Dis 6(4):349–361. https://doi.org/10.3233/JHD-170269

99. Byrne LM, Rodrigues FB, Johnson EB, Wijeratne PA, De Vita E, Alexander DC, Palermo G, Czech C, Schobel S, Scahill RI, Heslegrave A, Zetterberg H, Wild EJ (2018) Evaluation of mutant huntingtin and neurofilament proteins as potential markers in Huntington's disease. Sci Transl Med 10(458): eaat7108.   https://doi.org/10.1126/scitranslmed.aat7108

100. Hoffman PN, Lasek RJ (1975) The slow component of axonal transport. Identification of major structural polypeptides of the axon and their generality among mammalian neurons. J Cell Biol 66(2):351–366. https://doi.org/10.1083/jcb.66.2.351

101. Fuchs E, Cleveland DW (1998) A structural scaffolding of intermediate filaments in health and disease. Science 279(5350):514–519. https://doi.org/10.1126/science.279.5350.514

102. Petzold A (2005) Neurofilament phosphoforms: surrogate markers for axonal injury, degeneration and loss. Neurol Sci 233 (1-2):183–198. https://doi.org/10.1016/j.jns.2005.03.015

103. Petzold A, Altintas A, Andreoni L, Bartos A, Berthele A, Blankenstein MA, Buee L, Castellazzi M, Cepok S, Comabella M, Constantinescu CS, Deisenhammer F, Deniz G, Erten G, Espiño M, Fainardi E, Franciotta D, Freedman MS, Giedraitis V, Gilhus NE, Giovannoni G, Glabinski A, Grieb P, Hartung HP, Hemmer B, Herukka SK, Hintzen R, Ingelsson M, Jackson S, Jacobsen S, Jafari N, Jalosinski M, Jarius S, Kapaki E, Kieseier BC, Koel-Simmelink MJ, Kornhuber J, Kuhle J, Kurzepa J, Lalive PH, Lannfelt L, Lehmensiek V, Lewczuk P, Livrea P, Marnetto F, Martino D, Menge T, Norgren N, Papuć E, Paraskevas GP, Pirttilä T, Rajda C, Rejdak K, Ricny J, Ripova D, Rosengren L, Ruggieri M, Schraen S, Shaw G, Sindic C, Siva A, Stigbrand T, Stonebridge I, Topcular B,

Trojano M, Tumani H, Twaalfhoven HA, Vécsei L, Van Pesch V, Vanderstichele H, Vedeler C, Verbeek MM, Villar LM, Weissert R, Wildemann B, Yang C, Yao K, Teunissen CE (2010) Neurofilament ELISA validation. J Immunol Methods 352 (1-2):23–31. https://doi.org/10.1016/j.jim.2009.09.014

104. Gaiottino J, Norgren N, Dobson R, Topping J, Nissim A, Malaspina A, Bestwick JP, Monsch AU, Regeniter A, Lindberg RL, Kappos L, Leppert D, Petzold A, Giovannoni G, Kuhle J (2013) Increased neurofilament light chain blood levels in neurodegenerative neurological diseases. PLoS One 8(9):e75091. https://doi.org/10.1371/journal.pone.0075091

105. Kuhle J, Barro C, Andreasson U, Derfuss T, Lindberg R, Sandelius Å, Liman V, Norgren N, Blennow K, Zetterberg H (2016) Comparison of three analytical platforms for quantification of the neurofilament light chain in blood samples: ELISA, electrochemiluminescence immunoassay and Simoa. Clin Chem Lab Med 54(10):1655–1661. https://doi.org/10.1515/cclm-2015-1195

106. Gisslén M, Price RW, Andreasson U, Norgren N, Nilsson S, Hagberg L, Fuchs D, Spudich S, Blennow K, Zetterberg H (2015) Plasma concentration of the neurofilament light protein (NFL) is a biomarker of CNS injury in HIV infection: a cross-sectional study. EBioMedicine 3:135–140. https://doi.org/10.1016/j.ebiom.2015.11.036

107. Lewczuk P, Ermann N, Andreasson U, Schultheis C, Podhorna J, Spitzer P, Maler JM, Kornhuber J, Blennow K, Zetterberg H (2018) Plasma neurofilament light as a potential biomarker of neurodegeneration in Alzheimer's disease. Alzheimers Res Ther 10 (1):71. https://doi.org/10.1186/s13195-018-0404-9

108. Sampedro F, Pérez-González R, Martínez-Horta S, Marín-Lahoz J, Pagonabarraga J, Kulisevsky J (2020) Serum neurofilament light chain levels reflect cortical neurodegeneration in de novo Parkinson's disease. Parkinsonism Relat Disord 74:43–49. https://doi.org/10.1016/j.parkreldis.2020.04.009

109. Cai L, Huang J (2018) Neurofilament light chain as a biological marker for multiple sclerosis: a meta-analysis study. Neuropsychiatr Dis Treat 14:2241–2254. https://doi.org/10.2147/NDT.S173280

110. Kuhle J, Kropshofer H, Haering DA, Kundu U, Meinert R, Barro C, Dahlke F, Tomic D, Leppert D, Kappos L (2019) Blood neurofilament light chain as a biomarker of MS disease activity and treatment response. Neurology 92(10):e1007–e1015. https://doi.org/10.1212/WNL.0000000000007032

111. Olsson B, Alberg L, Cullen NC, Michael E, Wahlgren L, Kroksmark AK, Rostasy K, Blennow K, Zetterberg H, Tulinius M (2019) NFL is a marker of treatment response in children with SMA treated with nusinersen. J Neurol 266(9):2129–2136. https://doi.org/10.1007/s00415-019-09389-8

112. Byrne LM, Rodrigues FB, Blennow K, Durr A, Leavitt BR, Roos RAC, Scahill RI, Tabrizi SJ, Zetterberg H, Langbehn D, Wild EJ (2017) Neurofilament light protein in blood as a potential biomarker of neurodegeneration in Huntington's disease: a retrospective cohort analysis. Lancet Neurol 16 (8):601–609. https://doi.org/10.1016/S1474-4422(17)30124-2

113. Niemelä V, Landtblom AM, Blennow K, Sundblom J (2017) Tau or neurofilament light—which is the more suitable biomarker for Huntington's disease? PLoS One 12(2): e0172762. https://doi.org/10.1371/journal.pone.0172762

114. Constantinescu R, Romer M, Oakes D, Rosengren L, Kieburtz K (2009) Levels of the light subunit of neurofilament triplet protein in cerebrospinal fluid in Huntington's disease. Parkinsonism Relat Disord 15 (3):245–248. https://doi.org/10.1016/j.parkreldis.2008.05.012

115. Vinther-Jensen T, Börnsen L, Budtz-Jørgensen E, Ammitzbøll C, Larsen IU, Hjermind LE, Sellebjerg F, Nielsen JE (2016) Selected CSF biomarkers indicate no evidence of early neuroinflammation in Huntington disease. Neurol Neuroimmunol Neuroinflamm 3(6):e287. https://doi.org/10.1212/NXI.0000000000000287

116. Rodrigues FB, Byrne LM, McColgan P, Robertson N, Tabrizi SJ, Zetterberg H, Wild EJ (2016) Cerebrospinal fluid inflammatory biomarkers reflect clinical severity in Huntington's disease. PLoS One 11(9): e0163479. https://doi.org/10.1371/journal.pone.0163479

117. Johnson EB, Byrne LM, Gregory S, Rodrigues FB, Blennow K, Durr A, Leavitt BR, Roos RA, Zetterberg H, Tabrizi SJ, Scahill RI, Wild EJ, TRACK-HD Study Group (2018) Neurofilament light protein in blood predicts regional atrophy in Huntington disease. Neurology 90(8):e717–e723. https://doi.org/10.1212/WNL.0000000000005005

118. Rodrigues FB, Byrne LM, Tortelli R, Johnson EB, Wijeratne PA, Arridge M, De Vita E, Ghazaleh N, Houghton R, Furby H, Alexander DC, Tabrizi SJ, Schobel S, Scahill RI, Heslegrave A, Zetterberg H, Wild EJ (2020) Longitudinal dynamics of mutant huntingtin and neurofilament light in Huntington's disease: the prospective HD-CSF study. (article in pre-print). https://doi.org/10.1101/2020.03.31.20045260

119. Tabrizi SJ, Leavitt BR, Landwehrmeyer GB, Wild EJ, Saft C, Barker RA, Blair NF, Craufurd D, Priller J, Rickards H, Rosser A, Kordasiewicz HB, Czech C, Swayze EE, Norris DA, Baumann T, Gerlach I, Schobel SA, Paz E, Smith AV, Bennett CF, Lane RM, Phase 1–2a IONIS-HTTRx Study Site Teams (2019) Targeting huntingtin expression in patients with Huntington's disease. N Engl J Med 380(24):2307–2316. https://doi.org/10.1056/NEJMoa1900907

120. Ventimiglia R, Mather PE, Jones BE, Lindsay RM (1995) The neurotrophins BDNF, NT-3 and NT-4/5 promote survival and morphological and biochemical differentiation of striatal neurons in vitro. Eur J Neurosci 7 (2):213–222. https://doi.org/10.1111/j.1460-9568.1995.tb01057.x

121. Canals JM, Checa N, Marco S, Akerud P, Michels A, Pérez-Navarro E, Tolosa E, Arenas E, Alberch J (2001) Expression of brain-derived neurotrophic factor in cortical neurons is regulated by striatal target area. J Neurosci 21(1):117–124. https://doi.org/10.1523/JNEUROSCI.21-01-00117.2001

122. Gauthier LR, Charrin BC, Borrell-Pagès M, Dompierre JP, Rangone H, Cordelières FP, De Mey J, MacDonald ME, Lessmann V, Humbert S, Saudou F (2004) Huntingtin controls neurotrophic support and survival of neurons by enhancing BDNF vesicular transport along microtubules. Cell 118 (1):127–138. https://doi.org/10.1016/j.cell.2004.06.018

123. Tasset I, Sánchez-López F, Agüera E, Fernández-Bolaños R, Sánchez FM, Cruz-Guerrero-A, Gascón-Luna F, Túnez I (2012) NGF and nitrosative stress in patients with Huntington's disease. J Neurol Sci 315 (1-2):133–136. https://doi.org/10.1016/j.jns.2011.12.014

124. Ciammola A, Sassone J, Cannella M, Calza S, Poletti B, Frati L, Squitieri F, Silani V (2007) Low brain-derived neurotrophic factor (BDNF) levels in serum of Huntington's disease patients. Am J Med Genet 144B (4):574–577. https://doi.org/10.1002/ajmg.b.30501

125. Squitieri F, Cannella M, Simonelli M, Sassone J, Martino T, Venditti E, Ciammola A, Colonnese C, Frati L, Ciarmiello A (2009) Distinct brain volume changes correlating with clinical stage, disease progression rate, mutation size, and age at onset prediction as early biomarkers of brain atrophy in Huntington's disease. CNS Neurosci Ther 15(1):1–11. https://doi.org/10.1111/j.1755-5949.2008.00068.x

126. Zuccato C, Marullo M, Vitali B, Tarditi A, Mariotti C, Valenza M, Lahiri N, Wild EJ, Sassone J, Ciammola A, Bachoud-Lèvi AC, Tabrizi SJ, Di Donato S, Cattaneo E (2011) Brain-derived neurotrophic factor in patients with Huntington's disease. PLoS One 6(8): e22966. https://doi.org/10.1371/journal.pone.0022966

127. Wang R, Ross CA, Cai H, Cong WN, Daimon CM, Carlson OD, Egan JM, Siddiqui S, Maudsley S, Martin B (2014) Metabolic and hormonal signatures in pre-manifest and manifest Huntington's disease patients. Front Physiol 5:231. https://doi.org/10.3389/fphys.2014.00231

128. Gutierrez A, Corey-Bloom J, Thomas EA, Desplats P (2020) Evaluation of biochemical and epigenetic measures of peripheral brain-derived neurotrophic factor (BDNF) as a biomarker in Huntington's disease patients. Front Mol Neurosci 12:335. https://doi.org/10.3389/fnmol.2019.00335

129. Ou ZA, Byrne LM, Rodrigues FB, Tortelli R, Johnson EB, Foiani MS, Arridge M, De Vita E, Scahill RI, Heslegrave A, Zetterberg H, Wild EJ (2019) Brain-derived neurotrophic factor in cerebrospinal fluid and plasma as a potential biomarker for Huntington's disease. [Abstract]. Neurotherapeutics 16(4):1373–1374

130. Zigmond AS, Snaith RP (1983) The hospital anxiety and depression scale. Acta Psychiatr Scand 67(6):361–370. https://doi.org/10.1111/j.1600-0447.1983.tb09716.x

131. Tirassa P, Iannitelli A, Sornelli F, Cirulli F, Mazza M, Calza A, Alleva E, Branchi I, Aloe L, Bersani G, Pacitti F (2012) Daily serum and salivary BDNF levels correlate with morning-evening personality type in women and are affected by light therapy. Riv Psichiatr 47(6):527–534. https://doi.org/10.1708/1178.13059

132. Moreira A, Aoki MS, de Arruda AFS, Machado DGDS, Elsangedy HM, Okano AH (2018) Salivary BDNF and cortisol responses during high-intensity exercise and official basketball matches in sedentary individuals and elite players. J Hum Kinet

65:139–149. https://doi.org/10.2478/hukin-2018-0040

133. Tsukinoki K, Saruta J, Sasaguri K, Miyoshi Y, Jinbu Y, Kusama M, Sato S, Watanabe Y (2006) Immobilization stress induces BDNF in rat submandibular glands. J Dent Res 85 (9):844–848. https://doi.org/10.1177/154405910608500913

134. Mitoma M, Yoshimura R, Sugita A, Umene W, Hori H, Nakano H, Ueda N, Nakamura J (2008) Stress at work alters serum brain-derived neurotrophic factor (BDNF) levels and plasma 3-methoxy-4-hydroxyphenylglycol (MHPG) levels in healthy volunteers: BDNF and MHPG as possible biological markers of mental stress? Prog Neuropsychopharmacol Biol Psychiatry 32(3):679–685. https://doi.org/10.1016/j.pnpbp.2007.11.011

135. Lommatzsch M, Niewerth A, Klotz J, Schulte-Herbrüggen O, Zingler C, Schuff-Werner P, Virchow JC (2007) Platelet and plasma BDNF in lower respiratory tract infections of the adult. Respir Med 101 (7):1493–1499. https://doi.org/10.1016/j.rmed.2007.01.003

136. Karege F, Perret G, Bondolfi G, Schwald M, Bertschy G, Aubry JM (2002) Decreased serum brain-derived neurotrophic factor levels in major depressed patients. Psychiatry Res 109(2):143–148. https://doi.org/10.1016/s0165-1781(02)00005-7

137. Kobayashi K, Shimizu E, Hashimoto K, Mitsumori M, Koike K, Okamura N, Koizumi H, Ohgake S, Matsuzawa D, Zhang L, Nakazato M, Iyo M (2005) Serum brain-derived neurotrophic factor (BDNF) levels in patients with panic disorder: as a biological predictor of response to group cognitive behavioral therapy. Prog Neuropsychopharmacol Biol Psychiatry 29(5):658–663. https://doi.org/10.1016/j.pnpbp.2005.04.010

138. Spillantini MG, Goedert M (2013) Tau pathology and neurodegeneration. Lancet Neurol 12(6):609–622. https://doi.org/10.1016/S1474-4422(13)70090-5

139. Fernández-Nogales M, Cabrera JR, Santos-Galindo M, Hoozemans JJ, Ferrer I, Rozemuller AJ, Hernández F, Avila J, Lucas JJ (2014) Huntington's disease is a four-repeat tauopathy with tau nuclear rods. Nat Med 20 (8):881–885. https://doi.org/10.1038/nm.3617

140. Liu P, Smith BR, Huang ES, Mahesh A, Vonsattel JPG, Petersen AJ, Gomez-Pastor R, Ashe KH (2019) A soluble truncated tau species related to cognitive dysfunction and caspase-2 is elevated in the brain of Huntington's disease patients. Acta Neuropathol Commun 7(1):111. https://doi.org/10.1186/s40478-019-0764-9

141. Rodrigues FB, Byrne L, McColgan P, Robertson N, Tabrizi SJ, Leavitt BR, Zetterberg H, Wild EJ (2016) Cerebrospinal fluid total tau concentration predicts clinical phenotype in Huntington's disease. J Neurochem 139:22–25. https://doi.org/10.1111/jnc.13719

142. Constantinescu R, Romer M, Zetterberg H, Rosengren L, Kieburtz K (2011) Increased levels of total tau protein in the cerebrospinal fluid in Huntington's disease. Parkinsonism Relat Disord 17(9):714–715. https://doi.org/10.1016/j.parkreldis.2011.06.010

143. Hanna J, Hossain GS, Kocerha J (2019) The potential for microRNA therapeutics and clinical research. Front Genet 10:478. https://doi.org/10.3389/fgene.2019.00478

144. Bartel DP (2004) MicroRNAs: genomics, biogenesis, mechanism, and function. Cell 116(2):281–297. https://doi.org/10.1016/s0092-8674(04)00045-5

145. Johnson R, Zuccato C, Belyaev ND, Guest DJ, Cattaneo E, Buckley NJ (2008) A microRNA-based gene dysregulation pathway in Huntington's disease. Neurobiol Dis 29(3):438–445. https://doi.org/10.1016/j.nbd.2007.11.001

146. Hoss AG, Labadorf A, Latourelle JC, Kartha VK, Hadzi TC, Gusella JF, MacDonald ME, Chen JF, Akbarian S, Weng Z, Vonsattel JP, Myers RH (2015) miR-10b-5p expression in Huntington's disease brain relates to age of onset and the extent of striatal involvement. BMC Med Genomics 8:10. https://doi.org/10.1186/s12920-015-0083-3

147. Salzman DW (2019) Small RNA biomarkers for the early detection and monitoring of Huntington's disease. In: 14th annual HD therapeutics conference, Palm Springs, California

148. Bithell A, Johnson R, Buckley NJ (2009) Transcriptional dysregulation of coding and non-coding genes in cellular models of Huntington's disease. Biochem Soc Trans 37 (Pt 6):1270–1275. https://doi.org/10.1042/BST0371270

149. Moyzis RK, Buckingham JM, Cram LS, Dani M, Deaven LL, Jones MD, Meyne J, Ratliff RL, Wu JR (1988) A highly conserved repetitive DNA sequence, (TTAGGG)n, present at the telomeres of human chromosomes. Proc Natl Acad Sci U S A 85(18):6622–6626. https://doi.org/10.1073/pnas.85.18.6622

150. Cawthon RM, Smith KR, O'Brien E, Sivatchenko A, Kerber RA (2003) Association between telomere length in blood and mortality in people aged 60 years or older. Lancet 361(9355):393–395. https://doi.org/10.1016/S0140-6736(03)12384-7

151. Scarabino D, Broggio E, Gambina G, Corbo RM (2017) Leukocyte telomere length in mild cognitive impairment and Alzheimer's disease patients. Exp Gerontol 98:143–147. https://doi.org/10.1016/j.exger.2017.08.025

152. Martin-Ruiz C, Williams-Gray CH, Yarnall AJ, Boucher JJ, Lawson RA, Wijeyekoon RS, Barker RA, Kolenda C, Parker C, Burn DJ, Von Zglinicki T, Saretzki G (2020) Senescence and inflammatory markers for predicting clinical progression in Parkinson's disease: the ICICLE-PD Study. J Parkinsons Dis 10 (1):193–206. https://doi.org/10.3233/JPD-191724

153. Erusalimsky JD (2020) Oxidative stress, telomeres and cellular senescence: what non-drug interventions might break the link? Free Radic Biol Med 150:87–95. https://doi.org/10.1016/j.freeradbiomed.2020.02.008

154. Rai SN, Singh BK, Rathore AS, Zahra W, Keswani C, Birla H, Singh SS, Dilnashin H, Singh SP (2019) Quality control in Huntington's disease: a therapeutic target. Neurotox Res 36(3):612–626. https://doi.org/10.1007/s12640-019-00087-x

155. Scarabino D, Veneziano L, Peconi M, Frontali M, Mantuano E, Corbo RM (2019) Leukocyte telomere shortening in Huntington's disease. J Neurol Sci 396:2529. https://doi.org/10.1016/j.jns.2018.10.024

156. Castaldo I, De Rosa M, Romano A, Zuchegna C, Squitieri F, Mechelli R, Peluso S, Borrelli C, Del Mondo A, Salvatore E, Vescovi LA, Migliore S, De Michele G, Ristori G, Romano S, Avvedimento EV, Porcellini A (2019) DNA damage signatures in peripheral blood cells as biomarkers in prodromal Huntington disease. Ann Neurol 85:296–301. https://doi.org/10.1002/ana.25393

157. Kota LN, Bharath S, Purushottam M, Moily NS, Sivakumar PT, Varghese M, Pal PK, Jain S (2015) Reduced telomere length in neurodegenerative disorders may suggest shared biology. J Neuropsychiatry Clin Neurosci 27:e92–e96. https://doi.org/10.1176/appi.neuropsych.13100240

158. Bennett CF, Swayze EE (2010) RNA targeting therapeutics: molecular mechanisms of antisense oligonucleotides as a therapeutic platform. Annu Rev Pharmacol Toxicol 50:259–293. https://doi.org/10.1146/annurev.pharmtox.010909.105654

159. Smith AV, Tabrizi SJ (2020) Therapeutic antisense targeting of huntingtin. DNA Cell Biol 39(2):154–158. https://doi.org/10.1089/dna.2019.5188

160. Kordasiewicz HB, Stanek LM, Wancewicz EV, Mazur C, McAlonis MM, Pytel KA, Artates JW, Weiss A, Cheng SH, Shihabuddin LS, Hung G, Bennett CF, Cleveland DW (2012) Sustained therapeutic reversal of Huntington's disease by transient repression of huntingtin synthesis. Neuron 74 (6):1031–1044. https://doi.org/10.1016/j.neuron.2012.05.009

161. Stanek LM, Yang W, Angus S, Sardi PS, Hayden MR, Hung GH, Bennett CF, Cheng SH, Shihabuddin LS (2013) Antisense oligonucleotide-mediated correction of transcriptional dysregulation is correlated with behavioral benefits in the YAC128 mouse model of Huntington's disease. J Huntingtons Dis 2:217–228. https://doi.org/10.3233/JHD-130057

162. Schobel S (2020) Preliminary results from a 15-month open-label extension study investigating tominersen (RG6042) huntingtin protein antisense oligonucleotide in adults with manifest Huntington's disease. In: 15th Annual Huntington's disease therapeutics conference, Palm Springs, California

Chapter 11

Biofluid Biomarkers of Amyotrophic Lateral Sclerosis

Cory J. Holdom, Frederik J. Steyn, Robert D. Henderson, Pamela A. McCombe, Mary-Louise Rogers, and Shyuan T. Ngo

Abstract

Amyotrophic lateral sclerosis (ALS) is an adult-onset degenerative disease that is characterized by the progressive, irreversible loss of upper and lower motor neurons. It is a highly heterogeneous disease and variability in age of onset, site of onset, rate of disease progression, and survival between individuals present significant challenges for diagnosis and clinical care. Research into understanding the cause for ALS, and the clinical management of the disease is limited in part due to the absence of specific, sensitive biomarkers for the disease. To date, studies aimed at identifying reliable and specific biomarkers for ALS have revealed blood, cerebrospinal fluid (CSF), and urine markers as being useful. Here, we summarize some of the most promising fluid biomarkers identified to date, and discuss their proposed utility for improving our approach to ALS diagnosis, care, and research. Overall, while most studies into biomarkers for ALS explore single-marker utility, accurate profiling of individuals with ALS is likely to require a panel of complementary biomarkers providing insight into multiple aspects of the disease.

Keywords Amyotrophic lateral sclerosis, Biomarkers, Neurofilaments, Inflammation, Metabolism, Oxidative stress, p75ECD

1 Amyotrophic Lateral Sclerosis

Amyotrophic lateral sclerosis (ALS) is an adult-onset degenerative disease. ALS is characterized by the progressive, irreversible loss of upper and lower motor neurons. Thus, patients present with a combination of symptoms associated with loss of these neurons [1]. Primary lateral sclerosis (PLS) and progressive muscle atrophy (PMA) are related diseases, with loss of upper motor neurons and lower motor neurons, respectively [1].

The majority of patients have onset of weakness in the limbs. Bulbar dysfunction is observed in about 30% of patients at initial presentation, with dysphagia and dysarthria increasing as the disease progresses [1, 2]. Approximately 5% of patients present with involvement of respiratory muscles [3, 4]. Some patients with ALS also experience extramotor abnormalities [5], including changes in

Philip V. Peplow, Bridget Martinez and Thomas A. Gennarelli (eds.), *Neurodegenerative Diseases Biomarkers: Towards Translating Research to Clinical Practice*, Neuromethods, vol. 173, https://doi.org/10.1007/978-1-0716-1712-0_11,

cognition [6–9], sleep [10–12], sensory disturbance [13, 14], autonomic dysfunction [15–17], and bladder [18–20] and bowel dysfunction [21, 22]. Reports on the incidence of cognitive changes in ALS (these include behavioral, cognitive, and language deficits to varying degrees) vary considerably. Many patients have mild cognitive dysfunction [23], and up to 15% of patients meet criteria for diagnosis of concomitant frontotemporal dementia (FTD) [24]. Similarly, up to 10–15% of FTD patients may also have ALS [25, 26]. Onset of motor symptoms is typically asymmetrical [27]; however, evidence from patients with contaminant ALS and FTD suggests that disease might spread in a prion-like manner [28–30], beginning in the motor neocortex, progressing to the spinal cord and brainstem, then the frontal–parietal regions and finally the temporal lobes of the cortex [31]. Following symptom onset, the mean life expectancy for patients with ALS is 3–5 years; however, some individuals live for longer periods with little change in disease severity, while others have much shorter survival [32–34].

Some people with ALS are members of families with multiple affected individuals. In such patients, genetic variants have been reported. Commonly reported genes implicated in ALS include *Cu/Zn superoxide dismutase 1* (*SOD1*) [35], *TAR DNA-binding protein* (*TARDBP*) [36], *chromosome 9 open reading frame 72* (*C9orf72*) [37, 38], and *fused-in-sarcoma* (*FUS*) [39]. The other 90% of ALS patients have no family history [1]; however, individuals with sporadic (sALS) disease can also have variants in genes associated with familial ALS (fALS) [40]. These genetic causes may also explain the overlap between ALS and other neurodegenerative diseases. For example, links between ALS and FTD are reinforced by the *C9orf72* hexanucleotide repeat expansion [41].

ALS is a highly heterogeneous disease. Significant variability in age of onset, site of onset, rate of disease progression, and survival between individuals poses challenges for diagnosis and clinical care [42]. Diagnosis of ALS is based on clinical assessment and exclusion of disease mimics [43]. After a confirmed diagnosis, prognosis is highly variable among patients. Biomarkers that can detect physiological and/or pathological changes associated with disease offer potential to shorten diagnostic delay and improve specificity of diagnoses [44], while also helping multidisciplinary teams to refine disease management [45].

Research aimed at identifying reliable and specific biomarkers for ALS has revealed blood, cerebrospinal fluid (CSF) and urine markers as being useful (Fig. 1). In this chapter, we summarize some of the most promising fluid biomarkers identified to date, and discuss their proposed utility for improving our approach to ALS diagnosis, care, and research.

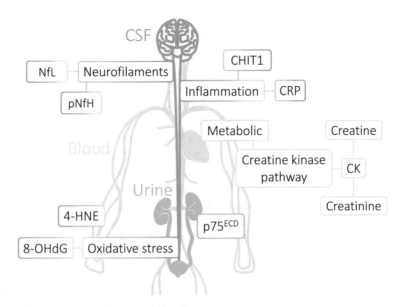

Fig. 1 Biofluid biomarkers of amyotrophic lateral sclerosis (ALS). Biomarkers for ALS have been identified in blood (blue), cerebrospinal fluid (CSF; dark green) and urine (light green). In the blood and CSF, the most promising fluid biomarkers identified to date include neurofilaments (neurofilament light (NfL) and phosphorylated neurofilament heavy chain/subunit (pNfH)), markers of inflammation (chitinase-1-like protein (CHIT-1) and C-reactive protein (CRP)), and metabolism associated changes in the creatine kinase pathway (creatine kinase (CK), creatine, and creatinine). Of these blood and CSF biomarkers, pNfH has the highest utility as a prediagnostic biomarker for ALS, but also as a biomarker for changes in disease progression. The expression of the extracellular domain of the low-affinity nerve growth factor receptor (p75^{ECD}) in urine is emerging as a very promising biomarker for ALS. Markers of oxidative stress, including 4-hydroxy-2-nonenal (4-HNE) and 8-hydroxy-2'-deoxyguanosine (8-OHdG), have also been identified blood, CSF, and urine, but their utility as a specific biomarker for ALS is relatively low

2 Cerebrospinal Fluid and Blood

Cerebrospinal fluid (CSF) is an ideal source of fluid-based ALS biomarkers. The direct contact of CSF with neuronal tissues [46] is advantageous as it results in CSF having a higher concentration of factors associated with neuronal function and/or death, and a lower concentration of proteins [44]. However, CSF can only be sampled via lumbar puncture, a procedure which cannot be readily performed on all patients with ALS. Unlike CSF, blood is distal to the site of disease in ALS and it contains a lower proportion of disease-specific markers [47]. However, sampling of peripheral blood is minimally invasive and is a routine procedure. While, in healthy individuals, CSF and blood are considered distinct compartments due to active shuttling of protein out of the CSF, proteomic profiling has shown that patients with ALS experience an influx of CSF factors into blood due to disruptions in the blood–brain barrier (BBB) and blood–spinal cord barrier (BSCB) [48, 49]. As such, assessment of blood is able to provide some insight into central nervous system (CNS) derived biomarkers in patients with ALS, as well as biomarkers from all peripheral tissues in the body.

Due to the increased exchange of factors between blood and CSF in patients with ALS, a number of biomarkers are conserved between the two fluids. While many blood- and CSF-based biomarkers have been investigated for their utility in ALS diagnosis and prognosis [50], the biomarkers we discuss span four domains: neurofilaments, and markers of inflammation, metabolism and oxidative stress.

2.1 Neurofilaments

Neurofilaments are structural components of the neuronal cytoskeleton. These components are released into the circulation following neuronal damage. Two classes of neurofilaments that are commonly observed in the blood and CSF of patients with ALS are neurofilament light (NfL) and phosphorylated neurofilament heavy chain/subunit (pNfH).

Levels of NfL and pNfH are consistently reported to be elevated in the CSF and blood/serum of patients with ALS [51–71] (Table 1). From a diagnostic viewpoint, studies that have compared the utility of neurofilaments as a biomarker for distinguishing ALS from disease mimics have shown that pNfH is more sensitive than NfL [63, 67, 70]. While these findings are supported, in part, by observations that levels of NfL and pNfH are increased in the CSF and/or blood of early symptomatic patients with ALS [68, 72], a study by Weydt and colleagues in 2016 reported that levels of NfL and pNfH were indistinguishable between controls (non-ALS mutation carriers) and asymptomatic individuals with known ALS mutations [69]. The lack of an increase in NfL or pNfH in the study by Weydt and colleagues is likely due to latency between sample collection and diagnosis; these samples were, on average, collected over 2 years before symptom onset [69]. Notably, two recent studies that aimed to determine the presymptomatic levels of serum pNfH and NfL have reported increased levels of these neurofilaments prior to the onset of motor symptoms and diagnosis of ALS in both sALS individuals [73] and asymptomatic individuals with fALS-linked mutations [73, 74]. Of note, both De Schaepdryver and colleagues [73], and Benatar and colleagues [74], found a steady increase in neurofilaments 12–24 months before the onset of symptoms. In each study, the prediagnostic increase in neurofilaments within a subset of patients was initially within normal ranges, exceeded this near symptom onset, and continued to increase for up to 6 months after disease onset [73, 74]. These studies, in conjunction with the normal neurofilament measures observed by

Table 1
Evidence for the use of neurofilament as a biomarker for disease progression in patients with ALS

Study [reference]	Design	Analytical technique	Outcomes
Benatar et al. (2020) [90]	• 229 ALS • 20 PLS • 11 PMA	• Quanterix Simoa • Iron Horse Diagnostics assay • Euroimmun CE marked ELISA	• Higher serum NfL in ALS, compared to PMA and PLS • High correlation between serum and CSF NfL, but not pNfH • Serum NfL (but not pNfH) predicts future decline in ALSFRS-R and survival • Initial serum NfL measures decreased simulated trial sample size by ~8% • Use of NfL and pNfH as outcome measures greatly reduces needed sample sizes
Boylan et al. (2009) [71]	• 20 ALS • 20 healthy control (HC) • Longitudinal	• Sandwich ELISA	• Plasma pNfH increased in ALS • Higher pNfH associated with higher ΔFRS • pNfH stable in patients • Results were also reflected in a SOD1^{G93A} mouse model
Boylan et al. (2012) [55]	• 63 ALS • Longitudinal	• Sandwich ELISA	• CSF pNfH correlated with ΔFRS in 4-month and 12-month follow-up • Plasma pNfH correlated with ΔFRS in 4-month, but not 12-month, follow-up • Plasma and serum pNfH associated with survival after 12 months • Plasma, serum, and CSF pNF-H did not differ between site of onset
Brettschneider et al. (2006) [65]	• 69 ALS • 73 Alzheimer's disease (AD) • 33 HC	• Sandwich ELISA	• CSF NfH increased in ALS, relative to AD and HC • CSF NfH higher in upper motor neuron dominant disease • CSF NfH higher in more rapid progression
Chen et al. (2016) [51]	• 40 sALS • 40 neurodegenerative control (NDC)	• Sandwich ELISA	• CSF pNfH correlates with ΔFRS • CSF pNfH elevated in ALS compared to NDC • CSF pNfH independently identifies ALS with 97% sensitivity and 84% specificity

(continued)

Table 1
(continued)

Study [reference]	Design	Analytical technique	Outcomes
De Schaepdryver et al. (2018) [63]	• 85 ALS • 31 ALS mimics • 215 neurologic controls	• Sandwich ELISA	• CSF and serum pNfH correlated • CSF pNfH more accurately identifies ALS than serum • Serum pNfH inversely correlated with survival • CSF and serum pNfH correlate with ΔFRS
Feneberg et al. (2017) [68]	• 189 ALS • 21 non-ALS MND • 27 MND mimics • 65 NDC • Longitudinal, Multicenter (8 European centers for collection, 1 center for analysis)	• Sandwich ELISA (CSF) • Simoa (serum)	• CSF NfL and pNfH, and serum NfL were increased in ALS (serum pNfH not assessed) • CSF NfL and pNfH, and serum NfL identified ALS from NDC and MND mimics • CSF NfL and pNfH, and serum NfL were not different between individuals initially diagnosed with ALS, and individuals later diagnosed • No consistent NF change observed over disease or between centers
Gaiottino et al. (2013) [61]	• 46 ALS • 20 Alzheimer's disease (AD) • 19 Guillain–Barre syndrome (GBS) • 68 neurological controls	• ECL	• CSF NfL increased in ALS, AD, and GBS, relative to neurological controls • CSF and serum NFL correlates • ALS serum NfL: sensitivity 91%, specificity 91%
Gendron et al. (2018) [54]	• 135 C9orf72 carriers (107 ALS/FTD) • 70 non-C9orf72 ALS/FTD • 37 HC • Longitudinal	• Meso Scale Discovery immunoassay	• CSF pNfH higher in C9orf72 ALS/FTD than asymptomatic carriers (sensitivity: 99%, specificity 96%) • CSF pNfH higher in C9orf72 ALS and ALS-FTD than pure FTD • CSF pNfH higher in non-C9orf72 ALS than HC and FTD • CSF pNfH did not change over time in all ALS cases

(continued)

Table 1
(continued)

Study [reference]	Design	Analytical technique	Outcomes
Kuhle et al. (2010) [64]	• 50 ALS • 95 multiple sclerosis (MS) • 20 mild cognitive impairment (MCI)/AD • 20 GBS • 20 subarachnoid hemorrhage (SAH)	• ECL	• Neurodegenerative patients had higher CSF NfH than controls
Lehnert et al. (2014) [53]	• 30 ALS • 30 HC • Multicenter • 6 European sites for collection • 1 site for processing	• Sandwich ELISA	• CSF pNfH increased in ALS in all sites • Different absolute pNfH between sites
Li et al. (2016) [79]	• 51 sALS • 12 multiple system atrophy (MSA) • 30 HC	• Sandwich ELISA	• Plasma and CSF pNfH are directly correlated • Plasma and CSF pNfH elevated in ALS compared to MSA and HC • Plasma: 80% sensitivity and 74% specificity • CSF: 82% sensitivity, 74% specificity
Li et al. (2018) [67]	• 85 ALS • 32 NDC	• Sandwich ELISA	• CSF pNfH and NfL higher in ALS • CSF pNfH inversely correlated with ALS disease duration • pNfH: sensitivity 100%, specificity 69% • NfL: sensitivity 96%, specificity 56%
Lu et al. (2015) [57]	• sALS: 103 London (plasma), 64 Oxford (serum, 38 with CSF) • HC: 42 London (plasma), 36 Oxford (serum, 20 with CSF) • Longitudinal	• ECL	• CSF NfL correlated with serum • Blood NfL increased in ALS • Blood NfL predicts survival • Blood NfL measures were stable over time • CSF: 97% sensitivity, 95% specificity • Serum: 89% sensitivity, 75% specificity • Plasma: 90% sensitivity, 71% specificity

(continued)

Table 1
(continued)

Study [reference]	Design	Analytical technique	Outcomes
McCombe et al. (2015) [52]	• 98 ALS • 59 HC • Longitudinal	• Sandwich ELISA	• Serum pNfH elevated in ALS • Serum pNfH increase in early stages of ALS • Initial serum pNfH lower in long-surviving individuals • Serum ΔpNfH correlated with survival
Menke et al. (2015) [57]	• ALS: 25 CSF, 40 serum • HC: 17 CSF, 25 serum	• Sandwich ELISA	• NfL increased in ALS (CSF and serum) • CSF NfL correlated with ΔFRS • Increased NfL associated with decreased corticospinal tract fractional anisotropy and increased radial diffusivity
Oeckl et al. (2016) [56]	• 75 ALS • 75 neurologic controls • Multicenter (15 European site collection, 2 site analysis)	• Sandwich ELISA	• CSF pNfH increased in 10/15 sites • CSF NfL increased in 5/12 sites • pNfH: sensitivity 79%, specificity 93% • NfL: sensitivity 79%, specificity 86%
Poesen et al. (2017) [70]	• 220 ALS • 50 ALS mimics • 316 NDC • Longitudinal	• Sandwich ELISA	• CSF pNfH and NfL correlated with ΔFRS, FVC and diagnostic delay, and number sites with both UMN and LMN involvement • CSF pNfH distinguished ALS from mimics with sensitivity 91%, specificity 88%
Reijin et al. (2009) [66]	• 32 ALS • 26 ALS mimics	• Sandwich ELISA	• CSF NfL and pNfH higher in ALS than disease mimics • Subtyping ALS did not show differences in NfL or pNfH • pNfH: sensitivity 72%, specificity 80% • NfL: sensitivity 75%, specificity 79%
Steinacker et al. (2015) [69]	• 253 ALS and PLS (20 fALS) • 85 MND mimics • 107 NDC	• Sandwich ELISA	• CSF NfL and pNfH increased in ALS, relative to mimics and NDC • CSF NfL and pNfH elevated early in disease • CSF NfL and pNfH correlated with ΔFRS and survival • NfL: 77% sensitivity, 85% specificity • pNfH: 83% sensitivity, 77% specificity

(continued)

Table 1
(continued)

Study [reference]	Design	Analytical technique	Outcomes
Tortelli et al. (2012) [59]	• 37 sALS • 46 NDC	• Sandwich ELISA	• CSF NfL higher in ALS than NDC • CSF NfL correlated with shorter diagnosis and ΔFRS • CSF NfL inversely correlated with ALSFRS-R • 78% sensitivity, 73% specificity
Tortelli et al. (2015) [58]	• 37 sALS	• Sandwich ELISA	• CSF NfL associated with faster progression (time to generalization) • High CSF NfL predicts shorter survival
Zetterberg et al. (2007) [62]	• 60 sALS • 19 fALS • 206 NDC • 40 HC	• Sandwich ELISA	• CSF NfL higher in ALS than NDC and HC • CSF NfL inversely correlated with disease duration • CSF NfL lower in mSOD1 ALS than wtSOD1

AD Alzheimer's disease, *ALS* amyotrophic lateral sclerosis, *ALSFRS-R* ALS functional rating scale—revised, *fALS* familial ALS, *FTD* frontotemporal dementia, *FVC* forced vital capacity, *GBS* Guillain–Barre syndrome, *HC* healthy control, *MCI* mild cognitive impairment, *MND* motor neuron disease, *MS* multiple sclerosis, *MSA* multiple system atrophy, *NDC* neurodegenerative control, *NfL* neurofilament light, *PLS* primary lateral sclerosis, *PMA* progressive muscle atrophy, *pNfH* phosphorylated neurofilament heavy, *SAH* subarachnoid hemorrhage, *sALS* sporadic ALS

Weydt and colleagues, highlight that the increase in neurofilaments is likely to be directly related to the active neurodegenerative processes of ALS. Moreover, they also suggest that the quantification of neurofilaments may be able to inform early diagnosis as well as the staging and the timing of treatments. In this regard, monitoring of levels of serum neurofilaments in asymptomatic individuals with a known ALS gene mutation or family history may provide crucial information to determine whether neurofilaments could be included in official ALS diagnostic criteria [75]. Alongside specifying diagnoses, neurofilaments may have value for improving clinical staging of people with ALS, and to better-inform the intervention windows for potential therapeutics.

As a prognostic biomarker, neurofilament expression has been shown to relate to clinical outcomes. Circulating levels of plasma and serum NfL are increased in patients with ALS; higher levels of NfL are associated with more rapid disease progression, and are predictive for survival [57, 76, 77]. Similarly, in a retrospective analysis of serum collected from individuals who later developed ALS, prediagnostic serum pNfH has been reported to predict shorter survival in individuals with fALS with known genetic risk factors, as well as sALS patients with no identified disease-linked mutations [73]. Benatar and colleagues observed presymptomatic neurofilament measures to be predictive for postdiagnostic disease progression rate, as well as survival [74]. A systemic review incorporating 20 studies exploring the effectiveness of neurofilaments as diagnostic and prognostic biomarkers for ALS found that both NfL and pNfH in CSF and blood were predictive for survival and future rate of disease progression [77]. Levels of NfL and pNfH have also been reported to be relatively stable, leading to the suggestion that neurofilaments can perform effectively as a biomarker of ALS throughout the course of disease [57, 71]. However, at the late stage of disease, when there has been severe loss of motor neurons, there will be little further degeneration and levels might decline. Overall, current studies that routinely report the use of neurofilaments for predicting the progression of clinical features of disease indicate that pNfH and NfL are reliable, and clinically relevant, biomarkers for ALS. However, larger, longitudinal, multicohort studies need to be undertaken to verify the true prognostic value of neurofilaments.

Notwithstanding the general consensus that neurofilaments are considered to be a useful biomarker for ALS [78], there are still caveats that limit their use. Firstly, given that neurofilaments may only provide about 60% specificity for ALS, the assessment of neurofilaments alone is unlikely to allow for the elimination of disease mimics during the process of an ALS diagnosis [57, 59, 67, 69, 79]. Secondly, there are no standardized methods for quantifying neurofilaments in the CSF and blood. Early approaches for detecting neurofilaments included immunoblotting [80–82] and enzyme-linked immunosorbent assays (ELISA) [66, 83–86]. Historically, these approaches have proved to be robust for detecting neurofilaments in CSF, but their lower sensitivity for detecting neurofilaments in blood limits the degree to which ALS can be distinguished from disease mimics and healthy controls [66, 79]. More recent techniques, including electrochemiluminescence (ECL) [87, 88] and single-molecule array (Simoa) [68, 76, 88], are promising as they have been shown to be sensitive enough to track the variation in neurofilaments over time within control individuals [89]. Congruent with this, by using Simoa to assess the levels of serum NfL in patients with ALS, Feneberg and colleagues were able to show that serum NfL was higher in patients with ALS when compared to nondiseased controls. Measures were sensitive/accurate enough to separate patients with ALS from nondiseased controls and motor neuron disease (MND) mimics [68]. Moreover, through the use of Simoa, a recent study by Benatar and colleagues has shown that serum NfL is predictive of future decline of the revised ALS functional rating scale (ALSFRS-R) [90]. In order to demonstrate replicability of published results, and to further validate and/or confirm current observations that support the use of neurofilaments as a biomarker for ALS, a gold-standard approach for quantification needs to be established.

2.2 Markers of Inflammation

Neuroinflammation is commonly reported to have both positive and negative effects in neurodegenerative diseases including Alzheimer's disease (AD) [91, 92], Parkinson's disease (PD) [93, 94], and multiple sclerosis (MS) [95, 96]. Neuroinflammation is also present in ALS, and is evidenced by infiltration of peripheral immune cells into CNS tissues, alongside increased presence of activated astrocytes and microglia [97, 98], and invasion of peripheral monocytes into the CSF [99]. In line with this, numerous candidate inflammatory proteins have been detected in the blood and CSF of patients with ALS through the use of liquid/gas chromatography [100–110]; an invaluable tool for unbiased screening of fluids. However, there has been low replicability between studies [111]. While the expression of markers of inflammation such as interleukin-6 (IL-6) [112–114], interleukin-8 (IL-8) [113, 115], tumor necrosis factor alpha (TNFa) [113, 114, 116], lipocalin-2 [113, 117], and complement [118–120] have consistently been found to be altered in ALS, two of the most promising neuroinflammatory proteins identified to date include chitotriosidase/chitinase 1 (CHIT1) [51, 108, 121–124] and C-reactive protein (CRP) [105, 125–128] (Table 2). Given that inflammation appears to be a key process in the progression of disease, there has been significant interest in the development of therapeutics to modulate and/or dampen inflammatory processes in ALS. For example, NP001 (sodium chlorite) (NCT02794857), a regulator of activated monocytes, has been shown to slow disease progression in patients with increased inflammation [129]. Dimethyl fumarate, a compound used in relapsing-remitting MS to promote effector T cells and inhibit production of regulatory T cells, is in Phase II trials in ALS (ACTRN12618000534280) for slowing disease progression, and extending survival [130]. Mecasin is a drug shown to improve motor symptoms in a small Korean cohort of patients, and is currently in Phase IIA (KCT0001984) to determine effective dosage [131]. The CD14 inhibitor, IC14 has recently been incorporated into a platform trial (NCT03474263) to assess the impact of immunosuppression in ALS [132]. Due to the significant focus on the targeting of neuroinflammation in ALS, the development of specific, insightful biomarkers for informing the degree/severity of inflammation with respect to the clinical features of disease have also been an area of intense research interest.

Table 2
Evidence for the use of chitotriosidase/chitinase 1 and C-reactive protein as biomarkers for disease progression in patients with ALS

Study [reference]	Design	Analytical technique	Outcomes
Varghese (2013) [108]	• 26 ALS • 23 non–neurodegenerative disease controls (NNDC)	• Liquid chromatography-mass spectrometry (LC-MS) • Sandwich enzyme-linked immunosorbent assay (ELISA)	• First study to show an increase in chitinase-1 (CHIT1) in ALS CSF, compared to NNDC • Chitinase-3-like protein 1 (CHI3L1) and chitinase-3-like protein 2 (CHI3L2) were also increased
Steinacker (2018) [122]	• 60 ALS • 46 ALS mimics • 25 healthy control (HC) • 25 Alzheimer's disease (AD) • 15 Parkinson's disease (PD) • 15 CJD • 80 frontotemporal dementia (FTD) • Longitudinal	• Sandwich ELISA • Immunohistochemistry (IHC) • Immunofluorescence (IF)	• CSF CHIT1 increased in ALS compared to disease mimics and HC • CSF CHIT1 increased in fast, compared to slow-progressing ALS • CSF CHIT1 not increased in slow-progressing ALS, compared to HC • CSF CHIT1 increased in ALS compared to PD, FTD and AD, but not CJD • Serum CHIT1 unchanged in ALS compared to other groups • CHIT1 correlated with disease duration and ALSFRS-R in fast-progressing ALS, but not with survival • Sensitivity 87% and specificity 84%, compared to HC • Sensitivity 82% and specificity 51%, compared to disease controls • No change in CFS CHIT1 after 6 months in fast or slow ALS

(continued)

Table 2
(continued)

Study [reference]	Design	Analytical technique	Outcomes
Thompson (2018) [123]	• 43 ALS • 6 PLS • 12 ALS mimics • 20 HC • 20 PD	• LC-MS/MS	• CSF CHIT1, CHI3L1 and CHI3L2 increased in ALS • CSF CHI3L1 also increased in PLS • ALS CHIT1, CHI3L1 and CHI3L2 correlated with change in ALSFRS-R, and pNfH • CHI3L1 increased over time • CHIT1 negatively associated with survival
Palgiardini (2015) [121]	• 75 ALS • 106 HC	• Dried blood spots	• Blood CHIT1 increased in ALS • Blood CHIT 1 higher in rapid progressing disease
Vu (2020) [124]	• 118 ALS • 17 neurological disease controls (NDC) • 24 HC	• Sandwich ELISA • IHC	• CSF CHIT1 increased in ALS compared to NDC and HC • CSF CHI3L1 increased in ALS and NDC, compared to HC • CHI3L1 and CHIT1 correlate well with pNfH • CHIT-1 and CHI3L1 increased in fast-progressing ALS • Increased astrocytic CHI3L1 expression in ALS
Chen (2016) [51]	• 40 sALS • 40 NDC	• Sandwich ELISA	• CSF CHIT-1 increased in sALS • Sensitivity: 83.8%, specificity: 81.1% • Increases to sensitivity: 83.8%, specificity: 91.9% with comeasures of pNfH and CHIT-1

(continued)

Table 2
(continued)

Study [reference]	Design	Analytical technique	Outcomes
Keizman (2009) [126]	• 80 ALS • 80 HC • Longitudinal	• Unstated	• First study to assess systemic low-grade inflammation in ALS • Blood CRP inversely proportional with ALSFRS-R • Inversely proportional relation between within-subject change in ALSFRS-R and CRP
Sanjak (2018) [128]	• 48 ALS no respiratory support • 23 ALS on NIV • 16 ALS on TIV	• Unstated	• Blood CRP increased in ALS patients on respiratory support • Blood CRP weakly correlated with ALSFRS-R • Blood CRP reduced after Riluzole administration
Lunetta (2017) [127]	• 520 ALS • 122 ALS (independent cohort) • Longitudinal	• Unstated	• Cross-sectional serum CRP correlates with ALSFRS-R • Serum CRP correlates with survival • Post-hoc inclusion of CRP in analysis of NP001 showed a significant dose-response improvement in functional scores in treatment group
De Schaepdryver (2020) [125]	• 383 ALS • Multicenter	• Immunoturbidimetry	• Serum CRP proportional to rate of disease progression • CRP did not associate with survival
Ryberg (2010) [105]	• 100 ALS • 100 NDC • 41 HC	• SELDI-MS	• CSF CRP increased in ALS, relative to all groups

AD Alzheimer's disease, *ALS* amyotrophic lateral sclerosis, *ALSFRS-R* ALS functional rating scale—revised, *CJD* Creutzfeldt–Jakob disease, *FTD* frontotemporal dementia, *NDC* neurodegenerative control, *NIV* noninvasive ventilation, *NNDC* nonneurodegenerative disease control, *PD* Parkinson's disease, *PLS* primary lateral sclerosis, *pNfH* phosphorylated neurofilament heavy, *sALS* sporadic ALS, *TIV* tracheostomy-invasive ventilation

2.2.1 Chitinase-1-Like Protein, CHIT1

CHIT1 is a hydrolase that binds and degrades chitin [133], a polysaccharide abundant within arthropods, as well as fungi [134]. This enzyme, as well as its related chitinase-3-like proteins 1 and 2 (CHI3L1/CHI3L2), is well-understood to be upregulated by activated macrophages in response to pathogenic infection [135]. The presence of these proteins in ALS is evidence for immune activation, which is a negative factor for survival [136]. There is also some evidence that chitinase and chitinase-like proteins may have a role in breaking down the extracellular matrix (ECM) [137–139]. Remodeling of the extracellular matrix has been proposed in ALS; changes in proteins responsible for ECM homeostasis are commonly observed in patients [140], and, in animal models, disruption of perineuronal nets has been reported [141]. Reparation of these perineuronal nets improves symptoms in animal models of ALS [141], although there is limited evidence of improvements in humans [142]. As a biomarker for neuroinflammation, chitinases may provide insight into understanding disease mechanisms, while contributing to the assessment of the efficacy of potential interventions [143].

CHIT1 was first identified in 2013 as a potential biomarker of neuroinflammation in ALS [108]. In this study, Varghese and colleagues used a combination of LC-MS and ELISA to investigate the expression of proteins in the CSF of patients with ALS, as compared to patients with tension headaches. They found that the expression of CHIT1 and CHI3L2 was increased by 17-fold and 2-fold, respectively, when compared to the tension headache cohort, and that CHIT1 activity was increased by 10-fold. They also showed that microglia exposed to CSF from patients with ALS exhibited a significant increase in the expression of CHIT1 [108]. These findings have now been well-supported [51, 108, 121–124], and two recent studies have highlighted a role for CSF CHIT1 as a biomarker for neuroinflammation in ALS, as well as disease stratification [122, 123]. In 2018, Steinacker and colleagues measured levels of CHIT1 in the CSF and serum of 60 patients with ALS, and compared this to ALS mimics, neurodegenerative controls, and healthy controls [122]. Of interest, levels of CHIT1 in the CSF were elevated in fast, but not slow, progressing patients with ALS; however, there was no overall change in the expression levels of serum CHIT1. Despite the increased CHIT1 production in fast-progressing ALS, Steinacker and colleagues did not observe a further change in CSF CHIT1 after 6 months in fast- or slow-progressing ALS, indicating that CHIT1 may function as a stable biomarker of fast-progressing ALS [122]. Although there was a relationship with disease duration, there was no association between circulating CHIT1 and survival outcomes. Interestingly, the increase in CHIT1 observed in ALS exceeded the levels of CHIT1 observed in AD, PD, and FTD patients; however, it was no different to Creutzfeldt–Jakob disease (CJD) patients. Although

CHIT1 was able to reasonably discriminate patients with ALS from healthy controls, there was only about 50% specificity when compared to neurodegenerative controls [122], providing some evidence for shared neuroinflammatory mechanisms between these diseases.

In another 2018 paper, Thompson and colleagues published an LC-MS/MS study comparing CSF in ALS, PLS, and healthy and disease control individuals [123]. Levels of CHIT1, CHI3L1, and CHI3L2 were found to be increased in ALS patients when compared to disease mimics and controls, however only CHI3L1 was also increased in patients with PLS [123]. They also reported that CHIT1, CHI3L1, and CHI3L2 were correlated with ALSFRS-R (a functional rating scale used to define progressive worsening of disability in patients across the motor neuron disease spectrum [144]), as well as pNfH, indicating potential for the chitinase enzymes to serve as a biomarker for disease progression. Congruent with this, CHI3L1 was found to be increased in follow-up measures within the ALS cohort. Contrasting the result in a study by Steinacker, CHIT1 (but not CHI3L1 or CHI3L2) was negatively associated with survival [123]. As CSF CHIT1 is primarily produced by resident microglia [108], and a primary event in most ALS cases is extensive microgliosis [136], an increase in circulating CHIT1 is likely a strong indicator for elevated microglial activation, which itself is a negative prognostic factor for ALS [145–147]. Furthermore, individuals who display increased inflammation within the CNS having faster disease progression and shorter survival [148], and cell culture [149] and mouse [150] experiments showing a feed-forward toxic effect of inflammatory CSF on healthy cells further highlight the importance of immune factors in the pathogenesis of ALS. In a more recent study, Vu et al., reported increased levels of CHIT1 and CHI3L1 in the CSF, as well as an abundance of CHI3L1 in post-mortem white-matter astrocytes of patients with ALS [124]. CHI3L1 transcription within reactive astrocytes is upregulated during various insults including cancers and neurodegenerative disease [151]. Together, monitoring CHIT1 and CHI3L1 in patients with ALS may provide objective measures for disease-driven gliosis, allowing more accurate assessment of patient status, and the effect of treatments that target neuroinflammation.

2.2.2 C-Reactive Protein, CRP

C-reactive protein (CRP) is secreted by hepatocytes [152], adipocytes [153], macrophages [154], endothelium [155], and smooth muscle [156] in a universal response to inflammation to drive activation of innate immune responses [152, 157]. Specifically, CRP is released as a pentamer that monomerizes at sites of inflammation. Actions of CRP include serving as an opsonin [158] and activating complement via the classical pathway [159]. This is important in ALS, as various complement proteins have been proposed to contribute to disease in both patients with ALS [160–162] and animal models of ALS [118, 163, 164]. Indeed, autopsied spinal cords and motor cortices from patients with ALS have increased C1q, C3 and C5b-9 within neurons, microglia and astrocytes [162, 165]. Motor pathways have also been reported to have increased C3d and C4d within oligodendrocytes, as well as increased C3 and C4 receptor expression in surrounding tissue [166]. Moreover, elevated plasma C5a and C5b-9 has been observed in patients with ALS [167]. In SOD1 mice, studies of complements C1q, C3 and C4 have shown no change in disease progression or survival [168, 169]. Knockout and inhibition of the downstream C5aR1 improves motor ability and survival, suggesting a further pathogenic role for the complement pathway in ALS [170, 171].

In 2009, a longitudinal investigation by Keizman and colleagues explored the role of low-grade inflammation in ALS [126]. Blood CRP, erythrocyte sedimentation rate (ESR) and fibrinogen were identified as being increased in ALS. These factors were inversely associated with ALSFRS-R and were also found to be increased as the disease progressed. This was the first published investigation to report changes in low-grade immune function, and an elevation in CRP, in ALS. However, the process for quantifying the immune factors and the accuracy for CRP to discriminate ALS were not reported [126]. A subsequent study by Ryberg and colleagues reported an increase in CSF CRP in ALS subjects, compared to neurodegenerative controls and healthy individuals [105]. Corroborating these findings, a 2020 study by De Schaepdryver and colleagues found serum CRP to be elevated in patients with ALS, and inversely proportional to rate of disease progression [125]. Collectively, these studies suggest that CRP could serve as an indicator of ALS neuroinflammatory status. However, further research is needed to determine whether CRP can be used as a biomarker of disease progression in ALS.

As well as potentially serving as a biomarker for the disease status of an individual, CRP could provide a meaningful outcome measure for assessing the efficacy of ALS treatments. In a 2018 study published by Sanjak and colleagues [128], CRP was shown to be elevated in patients who had progressed to noninvasive ventilation (NIV) or complete tracheostomy. Importantly, CRP was found to decline in patients following the commencement of Riluzole [128]. Interestingly, while Riluzole is indicated in ALS for management of glutamate-driven hyperexcitability [172], a second mechanism of action via the suppression of immune processes has been proposed in MS and spinal cord injury [173–175]. While the outcomes from the investigation by Sanjak and colleagues suggest that utility of CRP as a biomarker for disease progression might be limited to pretherapeutic use, it also provides justification for the use of CRP as an outcome measure for neuroinflammatory-regulating therapeutics. Indeed, in a large clinical trial of NP001 in an Italian cohort of patients with ALS ($n = 510$), serum CRP was shown to be inversely correlated with patient functional scores, and associated with shorter survival [127]. In this same study, a significant positive treatment effect of NP001 on ALSFRS-R was observed in patients with ALS who had elevated serum CRP, but not in patients with normal CRP [127]. In a subsequent multicenter study, however, De Schaepdryver and colleagues did not observe a relationship between serum CRP and survival [125]. As such, further explorations are needed to determine the efficacy for CRP as a biomarker for survival in patients with ALS.

While there appears to be some merit in using CRP as a biomarker for assessing the impact of trial drugs on outcome measures that relate to disease progression, current publications that explore the role of CRP as a biomarker for ALS are limited due to difficulties associated with accurate quantification of the protein. Assessment of CRP involves separation of free CRP from its bound forms before quantification. CRP is traditionally assessed using relatively imprecise quantitative nephelometry or semiquantitative latex agglutination [176, 177]. Within ALS studies, published techniques include immunoturbidimetry and surface-enhanced laser desorption/ionization mass spectrometry (SELDI-MS) [105, 125]. Immunoturbidimetry is rapid and low-cost, potentially allowing for simple high-throughput assessment of CRP measures in ALS plasma or CSF [178]. By contrast, SELDI-MS is highly precise but relatively expensive, allowing for accurate reporting in ALS research but reduced utility in a clinical setting [179]. At present, no technique has been consistently robust for assessing CRP in ALS. As such, the development of a reproducible approach for quantifying CRP will be crucial for determining whether CRP is a reliable biomarker for ALS.

2.3 Markers of Metabolism

Disruption of cellular and systemic metabolism is increasingly being recognized as a common feature of ALS. It is yet to be determined whether metabolic changes contribute to the causation of ALS; however, metabolic dysregulation has been repeatedly linked to worsened disease progression. On a systemic level, patients with ALS tend to have a lower prediagnostic BMI [180, 181], and this is potentially worsened through loss of appetite [182, 183] and weight loss as the disease progresses [181, 183–185]. Patients with ALS have also been reported to exhibit nondiabetic glucose intolerance [186, 187], hyperlipidemia [188, 189], and hypermetabolism [190–194].

Studies into cellular metabolism in ALS are early and under-explored; however, current findings in animal and cell models suggest that significant metabolic alterations exist. Evidence for decreased glucose utilization in neurons has primarily been obtained through studies that have traced radiolabeled glucose and acetate in the cortex and spinal cords of SOD1^{G93A} mice. In the cortex, decreased glucose metabolism products, including reduced pyruvate availability, decreased glucose-6-phosphate hydrogenase activity, and increased consumption of alternative tricarboxylic acid (TCA) cycle substrates were observed. In the spinal cord, the pentose phosphate pathway intermediate ribose 5-phosphate was decreased, alongside increased glucose-6-phosphate. Overall, these findings are indicative of significant metabolic perturbations in a mouse model of ALS [195, 196].

Extending on observations in the mouse, a number of studies have identified metabolic perturbations in human derived disease-relevant cells. In 2018, RNA sequencing of induced pluripotent stem cell (iPSC)-derived motor neurons from sALS and fALS individuals led to the identification of alterations in gene expression that were indicative of cellular metabolic alterations [197]. In a more recent study by Hor and colleagues, iPSC-derived motor neurons from sALS and fALS individuals exhibited mitochondrial hyperacetylation, which could be reversed upon upregulation of SIRT3, a mitochondrial deacetylase [198]. In contrast to these metabolic phenotypes observed by Fujimora et al. and Hor et al., a study of iPSC-derived FUS motor neurons by Guo and colleagues did not identify any alterations in motor neuron metabolism [199]. This is a curious observation, as motor neurons are traditionally considered the primary site of pathology for ALS. However, astrocytes are known to play a key role in regulating neuronal metabolism and have been proposed to exert a non–cell autonomous mechanism for disease pathology. Recent work by Allen and colleagues has shown that the selective restriction of substrates available for energy utilization within fibroblast-derived astrocytes from C9orf72 and sALS individuals results in reduced cellular metabolic flexibility [200], and this may increase the vulnerability of motor neurons in ALS.

It is important to note that altered cellular metabolism in ALS extends beyond the CNS. Metabolic perturbations have been identified in the skeletal muscle of mouse models of ALS, and altered expression or glucose and fatty acid metabolism genes in skeletal muscle biopsies from ALS patients has also been reported [201, 202]. These findings have been recently supported by the observation that myotubes derived from patients with ALS exhibit increased fatty acid oxidation when compared to myotubes generated from healthy control individuals, and that an increase in fatty acid oxidation is associated with slower progressing disease [203]. Collectively, these studies highlight a role for non–cell autonomous pathology in ALS, with metabolic alterations in neurons, glial cells, and skeletal muscle potentially modifying neurodegenerative mechanisms. By developing a better understanding of metabolic dysregulation in the disease, there is potential to identify an effective avenue for disease-modifying therapeutics. However, specific biomarkers are required to improve our ability to quantify metabolic processes impacted by ALS, and the effect of potential therapeutics on these pathways.

To date, extensive research into identifying metabolic biomarkers has not been undertaken, and there is no single blood biomarker to inform metabolic derangements specific to ALS [204]. Alongside changes in body weight and composition, changes in mitochondrial function and glucose utilization have been directly observed in tissues derived from ALS patients. Investigation of biochemical pathways, although of significant research interest, is not currently part of routine clinical evaluation, as fluid biomarkers have not yet been established to effectively track key metabolic aspects of the disease. There has, however, been some progress in imaging and electrophysiological approaches [205–208]. Of all the fluid biomarker candidates identified to date, the creatine kinase (CK) pathway has shown some potential for use in ALS as a biomarker for disease progression, and more specifically, as a measure for change in fat-free mass content [204].

2.3.1   The Creatine Kinase Pathway

Creatinine kinase (CK) and its metabolites, creatine and creatinine, have been proposed as biomarkers for change in skeletal muscle mass in patients with ALS [209–213]. CK enzymatically phosphorylates creatine to buffer phosphate and maintain ATP reserves during periods of increased energy utilization [214]. This is of particular importance in neurons and muscle as these tissues consume the bulk of energy within humans [215–218]. Phosphorylated creatine can stochastically cyclize, releasing the phosphate to form the stable waste product, creatinine, which is then excreted into urine. In this way, cells with increased energy utilization have decreased free creatine, increased CK turnover and creatinine production.

In patients with ALS, and in mouse models of ALS, CK has been found to be upregulated before the onset of motor symptoms, or loss of lean mass [212]. While existing measures do not directly compare CK with lean mass in patients with ALS, there is evidence that these measures are correlated. For example, after symptom onset, retrospective analysis of levels of circulating CK was found to be proportional to estimated lean muscle mass in patients with MND, and changes in CK decreased alongside declines of estimated muscle mass in these patients [219]. Given that levels of CK are indicative of the degree of ATP use, this observation is in line with studies in ALS that have demonstrated increased systemic energy use in both patients with ALS [190, 192–194], and mouse models of the disease [220]; an increase in CK production may occur as a consequence of increased energy utilization within tissues affected by the disease. This notion is further supported by observations by Szelechowski and colleagues demonstrating a decline in oxidative phosphorylation and increased fatty acid oxidation in presymptomatic ALS mice and in skin fibroblasts from humans with ALS [221].

Circulating levels of creatinine, a product of CK turnover, have been found to be proportional to estimates of lean mass in both patients with ALS [211, 219] and healthy non-ALS study participants [222, 223]. Cross-sectional creatinine measures have also been found to be an independent predictor for survival, wherein higher levels of creatinine were found to be associated with longer survival, and milder motor symptoms [211]. These observations were corroborated in a large ($n = 1241$) retrospective study by van Eijk and colleagues, who found that patients with ALS had decreasing muscle strength and plasma creatinine over the course of their disease [224]. Within this investigation, plasma creatinine was found to be a more consistent measure for disease progression than ALSFRS-R in long (>18 months) duration clinical trials [224].

Interestingly, levels of both circulating CK and creatinine have been shown to decline as disease progresses [211, 219, 224], and this may be due to a reduction in energy requirements as the atrophy of energy-intensive muscle fibers and death of motor neurons worsens throughout the course of disease [1]. The consistent performance for CK and creatinine as measures of lean mass, and as a measure of disease progression in ALS, highlights their utility as physiological biomarkers in ALS. Whether these can be used as reliable and informative biomarkers of metabolism in ALS, however, remains to be determined. Regardless, to perform effectively as a biomarker in the clinical setting, levels of CK and creatinine must be able to be assessed via simple and effective assays that are highly reproducible. These factors are routinely assayed when assessing kidney function, and so this is possible. While offering opportunities to conduct serial measures of CK and creatinine in patients with ALS with minimal difficulty, these measurements could also offer new avenues to study the changes in CK and creatinine over the course of the disease, thus allowing for better understanding of ALS pathology.

3   Urine

Urine has classically been used as a source of biomarkers for renal function [225]. In recent years, there has been an increasing interest in its use as a source of biomarkers for neurodegenerative disease [226, 227]. Although the sampling of urine is minimally invasive and relatively cheap, it has not been extensively investigated as a source of biomarkers in ALS. There has been recent success in the identification of the extracellular domain of nerve growth factor ($p75^{ECD}$) in urine as a biomarker for ALS. Moreover, its reliability as a biomarker within the clinical setting is supported by the fact that it shows minimal diurnal variation, and that it is highly stable at room temperature and when refrigerated [228].

3.1 Low-Affinity Nerve Growth Factor Receptor, p75^{ECD}

p75 (also called low-affinity nerve growth factor receptor) is a neurotrophic receptor implicated in driving neuronal apoptosis [229]. In healthy individuals, p75 is highly expressed during neuronal development, but downregulated in adult life [230]. Reexpression occurs following neuronal injury [231]. The expression of the receptor has been shown to be increased in motor neurons of patients with ALS and in murine models of ALS [232, 233]. Upon binding of nerve growth factor to p75, the extracellular domain ($p75^{ECD}$) is cleaved and released into the circulation [234], and subsequently shuttled into urine [235].

The $p75^{ECD}$ has been shown to be increased in the urine of patients with ALS, reflecting damage to motor neurons [228, 236, 237]. In 2014, a study by Shepheard and colleagues reported that urinary $p75^{ECD}$ was able to accurately distinguish ALS from other non-ALS neurodegenerative controls [237]. This observation was recapitulated in a following 2017 study by Jia and colleagues [236]. These findings suggest that $p75^{ECD}$ could be a specific and accurate biomarker for ALS where a diagnosis is otherwise uncertain.

Urinary $p75^{ECD}$ has proven to be a promising prognostic marker for ALS. Urinary $p75^{ECD}$ has been shown to be higher in ALS patients with rapidly progressing disease when compared to those with slow-progressing disease [228, 236]. Indeed, the concentration of $p75^{ECD}$ in urine increases alongside increased disease progression, and decreased ALSFRS-R [228, 236]. Overall higher levels of urinary $p75^{ECD}$ are predictive of shorter survival [228]. Unlike the biomarkers mentioned above, $p75^{ECD}$ consistently shows a close association with ALSFRS-R at baseline, and over the course of the disease, indicating that $p75^{ECD}$ serves as an accurate longitudinal biomarker for ALS progression [228].

$p75^{ECD}$ is able to be simply and accurately quantified in a standard laboratory setting with the use of sandwich ELISA [228, 236]. This, in combination with the studies outlined above deliver a strong rationale for using urinary $p75^{ECD}$ as a biomarker that will provide information about the efficacy of compounds that have entered clinical trials for ALS. Indeed, $p75^{ECD}$ has been included as an outcome measure in recent ALS clinical trials (NCT03487263, NCT02868580, ACTRN12618000534280, NCT03792490). In the TEALS trial, assessing the efficacy of Tecfidera in ALS (ACTRN12618000534280), urinary $p75^{ECD}$ was included as a secondary outcome measure [130]. Similarly, in a current phase 2 trial of Fasudil (NCT03792490), urinary p^{75ECD}

has been included as a longitudinal biomarker [238]. Notably, in the phase 2a Lighthouse trial (NCT02868580), $p75^{ECD}$ was found to acutely increase in patients with ALS, but subsequently decrease, after administration of Triumeq (dolutegravir 50mg, abacavir 600mg, lamivudine 300mg), potentially mirroring reductions in neuroinflammation [239]. Based on current data, urinary $p75^{ECD}$ along with serum NfL, has been included in the HEALY ALS platform trial as an outcome measure for multiple drug candidates (NCT04297683). These current trials will be important for further determining the validity for the continued use of $p75^{ECD}$ as a biomarker for ALS.

4 Biomarkers of Oxidative Stress in Blood, CSF, and Urine

Healthy maintenance of cells and tissues requires a highly regulated balance between pro-and antioxidative processes [240]. A common feature of many diseases is an accumulative disruption in this equilibrium, leading to increased oxidative stress within cells. Increased production of reactive oxygen species within neurons and glia has been observed in numerous neurological diseases including Alzheimer's disease [241], frontotemporal dementia [242, 243], Huntington's disease [244], Lewy body dementia [245, 246], and ALS [247–249]. Of note, SOD1, the first identified genetic contributor to ALS [35], plays a critical role in attenuating oxidative stress [250]. Many antioxidant-based therapies have shown strong effects in mutant SOD1 mice, but these benefits have not readily translated into clinical outcomes [251]. Regardless, markers of oxidative stress have been proposed as candidate biomarkers for ALS.

4-hydroxy-2-nonenal (4-HNE) is an amphiphilic lipid peroxidation product [252] that has been demonstrated to modify the activity of membrane enzymes such as choline acetyltransferase [253, 254], as well as inhibit glucose and glutamate membrane transport in cultured motor neuron cells [253]. In a postmortem investigation by Pedersen and colleagues, 4-HNE was shown to be elevated in lower motor neurons, as well as astrocytes of sALS individuals [255]. These findings are supported by observations from Simpson and colleagues, who reported that sALS patients had abnormally high serum 4-HNE levels relative to healthy and neurodegenerative controls. Interestingly, it was also shown that serum 4-HNE was increasingly elevated in sALS, but not fALS, patients over the course of their disease, suggesting that 4-HNE could be used as a marker of disease progression [256]. In line with this, levels of plasma 4-HNE have been shown to be predictive of the decline in functional capacity of ALS patients (as determined by a decline in ALSFRS-R), with 4-HNE being higher in patients with rapid-progressing disease [257]. As a protein-bound adduct, 4-HNE can be simply quantified using a standard competitive

ELISA format [256, 257]. It is able to be assessed under normal laboratory conditions, improving its potential as a biomarker. While this is an effective measure of total 4-HNE, it does not provide an understanding of which species 4-HNE is bound to. Mass-spectrometry could be used to identify which specific molecules have conjugated 4-HNE [258–262]. While 4-HNE is presently an effective measure of total lipid peroxidation in ALS, identifying which molecular pathways are altered through this peroxidation process can improve the ability for future therapeutic design.

Increased DNA oxidation has also been observed in ALS. 8-hydroxy-2'-deoxyguanosine (8-OHdG) is released from mitochondria as a product of mtDNA oxidation [263]. It is well-established as a biomarker for oxidative stress in many diseases, including cancer [264, 265]. In ALS, 8-OHdG has been shown to be elevated in serum [266], CSF [267] and plasma [257, 267], reflecting increased mtDNA oxidative stress [257, 266]. As the disease worsens, the amount of 8-OHdG detectable in circulation increases [257], mirroring progressive mitochondrial damage observed in ALS [268]. Collectively, these findings suggest that 8-OHdG may be an effective surrogate measure of the cumulative DNA fragmentation observed in ALS, and provide a rationale for exploring 8-OHdG as a prognostic biomarker for the disease. 8-OHdG has also been shown to be increased in the urine of sALS patients [269]. In a 2000 study, urinary 8-OHdG was reported to be increased in patients with ALS relative to controls, with levels of 8-OHdG further increasing over the course of the disease [267]. Moreover, urinary 8-OHdG was inversely correlated with ALSFRS and forced vital capacity (FVC) [267]. The consistent relationship between blood and urinary 8-OHdG suggest that either fluid can be sampled cross-sectionally or longitudinally to provide insight into mtDNA oxidative stress in ALS. 8-OHdG is routinely quantified using both liquid chromatography/electrochemistry [267, 269] and ELISA [257, 269], indicating that the assessment of 8-OHdG in ALS can be easily incorporated in the laboratory setting.

While no individual biomarker for oxidative stress has been validated in ALS, consistent observations of increased oxidative stress markers in plasma, serum, and urine underscore the utility for a panel of informative oxidative stress biomarkers to provide insight into DNA, protein and lipid damage accrued from the disease [266]. Further studies need to be conducted to determine the clinical utility for oxidative stress biomarkers in ALS.

5 Conclusions

Research into understanding the cause of ALS, and the clinical management of the disease is limited in part due to the absence of specific, sensitive biomarkers for the disease [44]. In a recent consensus conference sponsored by the International Federation of Clinical Neurophysiology, the World Federation of Neurology, the ALS Association, and the MND Association, it was noted that the assessment of biomarkers should be incorporated into new diagnostic criteria for ALS [270]. Individuals with ALS have a 14-month median latency between motor symptom onset and diagnosis [271]. Alongside this delay, prospective large-cohort studies have identified numerous physiological alterations preceding the onset of motor symptoms. This delay between ALS onset and diagnosis introduces difficulty with managing the disease as most physiological damage has occurred before a patient approaches their primary care physician, limiting the ability for therapeutic intervention to improve disease outcomes [1]. Likely as a result to these complications, to date, few diagnostic biomarkers for ALS have been identified. Alongside traditional El Escorial criteria, current diagnosis includes noninvasive measures such as electrophysiology and imaging of tissue [271]; however, direct sampling of biological tissues has potential to improve specificity and shorten diagnostic delay. Table 3 summarizes the fluid biomarkers discussed within this chapter and their evidence for improving ALS diagnoses. Of note, only urinary p75^{ECD} has been evidenced to have consistently improved diagnostic utility in ALS. Furthermore, specific biomarkers for ALS diagnoses will enable better patient stratification, allowing for better patient-oriented management and therapies [50].

At present, most studies into biomarkers for ALS explore single marker utility. Accurate profiling of individuals with ALS is likely to require a panel of complementary biomarkers providing insight into multiple aspects of the disease. Early explorations have shown that a combinatorial assessment of promising neurodegenerative biomarkers such as neurofilament, alongside inflammatory biomarkers provides high sensitivity and specificity for ALS diagnosis, as well as preserving the prognostic benefit of the individual biomarkers [272]. Inclusion of neurofilaments, CHIT1, CRP, p75^{ECD}, and peroxidation products in a general fluid biomarker panel for ALS has potential to provide insight into many physiological processes impacted by the disease. This, in turn, may allow for an improved understanding of the disease while informing disease management strategies. However, development of such a panel will require extensive experimental validation of each biomarker individually, and in combination with other biomarkers.

Table 3
Summary of diagnostic biomarkers in ALS

	Biomarker	Tissue	Findings	Specificity	Replicability	Prediagnostic	Change with disease progression	Score
Neurofilaments	Neurofilament light chain, NfL	Blood and CSF	Increased in ALS compared to healthy individuals and neurodegenerative controls. Correlated with rate of disease progression, and survival. Product of neuronal degradation	Medium	Medium	No	No	1
	Phosphorylated neurofilament heavy chain, pNfH	Blood and CSF	Increased in ALS compared to healthy individuals and neurodegenerative controls. Increases in early stages of ALS. Correlated with rate of disease progression and survival. Product of neuronal degradation	Medium	High	Yes	Yes	3.5
Inflammation	Chitinase-like protein 1, CHIT-1		Increased in ALS compared to healthy controls and disease mimics, and correlates with rate of disease progression. Marker for neuroinflammation.	Medium	High	Unknown	No	1.5
	C-reactive protein, CRP		Increased in ALS, compared to healthy and neurodegenerative controls. Proportional to rate of disease progression. Marker for neuroinflammation	Unknown	High	Unknown	Unknown	1
	Low-affinity nerve growth factor, p75[ECD]	Urine	Increased in ALS compared to healthy and neurodegenerative controls. Correlated with rate of disease progression and predictive	High	High	Unknown	Yes	3

Category	Biomarker	Biofluid	Description	Specificity	Replicability	Prediagnostic	Progression	Score
Metabolism	Creatine kinase pathway	Blood	CK increased before onset of ALS motor symptoms. Declines over time alongside creatinine mirroring loss of lean mass. Creatinine is a sensitive marker for disease progression in long-duration studies for survival. High specificity for ALS. Marker for neuronal repair	Unknown	High	Yes	Yes	3
Oxidative stress	4-hydroxy-2-nonenal, 4-HNE	Blood, CSF and urine	Increased in circulation over the course of sALS, and predictive for future ALSFRS-R. Measure of lipid peroxidation	Unknown	High	Unknown	Yes	2
	8-hydroxy-2′-deoxyguanosine, 8-OHdG		Increased over the course of ALS. Correlates with clinical measures of disease progression. Reflects mtDNA oxidation	Low	High	Unknown	Yes	2

Biomarkers scored on 4 criteria: specificity for ALS [high: 1, medium: 0.5, low/unknown: 0]; utility as a prediagnostic biomarker for ALS [yes: 1, no/unknown: 0]; change in biomarker alongside disease progression [yes: 1, no/unknown: 0]; replicability of findings between studies [high: 1, medium: 0.5, low/unknown: 0]. Score determined as sum of individual criteria. Higher scores indicate better performance as a diagnostic biomarker for ALS

4-HNE 4-hydroxy-2-nonenal, *8-OHdG* 8-hydroxy-2′-deoxyguanosine, *ALS* amyotrophic lateral sclerosis, *ALSFRS-R* ALS functional rating scale—revised, *CHIT-1* chitinase-like protein 1, *CK* creatine kinase, *CRP* C-reactive protein, *mtDNA* mitochondrial DNA, *NfL* neurofilament light chain, *p75ECD* low-affinity nerve growth factor extracellular domain, *pNfH* phosphorylated neurofilament heavy chain, *sALS* sporadic ALS

References

1. van Es MA, Hardiman O, Chio A, Al-Chalabi A, Pasterkamp RJ, Veldink JH, van den Berg LH (2017) Amyotrophic lateral sclerosis. Lancet 390(10107):2084–2098. https://doi.org/10.1016/S0140-6736(17) 31287-4

2. Clarke JL, Jackson JH (1867) On a case of muscular atrophy, with disease of the spinal cord and medulla oblongata. Med Chir Trans 50:489–498.481. https://doi.org/10.1177/095952876705000122

3. Hardiman O, van den Berg LH, Kiernan MC (2011) Clinical diagnosis and management of amyotrophic lateral sclerosis. Nat Rev Neurol 7(11):639–649. https://doi.org/10.1038/nrneurol.2011.153

4. Kiernan MC, Vucic S, Cheah BC, Turner MR, Eisen A, Hardiman O, Burrell JR, Zoing MC (2011) Amyotrophic lateral sclerosis. Lancet 377(9769):942–955. https://doi.org/10.1016/S0140-6736(10)61156-7

5. McCombe PA, Wray NR, Henderson RD (2017) Extra-motor abnormalities in amyotrophic lateral sclerosis: another layer of heterogeneity. Expert Rev Neurother 17(6):561–577. https://doi.org/10.1080/14737175.2017.1273772

6. Ahmed RM, Caga J, Devenney E, Hsieh S, Bartley L, Highton-Williamson E, Ramsey E, Zoing M, Halliday GM, Piguet O, Hodges JR, Kiernan MC (2016) Cognition and eating behavior in amyotrophic lateral sclerosis: effect on survival. J Neurol 263(8):1593–1603. https://doi.org/10.1007/s00415-016-8168-2

7. Stojkovic T, Stefanova E, Pekmezovic T, Peric S, Stevic Z (2016) Executive dysfunction and survival in patients with amyotrophic lateral sclerosis: preliminary report from a Serbian centre for motor neuron disease. Amyotroph Lateral Scler Frontotemporal Degener 17(7-8):543–547. https://doi.org/10.1080/21678421.2016.1211148

8. Hu WT, Shelnutt M, Wilson A, Yarab N, Kelly C, Grossman M, Libon DJ, Khan J, Lah JJ, Levey AI, Glass J (2013) Behavior matters—cognitive predictors of survival in amyotrophic lateral sclerosis. PLoS One 8(2):e57584. https://doi.org/10.1371/journal.pone.0057584

9. Caga J, Turner MR, Hsieh S, Ahmed RM, Devenney E, Ramsey E, Zoing MC, Mioshi E, Kiernan MC (2016) Apathy is associated with poor prognosis in amyotrophic lateral sclerosis. Eur J Neurol 23(5):891–897. https://doi.org/10.1111/ene.12959

10. Lo Coco D, Mattaliano P, Spataro R, Mattaliano A, La Bella V (2011) Sleep-wake disturbances in patients with amyotrophic lateral sclerosis. J Neurol Neurosurg Psychiatry 82(8):839–842. https://doi.org/10.1136/jnnp.2010.228007

11. Lo MH, Lin CL, Chuang E, Chuang TY, Kao CH (2017) Association of dementia in patients with benign paroxysmal positional vertigo. Acta Neurol Scand 135(2):197–203. https://doi.org/10.1111/ane.12581

12. Atalaia A, Carvalho MD, Evangelista T, Pinto A (2007) Sleep characteristics of amyotrophic lateral sclerosis in patients with preserved diaphragmatic function. Amyotroph Lateral Scler 8(2):101–105. https://doi.org/10.1080/17482960601029883

13. Gregory R, Mills K, Donaghy M (1993) Progressive sensory nerve dysfunction in amyotrophic lateral sclerosis: a prospective clinical and neurophysiological study. J Neurol 240(5):309–314. https://doi.org/10.1007/BF00838169

14. Ohashi N, Nonami J, Kodaira M, Yoshida K, Sekijima Y (2020) Taste disorder in facial onset sensory and motor neuronopathy: a case report. BMC Neurol 20(1):71. https://doi.org/10.1186/s12883-020-01639-x

15. Piccione EA, Sletten DM, Staff NP, Low PA (2015) Autonomic system and amyotrophic lateral sclerosis. Muscle Nerve 51(5):676–679. https://doi.org/10.1002/mus.24457

16. Oey PL, Vos PE, Wieneke GH, Wokke JH, Blankestijn PJ, Karemaker JM (2002) Subtle involvement of the sympathetic nervous system in amyotrophic lateral sclerosis. Muscle Nerve 25(3):402–408. https://doi.org/10.1002/mus.10049

17. Hu F, Jin J, Qu Q, Dang J (2016) Sympathetic skin response in amyotrophic lateral sclerosis. J Clin Neurophysiol 33(1):60–65. https://doi.org/10.1097/wnp.0000000000000226

18. de Carvalho MLL, Motta R, Battaglia MA, Brichetto G (2011) Urinary disorders in amyotrophic lateral sclerosis subjects. Amyotroph Lateral Scler 12(5):352–355. https://doi.org/10.3109/17482968.2011.574141

19. Nübling GS, Mie E, Bauer RM, Hensler M, Lorenzl S, Hapfelmeier A, Irwin DE, Borasio GD, Winkler AS (2014) Increased prevalence of bladder and intestinal dysfunction in

amyotrophic lateral sclerosis. Amyotroph Lateral Scler Frontotemporal Degener 15 (3-4):174–179. https://doi.org/10.3109/21678421.2013.868001

20. Arlandis S, Vázquez-Costa JF, Martínez-Cuenca E, Sevilla T, Boronat F, Broseta E (2017) Urodynamic findings in amyotrophic lateral sclerosis patients with lower urinary tract symptoms: results from a pilot study. NeurourolUrodyn 36(3):626–631. https://doi.org/10.1002/nau.22976

21. Toepfer M, Folwaczny C, Lochmüller H, Schroeder M, Riepl RL, Pongratz D, Müller-Felber W (1999) Noninvasive (13)C-octanoic acid breath test shows delayed gastric emptying in patients with amyotrophic lateral sclerosis. Digestion 60(6):567–571. https://doi.org/10.1159/000007708

22. Toepfer M, Schroeder M, Klauser A, Lochmüller H, Hirschmann M, Riepl RL, Pongratz D, Müller-Felber W (1997) Delayed colonic transit times in amyotrophic lateral sclerosis assessed with radio-opaque markers. Eur J Med Res 2(11):473–476

23. Xu Z, Alruwaili ARS, Henderson RD, McCombe PA (2017) Screening for cognitive and behavioural impairment in amyotrophic lateral sclerosis: frequency of abnormality and effect on survival. J Neurol Sci 376:16–23. https://doi.org/10.1016/j.jns.2017.02.061

24. Ringholz GM, Appel SH, Bradshaw M, Cooke NA, Mosnik DM, Schulz PE (2005) Prevalence and patterns of cognitive impairment in sporadic ALS. Neurology 65 (4):586–590. https://doi.org/10.1212/01.wnl.0000172911.39167.b6

25. Burrell JR, Kiernan MC, Vucic S, Hodges JR (2011) Motor neuron dysfunction in frontotemporal dementia. Brain 134 (Pt 9):2582–2594. https://doi.org/10.1093/brain/awr195

26. Lomen-Hoerth C, Anderson T, Miller B (2002) The overlap of amyotrophic lateral sclerosis and frontotemporal dementia. Neurology 59(7):1077–1079

27. Devine MS, Kiernan MC, Heggie S, McCombe PA, Henderson RD (2014) Study of motor asymmetry in ALS indicates an effect of limb dominance on onset and spread of weakness, and an important role for upper motor neurons. Amyotroph Lateral Scler Frontotemporal Degener 15(7-8):481–487. https://doi.org/10.3109/21678421.2014.906617

28. Braak H, Brettschneider J, Ludolph AC, Lee VM, Trojanowski JQ, Del Tredici K (2013) Amyotrophic lateral sclerosis—a model of corticofugal axonal spread. Nat Rev Neurol 9 (12):708–714. https://doi.org/10.1038/nrneurol.2013.221

29. Tan RH, Kril JJ, Fatima M, McGeachie A, McCann H, Shepherd C, Forrest SL, Affleck A, Kwok JB, Hodges JR, Kiernan MC, Halliday GM (2015) TDP-43 proteinopathies: pathological identification of brain regions differentiating clinical phenotypes. Brain 138(Pt 10):3110–3122. https://doi.org/10.1093/brain/awv220

30. Ludolph AC, Brettschneider J (2015) TDP-43 in amyotrophic lateral sclerosis—is it a prion disease? Eur J Neurol 22 (5):753–761. https://doi.org/10.1111/ene.12706

31. Brettschneider J, Del Tredici K, Toledo JB, Robinson JL, Irwin DJ, Grossman M, Suh E, Van Deerlin VM, Wood EM, Baek Y, Kwong L, Lee EB, Elman L, McCluskey L, Fang L, Feldengut S, Ludolph AC, Lee VM, Braak H, Trojanowski JQ (2013) Stages of pTDP-43 pathology in amyotrophic lateral sclerosis. Ann Neurol 74(1):20–38. https://doi.org/10.1002/ana.23937

32. Chiò A, Logroscino G, Hardiman O, Swingler R, Mitchell D, Beghi E, Traynor BG, Eurals C (2009) Prognostic factors in ALS: a critical review. Amyotroph Lateral Scler 10(5-6):310–323. https://doi.org/10.3109/17482960802566824

33. de Carvalho M, Swash M, Pinto S (2019) Diaphragmatic neurophysiology and respiratory markers in ALS. Front Neurol 10:143. https://doi.org/10.3389/fneur.2019.00143

34. McCombe PA, Garton FC, Katz M, Wray NR, Henderson RD (2020) What do we know about the variability in survival of patients with amyotrophic lateral sclerosis? Expert Rev Neurother 20(9):921–941. https://doi.org/10.1080/14737175.2020.1785873

35. Rosen DR, Siddique T, Patterson D, Figlewicz DA, Sapp P, Hentati A, Donaldson D, Goto J, O'Regan JP, Deng HX et al (1993) Mutations in Cu/Zn superoxide dismutase gene are associated with familial amyotrophic lateral sclerosis. Nature 362(6415):59–62. https://doi.org/10.1038/362059a0

36. Van Deerlin VM, Leverenz JB, Bekris LM, Bird TD, Yuan W, Elman LB, Clay D, Wood EM, Chen-Plotkin AS, Martinez-Lage M, Steinbart E, McCluskey L, Grossman M, Neumann M, Wu IL, Yang W-S, Kalb R, Galasko DR, Montine TJ, Trojanowski JQ, Lee VMY, Schellenberg GD, Yu C-E (2008) TARDBP mutations in amyotrophic lateral sclerosis with TDP-43 neuropathology: a

genetic and histopathological analysis. Lancet Neurol 7(5):409–416. https://doi.org/10.1016/S1474-4422(08)70071-1

37. DeJesus-Hernandez M, Mackenzie IR, Boeve BF, Boxer AL, Baker M, Rutherford NJ, Nicholson AM, Finch NCA, Flynn H, Adamson J, Kouri N, Wojtas A, Sengdy P, Hsiung G-YR, Karydas A, Seeley WW, Josephs KA, Coppola G, Geschwind DH, Wszolek ZK, Feldman H, Knopman DS, Petersen RC, Miller BL, Dickson DW, Boylan KB, Graff-Radford NR, Rademakers R (2011) Expanded GGGGCC hexanucleotide repeat in noncoding region of C9ORF72 causes Chromosome 9p-linked FTD and ALS. Neuron 72(2):245–256. https://doi.org/10.1016/j.neuron.2011.09.011

38. Renton AE, Majounie E, Waite A, Simón-Sánchez J, Rollinson S, Gibbs JR, Schymick JC, Laaksovirta H, van Swieten JC, Myllykangas L, Kalimo H, Paetau A, Abramzon Y, Remes AM, Kaganovich A, Scholz SW, Duckworth J, Ding J, Harmer DW, Hernandez DG, Johnson JO, Mok K, Ryten M, Trabzuni D, Guerreiro RJ, Orrell RW, Neal J, Murray A, Pearson J, Jansen IE, Sondervan D, Seelaar H, Blake D, Young K, Halliwell N, Callister JB, Toulson G, Richardson A, Gerhard A, Snowden J, Mann D, Neary D, Nalls MA, Peuralinna T, Jansson L, Isoviita V-M, Kaivorinne A-L, Hölttä-Vuori M, Ikonen E, Sulkava R, Benatar M, Wuu J, Chiò A, Restagno G, Borghero G, Sabatelli M, Heckerman D, Rogaeva E, Zinman L, Rothstein JD, Sendtner M, Drepper C, Eichler EE, Alkan C, Abdullaev Z, Pack SD, Dutra A, Pak E, Hardy J, Singleton A, Williams NM, Heutink P, Pickering-Brown S, Morris HR, Tienari PJ, Traynor BJ (2011) A Hexanucleotide Repeat Expansion in C9ORF72 Is the Cause of Chromosome 9p21-Linked ALS-FTD. Neuron 72(2):257–268. https://doi.org/10.1016/j.neuron.2011.09.010

39. Crozat A, Aman P, Mandahl N, Ron D (1993) Fusion of CHOP to a novel RNA-binding protein in human myxoid liposarcoma. Nature 363(6430):640–644. https://doi.org/10.1038/363640a0

40. Gibson SB, Downie JM, Tsetsou S, Feusier JE, Figueroa KP, Bromberg MB, Jorde LB, Pulst SM (2017) The evolving genetic risk for sporadic ALS. Neurology 89(3):226–233. https://doi.org/10.1212/WNL.0000000000004109

41. Hodges J (2012) Familial frontotemporal dementia and amyotrophic lateral sclerosis associated with the C9ORF72 hexanucleotide repeat. Brain 135(Pt 3):652–655. https://doi.org/10.1093/brain/aws033

42. Rosenfeld J, Strong MJ (2015) Challenges in the understanding and treatment of amyotrophic lateral sclerosis/motor neuron disease. Neurotherapeutics 12(2):317–325. https://doi.org/10.1007/s13311-014-0332-8

43. Ghasemi M (2016) Amyotrophic lateral sclerosis mimic syndromes. Iran J Neurol 15 (2):85–91

44. Verber NS, Shepheard SR, Sassani M, McDonough HE, Moore SA, Alix JJP, Wilkinson ID, Jenkins TM, Shaw PJ (2019) Biomarkers in motor neuron disease: a state of the art review. Front Neurol 10:291. https://doi.org/10.3389/fneur.2019.00291

45. Hogden A, Foley G, Henderson RD, James N, Aoun SM (2017) Amyotrophic lateral sclerosis: improving care with a multidisciplinary approach. J Multidiscip Healthc 10:205–215. https://doi.org/10.2147/JMDH.S134992

46. Tarasiuk J, Kułakowska A, Drozdowski W, Kornhuber J, Lewczuk P (2012) CSF markers in amyotrophic lateral sclerosis. J Neural Transm 119(7):747–757. https://doi.org/10.1007/s00702-012-0806-y

47. Brancia C, Noli B, Boido M, Boi A, Puddu R, Borghero G, Marrosu F, Bongioanni P, Orru S, Manconi B, D'Amato F, Messana I, Vincenzoni F, Vercelli A, Ferri GL, Cocco C (2016) VGF protein and its C-terminal derived peptides in amyotrophic lateral sclerosis: human and animal model studies. PLoS One 11(10):e0164689. https://doi.org/10.1371/journal.pone.0164689

48. Kakaroubas N, Brennan S, Keon M, Saksena NK (2019) Pathomechanisms of blood-brain barrier disruption in ALS. Neurosci J 2019:2537698–2537698. https://doi.org/10.1155/2019/2537698

49. Sasaki S (2015) Alterations of the blood-spinal cord barrier in sporadic amyotrophic lateral sclerosis. Neuropathology 35 (6):518–528. https://doi.org/10.1111/neup.12221

50. Vijayakumar UG, Milla V, Cynthia Stafford MY, Bjourson AJ, Duddy W, Duguez SM (2019) A systematic review of suggested molecular strata, biomarkers and their tissue sources in ALS. Front Neurol 10:400. https://doi.org/10.3389/fneur.2019.00400

51. Chen X, Chen Y, Wei Q, Ou R, Cao B, Zhao B, Shang HF (2016) Assessment of a multiple biomarker panel for diagnosis of amyotrophic lateral sclerosis. BMC Neurol 16:173. https://doi.org/10.1186/s12883-016-0689-x

52. McCombe PA, Pfluger C, Singh P, Lim CY, Airey C, Henderson RD (2015) Serial measurements of phosphorylated neurofilament-heavy in the serum of subjects with amyotrophic lateral sclerosis. J Neurol Sci 353 (1-2):122–129. https://doi.org/10.1016/j.jns.2015.04.032

53. Lehnert S, Costa J, de Carvalho M, Kirby J, Kuzma-Kozakiewicz M, Morelli C, Robberecht W, Shaw P, Silani V, Steinacker P, Tumani H, Van Damme P, Ludolph A, Otto M (2014) Multicentre quality control evaluation of different biomarker candidates for amyotrophic lateral sclerosis. Amyotroph Lateral Scler Frontotemporal Degener 15(5-6):344–350. https://doi.org/10.3109/21678421.2014.884592

54. Gendron TF, Group CONS, Daughrity LM, Heckman MG, Diehl NN, Wuu J, Miller TM, Pastor P, Trojanowski JQ, Grossman M, Berry JD, Hu WT, Ratti A, Benatar M, Silani V, Glass JD, Floeter MK, Jeromin A, Boylan KB, Petrucelli L (2017) Phosphorylated neurofilament heavy chain: a biomarker of survival for C9ORF72-associated amyotrophic lateral sclerosis. Ann Neurol 82(1):139–146. https://doi.org/10.1002/ana.24980

55. Boylan KB, Glass JD, Crook JE, Yang C, Thomas CS, Desaro P, Johnston A, Overstreet K, Kelly C, Polak M, Shaw G (2013) Phosphorylated neurofilament heavy subunit (pNF-H) in peripheral blood and CSF as a potential prognostic biomarker in amyotrophic lateral sclerosis. J Neurol Neurosurg Psychiatry 84(4):467–472. https://doi.org/10.1136/jnnp-2012-303768

56. Oeckl P, Jardel C, Salachas F, Lamari F, Andersen PM, Bowser R, de Carvalho M, Costa J, van Damme P, Gray E, Grosskreutz J, Hernandez-Barral M, Herukka SK, Huss A, Jeromin A, Kirby J, Kuzma-Kozakiewicz M, Amador Mdel M, Mora JS, Morelli C, Muckova P, Petri S, Poesen K, Rhode H, Rikardsson AK, Robberecht W, Rodriguez Mahillo AI, Shaw P, Silani V, Steinacker P, Turner MR, Tuzun E, Yetimler B, Ludolph AC, Otto M (2016) Multicenter validation of CSF neurofilaments as diagnostic biomarkers for ALS. Amyotroph Lateral Scler Frontotemporal Degener 17 (5–6):404–413. https://doi.org/10.3109/21678421.2016.1167913

57. Menke RA, Gray E, Lu CH, Kuhle J, Talbot K, Malaspina A, Turner MR (2015) CSF neurofilament light chain reflects corticospinal tract degeneration in ALS. Ann Clin Transl Neurol 2(7):748–755. https://doi.org/10.1002/acn3.212

58. Tortelli R, Copetti M, Ruggieri M, Cortese R, Capozzo R, Leo A, D'Errico E, Mastrapasqua M, Zoccolella S, Pellegrini F, Simone IL, Logroscino G (2015) Cerebrospinal fluid neurofilament light chain levels: marker of progression to generalized amyotrophic lateral sclerosis. Eur J Neurol 22 (1):215–218. https://doi.org/10.1111/ene.12421

59. Tortelli R, Ruggieri M, Cortese R, D'Errico E, Capozzo R, Leo A, Mastrapasqua M, Zoccolella S, Leante R, Livrea P, Logroscino G, Simone IL (2012) Elevated cerebrospinal fluid neurofilament light levels in patients with amyotrophic lateral sclerosis: a possible marker of disease severity and progression. Eur J Neurol 19 (12):1561–1567. https://doi.org/10.1111/j.1468-1331.2012.03777.x

60. Lu CH, Macdonald-Wallis C, Gray E, Pearce N, Petzold A, Norgren N, Giovannoni G, Fratta P, Sidle K, Fish M, Orrell R, Howard R, Talbot K, Greensmith L, Kuhle J, Turner MR, Malaspina A (2015) Neurofilament light chain: A prognostic biomarker in amyotrophic lateral sclerosis. Neurology 84(22):2247–2257. https://doi.org/10.1212/WNL.0000000000001642

61. Gaiottino J, Norgren N, Dobson R, Topping J, Nissim A, Malaspina A, Bestwick JP, Monsch AU, Regeniter A, Lindberg RL, Kappos L, Leppert D, Petzold A, Giovannoni G, Kuhle J (2013) Increased neurofilament light chain blood levels in neurodegenerative neurological diseases. PLoS One 8(9):e75091. https://doi.org/10.1371/journal.pone.0075091

62. Zetterberg H, Jacobsson J, Rosengren L, Blennow K, Andersen PM (2007) Cerebrospinal fluid neurofilament light levels in amyotrophic lateral sclerosis: impact of SOD1 genotype. Eur J Neurol 14(12):1329–1333. https://doi.org/10.1111/j.1468-1331.2007.01972.x

63. De Schaepdryver M, Jeromin A, Gille B, Claeys KG, Herbst V, Brix B, Van Damme P, Poesen K (2018) Comparison of elevated phosphorylated neurofilament heavy chains in serum and cerebrospinal fluid of patients with amyotrophic lateral sclerosis. J Neurol Neurosurg Psychiatry 89(4):367–373.

https://doi.org/10.1136/jnnp-2017-316605

64. Kuhle J, Regeniter A, Leppert D, Mehling M, Kappos L, Lindberg RL, Petzold A (2010) A highly sensitive electrochemiluminescence immunoassay for the neurofilament heavy chain protein. J Neuroimmunol 220 (1–2):114–119. https://doi.org/10.1016/j.jneuroim.2010.01.004

65. Brettschneider J, Petzold A, Sussmuth SD, Ludolph AC, Tumani H (2006) Axonal damage markers in cerebrospinal fluid are increased in ALS. Neurology 66 (6):852–856. https://doi.org/10.1212/01.wnl.0000203120.85850.54

66. Reijn TS, Abdo WF, Schelhaas HJ, Verbeek MM (2009) CSF neurofilament protein analysis in the differential diagnosis of ALS. J Neurol 256(4):615–619. https://doi.org/10.1007/s00415-009-0131-z

67. Li DW, Ren H, Jeromin A, Liu M, Shen D, Tai H, Ding Q, Li X, Cui L (2018) Diagnostic performance of neurofilaments in chinese patients with amyotrophic lateral sclerosis: a prospective study. Front Neurol 9:726. https://doi.org/10.3389/fneur.2018.00726

68. Feneberg E, Oeckl P, Steinacker P, Verde F, Barro C, Van Damme P, Gray E, Grosskreutz J, Jardel C, Kuhle J, Koerner S, Lamari F, Amador MDM, Mayer B, Morelli C, Muckova P, Petri S, Poesen K, Raaphorst J, Salachas F, Silani V, Stubendorff B, Turner MR, Verbeek MM, Weishaupt JH, Weydt P, Ludolph AC, Otto M (2018) Multicenter evaluation of neurofilaments in early symptom onset amyotrophic lateral sclerosis. Neurology 90(1):e22–e30. https://doi.org/10.1212/WNL.0000000000004761

69. Steinacker P, Feneberg E, Weishaupt J, Brettschneider J, Tumani H, Andersen PM, von Arnim CA, Bohm S, Kassubek J, Kubisch C, Lule D, Muller HP, Muche R, Pinkhardt E, Oeckl P, Rosenbohm A, Anderl-Straub S, Volk AE, Weydt P, Ludolph AC, Otto M (2016) Neurofilaments in the diagnosis of motoneuron diseases: a prospective study on 455 patients. J Neurol Neurosurg Psychiatry 87(1):12–20. https://doi.org/10.1136/jnnp-2015-311387

70. Poesen K, De Schaepdryver M, Stubendorff B, Gille B, Muckova P, Wendler S, Prell T, Ringer TM, Rhode H, Stevens O, Claeys KG, Couwelier G, D'Hondt A, Lamaire N, Tilkin P, Van Reijen D, Gourmaud S, Fedtke N, Heiling B, Rumpel M, Rodiger A, Gunkel A,

Witte OW, Paquet C, Vandenberghe R, Grosskreutz J, Van Damme P (2017) Neurofilament markers for ALS correlate with extent of upper and lower motor neuron disease. Neurology 88(24):2302–2309. https://doi.org/10.1212/WNL.0000000000004029

71. Boylan K, Yang C, Crook J, Overstreet K, Heckman M, Wang Y, Borchelt D, Shaw G (2009) Immunoreactivity of the phosphorylated axonal neurofilament H subunit (pNF-H) in blood of ALS model rodents and ALS patients: evaluation of blood pNF-H as a potential ALS biomarker. J Neurochem 111(5):1182–1191. https://doi.org/10.1111/j.1471-4159.2009.06386.x

72. Weydt P, Oeckl P, Huss A, Muller K, Volk AE, Kuhle J, Knehr A, Andersen PM, Prudlo J, Steinacker P, Weishaupt JH, Ludolph AC, Otto M (2016) Neurofilament levels as biomarkers in asymptomatic and symptomatic familial amyotrophic lateral sclerosis. Ann Neurol 79(1):152–158. https://doi.org/10.1002/ana.24552

73. De Schaepdryver M, Goossens J, De Meyer S, Jeromin A, Masrori P, Brix B, Claeys KG, Schaeverbeke J, Adamczuk K, Vandenberghe R, Van Damme P, Poesen K (2019) Serum neurofilament heavy chains as early marker of motor neuron degeneration. Ann Clin Transl Neurol 6(10):1971–1979. https://doi.org/10.1002/acn3.50890

74. Benatar M, Wuu J, Lombardi V, Jeromin A, Bowser R, Andersen PM, Malaspina A (2019) Neurofilaments in pre-symptomatic ALS and the impact of genotype. Amyotroph Lateral Scler Frontotemporal Degener 20 (7–8):538–548. https://doi.org/10.1080/21678421.2019.1646769

75. Gagliardi D, Meneri M, Saccomanno D, Bresolin N, Comi GP, Corti S (2019) Diagnostic and prognostic role of blood and cerebrospinal fluid and blood neurofilaments in amyotrophic lateral sclerosis: a review of the literature. Int J Mol Sci 20(17):4152. https://doi.org/10.3390/ijms20174152

76. Verde F, Steinacker P, Weishaupt JH, Kassubek J, Oeckl P, Halbgebauer S, Tumani H, von Arnim CAF, Dorst J, Feneberg E, Mayer B, Müller H-P, Gorges M, Rosenbohm A, Volk AE, Silani V, Ludolph AC, Otto M (2019) Neurofilament light chain in serum for the diagnosis of amyotrophic lateral sclerosis. J Neurol Neurosurg Psychiatry 90(2):157. https://doi.org/10.1136/jnnp-2018-318704

77. Xu Z, Henderson RD, David M, McCombe PA (2016) Neurofilaments as biomarkers for amyotrophic lateral sclerosis: a systematic

review and meta-analysis. PLoS One 11(10): e0164625. https://doi.org/10.1371/jour nal.pone.0164625

78. Poesen K, Van Damme P (2019) Diagnostic and prognostic performance of neurofilaments in ALS. Front Neurol 9:1167–1167. https://doi.org/10.3389/fneur.2018. 01167

79. Li S, Ren Y, Zhu W, Yang F, Zhang X, Huang X (2016) Phosphorylated neurofilament heavy chain levels in paired plasma and CSF of amyotrophic lateral sclerosis. J Neurol Sci 367:269–274. https://doi.org/10.1016/j. jns.2016.05.062

80. Huh JW, Laurer HL, Raghupathi R, Helfaer MA, Saatman KE (2002) Rapid loss and partial recovery of neurofilament immunostaining following focal brain injury in mice. Exp Neurol 175(1):198–208. https://doi.org/ 10.1006/exnr.2002.7880

81. Galvin JE, Nakamura M, McIntosh TK, Saatman KE, Sampathu D, Raghupathi R, Lee VM, Trojanowski JQ (2000) Neurofilament-rich intraneuronal inclusions exacerbate neurodegenerative sequelae of brain trauma in NFH/LacZ transgenic mice. Exp Neurol 165(1):77–89. https://doi.org/10.1006/ exnr.2000.7461

82. Gotow T (2000) Neurofilaments in health and disease. Med Electron Microsc 33 (4):173–199. https://doi.org/10.1007/ s007950000019

83. Petzold A, Keir G, Green AJ, Giovannoni G, Thompson EJ (2003) A specific ELISA for measuring neurofilament heavy chain phosphoforms. J Immunol Methods 278 (1-2):179–190. https://doi.org/10.1016/ s0022-1759(03)00189-3

84. Petzold A, Shaw G (2007) Comparison of two ELISA methods for measuring levels of the phosphorylated neurofilament heavy chain. J Immunol Methods 319(1-2):34–40. https://doi.org/10.1016/j.jim.2006.09. 021

85. Shaw G, Yang C, Ellis R, Anderson K, Parker Mickle J, Scheff S, Pike B, Anderson DK, Howland DR (2005) Hyperphosphorylated neurofilament NF-H is a serum biomarker of axonal injury. Biochem Biophys Res Commun 336(4):1268–1277. https://doi.org/10. 1016/j.bbrc.2005.08.252

86. Koel-Simmelink MJ, Vennegoor A, Killestein J, Blankenstein MA, Norgren N, Korth C, Teunissen CE (2014) The impact of pre-analytical variables on the stability of neurofilament proteins in CSF, determined by a novel validated SinglePlex Luminex assay and ELISA. J Immunol Methods 402

(1-2):43–49. https://doi.org/10.1016/j. jim.2013.11.008

87. Benatar M, Wuu J, Andersen PM, Lombardi V, Malaspina A (2018) Neurofilament light: a candidate biomarker of presymptomatic amyotrophic lateral sclerosis and phenoconversion. Ann Neurol 84 (1):130–139. https://doi.org/10.1002/ana. 25276

88. Kuhle J, Barro C, Andreasson U, Derfuss T, Lindberg R, Sandelius A, Liman V, Norgren N, Blennow K, Zetterberg H (2016) Comparison of three analytical platforms for quantification of the neurofilament light chain in blood samples: ELISA, electro-chemiluminescence immunoassay and Simoa. Clin Chem Lab Med 54(10):1655–1661. https://doi.org/10.1515/cclm-2015-1195

89. Khalil M, Teunissen CE, Otto M, Piehl F, Sormani MP, Gattringer T, Barro C, Kappos L, Comabella M, Fazekas F, Petzold A, Blennow K, Zetterberg H, Kuhle J (2018) Neurofilaments as biomarkers in neurological disorders. Nat Rev Neurol 14 (10):577–589. https://doi.org/10.1038/ s41582-018-0058-z

90. Benatar M, Zhang L, Wang L, Granit V, Statland J, Barohn R, Swenson A, Ravits J, Jackson C, Burns TM, Trivedi J, Pioro EP, Caress J, Katz J, McCauley JL, Rademakers R, Malaspina A, Ostrow LW, Wuu J (2020) Validation of serum neurofilaments as prognostic and potential pharmacodynamic biomarkers for ALS. Neurology 95 (1):e59–e69. https://doi.org/10.1212/ WNL.0000000000009559

91. Heneka MT, Carson MJ, El Khoury J, Landreth GE, Brosseron F, Feinstein DL, Jacobs AH, Wyss-Coray T, Vitorica J, Ransohoff RM, Herrup K, Frautschy SA, Finsen B, Brown GC, Verkhratsky A, Yamanaka K, Koistinaho J, Latz E, Halle A, Petzold GC, Town T, Morgan D, Shinohara ML, Perry VH, Holmes C, Bazan NG, Brooks DJ, Hunot S, Joseph B, Deigendesch N, Garaschuk O, Boddeke E, Dinarello CA, Breitner JC, Cole GM, Golenbock DT, Kummer MP (2015) Neuroinflammation in Alzheimer's disease. Lancet Neurol 14 (4):388–405. https://doi.org/10.1016/ s1474-4422(15)70016-5

92. Hampel H, Caraci F, Cuello AC, Caruso G, Nisticò R, Corbo M, Baldacci F, Toschi N, Garaci F, Chiesa PA, Verdooner SR, Akman-Anderson L, Hernández F, Ávila J, Emanuele E, Valenzuela PL, Lucía A, Watling M, Imbimbo BP, Vergallo A, Lista S (2020) A path toward precision medicine for

neuroinflammatory mechanisms in Alzheimer's disease. Front Immunol 11:456. https://doi.org/10.3389/fimmu.2020.00456

93. Rocha NP, de Miranda AS, Teixeira AL (2015) Insights into neuroinflammation in Parkinson's disease: from biomarkers to antiinflammatory based therapies. Biomed Res Int 2015:628192. https://doi.org/10.1155/2015/628192

94. He R, Yan X, Guo J, Xu Q, Tang B, Sun Q (2018) Recent advances in biomarkers for Parkinson's disease. Front Aging Neurosci 10:305. https://doi.org/10.3389/fnagi.2018.00305

95. Harris VK, Tuddenham JF, Sadiq SA (2017) Biomarkers of multiple sclerosis: current findings. Degener Neurol Neuromuscul Dis 7:19–29. https://doi.org/10.2147/dnnd.S98936

96. Harris VK, Sadiq SA (2009) Disease biomarkers in multiple sclerosis: potential for use in therapeutic decision making. Mol Diagn Ther 13(4):225–244. https://doi.org/10.1007/bf03256329

97. Hensley K, Abdel-Moaty H, Hunter J, Mhatre M, Mou S, Nguyen K, Potapova T, Pye QN, Qi M, Rice H, Stewart C, Stroukoff K, West M (2006) Primary glia expressing the G93A-SOD1 mutation present a neuroinflammatory phenotype and provide a cellular system for studies of glial inflammation. J Neuroinflammation 3:2. https://doi.org/10.1186/1742-2094-3-2

98. Qian K, Huang H, Peterson A, Hu B, Maragakis NJ, Ming GL, Chen H, Zhang SC (2017) Sporadic ALS astrocytes induce neuronal degeneration in vivo. Stem Cell Rep 8 (4):843–855. https://doi.org/10.1016/j.stemcr.2017.03.003

99. Zondler L, Müller K, Khalaji S, Bliederhäuser C, Ruf WP, Grozdanov V, Thiemann M, Fundel-Clemes K, Freischmidt A, Holzmann K, Strobel B, Weydt P, Witting A, Thal DR, Helferich AM, Hengerer B, Gottschalk KE, Hill O, Kluge M, Ludolph AC, Danzer KM, Weishaupt JH (2016) Peripheral monocytes are functionally altered and invade the CNS in ALS patients. Acta Neuropathol 132(3):391–411. https://doi.org/10.1007/s00401-016-1548-y

100. Ranganathan S, Williams E, Ganchev P, Gopalakrishnan V, Lacomis D, Urbinelli L, Newhall K, Cudkowicz ME, Brown RH Jr, Bowser R (2005) Proteomic profiling of cerebrospinal fluid identifies biomarkers for amyotrophic lateral sclerosis. J Neurochem 95(5):1461–1471. https://doi.org/10.1111/j.1471-4159.2005.03478.x

101. Pasinetti GM, Ungar LH, Lange DJ, Yemul S, Deng H, Yuan X, Brown RH, Cudkowicz ME, Newhall K, Peskind E, Marcus S, Ho L (2006) Identification of potential CSF biomarkers in ALS. Neurology 66 (8):1218–1222. https://doi.org/10.1212/01.wnl.0000203129.82104.07

102. Ranganathan S, Nicholl GCB, Henry S, Lutka F, Sathanoori R, Lacomis D, Bowser R (2007) Comparative proteomic profiling of cerebrospinal fluid between living and post mortem ALS and control subjects. Amyotroph Lateral Scler 8(6):373–379. https://doi.org/10.1080/17482960701549681

103. Brettschneider J, Mogel H, Lehmensiek V, Ahlert T, Süssmuth S, Ludolph AC, Tumani H (2008) Proteome analysis of cerebrospinal fluid in amyotrophic lateral sclerosis (ALS). Neurochem Res 33(11):2358–2363. https://doi.org/10.1007/s11064-008-9742-5

104. Brettschneider J, Lehmensiek V, Mogel H, Pfeifle M, Dorst J, Hendrich C, Ludolph AC, Tumani H (2010) Proteome analysis reveals candidate markers of disease progression in amyotrophic lateral sclerosis (ALS). Neurosci Lett 468(1):23–27. https://doi.org/10.1016/j.neulet.2009.10.053

105. Ryberg H, An J, Darko S, Lustgarten JL, Jaffa M, Gopalakrishnan V, Lacomis D, Cudkowicz M, Bowser R (2010) Discovery and verification of amyotrophic lateral sclerosis biomarkers by proteomics. Muscle Nerve 42(1):104–111. https://doi.org/10.1002/mus.21683

106. Mendonça DMF, Pizzati L, Mostacada K, de Martins SC, Higashi R, Sá LA, Neto VM, Chimelli L, Martinez AMB (2012) Neuroproteomics: an insight into ALS. Neurol Res 34(10):937–943. https://doi.org/10.1179/1743132812Y.0000000092

107. von Neuhoff N, Oumeraci T, Wolf T, Kollewe K, Bewerunge P, Neumann B, Brors B, Bufler J, Wurster U, Schlegelberger B, Dengler R, Zapatka M, Petri S (2012) Monitoring CSF proteome alterations in amyotrophic lateral sclerosis: obstacles and perspectives in translating a novel marker panel to the clinic. PLoS One 7(9):e44401. https://doi.org/10.1371/journal.pone.0044401

108. Varghese AM, Sharma A, Mishra P, Vijayalakshmi K, Harsha HC, Sathyaprabha TN, Bharath SM, Nalini A, Alladi PA, Raju TR (2013) Chitotriosidase—a putative biomarker for sporadic amyotrophic lateral

sclerosis. Clin Proteomics 10(1):19. https://doi.org/10.1186/1559-0275-10-19

109. Collins MA, An J, Hood BL, Conrads TP, Bowser RP (2015) Label-free LC-MS/MS proteomic analysis of cerebrospinal fluid identifies protein/pathway alterations and candidate biomarkers for amyotrophic lateral sclerosis. J Proteome Res 14 (11):4486–4501. https://doi.org/10.1021/acs.jproteome.5b00804

110. Chen Y, Liu XH, Wu JJ, Ren HM, Wang J, Ding ZT, Jiang YP (2016) Proteomic analysis of cerebrospinal fluid in amyotrophic lateral sclerosis. Exp Ther Med 11(6):2095–2106. https://doi.org/10.3892/etm.2016.3210

111. Barschke P, Oeckl P, Steinacker P, Ludolph A, Otto M (2017) Proteomic studies in the discovery of cerebrospinal fluid biomarkers for amyotrophic lateral sclerosis. Expert Rev Proteomics 14(9):769–777. https://doi.org/10.1080/14789450.2017.1365602

112. Chen Y, Xia K, Chen L, Fan D (2019) Increased Interleukin-6 levels in the astrocyte-derived exosomes of sporadic amyotrophic lateral sclerosis patients. Front Neurosci 13:574. https://doi.org/10.3389/fnins.2019.00574

113. Ngo ST, Steyn FJ, Huang L, Mantovani S, Pfluger CMM, Woodruff TM, O'Sullivan JD, Henderson RD, McCombe PA (2015) Altered expression of metabolic proteins and adipokines in patients with amyotrophic lateral sclerosis. J Neurol Sci 357(1):22–27. https://doi.org/10.1016/j.jns.2015.06.053

114. Moreau C, Devos D, Brunaud-Danel V, Defebvre L, Perez T, Destée A, Tonnel AB, Lassalle P, Just N (2005) Elevated IL-6 and TNF-alpha levels in patients with ALS: inflammation or hypoxia? Neurology 65 (12):1958–1960. https://doi.org/10.1212/01.wnl.0000188907.97339.76

115. Gonzalez-Garza MT, Martinez HR, Cruz-Vega DE, Hernandez-Torre M, Moreno-Cuevas JE (2018) Adipsin, MIP-1b, and IL-8 as CSF biomarker panels for ALS diagnosis. Dis Markers 2018:3023826. https://doi.org/10.1155/2018/3023826

116. Cereda C, Baiocchi C, Bongioanni P, Cova E, Guareschi S, Metelli MR, Rossi B, Sbalsi I, Cuccia MC, Ceroni M (2008) TNF and sTNFR1/2 plasma levels in ALS patients. J Neuroimmunol 194(1):123–131. https://doi.org/10.1016/j.jneuroim.2007.10.028

117. Petrozziello T, Mills AN, Farhan SMK, Mueller KA, Granucci EJ, Glajch KE, Chan J, Chew S, Berry JD, Sadri-Vakili G (2020) Lipocalin-2 is increased in amyotrophic lateral sclerosis. Muscle Nerve 62(2):272–283. https://doi.org/10.1002/mus.26911

118. Wang HA, Lee JD, Lee KM, Woodruff TM, Noakes PG (2017) Complement C5a-C5aR1 signalling drives skeletal muscle macrophage recruitment in the hSOD1G93A mouse model of amyotrophic lateral sclerosis. Skelet Muscle 7(1):10. https://doi.org/10.1186/s13395-017-0128-8

119. Tsuboi Y, Yamada T (1994) Increased concentration of C4d complement protein in CSF in amyotrophic lateral sclerosis. J Neurol Neurosurg Psychiatry 57(7):859–861. https://doi.org/10.1136/jnnp.57.7.859

120. Goldknopf IL, Sheta EA, Bryson J, Folsom B, Wilson C, Duty J, Yen AA, Appel SH (2006) Complement C3c and related protein biomarkers in amyotrophic lateral sclerosis and Parkinson's disease. Biochem Biophys Res Commun 342(4):1034–1039. https://doi.org/10.1016/j.bbrc.2006.02.051

121. Pagliardini V, Pagliardini S, Corrado L, Lucenti A, Panigati L, Bersano E, Servo S, Cantello R, D'Alfonso S, Mazzini L (2015) Chitotriosidase and lysosomal enzymes as potential biomarkers of disease progression in amyotrophic lateral sclerosis: a survey clinic-based study. J Neurol Sci 348 (1):245–250. https://doi.org/10.1016/j.jns.2014.12.016

122. Steinacker P, Verde F, Fang L, Feneberg E, Oeckl P, Roeber S, Anderl-Straub S, Danek A, Diehl-Schmid J, Fassbender K, Fliessbach K, Foerstl H, Giese A, Jahn H, Kassubek J, Kornhuber J, Landwehrmeyer GB, Lauer M, Pinkhardt EH, Prudlo J, Rosenbohm A, Schneider A, Schroeter ML, Tumani H, von Arnim CAF, Weishaupt J, Weydt P, Ludolph AC, Yilmazer Hanke D, Otto M (2018) Chitotriosidase (CHIT1) is increased in microglia and macrophages in spinal cord of amyotrophic lateral sclerosis and cerebrospinal fluid levels correlate with disease severity and progression. J Neurol Neurosurg Psychiatry 89(3):239–247. https://doi.org/10.1136/jnnp-2017-317138

123. Thompson AG, Gray E, Thézénas ML, Charles PD, Evetts S, Hu MT, Talbot K, Fischer R, Kessler BM, Turner MR (2018) Cerebrospinal fluid macrophage biomarkers in amyotrophic lateral sclerosis. Ann Neurol 83(2):258–268. https://doi.org/10.1002/ana.25143

124. Vu L, An J, Kovalik T, Gendron T, Petrucelli L, Bowser R (2020) Cross-sectional and longitudinal measures of chitinase proteins in amyotrophic lateral sclerosis and expression of CHI3L1 in activated astrocytes.

J Neurol Neurosurg Psychiatry 91 (4):350–358. https://doi.org/10.1136/jnnp-2019-321916

125. De Schaepdryver M, Lunetta C, Tarlarini C, Mosca L, Chio A, Van Damme P, Poesen K (2020) Neurofilament light chain and C reactive protein explored as predictors of survival in amyotrophic lateral sclerosis. J Neurol Neurosurg Psychiatry 91(4):436. https://doi.org/10.1136/jnnp-2019-322309

126. Keizman D, Rogowski O, Berliner S, Ish-Shalom M, Maimon N, Nefussy B, Artamonov I, Drory VE (2009) Low-grade systemic inflammation in patients with amyotrophic lateral sclerosis. Acta Neurol Scand 119(6):383–389. https://doi.org/10.1111/j.1600-0404.2008.01112.x

127. Lunetta C, Lizio A, Maestri E, Sansone VA, Mora G, Miller RG, Appel SH, Chio A (2017) Serum C-reactive protein as a prognostic biomarker in amyotrophic lateral sclerosis. JAMA Neurol 74(6):660–667. https://doi.org/10.1001/jamaneurol.2016.6179

128. Sanjak M, Bravver EK, Bockenek W, Williamson T, Lindblom SS, Dawson W, Johnson M, Lucas N, Lary C, Ranzinger L, Newell-Sturdivant A, Brandon N, Holsten S, Ward A, Hillberry R, Wright K, Rozario N, Brooks B (2018) C-reactive protein (CRP) is significantly higher in amyotrophic lateral sclerosis (ALS) patients on non-invasive ventilation (NIV) and tracheostomy-invasive ventilation (TIV) compared with ALS patients at intake clinic evaluation and decreases following Riluzole administration—is CRP potentially a biomarker for treatment responsiveness? Neurology 90 (15 Supplement):P4.447

129. Miller RG, Block G, Katz JS, Barohn RJ, Gopalakrishnan V, Cudkowicz M, Zhang JR, McGrath MS, Ludington E, Appel SH, Azhir A (2015) Randomized phase 2 trial of NP001-a novel immune regulator: safety and early efficacy in ALS. Neurol Neuroimmunol Neuroinflamm 2(3):e100. https://doi.org/10.1212/nxi.0000000000000100

130. Vucic S, Ryder J, Mekhael L, Rd H, Mathers S, Needham M, Dw S, Mc K (2020) Phase 2 randomized placebo controlled double blind study to assess the efficacy and safety of tecfidera in patients with amyotrophic lateral sclerosis (TEALS Study): study protocol clinical trial (SPIRIT Compliant). Medicine (Baltimore) 99(6):e18904. https://doi.org/10.1097/md.0000000000018904

131. Kim S, Kim JK, Son MJ, Kim D, Song B, Son I, Kang HW, Lee J, Kim S (2018) Mecasin treatment in patients with amyotrophic lateral sclerosis: study protocol for a randomized controlled trial. Trials 19(1):225. https://doi.org/10.1186/s13063-018-2557-z

132. Implicit B (2021) IC14 for treatment of amyotrophic lateral sclerosis. https://ClinicalTrials.gov/show/NCT03508453

133. Kanneganti M, Kamba A, Mizoguchi E (2012) Role of chitotriosidase (chitinase 1) under normal and disease conditions. J Epithel Biol Pharmacol 5:1–9. https://doi.org/10.2174/1875044301205010001

134. Elieh-Ali-Komi D, Hamblin MR (2016) Chitin and chitosan: production and application of versatile biomedical nanomaterials. Int J Adv Res (Indore) 4(3):411–427

135. Kzhyshkowska J, Gratchev A, Goerdt S (2007) Human chitinases and chitinase-like proteins as indicators for inflammation and cancer. Biomark Insights 2:128–146

136. Lewis C-A, Manning J, Rossi F, Krieger C (2012) The neuroinflammatory response in ALS: the roles of microglia and T cells. Neurol Res Int 2012:803701. https://doi.org/10.1155/2012/803701

137. Bigg HF, Wait R, Rowan AD, Cawston TE (2006) The mammalian chitinase-like lectin, YKL-40, binds specifically to type I collagen and modulates the rate of type I collagen fibril formation. J Biol Chem 281 (30):21082–21095. https://doi.org/10.1074/jbc.M601153200

138. Sun YJ, Chang NC, Hung SI, Chang AC, Chou CC, Hsiao CD (2001) The crystal structure of a novel mammalian lectin, Ym1, suggests a saccharide binding site. J Biol Chem 276(20):17507–17514. https://doi.org/10.1074/jbc.M010416200

139. Chang NC, Hung SI, Hwa KY, Kato I, Chen JE, Liu CH, Chang AC (2001) A macrophage protein, Ym1, transiently expressed during inflammation is a novel mammalian lectin. J Biol Chem 276(20):17497–17506. https://doi.org/10.1074/jbc.M010417200

140. Wong CO, Venkatachalam K (2019) Motor neurons from ALS patients with mutations in C9ORF72 and SOD1 exhibit distinct transcriptional landscapes. Hum Mol Genet 28 (16):2799–2810. https://doi.org/10.1093/hmg/ddz104

141. Forostyak S, Homola A, Turnovcova K, Svitil P, Jendelova P, Sykova E (2014) Intrathecal delivery of mesenchymal stromal cells protects the structure of altered perineuronal nets in SOD1 rats and amends the course of

ALS. Stem Cells 32(12):3163–3172. https://doi.org/10.1002/stem.1812

142. Petrou P, Gothelf Y, Argov Z, Gotkine M, Levy YS, Kassis I, Vaknin-Dembinsky A, Ben-Hur T, Offen D, Abramsky O, Melamed E, Karussis D (2016) Safety and clinical effects of mesenchymal stem cells secreting neurotrophic factor transplantation in patients with amyotrophic lateral sclerosis: results of phase 1/2 and 2a clinical trials. JAMA Neurol 73(3):337–344. https://doi.org/10.1001/jamaneurol.2015.4321

143. Goyal NA, Berry JD, Windebank A, Staff NP, Maragakis NJ, van den Berg LH, Genge A, Miller R, Baloh RH, Kern R, Gothelf Y, Lebovits C, Cudkowicz M (2020) Addressing heterogeneity in amyotrophic lateral sclerosis clinical trials. Muscle Nerve 62(2):156–166. https://doi.org/10.1002/mus.26801

144. Chio A, Hammond ER, Mora G, Bonito V, Filippini G (2015) Development and evaluation of a clinical staging system for amyotrophic lateral sclerosis. J Neurol Neurosurg Psychiatry 86(1):38–44. https://doi.org/10.1136/jnnp-2013-306589

145. Weydt P, Yuen EC, Ransom BR, Möller T (2004) Increased cytotoxic potential of microglia from ALS-transgenic mice. Glia 48(2):179–182. https://doi.org/10.1002/glia.20062

146. Beers DR, Henkel JS, Xiao Q, Zhao W, Wang J, Yen AA, Siklos L, McKercher SR, Appel SH (2006) Wild-type microglia extend survival in PU.1 knockout mice with familial amyotrophic lateral sclerosis. Proc Natl Acad Sci 103(43):16021. https://doi.org/10.1073/pnas.0607423103

147. Boillée S, Yamanaka K, Lobsiger CS, Copeland NG, Jenkins NA, Kassiotis G, Kollias G, Cleveland DW (2006) Onset and progression in inherited ALS determined by motor neurons and microglia. Science 312(5778):1389. https://doi.org/10.1126/science.1123511

148. Olesen MN, Wuolikainen A, Nilsson AC, Wirenfeldt M, Forsberg K, Madsen JS, Lillevang ST, Brandslund I, Andersen PM, Asgari N (2020) Inflammatory profiles relate to survival in subtypes of amyotrophic lateral sclerosis. Neurol Neuroimmunol Neuroinflamm 7(3):e697. https://doi.org/10.1212/NXI.0000000000000697

149. Mishra P-S, Vijayalakshmi K, Nalini A, Sathyaprabha TN, Kramer BW, Alladi PA, Raju TR (2017) Etiogenic factors present in the cerebrospinal fluid from amyotrophic lateral sclerosis patients induce predominantly pro-inflammatory responses in microglia. J Neuroinflammation 14(1):251. https://doi.org/10.1186/s12974-017-1028-x

150. Mishra PS, Boutej H, Soucy G, Bareil C, Kumar S, Picher-Martel V, Dupré N, Kriz J, Julien J-P (2020) Transmission of ALS pathogenesis by the cerebrospinal fluid. Acta Neuropathol Commun 8(1):65–65. https://doi.org/10.1186/s40478-020-00943-4

151. Bonneh-Barkay D, Wang G, Starkey A, Hamilton RL, Wiley CA (2010) In vivo CHI3L1 (YKL-40) expression in astrocytes in acute and chronic neurological diseases. J Neuroinflammation 7:34. https://doi.org/10.1186/1742-2094-7-34

152. Pepys MB, Hirschfield GM (2003) C-reactive protein: a critical update. J Clin Invest 111(12):1805–1812. https://doi.org/10.1172/jci18921

153. de Dios O, Gavela-Pérez T, Aguado-Roncero P, Pérez-Tejerizo G, Ricote M, González N, Garcés C, Soriano-Guillén L (2018) C-reactive protein expression in adipose tissue of children with acute appendicitis. Pediatr Res 84(4):564–567. https://doi.org/10.1038/s41390-018-0091-z

154. Devaraj S, Singh U, Jialal I (2009) The evolving role of C-reactive protein in atherothrombosis. Clin Chem 55(2):229–238. https://doi.org/10.1373/clinchem.2008.108886

155. Pasceri V, Willerson James T, Yeh Edward TH (2000) Direct proinflammatory effect of C-reactive protein on human endothelial cells. Circulation 102(18):2165–2168. https://doi.org/10.1161/01.CIR.102.18.2165

156. Calabró P, Willerson James T, Yeh Edward TH (2003) Inflammatory cytokines stimulated C-reactive protein production by human coronary artery smooth muscle cells. Circulation 108(16):1930–1932. https://doi.org/10.1161/01.CIR.0000096055.62724.C5

157. Gewurz H, Mold C, Siegel J, Fiedel B (1982) C-reactive protein and the acute phase response. Adv Intern Med 27:345–372

158. Janeway CA, Medzhitov R (2002) Innate immune recognition. Annu Rev Immunol 20(1):197–216. https://doi.org/10.1146/annurev.immunol.20.083001.084359

159. Agrawal A, Shrive AK, Greenhough TJ, Volanakis JE (2001) Topology and structure of the C1q-binding site on C-reactive protein. J Immunol 166(6):3998–4004. https://doi.org/10.4049/jimmunol.166.6.3998

160. Apostolski S, Nikolić J, Bugarski-Prokopljević C, Miletić V, Pavlović S,

Filipović S (1991) Serum and CSF immunological findings in ALS. Acta Neurol Scand 83 (2):96–98. https://doi.org/10.1111/j.1600-0404.1991.tb04656.x

161. Trbojević-Čepe M, Brinar V, Pauro M, Vogrinc Ž, Štambuk N (1998) Cerebrospinal fluid complement activation in neurological diseases. J Neurol Sci 154(2):173–181. https://doi.org/10.1016/S0022-510X(97)00225-6

162. Sta M, Sylva-Steenland RMR, Casula M, de Jong JMBV, Troost D, Aronica E, Baas F (2011) Innate and adaptive immunity in amyotrophic lateral sclerosis: evidence of complement activation. Neurobiol Dis 42 (3):211–220. https://doi.org/10.1016/j.nbd.2011.01.002

163. Woodruff TM, Costantini KJ, Crane JW, Atkin JD, Monk PN, Taylor SM, Noakes PG (2008) The complement factor C5a contributes to pathology in a rat model of amyotrophic lateral sclerosis. J Immunol 181 (12):8727. https://doi.org/10.4049/jimmunol.181.12.8727

164. Heurich B, el Idrissi NB, Donev RM, Petri S, Claus P, Neal J, Morgan BP, Ramaglia V (2011) Complement upregulation and activation on motor neurons and neuromuscular junction in the SOD1 G93A mouse model of familial amyotrophic lateral sclerosis. J Neuroimmunol 235(1):104–109. https://doi.org/10.1016/j.jneuroim.2011.03.011

165. Donnenfeld H, Kascsak RJ, Bartfeld H (1984) Deposits of IgG and C_3 in the spinal cord and motor cortex of ALS patients. J Neuroimmunol 6(1):51–57. https://doi.org/10.1016/0165-5728(84)90042-0

166. Kawamata T, Akiyama H, Yamada T, McGeer PL (1992) Immunologic reactions in amyotrophic lateral sclerosis brain and spinal cord tissue. Am J Pathol 140(3):691–707

167. Mantovani S, Gordon R, Macmaw JK, Pfluger CMM, Henderson RD, Noakes PG, McCombe PA, Woodruff TM (2014) Elevation of the terminal complement activation products C5a and C5b-9 in ALS patient blood. J Neuroimmunol 276(1):213–218. https://doi.org/10.1016/j.jneuroim.2014.09.005

168. Chiu IM, Phatnani H, Kuligowski M, Tapia JC, Carrasco MA, Zhang M, Maniatis T, Carroll MC (2009) Activation of innate and humoral immunity in the peripheral nervous system of ALS transgenic mice. Proc Natl Acad Sci 106(49):20960. https://doi.org/10.1073/pnas.0911405106

169. Lobsiger CS, Boillée S, Pozniak C, Khan AM, McAlonis-Downes M, Lewcock JW, Cleveland DW (2013) C1q induction and global complement pathway activation do not contribute to ALS toxicity in mutant SOD1 mice. Proc Natl Acad Sci 110(46):E4385. https://doi.org/10.1073/pnas.1318309110

170. Lee JD, Kumar V, Fung JNT, Ruitenberg MJ, Noakes PG, Woodruff TM (2017) Pharmacological inhibition of complement C5a-C5a1 receptor signalling ameliorates disease pathology in the hSOD1G93A mouse model of amyotrophic lateral sclerosis. Br J Pharmacol 174(8):689–699. https://doi.org/10.1111/bph.13730

171. Woodruff TM, Lee JD, Noakes PG (2014) Role for terminal complement activation in amyotrophic lateral sclerosis disease progression. Proc Natl Acad Sci 111(1):E3. https://doi.org/10.1073/pnas.1321248111

172. Hugon J (1996) Riluzole and ALS therapy. Wien Med Wochenschr 146(9–10):185–187

173. Liu BS, Ferreira R, Lively S, Schlichter LC (2013) Microglial SK3 and SK4 currents and activation state are modulated by the neuroprotective drug, riluzole. J Neuroimmune Pharmacol 8(1):227–237. https://doi.org/10.1007/s11481-012-9365-0

174. Gilgun-Sherki Y, Panet H, Melamed E, Offen D (2003) Riluzole suppresses experimental autoimmune encephalomyelitis: implications for the treatment of multiple sclerosis. Brain Res 989(2):196–204. https://doi.org/10.1016/s0006-8993(03)03343-2

175. Wu Y, Satkunendrarajah K, Teng Y, Chow DS, Buttigieg J, Fehlings MG (2013) Delayed post-injury administration of riluzole is neuroprotective in a preclinical rodent model of cervical spinal cord injury. J Neurotrauma 30(6):441–452. https://doi.org/10.1089/neu.2012.2622

176. Alhabbab R (2018) C-reactive protein (CRP) latex agglutination test, pp 59–62

177. Drieghe SA, Alsaadi H, Tugirimana PL, Delanghe JR (2014) A new high-sensitive nephelometric method for assaying serum C-reactive protein based on phosphocholine interaction. Clin Chem Lab Med 52 (6):861–867. https://doi.org/10.1515/cclm-2013-0669

178. Santos V, Guerreiro T, Suarez W, Faria R, Fatibello-Filho O (2011) Evaluation of turbidimetric and nephelometric techniques for analytical determination of N-Acetylcysteine and Thiamine in pharmaceutical formulations employing a lab-made portable microcontrolled turbidimeter and nephelometer. J Braz Chem Soc 22:1968–1978. https://doi.org/10.1590/S0103-50532011001000019

179. Seibert V, Wiesner A, Buschmann T, Meuer J (2004) Surface-enhanced laser desorption ionization time-of-flight mass spectrometry (SELDI TOF-MS) and ProteinChip technology in proteomics research. Pathol Res Pract 200(2):83–94. https://doi.org/10.1016/j.prp.2004.01.010

180. Nakken O, Meyer HE, Stigum H, Holmøy T (2019) High BMI is associated with low ALS risk: a population-based study. Neurology 93 (5):e424–e432. https://doi.org/10.1212/wnl.0000000000007861

181. Janse van Mantgem MR, van Eijk RPA, van der Burgh HK, Tan HHG, Westeneng H-J, van Es MA, Veldink JH, van den Berg LH (2020) Prognostic value of weight loss in patients with amyotrophic lateral sclerosis: a population-based study. J Neurol Neurosurg Psychiatry 91(8):867–875. https://doi.org/10.1136/jnnp-2020-322909

182. Holm T, Maier A, Wicks P, Lang D, Linke P, Münch C, Steinfurth L, Meyer R, Meyer T (2013) Severe loss of appetite in amyotrophic lateral sclerosis patients: online self-assessment study. Interact J Med Res 2(1): e8. https://doi.org/10.2196/ijmr.2463

183. Ngo ST, van Eijk RPA, Chachay V, van den Berg LH, McCombe PA, Henderson RD, Steyn FJ (2019) Loss of appetite is associated with a loss of weight and fat mass in patients with amyotrophic lateral sclerosis. Amyotroph Lateral Scler Frontotemporal Degener 20(7-8):497–505. https://doi.org/10.1080/21678421.2019.1621346

184. Nakayama Y, Shimizu T, Matsuda C, Haraguchi M, Hayashi K, Bokuda K, Nagao M, Kawata A, Ishikawa-Takata K, Isozaki E (2019) Body weight variation predicts disease progression after invasive ventilation in amyotrophic lateral sclerosis. Sci Rep 9 (1):12262. https://doi.org/10.1038/s41598-019-48831-9

185. Körner S, Hendricks M, Kollewe K, Zapf A, Dengler R, Silani V, Petri S (2013) Weight loss, dysphagia and supplement intake in patients with amyotrophic lateral sclerosis (ALS): impact on quality of life and therapeutic options. BMC Neurol 13(1):84. https://doi.org/10.1186/1471-2377-13-84

186. Pradat PF, Bruneteau G, Gordon PH, Dupuis L, Bonnefont-Rousselot D, Simon D, Salachas F, Corcia P, Frochot V, Lacorte JM, Jardel C, Coussieu C, Le Forestier N, Lacomblez L, Loeffler JP, Meininger V (2010) Impaired glucose tolerance in patients with amyotrophic lateral sclerosis. Amyotroph Lateral Scler 11(1-2):166–171. https://doi.org/10.3109/17482960902822960

187. Reyes ET, Perurena OH, Festoff BW, Jorgensen R, Moore WV (1984) Insulin resistance in amyotrophic lateral sclerosis. J Neurol Sci 63(3):317–324. https://doi.org/10.1016/0022-510X(84)90154-0

188. Dedic SI, Stevic Z, Dedic V, Stojanovic VR, Milicev M, Lavrnic D (2012) Is hyperlipidemia correlated with longer survival in patients with amyotrophic lateral sclerosis? Neurol Res 34(6):576–580. https://doi.org/10.1179/1743132812y.0000000049

189. Dupuis L, Corcia P, Fergani A, Gonzalez De Aguilar JL, Bonnefont-Rousselot D, Bittar R, Seilhean D, Hauw JJ, Lacomblez L, Loeffler JP, Meininger V (2008) Dyslipidemia is a protective factor in amyotrophic lateral sclerosis. Neurology 70(13):1004–1009. https://doi.org/10.1212/01.wnl.0000285080.70324.27

190. Steyn FJ, Ioannides ZA, van Eijk RPA, Heggie S, Thorpe KA, Ceslis A, Heshmat S, Henders AK, Wray NR, van den Berg LH, Henderson RD, McCombe PA, Ngo ST (2018) Hypermetabolism in ALS is associated with greater functional decline and shorter survival. J Neurol Neurosurg Psychiatry 89 (10):1016. https://doi.org/10.1136/jnnp-2017-317887

191. Jésus P, Fayemendy P, Nicol M, Lautrette G, Sourisseau H, Preux PM, Desport JC, Marin B, Couratier P (2018) Hypermetabolism is a deleterious prognostic factor in patients with amyotrophic lateral sclerosis. Eur J Neurol 25(1):97–104. https://doi.org/10.1111/ene.13468

192. Desport JC, Preux PM, Magy L, Boirie Y, Vallat JM, Beaufrère B, Couratier P (2001) Factors correlated with hypermetabolism in patients with amyotrophic lateral sclerosis. Am J Clin Nutr 74(3):328–334. https://doi.org/10.1093/ajcn/74.3.328

193. Desport JC, Torny F, Lacoste M, Preux PM, Couratier P (2005) Hypermetabolism in ALS: correlations with clinical and paraclinical parameters. Neurodegener Dis 2 (3-4):202–207. https://doi.org/10.1159/000089626

194. Bouteloup C, Desport JC, Clavelou P, Guy N, Derumeaux-Burel H, Ferrier A, Couratier P (2009) Hypermetabolism in ALS patients: an early and persistent phenomenon. J Neurol 256(8):1236–1242. https://doi.org/10.1007/s00415-009-5100-z

195. Tefera TW, Borges K (2019) Neuronal glucose metabolism is impaired while astrocytic TCA cycling is unaffected at symptomatic

stages in the hSOD1(G93A) mouse model of amyotrophic lateral sclerosis. J Cereb Blood Flow Metab 39(9):1710–1724. https://doi.org/10.1177/0271678x18764775

196. Tefera TW, Bartlett K, Tran SS, Hodson MP, Borges K (2019) Impaired pentose phosphate pathway in the spinal cord of the hSOD1 (G93A) mouse model of amyotrophic lateral sclerosis. Mol Neurobiol 56(8):5844–5855. https://doi.org/10.1007/s12035-019-1485-6

197. Fujimori K, Ishikawa M, Otomo A, Atsuta N, Nakamura R, Akiyama T, Hadano S, Aoki M, Saya H, Sobue G, Okano H (2018) Modeling sporadic ALS in iPSC-derived motor neurons identifies a potential therapeutic agent. Nat Med 24(10):1579–1589. https://doi.org/10.1038/s41591-018-0140-5

198. Hor JH, Santosa MM, Lim VJW, Xuan Ho B, Taylor A, Khong ZJ, Ravits J, Fan Y, Liou YC, Soh BS, Ng SY (2019) ALS motor neurons exhibit hallmark metabolic defects that are rescued by nicotinamide and SIRT3 activation. bioRxiv 713651. https://doi.org/10.1101/713651

199. Guo W, Naujock M, Fumagalli L, Vandoorne T, Baatsen P, Boon R, Ordovás L, Patel A, Welters M, Vanwelden T, Geens N, Tricot T, Benoy V, Steyaert J, Lefebvre-Omar C, Boesmans W, Jarpe M, Sterneckert J, Wegner F, Petri S, Bohl D, Vanden Berghe P, Robberecht W, Van Damme P, Verfaillie C, Van Den Bosch L (2017) HDAC6 inhibition reverses axonal transport defects in motor neurons derived from FUS-ALS patients. Nat Commun 8 (1):861. https://doi.org/10.1038/s41467-017-00911-y

200. Allen S (2020) Understanding metabolic flexibility: a potential key to unlocking metabolic therapies in amyotrophic lateral sclerosis? Neural Regen Res 15(9):1654–1655. https://doi.org/10.4103/1673-5374.276333

201. Palamiuc L, Schlagowski A, Ngo ST, Vernay A, Dirrig-Grosch S, Henriques A, Boutillier A-L, Zoll J, Echaniz-Laguna A, Loeffler J-P, René F (2015) A metabolic switch toward lipid use in glycolytic muscle is an early pathologic event in a mouse model of amyotrophic lateral sclerosis. EMBO Mol Med 7(5):526–546. https://doi.org/10.15252/emmm.201404433

202. Scaricamazza S, Salvatori I, Giacovazzo G, Loeffler JP, Renè F, Rosina M, Quessada C, Proietti D, Heil C, Rossi S, Battistini S, Giannini F, Volpi N, Steyn FJ, Ngo ST, Ferraro E, Madaro L, Coccurello R, Valle C,

Ferri A (2020) Skeletal-muscle metabolic reprogramming in ALS-SOD1(G93A) mice predates disease onset and is a promising therapeutic target. iScience 23(5):101087. https://doi.org/10.1016/j.isci.2020.101087

203. Steyn F, Kirk S, Tefera T, Xie T, Tracey T, Kelk D, Wimberger E, Garton F, Roberts L, Chapman S, Coombes J, Leevy W, Ferri A, Valle C, Rene F, Loeffler J-P, McCombe P, Henderson R, Ngo S (2020) Altered skeletal muscle glucose-fatty acid flux in amyotrophic lateral sclerosis (ALS)

204. Kirk SE, Tracey TJ, Steyn FJ, Ngo ST (2019) Biomarkers of metabolism in amyotrophic lateral sclerosis. Front Neurol 10:191–191. https://doi.org/10.3389/fneur.2019.00191

205. Pradat P-F, El Mendili M-M (2014) Neuroimaging to investigate multisystem involvement and provide biomarkers in amyotrophic lateral sclerosis. Biomed Res Int 2014:467560. https://doi.org/10.1155/2014/467560

206. Roubeau V, Blasco H, Maillot F, Corcia P, Praline J (2015) Nutritional assessment of amyotrophic lateral sclerosis in routine practice: value of weighing and bioelectrical impedance analysis. Muscle Nerve 51 (4):479–484. https://doi.org/10.1002/mus.24419

207. Jara JH, Sheets PL, Nigro MJ, Perić M, Brooks C, Heller DB, Martina M, Andjus PR, Ozdinler PH (2020) The electrophysiological determinants of corticospinal motor neuron vulnerability in ALS. Front Mol Neurosci 13:73. https://doi.org/10.3389/fnmol.2020.00073

208. Wirth AM, Khomenko A, Baldaranov D, Kobor I, Hsam O, Grimm T, Johannesen S, Bruun T-H, Schulte-Mattler W, Greenlee MW, Bogdahn U (2018) Combinatory biomarker use of cortical thickness, MUNIX, and ALSFRS-R at baseline and in longitudinal courses of individual patients with amyotrophic lateral sclerosis. Front Neurol 9:614. https://doi.org/10.3389/fneur.2018.00614

209. Bereman MS, Kirkwood KI, Sabaretnam T, Furlong S, Rowe DB, Guillemin GJ, Mellinger AL, Muddiman DC (2020) Metabolite profiling reveals predictive biomarkers and the absence of β-Methyl Amino-l-alanine in plasma from individuals diagnosed with amyotrophic lateral sclerosis. J Proteome Res 19(8):3276–3285. https://doi.org/10.1021/acs.jproteome.0c00216

210. Chen X, Guo X, Huang R, Zheng Z, Chen Y, Shang HF (2014) An exploratory study of serum creatinine levels in patients with amyotrophic lateral sclerosis. Neurol Sci 35 (10):1591–1597. https://doi.org/10.1007/s10072-014-1807-4

211. Chiò A, Calvo A, Bovio G, Canosa A, Bertuzzo D, Galmozzi F, Cugnasco P, Clerico M, De Mercanti S, Bersano E, Cammarosano S, Ilardi A, Manera U, Moglia C, Sideri R, Marinou K, Bottacchi E, Pisano F, Cantello R, Mazzini L, Mora G (2014) Amyotrophic lateral sclerosis outcome measures and the role of albumin and creatinine: a population-based study. JAMA Neurol 71(9):1134–1142.   https://doi.org/10.1001/jamaneurol.2014.1129

212. Ito D, Hashizume A, Hijikata Y, Yamada S, Iguchi Y, Iida M, Kishimoto Y, Moriyoshi H, Hirakawa A, Katsuno M (2019) Elevated serum creatine kinase in the early stage of sporadic amyotrophic lateral sclerosis. J Neurol 266(12):2952–2961. https://doi.org/10.1007/s00415-019-09507-6

213. Tai H, Cui L, Guan Y, Liu M, Li X, Shen D, Li D, Cui B, Fang J, Ding Q, Zhang K, Liu S (2017) Correlation of creatine kinase levels with clinical features and survival in amyotrophic lateral sclerosis. Front Neurol 8:322. https://doi.org/10.3389/fneur.2017.00322

214. McLeish MJ, Kenyon GL (2005) Relating structure to mechanism in creatine kinase. Crit Rev Biochem Mol Biol 40(1):1–20. https://doi.org/10.1080/10409230590918577

215. Huber K, Petzold J, Rehfeldt C, Ender K, Fiedler I (2007) Muscle energy metabolism: structural and functional features in different types of porcine striated muscles. J Muscle Res Cell Motil 28(4-5):249–258. https://doi.org/10.1007/s10974-007-9123-8

216. Hultman E, Greenhaff PL (1991) Skeletal muscle energy metabolism and fatigue during intense exercise in man. Sci Prog 75(298 Pt 3-4):361–370

217. Le Masson G, Przedborski S, Abbott LF (2014) A computational model of motor neuron degeneration. Neuron 83(4):975–988. https://doi.org/10.1016/j.neuron.2014.07.001

218. Magistretti PJ, Allaman I (2015) A cellular perspective on brain energy metabolism and functional imaging. Neuron 86(4):883–901. https://doi.org/10.1016/j.neuron.2015.03.035

219. Rafiq MK, Lee E, Bradburn M, McDermott CJ, Shaw PJ (2016) Creatine kinase enzyme level correlates positively with serum creatinine and lean body mass, and is a prognostic factor for survival in amyotrophic lateral sclerosis. Eur J Neurol 23(6):1071–1078. https://doi.org/10.1111/ene.12995

220. Dupuis L, Oudart H, René F, de Aguilar J-LG, Loeffler J-P (2004) Evidence for defective energy homeostasis in amyotrophic lateral sclerosis: Benefit of a high-energy diet in a transgenic mouse model. Proc Natl Acad Sci U S A 101(30):11159. https://doi.org/10.1073/pnas.0402026101

221. Szelechowski M, Amoedo N, Obre E, Léger C, Allard L, Bonneu M, Claverol S, Lacombe D, Oliet S, Chevallier S, Le Masson G, Rossignol R (2018) Metabolic reprogramming in amyotrophic lateral sclerosis. Sci Rep 8(1):3953. https://doi.org/10.1038/s41598-018-22318-5

222. Schutte JE, Longhurst JC, Gaffney FA, Bastian BC, Blomqvist CG (1981) Total plasma creatinine: an accurate measure of total striated muscle mass. J Appl Physiol Respir Environ Exerc Physiol 51(3):762–766. https://doi.org/10.1152/jappl.1981.51.3.762

223. Baxmann AC, Ahmed MS, Marques NC, Menon VB, Pereira AB, Kirsztajn GM, Heilberg IP (2008) Influence of muscle mass and physical activity on serum and urinary creatinine and serum cystatin C. Clin J Am Soc Nephrol 3(2):348–354. https://doi.org/10.2215/cjn.02870707

224. van Eijk RPA, Eijkemans MJC, Ferguson TA, Nikolakopoulos S, Veldink JH, van den Berg LH (2018) Monitoring disease progression with plasma creatinine in amyotrophic lateral sclerosis clinical trials. J Neurol Neurosurg Psychiatry 89(2):156. https://doi.org/10.1136/jnnp-2017-317077

225. Harpole M, Davis J, Espina V (2016) Current state of the art for enhancing urine biomarker discovery. Expert Rev Proteomics 13 (6):609–626.   https://doi.org/10.1080/14789450.2016.1190651

226. Yao F, Hong X, Li S, Zhang Y, Zhao Q, Du W, Wang Y, Ni J (2018) Urine-based biomarkers for alzheimer's disease identified through coupling computational and experimental methods. J Alzheimers Dis 65 (2):421–431. https://doi.org/10.3233/jad-180261

227. Watanabe Y, Hirao Y, Kasuga K, Tokutake T, Semizu Y, Kitamura K, Ikeuchi T, Nakamura K, Yamamoto T (2019) Molecular network analysis of the urinary proteome of Alzheimer's disease patients. Demen Geriatr Cognit Disord Extra 9(1):53–65. https://doi.org/10.1159/000496100

228. Shepheard SR, Wuu J, Cardoso M, Wiklendt L, Dinning PG, Chataway T, Schultz D, Benatar M, Rogers ML (2017) Urinary p75(ECD): a prognostic, disease progression, and pharmacodynamic biomarker in ALS. Neurology 88(12):1137–1143. https://doi.org/10.1212/WNL. 0000000000003741

229. Ibanez CF, Simi A (2012) p75 neurotrophin receptor signaling in nervous system injury and degeneration: paradox and opportunity. Trends Neurosci 35(7):431–440. https://doi.org/10.1016/j.tins.2012.03.007

230. Yan Q, Johnson EM (1988) An immunohistochemical study of the nerve growth factor receptor in developing rats. J Neurosci 8 (9):3481. https://doi.org/10.1523/JNEUROSCI.08-09-03481.1988

231. Ferri CC, Moore FA, Bisby MA (1998) Effects of facial nerve injury on mouse motoneurons lacking the p75 low-affinity neurotrophin receptor. J Neurobiol 34(1):1–9. https://doi.org/10.1002/(SICI)1097-4695 (199801)34:1<1::AID-NEU1>3.0.CO;2-C

232. Lowry KS, Murray SS, McLean CA, Talman P, Mathers S, Lopes EC, Cheema SS (2001) A potential role for the p75 low-affinity neurotrophin receptor in spinal motor neuron degeneration in murine and human amyotrophic lateral sclerosis. Amyotroph Lateral Scler Other Motor Neuron Disord 2(3):127–134. https://doi.org/10.1080/146608201753275463

233. Copray JC, Jaarsma D, Küst BM, Bruggeman RW, Mantingh I, Brouwer N, Boddeke HW (2003) Expression of the low affinity neurotrophin receptor p75 in spinal motoneurons in a transgenic mouse model for amyotrophic lateral sclerosis. Neuroscience 116 (3):685–694. https://doi.org/10.1016/s0306-4522(02)00755-8

234. Kenchappa RS, Tep C, Korade Z, Urra S, Bronfman FC, Yoon SO, Carter BD (2010) p75 neurotrophin receptor-mediated apoptosis in sympathetic neurons involves a biphasic activation of JNK and up-regulation of tumor necrosis factor-alpha-converting enzyme/ADAM17. J Biol Chem 285 (26):20358–20368. https://doi.org/10.1074/jbc.M109.082834

235. DiStefano PS, Clagett-Dame M, Chelsea DM, Loy R (1991) Developmental regulation of human truncated nerve growth factor receptor. Ann Neurol 29(1):13–20. https://doi.org/10.1002/ana.410290105

236. Jia R, Shepheard S, Jin J, Hu F, Zhao X, Xue L, Xiang L, Qi H, Qu Q, Guo F, Rogers ML, Dang J (2017) Urinary extracellular domain of neurotrophin receptor p75 as a biomarker for amyotrophic lateral sclerosis in a Chinese cohort. Sci Rep 7(1):5127. https://doi.org/10.1038/s41598-017-05430-w

237. Shepheard SR, Chataway T, Schultz DW, Rush RA, Rogers ML (2014) The extracellular domain of neurotrophin receptor p75 as a candidate biomarker for amyotrophic lateral sclerosis. PLoS One 9(1):e87398. https://doi.org/10.1371/journal.pone.0087398

238. Lingor P, Weber M, Camu W, Friede T, Hilgers R, Leha A, Neuwirth C, Günther R, Benatar M, Kuzma-Kozakiewicz M, Bidner H, Blankenstein C, Frontini R, Ludolph A, Koch JC (2019) ROCK-ALS: protocol for a randomized, placebo-controlled, double-blind phase iia trial of safety, tolerability and efficacy of the rho kinase (ROCK) inhibitor Fasudil in amyotrophic lateral sclerosis. Front Neurol 10:293. https://doi.org/10.3389/fneur.2019.00293

239. Gold J, Rowe DB, Kiernan MC, Vucic S, Mathers S, van Eijk RPA, Nath A, Garcia Montojo M, Norato G, Santamaria UA, Rogers ML, Malaspina A, Lombardi V, Mehta PR, Westeneng HJ, van den Berg LH, Al-Chalabi A (2019) Safety and tolerability of Triumeq in amyotrophic lateral sclerosis: the Lighthouse trial. Amyotroph Lateral Scler Frontotemporal Degener 20 (7-8):595–604. https://doi.org/10.1080/21678421.2019.1632899

240. Rahal A, Kumar A, Singh V, Yadav B, Tiwari R, Chakraborty S, Dhama K (2014) Oxidative stress, prooxidants, and antioxidants: the interplay. Biomed Res Int 2014:761264. https://doi.org/10.1155/2014/761264

241. Tönnies E, Trushina E (2017) Oxidative stress, synaptic dysfunction, and Alzheimer's disease. J Alzheimers Dis 57(4):1105–1121. https://doi.org/10.3233/jad-161088

242. Phan K, He Y, Pickford R, Bhatia S, Katzeff JS, Hodges JR, Piguet O, Halliday GM, Kim WS (2020) Uncovering pathophysiological changes in frontotemporal dementia using serum lipids. Sci Rep 10(1):3640. https://doi.org/10.1038/s41598-020-60457-w

243. Nascimento C, Nunes VP, Diehl Rodriguez R, Takada L, Suemoto CK, Grinberg LT, Nitrini R, Lafer B (2019) A review on shared clinical and molecular mechanisms between bipolar disorder and frontotemporal dementia. Prog Neuropsychopharmacol Biol Psychiatry 93:269–283. https://doi.org/10.1016/j.pnpbp.2019.04.008

244. Kumar A, Ratan RR (2016) Oxidative stress and Huntington's disease: the good, the bad, and the ugly. J Huntingtons Dis 5 (3):217–237. https://doi.org/10.3233/jhd-160205

245. Dalfó E, Portero-Otín M, Ayala V, Martínez A, Pamplona R, Ferrer I (2005) Evidence of oxidative stress in the neocortex in incidental Lewy body disease. J Neuropathol Exp Neurol 64(9):816–830. https://doi.org/10.1097/01.jnen.0000179050.54522.5a

246. Mao P (2013) Oxidative stress and its clinical applications in dementia. J Neurodegener Dis 2013:319898. https://doi.org/10.1155/2013/319898

247. Pollari E, Goldsteins G, Bart G, Koistinaho J, Giniatullin R (2014) The role of oxidative stress in degeneration of the neuromuscular junction in amyotrophic lateral sclerosis. Front Cell Neurosci 8:131. https://doi.org/10.3389/fncel.2014.00131

248. Carrì MT, Valle C, Bozzo F, Cozzolino M (2015) Oxidative stress and mitochondrial damage: importance in non-SOD1 ALS. Front Cell Neurosci 9:41. https://doi.org/10.3389/fncel.2015.00041

249. Barber SC, Mead RJ, Shaw PJ (2006) Oxidative stress in ALS: a mechanism of neurodegeneration and a therapeutic target. Biochim Biophys Acta Mol Basis Dis 1762 (11):1051–1067. https://doi.org/10.1016/j.bbadis.2006.03.008

250. Pansarasa O, Bordoni M, Diamanti L, Sproviero D, Gagliardi S, Cereda C (2018) SOD1 in amyotrophic lateral sclerosis: "Ambivalent" behavior connected to the disease. Int J Mol Sci 19(5). https://doi.org/10.3390/ijms19051345

251. Benatar M (2007) Lost in translation: treatment trials in the SOD1 mouse and in human ALS. Neurobiol Dis 26(1):1–13. https://doi.org/10.1016/j.nbd.2006.12.015

252. Benedetti A, Comporti M, Esterbauer H (1980) Identification of 4-hydroxynonenal as a cytotoxic product originating from the peroxidation of liver microsomal lipids. Biochim Biophys Acta Lipids Lipid Metab 620 (2):281–296. https://doi.org/10.1016/0005-2760(80)90209-X

253. Pedersen WA, Cashman NR, Mattson MP (1999) The lipid peroxidation product 4-hydroxynonenal impairs glutamate and glucose transport and choline acetyltransferase activity in NSC-19 motor neuron cells. Exp Neurol 155(1):1–10. https://doi.org/10.1006/exnr.1998.6890

254. Bruce-Keller AJ, Li YJ, Lovell MA, Kraemer PJ, Gary DS, Brown RR, Markesbery WR, Mattson MP (1998) 4-Hydroxynonenal, a product of lipid peroxidation, damages cholinergic neurons and impairs visuospatial memory in rats. J Neuropathol Exp Neurol 57(3):257–267. https://doi.org/10.1097/00005072-199803000-00007

255. Pedersen WA, Fu W, Keller JN, Markesbery WR, Appel S, Smith RG, Kasarskis E, Mattson MP (1998) Protein modification by the lipid peroxidation product 4-hydroxynonenal in the spinal cords of amyotrophic lateral sclerosis patients. Ann Neurol 44(5):819–824. https://doi.org/10.1002/ana.410440518

256. Simpson EP, Henry YK, Henkel JS, Smith RG, Appel SH (2004) Increased lipid peroxidation in sera of ALS patients. Neurology 62 (10):1758. https://doi.org/10.1212/WNL.62.10.1758

257. Devos D, Moreau C, Kyheng M, Garçon G, Rolland AS, Blasco H, Gelé P, Timothée Lenglet T, Veyrat-Durebex C, Corcia P, Dutheil M, Bede P, Jeromin A, Oeckl P, Otto M, Meininger V, Danel-Brunaud V, Devedjian JC, Duce JA, Pradat PF (2019) A ferroptosis-based panel of prognostic biomarkers for amyotrophic lateral sclerosis. Sci Rep 9(1):2918. https://doi.org/10.1038/s41598-019-39739-5

258. Tang X, Sayre LM, Tochtrop GP (2011) A mass spectrometric analysis of 4-hydroxy-2-(E)-nonenal modification of cytochrome c. J Mass Spectrom 46(3):290–297. https://doi.org/10.1002/jms.1890

259. Aldini G, Gamberoni L, Orioli M, Beretta G, Regazzoni L, Maffei Facino R, Carini M (2006) Mass spectrometric characterization of covalent modification of human serum albumin by 4-hydroxy-trans-2-nonenal. J Mass Spectrom 41(9):1149–1161. https://doi.org/10.1002/jms.1067

260. Ethen CM, Reilly C, Feng X, Olsen TW, Ferrington DA (2007) Age-related macular degeneration and retinal protein modification by 4-hydroxy-2-nonenal. Invest Ophthalmol Vis Sci 48(8):3469–3479. https://doi.org/10.1167/iovs.06-1058

261. Rauniyar N, Stevens SM, Prokai-Tatrai K, Prokai L (2009) Characterization of 4-hydroxy-2-nonenal-modified peptides by liquid chromatography-tandem mass spectrometry using data-dependent acquisition: neutral loss-driven MS3 versus neutral loss-driven electron capture dissociation. Anal Chem 81(2):782–789. https://doi.org/10.1021/ac802015m

262. Nadkarni DV, Sayre LM (1995) Structural definition of early lysine and histidine adduction chemistry of 4-hydroxynonenal. Chem Res Toxicol 8(2):284–291. https://doi.org/10.1021/tx00044a014

263. Richter C (1995) Oxidative damage to mitochondrial DNA and its relationship to ageing. Int J Biochem Cell Biol 27(7):647–653. https://doi.org/10.1016/1357-2725(95)00025-K

264. Valavanidis A, Vlachogianni T, Fiotakis C (2009) 8-hydroxy-2' -deoxyguanosine (8-OHdG): a critical biomarker of oxidative stress and carcinogenesis. J Environ Sci Health C Environ Carcinog Ecotoxicol Rev 27(2):120–139. https://doi.org/10.1080/10590500902885684

265. Suzuki S, Hinokio Y, Komatu K, Ohtomo M, Onoda M, Hirai S, Hirai M, Hirai A, Chiba M, Kasuga S, Akai H, Toyota T (1999) Oxidative damage to mitochondrial DNA and its relationship to diabetic complications. Diabetes Res Clin Pract 45(2):161–168. https://doi.org/10.1016/S0168-8227(99)00046-7

266. Blasco H, Garcon G, Patin F, Veyrat-Durebex C, Boyer J, Devos D, Vourc'h P, Andres CR, Corcia P (2017) Panel of oxidative stress and inflammatory biomarkers in ALS: a Pilot Study. Can J Neurol Sci 44(1):90–95. https://doi.org/10.1017/cjn.2016.284

267. Bogdanov M, Brown RH, Matson W, Smart R, Hayden D, O'Donnell H, Flint Beal M, Cudkowicz M (2000) Increased oxidative damage to DNA in ALS patients. Free Radic Biol Med 29(7):652–658. https://doi.org/10.1016/S0891-5849(00)00349-X

268. Smith EF, Shaw PJ, De Vos KJ (2019) The role of mitochondria in amyotrophic lateral sclerosis. Neurosci Lett 710:132933. https://doi.org/10.1016/j.neulet.2017.06.052

269. Mitsumoto H, Santella RM, Liu X, Bogdanov M, Zipprich J, Wu H-C, Mahata J, Kilty M, Bednarz K, Bell D, Gordon PH, Hornig M, Mehrazin M, Naini A, Flint Beal M, Factor-Litvak P (2008) Oxidative stress biomarkers in sporadic ALS. Amyotroph Lateral Scler 9(3):177–183. https://doi.org/10.1080/17482960801933942

270. Shefner JM, Al-Chalabi A, Baker MR, Cui L-Y, de Carvalho M, Eisen A, Grosskreutz J, Hardiman O, Henderson R, Matamala JM, Mitsumoto H, Paulus W, Simon N, Swash M, Talbot K, Turner MR, Ugawa Y, van den Berg LH, Verdugo R, Vucic S, Kaji R, Burke D, Kiernan MC (2020) A proposal for new diagnostic criteria for ALS. Clin Neurophysiol. https://doi.org/10.1016/j.clinph.2020.04.005

271. Chiò A (1999) ISIS Survey: an international study on the diagnostic process and its implications in amyotrophic lateral sclerosis. J Neurol 246(Suppl 3):Iii1–Iii5. https://doi.org/10.1007/bf03161081

272. Ganesalingam J, An J, Shaw CE, Shaw G, Lacomis D, Bowser R (2011) Combination of neurofilament heavy chain and complement C3 as CSF biomarkers for ALS. J Neurochem 117(3):528–537. https://doi.org/10.1111/j.1471-4159.2011.07224.x

MicroRNAs in Blood Serum, Blood Plasma, and Cerebrospinal Fluid as Diagnostic Biomarkers of Alzheimer's Disease

Philip V. Peplow and Bridget Martinez

Abstract

MicroRNAs (miRNAs) are a family of small, genome-encoded endogenous RNAs that are transcribed but not translated into proteins. They serve essential roles in virtually every aspect of brain function, including neurogenesis, neural development, and cellular responses leading to changes in synaptic plasticity. They are implicated in neurodegeneration and other neurological disorders. Complex interplay among multiple pathways including excitotoxicity, mitochondrial dysfunction, ionic imbalance, oxidative stress, and inflammation are involved in the mechanism of CNS degenerative disease including Alzheimer's disease (AD). Circulating miRNAs are largely stable in blood and may serve as diagnostic markers for CNS injury. This chapter presents the findings in human studies using blood serum, blood plasma, and CSF which indicate that individual or combinations of miRNAs can serve as important biomarkers in distinguishing AD and mild cognitive impairment (MCI) patients from controls. Appropriate miRNAs can differentiate between AD and MCI, and between mild and moderate–severe AD, and predict the progression of MCI to AD. They could potentially replace or be combined with the molecular markers of AD currently used in clinical practice.

Keywords Alzheimer's disease, Blood plasma, Blood serum, Cerebrospinal fluid, Diagnostic biomarkers, Disease progression, microRNA, Mild cognitive impairment

---

1 Introduction

Alzheimer's disease (AD), the most common age-related, progressive neurodegenerative disease, affects currently more than 35 million people globally and is forecast to affect 1% of all people world-wide by 2050 [1–3]. AD is characterized by memory loss and cognitive decline and accounts for most cases of dementia in the elderly. The cost and societal impact of AD are increased by the disability it causes, and the high levels of daily, supervised care and assistance needed. Diagnosis of probable AD is based on clinical examination, magnetic resonance imaging, positron emission

Philip V. Peplow, Bridget Martinez and Thomas A. Gennarelli (eds.), *Neurodegenerative Diseases Biomarkers: Towards Translating Research to Clinical Practice*, Neuromethods, vol. 173, https://doi.org/10.1007/978-1-0716-1712-0_12,
© Springer Science+Business Media, LLC, part of Springer Nature 2022

tomography (PET), cerebrospinal fluid (CSF) assays, and neuro-psychological testing which includes cognitive performance [4, 5]. Current health systems have limited capacity to select those individuals likely to have AD pathology in order to confirm the diagnosis with available cerebrospinal fluid and imaging biomarkers at memory clinics. To lower the incidence and burden of AD requires being able to detect the disease at an early stage so that treatments can be initiated to reduce its severity and retard its progression. In addition, in individuals at risk of developing AD, being able to detect the disease at the preonset stage (asymptomatic) and begin treatment would also have a considerable impact on disease incidence and health system burden, for example, detecting mild cognitive impairment (MCI) which may subsequently progress to AD [6].

2   Early- and Late-Onset AD and MCI

There are two major forms of AD, early-onset (familial) and late-onset (sporadic). Early-onset AD is rare, whereas late-onset AD accounts for >90% of cases [7]. Early-onset AD is defined as AD with onset at age 30 to 65 years, while late-onset AD is AD with onset at age > 65 years [8]. A combination of genetic (70%) and environmental factors (30%) is involved in the etiology of AD [9]. Many genes have been shown to be involved in the development of late-onset AD including ABCA7, APOE, BIN1 [10]. No drugs or other therapeutic agents are presently available to delay the progression of AD, and no biomakers have yet been confirmed for the early detection of AD before the onset of irreversible neurological damage [11]. Cellular and molecular changes occurring in the brains of individuals with AD include neuronal and synaptic loss, mitochondrial damage, production and accumulation of β-amyloid peptide (Aβ) and hyperphosphorylated tau protein, decrease of acetylcholine neurotransmitter, inflammation, and oxidative stress. Characteristic of AD is aggregation of Aβ peptide in extracellular plaques and the hyperphosphorylated tau protein in intracellular neurofibrillary tangles. Most research studies have been performed with blood or brain tissue samples (postmortem) at late-stage AD, and identifying molecular biomarkers characteristic of the early-stage AD is a major challenge. Individuals with MCI almost always have the neuropathologic features of AD [12–14] and could provide important information. About 50% of MCI patients progress to AD [15].

The currently available biomarkers of AD are detected by measurement of amyloid beta (Aβ) peptide and tau protein levels in the cerebrospinal fluid (CSF) [16], brain imaging using PET to detect Aβ deposits [17], or postmortem gross specimen analysis and histology of brain sections [18]. However, the high costs and invasiveness of these methods, which require skilled expertise to perform and interpret or are time-consuming, restrict their application to only a small number of cases [19–22]. The Aβ42 isoform and tau protein levels in CSF and particularly the ratio of tau/Aβ42 and phospho-tau/Aβ have been used to predict the risk of progressing from MCI/very mild dementia to AD [23, 24] and to identify MCI patients diagnosed with probable early AD [25]. However, the biomarkers of Aβ and tau do not provide differentiating features between early- and late-onset AD [26]. Also, significant gender and race disparities in the CSF biomarkers Aβ42 and tau levels have been reported both in healthy subjects and patients with AD [27, 28]. CSF may not easily be collected by lumbar puncture in the elderly owing to lumbar disc degeneration with narrowing of the intervertebral spaces. In addition, blood contamination occurs in up to 20% of CSF samples collected by lumbar puncture [29] and may be a confounding factor affecting Aβ42 levels [30] and other markers. In addition, hemolysis may occur. Blood serum or plasma is an appealing source of biomarkers as it can be collected with minimal discomfort to the patient, thereby encouraging greater compliance in clinical trials and frequent testing, and hemolysis of any samples can easily be detected visually or by measurement of absorbance at 414/375 nm (samples with 414/375 ratio > 1.4 are considered hemolyzed; see ref. 31). Whole blood is considered a less suitable source for measurement of biomarkers as hemolysis is not readily detected [32]; it is collected in heparinized tubes and requires centrifugation of part of the sample to determine whether there is hemolysis. Also freezing of whole blood for storage purposes is not recommended as it produces cell damage.

3 Cognitive Decline and AD Pathology

Within AD patients there is a range of disease states from mild dementia to severe impairment. The Mini-Mental State Examination (MMSE) is commonly used to estimate the extent of cognitive decline and scores range from 0–30. Probable AD patients can be classified as having mild (MMSE scores 21–24), moderate (MMSE scores 10–20), or severe dementia (MMSE scores 0–9), while normal subjects have MMSE scores 25–30 [33]. However, the MMSE does not have sufficient sensitivity to detect cognitive decline at the MCI stage (MMSE mean score 27.0, see ref. 34), and therefore other cognitive assessment examinations have to be used. In the Montreal Cognitive Assessment (MoCA) examination, scores range from 0–30 and a score of ≥27 is considered normal. Normal subjects had a mean MoCA score of 27.4, MCI individuals a mean MoCA score 22.1 (range 17–23), and probable AD patients a mean MoCA score 16.2 (range 0–24) [34]. Identifying biomarkers capable of distinguishing MCI from the normal elderly and probable AD cases, as well the transition between the stages of mild, moderate, and severe cognitive impairment in AD, is an important need in indicating dysregulated signaling pathways and suggesting possible potential treatment strategies.

The depostion of hyperphosphorylated tau protein within select neuronal types in specific nuclei or areas is used in the histomicroscopic evaluation of AD-related pathology, and the staging of AD-related neurofibrillary pathology has been described [18, 35]. Six stages have been recognized in the progression of AD and they are usually grouped under stages I–II, III–IV, V–VI (see Fig. 1) with the major characteristics being as follows.

- Stage I: lesions develop in the transentorhinal region. Subcortical nuclei (viz. locus coeruleus, magnocellular nuclei of the basal forebrain) occasionally exhibit the earliest changes in the absence of cortical involvement. The transentorhinal region is the first site in the cerebral cortex to be involved. The entorhinal region proper is not or minimally involved.

- Stage II: lesions extend into the entorhinal region.

- Stage III: lesions extend from the transentorhinal region to the neocortex of the fusiform and lingual gyri and then diminish markedly beyond this region.

- Stage IV: lesions progress more widely into neocortical association areas.

- Stage V: lesions appear in previously uninvolved areas and extend widely into the first temporal convolution, and into high order association areas of frontal, parietal, and occipital neocortex (peristriate region).

- Stage VI: lesions reach the secondary and primary neocortical areas and extend into the striate region of the occipital lobe.

Areas of the brain affected by AD are shown in Fig. 2.

4   MicroRNAs in AD and MCI

MicroRNAs (miRNAs) are abundant, endogenous, small noncoding RNAs that act as important posttranscriptional regulators of gene expression by binding to the 3′-untranslated region (3′-UTR) of their target mRNAs, thereby interfering with gene regulation and translation or causing destabilization or preferential cleavage of target mRNAs [11, 36, 37]. They have been detected in blood, CSF, saliva and urine, and also in blood cells such as mononuclear cells and erythrocytes. Altered expression of certain microRNA molecules suggests that they could have a regulatory role in various diseases and disorders [38]. In animals, miRNAs are synthesized from primary microRNAs (pri-miRNA) in two stages by the action

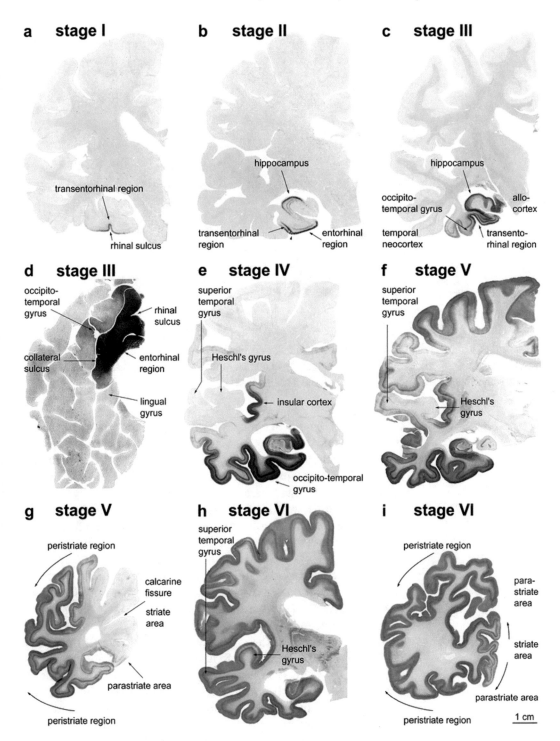

Fig. 1 Stages I–VI of cortical neurofibrillary pathology in 100 μm polyethylene glycol-embedded hemisphere sections immunostained for hyperphosphorylated tau (AT8, Innogenetics). Reproduced from [18]

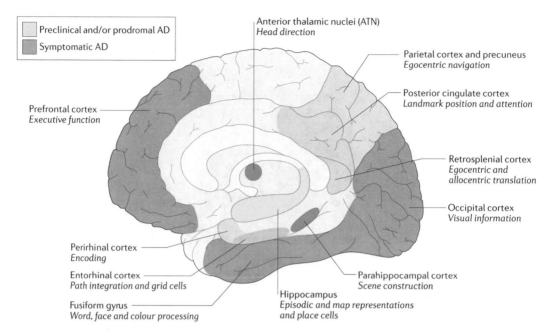

Fig. 2 Anatomical illustration of AD-related neuropathological changes. Yellow areas are brain areas affected by Alzheimer disease (AD) pathophysiology in preclinical and prodromal stages of the disease; red areas are brain areas affected in symptomatic stages of the disease. Reproduced from [65]

two RNase III-type enzymes: Drosha in the nucleus and Dicer in the cytoplasm. Within the nucleus, the pri-miRNA is processed by Drosha to generate a precursor microRNA (pre-miRNA). Exportin-5 in the cytoplasm mediates the transport of pre-miRNAs to the cytoplasm. In the cytoplasm, they are further processed by Dicer to generate mature miRNAs of approximately 22 nucleotides in length. The mature microRNAs are then loaded onto an Argonaute protein to produce the effector RNA-induced silencing complex (RISC) [39].

As mentioned above, miRNAs are a family of small, genome-encoded endogenous RNAs that are transcribed but not translated into proteins. They serve essential roles in virtually every aspect of brain function, including neurogenesis, neural development, and cellular responses leading to changes in synaptic plasticity [40]. They are also implicated in neurodegeneration and neurological disorders [38]. Complex interplay among multiple pathways including excitotoxicity, mitochondrial dysfunction, ionic imbalance, oxidative stress, and inflammation are involved in the mechanism of central nervous system (CNS) degenerative disease. Studies targeting miRNAs for therapies in cancer have established that circulating miRNAs are largely stable in blood [41] indicating that they likely represent a sampling of cellular activity, a route of intercellular communication, and as diagnostic and prognostic biomarkers for CNS injury [42]. It has been shown that serum or plasma miRNAs are indicative of the disease state [43]. Much interest has developed recently in the use of circulating cell-free miRNAs as novel markers in the clinical diagnosis of disease especially in cancer [44].

Circulating microRNAs within blood may be measured and used as minimally invasive diagnostic biomarkers. They may facilitate the early detection of AD and potentially the continual monitoring of disease progression and allow therapeutic interventions to be evaluated. The aim is to identify microRNA biomarkers of high sensitivity and specificity in early-stage AD, and paired microRNA analysis or using a panel of microRNAs may be an important method to accomplish this. We have performed a PubMed literature search of research articles published January 2016-December 2019 on microRNAs in AD and also MCI when included as prodromal or early-stage AD. This is to provide further information on microRNA dysregulation in AD. We have included studies using blood serum, blood plasma, and CSF. As mentioned previously, CSF is not an appropriate sample for screening and routine testing due to its high costs and invasiveness of sample collection. Blood-based biomarkers for AD screening and routine testing are far more suitable [45] and we have chosen to include studies using blood serum, and blood plasma, and to see whether the microRNA profile is different for blood serum and blood plasma compared to CSF. Reviews of microRNA profiles in AD have recently been presented [46–49].

4.1 MicroRNAs in Blood Serum

The miRNAs shown to have altered expression in blood serum of AD and MCI patients are summarized in Table 1.

There were seven studies and where reported all had recruited males and females in the study groups, but there was disparity in the proportions of males and females in the groups. Some of the studies had used large sized groups; for example, Yang et al. [50] analyzed blood serum from 107 AD patients, 101 MCI patients and 50 other neurologic controls, while in others the group sizes were small; for example, Kumar et al. [51] recruited 11 AD patients, 20 MCI patients and 18 normal controls. All the studies had used PCR as the method of analysis of miRNAs. There was limited overlap in the findings of the various studies. Where there was overlap, upregulation of miR-106-3p was observed in AD patients by Wu et al. [52] and Guo et al. [53], and downregulation of miR-22-3p in AD patients by Guo et al. [53] and Denk et al. [54].

Yang et al. [50] found that expression levels of miR-135a were significantly increased in AD and MCI compared to controls, but there was no significant difference between AD and MCI. Serum exosomal miR-193b expression was significantly decreased in AD and MCI compared with controls, and was significantly lower in AD than in MCI. Serum exosomal miR-384 was significantly

Table 1
Dysregulated microRNAs in blood serum of AD and MCI patients compared to controls

Reference	Subjects and age	Sample	Method of analysis	Increased expression of miRNAs	Decreased expression of miRNAs
Yang et al. [50]	AD, 41 M/66F, mean age 74.2 years; MCI, 42 M/59F, 61.6 years, ONC, 23 M/27F, 66.7	Serum exosomes	RT-qPCR	Exosomal miR-135a in AD and MCI but no difference between AD and MCI. Exosomal miR-384 in AD and MCI and higher in AD than MCI	Exosomal miR-193b in AD and MCI and lower in AD than MCI
Denk et al. [54]	AD, 47, mean age 65 years, MMSE 21; NC, 38, 64 years, NA	Serum	RT-qPCR	miR-103a-3p, −142-3p, −20a-5p, −29b-3p, −7b-5p, −7g-5p, −106a-5p, 106b-5p, −18b-5p, 223-3p, −26a-5p, −26b-5p, −301a-3p, −30b-5p in AD	miR-132-3p, −146a-5p, −15a-5p, −22-3p, −320a, −320b, −92a-3p, −1246 in AD
Wu et al. [52]	AD, 32 M/23F, mean age 72.1 years, MMSE mean score 12.8; NC, 30 M/20F, 73.0 years, MMSE 28.9	Serum	RT-qPCR	miR-146a-5p, −106b-3p, −195-5p, −20b-5p, −497-5p in AD	miR-125b-3p, −29c-3p, −93-5p, −19b-3p in AD
Kumar et al. [51]	AD, 6 M/5F, 76.3 years, MMSE 22.8; MCI, 10 M/10F, 74.1 years, MMSE 25.1; NC, 6 M/12F, 74.6 years, MMSE 27.3	Serum	RT-qPCR for serum and also validated using postmortem brain samples	miR-455-3p in AD compared to MCI and NC. miR-4668-5p in MCI but not in AD. miR-455-3p and miR-3613-3p in frontal cortex for AD Braak stage V, and miR-4674 in AD Braak stage VI	
Hara et al. [55]	Validation set: AD, 13 M/23F, 74.7 years, MMSE 19.3; NC, 4 M/18F, 73.7 years, MMSE 29.3	Serum	RT-qPCR	miR-26b-5p in AD	miR-501-3p in AD

Gou et al. [53]	*Validation set:* Mild AD, 16 M/15F, mean age 71.4 years, MMSE 24.1, CDR 0.5 or 1; moderate AD, 28 M/24F, 72.5 years, MMSE 15.5, CDR 2; severe AD, 17 M/21F, 74.1 years, MMSE 7.0, CDR 3; NC, 45 M/41F, 68.0 years, MMSE 28.9, CDR 0	Serum from fasted patients	RT-qPCR	miR-106b-3p, −6119-5p, −1246, −660-5p in AD	miR-26a-5p, −181c-3p, −126-5p, −22-3p -148b-5p, −144-5p in AD
Jia et al. [56]	AD, 62 M/22F, 81.4 years, MMSE 18.3; NC, 41 M/21F, 76.2 years, MMSE ND	Serum	PCR screening	miR-519 in AD	miR-29, −125b, −223 in AD

AD Alzheimer's disease, *MCI* mild cognitive impairment, *NC* normal controls, *MMSE* Mini-Mental State Examination, *CDR* clinical dementia rating, *RT-qPCR* real-time quantitative polymerase chain reaction, *ND* not determined

increased in AD and MCI compared with controls, and was significantly higher in AD than in MCI. Receiver operating curve analysis indicated that with appropriate cut-off values each of these miRNAs had very high sensitivity/specificity in discriminating AD patients from controls. MiR-135a had the highest diagnostic accuracy among the three miRNAs in discriminating MCI patients from control subjects (AUC 0.981). AUC values for the combinations miR-135a/miR-193b, miR-135a/miR-384, miR-193b/miR-384, and miR-135a/miR-193b/miR-384 were 0.994, 0.983, 0.905, and 0.995, respectively. Exosomal miR-384 in the serum increased the risk of MCI while exosomal miR-193b in the serum decreased the risk of MCI. A study by Denk et al. [54] indicated each of miR-20a-5p, miR-29b-3p, miR-26b-5p, and miR-320a to be a good test to distinguish AD from NC (AUC 0.85, specificity 92%; AUC 0.83, sensitivity 93%; AUC 0.97, sensitivity 89%, specificity 89%; AUC 0.90, sensitivity 83%, specificity 90%, respectively). A small scale study by Kumar et al. [51] found a significant increase in serum expression level of miR-455-3p in AD compared to MCI and NC. Expression of serum miR-4668-5p was significantly increased in MCI compared with NC, but was not elevated in AD. Serum miR-455-3p was a fair test to distinguish AD from NC (AUC 0.79). Hara et al. [55] found that miR-501-3p was decreased in AD compared with NC and correlated with MMSE scores. In brains of patients in the screening/discovery set, miR-501-3p was markedly upregulated in AD compared with NC and positively correlated with disease progression as indicated by Braak staging. Serum miR-501-3p was a good test to distinguish AD from NC (AUC 0.82, sensitivity 53%, specificity 100%). Guo et al. [53] showed that miR-22-3p had the highest sensitivity 82% and specificity 71% for differentiating AD from NC. The combination of miR-26a-5p/miR-181c-3p/miR-22-3p/miR-148b-5p/miR-106b-3p/miR-6119-5p/miR-660-5p had AUC 0.986, sensitivity 82% and specificity 71%. MiR-26a-5p, miR-181c-3p, miR-126-5p,miR-22-3p, miR-148b-5p, miR-106b-3p were differentially expressed at "mild" stage of AD; miR-6119-5p, miR-1246 were differentially expressed at "moderate" stage of AD; and miR-660 was differentially expressed at "severe" stage of AD. MiR-26a-5p was significantly different in severe AD compared to mild AD; miR-6119-5p was significantly different in moderate and severe AD compared to mild AD; while miR-1246 and miR-660-5p were significantly different in severe AD compared to mild and moderate AD stages. In addition, miR-26a-5p, miR-181c-3p, miR-126-5p, miR-22-3p, and miR-148b-5p were positively correlated to MMSE score, while miR-106b-3p, miR-6119-5p, miR-1246, and miR-660-5p were negatively correlated to MMSE score. A large scale study by Jia et al. [56] showed a significant decrease in miR-29, miR-125b, and miR-223, and an increase in miR-519 in AD compared with NC. Serum miR-223 was positively correlated with MMSE scores and might be a biomarker of AD severity. MiR-223 was a fair test to distinguish AD from NC (AUC 0.786). Also the combination miR-223 and miR-125b gave improved sensitivity/specificity (AUC 0.879) than miR-223 or miR-125b alone.

4.2 MicroRNAs in Blood Plasma

The miRNAs shown to have altered expression in blood plasma of AD and MCI patients are summarized in Table 2.

Four studies had used both males and females but there was disparity in the proportions of males and females in the groups. Siedlecki-Wullch et al. [31] analyzed blood plasma from 56 AD patients, 25 MCI patients and 14 NC, but in others the group sizes were small. In all the studies, PCR was the method of analysis of miRNAs. There was limited overlap in the findings of the various studies. Where there was overlap, upregulation of miR-181-5p was observed in AD and MCI [31, 57], and was higher in AD than in MCI [57].

Siedlecki-Wullch et al. [31] found the expression levels of plasma miR-92a-3p. miR-181c-5p, and miR-210-3p in AD patients was significantly increased compared with NC. A significant increase in expression was also found in miR-181c-5p and miR-210-3p, and an increasing trend for miR-92a-3p, in MCI patients. AUC values for miR-92a-3p, miR-181c-5p, and miR-210-3p were 0.70, 0.77, and 0.80, respectively, when AD subjects were compared with NC. MiR-92a-3p had 47% sensitivity and 93% specificity; miR-181c-5p had 70% sensitivity and 86% specificity; miR-210-3p had 81% sensitivity and 71% specificity. Comparing MCI subjects with NC, AUC values for miR-92a-3p, miR-181c-5p, and miR-210-3p were 0.67, 0.84, and 0.74, respectively; 50%, 85%, and 77% sensitivity, respectively, and 93%, 86%, and 71% specificity, respectively. Using a combination of miR-92a-3p, miR-181c-5p and miR-210-3p gave an AUC value 0.855, 93% sensitivity, and 71% specificity for AD patients compared with NC. With this combination, an AUC value 0.893, sensitivity 85% and specificity 86% was obtained for distinguishing MCI subjects from NC. This combination therefore served as a good test to distinguish AD patients and MCI subjects from controls. A small scale study by Nagaraj et al. [25] found there were significant increases in expression levels of miR-486-5p, miR-483-5p, miR-502-3p, and miR-200a-3p in AD and MCI compared to NC. MiR-483-5p and miR-502-3p were good tests to distinguish AD from NC, and MCI from NC (AUC >0.9, specificity and sensitivity >80%). Zirnheld et al. [57] observed that miR-34c was more highly expressed in moderate–severe and mild AD than in MCI or NC, but with no significant difference between moderate–severe and mild AD or between MCI and NC. In addition, miR-181c was more abundant in moderate–severe and mild AD than MCI or NC, but no difference between moderate–severe and

Table 2
Dysregulated microRNAs in blood plasma of AD and MCI patients compared to controls

Reference	Subjects and age	Sample	Method of analysis	Increased expression of miRNAs	Decreased expression of miRNAs
Siedlecki-Wullich et al. [31]	AD, 15 M/41F, 77.8 years, global dementia scale 4.64, MMSE 16.1; MCI 10 M/15F, 72.0 years, 3.15, 26.9; NC, 7 M/7F, 68.3 years, 2.07, 29.2	Plasma	RT-qPCR		miR-92a-3p. −181c-5p, and −210-3p in AD. miR-181c-5p, −210-3p, and an increasing trend for miR-92a-3p in MCI
Nagaraj et al. [25]	AD, 10 M/10F, 69.7 years, MMSE 20.5, CSF Aβ42 390 pg/mL, t-tau 767 pg/mL; MCI, 7 M/8F, 65.1 years, MMSE 25.9, CSF Aβ42 627 pg/mL, t-tau 450 pg/mL; NC, 15, age-matched	Plasma	RT-qPCR		miR-486-5p, −483-5p, −502-3p and −200a-3p in AD and MCI
Zirnheld et al. [57]	Mild AD, 9 M/7F, 78.2 years, MMSE 23.0; moderate–severe AD, 6 M/14F, 77.7 years, MMSE 13.3; MCI, 19 M/15F, mean age 78.9 years, MoCA mean score 20.7; NC, 6 M/31F, 75.3 years, MMSE 28.7	Plasma	RT-qPCR	miR-34c in mild and moderate–severe AD higher than in MCI or NC, but no significant difference between mild and moderate–severe AD. miR-181c and miR-411 higher in mild and moderate–severe AD than MCI or NC, but no significant difference between mild and moderate–severe AD or between MCI and NC	
Kayano et al. [58]	MCI, 11 M/12F, 72.8 years, MMSE 24.3; NC, 12 M/18F, 70.4 years, MMSE 28.6	Plasma	RT-qPCR screening	Differential correlation analysis was applied to the data set with 85 miRNAs. The 20 pairs of miRNAs which had the difference of correlation coefficients [r₁–r₂] >0.8 were selected as biomarkers that distinguish MCI from NC	

MoCA Montreal cognitive assessment

mild AD or between MCI and NC. MiR-411 was more highly expressed in moderate–severe and mild AD than in MCI or NC, but no difference between moderate–severe and mild AD or between MCI and NC. MiR-34c was a fair test to differentiate moderate–severe AD from MCI (AUC 0.78), a good test to distinguish moderate–severe AD from NC (AUC 0.82), and a poor test to distinguish MCI from NC (AUC 0.68). Similarly, miR-181c differentiated moderate–severe AD from both MCI and NC (AUC 0.86, 0.79, respectively) but was a poor test to distinguish between mild AD and MCI or NC (AUC 0.63, 0.60, respectively). MiR-411 was a good test to distinguish moderate–severe AD from MCI and NC (AUC 0.92, 0.99, respectively, and sensitivity 77%, 77% respectively) and for distinguishing mild AD from MCI and NC (AUC 0.93, 0.98, respectively, and sensitivity 79%, 93%, respectively). Thus, miR-411 was a more powerful biomarker than miR-34c or miR-181c to distinguish MCI from mild, moderate, and severe AD. Kayano et al. [58] using differential corelation analysis identified 20 pairs of miRNAs that distinguished MCI from NC. Two pairs miR-191/miR-101 and miR-103/miR-222 had the highest AUC 0.962, and were good tests to distinguish MCI from NC. Other microRNA pairs that included miR-191, miR-125b, and miR-590-5p also had high AUC > 0.95. Differential correlation analysis detected much different and more sensitive biomarkers for MCI than t-test (mean AUC 0.800).

4.3 MicroRNAs in Cerebrospinal Fluid

The miRNAs shown to have altered expression in CSF of AD and MCI patients are summarized in Table 3.

There were nine studies and where reported all had recruited males and females in the study groups, but there was disparity in the proportions of males and females in the groups. Wiedrick et al. [59] had analysed miRNAs in CSF of 47 AD patients and 71 NC, but in most of the other studies the group sizes were relatively small. In all the studies except one PCR was the method of analysis. There was limited overlap in the findings of the various studies. Where there was overlap, upregulation of miR-30a-5p was shown in AD patients [54, 60] and downregulation of miR-16-5p in AD [22, 26] and miR-146a in AD [22, 61].

Wiedrick et al. [59] found increased expression in CSF samples of miR-193a-5p, miR-597-5p, miR-195-5p, miR-484, miR-584-5p, miR-331-3p in AD compared to NC. Combinations of 7 miRNAs (mean AUC 0.796) were incrementally better at identifying AD samples than combinations that consisted of fewer miRNAs. Those combinations that contained as few as four miRNAs had a mean AUC > 0.72. Some combinations of six and seven miRNAs had AUC > 0.80. The highest ranked miRNAs tended to be, but were not always, the most important contributors to the multi-marker models. Higher ranked miRNAs such as miR-193a-5p had expression levels that correlated more strongly with MMSE scores

Table 3
Dysregulated microRNAs in cerebrospinal fluid of AD and MCI patients compared to controls

Reference	Subjects and age	Sample	Method of analysis	Increased expression of miRNAs	Decreased expression of miRNAs
Wiedrick et al. [59]	AD, 26 M/21F, mean age 73.1 years, MMSE 22.1, APOE ε4 (positive, %) 64.4, CSF Aβ42 306 pg/mL, t-tau 689 pg/mL; NC, 24 M/47F, 72.7 years, MMSE 29.1, APOE ε4 37.9, CSF Aβ42 479 pg/mL, t-tau 392 pg/mL	CSF centrifuged	qPCR	miR-193a-5p, −597-5p, −195-5p, −484, −584-5p in AD	
Jain et al. [60]	AD, 42; MCI, 17; HC + ONC, 82	CSF exosome-enriched	Next-generation sequencing for screening	miR-27a-3p, −30a-5p, −34c-3p in AD	
McKeever et al. [26]	EOAD, 7 M/10F, mean age of onset 56.8 ± 4.9 years, duration of disease 3.9 ± 2.3 years, CSF obtained at mean age 60.9 ± 4.6 years; LOAD, 8 M/5F, CSF obtained at mean age 75.5 ± 4.6 years; NC, 5 M/7F, CSF obtained at mean age 66.5 ± 7.7 years	Exosomes isolated from CSF	RT-qPCR	miR-125b-5p in both EOAD and LOAD	miR-16-5p in EOAD but not LOAD. miR-451a and −605-5p in both EOAD and LOAD
Liu et al. [62]	AD, 8 M/9F, mean age 74.1 years, APOE ε4 (positive, %) 76.5; MCI, 19 M/17F, 72.4 years, 55.6; SMC, 21 M/20F, 71.6 years, 17.1	CSF centrifuged, cell pellet sorted into lymphocyte populations by flow cytometry	RT-PCR screening	Let-7b higher in AD and MCI compared to SMC. Let-7b higher in CD4+ lymphocyte population from AD and MCI than SMC, while let-7b expression in CD8+ T lymphocytes, B lymphocytes and NK cells did not significantly differ	

Reference	Demographics	Sample	Method		
Denk et al. [54]	AD, 48, mean age 65 years, MMSE 21; NC, 44, 64 years, NA	CSF	RT-qPCR	miR-24-3p, −99b-5p, −124-3p, −125a-5p, −223-3p, −140-3p, −30a-5p, −22-3p in AD	miR-15a-5p in AD
Riancho et al. [63]	AD, 6 M/22F, mean age 67.7 years; NC, 10 M/18F, 70.2 years	CSF, raw and exosome enriched	RT-qPCR	miR-9-5p and miR-598 detected in 70% and 80% of AD exosomes compared to 30% and 50% of NC exosomes, respectively	miR-9-5p and miR-598 not detected in any CSF samples of AD, whereas present in 50% and 72% of NC samples, respectively
Lusardi et al. [22]	AD, 30 M/20F, 69.5 years, MMSE 18.3; NC, 23 M/26F, 67.8 years, MMSE 29.2	CSF	RT-qPCR	miR-378a-3p, −1291, −597-5p, −520b, −603, −202-3p, −519b-3p, −484 in AD	miR-143-3p, −142-3p, −328-3p, −193a-5p, −30a-3p, −19b-3p, −30d-5p, −340-5p, −140-5p, −125b-5p, −26b-5p, −16-5p, −146a-5p, −29a-3p, −195-5p, −15b-5p, −223-3p, −584-5p, −145-3p, −24-3p, −532-5p, −28-3p, −146b-5p, −27b-3p, −331-3p, −145-5p, 590-5p, −365a-3p in AD
Müller et al. [61]	AD, 2 M6/31F, 72.9 years, MMSE 20.0; MCI 15 M/22F, 73.1 years, MMSE 24.8; NC, 21 M/19F, 61.4 years, MMSE 27.2	CSF	qPCR screening		miR-146a
Müller et al. [64]	AD, 8 M/10F, 70.4 years, MMSE 19.7; NC, 8 M/12F, 69.2 years	CSF	qPCR screening	miR-29a in AD	

HC healthy controls, ONC other neurological disease controls, EOAD early-onset AD, LOAD late-onset AD, SMC subjective memory complaints

than lower ranked miRNAs such as miR-331-3p, suggesting that these miRNAs might also be able to signal disease progression in AD patients. Jain et al. [60] using next-generation sequencing to analyze exosome-enriched CSF samples showed that miR-27a-3p, miR-30a-5p, and miR-34c-3p were increased in AD patients. In addition, piwi-interacting RNA (piR)-019949 and piR-020364 were increased while piR-019324 was decreased in AD patients. The miRNA-piRNA signature was able to distinguish AD dementia patients from controls with an AUC of 0.83 that was similar to that of p-tau and Aβ42/40 ratio. Furthermore, CSF exosomes were obtained from individuals that were diagnosed with MCI and subjected to CSF collection between 2003 and 2004. Ten years later, 6 individuals had progressed to AD dementia, while 11 patients had developed stable MCI. When the piRNA signature was combined with measures of p-tau and Aβ42/40 ratio, it was able to predict conversion from MCI to AD dementia with an AUC 0.96. In another study of exosomes isolated from CSF, McKeever et al. [26] found a significant decrease in expression of miR-16-5p in early-onset AD but not late-onset AD, a significant increase in miR-125b-5p in both early-onset AD and late-onset AD, a significant robust decrease in miR-451a in both early-onset AD and late-onset AD, and a significant decrease in miR-605-5p in early-onset AD and late-onset AD compared to NC. There was a significant decrease in miR-16-5p in early-onset AD compared to late-onset AD, but no significant difference for miR-125-5p, miR-451a, and miR-605-5p. For distinguishing early-onset AD from NC, miR-16-5p gave AUC 0.760, miR-125b-5p AUC 0.723, miR-605-5p AUC 0.706, and miR-451a AUC 0.951 Moreover, for distinguishing late-onset AD from NC, miR-125b-5p gave AUC 0.785, miR-605-5p AUC 0.765, and for miR-451a AUC 0.847. Combining all four miRNAs resulted in an AUC 0.976 in distinguishing early-onset AD from NC. Combining the two best performing miRNAs at distinguishing early-onset AD from NC, miR-451a and miR-16-5p, gave AUC 0.946. For late-onset AD versus NC, combining the three miRNAs miR-125b-5p, miR-451a, miR-605-5p gave AUC 0.847. Liu et al. [62] collected CSF from AD, MCI and SMC (subjective memory complaints) patients which was centrifuged and cell pellet sorted into lymphocyte populations by flow cytometry. The relative expression of let-7b in CSF and cells was significantly increased in AD and MCI compared with SMC. Total number of CSF lymphocytes and ratio of CD4$^+$ lymphocytes in AD and MCI were significantly higher than in SMC. Let-7b expression in CD4$^+$ lymphocytes from AD and MCI was significantly higher than SMC, while let-7b expression in CD8$^+$ T lymphocytes, B lymphocytes and NK cells was not significantly different. Adding let-7b to Aβ40 and Aβ42 as predictive parameters increased the probability of predicting AD to 89.7% with AUC 0.93. Denk et al. [54] showed the expression level in AD of miR-24-3p, miR-99b-5p, miR-124-3p, miR-125a-5p, miR-223-3p, miR-140-3p, miR-30a-5p, miR-22-3p was significantly increased, while that of miR-15a-5p was significantly decreased, compared to NC. MiR-125a-5p had AUC 0.84 and 74% sensitivity and 82% specificity in discriminating AD cases from NC. In a validation phase study with CSF samples from AD and NC patients, Riancho et al. [63] found that miR-9-5p was not detected in any of AD samples but was present in 9 (50%) of NC samples. Similarly, miR-598 was absent in all AD samples but present in 13 (72%) of NC samples. MiR-134 was detected in none of AD samples but present in 2 (11%) of NC samples. An opposite pattern was observed in exosome-enriched CSF samples. MiR-9-5p was detected in 7 (70%) of AD and 3 (30%) of NC samples. MiR-598 was detected in 8 (80%) of AD and 5 (50%) of NC samples. MiR-134 was found in 2 (20%) of AD and 2 (20%) of NC samples. The tendency of increased miRNA expression in exosome-enriched CSF samples from AD patients compared with NC was not statistically significant. Lusardi et al. [22] found that replicated microRNAs increased in AD were miR-378a-3p, miR-1291, miR-597-5p, and those decreased in AD were miR-143-3p, miR-142-3p, miR-328-3p, miR-193a-5p, miR-30a-3p, miR-19b-3p, miR-30d-5p, miR-340-5p, miR-140-5p, miR-125b-5p, miR-26b-5p, miR-16-5p, miR-146a-5p, miR-29a-3p, miR-195-5p, miR-15b-5p, miR-223-3p. Candidate miRNAs not retested increased in AD were miR-520b, miR-603, miR-202-3p, miR-519b-3p, miR-484, while those decreased in AD were miR-584-5p, miR-145-3p, miR-24-3p, miR-532-5p, miR-28-3p, miR-146b-5p, miR-27b-3p, miR-331-3p, miR-145-5p, miR-590-5p, miR-365a-3p. The top performing linear combinations of 3 or 4 microRNAs were a fair test to distinguish AD from NC (AUC 0.80–0.82). Müller et al. [62] used large group sizes of AD, MCI and NC individuals to perform qPCR screening of microRNAs in CSF samples collected at three different centers. Importantly, no differences in miRNAs levels were found between AD, MCI and NC after correcting for confounding factors that included age, gender, sample storage time, centrifugation status. In a separate study, Müller et al. [64] showed using CSF samples from AD and NC subjects with low concentrations of erythrocytes and leukocytes that miR-29a expression was increased in AD compared with NC and differentiated AD from NC with a sensitivity of 78% and specificity 60%.

5 Conclusion

The data presented in these studies using blood serum, blood plasma and CSF indicate that individual or combinations of miRNAs can serve as important biomarkers in distinguishing AD and

Table 4
Dysregulated microRNAs of AD and MCI patients compared to controls

Sample analyzed	Increased expression of miRNAs	Decreased expression of miRNAs
Blood serum	AD patients: miR-103a-3p, −142-3p, −20a-5p, −29b-3p, −7b-5p, −7 g-5p, −106a-5p, 106b-5p, −18b-5p, 223-3p, −26a-5p, −26b-5p, −301a-3p, −30b-5p, −146a-5p, −106b-3p, −195-5p, −20b-5p, −497-5p, −455-3p, −6119-5p, −1246, −660-5p, −519 MCI patients: miR-4668-5p	AD patients: miR-132-3p, −146a-5p, −15a-5p, −22-3p, −320a, −320b, −92a-3p, −1246, −125b-3p, −29c-3p, −93-5p, −19b-3p, −501-3p, −26a-5p, −181c-3p, −126-5p, −148b-5p, −144-5p, −125b, −223
Blood plasma	AD patients: miR-: −92a-3p, −181c-5p, −210-3p, −486-5p, −483-5p, −502-3p, −200a-3p, −34c, −411 MCI patients: miR-181c-5p, −210-3p, −486-5p, −483-5p, −502-3p, −200a-3p	
CSF	AD patients: miR-193a-5p, −597-5p, −195-5p, −484, −584-5p, −24-3p, −99b-5p, −124-3p, −125a-5p, −223-3p, −140-3p, −30a-5p, −22-3p, −378a-3p, −1291, −520b, −603, −202-3p, −519b-3p, let-7b MCI patients: Let-7b	AD patients: miR-15a-5p, −143-3p, −142-3p, −328-3p, -193a-5p, −30a-3p, −19b-3p, −30d-5p, −340-5p, −140-5p, −125b-5p, −26b-5p, −16-5p, −146a-5p, −29a-3p, −195-5p, −15b-5p, −223-3p, −584-5p, −145-3p, −24-3p, −532-5p, −28-3p, −146b-5p, −27b-3p, −331-3p, −145-5p, −590-5p, −365a-3p, −146a

This table does not include miRNA expression measured in exosomes from serum and CSF, exosome-enriched CSF, or CSF lymphocytes

MCI patients from controls. Appropriate miRNAs can differentiate between AD and MCI, and between mild and moderate–severe AD, and predict the progression of MCI to AD. They could potentially replace or be combined with the molecular markers of AD currently used in clinical practice. The miRNAs found to be dysregulated in AD are summarized in Table 4.

These findings need to be replicated in large-scale studies in other research centers. The information indicates that miR-519 (up), miR-15a-5p, miR-19b-3p, miR-125b (down) are the only miRNAs dysregulated in the same direction in blood serum and CSF, otherwise the microRNA patterns for blood serum and CSF are different which has also been reported by Denk et al. [54]. The miRNA patterns of blood plasma and CSF were dissimilar, although the miRNA profiling of blood plasma is less extensive. Also there are several inconsistencies with certain miRNAs having been found to have increased and decreased expression in the same biofluid type. Importantly using differential correlation analysis, Kayano et al. [58] identified pairs of microRNAs in blood plasma that distinguish MCI from NC. Differential correlation analysis detected different and more sensitive MCI biomarkers compared to t-test. Also the highest AUC value of any four microRNAs was less than the highest in the two-pair approach. An issue with analyzing CSF samples is that the expression levels of miRNAs may be different in centrifuged versus noncentrifuged samples [61]. Although not included in this chapter, specific miRNAs could be important therapeutic targets in experimental animal studies of AD and MCI and in clinical trials with miRNA mimics or inhibitors.

References

1. Brookmeyer R, Johnson E, Ziegler-Graham K, Arrighi HM (2007) Forecasting the global burden of Alzheimer's disease. Alzheimers Dement 3(3):186–191. https://doi.org/10.1016/j.jalz.2007.04.381

2. Querfurth HW, LaFerla FM (2010) Alzheimer's disease. N Engl J Med 362(4):329–344. https://doi.org/10.1056/NEJMra0909142

3. The Alzheimer's Association (2011) 2011 Alzheimer's disease facts and figures. Alzheimers Dement 7(2):208–244. https://doi.org/10.1016/j.jalz.2011.02.004

4. McKhann G, Drachman D, Folstein M, Katzman R, Price D, Stadlan EM (1984) Clinical diagnosis of Alzheimer's disease. Neurology 34(7):939–944. https://doi.org/10.1212/WNL.34.7.939

5. McKhann GM, Knopman DS, Chertkow H, Hyman BT, Jack CR Jr, Kawas CH, Klunk WE, Koroshetz WJ, Manly JJ, Mayeux R, Mohs RC, Morris JC, Rossor MN, Scheltens P, Carrillo MC, Thies B, Weintraub S, Phelps CH (2011) The diagnosis of dementia due to Alzheimer's disease: recommendations from the National Institute on Aging-Alzheimer's association workgroups on diagnostic guidelines for Alzheimer's disease. Alzheimers Dement 7(3):263–269. https://doi.org/10.1016/j.jalz.2011.03.005

6. Hu CJ, Octave JN (2019) Editorial: risk factors and outcome predicating biomarker of neurodegenerative diseases. Front Neurol 10:45. https://doi.org/10.3389/fneur.2019.00045

7. Bali J, Gheinani AH, Zurbriggen S, Rajendran L (2012) Role of genes linked to sporadic Alzheimer's disease risk in the production of β-amyloid peptides. Proc Natl Acad Sci U S A 109(38):15307–15311. https://doi.org/10.1073/pnas.1201632109

8. Piaceri I, Nacmias B, Sorbi S (2013) Genetics of familial and sporadic Alzheimer's disease. Front Biosci E5:167–177. https://doi.org/10.2741/e605

9. Dorszewska J, Prendecki M, Oczkowska A, Dezor M, Kozubski W (2016) Molecular basis of familial and sporadic Alzheimer's disease. Curr Alzheimer Res 13(9):952–963. https://doi.org/10.2174/1567205013666160314150501

10. Barber RC (2012) The genetics of Alzheimer's disease. Scientifica 2012:246210. https://doi.org/10.6064/2012/246210

11. Reddy PH, Tonk S, Kumar S, Vijayan M, Kandimalla R, Kuruva CS, Reddy AP (2017) A critical evaluation of neuroprotective and neurodegenerative microRNAs in Alzheimer's disease. Biochem Biophys Res Commun 483(4):1156–1165. https://doi.org/10.1016/j.bbrc.2016.08.067

12. Morris JC, Storandt M, Miller JP, McKeel DW, Price JL, Rubin EH, Berg L (2001) Mild cognitive impairment represents early-stage Alzheimer disease. Arch Neurol 58(3):397–405. https://doi.org/10.1001/archneur.58.3.397

13. Morris JC, Cummings J (2005) Mild cognitive impairment (MCI) represents early-stage Alzheimer's disease. J Alzheimers Dis 7(3):235–239. https://doi.org/10.3233/jad-2005-7306

14. Garcia-Ptacek S, Eriksdotter M, Jelic V, Porta-Etessam J, Kåreholt I, Manzano Palomo S (2016) Subjective cognitive impairment: towards early identification of Alzheimer disease. Neurologia 31(8):562–571. https://doi.org/10.1016/j.nrl.2013.02.007

15. Sewell MC et al (2010) Neuropsychology in the diagnosis and treatment of dementia. In: Fillit HM, Rockwood K, Woodhouse K (eds) Brocklehurst's textbook of geriatric medicine and gerontology, 7th edn. Elsevier, Amsterdam, pp 402–410

16. Mattsson N, Zetterberg H, Hansson O, Andreasen N, Parnetti L, Jonsson M, Herukka SK, van der Flier WM, Blankenstein MA, Ewers M, Rich K, Kaiser E, Verbeek M, Tsolaki M, Mulugeta E, Rosén E, Aarsland D, Visser PJ, Schröder J, Marcusson J, de Leon M, Hampel H, Scheltens P, Pirttilä T, Wallin A, Jönhagen ME, Minthon L, Winblad B, Blennow K (2009) CSF biomarkers and incipient Alzheimer disease in patients with mild cognitive impairment. JAMA 302(4):385–393. https://doi.org/10.1001/jama.2009.1064

17. Vlassenko AG, Benzinger TL, Morris JC (2012) PET amyloid-beta imaging in preclinical Alzheimer's disease. Biochim Biophys Acta 1822(3):370–379. https://doi.org/10.1016/j.bbadis.2011.11.005

18. Braak H, Alafuzoff I, Arzberger T, Kretzschmar H, Del Tredici K (2006) Staging of Alzheimer disease-associated neurofibrillary pathology using paraffin sections and immunocytochemistry. Acta Neuropathol 112 (4):389–404. https://doi.org/10.1007/s00401-006-0127-z

19. Mintun MA, Larossa GN, Sheline YI, Dence CS, Lee SY, Mach RH, Klunk WE, Mathis CA, DeKosky ST, Morris JC (2006) [^{11}C]PIB in a nondemented population: potential antecedent marker of Alzheimer disease. Neurology 67(3):446–452. https://doi.org/10.1212/01.wnl.0000228230.26044.a4

20. Villemagne VL, Burnham S, Bourgeat P, Brown B, Ellis KA, Salvado O, Szoeke C, Macaulay SL, Martins R, Maruff P, Ames D, Rowe CC, Masters CL (2013) Amyloid β deposition, neurodegeneration, and cognitive decline in sporadic Alzheimer's disease: a prospective cohort study. Lancet Neurol 12 (4):357–367. https://doi.org/10.1016/S1474-4422(13)70044-9

21. Schneider P, Hampel H, Buerger K (2009) Biological marker candidates of Alzheimer's disease in blood, plasma, and serum. CNS Neurosci Ther 15(4):358–374. https://doi.org/10.1111/j.1755-5949.2009.00104.x

22. Lusardi TA, Phillips JI, Wiedrick JT, Harrington CA, Lind B, Lapidus JA, Quinn JF, Saugstad JA (2017) MicroRNAs in human cerebrospinal fluid as biomarkers for Alzheimer's disease. J Alzheimers Dis 55 (3):1223–1233. https://doi.org/10.3233/JAD-160835

23. Holzman DM (2011) CSF biomarkers for Alzheimer's disease: current utility and potential future use. Neurobiol Aging 32(Suppl 1):S4–S9. https://doi.org/10.1016/j.neurobiolaging.2011.09.003

24. Fagan AM, Perrin RJ (2012) Upcoming candidate cerebrospinal fluid biomarkers of Alzheimer's disease. Biomark Med 6(4):455–476. https://doi.org/10.2217/bmm.12.42

25. Nagaraj S, Laskowska-Kaszub K, Dębski KJ, Wojsiat J, Dąbrowski M, Gabryelewicz T, Kuźnicki J, Wojda U (2017) Profile of 6 microRNA in blood plasma distinguish early stage Alzheimer's disease patients from non-demented subjects. Oncotarget 8 (10):16122–16143. https://doi.org/10.18632/oncotarget.15109

26. McKeever PM, Schneider R, Taghdiri F, Weichert A, Multani N, Brown RA, Boxer AL, Karydas A, Miller B, Robertson J, Tartaglia MC (2018) MicroRNA expression levels are altered in the cerebrospinal fluid of patients with young-onset Alzheimer's disease. Mol Neurobiol 55(12):8826–8841. https://doi.org/10.1007/s12035-018-1032-x

27. Koran ME, Wagener M, Hohman TJ (2017) Sex differences in the association between AD biomarkers and cognitive decline. Brain Imaging Behav 11(1):205–213. https://doi.org/10.1007/s11682-016-9523-8

28. Morris JC, Schindler SE, McCue LM, Moulder KL, Benzinger TLS, Cruchaga C, Fagan AM, Grant E, Gordon BA, Holtzman DM, Xiong C (2019) Assessment of racial disparities in biomarkers for Alzheimer disease. JAMA Neurol 76(3):264–273. https://doi.org/10.1001/jamaneurol.2018.4249

29. Aasebø E, Opsahl JA, Bjørlykke Y, Myhr KM, Kroksveen AC, Berven FS (2014) Effects of blood contamination and the rostro-caudal gradient on the human cerebrospinal fluid proteome. PLoS One 9(3):e90429. https://doi.org/10.1371/journal.pone.0090429

30. Bjerke M, Portelius E, Minthon L, Wallin A, Anckarsäter H, Anckarsäter R, Andreasen N, Zetterberg H, Andreasson U, Blennow K (2010) Confounding factors influencing amyloid Beta concentration in cerebrospinal fluid. Int J Alzheimers Dis 2010:986310. https://doi.org/10.4061/2010/986310

31. Siedlecki-Wullich D, Català-Solsona J, Fábregas C, Hernández I, Clarimon J, Lleó A, Boada M, Saura CA, Rodríguez-Alvarez J, Miñano-Molina AJ (2019) Altered microRNAs related to synaptic function as potential plasma biomarkers for Alzheimer's disease. Alzheimers Res Ther 11(1):46. https://doi.org/10.1186/s13195-019-0501-4

32. Wilson A, Sweeney M, Lynch PL, O'Kane MJ (2018) Hemolysis rates in whole blood samples for blood gas/electrolyte analysis by point-of-care testing. J Appl Lab Med 3(1):144–145. https://doi.org/10.1373/jalm.2018.026427

33. Galasko D (1998) An integrated approach to the management of Alzheimer's disease: assessing cognition, function and behavior. Eur J Neurol 5(S4):S9–S17. https://doi.org/10.1111/j.1468-1331.1998.tb00444.x

34. Nasreddine ZS, Phillips NA, Bédirian V, Charbonneau S, Whitehead V, Collin I, Cummings JL, Chertkow H (2005) The Montreal cognitive assessment, MoCA: a brief screening tool for mild cognitive impairment. J Am Geriatr Soc 53(4):695–699. https://doi.org/10.1111/j.1532-5415.2005.53221.x

35. Braak H, Braak E (1991) Neuropathological stageing of Alzheimer-related changes. Acta Neuropathol 82(4):239–259. https://doi.org/10.1007/bf00308809

36. Guo H, Ingolia NT, Weissman JS, Bartel DP (2010) Mammalian microRNAs predominantly act to decrease target mRNA levels. Nature 466:835–840. https://doi.org/10.1038/nature09267

37. Ha M, Kim VN (2014) Regulation of microRNA biogenesis. Nat Rev Mol Cell Biol 15 (8):509–524. https://doi.org/10.1038/nrm3838

38. Peplow PV, Martinez B, Calin GA, Esquela-Kerscher A (eds) (2019) MicroRNAs in diseases and disorders. Royal Society of Chemistry, London. ISSN:2041-3203, ISSN:2041–3211

39. Wahid F, Shehzad A, Khan T, Kim YY (2010) MicroRNAs: synthesis, mechanism, function, and recent clinical trials. Biochim Biophys Acta 1803(11):1231–1243. https://doi.org/10.1016/j.bbamcr.2010.06.013

40. Ye Y, Xu H, Su X, He X (2016) Role of microRNA in governing synaptic plasticity. Neural Plast 2019:4959523. https://doi.org/10.1155/2016/4959523

41. Mitchell PS, Parkin RK, Kroh EM, Fritz BR, Wyman SK, Pogosova-Agadjanyan EL, Peterson A, Noteboom J, O'Briant KC, Allen A, Lin DW, Urban N, Drescher CW, Knudsen BS, Stirewalt DL, Gentleman R, Vessella RL, Nelson PS, Martin DB, Tewari M (2008) Circulating microRNAs as stable blood-based biomarkers for cancer detection. Proc Natl Acad Sci U S A 105 (30):10513–10518. https://doi.org/10.1073/pnas.0804549105

42. Stary CM, Bell JD, Cho JE, Giffard RG (2018) Identification of microRNAs as targets for treatment of ischemic stroke. In: Peplow PV, Dambinova SA, Gennarelli TA, Martinez B (eds) Acute brain impairment. Royal Society of Chemistry, London, pp 105–127. ISBN-13: 978-1782629504, ISBN-10: 1782629505

43. Chen X, Ba Y, Ma L, Cai X, Yin Y, Wang K, Guo J, Zhang Y, Chen J, Guo X, Li Q, Li X, Wang W, Zhang Y, Wang J, Jiang X, Xiang Y, Xu C, Zheng P, Zhang J, Li R, Zhang H, Shang X, Gong T, Ning G, Wang J, Zen K, Zhang J, Zhang CY (2008) Characterization of microRNAs in serum: a novel class of biomarkers for diagnosis of cancer and other diseases. Cell Res 18(10):997–1006. https://doi.org/10.1038/cr.2008.282

44. Ho AS, Huang X, Cao H, Christman-Skieller C, Bennewith K, Le QT, Koong AC (2010) Circulating miR-210 as a novel hypoxia marker in pancreatic cancer. Transl Oncol 3 (2):109–113. https://doi.org/10.1593/tlo.09256

45. Kim HJ, Park KW, Kim TE, Im JY, Shin HS, Kim S, Lee DH, Ye BS, Kim JH, Kim EJ, Park KH, Han HJ, Jeong JH, Choi SH, Park SA (2015) Elevation of the plasma Aβ40/Aβ42 ratio as a diagnostic marker of sporadic early-onset Alzheimer's disease. J Alzheimers Dis 48 (4):1043–1050. https://doi.org/10.3233/JAD-143018

46. Mendes-Silva AP, Pereira KS, Tolentino-Araujo GT, Nicolau Ede S, Silva-Ferreira CM, Teixeira AL, Diniz BS (2016) Shared biologic pathways between Alzheimer's disease and major depression: a systematic review of microRNA expression studies. Am J Geriatr Psychiatry 24 (10):903–912. https://doi.org/10.1016/j.jagp.2016.07.017

47. Van Giau V, An SS (2016) Emergence of exosomal miRNAs as a diagnostic biomarker for Alzheimer's disease. J Neurol Sci 360:141–152. https://doi.org/10.1016/j.jns.2015.12.005

48. Wu HZ, Ong KL, Seeher K, Armstrong NJ, Thalamuthu A, Brodaty H, Sachdev P, Mather K (2016) Circulating microRNAs as biomarkers of Alzheimer's disease: a systematic review. J Alzheimers Dis 49(3):755–766. https://doi.org/10.3233/JAD-150619

49. Martinez B, Peplow PV (2019) MicroRNAs as diagnostic and therapeutic tools for Alzheimer's disease: advances and limitations. Neural Regen Res 14(2):242–255. https://doi.org/10.4103/1673-5374.244784

50. Yang TT, Liu CG, Gao SC, Zhang Y, Wang PC (2018) The serum exosome derived microRNA-135a, −193b, and −384 were potential Alzheimer's disease biomarkers.

Biomed Environ Sci 31(2):87–96. https://doi.org/10.3967/bes2018.011

51. Kumar S, Vijayan M, Reddy PH (2017) MicroRNA-455-3p as a potential peripheral biomarker for Alzheimer's disease. Hum Mol Genet 26(19):3808–3822. https://doi.org/10.1093/hmg/ddx267

52. Wu Y, Xu J, Xu J, Cheng J, Jiao D, Zhou C, Dai Y, Chen Q (2017) Lower serum levels of miR-29c-3p and miR-19b-3p as biomarkers for Alzheimer's disease. Toihoku J Exp Med 242(2):129–136. https://doi.org/10.1620/tjem.242.129

53. Guo R, Fan G, Zhang J, Wu C, Du Y, Ye H, Li Z, Wang L, Zhang Z, Zhang L, Zhao Y, Lu Z (2017) A 9-microRNA signature in serum serves as a noninvasive biomarker in early diagnosis of Alzheimer's disease. J Alzheimers Dis 60(4):1365–1377. https://doi.org/10.3233/JAD-170343

54. Denk J, Oberhauser F, Kornhuber J, Wiltfang J, Fassbender K, Schroeter ML, Volk AE, Diehl-Schmid J, Prudlo J, Danek A, Landwehrmeyer B, Lauer M, Otto M, Jahn H (2018) Specific serum and CSF microRNA profiles distinguish sporadic behavioural variant of frontotemporal dementia compare with Alzheimers patients and cognitively healthy controls. PLoS One 13(5):e0197329. https://doi.org/10.1371/journal.pone.0197329

55. Hara N, Kikuchi M, Miyashita A, Hatsuta H, Saito Y, Kasuga K, Murayama S, Ikeuchi T, Kuwano R (2017) Serum microRNA miR-501-3p as a potential biomarker related to the progression of Alzheimer's disease. Acta Neuropathol Commun 5(1):10. https://doi.org/10.1186/s40478-017-0414-z

56. Jia LH, Liu YN (2016) Downregulated serum miR-223 serves as biomarker in Alzheimer's disease. Cell Biochem Funct 34(4):233–237. https://doi.org/10.1002/cbf.3184

57. Zirnheld AL, Shetty V, Chertkow H, Schipper HM, Wang E (2016) Distinguishing mild cognitive impairment from Alzheimer's disease by increased expression of key circulating microRNAs. Curr Neurobiol 7(2):38–50

58. Kayano M, Higaki S, Satoh JI, Matsumoto K, Matsubara E, Takikawa O, Niida S (2016) Plasma microRNA biomarker detection for mild cognitive impairment using differential correlation analysis. Biomark Res 4:22. https://doi.org/10.1186/s40364-016-0076-1

59. Wiedrick JT, Phillips JI, Lusardi TA, McFarland TJ, Lind B, Sandau US, Harrington CA, Lapidus JA, Galasko DR, Quinn JF, Saugstad JA (2019) Validation of microRNA biomarkers for Alzheimer's disease in human cerebrospinal fluid. J Alzheimers Dis 67(3):875–891. https://doi.org/10.3233/JAD-180539

60. Jain G, Stuendl A, Rao P, Berulava T, Pena Centeno T, Kaurani L, Burkhardt S, Delalle I, Kornhuber J, Hüll M, Maier W, Peters O, Esselmann H, Schulte C, Deuschle C, Synofzik M, Wiltfang J, Mollenhauer B, Maetzler W, Schneider A, Fischer A (2019) A combined miRNA-piRNA signature to detect Alzheimer's disease. Transl Psychiatry 9(1):250. https://doi.org/10.1038/s41398-019-0579-2

61. Müller M, Kuiperij HB, Versleijen AA, Chiasserini D, Farotti L, Baschieri F, Parnetti L, Struyfs H, De Roeck N, Luyckx J, Engelborghs S, Claassen JA, Verbeek MM (2016) Validation of microRNAs in cerebrospinal fluid as biomarkers for different forms of dementia in a multicenter study. J Alzheimers Dis 52(4):1321–1333. https://doi.org/10.3233/JAD-160038

62. Liu Y, He X, Li Y, Wang T (2018) Cerebrospinal fluid CD4+ T lymphocyte-derived miRNA-let-7b can enhances the diagnostic performance of Alzheimer's disease biomarkers. Biochem Biophys Res Commun 495(1):1144–1150. https://doi.org/10.1016/j.bbrc.2017.11.122

63. Riancho J, Vázquez-Higuera JL, Pozueta A, Lage C, Kazimierczak M, Bravo M, Calero M, Gonalezález A, Rodríguez E, Lleó A, Sánchez-Juan P (2017) MicroRNA profiles in patients with Alzheimer's disease: analysis of miR-9-5p and miR-598 in raw and exosome enriched cerebrospinal fluid samples. J Alzheimers Dis 57(2):483–491. https://doi.org/10.3233/JAD-161179

64. Müller M, Jäkel L, Bruinsma IB, Claassen JA, Kuiperij HB, Verbeek MM (2016) MicroRNA-29a is a candidate biomarker for Alzheimer's disease in cell-free cerebrospinal fluid. Mol Neurobiol 53(5):2894–2899. https://doi.org/10.1007/s12035-015-9156-8

65. Coughlan G, Laczó J, Hort J, Minihane AM, Hornberger M (2018) Spatial navigation deficits - overlooked cognitive marker for preclinical Alzheimer disease? Nat Rev Neurol 14:496–506

Chapter 13

Tau Biomarkers for Long-Term Effects of Neurotrauma: Technology Versus the Null Hypothesis

Rudy J. Castellani

Abstract

The broadly accepted concept that neurotrauma causes progressive neurodegenerative proteinopathy originated in boxers of the late nineteenth and early twentieth centuries. The protracted and extensive neurotrauma in this era was unprecedented and led to an enigmatic condition known as punch drunk. As medical science struggled to understand the precise nature of the neurological deficits in punch drunk boxers, a limited number of neuropathological studies identified, among other things, neurofibrillary pathology, now understood as an accumulation of phosphorylated tau. The apparent overlap with Alzheimer's disease pathology led to the hypothesis that punch drunk syndrome was neurodegenerative in nature. A careful appraisal of the historic case material, however, argues against a neurodegenerative disease and in favor of a largely stationary condition caused by traumatic brain injury. Moreover, a number of large-scale epidemiological studies have failed to identify a causal link between neurotrauma and neurodegenerative disease, and in aggregate argue against such a link. The neurotrauma–neurodegenerative proteinopathy hypothesis continues to be pursued nevertheless with sophisticated technology, including highly sensitive biomarker analyses for various tau protein species or tau surrogates. CSF and plasma studies for phosphorylated tau and total tau have so far shown marginal differences between neurotrauma exposure and control, with unclear clinical significance. PET scan data looking at putative tau ligands have also shown differences between neurotrauma and control in cross sectional designs, again with unclear clinical significance. While studies will continue and more data will likely be assembled, it may be important to consider that the null hypothesis, that is, that neurotrauma does not lead to neurodegenerative proteinopathy, remains applicable. This may help guide the interpretation of biomarker studies and provide avenues for novel studies, including therapeutic constructs unrelated to tau.

Keywords Tau protein, Phosphorylated tau, Punch Drunk syndrome, PET scan, CSF

1 Introduction: The Origin of the Neurotrauma–Neurodegenerative Disease Narrative

In order to critically examine the current hypothesis that neurotrauma is linked to neurodegenerative disease through biomarkers, it is important to look at the origin of the hypothesis in early twentieth century boxers. In doing so, considerable doubt

Philip V. Peplow, Bridget Martinez and Thomas A. Gennarelli (eds.), *Neurodegenerative Diseases Biomarkers: Towards Translating Research to Clinical Practice*, Neuromethods, vol. 173, https://doi.org/10.1007/978-1-0716-1712-0_13,
© Springer Science+Business Media, LLC, part of Springer Nature 2022

emerges, despite the attractiveness of the hypothesis and the many avenues of study that the hypothesis generates.

Boxing arose from the bare knuckle era from the late nineteenth century and made its way into the mainstream, due mainly to rules of engagement that made fighting palatable for a broader segment of society. A turning point in the United States was the fight between John L. Sullivan and "Gentleman Jim" Corbett [1]. This was a widely promoted fight that circumvented existing law by staging the fight in a private club. It was governed by the so-called Marquess of Queensbury rules (use of padded gloves, 3-min rounds with 1 min of rest, no wrestling, 10 s for a knockout, etc.), and was the first heavyweight championship fight to do so. The fight was held in New Orleans but was followed as far away as New York, San Francisco, Boston, and other major cities, and brought in substantial tourism dollars to the city and state. Other cities would soon adopt the New Orleans model and more widespread legalization of boxing followed, thus creating the "model experiment" [2] in neurotrauma by virtue of widespread and largely unregulated fighting.

In the early twentieth century, boxing had essentially no medical oversight. There was little interest in matching evenly skilled athletes, no inclination to stop fights until one of the participants was beaten into unconsciousness, and no mandatory exclusion times. Regulations in various states and municipalities differed, such that 45-round fights were sometimes promoted to produce a "fight to the finish," which was sought by boxers and spectators alike [3]. Many hundreds of fights during the course of a career was common [4]. Boxing in booths or "blood tubs" at fairs [5] in which boxers would "take on all comers" as often as 30 or 40 times a day was embedded in the entertainment culture. All told, boxer exposure to neurotrauma in this era was extreme. As an example, one individual who held the lightweight championship between 1910 and 1912, deteriorated to the point of legal mental incompetence while still boxing (Fig. 1). He was institutionalized by 1917 (at age 28), and lived out his years as a ward of the state, dying of heart disease at age 67 [3]. Such tragic cases have no parallel in modern day sport, football or otherwise, and are now a relic of history.

With the above in mind, it is not surprising that a condition called "punch drunk" (or "dementia pugilistica" from 1937 forward [6]) would appear in the medical literature in the early twentieth century [7]. The original article was essentially a communication of a well-known problem in boxing circles to the medical and scientific community. Neurological deficits were plainly apparent to lay people, which is quite different from modern day chronic traumatic encephalopathy (CTE), in which consensus groups of international experts have failed to identify specific

Fig. 1 Adolphus Wolgast (1888–1955) was the lightweight champion from 1910–1912, achieving the title in a 40 round conflagration that was among the bloodiest in history. Mr. Wolgast's skills declined after 1912, though he continued to fight (and perform badly). By 1917, while still fighting, a court declared him incompetent to manage his own affairs. He was ultimately institutionalized and lived out his years as a ward of the state. This case typifies punch drunk syndrome as it was classically defined. Indeed, the initials "AW" and "asylum" appear in a table in Martland's original paper. Although autopsy examination is not available, this case is informative. The fact that he had a stationary, albeit severe, neurological impairment after he stopped fighting, indicates brain damage from boxing-related neurotrauma, and in no way resembles neurodegenerative disease. Brain examination may well have revealed some extent of neurofibrillary change, but such a change would have been epiphenomenal to the more relevant tissue damage from trauma per se, and the resulting neurological deficits

clinical signs or sets of clinical signs. Neurological deficits attributable to trauma are typically absent in modern case series [8].

Importantly, punch drunk as originally described was a purely clinical condition (modern day CTE is purely pathological, or, more precisely, immunohistochemical [9]) characterized by diverse neurological signs such as slurring dysarthria, asymmetrical hyperreflexia, Parkinsonian-like features, ataxia, dragging of extremities, euphoria, memory loss, and fatuous dementia, reflecting multifocal traumatic brain injuries [6, 7, 10, 11].

2 Neuropathology as a Biomarker

Neuropathological examination could be considered an early form of biomarker, insofar as it is an assessment for structural brain changes as a corollary to a clinical disease. The legitimacy of the clinical disease is obviously an a priori requirement. In this respect, it is interesting that the first autopsy description of a boxer with neurological deterioration (which did not appear in the literature until 1954 [12]), was actually a case of genetic, early-onset Alzheimer's disease (AD) having nothing to do with boxing (reviewed in [13]). This would be the first but not the only erroneous interpretation of autopsy neuropathology in an index case (see below [14]). In both instances, neurofibrillary pathology or phosphorylated tau (p-tau) would be a component of the pathology, but in neither instance is there a legitimate clinicopathologic entity attributable to neurotrauma.

Corsellis in 1973 reported the largest autopsies series (14 cases) of classical punch drunk syndrome, all of whom fought between 1900 and 1940 [15]. Classical punch drunk syndrome is of historical relevance only; a series larger than this is no longer possible. The cases are simply too few. As evidence of this fact, researchers have gone back to the original paraffin blocks from the Corsellis series as many as four times since [16–19], including as recently as 2018 [19].

The Corsellis series showed heterogeneous pathology, encompassing damage to the septum pellucidum, damage to the cerebellum, damage to the substantia nigra, and neurofibrillary tangles, mainly in the temporal lobes. The most recent reexamination of this series in 2018 suggests that only 50% of the Corsellis cases meet modern consensus criteria for CTE [19], which is remarkable in light of the extensive neurotrauma of the era, compared to recent studies showing much higher percentages of CTE in athletes with far less neurotrauma exposure [20].

The original series described a number of other lesions (Table 1), including encephalomalacia from trauma, hippocampal sclerosis, cerebrovascular disease, neurosyphilis, and chronic alcoholism in about half of the cases [15]. Moreover, several individuals in this series are now believed to have had neurodegenerative diseases (AD, Lewy body dementia, progressive supranuclear palsy [19]) of unknown cause. Many also have aging-related tau astrogliopathy [21]. Disease duration ranged to greater than 40 years, which is unlike typical neurodegenerative disease. All told, this largest (but still very small) autopsy series of punch drunk

Table 1
Neuropathological findings in punch drunk syndrome unrelated to tauopathy

Septal fenestrations
Cerebellar sclerosis
Tabes dorsalis
Cerebral contusion
Lacunar infarcts
Flattened, detached fornices
Substantia nigra neuron loss
Atrophic, gliotic mamillary bodies
Focal midbrain cystic encephalomalacia
Gliosis of inferior olivary nucleus
Thinning of the corpus callosum
Leptomeningeal thickening
Thalamic/hypothalamic gliosis

syndrome demonstrates heterogeneous and severe neuropathology along with some misinterpretations. The role of tau aggregates as a biomarker in this context thus confronts a number of challenges.

3 Does Tau as a Biomarker Link Traumatic Brain Injury and Neurodegenerative Disease?

Investigations into tau protein analyses as biomarkers come with the presumption that those analyses monitor neurodegenerative disease, perhaps even as a driver of disease. It is important for investigators to keep in mind, however, that the etiologies for all major nosological categories of neurodegenerative disease (e.g., AD, Parkinson's disease, amyotrophic lateral sclerosis, frontotemporal dementia) are unknown, apart from rare subtypes with pathogenic mutation (i.e., autosomal dominant disease); even then, the precise molecular mechanism, or trigger, for neurodegeneration is unknown. In the case of AD, in which the amyloid cascade has been under study for 30 years at the cost of billions of dollars, and has achieved proof of concept (therapeutic reduction in amyloid [22]), the inability to modify the disease trajectory speaks to the collective ignorance on the subject of etiology. Factors responsible for initiating progressive deterioration at the molecular level remain a mystery. By extension, the notion that traumatic brain injury (TBI) causes neurodegenerative disease is, strictly speaking, hypothetical, with a number of factors complicating that hypothesis (Table 2).

Table 2
Factors complicating the neurotrauma–neurodegenerative proteinopathy concept

Stationary neurological disease in classical punch drunk syndrome
Stationary neurological disease following severe traumatic brain injury
No neurological signs in modern cases, unless coexisting neurodegenerative disease
Lack of clinical disease overall in modern cases
Benignity of tau with age
Small effect sizes in large scale epidemiological studies (no evidence of causality)

The term "neurodegenerative disease" should therefore not be used casually with respect to punch drunk syndrome or modern day CTE, although this tends to be the case. The terminology itself, such as "dementia pugilistica" or "tauopathy," tends to imply neurodegeneration, which would be overstated in both instances. Neurodegenerative diseases are invariably progressive and fatal conditions. Punch Drunk and modern CTE lack these features. Neurodegenerative diseases invariably result in consistent neurological deficits and consistent neuropathological changes across subtypes. Punch Drunk syndrome and modern CTE lack these features as well. Nevertheless, the Corsellis series suggested the neurofibrillary tangle (and therefore tau) as a primary pathological biomarker of punch drunk syndrome. Because the neurofibrillary tangle overlapped with AD pathology, a neurodegenerative mechanism was tacitly accepted, and continues to the present day. The reality, in retrospect, is that neurofibrillary change was one of a number of findings in a now historical condition confounded by severe structural brain injury, coincidental neurodegenerative disease, chronic alcoholism, and neurovascular disease.

Importantly, clinical studies argue strongly against a neurodegenerative mechanism as an explanation for the clinical presentation in punch drunk syndrome. Martland noted in his original punch drunk series that "Many cases remain mild in nature and do not progress beyond this point." [7] In his study of about 50 professional boxers, Winterstein noted "Generally the symptoms do not increase after the repeated injuries are stopped—conversely there are remissions and recoveries if the boxer gives up fighting at an early stage." [2] In 1969, Roberts reported the only large scale epidemiological study available on the topic, where he noted a 6% prevalence of punch drunk in boxers, and an additional 11% with focal neurological deficits that "were in no way disabling" [23]. With regard to progression, Roberts noted "There is a good deal of evidence in the present study to suggest that in most cases the condition remains stationary when the individual has stopped boxing, and indeed there are excellent independent accounts for a few of undoubted improvement after their retirement." [23] Such observations categorically exclude neurodegenerative disease, and are more in line with the clinical course following moderate-to-severe traumatic brain injury [24].

In the mid-1980s, with the advent of modern molecular biology, tau protein was recognized as the major protein component of neurofibrillary tangles [25]. Immunodiagnostic techniques with high sensitivity and specificity followed shortly thereafter. (The biology and metabolic processing of tau protein is reviewed in depth elsewhere [26, 27].) The human brain has since been studied extensively via p-tau immunohistochemistry, but the challenges in clinicopathologic correlation and in discerning the meaning of p-tau in brain continue [27]. It is known, for example, that accumulation of tau aggregates is an inevitable consequence of age, starting as early as childhood [28]. Such entities as primary age-related tauopathy (PART) [29] and aging-related tau astrogliopathy (ARTAG) [21]—that is, age-related rather than disease-related accumulations of p-tau—are evidence of the on-going challenges.

4 Lack of a Clinical Phenotype Limits the Tau Biomarker Concept

With the disappearance of punch drunk from boxing because of reduced exposure, and the emergence of highly sensitive immunohistochemistry in the late twentieth century, sport-associated manifestations of disease transitioned from a purely clinical syndrome to a change identified *exclusively* at autopsy. This rather dramatic paradigm shift also placed the diagnosis squarely within the province of neuropathological examination. Lacking a clinical requirement and a lower threshold for the extent of tauopathy [9], sport-implicated neuropathology has become both extremely common [8, 20] and subject to pattern interpretation. For unclear reasons, the CTE pattern (p-tau aggregates in neurons, astrocytes, and cell processes around small vessels in an irregular pattern at the depths of the cortical sulci) was deemed "pathognomonic" for a neurological state (encephalopathy) and a mechanism (trauma) in a consensus paper [9]. Regardless, the lack of a definable clinical disease precludes tau or any other measured quantity as a biomarker. More research is needed.

In light of the above, it is perhaps not surprising that a neurotrauma–neurodegenerative disease narrative emerged in football, but more remarkable are the often sparse nature of the findings and the overlap with the aging process. The depicted immunohistochemical neuropathology in the index case, for example, could potentially be found in any 50-year-old person [14].

5   Establishing a Causal Link Between TBI and Neurodegenerative Disease

Because TBI is a form of environmental exposure, causal assertions that TBI causes neurodegenerative disease require robust and well-designed, large-scale epidemiological study. An abundance of cross-sectional studies in the sport literature are interesting along these lines but are "hypothesis-generating." The Roberts study noted above is somewhat different in that it obtained a random sample of all professional boxers registered in the United Kingdom at that time and accounted for athletes who died or did not participate [23], although it was not longitudinal and more precisely looked at occupation rather than TBI per se. The study was compelling in its lack of evidence that boxing at the extremity of neurotrauma caused neurodegenerative disease. The most severely affected individuals had stationary disease for the most part, while a few improved; such characteristics run directly counter to neurodegenerative disease as noted previously. To the extent that a small subset in the Roberts series showed progressive disease, coincidental neurological disease such as Parkinson disease or AD, or the overlay of the aging process could not be excluded.

On the other hand, it seemed clear enough, given the 6% prevalence of punch drunk and the 17% prevalence of neurological signs not attributable to known conditions, that boxing exposure in this era *caused* neurological injury in some participants. That injury, however, was more in line with parenchymal brain damage from trauma per se, rather than a progressive neurodegenerative proteinopathy. In this sense, it is important to separate "traumatic dementia" from "neurodegenerative dementia". The former has an abundance of evidence to support cause (traumatic brain injury), the latter is of unknown cause, with trauma conferring only a marginal relative risk. The utility of protein biomarkers as a surrogate for neurodegenerative disease caused by trauma thus appears limited based on the Roberts study, and becomes even more so with modern contact sport athletes in whom neurotrauma exposure is far less.

One study using vital statistics have found a marginal increase in risk of AD and amyotrophic lateral sclerosis in National Football League athletes appearing on death certificates [30]. Such studies, however, are of an ecological nature with no causal ramifications with respect to TBI, and leave a number of confounders (e.g., genetics, age, access to health care) unaddressed. The same data set showed lower all-cause mortality, less cancer, less cardiovascular disease, and fewer suicides in the professional athletes compared to men in the general population [31]. To date, there are no studies in the contact sport literature with a study design sufficient to address the role of TBI from sport as a cause for subsequent neurodegenerative disease.

A number of studies have examined the relationship between TBI in general (apart from contact sport) and neurological diseases. In a systematic review, Godbolt et al. in 2014 [32] concluded that there was lack of risk for dementia after mild TBI. It was noteworthy that only eight studies met inclusion criteria (systematic reviews, meta-analyses, randomized controlled trials, case control studies, cohort studies, 30 or more cases). Moreover, only one study directly considered the risk of dementia after mild TBI and found that a history of mild TBI was not associated with a future diagnosis of dementia [33].

A case-control study by Plassman et al. [34] demonstrated a twofold and fourfold increased risk of AD specifically (as determined by clinical consensus criteria) following moderate and severe TBI respectively, but found no risk for AD following mild TBI. The demonstration of dose dependency has been influential and is generally accepted as evidence in favor of a relationship between TBI and AD specifically. However, a subset of people with TBI may suffer dementia or diminished cognitive reserve on the basis of structural damage to the brain. Moderate and severe TBI confer significant morbidity and mortality, so it is plausible that some percentage of such people have sufficient brain damage, or depleted reserve from brain damage, to increase the likelihood of a dementia diagnosis and even an Alzheimer's disease diagnosis clinically. Longitudinal studies demonstrating exposure, latency, progressive deterioration, and neuropathology are needed.

A number of large scale epidemiological studies have been published since the systematic review by Godbolt, examining TBI as an exposure and various outcomes including dementia in general [35–40], Alzheimer's disease [41], early onset dementia [37, 42, 43], and Parkinson's disease [44–46]. The studies overall are heterogeneous in definitions of TBI exposure and in outcome definitions both between and within studies. In the many studies examining dementia as an outcome, a number of International Classification of Diseases (ICD) codes are used, which include both neurodegenerative and nonneurodegenerative (e.g., alcohol-related, vascular) forms of dementia, which generally precludes comment on specific forms of dementia that might be associated with a specific neurodegenerative proteinopathy. Interestingly, regardless of definitions of exposure and outcome, the studies consistently show a significant but small effect size, with hazard ratios generally less than 2, which could be explained by any number of confounders. Reverse causality (i.e., TBI occurring as a result of a prodromal neurological condition) is a common finding across studies for example, evident in the consistent demonstration of a decreasing risk of the neurological outcome under study, the greater the time interval between TBI and outcome diagnosis [36, 39, 41, 43].

Because of the size of such studies, with subjects numbering in the thousands or more, neuropathological data are precluded, so the issue of TBI causing structural brain damage in some cases, that then leads to a given neurological outcome (e.g., parkinsonism, dementia), rather than progressive neurodegenerative proteinopathy, remains an open question. Overall, the epidemiological data to date do not permit a causal relationship between TBI and various neurodegenerative disease outcomes, but do raise a number of confounders, emphasizing again the hypothetical nature of biomarkers of TBI-associated tauopathy.

6   Tau as a Fluid Biomarker

The evidence base with respect to tau as a CSF or plasma biomarker is derived from studies in AD (reviewed in [47]). Both CSF total-tau (t-tau) and p-tau are used for the diagnosis of AD [48, 49] along with CSF amyloid-β1–42 (Aβ42). CSF t-tau and p-tau are higher in AD compared to control. In longitudinal studies CSF tau may predict disease progression in mild cognitive impairment (MCI) as well as nondemented control subjects [47]. CSF p-tau seems to be more specific than t-tau and may better discriminate AD from non-AD neurodegenerative disease [47]. In general, combined CSF tau and Aβ42, or the tau/Aβ42 ratio has better diagnostic value than individual biomarkers alone [50, 51]. A recent study noted a signature of p-tau217 and p-tau181 along with amyloid-β reduction (indirect evidence of amyloid-β "aggregation") in familial AD (*PSEN1*, *PSEN2*, or *APP*) as early as two decades prior to the widespread appearance of p-tau205 and total tau [52] (indirect evidence of tau aggregation). The latter corresponds more closely to neuronal loss and hypometabolism in symptomatic AD. These data suggest that amyloid-β aggregation precedes tau aggregation in disease, and may support the link between solubility and toxicity of hallmark proteins.

In the past few years, highly sensitive assays have demonstrated some utility in plasma-based tau, with higher plasma t-tau measurements in AD versus MCI and healthy controls. A large meta-analysis also showed an association between plasma t-tau levels and AD [53]. Plasma t-tau concentration via single-molecule array (Simoa) has shown overlap with control samples, but may have some predictive value in longitudinal study [47]. Plasma p-tau has also shown promise in AD using a novel enhanced chemiluminescence immunoassay, with some discrimination of AD and frontotemporal lobar degeneration [54].

The neurotrauma–tauopathy issue is considerably different from AD. Clinical AD is always associated with an abundance of

Aβ, which appears in aggregated from in advance of p-tau. P-tau in the hypothetical neurotrauma construct is much sparser and is unassociated with Aβ, unless it coexists with AD. The extent and the pathophysiologic basis for its appearance remain obscure. Neurotrauma-related p-tau also may co-occur with age-related tauopathy, while the aggregates themselves appear to be comprised of the same molecular p-tau phenotype of both AD and aging-related tau [55]. Identifying a discriminating concentration of tau in a fluid specimen in this setting presents an obvious challenge.

That said, a recent cross sectional study of p-tau181 and t-tau in athletes compared to controls suggested higher t-tau in athletes (presenting with multiple concussions), but no difference in p-tau181 [56]. A study in National Football League athletes with similar design showed comparable results, with a marginal increase in total tau and no differences in p-tau181 [57]. Another study noted racial differences in the p-tau181/total tau ratio in CSF [58], emphasizing variability in the population and the importance of rigorous control groups.

Overall the CSF-tau data suggest marginal differences between athletes and controls that need to be further explored for physiological relevance, and are in any event confounded by potentially coexisting age-related and/or AD-related tau. CSF tau analyses as evidence for a neurotrauma–neurodegenerative disease construct appears limited so far.

7 Selected Positron Emission Tomography (PET) Studies Using Putative Tau Ligands

Tau ligands or radiotracers comprise an array of heterocyclic aromatic organic compounds, the chemistry of which is reviewed elsewhere [59] (Fig. 2). They are lipophilic molecules with a low molecular weight by necessity, that is, capable of crossing the blood-brain barrier. The compounds generally share affinity for β-sheet secondary structure, which is much less specific than antibody probes to specific p-tau epitopes. The radiotracers are only specific for p-tau insofar as there is broad overlap between the predominant PET signals and the brain regions with known selective vulnerability to p-tau in autopsy specimens. In vitro binding assays may also be informative [60], but the precise ligand at the molecular level in any PET signal is unclear. Of note is that PET signals with tau ligands are disrupted by chemical denaturants such as formic acid [61], whereas p-tau immunohistochemistry is enhanced by denaturants. The differences between p-tau by brain tissue immunohistochemistry and tau signals by PET scan may be therefore be considerable and possibly even be mutually exclusive at the molecular level.

With the above caveats, $[^{18}F]$flortaucipir has shown promise and is the first and only tau ligand approved for experimental use by

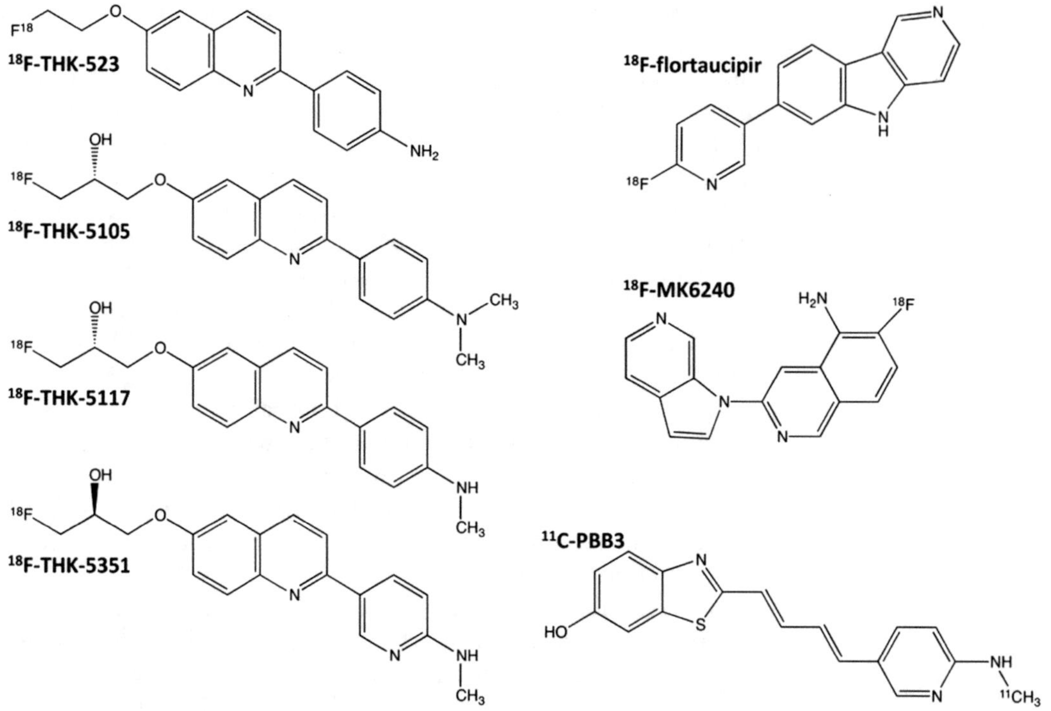

Fig. 2 Chemical classes of some tau PET ligands [66]

the Food and Drug Administration in the United States. Several recent studies are of note with respect to the subject under discussion.

Takahata et al. studied 30 TBI patients and 16 controls using the tau tracer ^{11}C-pyridinyl-butadienyl-benzothiazole 3 (^{11}C-PBB3) [62]. The TBI patients consisted of individuals with severe TBI (loss of consciousness greater than 24 h) or professional contact sport athletes. The rationale for considering these two groups under a single overall category of TBI was not detailed. The authors found that patients with TBI had higher ^{11}C-PBB3 binding capacities in the neocortical gray and white matter segments than healthy control subjects. TBI patients with traumatic encephalopathy syndrome had higher ^{11}C-PBB3 binding capacity in white matter than those without traumatic encephalopathy syndrome. ^{11}C-PBB3 binding capacity in white matter also correlated with the severity of psychosis. Noteworthy in this study is the discussion of functional disorders (traumatic encephalopathy syndrome, psychosis) with respect to tau, rather than neurodegenerative disease.

Mantyh and colleagues published a case report showing a "modest" correlation between [^{18}F]flortaucipir PET and post-mortem p-tau immunohistochemistry in a 72-year-old decedent and former National Football League athlete

[63]. Remarkable in this study is the extent of the PET labeling, which was present throughout the cerebrum. The extent of tau pathology is unclear from the description, but appears to be substantially less than the tau neuroimaging. The report also describes an absence of amyloid-β beta and "flamed shaped neurofibrillary tangles typical of Alzheimer's disease (AD)," and Braak stage III neurofibrillary change, which would be consistent with so-called primary age-related tauopathy. The report describes mild Parkinsonism, mild impairment in attention and executive functioning, and absence of synuclein reactivity, although nigral changes are not described. The basis for the new-onset seizure disorder was not clear from the report. Overall, the case appears to be a complex mixture of age-related pathology with features of CTE, but with limited evidence for a major neurocognitive disorder and a tau PET scan that outstrips the depicted cortical p-tau.

Stern et al. described tau PET (also [^{18}F]flortaucipir) in 26 former National Football League athletes and 31 control subjects [64]. The authors reported higher uptake among former players compared to controls. Three brain regions were specifically highlighted: bilateral superior frontal; bilateral medial temporal; and left parietal. They found no association between tau deposition and scores on cognitive and neuropsychiatric tests. This study highlights the difficulties in correlating PET signal uptake with clinical signs. The cross-sectional design (similar to Takahata et al.) precludes risk analyses or distinctions between cause and effect.

Interestingly, tau PET uptake appeared to be an insensitive measure of p-tau immunolabeling in a recent study, in that tauopathy up to Braak stage IV was not reliably detected by the ^{18}F-flortaucipir tracer [65]. This tends to suggest that the lesions described in contact sport athletes, which are often sparse, may not be detectable by this technique.

8   Conclusions

The potential utility of various tau species as biomarkers for trauma-associated neurodegenerative proteinopathy relies on the existence of such a mechanism in vivo. To date, the evidence is limited. The most compelling human construct of the early twentieth century, purportedly in support of such a mechanism, has a number of aspects that argue against typical neurodegenerative disease. A more modern sport-implicated central nervous system process differs considerably from the classical disease state, and suffers from the absence of a clinical correlate, rendering the concept of a biomarker moot. Given these deficiencies, biomarker studies related to tau, in particular plasma and CSF studies for tau and PET scanning of tau surrogates, have predictably shown marginal results, and will likely continue to do so.

In short, an asymmetry seems to have developed between the sophistication of biomarker analyses which is immense, and the empirical evidence in support of the underlying hypothesis which is limited. Greater recognition of this asymmetry may be needed to slow the growth of bewildering biomarker superstructure that, in the end, will not overcome the null hypothesis.

References

1. Rodriguez RG (2009) The history of boxing regulations. The regulation of boxing. McFarland, Jefferson, NC, pp 30–31
2. Winterstein CE (1937) Head injuries attributable to boxers. Lancet 230:719–722
3. Lang A (2012) The Nelson-Wolgast fight and the San Francisco boxing scene, 1900–1914. McFarland, Jefferson, NC
4. Zazryn T (2007) The evidence for chronic traumatic encephalopathy in boxing. Sports Med 37(6):467–476
5. Critchley M (1957) Medical aspects of boxing, particularly from a neurological standpoint. Br Med J 1:357–362. https://doi.org/10.1136/bmj.1.5015.357
6. Millspaugh J (1937) Dementia pugilistica. US Nav Med Bull 35:297–361
7. Martland HS (1928) Punch drunk. JAMA 91:1103–1107. https://doi.org/10.1001/jama.1928.02700150029009
8. McKee AC, Stein TD, Nowinski CJ, Stern RA, Daneshvar DH, Alvarez VE et al (2013) The spectrum of disease in chronic traumatic encephalopathy. Brain 136:43–64. https://doi.org/10.1093/brain/aws307
9. McKee AC, Cairns NJ, Dickson DW, Folkerth RD, Keene CD, Litvan I et al (2016) The first NINDS/NIBIB consensus meeting to define neuropathological criteria for the diagnosis of chronic traumatic encephalopathy. Acta Neuropathol 131(1):75–86. https://doi.org/10.1007/s00401-015-1515-z
10. Parker HL (1934) Traumatic encephalopathy ('punch drunk') of professional pugilists. J Neurol Psychopathol 15(57):20–28
11. Critchley M (1949) Punch-drunk syndromes: the chronic traumatic encephalopathy of boxers. In: Maloine (ed) Hommage a Clovis Vincent. Imprimerie Alascienne, Strasbourg, pp 131–145
12. Brandenburg W, Hallervorden J (1954) Dementia pugilistica mit anatomischem Befund. Virchows Arch 325S:680–709
13. Castellani RJ, Perry G (2017) Dementia pugilistica revisited. J Alzheimers Dis 60(4):1209–1221. https://doi.org/10.3233/Jad-170669
14. Omalu BI, DeKosky ST, Minster RL, Kamboh MI, Hamilton RL, Wecht CH (2005) Chronic traumatic encephalopathy in a National Football League player. Neurosurgery 57(1):128–133. https://doi.org/10.1227/01.Neu.0000163407.92769.Ed
15. Corsellis JA, Bruton CJ, Freeman BD (1973) Aftermath of boxing. Psychol Med 3(3):270–303. https://doi.org/10.1017/S0033291700049588
16. Roberts GW (1988) Immunocytochemistry of neurofibrillary tangles in dementia Pugilistica and Alzheimers-disease—evidence for common genesis. Lancet 2(8626–7):1456–1458
17. Roberts GW, Allsop D, Bruton C (1990) The occult aftermath of boxing. J Neurol Neurosurg Psychiatry 53(5):373–378. https://doi.org/10.1136/jnnp.53.5.373
18. Tokuda T, Ikeda S, Yanagisawa N, Ihara Y, Glenner GG (1991) Reexamination of ex-boxers brains using immunohistochemistry with antibodies to amyloid beta-protein and tau protein. Acta Neuropathol 82(4):280–285. https://doi.org/10.1007/Bf00308813
19. Goldfinger MH, Ling H, Tilley BS, Liu AKL, Davey K, Holton JL et al (2018) The aftermath of boxing revisited: identifying chronic traumatic encephalopathy pathology in the original Corsellis boxer series. Acta Neuropathol 136(6):973–974. https://doi.org/10.1007/s00401-018-1926-8
20. Mez J, Daneshvar DH, Kiernan PT, Abdolmohammadi B, Alvarez VE, Huber BR et al (2017) Clinicopathological evaluation of chronic traumatic encephalopathy in players of American football. JAMA 318(4):360–370. https://doi.org/10.1001/jama.2017.8334
21. Kovacs GG, Ferrer I, Grinberg LT, Alafuzoff I, Attems J, Budka H et al (2016) Aging-related tau astrogliopathy (ARTAG): harmonized evaluation strategy. Acta Neuropathol 131(1):87–102. https://doi.org/10.1007/s00401-015-1509-x

22. Karran E, Hardy J (2014) A critique of the drug discovery and phase 3 clinical programs targeting the amyloid hypothesis for Alzheimer disease. Ann Neurol 76(2):185–205. https://doi.org/10.1002/ana.24188

23. Roberts A (1969) Brain damage in boxers: a study of prevalence of traumatic encephalopathy among ex-professional boxers. Pitman Medical and Scientific Publishing, London

24. Himanen L, Portin R, Isoniemi H, Helenius H, Kurki T, Tenovuo O (2006) Longitudinal cognitive changes in traumatic brain injury—A 30-year follow-up study. Neurology 66(2):187–192. https://doi.org/10.1212/01.wnl.0000194264.60150.d3

25. Grundke-Iqbal I, Iqbal K, Quinlan M, Tung YC, Zaidi MS, Wisniewski HM (1986) Microtubule-associated protein tau. A component of Alzheimer paired helical filaments. J Biol Chem 261(13):6084–6089

26. Arendt T, Stieler JT, Holzer M (2016) Tau and tauopathies. Brain Res Bull 126 (Pt 3):238–292. https://doi.org/10.1016/j.brainresbull.2016.08.018

27. Castellani RJ, Perry G (2019) Tau biology, Tauopathy, traumatic brain injury, and diagnostic challenges. J Alzheimers Dis 67 (2):447–467. https://doi.org/10.3233/Jad-180721

28. Braak H, Thal DR, Ghebremedhin E, Del Tredici K (2011) Stages of the pathologic process in Alzheimer disease: age categories from 1 to 100 years. J Neuropathol Exp Neurol 70 (11):960–969. https://doi.org/10.1097/NEN.0b013e318232a379

29. Crary JF, Trojanowski JQ, Schneider JA, Abisambra JF, Abner EL, Alafuzoff I et al (2014) Primary age-related tauopathy (PART): a common pathology associated with human aging. Acta Neuropathol 128(6):755–766. https://doi.org/10.1007/s00401-014-1349-0

30. Lehman EJ, Hein MJ, Baron SL, Gersic CM (2012) Neurodegenerative causes of death among retired National Football League players. Neurology 79(19):1970–1974. https://doi.org/10.1212/WNL.0b013e31826daf50

31. Baron SL, Hein MJ, Lehman E, Gersic CM (2012) Body mass index, playing position, race, and the cardiovascular mortality of retired professional football players. Am J Cardiol 109 (6):889–896. https://doi.org/10.1016/j.amjcard.2011.10.050

32. Godbolt A, Cancelliere C, Hincapie C, Marras C, Boyle E, Kristman V et al (2014) A systematic review of the risk of dementia and chronic cognitive impairment after mild traumatic brain injury. Results of the International Collaboration on MTBI Prognosis (ICoMP). Brain Inj 28(5–6):729

33. Helmes E, Ostbye T, Steenhuis RE (2011) Incremental contribution of reported previous head injury to the prediction of diagnosis and cognitive functioning in older adults. Brain Inj 25(4):338–347. https://doi.org/10.3109/02699052.2011.556104

34. Plassman BL, Havlik RJ, Steffens DC, Helms MJ, Newman TN, Drosdick D et al (2000) Documented head injury in early adulthood and risk of Alzheimer's disease and other dementias. Neurology 55(8):1158–1166. https://doi.org/10.1212/Wnl.55.8.1158

35. Lee YK, Hou SW, Lee CC, Hsu CY, Huang YS, Su YC (2013) Increased risk of dementia in patients with mild traumatic brain injury: a nationwide cohort study. PLoS One 8(5): e62422. https://doi.org/10.1371/journal.pone.0062422

36. Gardner RC, Burke JF, Nettiksimmons J, Kaup A, Barnes DE, Yaffe K (2014) Dementia risk after traumatic brain injury vs nonbrain trauma the role of age and severity. JAMA Neurol 71(12):1490–1497. https://doi.org/10.1001/jamaneurol.2014.2668

37. Nordstrom P, Michaelsson K, Gustafson Y, Nordstrom A (2014) Traumatic brain injury and young onset dementia: a nationwide cohort study. Ann Neurol 75(3):374–381. https://doi.org/10.1002/ana.24101

38. Barnes DE, Byers AL, Gardner RC, Seal KH, Boscardin WJ, Yaffe K (2018) Association of Mild Traumatic Brain Injury with and without Loss of consciousness with dementia in US military veterans. JAMA Neurol 75 (9):1055–1061. https://doi.org/10.1001/jamaneurol.2018.0815

39. Nordstrom A, Nordstrom P (2018) Traumatic brain injury and the risk of dementia diagnosis: A nationwide cohort study. PLoS Med 15(1): e100249. https://doi.org/10.1371/journal.pmed.1002496

40. Yang JR, Kuo CF, Chung TT, Liao HT (2019) Increased risk of dementia in patients with craniofacial trauma: a Nationwide population-based cohort study. World Neurosurg 125: E563–EE74. https://doi.org/10.1016/j.wneu.2019.01.133

41. Tolppanen AM, Taipale H, Hartikainen S (2017) Head or brain injuries and Alzheimer's disease: a nested case-control register study. Alzheimers Dement 13(12):1371–1379. https://doi.org/10.1016/j.jalz.2017.04.010

42. Cations M, Draper B, Low LF, Radford K, Trollor J, Brodaty H et al (2018) Non-genetic

risk factors for degenerative and vascular young onset dementia: results from the INSPIRED and KGOW studies. J Alzheimers Dis 62(4):1747–1758. https://doi.org/10.3233/Jad-171027

43. Fann JR, Ribe AR, Pedersen HS, Fenger-Gron M, Christensen J, Benros ME et al (2018) Long-term risk of dementia among people with traumatic brain injury in Denmark: a population-based observational cohort study. Lancet Psychiatry 5(5):424–431. https://doi.org/10.1016/S2215-0366(18)30065-8

44. Fang F, Chen HL, Feldman AL, Kamel F, Ye WM, Wirdefeldt K (2012) Head injury and Parkinson's disease: a population-based study. Mov Disord 27(13):1632–1635. https://doi.org/10.1002/mds.25143

45. Gardner RC, Burke JF, Nettiksimmons J, Goldman S, Tanner CM, Yaffe K (2015) Traumatic brain injury in later life increases risk for Parkinson disease. Ann Neurol 77(6):987–995. https://doi.org/10.1002/ana.24396

46. Gardner RC, Byers AL, Barnes DE, Li YX, Boscardin J, Yaffe K (2018) Mild TBI and risk of Parkinson disease a chronic effects of Neurotrauma consortium study. Neurology 90(20):E1771–E17E9. https://doi.org/10.1212/Wnl.0000000000005522

47. Del Prete E, Beatino MF, Campese N, Giampietri L, Siciliano G, Ceravolo R et al (2020) Fluid candidate biomarkers for Alzheimer's disease: a precision medicine approach. J Pers Med 10(4):221. https://doi.org/10.3390/jpm10040221

48. Jack CR, Albert MS, Knopman DS, McKhann GM, Sperling RA, Carrillo MC et al (2011) Introduction to the recommendations from the National Institute on Aging-Alzheimer's Association workgroups on diagnostic guidelines for Alzheimer's disease. Alzheimers Dement 7(3):257–262. https://doi.org/10.1016/j.jalz.2011.03.004

49. Mattsson N, Zetterberg H, Janelidze S, Insel PS, Andreasson U, Stomrud E et al (2016) Plasma tau in Alzheimer disease. Neurology 87(17):1827–1835. https://doi.org/10.1212/Wnl.0000000000003246

50. Tapiola T, Alafuzoff I, Herukka SK, Parkkinen L, Hartikainen P, Soininen H et al (2009) Cerebrospinal fluid beta-Amyloid 42 and Tau proteins as biomarkers of Alzheimer-type pathologic changes in the brain. Arch Neurol 66(3):382–389. https://doi.org/10.1001/archneurol.2008.596

51. Dhiman K, Blennow K, Zetterberg H, Martins RN, Gupta VB (2019) Cerebrospinal fluid biomarkers for understanding multiple aspects of Alzheimer's disease pathogenesis. Cell Mol Life Sci 76(10):1833–1863. https://doi.org/10.1007/s00018-019-03040-5

52. Barthelemy NR, Li Y, Joseph-Mathurin N, Gordon BA, Hassenstab J, Benzinger TLS et al (2020) A soluble phosphorylated tau signature links tau, amyloid and the evolution of stages of dominantly inherited Alzheimer's disease. Nat Med 26(3):398. https://doi.org/10.1038/s41591-020-0781-z

53. Lue LF, Sabbagh MN, Chiu MJ, Jing NM, Snyder NL, Schmitz C et al (2017) Plasma levels of A beta 42 and Tau identified probable Alzheimer's dementia: findings in two cohorts. Front Aging Neurosci 9:226. https://doi.org/10.3389/fnagi.2017.00226

54. Ashton NJ, Hye A, Rajkumar AP, Leuzy A, Snowden S, Suarez-Calvet M et al (2020) An update on blood-based biomarkers for non-Alzheimer neurodegenerative disorders. Nat Rev Neurol 16(5):265–284. https://doi.org/10.1038/s41582-020-0348-0

55. Arena JD, Smith DH, Lee EB, Gibbons GS, Irwin DJ, Robinson JL et al (2020) Tau immunophenotypes in chronic traumatic encephalopathy recapitulate those of ageing and Alzheimer's disease. Brain 143:1572–1587. https://doi.org/10.1093/brain/awaa071

56. Taghdiri F, Multani N, Tarazi A, Naeimi SA, Khodadadi M, Esopenko C et al (2019) Elevated cerebrospinal fluid total tau in former professional athletes with multiple concussions. Neurology 92(23):E2717–E2E26. https://doi.org/10.1212/Wnl.0000000000007608

57. Alosco ML, Tripodis Y, Fritts NG, Heslegrave A, Baugh CM, Conneely S et al (2018) Cerebrospinal fluid tau, a beta, and sTREM2 in former National Football League Players: modeling the relationship between repetitive head impacts, microglial activation, and neurodegeneration. Alzheimers Dement 14(9):1159–1170. https://doi.org/10.1016/j.jalz.2018.05.004

58. Alosco ML, Tripodis Y, Koerte IK, Jackson JD, Chua AS, Mariani M et al (2019) Interactive effects of racial identity and repetitive head impacts on cognitive function, structural MRI-derived volumetric measures, and cerebrospinal fluid Tau and A beta. Front Hum Neurosci 13:440. https://doi.org/10.3389/fnhum.2019.00440

59. Choi Y, Ha S, Lee YS, Kim YK, Lee DS, Kim DJ (2018) Development of tau PET imaging ligands and their utility in preclinical and clinical studies. Nucl Med Mol Imaging 52(1):24–30. https://doi.org/10.1007/s13139-017-0484-7

60. Ono M, Sahara N, Kumata K, Ji B, Ni RQ, Koga S et al (2017) Distinct binding of PET ligands PBB3 and AV-1451 to tau fibril strains in neurodegenerative tauopathies. Brain 140:764–780. https://doi.org/10.1093/brain/aww339

61. Harada R, Okamura N, Furumoto S, Tago T, Yanai K, Arai H et al (2016) Characteristics of Tau and its ligands in PET imaging. Biomolecules 6(1). https://doi.org/10.3390/biom6010007

62. Takahata K, Kimura Y, Sahara N, Koga S, Shimada H, Ichise M et al (2019) PET-detectable tau pathology correlates with long-term neuropsychiatric outcomes in patients with traumatic brain injury. Brain 142:3265–3279. https://doi.org/10.1093/brain/awz238

63. Mantyh WG, Spina S, Lee A, Iaccarino L, Soleimani-Meigooni D, Tsoy E et al (2020) Tau positron emission tomographic findings in a former US football player with

pathologically confirmed chronic traumatic encephalopathy. JAMA Neurol 77 (4):517–521. https://doi.org/10.1001/jamaneurol.2019.4509

64. Stern RA, Adler CH, Chen KW, Navitsky M, Luo J, Dodick DW et al (2019) Tau positron-emission tomography in former National Football League Players. N Engl J Med 380 (18):1716–1725. https://doi.org/10.1056/NEJMoa1900757

65. Soleimani-Meigooni DN, Iaccarino L, La Joie R, Baker S, Bourakova V, Boxer AL et al (2020) 18F-flortaucipir PET to autopsy comparisons in Alzheimer's disease and other neurodegenerative diseases. Brain 143 (11):3477–3494. https://doi.org/10.1093/brain/awaa276

66. Lyoo CH, Cho H, Choi JY, Ryu YH, Lee MS (2018) Tau positron emission tomography imaging in degenerative parkinsonisms. J Mov Disord 11(1):1–12. https://doi.org/10.14802/jmd

Chapter 14

Impedimetric Immunosensing for Neuroinflammatory Biomarker Profiling

Andrea Cruz, Catarina M. Abreu, Paulo P. Freitas, and Inês Mendes Pinto

Abstract

Neuroinflammation is a hallmark of several neurodegenerative diseases including multiple sclerosis, Alzheimer's, and Parkinson's disease. Recent evidence shows that chronic activation of the immune system could be used to pinpoint the onset and progression of these diseases in concomitant analysis with disease-specific signatures. Tumor necrosis factor alpha (TNFα) a pleiotropic proinflammatory cytokine with distinct functions in homeostasis and disease pathogenesis, has been recognized as a potential biomarker for several neurological disorders. In the context of neuroinflammation and neurodegeneration, here we describe a method for quantitative detection of TNFα, in clinically relevant body fluids, based on an ultrasensitive impedimetric system.

Keywords Neurodegenerative, Neuroinflammation, Biomarkers, Impedimetric immunosensors

1 Introduction

Inflammation represents a natural protective mechanism of cells against harmful and inflicted stimuli [1]. In the central nervous system (CNS), neuroinflammatory responses are exclusively mediated by resident glial cells (microglia and astrocytes) and eventually by infiltrated macrophages and lymphocytes from the peripheral immune system upon blood–brain barrier breakdown [2]. Microglial cells are responsible for the maintenance of CNS homeostasis while scavenging for apoptotic cells, monitoring ongoing synaptic activity and providing trophic support for neurons [3]. Microglia react to a pathological stimulus by activating several proinflammatory molecular cascades that subsequently allow for the recruitment of other immune cells to restore homeostasis

Philip V. Peplow, Bridget Martinez and Thomas A. Gennarelli (eds.), *Neurodegenerative Diseases Biomarkers: Towards Translating Research to Clinical Practice*, Neuromethods, vol. 173, https://doi.org/10.1007/978-1-0716-1712-0_14,
© Springer Science+Business Media, LLC, part of Springer Nature 2022

Table 1
Biomarkers of neuroinflammation[a]

Disorder	Biological fluid	Inflammatory biomarker	Ref.
MS	Serum	TNFα, IL-1β, RANKL, IL-17, PTX3, IL-10	[7]
	Serum/CSF	OPN	[8]
	CSF	CXCL13, IL-23, IL-17, CXCL10, TNFα, TGF-β	[9]
	CSF	CHIT1, MCP-1, GFAP, CHI3L1	[10]
	CSF	sTREM2	[11]
PD	Serum	IFNγ, IL-1β, IL-2, IL-3, IL-10, MIF, TNFα	[12]
	Serum	MIP-1β, MCP-1, IL-8	[13]
	CSF	β2-microglobulin, IL-8, IL-6, TNFα, CHI3L1	[14]
AD	Serum	CHI3L1	[15]
	Serum	IL-8	[16]
	Serum	FGF-1, IL-1β, IL-10, IL-11, IL-18	[17]
	Serum	IL-3, MCP-1, RANTES, sIL-6R, TGF-β1	[18]
	Serum	sCD40L	[19]
	CSF	CHI3L1, MCP-1	[15]
	CSF	IL-15, sFLT-1, sICAM-1	[16]
	CSF	MIP-1β, MIP-3β, sIL-6R	[18]
	CSF	IL-1β	[20]
	CSF	TDP-43	[21]

[a]Table 1 is not intended to be an exhaustive list, only biomarkers characterized in human subjects were included. *AD* Alzheimer's disease, *CHI3L1* chitinase-3-like protein 1, *CHIT1* chitotriosidase, *CXCL-10* C-X-C Motif Chemokine Ligand 10, *CXCL13* C-X-C motif ligand 13, *FGF-1* fibroblast growth factor 1, *GFAP* glial fibrillary acidic protein, *IL-1β* Interleukin 1 beta, *IL-2* Interleukin 2, *IL-3* Interleukin-3, *IL-6* Interleukin 6, *IL-8* Interleukin 8, *IL-10* Interleukin 10, *IL-11* Interleukin 11, *IL-15* Interleukin 15, *IL-17* Interleukin 17, *IL-23* Interleukin 23, *IFNγ* Interferon gamma, *MCP-1* monocyte chemoattractant protein-1, *MIF* Macrophage migration inhibitory factor, *MIP-1β* Macrophage Inflammatory Proteins 1 beta, *MIP-3β* Macrophage Inflammatory Protein 3 beta, *MS* Multiple Sclerosis, *OPN* osteopontin, *PD* Parkinson's Disease, *PTX3* pentraxin 3, *RANKL* receptor activator of nuclear factor kappa-B ligand, *RANTES* Regulated on Activation Normal T Cell Expressed and Secreted, *sCD40L* soluble CD40 ligand, *sFLT-1* soluble fms-like tyrosine kinase 1, *sICAM-1* soluble intercellular adhesion molecule 1, *sIL-6R* Soluble Interleukin-6 receptor, *TDP-43* TAR DNA-binding protein 43, *TNFα* tumor necrosis factor alpha, *sTREM2* secreted form of the triggering receptor expressed on myeloid cells 2, *TGF-β1* transforming growth factor beta 1

[4]. Recent studies have reported the presence of several cytokine inflammatory mediators in the cerebrospinal fluid (CSF), tears and blood of Alzheimer's, Parkinson's, and multiple sclerosis patients [5, 6] (Table 1), suggesting a potential role of abnormal cytokine production in the prognosis of neurodegeneration.

Potential biomarkers for neuroinflammatory disease diagnosis, prognosis, and treatment responses, in particular those measurable in non- or minimally invasive body fluids (e.g., blood, tears, saliva), remain poorly characterized. However, several studies have correlated the levels of inflammatory biomarkers in CSF and plasma with diseases prognosis and treatment responses in different clinical trials for MS [22, 23], PD [24, 25], and AD [26, 27].

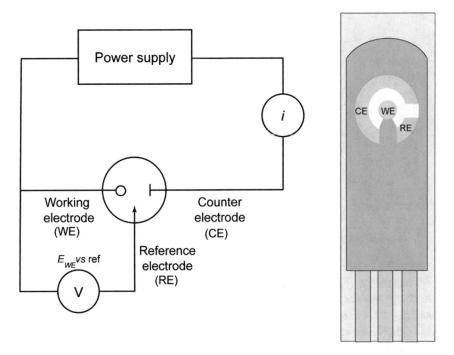

Fig. 1 Three electrode electrochemical system and schematic representation of a gold screen-printed electrode

Although CSF is considered the reference sample for CNS disorders, the biochemical profiling of more accessible biological fluids is gaining attention. Due to its noninvasive nature, tears represent an attractive alternative tissue source for neuroinflammatory/neurodegenerative disease diagnosis and progression monitoring, with some studies reporting high levels of tumor necrosis factor alpha (TNFα) in tear fluids of Parkinson's patients [6]. Quantitative biomarker detection is conventionally performed by enzyme-linked immunosorbent assays (ELISAs); however, due to its high limit of detection (LOD), volume requirements, time of acquisition and advanced technical expertise it is not compatible with point-of-care (POC) testing; thus, miniaturized and easy-of-use biosensing technologies have been recently explored to overcome the limitations of conventional methodologies.

In this framework, impedimetric biosensors are seen as promising analytical devices able to convert biochemical interactions between a receptor (e.g., antibody, enzyme, and DNA) and a biomarker (e.g., antigen, enzyme substrate, and DNA) into a measurable electrical signal proportional to the biomarker concentration. This analysis is typically performed by a three-electrode system: a working electrode (WE), the transduction element of the system, where the antibody–biomarker interactions occur; a

a   **b**

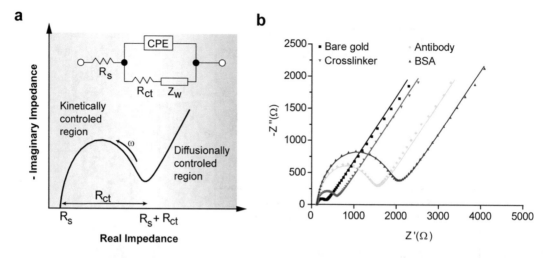

Fig. 2 Schematic representation of the detection mechanism based on Electrochemical Impedance Spectroscopy (EIS). (**a**) Impedance (Z) data represented as a Nyquist plot and corresponding equivalent circuit.
Resistance to charge transfer (R_{ct}); solution resistance (R_s); constant phase element (CPE) (**b**) Nyquist plot
of the consecutive functionalization steps by the representation (dots) of experimental values of the imaginary
(Z') and real (Z) impedance values and the values obtained upon the fitting (lines) to the Randles equivalent
circuit [R_s(CPE[$R_{ct}Z_W$])]

reference electrode (RE) that is connected to the WE to establish a
known and stable potential; and a counterelectrode (CE) that
establishes the connection to an ionic conducting solution, an
electrolyte, so that current can be applied to the WE [28] (Fig. 1).

Electrochemical impedance spectroscopy (EIS), is a sensitive
methodology that monitors variations in electrical properties arising from interfacial interactions, such as biorecognition events, on
surface-modified electrodes. The impedance (Z) can be determined
using the Ohm's law and applying a sinusoidal perturbation to the
system and measuring the output current [29]. The complex value
of impedance can be graphically represented in Nyquist plots,
where the imaginary (Z') and real (Z) impedance values are plotted
in the y and x-axes, respectively, and each point is the impedance for
a specific angular frequency (Fig. 2a). Impedance data is posteriorly
fitted to electrical equivalent circuits to understand how the different electrical components contribute to the overall impedance. The
Randles equivalent circuit is one of the simplest and most widely
used equivalent circuits in EIS [30], it considers electrical components resulting from (1) the diffusion of the redox probe—R_s and
Z_W, accounting for the solution resistance and Warburg impedance,
respectively—which do not affect electron transfer at the surface of
the electrode and (2) electrical components highly sensitive to
interfacial properties of the WE—R_{ct} and C_{dl}, representative of
the resistance to charge transfer and double layer capacitance
(Fig. 2a).

To achieve biomarker-specific detection, the WE are chemically modified to allow the attachment of a biorecognition element (e.g., an antibody) that will capture target biomarker and alter the conductivity, and hence the impedance of the system. The consecutive addition of resistive layers, to the WE surface, causes a surface blocking, and increases the impedance of the system, which appears as a larger semi-circle in the Nyquist plot (Fig. 2b).

Even though impedimetric measurements can be performed in the absence of electrolyte solutions, the use of redox probes such as the ferricyanide couple ($Fe(CN)_6^{3-/4-}$), are often used to enhance the charge-transfer resistance (R_{ct}) signal. In these systems, when a clinical sample is applied to the WE, an increase in the R_{ct} can be used as a direct measurement of the biorecognition event (Fig. 3).

In this chapter, we present a detailed protocol [31] for the functionalization of gold electrochemical biosensors for the detection of TNFα in cerebrospinal fluid, blood serum and tears. Importantly, this protocol [31] can be adapted for the quantitative detection of other biomarkers [32].

2 Materials

2.1 Reagents

1. Phosphate buffer saline buffer (PBS, 10 mM pH 7.4) tablets.
2. Glycerol.
3. Bovine serum albumin (BSA).
4. Potassium ferricyanide (III), $K_3Fe(CN)_6$.
5. Potassium hexacyanoferrate (II) trihydrate, $K_4Fe(CN)_6 \cdot 3H_2O$.
6. 10 mM Phosphate buffer (PB).
7. Sulfosuccinimidyl 6-(3'-(2-pyridyldithio)propionamido)hexanoate crosslinker (sulfo-LC-SPDP) (see **Note 1**).
8. Anti-TNFα monoclonal coating antibody.
9. TNFα recombinant protein standards.
10. Recombinant interferon gamma, IFNγ.
11. Recombinant interleukin-4, IL-4.
12. Isopropanol.
13. Milli-Q water.
14. 0.9% NaCl (physiological serum).
15. Precision wipes (Kimtech, Kimberly-Clark).

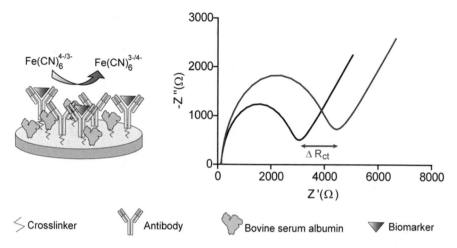

Fig. 3 Representation of the clinical sample incubation and corresponding increase in the resistance to charge transfer in the Nyquist plot

2.2 Equipment

1. Potentiostat/galvanostat (PGSTAT302N, Autolab, Metrohm).

2. Frequency Response Analysis module (FRA32M, Autolab, Metrohm).

3. Three-electrode gold screen-printed electrode (C223AT, DropSens, Metrohm).

4. Screen-printed electrode (SPE) connector interface with the potentiostat (DRP-DSC4MM, DropSens, Metrohm).

3 Methods

3.1 Solution Preparation

1. **5.0 mM $[Fe(CN)_6]^{3-/4-}$ electrolyte solution**
   Prepare 25 mL of electrolyte solution by dissolving 0.0528 g of potassium hexacyanoferrate (II) trihydrate and 0.0412 g of potassium ferricyanide (III) in 25 mL of PBS 10 mM pH 7.4. (*see* **Note 2**).

2. **10 mg/mL Sulfo-LC-SPDP crosslinking solution**
   Prepare the crosslinker Sulfo-LC-SPDP at a concentration of 10 mg/mL in 10 mM PB, pH 7.4 with 5% glycerol (*see* **Note 3**).

3. **Anti-TNFα antibody solution**
   Prepare the antibody solution by diluting to 0.25 mg/mL in 10 mM PB, pH 7.4 with 5% glycerol (*see* **Note 4**).

4. **BSA blocking solution**
   Dissolve 0.0005 g of BSA in 1 mL of 10 mM PB pH 7.4 with 5% glycerol (*see* **Note 5**).

3.2 Immunosensor Functionalization

The biorecognition of a specific biomolecule is achieved through the immobilization of its bioreceptor on the WE surface. One of the most common strategies for covalent antibody immobilization on metallic surfaces, such as gold, silver and platinum WE electrodes, consists of the formation of self-assembled monolayers (SAMs) via thiol (R-SH), sulfide (R-S-R), or disulfide (R-S-S-R) functional groups [32, 33]. Surface modification allows for the introduction of reactive groups such as succinimidyl esters (NHS ester) able to react with the primary amines on monoclonal antibodies to achieve covalent binding [34, 35].

1. Clean electrodes with isopropanol and milli-Q water. Dry gently with a nitrogen spray gun.

2. Add 100 µL of electrolyte solution and measure the impedance of the system at a fixed potential of +0.125 V, using a sinusoidal perturbation with 5 mV amplitude at frequency range of 100,000–0.1 Hz. Impedance data is fitted to a Randles equivalent circuit $[R_s(CPE[R_{ct}Z_W])]$ using the Nova Software (see **Note 6**).

3. Remove the screen-printed electrode from the connector, rinse with Milli-Q water and dry gently with the nitrogen spray gun.

4. Add 1 µL of the sulfo-LC-SPDP solution and incubate for 20 min in a dark and humid chamber (see **Note 7**).

5. Remove the drop of the sulfo-LC-SPDP solution with a clean precision wipe.

6. Rinse with 10 mM phosphate buffer and measure the impedance of the system according to step 2 (see **Note 8**).

7. Remove the SPE from the connector, rinse with milli-Q water and dry gently with the nitrogen spray gun.

8. Add 1 µL of 0.25 mg/mL anti-TNFα antibody and incubate overnight at 4 °C in a humid chamber.

9. Remove the drop of the coating antibody solution with a clean precision wipe.

10. Rinse with 10 mM phosphate buffer and measure the impedance of the system according to step 2.

11. Remove the SPE from the connector, rinse with milli-Q water and dry gently with the nitrogen spray gun.

12. Incubate 1 µL of 1% BSA blocking solution in the WE of the SPE for 30 min at room temperature in a humid chamber (see **Note 9**).

13. Remove the drop of the blocking solution with a clean precision wipe.

14. Rinse with 10 mM phosphate buffer and measure the impedance of the system according to step 2.

15. Remove the SPE from the connector, rinse with milli-Q water, and dry gently with the nitrogen spray gun.

3.3 Immunosensor Calibration

Calibration of the immunosensor system requires the preparation of increasing concentrations of recombinant protein solutions of interest in an appropriate sample matrix followed by incubation on the WE. Specifically for TNFα analysis in tear samples, 0.9% NaCl should be used to prepare the recombinant protein standard solutions as current clinical practice consists in the recovery of tear proteins from diagnostic ophthalmic strips with 0.9% NaCl [36]. The same matrix can be used for biomarker detection in CSF and blood serum as sample dilution in 0.9% NaCl is necessary to improve biomarker–antibody interaction at the WE surface.

1. Add 1μL of 0.9% NaCl to the SPE in the absence of any recombinant protein to evaluate the effect of sample matrix on the EIS measurement. Incubate for 90 min at room temperature in a humidified chamber (*see* **Note 10**).

2. Wash the SPE with milli-Q water and measure EIS according to Subheading 3.2, **step 2**.

3. Remove the SPE from the connector, rinse with milli-Q water and dry gently with the nitrogen gun.

4. Incubate the lowest concentration of recombinant protein solution for 90 min at room temperature in a humidified chamber.

5. Wash the SPE with milli-Q water and measure EIS according to Subheading 3.2, **step 2**.

6. Remove the SPE from the connector, rinse with Milli-Q water and dry gently with the nitrogen gun.

7. Extract the R_{ct} values from the Nyquist plot through the Randles equivalent circuit.

8. Repeat steps 4 to 7 until all the desired protein concentrations are tested (see **Note 11**).

9. Determine the calibration curve for the biomarker of interest by establishing a linear regression between ΔR_{ct} and log[biomarker], where ΔR_{ct} corresponds to R_{ct} (biomarker) $-R_{ct}$ (BSA) and log[biomarker] is the base 10 logarithm of the concentration of the biomarker (see **Note 12**). The LOD of the immunosensor is calculated as the biomarker concentration in the respective calibration curve at which the ΔR_{ct} signal corresponds to the following.

$$\Delta R_{ct} = 3 \times \text{standard deviation}_{(blank)} + \Delta R_{ct\,(blank)}$$

3.4 Immunosensor Specificity Analysis

The analysis of the immunosensor selectivity for detection of targeted biomarkers is fundamental to ensure that nonspecific signals remain below the sensor's LOD or its linear range. In this framework, increasing concentrations of interfering molecules expected to be present in the clinical sample should be tested. For the TNFα detection in tear samples, other cytokines such as interferon gamma (IFNγ) and interleukin-4 (IL-4) could be used.

1. Incubate the lowest concentration of recombinant protein solution for 90 min at room temperature in a humidified chamber (*see* **Note 13**).

2. Wash the SPE with milli-Q water and measure EIS according to Subheading 3.2, **step 2**.

3. Remove the SPE from the connector, rinse with milli-Q water and dry gently with the nitrogen gun.

4. Repeat steps 1 to 3 until all the desired protein concentrations are tested.

3.5 Quantitative Detection of TNFα in Clinical Samples

To show the potential of the technology toward biomarker detection in clinical samples, the clinical sample is tested on the immunosensor functionalized for the biomarker of interest, in this case, TNFα. For tear samples, no sample dilution is required, however biomarker detection for more complex matrixes such as in CSF and blood serum a sample dilution (1:4) is required [31].

1. Incubate the sample for 90 min at room temperature in a humidified chamber (*see* **Note 13**).

2. Wash the SPE with milli-Q water and measure EIS according to Subheading 3.2, **step 2**.

3. Remove the SPE from the connector, rinse with milli-Q water, and dry gently with the nitrogen spray gun.

4. Extract the R_{ct} values from the Nyquist plot through the Randles equivalent circuit and extrapolate the results based on the calibration curve defined in Subheading 3.3.

4 Notes

1. Store desiccated chemicals at −20 °C (humidity sensitive).

2. Always keep 5.0 mM $[Fe(CN)6]^{3-/4-}$ electrolyte solution protected from light (e.g., cover with an aluminum foil), store at 4 °C, and use within a week for experiments.

3. The solution must be prepared at the time of functionalization; the 10 mM PB, pH 7.4 with 5% glycerol can be prepared in advance and stored in ambient conditions.

4. Concentration of the antibody solution may have to be optimized for a different antibody.

5. The blocking solution should be prepared and stored in small aliquots (20μL) at −20 °C.

6. The optimal frequency range can be different based on the biorecognition probes and electrodes (e.g., graphene/gold) in use. The electrolyte solution volume should cover the three-electrode system.

7. Create a humid chamber to minimize droplet evaporation; make sure the solution only covers the WE. The formation of packed SAMs of alkanethiols will be formed within 1 h, with shorter times possible for more concentrated solutions [33].

8. Alternative strategy includes the immobilization of the antibody to the WE via carboxylic groups on the C-terminals, aspartate or glutamate amino acid residues present side chains of the Fc region of the antibody [34]. The carboxylic groups can be converted to amine reactive NHS esters through the EDC (1-ethyl-3-(3-dimethylaminopropyl)carbodiimide)/NHS chemistry to establish a stable amine bond [35].

9. Other blocking solutions with amine functional groups, such as ethanolamine, can be used.

10. This measurement is necessary to define the immunosensor LOD. Other incubation times should be tested for other antibody–protein biomarker interactions.

11. Preform signal acquisitions until saturation or at the desired dynamic range. In the conditions described in the protocol, the linear range is between 1 pg/mL and 25 pg/mL.

12. Other approaches might be considered, namely, the normalization of the $R_{ct(biomarker)}$ based on the $R_{ct(BSA)}$.

$$\frac{R_{ct\,(biomarker)} - R_{ct\,(BSA)}}{R_{ct\,(BSA)}}$$

13. The time used for sample incubation will be the same as that used for the calibration curve

Acknowledgments

I.M.P. and A.C. acknowledge the financial support from the Marie Curie COFUND Programme "NanoTRAINforGrowth," from the European Union's Seventh Framework Programme for research, technological development and demonstration under Grant Agreement No. 600375. This chapter is a result of the project advancing cancer research: from basic knowledge to application (NORTE-01-0145-FEDER-000029), cofinanced by Norte Portugal Regional Operational Programme (NORTE, 2020), under the PORTUGAL 2020 Partnership Agreement, through the European Regional Development Fund, European Union. This work was partially financed by N2020 project: RHAQ/COLAB NORTE-06-3559-FSE-000044 and FCT (Portuguese Foundation for Science and Technology).

References

1. Medzhitov R (2008) Origin and physiological roles of inflammation. Nature 454:428–435. https://doi.org/10.1038/nature07201
2. DiSabato DJ, Quan N, Godbout JP (2016) Neuroinflammation: the devil is in the details. J Neurochem 139:136–153. https://doi.org/10.1111/jnc.13607
3. Norris GT, Kipnis J (2019) Immune cells and CNS physiology: microglia and beyond. J Exp Med 216:60–70. https://doi.org/10.1084/jem.20180199
4. Voet S, Prinz M, van Loo G (2019) Microglia in central nervous system inflammation and multiple sclerosis pathology. Trends Mol Med 25:112–123. https://doi.org/10.1016/j.molmed.2018.11.005
5. Abreu CM, Soares-dos-Reis R, Melo PN, Relvas JB, Guimarães J, Sá MJ, Cruz AP, Mendes Pinto I (2018) Emerging biosensing technologies for neuroinflammatory and neurodegenerative disease diagnostics. Front Mol Neurosci 11:164. https://doi.org/10.3389/fnmol.2018.00164
6. Çomoğlu SS, Güven H, Acar M, Öztürk G, Koçer B (2013) Tear levels of tumor necrosis factor-alpha in patients with Parkinson's disease. Neurosci Lett 553:63–67. https://doi.org/10.1016/j.neulet.2013.08.019
7. D'Ambrosio A, Pontecorvo S, Colasanti T, Zamboni S, Francia A, Margutti P (2015) Peripheral blood biomarkers in multiple sclerosis. Autoimmun Rev 14:1097–1110. https://doi.org/10.1016/j.autrev.2015.07.014
8. Housley WJ, Pitt D, Hafler DA (2015) Biomarkers in multiple sclerosis. Clin Immunol 161:51–58. https://doi.org/10.1016/j.clim.2015.06.015
9. Kothur K, Wienholt L, Brilot F, Dale RC (2016) CSF cytokines/chemokines as biomarkers in neuroinflammatory CNS disorders: a systematic review. Cytokine 77:227–237. https://doi.org/10.1016/j.cyto.2015.10.001
10. Novakova L, Axelsson M, Khademi M, Zetterberg H, Blennow K, Malmeström C, Piehl F, Olsson T, Lycke J (2017) Cerebrospinal fluid biomarkers as a measure of disease activity and treatment efficacy in relapsing-remitting multiple sclerosis. J Neurochem 141:296–304. https://doi.org/10.1111/jnc.13881
11. Zetterberg H (2017) Fluid biomarkers for microglial activation and axonal injury in multiple sclerosis. Acta Neurol Scand 136:15–17. https://doi.org/10.1111/ane.12845
12. Rocha NP, De Miranda AS, Teixeira AL (2015) Insights into neuroinflammation in Parkinson's disease: from biomarkers to anti-inflammatory based therapies. Biomed Res Int 2015:1–12. https://doi.org/10.1155/2015/628192
13. Brockmann K, Schulte C, Schneiderhan-Marra N, Apel A, Pont-Sunyer C, Vilas D, Ruiz-Martinez J, Langkamp M, Corvol J-C, Cormier F, Knorpp T, Joos TO, Bernard A, Gasser T, Marras C, Schüle B, Aasly JO, Foroud T, Marti-Masso JF, Brice A, Tolosa E, Berg D, Maetzler W (2017) Inflammatory profile discriminates clinical subtypes in LRRK2-associated Parkinson's disease. Eur J Neurol 24:427–4e6. https://doi.org/10.1111/ene.13223
14. Andersen AD, Binzer M, Stenager E, Gramsbergen JB (2017) Cerebrospinal fluid biomarkers for Parkinson's disease—a systematic review. Acta Neurol Scand 135:34–56. https://doi.org/10.1111/ane.12590
15. Olsson B, Lautner R, Andreasson U, Öhrfelt A, Portelius E, Bjerke M, Hölttä M, Rosén C, Olsson C, Strobel G, Wu E, Dakin K, Petzold M, Blennow K, Zetterberg H (2016) CSF and blood biomarkers for the diagnosis of Alzheimer's disease: a systematic review and meta-analysis. Lancet Neurol 15:673–684. https://doi.org/10.1016/S1474-4422(16)00070-3
16. Popp J, Oikonomidi A, Tautvydaitė D, Dayon L, Bacher M, Migliavacca E, Henry H, Kirkland R, Severin I, Wojcik J, Bowman GL (2017) Markers of neuroinflammation

associated with Alzheimer's disease pathology in older adults. Brain Behav Immun 62:203–211. https://doi.org/10.1016/j.bbi.2017.01.020

17. Brosseron F, Krauthausen M, Kummer M, Heneka MT (2014) Body fluid cytokine levels in mild cognitive impairment and Alzheimer's disease: a comparative overview. Mol Neurobiol 50:534–544. https://doi.org/10.1007/s12035-014-8657-1

18. Delaby C, Gabelle A, Blum D, Schraen-Maschke S, Moulinier A, Boulanghien J, Séverac D, Buée L, Rème T, Lehmann S (2015) Central nervous system and peripheral inflammatory processes in Alzheimer's disease: biomarker profiling approach. Front Neurol 6:1–11. https://doi.org/10.3389/fneur.2015.00181

19. Yu S, Liu Y-P, Liu Y-H, Jiao S-S, Liu L, Wang Y-J, Fu W-L (2016) Diagnostic utility of VEGF and soluble CD40L levels in serum of Alzheimer's patients. Clin Chim Acta 453:154–159. https://doi.org/10.1016/j.cca.2015.12.018

20. Hesse R, Wahler A, Gummert P, Kirschmer S, Otto M, Tumani H, Lewerenz J, Schnack C, von Arnim CAF (2016) Decreased IL-8 levels in CSF and serum of AD patients and negative correlation of MMSE and IL-1β. BMC Neurol 16:185. https://doi.org/10.1186/s12883-016-0707-z

21. Majumder V, Gregory JM, Barria MA, Green A, Pal S (2018) TDP-43 as a potential biomarker for amyotrophic lateral sclerosis: a systematic review and meta-analysis. BMC Neurol 18:90. https://doi.org/10.1186/s12883-018-1091-7

22. Huang J, Khademi M, Fugger L, Lindhe Ö, Novakova L, Axelsson M, Malmeström C, Constantinescu C, Lycke J, Piehl F, Olsson T, Kockum I (2020) Inflammation-related plasma and CSF biomarkers for multiple sclerosis. Proc Natl Acad Sci 117:12952–12960. https://doi.org/10.1073/pnas.1912839117

23. Stilund M, Gjelstrup MC, Petersen T, Møller HJ, Rasmussen PV, Christensen T (2015) Biomarkers of inflammation and axonal degeneration/damage in patients with newly diagnosed multiple sclerosis: contributions of the soluble CD163 CSF/serum ratio to a biomarker panel. PLoS One 10:e0119681. https://doi.org/10.1371/journal.pone.0119681

24. Santaella A, Kuiperij HB, van Rumund A, Esselink RAJ, van Gool AJ, Bloem BR, Verbeek MM (2020) Inflammation biomarker discovery in Parkinson's disease and atypical parkinsonisms. BMC Neurol 20:26. https://doi.org/10.1186/s12883-020-1608-8

25. Hall S, Janelidze S, Surova Y, Widner H, Zetterberg H, Hansson O (2018) Cerebrospinal fluid concentrations of inflammatory markers in Parkinson's disease and atypical parkinsonian disorders. Sci Rep 8:13276. https://doi.org/10.1038/s41598-018-31517-z

26. Morgan AR, Touchard S, Leckey C, O'Hagan C, Nevado-Holgado AJ, Barkhof F, Bertram L, Blin O, Bos I, Dobricic V, Engelborghs S, Frisoni G, Frölich L, Gabel S, Johannsen P, Kettunen P, Kłoszewska I, Legido-Quigley C, Lleó A, Martinez-Lage P, Mecocci P, Meersmans K, Molinuevo JL, Peyratout G, Popp J, Richardson J, Sala I, Scheltens P, Streffer J, Soininen H, Tainta-Cuezva M, Teunissen C, Tsolaki M, Vandenberghe R, Visser PJ, Vos S, Wahlund L-O, Wallin A, Westwood S, Zetterberg H, Lovestone S, Morgan BP (2019) Inflammatory biomarkers in Alzheimer's disease plasma. Alzheimers Dement 15:776–787. https://doi.org/10.1016/j.jalz.2019.03.007

27. Huang L-K, Chao S-P, Hu C-J (2020) Clinical trials of new drugs for Alzheimer disease. J Biomed Sci 27:18. https://doi.org/10.1186/s12929-019-0609-7

28. Bard AJ, Faulkner LR (2001) In: Harris D, Swain E (eds) Electrochemical methods: fundamentals and applications, 2nd edn. Wiley, New York. ISBN: 978-0-471-04372-0

29. Grieshaber D, MacKenzie R, Vörös J, Reimhult E, Grieshaber D, MacKenzie R, Vörös J, Reimhult E (2008) Electrochemical biosensors—sensor principles and architectures. Sensors 8:1400–1458. https://doi.org/10.3390/s80314000

30. Randles JEB (1947) Kinetics of rapid electrode reactions. Faraday Discuss 1:11–19. https://doi.org/10.1039/DF9470100011

31. Cruz A, Queirós R, Abreu CM, Barata C, Fernandes R, Silva R, Ambrosio AF, Soares dos Reis R, Guimarães J, Sá MJ, Relvas JB, Freitas PP, Pinto IM (2019) Electrochemical immunosensor for TNFα-mediated inflammatory disease screening. ACS Chem Nerosci 10(6):2676–2682. https://doi.org/10.1021/acschemneuro.9b00036

32. Abreu CM, Thomas V, Knaggs P, Bunkheila A, Cruz A, Teixeira SR, Alpuim P, Francis LW, Gebril A, Ibrahim A, Margarit L, Gonzalez D, Freitas PP, Conlan RS, Mendes Pinto I (2020) Non-invasive molecular assessment of human embryo development and implantation potential. Biosens Bioelectron 157:112144. https://doi.org/10.1016/j.bios.2020.112144

33. Wink T, van Zuilen SJ, Bult A, van Bennekom WP (1997) Self-assembled monolayers for

biosensors. Analyst 122:43R–50R. https://doi.org/10.1039/a606964i

34. Welch NG, Scoble JA, Muir BW, Pigram PJ (2017) Orientation and characterization of immobilized antibodies for improved immunoassays (review). Biointerphases 12:02D301. https://doi.org/10.1116/1.4978435

35. Hermanson GT (2013) In: Audet J, Preap M (eds) Bioconjugate techniques, 3rd edn. Elsevier, London. ISBN: 9780123822390

36. Denisin AK, Karns K, Herr AE (2012) Post-collection processing of Schirmer strip-collected human tear fluid impacts protein content. Analyst 137:5088. https://doi.org/10.1039/c2an35821b

Chapter 15

Aptamer Detection of Neurodegenerative Disease Biomarkers

Hui Xi and Yang Zhang

Abstract

Neurodegenerative disease is a kind of disease caused by the degeneration of neurons and myelin sheaths. With the aging of the global population, the increase of the incidence of neurodegenerative diseases, such as Alzheimer's disease (AD), Parkinson's disease (PD), and Huntington's disease (HD), has become a great challenge for society at large. Most of these diseases relate to misfolded proteins in the central nervous system, such as: amyloid-ß, tau, α-synuclein, huntingtin, and prion proteins. Additionally, there is a large amount of evidence linking vascular dysfunction and vascular risk factors with the pathogenesis of neurodegenerative diseases, such as vascular dementia (VAD). It is also possible to detect neurodegenerative diseases by detecting angiogenesis biomarkers. As biomarkers of neurodegenerative diseases, detecting these proteins can potentially allow for the slowing down or prevent the onset of these diseases. Aptamers can be used to detect these neurodegenerative disease biomarkers. Aptamer, a small oligonucleotide sequence, is screened in vitro, which can bind to corresponding protein markers with high affinity and specificity. As an ideal biorecognition element for neurodegenerative disease detection, the chapter summarizes recent advances on aptamers and their application in the detection of neurodegenerative diseases.

Keywords Aptamers, Neurodegenerative diseases, Biomarkers, Biosensors

1 Introduction

Neurodegenerative disease (NDs) is a chronic disease, which is characterized by loss of a large number of neurons, which leads to progressive disability, cognitive dysfunction, and possibly death [1]. With the ageing population, this kind of disease is increasing dramatically. Many countries have invested a lot of resources into the research and treatment of neurodegenerative diseases. However, the increase of the incidence of neurodegenerative diseases is still a huge challenge for society [2]. Neurodegenerative diseases can be divided into either acute neurodegenerative diseases or chronic neurodegenerative diseases. The former includes strokes and traumatic brain injuries, the latter mainly includes Alzheimer's disease (AD), Parkinson's disease (PD), and Huntington's disease

Philip V. Peplow, Bridget Martinez and Thomas A. Gennarelli (eds.), *Neurodegenerative Diseases Biomarkers: Towards Translating Research to Clinical Practice*, Neuromethods, vol. 173, https://doi.org/10.1007/978-1-0716-1712-0_15,
© Springer Science+Business Media, LLC, part of Springer Nature 2022

(HD) [1, 2]. Although the pathological location and etiology of these diseases are different, their common feature is characterized by the gradual appearance and diffusion of the misfolded proteins accumulated in the brain [3]. These proteins are easy to self-assemble into small aggregates, which are called oligomers, and can sometimes absorb additional monomers and extend through the fiber structure. The accumulation of these can form larger aggregates, which can be either intracellular or extracellular. It is unclear how the aggregation of the proteins could produce toxicity to neurons [4]. Biomarkers refer to biochemical indicators that can be used to indicate changes in the structure or function of organs, tissue, cells and subcellular. Compared with their normal condition, the content, structure and number of biomarkers may change according to the progression of disease [5], and this can be used to detect the disease. Previous work showed that the biomarkers of Alzheimer's disease (AD) are amyloid-ß (Aβ), tau, α-synuclein (α-syn), and TAR DNA/RNA binding protein 43 (TDP-43) [6–8]; the biomarker of Parkinson's disease (PD) is α-syn [9]; the biomarker of prion diseases is prion [10]; and so on. These biomarkers are closely related to the pathogenesis of neurodegenerative diseases.

Although neural cell dysfunction and death are the basis of clinical symptoms of NDs, the mechanisms that promote this pathology remain to be determined. More and more evidences show that cardiovascular diseases are closely related to neurodegenerative diseases such as AD [1]. People with heart disease have a high rate of developing NDs. In the brain of nondementia patients with heart disease, the incidence of AD-like Aβ deposits in neurons is increased [6]. However, the incidence of neurodegenerative diseases can also lead to the growth of cardiovascular diseases [11]. Several studies have shown that there is a high correlation between cardiovascular mortality and AD, as well as between hypertension, hyperlipidemia, diabetes, and dementia. Vascular components are related to the pathogenesis of neuronal degeneration [12]. In NDs, there are many structural and functional abnormalities of cerebral microvasculature, such as vascular distortion, torsion, bending and circulation, which lead to the decrease of vascular density [11]. Grammas et al. [13] believe that hypoperfusion/hypoxia can contribute to the pathogenesis of NDs. Hypoxia-inducible factors (HIF) increase in the microcirculation of AD patients. Cerebral hypoxia is a powerful mechanism to stimulate vascular activation and angiogenesis. Experimental results show that the microvessels isolated from the brain of AD patients express a large number of angiogenic proteins. Although there is no study on the relationship between angiogenesis and NDs, the evidence suggests that there is a relationship between angiogenesis biomarkers and their target ligands, which is beneficial to the prevention and treatment of neurodegenerative diseases. There are promoting factors and inhibiting factors for angiogenesis, both of which can be understood as the natural defense mechanism to help restore the oxygen and nutrition supply of the affected brain tissue, as angiogenesis provides nutrients for new neuron formation and nerve cells. Matrix metalloproteinase (MMPs) and vascular endothelial growth factor (VEGF), stimulating angiogenesis mediators, thrombospondins (TSP) and endostatin and antiangiogenesis mediators can be used as detectable biomarkers [14, 15].

A biosensor is a kind of detection instrument which converts biological substances into easily observed signals. Aptamer is a small oligonucleotide sequence or a short polypeptide screened in vitro, which can bind with corresponding ligands with high affinity and specificity [16–18]. The aptamer-based biosensor is a kind of biosensor using aptamers as its recognition element. With the combination of the known sequence of nucleic acid molecules fixed on the surface of the sensor or transducer and the target material, the molecule interaction events can be converted into a specific signal transformed by the transducer. It is common and intuitive to detect an electrical signal transformation. This kind of sensor becomes an electrochemical aptamer biosensor. An electrochemical aptamer biosensor has applications in cancer [19], pathology [20], toxicology [21], and other fields, such as gene expression [22], drug delivery [20], and molecular sequencing and diagnosis [19, 23]. Therefore, it can also be used to detect biomarkers of neurodegenerative diseases. The aptamer on the sensor can be modified to improve the sensitivity and reduce the detection limit.

Aptamer modification has been widely used in disease detection with biosensors. It can improve the sensitivity, selectivity, affinity, and other characteristics of biosensors, and overcome the shortcomings of traditional sensors [24]. This is because aptamers have a large specific surface area and a large number of receptor binding sites, with many spatial structural forms, and they can easily form hairpin, stem ring, clover, and quadrupole structures, which can be closely combined with the target substance based on van der Waals force and hydrogen bond action [25]. They can also distinguish between similar substances and improve specificity. Aptamers are oligonucleotide sequences with high affinity and specificity with the target substance selected from the oligonucleotide library by SELEX technology [26, 27]. Compared with the antibody antigen binding detection target, SELEX screening of aptamers is easy to synthesize, easy to modify; has good repeatability and stability; and is easy to store. Normally, an initial library of oligonucleotide sequences containing random and fixed primers regions was used. Sequences can be of either natural DNA or RNA chemistry, or synthetic nucleic acids. Aptamer can adapt and fold to recognize many nonnucleic acids, small molecules, heavy metal ions, proteins, and so on [28–31]. Different aptamer folding into different structures may have a different ability to recognize target molecules, which affects their sensitivity and specificity. The random region sequence of oligonucleotide normally contains 30–100 bases, which can form common structural motifs, such as hairpin, bulge, pseudoknot, and G-quartet [32–35].

This chapter summarizes several common neurodegenerative diseases and their biomarkers and also introduces the methods of detecting these biomarkers of NDs, focusing especially on the application of biosensors in this detection. It will also summarize the recent advance of biomarkers related to NDs, such as angiogenesis biomarkers and their corresponding aptamers and the detection of biosensors based on aptamers [36–39].

2 Detection Methods of Neurodegenerative Diseases

2.1 Current Detection Methods

In neurodegenerative diseases, the search for potential biomarkers is necessary. Mass spectrometry is the key technology for qualitative and quantitative identification of protein biomarkers [40]. Mass spectrometry is used to separate proteins by electrophoresis, chromatography and other separation technologies, and is usually combined with high performance liquid chromatography and gas chromatography [41]. Great achievements have been made in the understanding of the biomarkers of Alzheimer's disease (AD), Parkinson's disease (PD), and multiple sclerosis. Shi M et al. [42] reported that tau in phosphorylated saliva can be detected in patients with Alzheimer's disease by proteomic method, which shows that this method has the potential for detection and diagnosis.

The detection of neurodegenerative diseases has always been a challenge in the medical field, because the high complexity of brain structure changes with age and pathological history. Although imaging technology can directly observe the structure of brain [43, 44], developing effective technology to identify biomarkers is of great interest. A comparison of those common methods is summarized in Table 1.

2.2 Application of Electrochemical Biosensors in Neurodegenerative Diseases

To construct a biosensor for the analysis and detection of NDs biomarkers, selecting the appropriate biological receptor on the sensor and the transducer with high sensitivity and good selectivity of the NDs biomarkers is important. The common biological receptors for biomarker detection include aptamers and antibodies. Compared with antibodies, aptamers have many advantages. Many common detection methods are relatively expensive, time-consuming, inefficient, and less sensitive. Therefore, biosensors, especially electrochemical biosensors powered by aptamers, are faster, more sensitive, more specific, and more cost-effective [22, 23].

Table 1
Comparison of different detection methods for neurodegenerative diseases

Method	Advantages	Disadvantages	Ref.
Mass spectrometry	High accuracy and sensitivity	Expensive	[45]
Myocardial perfusion imaging	Visual stereoscopic image	Expensive	[46]
Positron emission computed tomography	Good sensitivity and security	Expensive Low scanning speed	[47]
Enzyme-linked immunosorbent assay	Easy to operate, fast, wide range of application, cheap	Expensive Poor spatial resolution	[48]
Western blot	Simple and efficient qualitative test	Insensitive to low level markers, false positives	[49]

In recent years, electrochemical biosensors have been widely used in food monitoring, environmental detection, medical research, and clinical diagnosis due to their advantage of having high sensitivity, a fast response, a low price, good biocompatibility, and miniaturization [40]. For the detection of biomarkers in NDs, electrochemical biosensors, as a low-cost and fast detection tool, provide a potential alternative to advanced biological analysis system for complex samples. An electrochemical biosensor is a kind of detection instrument which is sensitive to biological substances and converts biological concentrations into electrical signals. Biosensors consists of two parts: a biometric part and a transformation part. Enzymes, antibodies, nucleic acids, and other bioactive substances are used as recognition elements, and the molecular recognition part is used to identify the target that is to be tested, by causing some physical or chemical changes, which is the basis of the sensor's selective determination. An analysis tool or system composed of an appropriate transducer, such as an oxygen electrode, a photosensitive tube, a field-effect tube or a signal amplification device, which can transform the signal induced by biological activity into an electrical signal transducer, has the function of a receiver and a converter [25]. An electrochemical biosensor, as a low-cost and rapid method to detect neurodegenerative diseases, provides a potential choice for the biological analysis system of complex samples. Through the modification of the electrode combined with the appropriate aptamer, combined with the biomarkers of neurodegenerative diseases, the change in electrochemical signals can be detected (Fig. 1).

2.2.1 The Principle of Electrochemical Biosensor

A biologically active substance with a known sequence is fixed on the surface of the biosensor or converter as a probe. Specific binding of antibodies, enzymes, aptamers with target substances refers to the change of the electrochemical response signal of the

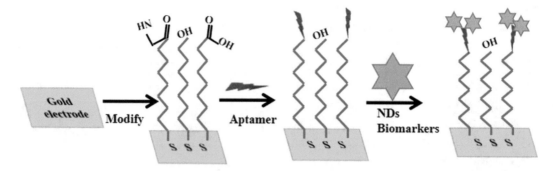

Fig. 1 A schematic diagram of NDS biomarkers detection. The gold electrode of the biosensor was modified with the active group. The aptamer targeting the corresponding ND's biomarkers were modified on the sensor. When biomarkers are measured in the solution, the distance or structure of the aptamer to the electrode will be changed, thus the capacitance, resistance or voltage will be changed, and the biomarkers of neurodegenerative diseases can be detected

indicator, the detection of the existence of the target gene, and the quantitative analysis of the target [16]. When the concentration of the target substance changes, the electrical signal of the indicator, after embedding the aptamer and the target binding structure, has a certain relationship with the amount and the concentration of the measured substance within a certain range, so the quantitative analysis of the substance to be measured can be carried out. So, the strength of the electrochemical response signal can be attributed to the redox reaction of the indicator molecule in the binding target substance, and the current corresponding to the potential changes in the redox reaction; and additionally, the contact area and distance between the electroactive substance and the electrode change after the binding with the target, resulting in a change of the electrical signal [18–20].

2.2.2 Aptamer-Based Electrochemical Biosensor

The preparation of an electrochemical sensor based on aptamer relies on the combination of probe and electrode. Taking an electrochemical DNA sensor as an example, in order to make the DNA and the electrode closely linked, stable, and to maintain their current activity and meet the sensitivity requirements, the DNA biosensor can be constructed by different methods, and a DNA probe can be modified on the electrode [50]. Adsorption method: due to the interaction between a negatively charged phosphate group of DNA probe and a positively charged electrode, there is direct adsorption, such as chitosan formation [51]. Biotin–avidin method: biotin is modified on the electrode surface, avidin is modified on a DNA probe and the DNA is modified on the electrode through a biotin avidin interaction [52]. Chemical bonding: the probe molecule and electrode interact through an amide bond, ether bond or covalent bond to the biosensor [53]. Self-assembly method: Au is the base electrode, and a sulfhydryl group is modified on the DNA to form an Au-S bond of DNA, and self-assembled on the electrode. When based on a DNA biosensor, target molecules can be selectively identified, and electrochemical detection can be performed by combining the hybridization indicator of a recognition signal [54]. The response of electrical signals can be detected by the change of current, the potential, conductance, resistance and capacitance. The indicator here can be a hybrid indicator of electrochemical activity as an identifier, or an electrochemical activity functional group modified on the oligonucleotide sequence [55]. This includes metal complexes, such as Fe, Ru and other metals, which form complexes with bipyridine and other substances, because the changes of metal ions in the center site can be matched to the change in electrical signals. A dye hybridization indicator, acting on a DNA molecule, changes its spatial structure [56–59].

Most of the aptamer-based sensors are based on single strand DNA or RNA. Natural nucleic acids do not have functional groups that can produce absorption in visible regions, fluorescence or magnetic or electrochemical signals [58]. Therefore, to make the aptamer into a sensor, an external signal label needs to be applied. In order to achieve this goal, many organic fluorescent chromophores and electrochemical active tags have been used [60–62]. The development of electrochemical DNA biosensors can be used to detect both unmarked and labeled DNA targets. Therefore, it can be divided into labeled electrochemical biosensors and unmarked electrochemical biosensors [63, 64]. The labeling step improves the sensitivity and selectivity of measurements, but also increases the time, complexity and cost of each measurement. The labeled electrochemical biosensor utilizes the electrochemical activity of the probe as an indicator [65]. When the probe is hybridized with the target molecule, the structure of the modified probe molecule will change, and then the distance between the probe molecule and the electrode surface will change too, which generates different signals which can then be detected. The commonly used markers are ferrocene (Fc), methylene blue (MB) redox markers, enzyme markers and nano material markers [60, 66–68]. It is reported that the size of the electrical signal can be changed according to the presence of the target, in order to achieve the target detection. The detection limit of the sensor is low and the sensitivity is high. The principle of the sensor being based on the conformational change of the MB marker molecule is similar. After binding to the target molecule, the MB marker which was previously far away from the electrode surface is closer to the electrode surface, so it can generate electrical signals and improve the sensitivity of the sensor [69]. Compared with the labeled DNA biosensor, the unmarked electrochemical DNA biosensor does not need to label the electroactive substance on the nucleic acid probe.

According to the different affinity before and after the hybridization of an indicator with the target, the amount of each indicator in the system is different before and after the hybridization, and differing electrochemical signals are generated to detect the target molecule [70–72]. Han et al. [69] first developed a simple, tagless, amplification-free electrochemical biosensor, which uses methylene blue (MB) as the indicator of hybrid redox, and uses a DNA nanostructure probe to detect miRNA. Under the condition of low ion strengthening, high specific hybridization was carried out to improve the sensitivity. In order to improve the sensitivity of the sensor and reduce the detection limit, we can modify the aptamer and amplify the signal of the detector.

3   Aptamers in Neurodegenerative Diseases

The selection of aptamers for neurodegenerative diseases can be used in basic research to better understand the development of these diseases. Moreover, they can also be used to develop new diagnoses or treatments. Suitable ligands can be selected according to a variety of targets from small compounds (such as vitamins, minerals, or amino acids) to large molecules (nucleic acid structures, peptides, or proteins). Generally, aptamers can be detected in the range of nM to pM, but for small molecules, their affinity is often lower than the μM range [73]. Aptamers can be used in treatment, purification processes and diagnosis, and also in biosensor detection.

Compared with other bioactive molecules, aptamers have a wide range of targets, including tissues, cells, viruses, proteins, and toxins [36–38]. Moreover, aptamers are easy to label with fluorescence without their activity being affected. Therefore, aptamers can be easily synthesized to obtain corresponding nucleic acid sequences, the combination of which can then be used in the application of sensors [39]. Aptamers have many advantages, such as high affinity and specificity, nontoxicity, and nonimmunity (Table 2).

3.1   Aptamer and SELEX

An aptamer is a small oligonucleotide sequence or short polypeptide obtained through in vitro screening. These single chain nucleotides or peptides can specifically recognize and bind a large number of target molecules, such as gold ions, organic dyes and amino acids [28–30], antibodies [31], proteins [32, 33], and even entire cell [34–36], due to various intermolecular forces, such as van der Waals force, hydrogen bonds, and base accumulation. Aptamers were originally thought to be DNA or RNA fragments. They were originally developed almost simultaneously by two research groups in the 1990s [21, 40]. The application of the chemical modification of aptamers in analytical chemistry and

Table 2
Aptamer vs. antibody. A detailed comparison between aptamer and antibody

Aptamer	Antibody
It is easy to synthesize in vitro and easy to produce	Produced in vivo and with a long production time
Strong specificity and good repeatability	Affinity depends on the antigens
Low cost and easy to modify	Expensive, and hard to modify
Small size, meaning a more effective access to biological compartment	Large size, and the antibody is blocked from entering the biological compartment
Nonimmunization	Immunization

enzyme chemistry provides a research platform for chemical biology and biomedicine. Over the past few decades, more than 2000 aptamers have been developed for about 141 target ligands [74]. Studies have shown that aptamers bind to proteins to prevent their function and accumulation and they generate different signals for qualitative and quantitative analysis of proteins. Scientists are trying to use aptamers to treat diseases or detect pathological processes [75].

The library sequences that are not combined with the target are removed. The target binding sequences obtained by heating are used as the template for PCR amplification (or RT-PCR for RNA Library), and the library's next cycle is prepared. After 8–20 cycles of continuous selections, the oligonucleotide sequences with high specificity and high affinity for the target are obtained. Then, after cloning, the corresponding nucleic acid sequence is obtained. The α-synuclein (α-syn) of the AD biomarker and the alpha synuclein (α-syn) of the PD biomarker can be screened by SELEX. Since the classical SELEX aptamer selection method has been developed, various improvements and modifications have been introduced to shorten the selection time and improve the binding affinity [38, 39]. These improved SELEX methods are as follows: cell SELEX, immunoprecipitation coupled SELEX (IP-SELEX), capillary electrophoresis SELEX (CE- SELEX), capture SELEX and atomic force microscope SELEX (AFM-SELEX), and so on [26].

3.2 Characteristics and Optimization of Aptamers

An aptamer has shorter detection cycle, lower detection limit, high affinity and strong specificity. Due to the large surface area of an aptamer and a large number of receptor binding sites, a unique spatial structure recognition target is formed. Compared with antibody detection, the aptamer obtained by SELEX is easy to synthesize in vitro, with good repeatability, stability, and in vitro modification. Compared with molecular biology, it is simple and easy to operate, with low detection costs, and a wide range of aptamer targets, including tissues, cells, viruses, and proteins. [32, 33]. Aptamers are easy to label with fluorescence without any side effect, so they are easily integrated into other detection technologies. Despite the advantages of aptamers, they still have some obvious limitations, which may hinder their clinical application. The natural unmodified aptamer is unstable, especially for RNA aptamers. These may be degraded by nuclease in the blood and have shorter span of effective activity, which limits their application in practice. Most aptamers have a molecular weight between 5 and 10 kDa, so they can rapidly penetrate into the target tissue or tumor, and they can also be removed by the body without having interacted with their target at all [34, 35].

In order to overcome these obstacles, the natural aptamer can be modified in vitro to improve its stability and specificity. The aptamers can be chemically modified or conjugated with nanoparticles. Some functional groups such as amino group, carboxyl group and hydroxyl group can improve the stability of aptamer. Rhie A et al. [76] have shown that 2′-fluoro-RNA modified aptamers preferentially bind to prions disease-related conformations and inhibit transformation, which in turn shows that the stability of aptamers has been significantly improved. Similarly, 4-′-thiopyrimidines and 2′-amino pyrimidines can also be modified on aptamers. 3′- and 5′-ligands are often modified, such as in the case of biotin streptavidin and cholesterol. Additionally, the conjugated modification with nanoparticles can also improve the stability of aptamers. Nano materials have many properties, such as a small size, a large specific surface area, excellent electronic performance, electrocatalytic activity, and good biocompatibility, alongside their good physical and chemical properties, which have a promising application in ultrasensitive electrochemical biosensors. Nanoparticles, such as liposomes [77], quantum dots [78], aggregation-induced emission (AIE) [79], and virus-based vectors [80] improve the affinity of aptamers. To improve the sensitivity of electrochemical sensors, Park et al. [81] used a sandwich of nano gold to improve the detection sensitivity of the target DNA to 5×10^{-13} M. DNA probes are placed between the separated microelectrodes to separate the positive and negative hybridization signals according to their circuit resistance between the electrons. Due to the large surface of the electrode and the amount of oxidized gold atoms of each nano particle, the resistance is regulated by the gold nano particle that reacts to the target, so the detection target molecules with a high sensitivity can be detected. Farokhzad et al. [82] showed that conjugating of nanoparticles to aptamers can improve the affinity between aptamers and their targets when targeting prostate cancer cells. In order to avoid the degradation of aptamers before the detection, carriers such as polycation [28], liposome [77], chitosan [22], peptide [32], and quantum dots [78] can protect aptamers from degradation, to improve the stability of aptamers.

3.3 Application of Aptamers in Neurodegenerative Diseases

The prevalence of neurodegenerative diseases has attracted worldwide attention, so using aptamers to detect the biomarkers of neurodegenerative diseases has attracted the attention of many scholars. Neurodegenerative diseases are characterized by the accumulation of misfolded proteins in the central nervous system (CNS). These diseases, including cognitive disorders such as AD, PD, HD, and MS, can cause strokes in severe cases, and a stroke is harmful to people's health. Aptamers can be combined with these targets to prevent the accumulation of misfolded proteins or reduce the negative effects, in order to further achieve the purpose of the detection and treatment of diseases [76].

3.3.1 Aptamers for Alzheimer's Disease Biomarkers

Alzheimer's disease (AD) is one of the main neurodegenerative diseases, and is characterized by memory regression, cognitive impairment, gradual loss of activities in daily life, and an evolution into dementia and death. With an increasingly older population, AD has become a serious public health problem in the world [8]. AD is a slow and persistent disease. The first symptom of cognitive dysfunction is not obvious, so it is difficult to identify the pathological process only by the clinical manifestations. For this reason, biomarkers are used for the early detection of AD. AD is characterized by amyloid plaques and neurofibrillary tangles in the brain, which contain the proteins amyloid-β (Aβ), peptides, and hyperphosphorylated tau [6–8]. The misfolding of these proteins is toxic to neurons, which leads to cell death and neurodegeneration by way of destroying the normal function of cells. This induces oxidative stress and causes inflammation. But the effects of a more toxic misfolded monomer, oligomer or final aggregate remains to be studied. At present, there are three biomarkers used for the diagnosis of AD in human cerebrospinal fluid and blood: tau protein, Aβ and apolipoprotein E4 (APOE4) [6, 7, 83]. According to the structure of these biomarkers, we can choose the best biological receptor and bioassay in order to improve sensitivity and selectivity, and link them with biosensors for detection.

Tau protein is a microtubule related protein that forms intracellular aggregation in several neurodegenerative diseases. In AD [84], the abnormal phosphorylation and modification of the tau protein are the characteristics of its pathology. Both over phosphorylated tau and oligomeric tau are attracted to synaptic loss, which initiates the misfolding of the tau protein in neurons. Different types of tau control different areas, such as p-tau181, which can highlight the difference between AD and dementia in a Lewy body. P-tau231 can also differentiate between AD and frontotemporal dementia (FTD) [85]. One of the characteristics of AD is the existence of an Aβ viscous plaque, which forms around neurons.

Since the aggregation of peptide Aβ is a marker of AD, many studies have established aptamers for this peptide. The monomeric form of Aβ does not have neurotoxicity, but oligomers and fibers formed through the complex process of nucleation dependence show neurotoxicity and prevent long-term enhancement in the affecting of synapses. Associated with AD are neurofibrillary tangle, Aβ plaque and insoluble hydrophobic Aβ aggregation. Amyloid precursor protein (APP) is a type I transmembrane monosaccharide protein, which is located on chromosome 21q and consists of 695–770 amino acid residues [84]. APP is widely expressed in almost all neurons and nonneuron tissues. There are two mechanisms of APP degradation: one is the nonamyloid protein generation pathway, in which the neurons located in an Aβ sequence contain proteins with metalloproteinase domain for degradation; the other is the amyloid protein generation pathway, which forms toxic oligomeric species and plaques through aggregation. Human apolipoprotein E (APOE) is a risk factor of AD. It mediates the binding of lipoproteins to low density lipoprotein receptors in peripherals and the central nervous system (CNS). APOE stimulates the transcription factor AP-1, which improves the transcription of APP and improves the level of Aβ. The most common APOE subtypes are APOE2, APOE3, and APOE4. APOE4 transports cholesterol; APOE4 [83] isomers have different transporting efficiency. The APOE4 allele is related to cholinergic dysfunction due to an increased amyloid burden, and APOE4 is related to an increased risk of AD.

Although the mechanism of AD is not clear, its unique feature is the accumulation of misfolded proteins in CNS. Therefore, it is necessary to develop possible measures to reduce or prevent the occurrence of AD. To do this it is necessary to find suitable aptamers to bind to these target proteins, and to detect AD early enough to prevent their accumulation, so as to prevent the progress of AD. The specific aptamers of the tau protein have been developed. These aptamers inhibit the oligomerization tendency in tau, but do not inhibit the oligomerization tendency in the tau disease, which affects the half-life of the tau. Tau aptamers also significantly reduce the neurotoxicity of tau oligomer mediated hippocampal neurons and the synthesis of dendritic spine loss. In recent years, many studies on the aptamers of tau-381, tau-410 and tau-441 have been reported [86–90]. Since the aggregation of peptide Aβ is a marker of AD, many studies have been done to establish aptamers to fight against this disease.

The aptamers for these proteins can be incorporated into fluorescently labeled biosensors, and electrochemical biosensors. Electrochemical biosensors can convert chemical signals into measurable electrical signals by using potential, ampere and impedance sensors. Cyclic voltammetry (CV), differential pulse voltammetry (DPV) and electrochemical impedance spectroscopy (EIS) [90–92] are the most commonly used electroanalytical techniques for the detection of AD biomarkers. Esteves Villanueva Jo [88] and others developed a biosensor based on electrochemical analysis to detect the misfolding of the tau protein. Because tau binding causes protein conformation changes, and the changes can be transferred to a change in resistance, which can be recorded by the Au electrode, which can be used to detect AD. The sensor has a detection range of 0.2 to 1.0 μM tau. Scarlet et al. [93] reported that the electrochemical sensor was used to detect tau-441, and that the detection limit was 0.03 pM, with high sensitivity and low detection limit: aptamers for Alzheimer's disease biomarkers (Table 3). Because these aptamers are specific for oligomers, they are used to recognize Aβ oligomers, prevent Aβ toxicity, and diagnose AD.

3.3.2 Aptamers for Parkinson's Disease Biomarkers

Parkinson's disease (PD) is the second most common neurodegenerative disease, and is most common in people over 60 years old. PD is mainly defined by the symptoms of dyskinesia, such as resting tremor, bradykinesia, postural instability, and stiffness. The cause of PD symptoms is the aggregation of α-synuclein misfolding in the neurons within cytoplasmic Lewy bodies (LBS) and the decrease of dopamine caused by the death of dopaminergic neurons in the substantia nigra [94]. The aggregation of α-synuclein, especially oligomer, has a negative effect in neurons, causing damage to the degradation system, mitochondrial damage, oxidative stress, and toxicity. In other neurodegenerative diseases, the aggregation of α-synuclein in Lewy bodies was also found, such as Lewy body dementia and multiple system atrophy. Therefore, aptamers against α-synuclein are potential tools for the diagnosis or treatment of PD. Ikebukuro and his team [95] proposed the first aptamer for a α-synuclein oligomer, called "M5–15." It can reduce the protein of an α-syn amyloid fiber and a soluble oligomer in vitro, and can be used for the detection and treatment of PD. DA is related to the signal transduction of neurons, and the loss of DA will lead to the relaxation of the basal ganglia and dyskinesia. In recent years, aptamer biosensors for the detection of biomarkers in Parkinson's disease have been developed (Table 4).

3.3.3 Aptamers for Huntington's Disease Biomarkers.

Huntington's disease (HD) is an autosomal dominant genetic disease, which leads to the degeneration of the nervous system and causes serious motor, cognitive and mental disorders. HD is one of the major inherited neuronal tissue diseases. It is speculated that about 5 to 10 of every 100,000 people are affected by the disease. This disease is related to the mutant form of huntingtin (HTT), which originates from the increase of glutamine codon (CAG) in the gene. These mutations lead to the production of a

Table 3
Aptamers for Alzheimer's disease biomarkers

Target	Aptamer sequences	Detection method	Linear range	Limit of detection	Ref.
Tau-381	ssDNA: 5′-GCGGGAGCGTGGCAGG-3′	Fluorescence	–	–	[86]
Tau-410	dsDNA: 5′-CTTCTGCCCGCCTCCTTCC-3′ 3′-GAAGACGGGCGGAGGAAGG-5′	Fluorescence	–	–	[86]
Tau-441	RNA: 5′-CCGUGUCUUCGUGAGGUCGGUGUCGGCUUGGC AGAAAGGG-3′	CV and EIS	0.2 to 1.0 μM	0.2 μM	[87, 88]
Aβ	DNA: 5′-HS-GCCTGTGTTGGGGGGGTGCG-3′	CV and EIS	5 pM–2 nM	3 pM	[89]
Aβ1–40	RNA: 5′-UUUACCGUAAGGCCUGUCUUCGUUUGACAGCGGC UUGUUGACCCUCACACUUUGUACCUGCUGCCA-3′	CV	0.02–1.50 nM	10 pM	[90]
Aβ40	DNA: 5′GCGTAATACGACTCACTATAGGGCGGGGAATTCGAGCTC GGTACCTTTACCGTAAGGCCTGTCTTCGTTTGACAGCGGC TTGTTGACCCTCACACTTTGTACCTGCTGCCAACTGCAGGC ATGCAA GCTTGG-3′	Optical	–	–	[91]
APOE4	–	CV	1.0–10,000 ng/ mL	0.3 ng/mL	[92]

Table 4
Aptamers for Parkinson's disease biomarkers

Target	Aptamer	Detection method	K_d	Ref.
DA	DNA: 5'-GTCTCTGTGTGCGCCAGAGAACAC TGGGGCAGATATGGGCCAGCAC AGAATGAGGCCC-3'	Electrochemical	2.8 μM	[99]
	RNA: 5'-GGGAAUUCCGCGUGUGCGCCGC GGAAGAGGGAAUAUAGAGGCCAGCAC AUAGUGAGGCCCUCCUCCC-3'	Electrochemical	–	[97]
	RNA: 5'-GUCUCUGUGUGCGCCAGAGAACAC UGGGGCAGAUAUGGGCCAGCAC AGAAUGAGGCCC-3'	Fluorescence	1.6 μM	[95]
α-Synuclein oligomer	DNA: 5'-ATAGTCCCATCATTC ATTGTATGGTACGGCGCGGTGGCGGGTGC GTGGAGATATTAGCAAGTGTCA-3'	Fluorescence	–	[99]
	DNA: 5'-GCCTGTGGTGTTGGGGCGGGTGCG-3'	Fluorescence	68 nM	[98]

misfolded protein, which can gather together. Therefore, it is vital to find an aptamer for the detection and treatment of HD. Skogen et al. reported an aptamer, 20-mG-rich oligonucleotide, which can effectively inhibit huntingtin protein aggregation [101, 102].

3.3.4 Aptamers for Other Neurodegenerative Diseases Biomarkers

Several aptamers have been developed for prionopathies, also known as transmissible spongiform encephalopathies (TSE) [103]. TSE is a neurodegenerative disease that affects human beings and other mammals. The pathological feature of TSE is a normal cell prion protein (PrP^C), an isomer rich in α-helix transformed into an abnormal PrP^{Sc} subtype and an isomer rich in β-sheet accumulates in the brain. The process of the prion damaging the host is not clear, but some experiments show that PrP^{Sc} plays a role in neuronal dysfunction. Therefore, it is necessary to select a large number of aptamers for the detection of PrP^c and PrP^{Sc}. In recent years, a wide range of research has been carried out to study the pathology of different prion proteins and to establish suitable aptamer sensors that can detect different kinds and different conformational proteins. Rhie A et al. [104] selected SAF-93, fluoro-modified RNA oligomers as two suitable aptamers for PrP^{Sc} detection. The affinity of these aptamers to PrP^{Sc} is ten times higher than that of PrP^c, because there are two specific heparin binding sites in a PrP^{Sc} molecule. In addition, SAF-93 can convert PrP^c to PrP^{Sc} for detection and treatment. Some DNA aptamers have been used in further studies to develop the detection of PrP^c and PrP^{Sc} using fluorescence, electromagnetic luminescence [105], surface plasmon resonance [106] or electrochemical sensors. Multiple sclerosis (MS) is an autoimmune inflammatory disease of the central nervous system. Its pathological feature is that the insulating myelin sheath of the nerve axon is destroyed locally. At present, there is no therapy to prevent the progress or induce the repair of patients with vision loss, gait and cognitive impairment. Therefore, MS therapy is a major challenge. For example, Nastasijevic et al. [107] selected a 40 nucleotide DNA aptamer to use against a mouse's myelin sheath, and the aptamer was seen to promote the remyelination of CNS lesions in these mice.

Neurodegenerative diseases are characterized by the accumulation of misfolded proteins in the central nervous system, including AD, PD, TSE, HD, and MS, which are all "neurocentric." We recognize that neural cell dysfunction and death are the base clinical symptoms of NDs, but NDs are complex diseases in comparison to many other biological mechanisms. Neural restorative events include neurogenesis, synaptosomes and the formation of new blood vessels. More and more evidences show that angiogenesis is closely related to neurodegenerative diseases. Neurovascular abnormalities occur in many brain diseases and neurodegenerative diseases. The characteristics of angiogenic factors and processes are found in the brain of a ND patient. The whole genome expression profile of a brain with an ND has determined the significant increase in the regulation of genes that promote angiogenesis [86]. Further clinical views may come from the recently discovered vascular endothelial growth factor (VEGF) as a modification of amyotrophic lateral sclerosis (ALS) in neurodegenerative diseases. This can be called the "vascular hypothesis" of NDs, which considers NDs as vascular diseases with neurodegenerative consequences or "vascular lesions" [108]. In the process of physiological angiogenesis, the response to the brain's endothelial cells (ECs) is activated and the vasogenic medium is released to form new blood vessels, and then the activated ECs are closed by a feedback inhibition circuit signal. In NDs, because of continuous stimulation, activated ECs release a large number of inflammatory proteins and angiogenic mediators [106], many of which have a direct impact on neuronal activity (Fig. 2).

The process of angiogenesis is complex, including matrix decomposition, endothelial germination, endothelial proliferation, migration, tube formation, and the transfer of peripheral/smooth muscle cells (SMC) to mature vessels. Healthy blood vessel formation is strictly regulated, and disorders occur when in a pathological state [11]. The process of angiogenesis is present in various pathological states, including ischemic heart disease, peripheral arterial occlusive disease, brain tumor formation, cancer, age-related macular degeneration, neurological diseases, and neurodegenerative diseases, including Alzheimer's disease, amyotrophic lateral

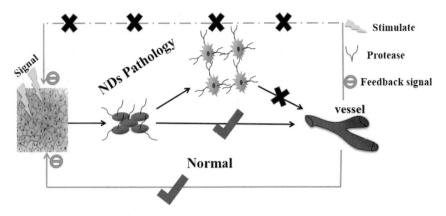

Fig. 2 Dysfunctional cells continue to express and release high levels of inflammatory and neurotoxic factors, leading to the decrease of microvascular density in the brain and the loss of feedback inhibition of endothelial cells. Those are the causes of neuronal dysfunction, death and neurodegenerative diseases

sclerosis, diabetic neuropathy, and stroke. The vascular reconstruction of the brain is either through the formation of new blood vessels from existing ones, or through the deposition of some endothelial progenitor cells in the process of angiogenesis. The formation of a blood–brain barrier (BBB) is an important step in the angiogenesis in the physiology of the central nervous system (CNS). BBB is regulated by a neurovascular unit (NVU) composed of endothelial cells, perivascular cells and vascular astrocytes. NVU plays a key role in preventing the entry of neurotoxicity and regulates the blood flow of the CNS. The central nervous system, consisting of the brain and spinal cord, has highly specialized blood vessels to meet the needs of a highly active metabolism and to protect sensitive neurons from toxic metabolites and exogenous substances. The tissue-specific function of CNS blood cells is regulated by the NVU, which is regulated by the dialogue between nerve cells and vascular cells [11, 109].

Angiogenic markers are mainly composed of angiogenic proteins, the most important of which are the vascular endothelial growth factor family (VEGF-A, VEGF-B, VEGF-C, VEGF-D), human angiopoietin receptor tyrosine kinase 2 (Tie-2), soluble fms-like tyrosine kinase receptor 1 (sFlt-1), placenta growth factor (PLGF) and basic fibroblast growth factor (bFGF). These are potential prognostic indicators for disease activity and in the detection of NDs. These biomarkers can be used to detect signs of angiogenesis. Angiopoietin seems to be one of the most important partners of VEGF. For example, VEGF has been shown to induce nerve growth, especially neuroprotection after ischemia or spinal cord injury [11]. Two of the most characteristic angiopoietins (Ang), Ang1 and Ang2, are ligands of the Tie2 receptor tyrosine kinase in endothelial and endothelial progenitor cells. VEGF induces endothelial cell migration and proliferation, and initiates angiogenesis. Conversely, Ang1-Tie2 interactions mediate neovas-cularization by recruiting endothelial cells (smooth muscle cells and peripheral cells) and establishing a basement membrane. However, in stationary mature vessels, Ang1 signaling inhibits endothelial cell activation and prevents angiogenesis. The experimental results show that the expression and activation of MMPs are the reason for the opening of the BBB [11, 109]. In a normal human brain, MMPs are either in a latent state (MMP-2) or at a very low level. Now, they have reached an undetectable expression level (MMP-3 and MMP-9). When lesions occur, the content of MMPs will be high and the expression will be abnormal [11, 108, 109].

Many structural and functional microvascular abnormalities have been found in AD. The loss of neuron cells in AD may be due to the pathological changes of the vascular structure, the decrease of cerebral blood flow and the change of oxygen utiliza-tion, which lead to the disturbance of cerebral microcirculation, resulting in vascular dementia and other diseases [105]. When compared with the age-matched control group, the cerebral micro-vascular abnormalities in AD have higher release levels, which includes nitric oxide (NO), thrombin, vascular endothelial growth and other factors (VEGF), angiopoietin-2 (Ang-2), tumor necrosis factor-α (TNF-α), transforming growth factor-β (TGF-β), interleukin-1 β, IL-6, IL-8, and matrix metalloproteinase (MMPs) in AD. In AD and other dementias, microvessel density decreased and a large number of atrophic vessels were observed. Age-related vascular wall degeneration prevents the brain from clearing the Aβ from the perivascular space, and it promotes the aggregation of Aβ, resulting in the decline of cognitive ability. The decrease of vascular density may cause AD, or vascular abnormality may cooperate with other pathogenic processes to increase this brain pathology and the neuronal cell deaths evident in AD. Therefore, it is necessary to select appropriate aptamers to detect angiogenesis-related proteins. Alice et al. [110] developed a multiplex sandwich aptamer (micro-array) gene chip for the detection of VEGF$_{165}$ and thrombin. A new sensor for VEGF$_{165}$ detection was described, which uses Vap7 and VEA5, two DNA aptamers that can recognize different sites of proteins. Because thrombin upregulates the VEGF expression, the simultaneous recognition of these two proteins may be helpful for the analysis of pathological biomarkers that are characterized by neovascularization. Vap7 is a binding aptamer of VEGF$_{165}$, folded into a G-quadruplex chain structure, while VEA5 has three stem ring structures. The KD value of VEA5 binding to VEGF$_{165}$ was 130 nM, and the KD value of Vap7 binding to VEGF$_{165}$ was 20 nM. There are related studies that show that the abnormal astrocytes in Huntington's disease inhibit vascular responses by aptamer detection [111]. HD astrocytes produce more VEGF, which can promote the angiogenesis of mice with HD and increase the density of cerebral blood vessels. Therefore, an aptamer for VEGF can potentially be used in the detection of HD with good sensitivity, precision, accuracy and robustness.

Biosensors based on aptamers are proving promising in the detection of angiogenesis biomarkers, such as vascular endothelial growth factor (VEGF). The downregulation and upregulation of VEGF were observed in the inhibition of neurologic disorders and neurodegenerations respectively. From a clinical point of view, VEGF has been used as a serum biomarker of various NDs [11]. In recent years, the research progress of aptamer-based biosensors for detection and quantification has been widely studied on optical and electrochemical platforms. The optical aptamer sensor has the advantages of being done in a real-time and quantitative way, with no need for amplification or advanced instruments. It is more convenient and the detection can be done without labelling. It is a good platform for VEGF detection. The detection can be seen by a change in chemiluminescence [112], fluorescence [113], colorimetry [114] and other optical signals generated by the combination of the aptamer and angiogenesis biomarkers. The sensor based on an electrochemical platform for VEGF detection needs a low volume sample, so it can be repeated easily and can be used for quantitative detection. Nonaka et al. [115] found that the DNA sensor made by a sandwich/pyrroloquinoline-quinone glucose dehydrogenase (PQQ-GDH)/enzyme signal amplification has a high sensitivity and its detection limit is 15 nM. Crulhas et al. [116] used the DNA sensor which was prepared by the self-assembly of thiol aptamer on gold coverage to detect the VEGF with a sensitivity of 0.15 ng/ml. The electrochemical aptamer sensor can be modified by different methods in order to improve its sensitivity. However, there may be false positive results in clinical application, and factors such as electrolyte and current may cause this. This is the main problem to overcome in the future in order to develop aptamer-based biosensors for neurodegenerative diseases.

4 Conclusions and Perspectives

Neurodegenerative diseases are associated with the degeneration of neurons and their myelin sheath, and the accumulation of misfolded proteins in the central nervous system. The chronic neurodegenerative diseases are Alzheimer's disease (AD), Parkinson's disease (PD), Lewy body dementia (LBD) and frontotemporal dementia (FTD), Huntington's disease (HD), amyotrophic lateral sclerosis (ALS), different types of spinocerebellar ataxia (SCA), and Pick's disease. There are also acute neurodegenerative diseases, such as ischemia and brain injury caused by strokes. Their incidence will continue to increase with the aging population. Therefore, there is an increasingly urgent need to find a way to diagnose neurodegenerative diseases. In addition to other molecular probes, aptamers should also contribute to the development of basic research on NDs, and they can also be used in detection and treatment. Aptamers are a kind of molecules that can be screened in vitro, targeting proteins, RNA, DNA fragments, and so on with similar functions to antibodies. They have clear advantages when compared with antibodies, including a smaller size, lower toxicity or nontoxicity, and a lower immunogenicity. Despite their advantages over antibodies, there are still obstacles to overcome in their clinical application, such as their rapid degradation and susceptibility to nuclease. Fortunately, various methods have been proposed to overcome these limitations, such as aptamer chemical modification and aptamer nanoparticle conjugation. This will improve the clinical application of aptamers, such as in the treatment of cancer, the eyes and inflammatory diseases. Many other aptamers are being developed, which indicates that aptamers will be more frequently used in the treatment of neurodegenerative diseases. However, although this chapter describes several aptamers for NDs, their number is still fewer compared with those for cancer or infectious diseases. It may be because that NDs pathology is more complex compared with other diseases, and aggregation proteins play a role in these diseases. The conformation of proteins is important to the pathology of disease, which is an ideal trait when detecting them with aptamers.

Therefore, it may be very helpful to develop more aptamers that can distinguish between different proteins and their conformation changes for NDs in the future. These aptamers can be integrated into biosensors to detect the biomarkers of neurodegenerative diseases in blood or cerebrospinal fluid.

References

1. Trojanowski JQ, Hampel H (2011) Neurodegenerative disease biomarkers: guideposts for disease prevention through early diagnosis and intervention. Prog Neorobiol 95 (4):491–495. https://doi.org/10.1016/j.pneurobio.2011.07.004
2. Qu J, Yu S, Zheng Y, Zheng Y, Yang H, Zhang J (2016) Aptamer and its applications in neurodegenerative diseases. Cell Mol Life Sci 74(4):683–695. https://doi.org/10.1007/s00018-016-2345-4
3. Gallucci M, Limbucci N, Catalucci A, Caulo M (2008) Neurodegenerative diseases. Radiol Clin North Am 46(4):799–817. https://doi.org/10.1016/j.rcl.2008.06.002
4. Haenseler W, Rajendran L (2019) Concise review: modelling neurodegenerative diseases with human pluripotent stem cell derived microglia. Stem Cells 37:724–730. https://doi.org/10.1002/stem.2995
5. Miller DH (2004) Biomarkers and surrogate outcomes in neurodegenerative disease: lessons from multiple sclerosis. NeuroRx 1 (2):284–294. https://doi.org/10.1602/neurorx.1.2.284
6. Bouvier-Müller A, Ducongé F (2018) Nucleic acid aptamers for neurodegenerative diseases. Biochimie 145:73–83. https://doi.org/10.1016/j.biochi.2017.10.026
7. Henley SM, Bates GP, Tabrizi SJ (2005) Biomarkers for neurodegenerative diseases. Curr Opin Neurol 18(6):698–705. https://doi.org/10.1097/01.wco.0000186842.51129.cb

8. Chen JA, Fears SC, Jasinska AJ, Huang A, Al-Sharif NB, Scheibel KE, Coppola G (2018) Neurodegenerative disease biomarkers Aβ1-40, Aβ1-42, tau, and p-tau181 in the vervet monkey cerebrospinal fluid: relation to normal aging, genetic influences, and cerebral amyloid angiopathy. Brain Behav 8 (2):e00903. https://doi.org/10.1002/brb3.903

9. Castrillon R, Acien A, Orozco-Arroyave JR, Morales A, Vargas JF, Vera-Rodriguez R, Villegas A (2019) Characterization of the handwriting skills as a biomarker for Parkinson's disease. In: 2019 14th IEEE international conference on Automatic Face & Gesture Recognition (FG 2019). https://doi.org/10.1109/FG.2019.8756508

10. Westergard L, Christensen HM, Harris DA (2007) The cellular prion protein (PrPC): its physiological function and role in disease. Biochim Biophys Acta 1772(6):629–644. https://doi.org/10.1016/j.bbadis.2007.02.011

11. Mashreghi M, Azarpara H, Bazaz MR, Jafari A, Masoudifar A, Mirzaei H, Jaafari MR (2017) Angiogenesis biomarkers and their targeting ligands as potential targets for tumor angiogenesis. J Cell Physiol 233 (4):2949–2965. https://doi.org/10.1002/jcp.26049

12. Vallon M, Chang J, Zhang H, Kuo CJ (2014) Developmental and pathological angiogenesis in the central nervous system. Cell Mol Life Sci 71(18):3489–3506. https://doi.org/10.1007/s00018-014-1625-0

13. Grammas P, Martinez J, Miller B (2011) Cerebral microvascular endothelium and the pathogenesis of neurodegenerative diseases. Expert Rev Mol Med 13:e19. https://doi.org/10.1017/S1462399411001918

14. Sosic A, Meneghello A, Antognoli A, Cretaio E, Gatto B (2013) Development of a multiplex Sandwich aptamer microarray for the detection of VEGF165 and thrombin. Sensors 13(10):13425–13438. https://doi.org/10.3390/s131013425

15. Cho H, Yeh EC, Sinha R, Laurence TA, Bearinger JP, Lee LP (2012) Single-step Nanoplasmonic VEGF165 Aptasensor for early cancer diagnosis. ACS Nano 6 (9):7607–7614. https://doi.org/10.1021/nn203833d

16. Drummond TG, Hill MG, Barton JK (2003) Electrochemical DNA sensors. Nat Biotechnol 21(10):1192–1199. https://doi.org/10.1038/nbt873

17. Cagnin S, Caraballo M, Guiducci C, Martini P, Ross M, SantaAna M, Lanfranchi G (2009) Overview of electrochemical DNA biosensors: new approaches to detect the expression of life. Sensors 9(4):3122–3148. https://doi.org/10.3390/s90403122

18. Chen M, Hou C, Huo D, Yang M, Fa H (2016) An ultrasensitive electrochemical DNA biosensor based on a copper oxide nanowires/single-walled carbon nanotubes nanocomposite. Appl Surf Sci 364:703–709. https://doi.org/10.1016/j.apsusc.2015.12.203

19. Phillips JA, Xu Y, Xia Z, Fan ZH, Tan W (2009) Enrichment of cancer cells using aptamers immobilized on a Microfluidic Channel. Anal Chem 81(3):1033–1039. https://doi.org/10.1021/ac802092j

20. Yuan Q, Liu Y, Ye C, Sun H, Dai D, Wei Q, Lin CT (2018) Highly stable and regenerative graphene–diamond hybrid electrochemical biosensor for fouling target dopamine detection. Biosens Bioelectron 111:117–123. https://doi.org/10.1016/j.bios.2018.04.006

21. Tuerk C, Gold L (1990) Systematic evolution of ligands by exponential enrichment: RNA ligands to bacteriophage T4 DNA polymerase. Science 249(4968):505–510. https://doi.org/10.1126/science.2200121

22. Famulok M, Hartig JS, Mayer G (2007) Functional aptamers and aptazymes in biotechnology, diagnostics, and therapy. Chem Rev 107:3715–3743. https://doi.org/10.1021/cr0306743

23. Song S, Wang L, Li J, Fan C, Zhao J (2008) Aptamer-based biosensors. TrAC Trends Anal Chem 27(2):108–117. https://doi.org/10.1016/j.trac.2007.12.004

24. Muller J, El-Maarri O, Oldenburg J, Potzsch B, Mayer G (2008) Monitoring the progression of the in vitro selection of nucleic acid aptamers by denaturing highperformance liquid chromatography. Anal Bioanal Chem 390:1033–1037. https://doi.org/10.1007/s00216-007-1699-8

25. Khan N, Maddaus A, Song E (2018) A low-cost inkjet-printed aptamer-based electrochemical biosensor for the selective detection of lysozyme. Biosensors 8(1):7. https://doi.org/10.3390/bios8010007

26. Zhang Y, Lai B, Juhas M (2019) Recent advances in aptamer discovery and applications. Molecules 24(5):941. https://doi.org/10.3390/molecules24050941

27. Zhuo Z, Yu Y, Wang M, Li J, Zhang Z, Liu J, Wu X, Lu A, Zhang G, Zhang B (2017) Recent advances in SELEX technology and aptamer applications in biomedicine. Int J

Mol Sci 18(10):2142. https://doi.org/10.3390/ijms18102142

28. Ruscito A, McConnell EM, Koudrina A, Velu R, Mattice C, Hunt V, DeRosa MC (2017) In vitro selection and characterization of DNA aptamers to a small molecule target. Curr Protoc Chem Biol:233–268. https://doi.org/10.1002/cpch.28

29. Alsager OA, Kumar S, Hodgkiss JM (2017) A lateral flow Aptasensor for small molecule targets exploiting adsorption and desorption interactions on Gold nanoparticles. Anal Chem 89(14):7416–7424. https://doi.org/10.1021/acs.analchem.7b00906

30. Peng Y, Huang H, Zhang Y, Kang C, Chen S, Song L, Zhong C (2018) A versatile MOF-based trap for heavy metal ion capture and dispersion. Nat Commun 9(1). https://doi.org/10.1038/s41467-017-02600-2

31. Kulbachinskiy AV (2007) Methods for selection of aptamers to protein targets. Biochem 72(13):1505–1518. https://doi.org/10.1134/s000629790713007x

32. Bonanni A, Pumera M (2011) Graphene platform for hairpin-DNA-based Impedimetric Genosensing. ACS Nano 5(3):2356–2361. https://doi.org/10.1021/nn200091p

33. Wang Y, Wang C, Bo H, Gao Q, Qi H, Zhang C (2013) Specific recognition of a single guanine bulge in dsDNA using a surface plasmon resonance sensor with immobilized 2-(2-aminoacetyl)amino-5,6,7-trimethyl-1,8-naphthyridine. Sensor Actuator B Chem 177:800–806. https://doi.org/10.1016/j.snb.2012.11.026

34. Lax P, Becerra AG, Soteras F, Cabello M, Doucet ME (2011) Effect of the arbuscular mycorrhizal fungus glomus intraradices on the false root-knot nematode Nacobbus aberrans in tomato plants. Biol Fertil Soils 47(5):591–597. https://doi.org/10.1007/s00374-010-0514-4

35. Gibriel A, Adel O (2017) Advances in ligase chain reaction and ligation-based amplifications for genotyping assays: detection and applications. Mutat Res/Rev Mutat Res 773:66–90. https://doi.org/10.1016/j.mrrev.2017.05.001

36. Kim M, Kim DM, Kim KS, Jung W, Kim DE (2018) Applications of cancer cell-specific aptamers in targeted delivery of anticancer therapeutic agents. Molecules 23(4):830. https://doi.org/10.3390/molecules23040830

37. Bai C, Lu Z, Jiang H, Yang Z, Liu X, Ding H, Shao N (2018) Aptamer selection and application in multivalent binding-based electrical impedance detection of inactivated H1N1 virus. Biosens Bioelectron 110:162–167. https://doi.org/10.1016/j.bios.2018.03.047

38. Li Y, Lee JS (2019) Recent developments in affinity-based selection of aptamers for binding disease-related protein targets. Chem Pap 2019:1–17. https://doi.org/10.1007/s11696-019-00842-6

39. Yang Q, Zhou L, Wu YX, Zhang K, Cao Y, Zhou Y, Gan N (2018) A two-dimensional metal–organic framework nanosheets-based fluorescence resonance energy transfer aptasensor with circular strand-replacement DNA polymerization target-triggered amplification strategy for homogenous detection of antibiotics. Anal Chem Acta 1020:1–8. https://doi.org/10.1016/j.aca.2018.02.058

40. Kuusisto E, Salminen A, Alafuzoff I (2001) Ubiquitin-binding protein p62 is present in neuronal and glial inclusions in human tauopathies and synucleinopathies. Neuroreport 12(10):2085–2090. https://doi.org/10.1097/00001756-200107200-00009

41. Davidsson P, Sjögren M (2005) The use of proteomics in biomarker discovery in neurodegenerative diseases. Dis Markers 21(2):81–92. https://doi.org/10.1155/2005/848676

42. Shi M, Sui YT, Peskind ER, Li G, Hwang H, Devic I, Ginghina C, Edgar JS, Pan C, Goodlett DR (2011) Salivary tau species are potential biomarkers of Alzheimer's disease. J Alzheimers Dis 27(2):299–305. https://doi.org/10.3233/JAD-2011-110731

43. Saini S, Arora K (2014) Study analysis on the different image segmentation techniques. IJICT 4(14):1445–1452

44. Suwalka I, Agrawal N (2016) Assessment of segmentation techniques for neurodegenerative disease detection. ICCPCT. https://doi.org/10.1109/ICCPCT.2016.7530187

45. Ngounou Wetie AG, Sokolowska I, Wormwood K, Beglinger K, Michel T, Thome J, Woods AG (2013) Mass spectrometry for the detection of potential psychiatric biomarkers. J Mol Psychiatr 1(1):8. https://doi.org/10.1186/2049-9256-1-8

46. Bois JP, Scott C, Chareonthaitawee P, Gibbons RJ, Rodriguez-Porcel M (2019) Phase analysis single-photon emission computed tomography (SPECT) myocardial perfusion imaging (MPI) detects dyssynchrony in myocardial scar and increases specificity of MPI. EJNMMI Res 9:11. https://doi.org/10.1186/s13550-019-0476-y

47. Barthel H, Schroeter ML, Hoffmann KT, Sabri O (2015) PET/MR in dementia and other neurodegenerative diseases. Semin Nucl Med 45(3):224–233. https://doi.org/10.1053/j.semnuclmed.2014.12.003

48. Goñi F, Martá-Ariza M, Peyser D, Herline K, Wisniewski T (2017) Production of monoclonal antibodies to pathologic β-sheet oligomeric conformers in neurodegenerative diseases. Sci Rep 7(1):9881. https://doi.org/10.1038/s41598-017-10393-z

49. Martins-Gomes C, Silva M (2018) Western blot methodologies for analysis of in vitro protein expression induced by teratogenic agents. Methods Mol Biol 1797:191–203. https://doi.org/10.1007/978-1-4939-7883-0_9

50. Yang J, Xu CQ, Nutiu R, Li Y (2004) Immobilized DNA biosensor based on evanescent wave long-period fiber gratings. In: Photonics North 2004: photonic applications in astronomy, biomedicine, imaging, materials processing, and education. https://doi.org/10.1117/12.567502

51. Mojiri A, Ohashi A, Ozaki N, Shoiful A, Kindaichi T (2018) Pollutant removal from synthetic aqueous solutions with a combined electrochemical oxidation and adsorption method. Int J Environ Res Public Health 15 (7):1443. https://doi.org/10.3390/ijerph15071443

52. Fu X, Huang R, Wang J, Chang B (2013) Sensitive electrochemical immunoassay of a biomarker based on biotin-avidin conjugated DNAzyme concatamer with signal tagging. RSC Adv 3(32):13451. https://doi.org/10.1039/C3RA41429A

53. Park H-S, Hwang S-J, Choy J-H (2001) Relationship between chemical bonding character and electrochemical performance in nickel-substituted lithium manganese oxides. J Phys Chem 105:4860–4866. https://doi.org/10.1021/jp010079+

54. Kong D, Liao F, Lin Y, Cheng L, Peng H, Zhang J, Fan H (2018) A homogenous electrochemical sensing DNA sensor by using bare au electrode based on potential-assisted chemisorption technique. Sensor Actuat B Chem 266:288–293. https://doi.org/10.1016/j.snb.2018.03.011

55. Zhao N, Pei SN, Qi J, Zeng Z, Iyer SP, Lin P, Zu Y (2015) Oligonucleotide aptamer-drug conjugates for targeted therapy of acute myeloid leukemia. Biomaterials 67:42–51. https://doi.org/10.1016/j.biomaterials.2015.07.025

56. Lucarelli F, Kicela A, Palchetti I, Marrazza G, Mascini M (2002) Electrochemical DNA biosensor for analysis of wastewater samples. Bioelectrochemistry 58(1):113–118. https://doi.org/10.1016/s1567-5394(02)00133-0

57. Mogha NK, Sahu V, Sharma RK, Masram DT (2018) Reduced graphene oxide nanoribbon immobilized gold nanoparticle based electrochemical DNA biosensor for the detection of Mycobacterium tuberculosis. J Mater Chem B 6(31):5181–5187. https://doi.org/10.1039/C8TB01604F

58. Jiang D, Ge P, Wang L, Jiang H, Yang M, Yuan L, Ju X (2019) A novel electrochemical mast cell-based paper biosensor for the rapid detection of milk allergen casein. Biosens Bioelectron. https://doi.org/10.1016/j.bios.2019.01.050

59. Rahman M, Heng LY, Futra D, Chiang CP, Rashid ZA, Ling TL (2017) A highly sensitive electrochemical DNA biosensor from acrylic-gold nano-composite for the determination of Arowana fish gender. Nanoscale Res Lett 12 (1):484. https://doi.org/10.1016/j.snb.2016.11.061

60. Zhang J, Qi H, Li Y, Yang J, Gao Q, Zhang C (2008) Electrogenerated chemiluminescence DNA biosensor based on hairpin DNA probe labeled with ruthenium complex. Anal Chem 80(8):2888–2894. https://doi.org/10.1021/ac701995g

61. Thiruppathiraja C, Kamatchiammal S, Adaikkappan P, Santhosh DJ, Alagar M (2011) Specific detection of mycobacterium sp. genomic DNA using dual labeled gold nanoparticle based electrochemical biosensor. Anal Biochem 417(1):73–79. https://doi.org/10.1016/j.ab.2011.05.034

62. Hu P, Liu N, Wu KY, Zhai LY, Xie BP, Sun B, Chen JX (2018) Successive and specific detection of Hg2+ and I– by a DNA@MOF biosensor: experimental and simulation studies. Inorg Chem 57(14):8382–8389. https://doi.org/10.1021/acs.inorgchem.8b01051

63. Kim DK, Kerman K, Saito M, Sathuluri RR, Endo T, Yamamura S, Tamiya E (2007) Label-free DNA biosensor based on localized surface Plasmon resonance coupled with interferometry. Anal Chem 79 (5):1855–1864. https://doi.org/10.1021/ac061909o

64. Zari N, Amine A, Ennaji MM (2009) Label-free DNA biosensor for electrochemical detection of short DNA sequences related to human papilloma virus. Anal Lett 42 (3):519–535. https://doi.org/10.1080/00032710802421897

65. Wang Z, Yang Y, Leng K, Li J, Zheng F, Shen G, Yu R (2008) A sequence-selective electrochemical DNA biosensor based on

HRP-Labeled probe for colorectal cancer DNA detection. Anal Lett 41(1):24–35. https://doi.org/10.1080/00032710701746873

66. Wang J, Shi A, Fang X, Han X, Zhang Y (2014) Ultrasensitive electrochemical super-sandwich DNA biosensor using a glassy carbon electrode modified with gold particle-decorated sheets of graphene oxide. Microchim Acta 181(9–10):935–940. https://doi.org/10.1007/s00604-014-1182-0

67. Chen X, Xie H, Seow ZY, Gao Z (2010) An ultrasensitive DNA biosensor based on enzyme-catalyzed deposition of cupric hexacyanoferrate nanoparticles. Biosens Bioelectron 25(6):1420–1426. https://doi.org/10.1016/j.bios.2009.10.041

68. Huang B, Liu J, Lai L, Yu F, Ying X, Ye BC, Li Y (2017) A free-standing electrochemical sensor based on graphene foam-carbon nanotube composite coupled with gold nanoparticles and its sensing application for electrochemical determination of dopamine and uric acid. J Electroanal Chem 801:129–134. https://doi.org/10.1016/j.jelechem.2017.07.029

69. Zamfir LG, Fortgang P, Farre C, Ripert M, De Crozals G, Jaffrezic-Renault N, Chaix C (2015) Synthesis and electroactivated addressing of ferrocenyl and azido-modified stem-loop oligonucleotides on an integrated electrochemical device. Electrochim Acta 164:62–70. https://doi.org/10.1016/j.electacta.2015.02.167

70. Zhang Y, Huang L (2012) Label-free electrochemical DNA biosensor based on a glassy carbon electrode modified with gold nanoparticles, polythionine, and graphene. Microchim Acta 176(3-4):463–470. https://doi.org/10.1007/s00604-011-0742-9

71. Yan F, Wang F, Chen Z (2011) Aptamer-based electrochemical biosensor for label-free voltammetric detection of thrombin and adenosine. Sensor Actuat B Chem 160(1):1380–1385. https://doi.org/10.1016/j.snb.2011.09.081

72. Lee JG, Yun K, Lim GS, Lee SE, Kim S, Park JK (2007) DNA biosensor based on the electrochemiluminescence of $Ru(bpy)_3^{2+}$ with DNA-binding intercalators. Bioelectrochemistry 70(2):228–234. https://doi.org/10.1016/j.bioelechem.2006.09.003

73. Sekhon SS, Ahn G, Park GY, Park DY, Lee SH, Ahn JY, Kim YH (2019) The role of aptamer loaded exosome complexes in the neurodegenerative diseases. Toxicol Environ Health Sci 11(2):85–93. https://doi.org/10.1007/s13530-019-0392-6

74. Cowperthwaite MC, Ellington AD (2008) Bioinformatic analysis of the contribution of primer sequences to aptamer structures. J Mol Evol 67(1):95–102. https://doi.org/10.1007/s00239-008-9130-4

75. Toh SY, Citartan M, Gopinath SC, Tang TH (2015) Aptamers as a replacement for antibodies in enzyme-linked immunosorbent assay. Biosens Bioelectron 64:392–403. https://doi.org/10.1016/j.bios.2014.09.026

76. Rhie A, Kirby L, Sayer N, Wellesley R, Disterer P, Sylvester I, Tahiri-Alaoui A (2003) Characterization of 2′-Fluoro-RNA aptamers that bind preferentially to disease-associated conformations of prion protein and inhibit conversion. J Biol Chem 278(41):39697–39705. https://doi.org/10.1074/jbc.M305297200

77. Li L, Hou J, Liu X, Guo Y, Wu Y, Zhang L, Yang Z (2014) Nucleolin-targeting liposomes guided by aptamer AS1411 for the delivery of siRNA for the treatment of malignant melanomas. Biomaterials 35(12):3840–3850. https://doi.org/10.1016/j.biomaterials.2014.01.019

78. Ma X, Du C, Zhang J, Shang M, Song W (2019) A system composed of vanadium (IV) disulfide quantum dots and molybdenum (IV) disulfide nanosheets for use in an aptamer-based fluorometric tetracycline assay. Microchim Acta 186(12). https://doi.org/10.1007/s00604-019-3983-7

79. Zhang S, Fan J, Wang Y, Li D, Jia X, Yuan Y, Cheng Y (2019) Tunable aggregation-induced circularly polarized luminescence of chiral AIEgens via the regulation of mono−/di-substituents of molecules or nanostructures of self-assemblies. Mater Chem Front 00:1–3. https://doi.org/10.1039/C9QM00358D

80. Kolb G, Reigadas S, Castanotto D, Faure A, Ventura M, Rossi JJ, Toulme JJ (2006) Endogenous expression of an Anti-TAR aptamer reduces HIV-1 replication. RNA Biol 3(4):150–156. https://doi.org/10.4161/rna.3.4.3811

81. Zhang Y, Wang M, Huang L (2012) Fabrication of a sensitive electrochemical biosensor for detection of DNA hybridization based on Gold nanoparticles/CuO Nanospindles modified glassy carbon electrode. Chin J Chem 30(1):167–172. https://doi.org/10.1002/cjoc.201180451

82. Xu X, Ho W, Zhang X, Bertrand N, Farokhzad O (2015) Cancer nanomedicine: from targeted delivery to combination therapy. Trends Mol Med 21(4):223–232. https://doi.org/10.1016/j.molmed.2015.01.001

83. Chen W, Jin F, Cao G, Mei R, Wang Y, Long P, Ge W (2018) ApoE4 may be a promising target for treatment of coronary heart disease and Alzheimer's disease. Curr Drug Targets 19(9):1038–1044. https://doi.org/10.2174/1389450119666180406112050

84. Chakrabarti S, Khemka VK, Banerjee A, Chatterjee G, Ganguly A, Biswas A (2015) Metabolic risk factors of sporadic Alzheimer's disease: implications in the pathology, pathogenesis and treatment. Aging Dis 6 (4):282–299. https://doi.org/10.14336/AD.2014.002

85. Davatzikos C, Resnick SM, Wu X, Parmpi P, Clark CM (2008) Individual patient diagnosis of AD and FTD via high-dimensional pattern classification of MRI. Neuroimage 41 (4):1220–1227. https://doi.org/10.1016/j.neuroimage.2008.03.050

86. Krylova SM, Musheev M, Nutiu R, Li Y, Lee G, Krylov SN (2005) Tau protein binds single-stranded DNA sequence specifically— the proof obtained in vitro with non-equilibrium capillary electrophoresis of equilibrium mixtures. FEBS Lett 579:1371–1375. https://doi.org/10.1016/j.febslet.2005.01.032

87. Kim JH, Kim E, Choi WH, Lee J, Lee JH, Lee H, Kim DE, Suh YH, Lee MJ (2016) Inhibitory RNA aptamers of tau oligomerization and their neuroprotective roles against proteotoxic stress. Mol Pharm 13:2039–2048. https://doi.org/10.1021/acs.molpharmaceut.6b00165

88. Esteves-Villanueva JO, Trzeciakiewicz H, Martic S (2014) A protein-based electrochemical biosensor for detection of tau protein, a neurodegenerative disease biomarker. Analyst 139:2823–2831. https://doi.org/10.1039/c4an00204k

89. Liu L, Xia N, Jiang M, Huang N, Guo S, Li S, Zhang S (2015) Electrochemical detection of amyloid-β oligomer with the signal amplification of alkaline phosphatase plus electrochemical- chemical-chemical redox cycling. J Electroanal Chem 754:40–45. https://doi.org/10.1016/j.jelechem.2015.06.017

90. Liu L, Zhao F, Ma F, Zhang L, Yang S, Xia N (2013) Electrochemical detection of beta-amyloid peptides on electrode covered with N-terminus-specific antibody based on electrocatalytic O2 reduction by Abeta(1-16)-heme-modified gold nanoparticles. Biosens Bioelectron 49:231–235. https://doi.org/10.1016/j.bios.2013.05.028

91. Rahimi F, Murakami K, Summers J L, Chen C H, Bitan G (2009) RNA aptamers generated against oligomeric Abeta40 recognize common amyloid aptatopes with low specificity but high sensitivity. PLoS one 4 (2009), e7694. https://doi.org/10.1371/journal.pone.0089901

92. Liu Y, Xu LP, Wang S, Yang W, Wen Y, Zhang X (2015) An ultrasensitive electrochemical immunosensor for apolipoprotein E4 based on fractal nanostructures and enzyme amplification. Biosens Bioelectron 71:396–400. https://doi.org/10.1016/j.bios.2015.04.068

93. Wang SX, Acha D, Shah AJ, Hills F, Roitt I, Demosthenous A, Bayford RH (2017) Detection of the tau protein in human serum by a sensitive four-electrode electrochemical biosensor. Biosens Bioelectron 92:482–488. https://doi.org/10.1016/j.bios.2016.10.077

94. Kamel F (2013) Paths from pesticides to Parkinson's. Science 341(6147):722–723. https://doi.org/10.1126/science.1243619

95. Alvarez-Martos I, Ferapontova EE (2016) Electrochemical label-free aptasensor for specific analysis of dopamine in serum in the presence of structurally related neurotransmitters. Anal Chem 88:3608e3616. https://doi.org/10.1021/acs.analchem.5b04207

96. Tsukakoshi K, Harada R, Sode K, Ikebukuro K (2010) Screening of DNA aptamer which binds to alpha-synuclein. Biotechnol Lett 32:643–648. https://doi.org/10.1007/s10529-010-0200-5

97. Mannironi C, Di A, Nardo FP, Tocchini-Valentini GP (1997) In vitro selection of dopamine RNA ligands. Biochem 36:9726e9734. https://doi.org/10.1021/bi9700633

98. Tsukakoshi K, Abe K, Sode K, Ikebukuro K (2012) Selection of DNA aptamers that recognize alpha-synuclein oligomers using a competitive screening method. Anal Chem 84:5542e5547. https://doi.org/10.1021/ac300330g

99. Walsh R, DeRosa MC (2009) Retention of function in the DNA homolog of the RNA dopamine aptamer. Biochem Biophys Res Commun 388(4):732–735. https://doi.org/10.1016/j.bbrc.2009.08.084

100. Weng CH, Huang CJ, Lee GB (2012) Screening of aptamers on microfluidic systems for clinical applications. Sensor 12 (7):9514–9529. https://doi.org/10.3390/s120709514

101. Munoz-Sanjuan I, Bates GP (2011) The importance of integrating basic and clinical research toward the development of new

therapies for Huntington disease. J Clin Investig 121(2):476–483. https://doi.org/10.1172/JCI45364

102. Skogen M, Roth J, Yerkes S, Parekh-Olmedo-H, Kmiec E (2006) Short G-rich oligonucleotides as a potential therapeutic for Huntington's disease. BMC Neurosci 7:65. https://doi.org/10.1186/1471-2202-7-65

103. Proske D, Gilch S, Wopfner F, Schatzl HM, Winnacker EL, Famulok M (2002) Prion-protein-specifific aptamer reduces PrPSc formation. ChemBioChem 3:717–725. https://doi.org/10.1002/1439-7633(20020802)3:8%3c717::AID-CBIC717%3e3.0.CO;2-C

104. Rhie A, Park WS, Choi MK, Kim JH, Ryu J, Ryu CH, Jung YS (2015) Genomic copy number variations characterize the prognosis of both P16-positive and P16-negative oropharyngeal squamous cell carcinoma after curative resection. Medicine 94(50):e2187. https://doi.org/10.1097/md.0000000000002187

105. Hossain MT, Shibata T, Kabashima T, Kai M (2010) Aptamer-mediated chemiluminescence detection of prion protein on a membrane using trimethoxyphenylglyoxal. Anal Sci Int J Jpn Soc Anal Chem 26:645e647. https://doi.org/10.2116/analsci.26.645

106. Miodek A, Poturnayova A, Snejdarkova M, Hianik T, Korri-Youssoufifi H (2013) Binding kinetics of human cellular prion detection by DNA aptamers immobilized on a conducting polypyrrole. Anal Bioanal Chem 405:2505–2514. https://doi.org/10.1007/s00216-012-6665-4

107. Nastasijevic B, Wright BR, Smestad J, Warrington AE, Rodriguez M, Maher LJ (2012) Remyelination induced by a DNA aptamer in a mouse model of multiple sclerosis. PLoS One 7(6):e39595. https://doi.org/10.1371/journal.pone.0039595

108. Buysschaert I, Carmeliet P, Dewerchin M (2007) Clinical and fundamental aspects of angiogenesis and ANTI-angiogenesis. Acta Clin Belg 62(3):162–169. https://doi.org/10.1179/acb.2007.027

109. Xiong Y, Mahmood A, Chopp M (2010) Angiogenesis, neurogenesis and brain recovery of function following injury. Curr Opin Investig Drugs 11:298–308. https://doi.org/10.1016/j.cct.2010.01.003

110. Sosic A, Meneghello A, Antognoli A, Cretaio E, Gatto B (2013) Development of a multiplex Sandwich aptamer microarray for the detection of VEGF165 and thrombin. Sensor 13(10):13425–13438. https://doi.org/10.3390/s131013425

111. Hsiao HY, Chen YC, Huang CH, Chen CC, Hsu YH, Chen HM, Chern Y (2015) Aberrant astrocytes impair vascular reactivity in Huntington disease. Ann Neurol 78(2):178–192. https://doi.org/10.1002/ana.24428

112. Kopra K, Syrjänpää M, Hänninen P, Härmä H (2014) Non-competitive aptamer-based quenching resonance energy transfer assay for homogeneous growth factor quantification. Analyst 139(8):2016. https://doi.org/10.1039/c3an01814h

113. Charbgoo F, Soltani F, Taghdisi SM, Abnous K, Ramezani M (2016) Nanoparticles application in high sensitive aptasensor design. TrAC Trend Anal Chem 85:85–97. https://doi.org/10.1016/j.trac.2016.08.008

114. Xu H, Kou F, Ye H, Wang Z, Huang S, Liu X, Chen G (2017) Highly sensitive antibody-aptamer sensor for vascular endothelial growth factor based on hybridization chain reaction and pH meter/indicator. Talanta 175:177–182. https://doi.org/10.1016/j.talanta.2017.04.073

115. Nonaka Y, Yoshida W, Abe K, Ferri S, Schulze H, Bachmann TT, Ikebukuro K (2012) Affinity improvement of a VEGF aptamer by in silico maturation for a sensitive VEGF-detection system. Anal Chem 85(2):1132–1137. https://doi.org/10.1021/ac303023d

116. Crulhas BP, Karpik AE, Delella FK, Castro GR, Pedrosa VA (2017) Electrochemical aptamer-based biosensor developed to monitor PSA and VEGF released by prostate cancer cells. Anal Bioanal Chem 409(29):6771–6780. https://doi.org/10.1007/s00216-017-0630-1

Possible Biomarkers for Frontotemporal Dementia and to Differentiate from Alzheimer's Disease and Amyotrophic Lateral Sclerosis

Donald M. R. Harker, Bridget Martinez, and Ruben K. Dagda

Abstract

Though not as well-known as Alzheimer's disease, yet with a prevalence of 15–22/100,000, and an incidence 2.7–4.1/100,000 cases per year, frontotemporal dementia is a grave, chronic neurodegenerative disorder with low life expectancy, a survival comparable to that of Alzheimer's disease, and a distressing clinical course for patients as well as family and caregivers. Given that some pathological features in many neurodegenerative diseases are convergent, coupled with the fact that treatment for one specific disease may worsen outcome if misdiagnosed for another, the need for sensitive and specific biomarkers has garnered the attention of physicians and neuroscientists alike. Here, we explore the clinical presentation of frontotemporal dementia with a focus on behavioral manifestations. We also discuss neuroanatomy and specific symptomology as a guiding diagnostic tool given that affected regions of the cortex are responsible for certain movements, motivation, reward processing, decision making, executive function, expression, language functions, social inhibition, and many of the complex social functions. Finally, we review the most commonly used biomarkers in the workup and assessment of neurodegenerative disorders, with a focus on both structural and biofluid markers.

Keywords Frontotemporal dementia, Magnetic resonance imaging, Behavioral variant, Frontal lobe, Neurodegeneration, α-Synuclein, Mitochondria, GABAergic/glutamatergic

1 Introduction to Neurodegeneration

Dementia, as its English translation of Latin *demens* suggests, "without mind" [1], has become a prevalent phrase describing neurodegenerative processes, many of which have overlapping symptomatology and present difficulties in accurate initial diagnosis. 5–15% of all cases of dementia are caused by frontotemporal lobar degeneration (FTLD), which encompasses various different diseases of the brain, all brought on by distinct etiologies but all affecting the prefrontal and anterior temporal lobes of both hemispheres [2]. The heterogeneous nature of the etiologies warrants

Philip V. Peplow, Bridget Martinez and Thomas A. Gennarelli (eds.), *Neurodegenerative Diseases Biomarkers: Towards Translating Research to Clinical Practice*, Neuromethods, vol. 173, https://doi.org/10.1007/978-1-0716-1712-0_16,
© Springer Science+Business Media, LLC, part of Springer Nature 2022

the need for novel and specific biomarkers. Although neuronal damage and cerebral dysfunction can span across multiple areas of the brain (most commonly the frontal and anterior temporal lobes of the cortex) presentation will vary significantly based on the precise areas affected [3–5]. Clinically, it is exceedingly difficult to distinguish between the varying subtypes of FTLD, and even more exacerbating is the overlap of disease with other forms of neurodegeneration, such as Alzheimer's disease (AD) and amyotrophic lateral sclerosis (ALS).

FTD is a brain degenerative disease that represents a broad clinical spectrum that was initially identified in 1892 when Arnold Pick first began to describe the disease. However, it was not until 1926 when the characteristic cytoplasmic inclusion bodies were associated with cerebral atrophy and the term "Pick's disease" was first coined [3, 4]. As the disease became more well understood, it became apparent that clinical presentation was a continuum/spectrum of symptomology and FTD became the moniker based on scientific criteria as an umbrella term specific to the spectrum covering multiple syndromes and subtypes such as Pick's disease, cortical basal degeneration, and progressive supranuclear palsy [5–7]. Incidence ranges are estimated between 15–22 per 100,000 cases per year though underestimation is a true possibility due to the difficulty of accurate diagnosis in the setting of overlapping symptomatology in neurodegenerative disease [8]. Unfortunately, even with such high incidence and prevalence, FTD may not be as well-known as diseases such as AD to the public at large, even though it is the second most common cause of dementia in populations over 65 years of age and the most common cause of early onset dementia in people <60 years of age [8]. Once thought to be more common in males, prevailing clinical research has shown a more even distribution of incidence between male and female, with some discrepancy noted between specific syndromes [8]. Neurodegeneration is generally not an acute process and chronic/continuous insults take time to accumulate and cause notable symptomology, which is why presentations earlier than 45 years of age are usually due to massive trauma or poisoning. In the case of true FTD, a family history of dementia is usually present but traceable autosomal dominant genetic inheritance is only described in 10–20% of patients [8]. Another poignant consideration in diseases involving neurodegeneration is the financial and emotional burden on caregivers, with studies suggesting the economic burden of FTD being twice that of AD [9]. Patient's average survival can range from 3 to 12 years among diseases subtypes requiring significant care and as disease progresses, patient personality and behaviors can be drastically altered causing substantial emotional trauma to the caregiver [8, 10–13]. Particularly interesting and unfortunate is the fact that caregivers report a higher level of emotional burden while caring for patients with FTD when compared to those with AD [10]. With these considerations in mind coupled with the diagnostic difficulties due to overlapping symptomatology, the need for specific, early alert testing and evidence-based treatments is clearly necessary [14].

2   Nomenclature and Classification

As previously mentioned, FTD is described as a spectrum of disease and is classically demarcated into three core clinical syndromes of FTD with other related movement disorders and aphasias considered under the same clinical umbrella [15–17]. The behavioral variant (bvFTD) is seen with the highest incidence while the other syndromes are versions of primary progressive aphasia (PPA), differentiated by symptoms relating to either semantics and fluency such as semantic variant PPA (sv-PPA), nonfluent PPA (nfv-PPA), and logopenic PPA (lv-PPA) [17, 18]. The bvFTD is characterized by psychiatric changes in the patient's core temperament and behavior as well as early signs of disease, especially when compared to other types of neurodegenerative disease [18]. It is defined by loss of empathy alongside deficits in memory, social compartment, insight, executive function, and abstract thought. Additionally, an affected patient's behavior will appear disinhibited, impulsive and include newly acquired apathy; these patients will also exhibit hypersexual behavior and a predilection for sweet/fatty foods. These psychiatric changes are associated with severe morbidity, though cognitive function is commonly preserved, the patient is often perceived unrecognizable by family and friends [19, 20]. What makes these pathological transitions more distressing to family and caregivers is patients' lack of insight into their own disinhibited and often dangerous attitude behavior [19, 20].

3   Anatomy and Disease Correlation

Specific symptomatology from disease results from degeneration of well understood circuits in the frontotemporal region of the brain due to atrophy of the connected regions and disruption of the pathways in between [21]. For example, the anterior cingulate cortex (ACC), a region of the medial prefrontal cortex required for motor control (including both the inhibition and excitation of motor responses) as well as empathy, impulse control, emotion, and decision-making is most commonly implicated in bvFTD; in addition to behavioral disturbances patients are afflicted with retardation of spontaneous action [18]. This region is also commonly implicated in goal-directed behaviors, including communication through speech [18]. Language has long been considered one of the most lateralized functions of the cerebral cortex. it is estimated that right-hand dominance equates to a 95–99% correlation with left cortical lateralization of language [18, 22]. However, while bvFTD can affect either hemisphere, studies have shown distinct subsyndromes presentations based on lateralization, adding additional difficulties in diagnosis [22, 23]. In patients with right sided lesions, for example, markedly decreased motivation may present initially, in extreme cases, to the point that patients may be drawn into complete sedentary activities for the majority of waking hours with indifference toward self-maintenance; meanwhile deficits in language may not be seen until late-stage progression [24].

Another cerebral structure commonly implicated in FTD is the orbitofrontal cortex (OFC) [25]. This region of the brain processes risk vs reward in behavior planning between the more primal subcortical structures of the brain and the evolved frontal cortices [25]. In an article by Moll et al., published in 2011 for example, it was shown that prosocial sentiments, such as guilt, pity, embarrassment, and empathy, were blunted in patients with degeneration of ventral medial and polar sectors of the prefrontal cortex regions, as well as the anterior temporal lobes of the brain [26–28]. Consequences of lesions to the OFC are best demonstrated in the infamous story of Phineas Gage, a railway worker who was accidentally struck through the skull with a piece of iron [23, 24]. Unfortunately, Mr. Gage developed many of the symptoms described above; his family describing a drastic change in personality in someone they once knew to be a responsible, hard-working, and compassionate man to an impulsive, apathetic, and hyperoral one [24, 29].

Outside of the core syndromes, other diseases included under the umbrella of FTD include corticobasal syndrome, progressive supranuclear palsy, and frontotemporal degeneration with amyotrophic lateral sclerosis (FTD/ALS) [15]. ALS is defined as a mixed upper and lower motor neuron deficit, commonly affecting the lower limbs. We include it here because although it is predominantly a motor disorder, ALS is associated with cognitive decline (in a pattern consistent with frontotemporal dementia) [30, 31]. Additionally, studies have shown that approximately 10% of ALS cases are inherited and the associated mutations include a hexanucleotide repeat on chromosome 9, which is associated with frontotemporal dementia [30], often hindering proper diagnosis. AD, is a neurodegenerative disease process is characterized by β-amyloid plaque deposition and neurofibrillary tangles of hyperphosphorylated tau [32]. AD is diagnosed based on clinical presentation coupled with fluid and imaging biomarkers; however, treatment is non–disease-modifying and aimed only at symptomatic improvement [32]. As previously mentioned, there exists substantial symptom overlap between AD and FTD; with the former presenting with predominance of memory and visuospatial deficits, normal neurological examination, and evidence of generalized brain atrophy on imaging [17].

4 Clinical Investigation and Assessment

While clinical presentation is the initial nidus for investigation, it is most useful when patients are first assessed by skilled clinicians with experience with FTD. Patients with true dementia from one pathology or the other, are usually brought to medical attention by concerned family members, meanwhile the patients themselves have little to no awareness of their cognitive decline. Initial assessment begins with a Mini-Mental State Examination (MMSE) or other similar self-surveying measures [33]. Lab work is also part of initial clinical workup in the setting of dementia, these include complete blood cell count (differential); electrolyte panel; renal, liver, and thyroid function tests as well as vitamin B12 levels to rule out common and reversible causes of cognitive impairment [34]. Clinical exam conclusions are determined by the regions of the cortex affected; degeneration of the medial/orbital frontal and anterior insula suggests bvFTD, meanwhile dominant hemisphere lateral frontal and precentral gyrus deterioration would point toward fluency related PPA, and semantic variants demonstrate anterior temporal disease. No current recommendations exist to screen populations without significant family history and markers for familial disease, a physical(neuro) exam and most importantly a high index of suspicion in a skilled clinician with a patient showing characteristic symptoms is the primary factor for further workup [34]. The recommendation to not screen further highlights the importance of exploring biomarkers in neurodegenerative diseases such as FTLD, as the advent of novel and unique biomarkers could potentially make screening part of regular checkups and early treatment would mean longer, more improved quality of life in susceptible patients, and be able to monitor progression of the disease and response to treatment.

Research directed at FTD and associated biomarkers has greatly transformed our understanding of the disease. Largely, a contributing factor is the development of new medical technologies coupled with increases in average life expectancy/increase prevalence of disease. While the pathology of aging and related corporeal degeneration is far from well understood, an acceptable pursuit would be to study the process of neurodegeneration as well develop diagnostic tests which could aim to provide earlier alert systems and treatment that may stave off the progression of symptoms of neurodegeneration or perhaps even reverse them in the future. It is therefore not surprising that currently diagnostics aimed at forecasting the fundamental pathology of FTD in vivo is the prime emphasis of investigation [35]. Diagnosis of FTD is clinically based on certain criteria of observation or history of behavioral changes, and cognitive tests, while imaging techniques play an analytical role mainly used for the elimination of focal pathology, such as tumors [35]. Unfortunately, most cases of FTD and its specific etiology are only definitively diagnosed at autopsy, with postmortem tissue demonstrating a combination of extreme degeneration of the frontal, insular or temporal cortices [35]. Many patients diagnosed with FTD were found to have AD at postmortem.

5   Genetic Biomarkers

The genetic predisposition of FTD is substantial, with 40% of patients having a family history of dementia and approximately 10–25% showing observable autosomal dominant patterns of heritability [35]. This association appears to be most common in the bvFTD subtype [35]. Genetic factors have been investigated and the currently accepted bona fide mutations include microtubule-associated protein tau (MAPT), progranulin (GRN), Optineurin (OPTN), TDP-43, and C90RF72 [36–38]. The first chromosome linked to FTD was chromosome 17, this genomic site is now known as FTDP-17 (P for parkinsonism) and most recently, serine-threonine-protein-kinase (TBK1) was discovered in 2015 and implicated in the genetic inheritance associated with FTD [37, 39]. While mutations in the TAR DNA binding protein of 43 kDa (TDP-43) is predominantly causative of familial ALS, a significant subset of FTD patients harbor mutations in this gene which exhibit lobar degeneration and ubiquitinated inclusions. This subtype of FTD is known as frontotemporal lobar degeneration with TDP (FTLD-TDP). Mutations in the OPTN, a gene, which is involved in autophagy of large protein aggregates and dysfunctional organelles and contributes to familial ALS, have been identified in a subset of FTD patients (4.8%) without motor involvement [40]. This subset of patients develop ALS before manifesting FTD symptoms with significant protein aggregation (in the absence of Tau aggregation) caused by TDP-43 [40]. Mutations in OPTN leads to a dysfunction in autophagy, specifically in the turnover of damaged mitochondria, vesicular trafficking, maintenance of the Golgi apparatus, and protein aggregation. MAPT mutations more consistently demonstrate focal and symmetrical temporal lobe atrophy and cause the tau protein to pathogenically combine and form fibrils within neurons. Meanwhile the hexanucleotide repeat expansion on chromosome 9, *C9ORF72 mutation* has been best documented in bvFTD and other subtypes related to motor neuron disease [37]. In terms of mutation prevalence, in a cohort of 95 cases of pathologically confirmed frontotemporal lobar degeneration, postmortem brain revealed that (51%) had

TDP-43 associated pathology, 42 (44%) revealed tau pathology and five (5%) revealed fused-in-sarcoma pathology [41]. Aβ metabolism in familial AD (FAD), has been linked to autosomal dominant FAD-causing mutations; implicated genes include APPswe, APParc, and PSEN1 H163Y, all associated with the disruption in APP processing [42]. While genetic research is a promising field in the earlier detection of neurodegenerative disease, this path poses difficulties as not all FTD appears to emerge from inherited genetic mutations. Additionally, uncertainties about penetrance make it difficult to interpret results (Table 1).

6 Neuroanatomical Biomarkers

Structural brain MRI is helpful in distinguishing FTD from AD, with a sensitivity of 55–94% and a specificity of 81–97% through the comparison of atrophy patterns. In patients with FTD, for example, MRI scans generally reveal frontal and temporal atrophy with relative sparing of the hippocampus [43]. In contrast, AD patients will exhibit bilateral hippocampal and medial temporal lobe atrophy [43]. The pattern of symmetry is also important, for example, asymmetrical involvement of the hemispheres is commonly noted in FTD patients but rare in patients with AD [43]. In addition, support vector machine approaches are now able to integrate T1 and DTI imaging results and aid in the differential diagnostic between FTLD and AD [44].

Structurally speaking, volumetric T1-weighted MRI has been used extensively in an effort to assess gray matter volume/atrophy of specific regions, in this case the frontal lobe [45, 46]. Postprocessing techniques include voxel-based morphometry studies, surface based morphometry methods and software such as FreeSurfer which enables the independent measurement of cortical area and thickness [47]. In bvFTD this imaging modality shows distinct atrophy of the temporal, frontal, as well as the insula and the anterior cingulate cortex. Meanwhile in the genetic forms of FTD, T1-weighted MRI demonstrates specific changes. For example, in the GRN-mutation associated FTD, studies demonstrate asymmetrical frontotemporal-parietal atrophy, as opposed to the C9ORF72 mutation, which demonstrates a symmetrical pattern of widespread atrophy targeting the thalamus and superior cerebellum [47]. Additionally, in patients with MAPT mutations, involvement is noted to begin in the hippocampus and amygdala, followed by the temporal lobe. Resting-state functional MRI (RS-fMRI) has emerged as another useful biomarker for diagnosis given its ability to measure functional connectivity between various areas of the brain based on signals dependent on blood-oxygen levels [48, 49]. Like T1-weighted MRI, RS-fMRI is a noninvasive procedure and in patients with FTD, RS-fMRI highlights decreased

Table 1
Summary table of findings, highlighting important biomarkers associated with neurodegenerative diseases

Biomarker	Description	Specificity
Genetic biomarkers		
Microtubule-associated protein tau (MAPT)	Mutations in this gene disrupt the normal binding of tau to tubulin; results in pathological deposits of hyperphosphorylated tau	Mutations seen in multiple neuropathologies; including frontotemporal dementia, Pick's disease, Alzheimer disease, argyrophilic grain disease, progressive supranuclear palsy, and corticobasal degeneration
Progranulin (GRN)	Marker for *GRN* loss of function mutation that is measurable in both CSF and plasma	100% sensitive and specific to *GRN* mutation carriers
Optineurin (OPTN)	Protein involved in inflammatory response, autophagy, Golgi maintenance, and vesicular transport	Associated with both ALS and glaucoma but also present within the inclusions in FTD, Alzheimer's disease and Huntington's disease
TDP-43	Predominantly causative of familial ALS; a significant subset of FTD patients harbor mutations in this gene which exhibit lobar degeneration and ubiquitinated inclusions; TDP (FTLD-TDP)	Specific to FTLD-TDP, should be noted that TDP co-pathology exists in other neurodegenerative brain disorders
C9ORF72	Pathologic expansion of a noncoding GGGGCC hexanucleotide repeat of the C9orf72 gene	Strongly associated with ALS and FTD
FTDP-17	Linked genetically to mutations in tau gene; an autosomal dominant neurodegenerative disorder	Both FTD and parkinsonism have been linked to chromosome 17 abnormalities
Serine-threonine-protein-kinase (TBK1)	Involved in innate immunity, inflammation, selective autophagy, oncogenesis, and cell death	Mutations are known to cause FTD and ALS (both sporadic and familial)
Amyloid-β precursor protein (aβpp), presenilin 1 (PSEN1), and presenilin 2 (PSEN2)	Causative genes identified in AD pathology; pattern of an autosomal dominant mode of inheritance	Strongly implicated in AD Early onset Alzheimer's disease (EOAD)
Neuroanatomical biomarkers		
T1-weighted MRI	Assess gray matter volume/atrophy	Distinguishing FTD from AD, with a specificity of 81–97%

(continued)

Table 1
(continued)

Biomarker	Description	Specificity
DTI	MRI technique that visualizes white matter	DTI imaging demonstrates involvement of the cingulum, superior cerebellar peduncles as well as the corpus collosum; involvement of these areas considered typical for FTD
Tau PET imaging	A measure of NFT burden, present in all tauopathies	Not specific to FTLD; present in all tauopathies
Support vector machine	Supervised machine learning models with associated learning algorithms; analyze data, integrate T1 and DTI imaging results for classification and regression analysis	Effective in distinguishing patients with AD from controls
Resting-state functional MRI (RS-fMRI)	Measure functional connectivity between various areas of the brain based on signals dependent on blood-oxygen levels (BOLD) signal, which is sensitive to spontaneous neural activity	Can detect disease-specific functional connectivity differences; useful in distinguishing the pathophysiology of AD and bvFTD
Biofluid biomarkers		
Aβ:Tau	Ratio could be useful in differential diagnosis of AD and FTD	Utility in differential diagnosis of AD and FTD with primary language disturbances, it cannot be used to distinguish AD from behavioral variant FTD
TDP-43 in CSF	In accordance with increased TDP-43 gene expression in FTLD, TDP-43 levels in CSF tend to be elevated in disease states	May increase in the early stage of ALS, correlating with early-stage TDP-43 pathology, also increased in FTLD (strong evidence that mechanisms of neurodegeneration in ALS are linked to pathologic TDP-43)
Neurofilament light chain (NfL)	Function is intrinsically structural, and they are involved in the modulation of response to stimuli through axonal diameter control and maintenance of structural integrity	Linear relationship with disease progression in patients with FTD; Low specificity (it also increases in other neurodegenerative disease such as ALS, AD
Progranulin	The nonsense or frameshift (most common) types of mutation usually lead to a protein insufficiency and measurable decreases in progranulin	95% specific GRN-related pathogenesis of FTD

(continued)

Table 1
(continued)

Biomarker	Description	Specificity
TDP-43 in blood	Uses ELISA to detect the presence, or increased amounts, of TDP-43 in plasma	It may help to distinguish those cases of FTLD with ubiquitin/TDP-43 pathology from those with tauopathy; may help distinguish patients with FTD from AD

connectivity between the anterior cingulate cortex and frontoinsula [48]. Mutation-specific findings include reduced left frontal connectivity in GRN mutations of FTD, and reduced connectivity in sensorimotor networks in FTD patients with C9ORF72 mutations [47].

Another imaging technique employed to assess structural integrity, of white matter in this instance, is diffusion tensor imaging, (DTI) [50]. Like MRI, this technique is noninvasive and works by measuring diffusion via the motion of water molecules to specifically measure the microstructural changes in white matter tracts and overall loss of neuronal connectivity with higher resolution [50]. Importantly, studies have highlighted that white matter involvement often occurs even before gray matter involvement, with results showing degeneration in frontal, temporal as well as insular regions [51]. Unique to bvFTD, white matter abnormalities can be seen in the uncinate fasciculus, cingulum bundle, and corpus collosum using DTI [52]. Just as with the MRI imaging modality, specific genetic mutations confer specific white matter degeneration patterns. For example, MAPT mutation associated-FTD has been associated with uncinate fasciculus, and parahippocampal cingulum involvement. In contrast, patients with C9ORF72 mutation-associated FTD, DTI imaging demonstrates involvement of the cingulum, superior cerebellar peduncles as well as the corpus collosum [47, 53].

PET scans have also been found useful and are implemented in the assessment of metabolic changes that precede atrophy in FTD with the use of ^{18}F-fluorodeoxyglucose as a tracer (FDG-PET) [54, 55]. With high specificity for bvFTD, FDG-PET reveals hypometabolism in the orbitofrontal cortex, basal ganglia, dorsolateral cortex, medial prefrontal cortex, and anterior temporal poles. FDG-PET offers a unique medical insight for patients that pose high risk given that studies have shown asymmetrical hypometabolism before symptoms present and even before the onset of gray matter atrophy. In patients with GRN mutations, FDG-PET studies show asymmetrical hypometabolism in the frontal and temporal areas; in contrast, patients with MAPT mutations have associated patterns of hypometabolism in the frontal, parietal and in the medial temporal lobe on FDG-PET [56]. Meanwhile in disease caused by C9ORF72 mutations, affected areas observed on FDG-PET include the basal ganglia, thalamus as well as the limbic system [57]. However, besides ^{18}F-fluorodeoxyglucose, other tracers have been used as diagnostic biomarkers, namely, amyloid and tau. Tau Pet (flortaucipir), specifically known as the ^{18}F-AV-1451 ligand, demonstrates an increased uptake in some FTD patients in the basal ganglia as well as the frontal and temporal regions of the brain [58]. Meanwhile, Pittsburg compound (PiB), also known as amyloid tracer, with its high sensitivity, has been used to differentiate between AD and FTD, as well varying types of FTD [59] (Table 1).

7 Biofluid Biomarkers

Unfortunately, because of the similar, often overlapping clinical symptomology between AD, and FTD, correct diagnosis is often delayed and/or incorrect. It is for this reason that many of the biofluid biomarkers that have been developed/studied, seek to primarily differentiate between these two neurological disorders [17, 60].

Given that cerebral spinal fluid (CSF) is produced by the choroid plexus in the lateral ventricles and travels through the third and fourth ventricles, coursing over the exterior surface of the brain and spinal cord, CSF contains important biomarkers in those afflicted with neurodegenerative disorders [61]. In CSF, the ratio of levels Aβ:tau can be used to distinguish FTD and AD; however, it should be noted that CSF can be difficult to collect in the elderly and involves an invasive procedure [62]. A major challenge for early and correct treatment of neurodegenerative disorders is an accurate diagnosis, especially in the differential diagnosis between AD and FTD given the poverty of specific biomarkers for FTD. However, although studies suggest that p-tau/Aß42 ratio could be useful in differential diagnosis of AD and FTD with primary language disturbances, it cannot be used to distinguish AD from behavioral variant FTD [62, 63]. For patients suffering from ALS, studies suggest that the levels of TDP-43 in CSF may increase in the early stage of ALS, correlating with early-stage TDP-43 pathology, as a promising biomarker for the early stage of ALS [64]. Moreover, neurofilament light chain (NfL) represents another helpful blood and CSF biomarker. NfL is uniquely useful as a biomarker given its linear relationship with disease progression in patients with FTD and therefore its utility in monitoring disease progression [65–67]. These intracellular filaments are located in the central and peripheral nervous system and found in several variations. Their function is intrinsically structural, and they are involved in the modulation of response to stimuli through axonal diameter control and maintenance of structural integrity [68]. Importantly, damage to neurons leads to leakage of measurable amounts of NfL into the CSF. Another benefit to NfL as a biomarker is that blood levels correlate with CSF concentration, making sample retrieval convenient and relatively noninvasive. However, given its low specificity (it also increases in other neurodegenerative disease such as ALS, AD) it should be assessed/interpreted in combination with other biomarkers/diagnostic data.

Another highly specific (95%) and sensitive (95%) blood biomarker is progranulin [69–71]. Levels of progranulin become important in specific GRN-related pathogenesis of FTD; this is because the nonsense or frameshift (most common) types of mutation usually lead to a protein insufficiency and measurable decreases in progranulin [72]. TDP-43 has also been shown to serve as useful biomarker; it is a nuclear DNA/RNA-binding protein, part of a group of RNA-binding proteins which we now understand to be associated with ALS/FTLD, including Fused in Sarcoma (FUS), heterogeneous nuclear ribonucleoprotein A1 (hnRNP A1), and heterogeneous nuclear ribonucleoprotein A2/B1 (hnRNP A2/B1) [73]. Cell death from processes such as aberrant phosphorylation ubiquitination results in elevated levels of TDP-43, with measurable levels found in both CSF as well as in blood. Unfortunately, results from controls measuring TDP-43 have shown substantial overlap. TDP-43 has been the subject of many studies given that it is a common feature in the more prevalent neurodegenerative diseases, including ALS, FTLD as well as AD first discovered in 2006 [74–76]. Its pathomechanism and marker as a neurotoxicity trigger has been studied in vitro and in vivo in over 2000 studies since then, with hopes of elucidating a common therapeutic target in neurodegeneration [76–78]. Given the invasive nature of CSF collection, the ideal biofluid biomarker, would be one that can discriminate between FTD and AD and involve a minimally invasive procedure such as collecting peripheral venous blood (Table 1).

8 Unique Biomarkers: Mitochondrial Respiration Assays in Muscle Biopsies and Fibroblasts

An involvement of mitochondrial dysfunction in FTD has been suggested recently, albeit these studies have been limited in a very small number of patients. TDP-43 and Optineurin mutants in familial ALS-FTD patients exhibit toxic gain of function by inhibiting mitochondrial function and structure. Specifically, TDP-43 and

OPTN can translocate and aggregate in mitochondria to promote mitochondrial dysfunction as evident by a decrease in oxidative phosphorylation, a decrease in oxygen consumption rates, decreased ATP synthesis, and mitochondrial fragmentation (decreased form factor and increased circularity of mitochondria) [79]. In addition to mutations in TDP-43 and OPTN, mutations in coiled-coil-helix coiled-coil-helix protein (CHCHD10), a protein localized to the intermembrane space and matrix that regulates mitochondrial structure, give rise to a subset of FTD that shows mitochondrial pathology beyond CNS pathology, given that FTD affects peripheral tissues in humans [80]. Therefore, the use of patient-derived fibroblasts and muscle biopsies not only provides valuable insight on how familial mutations in FTD-associated mutations in TDP-43 and other FTD-associated proteins contribute to mitochondrial dysfunction and bioenergetics alterations in humans but can be a diagnostic tool to measure the extent of bioenergetic dysfunction and disease severity in FTD. For instance, it has been well documented that muscle biopsies from FTD patients contain ragged fibers and lack cytochrome c oxidase and a loss of complex proteins [80]. In the same study, it was observed that patient-derived fibroblasts from FTD patients harboring CHCHD10 mutations showed swollen mitochondria with disrupted cristae, significantly decreased mitochondrial respiration and decreased complex-driven activities [80]. Another, recent study performed in primary human fibroblasts derived from three FTD patients harboring mutations in TDP-43 and C9ORF72 exhibited aberrant mitochondrial fragmentation, and altered transmembrane potential in the absence of significant alterations in mitochondrial respiration [81]. These studies involved obtaining small skin biopsies (1 mm punches) from patients, culturing fibroblasts for several weeks and performing both immunohistochemical assays in fixed cells to analyze for mitochondrial morphology coupled with assays that measured the level of mitochondrial-derived reactive oxygen species (ROS), and transmembrane potential changes in the mitochondria of fibroblasts in culture [81]. Although these techniques are laborious, can be technically challenging and protracted, performing mitochondrial function assays in patient-derived fibroblasts provides an opportunity to assess for bioenergetic deficiencies and the extent of mitochondrial dysfunction and structural aberrations by using peripheral tissues in FTD patients. However, it is worth noting that all of the aforementioned mitochondrial functional assays are not intended to be used as bone fide diagnostic markers for FTD but as prognostic tools that can be complemented with diagnostic markers to understand mitochondrial pathology and bioenergetic deficiencies in FTD. Hence, when all these assays are combined with other diagnostic markers mentioned above, these tools can be prognostic in terms of measuring disease progression, bioenergetic deficiencies, and efficacy of specific treatments for FTD.

9   Conclusion

Given the current state of the field and the overlapping nature of many of the biomarkers (i.e., poor specificity) we must rely on a multitude of modalities to accurately assess and diagnose the various potential causes of neurodegeneration. Both the diagnostic implementation of a combination of biofluid biomarkers coupled with imaging modalities and clinical history and presentation are all equally integral aspects that must currently be accounted for, as more specific and sensitive biomarkers are discovered. Already, the search for effective and affordable biomarkers of neurodegeneration has led to more unique avenues such the use of electroencephalographic (EEG) signals and a novel information-sharing method a as a promising alternative to the abovementioned biomarkers. However, even in such innovative studies, the need for multiple diagnostics tools is still critical [82]. Until then, a substantial amount of diagnostic weight lies upon clinical assessment of cognitive health deviation from a normal baseline.

References

1. Jellinger KA (2010) Should the word 'dementia' be forgotten? J Cell Mol Med 14 (10):2415–2416

2. Graff-Radford NR, Woodruff BK (2007) Frontotemporal dementia. Semin Neurol 27 (1):48–57

3. Bain HDC et al (2019) The role of lysosomes and autophagosomes in frontotemporal lobar degeneration. Neuropathol Appl Neurobiol 45 (3):244–261

4. Pick A (1892) Uber die Beziehungen der senilen Hirnatrophie zur Aphasie. Prag Med Wochenschr 17:165–167

5. Bergeron C, Davis A, Lang AE (1998) Corticobasal ganglionic degeneration and progressive supranuclear palsy presenting with cognitive decline. Brain Pathol 8(2):355–365

6. Kertesz A, Munoz D (2004) Relationship between frontotemporal dementia and corticobasal degeneration/progressive supranuclear palsy. Dement Geriatr Cogn Disord 17 (4):282–286

7. Kertesz A (2003) Pick complex: an integrative approach to frontotemporal dementia: primary progressive aphasia, corticobasal degeneration, and progressive supranuclear palsy. Neurologist 9(6):311–317

8. Onyike CU, Diehl-Schmid J (2013) The epidemiology of frontotemporal dementia. Int Rev Psychiatry 25(2):130–137

9. Galvin JE et al (2017) The social and economic burden of frontotemporal degeneration. Neurology 89(20):2049–2056

10. Kaizik C et al (2017) Factors underpinning caregiver burden in frontotemporal dementia differ in spouses and their children. J Alzheimers Dis 56(3):1109–1117

11. Mioshi E et al (2009) Factors underlying caregiver stress in frontotemporal dementia and Alzheimer's disease. Dement Geriatr Cogn Disord 27(1):76–81

12. Diehl-Schmid J et al (2013) Caregiver burden and needs in frontotemporal dementia. J Geriatr Psychiatry Neurol 26(4):221–229

13. Riedijk SR et al (2006) Caregiver burden, health-related quality of life and coping in dementia caregivers: a comparison of frontotemporal dementia and Alzheimer's disease. Dement Geriatr Cogn Disord 22 (5–6):405–412

14. Gossye H, Van Broeckhoven C, Engelborghs S (2019) The use of biomarkers and genetic screening to diagnose frontotemporal

dementia: evidence and clinical implications. Front Neurosci 13:757

15. Boeve BF (2007) Links between frontotemporal lobar degeneration, corticobasal degeneration, progressive supranuclear palsy, and amyotrophic lateral sclerosis. Alzheimer Dis Assoc Disord 21(4):S31–S38

16. Jalilianhasanpour R et al (2019) Functional connectivity in neurodegenerative disorders: Alzheimer's disease and frontotemporal dementia. Top Magn Reson Imaging 28 (6):317–324

17. Bang J, Spina S, Miller BL (2015) Frontotemporal dementia. Lancet 386 (10004):1672–1682

18. Lanata SC, Miller BL (2016) The behavioural variant frontotemporal dementia (bvFTD) syndrome in psychiatry. J Neurol Neurosurg Psychiatry 87(5):501–511

19. Lee GJ et al (2014) Neuroanatomical correlates of emotional blunting in behavioral variant frontotemporal dementia and early-onset Alzheimer's disease. J Alzheimers Dis 41 (3):793–800

20. Migliaccio R et al (2020) Cognitive and behavioural inhibition deficits in neurodegenerative dementias. Cortex 131:265–283

21. Daianu M et al (2015) Communication of brain network core connections altered in behavioral variant frontotemporal dementia but possibly preserved in early-onset Alzheimer's disease. Proc SPIE Int Soc Opt Eng 9413:941322

22. Nielsen JA et al (2013) An evaluation of the left-brain vs. right-brain hypothesis with resting state functional connectivity magnetic resonance imaging. PLoS One 8(8):e71275–e71275

23. Haas LF (2001) Phineas gage and the science of brain localisation. J Neurol Neurosurg Psychiatry 71(6):761

24. Torregrossa MM, Quinn JJ, Taylor JR (2008) Impulsivity, compulsivity, and habit: the role of orbitofrontal cortex revisited. Biol Psychiatry 63(3):253–255

25. Perry A et al (2016) The role of the orbitofrontal cortex in regulation of interpersonal space: evidence from frontal lesion and frontotemporal dementia patients. Soc Cogn Affect Neurosci 11(12):1894–1901

26. Moll J et al (2011) Impairment of prosocial sentiments is associated with frontopolar and septal damage in frontotemporal dementia. Neuroimage 54(2):1735–1742

27. Eslinger PJ, Damasio AR (1985) Severe disturbance of higher cognition after bilateral frontal lobe ablation: patient EVR. Neurology 35 (12):1731–1741

28. Edwards-Lee T et al (1997) The temporal variant of frontotemporal dementia. Brain 120 (Pt 6):1027–1040

29. Van Horn JD et al (2012) Mapping connectivity damage in the case of Phineas gage. PLoS One 7(5):e37454–e37454

30. Strong MJ et al (2017) Amyotrophic lateral sclerosis—frontotemporal spectrum disorder (ALS-FTSD): revised diagnostic criteria. In: Amyotrophic lateral sclerosis & frontotemporal degeneration, vol 18, pp 153–174

31. Woolley SC, Strong MJ (2015) Frontotemporal dysfunction and dementia in amyotrophic lateral sclerosis. Neurol Clin 33 (4):787–805

32. Kumar A et al (2021) Alzheimer disease. In: StatPearls. StatPearls Publishing Copyright © 2021, StatPearls Publishing LLC, Treasure Island, FL

33. Folstein MF, Folstein SE, McHugh PR (1975) "Mini-mental state". A practical method for grading the cognitive state of patients for the clinician. J Psychiatr Res 12(3):189–198

34. Scott KR, Barrett AM (2007) Dementia syndromes: evaluation and treatment. Expert Rev Neurother 7(4):407–422

35. Bott NT et al (2014) Frontotemporal dementia: diagnosis, deficits and management. Neurodegen Dis Manag 4(6):439–454

36. Pottier C et al (2018) Potential genetic modifiers of disease risk and age at onset in patients with frontotemporal lobar degeneration and GRN mutations: a genome-wide association study. Lancet Neurol 17(6):548–558

37. Paulson HL, Igo I (2011) Genetics of dementia. Semin Neurol 31(5):449–460

38. Feng S-M et al (2019) Novel mutation in optineurin causing aggressive ALS+/−frontotemporal dementia. Ann Clin Transl Neurol 6 (12):2377–2383

39. Gijselinck I et al (2015) Loss of TBK1 is a frequent cause of frontotemporal dementia in a Belgian cohort. Neurology 85 (24):2116–2125

40. Hu WT, Grossman M (2009) TDP-43 and frontotemporal dementia. Curr Neurol Neurosci Rep 9(5):353–358

41. Rohrer JD et al (2011) Clinical and neuroanatomical signatures of tissue pathology in frontotemporal lobar degeneration. Brain 134 (Pt 9):2565–2581

42. Thordardottir S et al (2017) The effects of different familial Alzheimer's disease mutations

on APP processing in vivo. Alzheimers Res Ther 9(1):9–9

43. Harper L et al (2016) MRI visual rating scales in the diagnosis of dementia: evaluation in 184 post-mortem confirmed cases. Brain 139 (Pt 4):1211–1225

44. Grossman M (2010) Biomarkers in frontotemporal lobar degeneration. Curr Opin Neurol 23 (6):643–648

45. Bruun M et al (2019) Detecting frontotemporal dementia syndromes using MRI biomarkers. NeuroImage Clin 22:101711–101711

46. Del Sole A, Malaspina S, Magenta Biasina A (2016) Magnetic resonance imaging and positron emission tomography in the diagnosis of neurodegenerative dementias. Funct Neurol 31(4):205–215

47. Meeter LH et al (2017) Imaging and fluid biomarkers in frontotemporal dementia. Nat Rev Neurol 13(7):406–419

48. Hohenfeld C, Werner CJ, Reetz K (2018) Resting-state connectivity in neurodegenerative disorders: is there potential for an imaging biomarker? Neuroimage Clin 18:849–870

49. Moguilner S et al (2018) Weighted symbolic dependence metric (wSDM) for fMRI resting-state connectivity: a multicentric validation for frontotemporal dementia. Sci Rep 8(1):11181

50. Mahoney CJ et al (2015) Longitudinal diffusion tensor imaging in frontotemporal dementia. Ann Neurol 77(1):33–46

51. Feis RA et al (2019) Multimodal MRI of grey matter, white matter, and functional connectivity in cognitively healthy mutation carriers at risk for frontotemporal dementia and Alzheimer's disease. BMC Neurol 19(1):343–343

52. Lu PH et al (2014) Regional differences in white matter breakdown between frontotemporal dementia and early-onset Alzheimer's disease. J Alzheimers Dis 39:261–269

53. Mahoney CJ et al (2014) Profiles of white matter tract pathology in frontotemporal dementia. Hum Brain Mapp 35(8):4163–4179

54. Tsai RM et al (2019) (18)F-flortaucipir (AV-1451) tau PET in frontotemporal dementia syndromes. Alzheimers Res Ther 11 (1):13–13

55. Martínez G et al (2017) 18F PET with florbetapir for the early diagnosis of Alzheimer's disease dementia and other dementias in people with mild cognitive impairment (MCI). Cochrane Database Syst Rev 11(11): CD012216–CD012216

56. Deters KD et al (2014) Cerebral hypometabolism and grey matter density in MAPT intron 10 +3 mutation carriers. Am J Neurodegener Dis 3(3):103–114

57. Cistaro A et al (2014) The metabolic signature of C9ORF72-related ALS: FDG PET comparison with nonmutated patients. Eur J Nucl Med Mol Imaging 41(5):844–852

58. Smith R et al (2016) 18F-AV-1451 tau PET imaging correlates strongly with tau neuropathology in MAPT mutation carriers. Brain 139 (9):2372–2379

59. Ishii K (2014) PET approaches for diagnosis of dementia. Am J Neuroradiol 35 (11):2030–2038

60. Rascovsky K et al (2011) Sensitivity of revised diagnostic criteria for the behavioural variant of frontotemporal dementia. Brain 134 (Pt 9):2456–2477

61. Lewczuk P et al (2018) Cerebrospinal fluid and blood biomarkers for neurodegenerative dementias: an update of the consensus of the task force on biological markers in psychiatry of the world Federation of Societies of biological psychiatry. World J Biol Psychiatry 19 (4):244–328

62. Casoli T et al (2019) Cerebrospinal fluid biomarkers and cognitive status in differential diagnosis of frontotemporal dementia and Alzheimer's disease. J Int Med Res 47 (10):4968–4980

63. Ritchie C et al (2014) Plasma and cerebrospinal fluid amyloid beta for the diagnosis of Alzheimer's disease dementia and other dementias in people with mild cognitive impairment (MCI). Cochrane Database Syst Rev 2014(6): CD008782

64. Kasai T et al (2009) Increased TDP-43 protein in cerebrospinal fluid of patients with amyotrophic lateral sclerosis. Acta Neuropathol 117 (1):55–62

65. Spotorno N et al (2020) Plasma neurofilament light protein correlates with diffusion tensor imaging metrics in frontotemporal dementia. PLoS One 15(10):e0236384–e0236384

66. Mattsson N et al (2017) Association of plasma neurofilament light with neurodegeneration in patients with Alzheimer disease. JAMA Neurol 74(5):557–566

67. Sjögren M et al (2000) Cytoskeleton proteins in CSF distinguish frontotemporal dementia from AD. Neurology 54(10):1960–1964

68. Olsson B et al (2019) Association of cerebrospinal fluid neurofilament light protein levels with cognition in patients with dementia, motor neuron disease, and movement disorders. JAMA Neurol 76(3):318–325

69. Kao AW et al (2017) Progranulin, lysosomal regulation and neurodegenerative disease. Nat Rev Neurosci 18(6):325–333

70. Kessenbrock K et al (2008) Proteinase 3 and neutrophil elastase enhance inflammation in mice by inactivating antiinflammatory progranulin. J Clin Invest 118(7):2438–2447

71. Tolkatchev D et al (2008) Structure dissection of human progranulin identifies well-folded granulin/epithelin modules with unique functional activities. Protein Sci 17(4):711–724

72. Nguyen AD et al (2018) Murine knockin model for progranulin-deficient frontotemporal dementia with nonsense-mediated mRNA decay. Proc Natl Acad Sci U S A 115 (12):E2849–E2858

73. Baloh RH (2012) How do the RNA-binding proteins TDP-43 and FUS relate to amyotrophic lateral sclerosis and frontotemporal degeneration, and to each other? Curr Opin Neurol 25(6):701–707

74. Riku Y et al (2014) Lower motor neuron involvement in TAR DNA-binding protein of 43 kDa-related frontotemporal lobar degeneration and amyotrophic lateral sclerosis. JAMA Neurol 71(2):172–179

75. Steinacker P, Barschke P, Otto M (2019) Biomarkers for diseases with TDP-43 pathology. Mol Cell Neurosci 97:43–59

76. Shenouda M et al (2018) Mechanisms associated with TDP-43 neurotoxicity in ALS/FTLD. Adv Neurobiol 20:239–263

77. Liu YC, Chiang PM, Tsai KJ (2013) Disease animal models of TDP-43 proteinopathy and their pre-clinical applications. Int J Mol Sci 14 (10):20079–20111

78. Huang C, Yan S, Zhang Z (2020) Maintaining the balance of TDP-43, mitochondria, and autophagy: a promising therapeutic strategy for neurodegenerative diseases. Transl Neurodegener 9(1):40

79. Huang C, Yan S, Zhang Z (2020) Maintaining the balance of TDP-43, mitochondria, and autophagy: a promising therapeutic strategy for neurodegenerative diseases. Transl Neurodegen 9(1):40

80. Bannwarth S et al (2014) A mitochondrial origin for frontotemporal dementia and amyotrophic lateral sclerosis through CHCHD10 involvement. Brain 137(Pt 8):2329–2345

81. Onesto E et al (2016) Gene-specific mitochondria dysfunctions in human TARDBP and C9ORF72 fibroblasts. Acta Neuropathol Commun 4(1):47

82. Dottori M et al (2017) Towards affordable biomarkers of frontotemporal dementia: a classification study via network's information sharing. Sci Rep 7(1):3822

Part III

Clinical Methods

Chapter 17

TSPO PET Imaging as a Biomarker of Neuroinflammation in Neurodegenerative Disorders

Eryn L. Werry, Fiona M. Bright, and Michael Kassiou

Abstract

Neuroinflammation is a hallmark feature across the spectrum of neurodegenerative disorders. Central to neuroinflammation is the activation of microglia and astrocytes. Activated microglia, and perhaps astrocytes, display an upregulation of TSPO in neuroinflammation and in neurodegenerative disease models, based on culture and animal studies. This indicates TSPO may be a biomarker for neuroinflammation, however, clinical use of TSPO-targeting positron emission tomography to monitor neuroinflammation in neurodegenerative disorders has been hindered by the presence of a TSPO polymorphism (A147T). TSPO ligands bind with lower affinity to A147T TSPO, restricting the clinical utility of this approach. This chapter reviews the ongoing efforts to produce ligands that bind highly to A147T TSPO. It also explores the question of how to interpret the TSPO PET signal, by examining which microglial phenotypes upregulate TSPO in neuroinflammation, and what other brain cell types might contribute to the TSPO PET signal.

Keywords Translocator protein, Neuroinflammation, Neurodegeneration, Microglia, Astrocytes

1 Neuroinflammation in Neurodegenerative Disorders

A critical obstacle for neuroscientists seeking to translate neurodegenerative research to the clinic is the extensive clinical, genetic, and pathological heterogeneity and overlap of neurodegenerative diseases, further complicating the complexity of these disorders. For clinicians, this also poses challenges for the accurate diagnosis of these disorders during life. Therefore, for both neuroscientists and clinicians, identifying sensitive and specific biomarkers of disease and development of disease-modifying therapeutics is crucial, yet has remained elusive to date. However, the trajectory of research targeted at neuroinflammatory mechanisms and pathways associated with the pathophysiology of neurodegeneration has been intensifying and is very much at the forefront of current research, potentially offering promising avenues for biomarker development.

Philip V. Peplow, Bridget Martinez and Thomas A. Gennarelli (eds.), *Neurodegenerative Diseases Biomarkers: Towards Translating Research to Clinical Practice*, Neuromethods, vol. 173, https://doi.org/10.1007/978-1-0716-1712-0_17,
© Springer Science+Business Media, LLC, part of Springer Nature 2022

Advancing age is a primary risk factor for chronic conditions including neurodegenerative diseases, as reviewed elsewhere [1]. There is evidence that the brain undergoes "inflamm-aging," acquiring a progressively heightened proinflammatory environment across the lifespan [2], which may contribute to the age-related risk of sporadic neurodegeneration. In addition to a heightened inflammatory environment with age, the immune system itself undergoes a gradual deterioration or remodeling termed "immunosenescence" and is responsible for the increased susceptibility of older individuals to diseases, particularly inflammatory age-related disease [3].

Despite different etiologies, neuroinflammation is considered a characteristic pathological feature across the spectrum of neurodegenerative diseases [4–7]. Neuroinflammation involves a complex multistage physiological response triggered by cell-damaging processes in the brain, which can include infection, toxins, autoimmunity, trauma, and responses to altered neuronal activity. Neuroinflammation is driven by the reactive morphology and function of resident central nervous system (CNS) innate immune glial cells, predominantly microglia and astrocytes and is accompanied by a complex array of interdependent inflammatory factors that impact the surrounding cells. The degree of neuroinflammation that exists across the spectrum of neurodegenerative disease varies, and is dependent on multiple factors including the period of time and course of its evolution, and the circumstances underlying the initial cause [8–10]. The neuroinflammatory response aims to mitigate the triggering factors by conjuring CNS immunity to defend from harm and restore homeostasis [9, 11].

Microglia, the primary CNS immunocompetent cells, are at the epicenter of the CNS immune response. Increasing evidence suggests that microglia and endothelial cells continue to change phenotypes with age and experience in humans [12, 13], driven by inflammaging processes [14]. Microglia enact diverse functions, surveying their surroundings with processes and adapting quickly to any changes in homeostasis by undergoing alterations to their morphology and functional state, and migrating to sites of injury to initiate tissue repair within the CNS [15–18]. Microglia are activated via microglial pattern recognition receptors (PRRs, including toll-like receptors (TLRs), pathogen-associated molecular patterns (PAMPs) and danger-associated molecular patterns (DAMPs)). In addition, they are also activated in response to glial production of a milieu of inflammatory factors including cytokines, chemokines, secondary messengers, and reactive oxygen species [4, 19–21]. This activation promotes an inflammatory response that further engages the innate immune system. Microglia and astrocytes function as both the target and source of inflammatory factors including pro- and anti-inflammatory cytokines and chemokines [4, 19, 22, 23]. Research has uncovered the heterogeneity of microglia and astrocytes, and their reactive responses are context- and regionally dependent [24–27]. Therefore, the inflammatory response produced may vary across different brain regions and cell populations affected and be dependent on the subtype of neurodegenerative disease. This suggests that specific neuroimmune or inflammatory phenotypes may exist across the various neurodegenerative diseases potentially providing targets for techniques such as in vivo monitoring that are specific to a distinct type of neurodegeneration.

Importantly, although an acute neuroinflammatory reaction to a dangerous stimulus can have protective effects, a chronic, uncontrolled neuroinflammatory response, constituted by the prolonged overactivation of microglia and astrocytes, is deleterious given the excessive and dysregulated production of proinflammatory factors. This highly damaging, chronic response prohibits neuronal repair and results in synaptic impairment, oxidative damage and mitochondrial dysfunction, collectively contributing to, or exacerbating, neurodegenerative processes [10, 22, 28]. A chronic unresolved inflammatory response within the CNS can also result in involvement of adaptive immunity, with the recruitment and infiltration of peripheral immune cells caused via disruption of the blood–brain barrier (BBB), which can further initiate neurodegenerative processes [29].

There is increasing evidence supporting an instrumental role of neuroinflammation in the pathogenesis and progression of most neurodegenerative diseases. Animal models of disease, particularly those modelling Alzheimer's disease (AD) and Parkinson's disease (PD), show activation of inflammatory processes that occur before neurodegeneration. In some reports, this occurs within a similar timeframe to the deposition of intracellular and intercellular protein aggregates [30–32]. Post-mortem studies assessing brain tissue from patients with neurodegenerative diseases consistently report increases in proinflammatory cytokines, activated microglia, and reactive astrocytes [4, 33]. In addition to brain tissue, post-mortem and clinical patient cohort studies across the spectrum of neurodegenerative disorders have also shown dysregulation of inflammatory factors in the periphery, including pro-inflammatory cytokines, chemokines, and complement in the blood, serum, and cerebrospinal fluid of a range of patients with different neurodegenerative diseases [34–38]. Indeed, the chronic neuroinflammatory response observed in neurodegenerative disorders in increasingly recognized to not only be restricted to neuronal pathology, but also involves multifaceted interactions with immunological mechanisms throughout both the central and peripheral nervous systems [22, 28, 39].

There has been much debate as to whether neuroinflammation is a causative or reactive process in the context of neurodegeneration [28, 40]. It has been suggested that the initial disease- and case-specific pathological insult (e.g., amyloid plaques, α-synuclein, tau, TDP-43 etc.) may induce an ongoing cytotoxic response, resulting in a secondary chronic neuroinflammatory event that overlaps with the timing of altered neuronal function in vulnerable areas [41, 42]. Alternatively, there is a body of evidence that neuroinflammation may play more of an initiating role in neurodegeneration. Indeed, immune activation has been suggested to be an early feature, as opposed to a late consequence of neurodegeneration, indicating an early role in disease pathogenesis. Supporting this are studies reporting neuroinflammation or microglial activation in the prodromal stages of AD [43, 44], PD [45–48] and frontotemporal dementia (FTD) [49]. Furthermore, induced pluripotent stem cell-derived microglia and astrocytes established from early onset familial dementia and amyotrophic lateral sclerosis (ALS) patients show aberrant activity, indicative of an underlying immune issue that preexists prior to exacerbating cytotoxicity events [50, 51]. In addition, the most compelling evidence to date for a significant causative role of neuroinflammatory and immune mechanisms in neurodegeneration comes from genome-wide association studies (GWAS) that have shown the expression of function and disease-causative mutations or associated polymorphisms in genes responsible for particular neurodegenerative diseases, also implicated in neuroinflammation [52–56]. A significant association between neurodegenerative diseases and human leukocyte antigen (HLA)-loci (immune system) has also been reported [57–59]. This suggests that the genetic background may influence the degree of inflammation in neurodegeneration [41, 42].

Irrespective of whether neuroinflammation is a causative initiator or reactive participant in neurodegenerative diseases, it is clear neuroinflammation is present early, continues throughout the degeneration, and plays an important role in progression. This suggests in vivo monitoring of neuroinflammation could be a useful tool to track progression of neurodegenerative diseases. Given neuroinflammation is a common feature across the spectrum of neurodegenerative diseases, development of neuroinflammatory tracers that can detect and track pathophysiological changes upstream of neuronal loss are highly desirable. Such tracers would minimize the need to develop separate tools for each disease and prevent the need to perform multiple imaging of patients with an unclear diagnosis. The ability to monitor the evolution of neuroinflammation with disease progression will assist in further understanding the extent to which it actively initiates neurodegeneration, or simply is a secondary response to initial neuronal death. In addition, identifying inflammatory processes as early as possible in the primary stages of neurodegeneration could enable early delivery of tailored therapeutics, potentially in the prodromal phase of disease [60]. Clinical trials assessing agents that modulate neuroinflammation are actively being conducted in diseases such as ALS

[61]. Monitoring neuroinflammation could assist in streamlining clinical trials by enabling disease progression to be tracked and provide information about therapeutic efficacy and patient response to trialed drugs [62]. Furthermore, such clinical trials would be advantageous in uncovering the underlying pathophysiology of neuroinflammatory and regulatory pathways that may be specific to distinct neurodegenerative diseases. To further pursue the prospects of in vivo monitoring of neuroinflammation, an adequate biomarker of neuroinflammation is essential. To date, the translocator protein (TSPO) has shown particular promise, as recently reviewed [63].

2 Establishment of TSPO as a Neuroinflammation Biomarker

The translocator protein (TSPO; 18 kDa) is found in the outer mitochondrial membrane of many different cells, but is particularly abundant in steroid-synthesizing cells, such as in the adrenal gland, gonads, and brain [64]. While there is some interest in targeting TSPO as a treatment for neurodegeneration and neuroinflammation [65–67], the majority of clinical interest in TSPO has focused on understanding and leveraging TSPO as a biomarker for neuroinflammation. TSPO is expressed at low levels in the noninflamed brain, and upregulates in disease-relevant areas in animal models of many neuroinflammatory conditions, including Alzheimer's disease (AD), stroke, brain injury, experimental autoimmune encephalitis, and epilepsy [68–80]. TSPO upregulation was abrogated in successful animal trials of therapeutics for Huntington's disease (HD) and AD, suggesting longitudinal changes in TSPO levels could be used to monitor treatment progress in clinical trials for these conditions [81, 82].

Given the strong evidence of neuroinflammation-related TSPO upregulation in preclinical studies, the ability of TSPO PET ligands to detect neuroinflammation was examined in humans suffering from neurodegenerative conditions. Many early clinical studies produced conflicting findings, with results dependent on the TSPO PET ligand used. When compared to healthy controls, increased central nervous system TSPO PET signal was seen in patients with motor neuron disease, AD, PD and prodromal HD [45, 83–89]. These studies used one of the first TSPO PET ligands, [^{11}C]PK 11195 (Fig. 1). In contrast, TSPO PET signal did not increase in patients with AD and multiple sclerosis when probed with second-generation ligands such as [^{11}C]-DPA-713, [^{18}F]-DPA-714, and [^{18}F]-FEDAA1106 [90–94] (Fig. 1), which were designed to mitigate the low brain permeability and high nonspecific binding of PK 11195 which limit its use as a PET imaging agent [95].

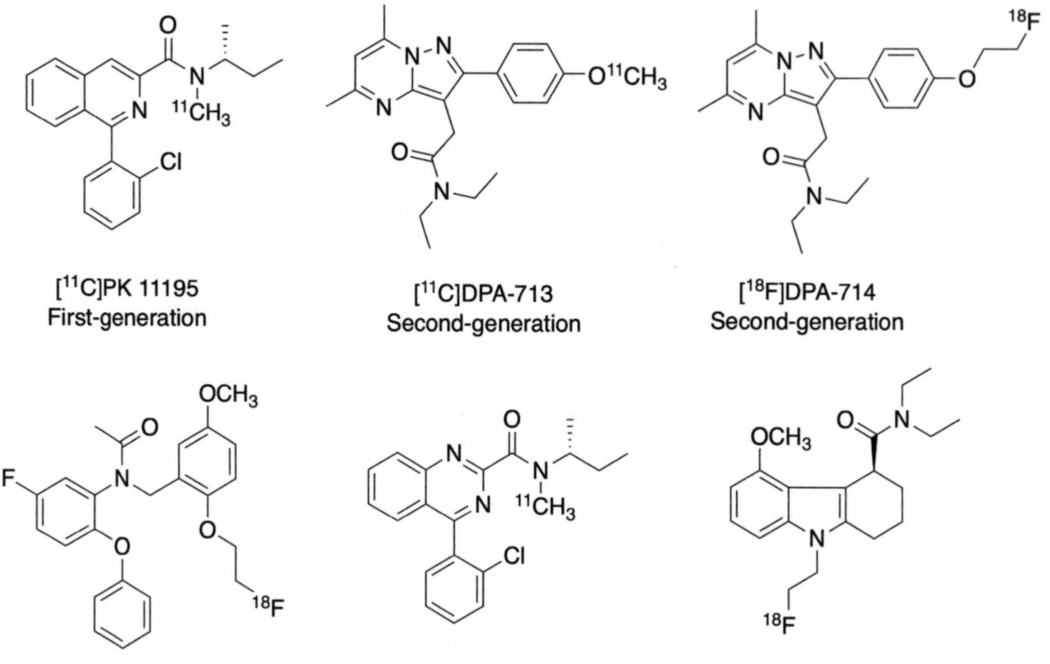

[^{11}C]PK 11195
First-generation

[^{11}C]DPA-713
Second-generation

[^{18}F]DPA-714
Second-generation

[^{18}F]FEDAA1106
Second-generation

[^{11}C]ER176
Third-generation

[^{18}F]GE-180
Third-generation

Fig. 1 Examples of TSPO ligands

Large standard deviations were seen in clinical studies using these second-generation ligands, reflecting unpredicted variability in PET signal between patients. To understand the source of this variance, ex vivo radioligand binding using these ligands was performed on tissue from human donors. Three different binding patterns were seen with second-generation ligands—high affinity binding occurred in tissue from ~50–65% of donors (high affinity binders; HABs), poor binding was seen in tissue from ~5–25% of donors (low affinity binders; LABs), and two-site or moderate binding was seen in the remaining tissue (~30%; mixed affinity binders; MABs) [96–98]. The degree of binding drop-off between HABs and LABs differed for each ligand, with DPA-713 and DAA1106 displaying ~fourfold decrease in affinity [96–98]. In contrast, PK 11195 showed equally high affinity in HAB and LAB tissue [96].

The presence of single nucleotide polymorphisms (SNPs) can be one cause of such variance. Genotyping LABs revealed they are homozygous for a missense SNP, rs6971 [99, 100], while MABs are heterozygous and HABs are homozygous for wild type TSPO. This polymorphism has a prevalence that roughly corresponds with the prevalence of LABs, being present in 30% of Caucasians, 25% of Africans, and to a lesser degree in other races [101]. Rs6971 leads to replacement of the nonpolar alanine with the polar threonine at amino acid 147 (A147T). Given this amino acid is within the binding pocket of the protein [102], this switch from nonpolar to polar may contribute to the reported reduction in affinity of second-generation TSPO ligands.

More recent clinical studies with second-generation TSPO ligands mitigate the influence of A147T TSPO by excluding recruitment of LABs or by stratifying for genotype when analyzing data. These studies more consistently report higher TSPO PET signal in the central nervous system of neuroinflammatory conditions compared to healthy controls (e.g., motor neuron disease, mild cognitive impairment, AD and multiple sclerosis) [43, 103–112]. Given that increased neuroinflammation-related TSPO PET signals are more replicable in these recent studies that account for SNP sensitivity, TSPO is a good target for development of neuroinflammation imaging agents.

3   Advancing the Clinical Potential of TSPO Imaging in Neuroinflammation

Despite the clinical potential of TSPO imaging in neuroinflammatory conditions suggested by these previous studies, the impact of this approach may be widened by development of TSPO PET ligands that do not show sensitivity for rs6971 and by understanding the cell types that TSPO is upregulated on in neuroinflammation.

3.1   New Ligands to Overcome rs6971 Sensitivity

Although stratifying for genotype has increased congruence in clinical studies using second-generation TSPO ligands [113], development of a high affinity and brain-permeable TSPO ligand that binds equally well in HAB and LAB tissue would increase the clinical potential of TSPO PET imaging. A number of newly disclosed ligands have improved upon the degree of A147T TSPO sensitivity seen with second-generation TSPO ligands. These "third-generation" TSPO ligands include [^{11}C]ER-176 and [^{18}F]-GE180.

[^{11}C]ER176 (Fig. 1), is a quinazoline analog of the first-generation PK 11195, designed to retain the structural elements involved in high affinity binding at A147T while mitigating its poor pharmacokinetic properties [114, 115]. This ligand bound with equal affinity to wild type (WT) and A147T TSPO in membranes prepared from human brain tissue, but yielded reduced binding potential in PET studies on healthy control LABs compared to HABs. Binding was sufficiently high, however, to detect a TSPO signal in both groups, suggesting genotyping is still necessary with [^{11}C]ER176, but that LABs will not generate a false negative and so can be included in sampling [114, 115]. Studies are ongoing to probe the utility of this ligand for detecting neuroinflammation in neurodegenerative patients.

Another third-generation TSPO ligand is [^{18}F]GE-180 (Fig. 1). Incorporation of a fluorine-18 label gives this ligand a much longer radioactive half-life compared to carbon-11 labelled compounds, allowing their use in imaging facilities that do not have their own cyclotron [116, 117]. Fluorine-18-labelled ligands also have a reduced risk of radiometabolite production because they are more metabolically stable than their respective ^{11}C-containing derivatives. In radioligand binding studies, GE-180 shows a fivefold reduction in affinity at A147T compared to WT TSPO [118]. Despite this, it generated a TSPO PET signal in a small study on multiple sclerosis patients, and this signal did not significantly differ between HABs, MABs and LABs [119]. There is debate about whether this reflects a lack of in vivo polymorphism sensitivity or could be due to either high variability in neuroinflammation within the sampled patients reducing statistical power to detect differences between groups, or low brain uptake resulting in poor signal to noise ratio [120–124]. Further studies have been proposed to address some of these alternate explanations and shed further light on the clinical utility of [^{18}F]GE-180 [120, 122–124].

In addition to the ongoing clinical validation of [^{11}C]ER176 and [^{18}F]GE-180, in vitro and in situ high throughput screening of ligands at A147T and WT TSPO is being conducted to develop a better understanding of the influence of ligand structure on differential affinity at the two TSPO forms [118, 125, 126]. This approach has already yielded a candidate that has shown brain permeability in animal models [118], although the clinical utility of these ligands is yet to be evaluated.

3.2 Understanding Brain Cell Types That Show Neuroinflammation-Induced TSPO Upregulation

Although it is clear that there is an upregulation of TSPO on microglia in neuroinflammation [127–133], it is not yet clear whether this upregulation represents destructive proinflammatory microglial functions or protective anti-inflammatory microglial functions. It is also not clear the extent to which upregulation occurs on other brain cell types. Answering these questions will be key to interpreting the functional implications of an upregulated TSPO PET signal.

3.2.1 Microglial Phenotypes

The major functions of microglia include maintaining homeostasis, regulating synaptogenesis, refining synapses and responding to pathogens and cell damage [134]. To carry out these functions, microglia need to sense and respond to challenges in the cellular environment. Given this, and the variety of challenges that microglia encounter across the lifespan, they do not comprise a homogenous population but demonstrate multiple phenotypes that may carry out constructive or destructive functions [135]. To understand what functional microglial state the TSPO signal in the brain represents will require investigating which microglial phenotypes upregulate TSPO in neuroinflammation.

Up until 2017, three main microglial phenotypes were recognized, based on morphology, the expression of a limited number of cell markers, and measurement of released cytokines. M_0 microglia describe a population of "resting" microglia that display a ramified morphology, with a primary role in maintaining homeostasis. When M_0 microglia became activated by pathogens or cellular damage, two activated microglial forms arise. The term "M_1 microglia" is used to describe a destructive phenotype that released proinflammatory cytokines such as interleukin-1β (IL-1β) and tumor necrosis factor-α (TNF-α). The term "M_2 microglia" is used to describe a protective phenotype that emerges after exposure to anti-inflammatory cytokines such as interleukin-4 (IL-4), IL-10, and IL-13 [136].

The majority of work probing the phenotypic expression of TSPO has been conducted using this $M_0/M_1/M_2$ phenotypic classification. In vitro and in vivo studies on murine subjects suggest proinflammatory M_1 microglia display an upregulation of TSPO that is not apparent in M_2 microglia [137, 138]. Human microglia often behave differently to murine microglia [139], and in human cells the phenotypes displaying upregulated TSPO levels in neuroinflammation are less clear. Three human studies have used proinflammatory stimuli to generate M_1 microglia, and each have reported conflicting results. An in vivo study found an increase in TSPO on M_1 microglia, and two in vitro studies found a decrease, and no change, in TSPO expression [138, 140, 141]. As human microglia quickly change their physiological properties during culturing [142, 143], these conflicting results may be partly explained by the culturing process.

The conflicting results in these human microglial studies may also be due to deficiencies in the $M_0/M_1/M_2$ paradigm. Allocation of phenotype within this triphenotype continuum depends on the presence of a small number of surface markers or released cytokines, after stimulation with stereotyped pro- or anti-inflammatory stimuli (e.g., lipopolysaccharide (LPS), IL-4). In the inflamed environment that accompanies neurodegeneration, however, a much more varied and dense spectrum of stimuli may influence microglia. These may include any combination of excitotoxic neurotransmitter levels, assemblies of aggregation-prone proteins, and a large number of pro- and anti-inflammatory cytokines. This suggests that the $M_0/M_1/M_2$ paradigm may be too simplistic to fully describe the variety of microglial phenotypes induced by these stimuli. In accordance with this, since 2017, more than 8 new microglial phenotypes have been discovered using genomic and proteomic techniques to identify subsets of microglia displaying signature networks of dysregulated genes and proteins in neuroinflammatory conditions [135, 144–146]. These are likely to be a small portion of the total number of microglial phenotypes in neurodegenerative diseases [135, 144–146]. Some of these newly described phenotypes include disease-associated microglia, microglia neurodegenerative phenotype, LPS-related transcriptomic signature microglia, interferon-related transcriptomic signature microglia, and proliferation-related transcriptomic signature microglia (see Fig. 2). TSPO is upregulated in three of these: LPS-signature microglia (identified from mouse data across 69 different conditions [144]), microglia neurodegenerative phenotype (described in mouse and human AD datasets), and disease-associated microglia (described in mouse and human AD datasets) [135, 144] (Fig. 2). Some of the newly described phenotypes, however, did not display an upregulation of TSPO, suggesting the TSPO PET signal might not signify every microglial phenotype enriched in neuroinflammation or neurodegenerative conditions [135]. An understanding of the functional capabilities of microglial phenotypes that express, and do not express, TSPO is growing and will lead to a greater understanding of the functional interpretation of the TSPO PET signal.

3.2.2 Astrocytes

As introduced earlier, microglia are not the only cells that play a role in neuroinflammation. Another type of glial cell, astrocytes, also contributes to the neuroinflammatory environment by releasing cytokines. Many studies examining the colocalization of TSPO with astrocytic markers such as glial-fibrillary acidic protein (GFAP) find no colocalization [129, 147–149], although a minority of papers report some overlap in staining [132, 133, 150–152]. These studies have all examined colocalization at different time points after the onset of neuroinflammation, and it is possible that astrocytic TSPO expression is low after initial induction of neuroinflammation, but upregulates over time, as is seen in a demyelination model [152]. Even if astrocytes contribute to the upregulation of TSPO seen in clinical PET studies, it is likely that any signal they contribute is reflective of neuroinflammatory processes, given the role of astrocytes in neuroinflammation.

3.2.3 Neurons

TSPO is expressed at low levels in neurons [106]. While an increase in TSPO levels is seen in dorsal root ganglia after injury to peripheral nerves [153], most studies find no TSPO upregulation in brain neurons during neuroinflammation [68, 154–157], although two studies challenge this finding. The first reported upregulation of TSPO that colocalized with the neuronal marker, NeuN, but only when TSPO was probed using a polyclonal antibody. No signal was found in the same paper using a monoclonal antibody, suggesting antibody specificity must be verified before use in TSPO localization studies [68]. A more recent study found a small (<50%) increase in hippocampal neuronal TSPO mRNA levels after physiological stimuli (exposure to a novel environment) [106], however

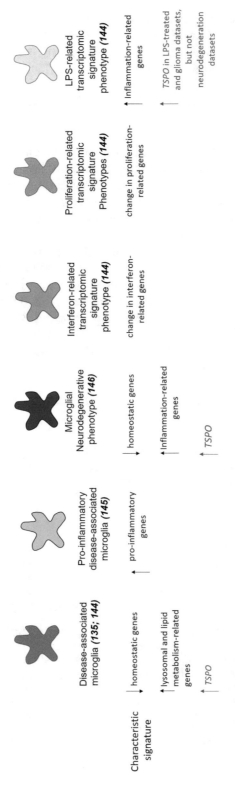

Fig. 2 Examples of new microglial phenotypes displaying signature gene or protein expression networks

this increase is low compared to that seen on microglia in neuroinflammation (7–20× upregulation; [138, 158]), suggesting the contribution of neuronal TSPO upregulation from physiological stimuli is not a major contributor to the TSPO PET signal in neuroinflammation.

4   Conclusion

Neuroinflammation is a characteristic feature of many disorders that involve neurodegeneration. TSPO is upregulated on microglia and astrocytes in neuroinflammation, and may provide a useful biomarker to identify neurodegenerative diseases in prodromal states and monitor disease evolution in clinical trials. Current second-generation TSPO PET radioligands allow detection of neuroinflammation in many subjects, but these ligands lose affinity in subjects that have the rs6971 TSPO polymorphism. An understanding of the functional meaning of the TSPO signal may also become possible through mapping TSPO expression levels on newly defined microglial phenotypes. Ongoing work to produce a TSPO PET ligand that binds equally highly to patients with both wild type and rs6971 TSPO will increase the clinical utility of this technique.

Funding

The authors research is supported by the National Health and Medical Research Council of Australia (NHMRC). MK is an NHMRC Principal Research Fellow (APP1154692).

References

1. Hou Y, Dan X, Babbar M, Wei Y, Hasselbalch SG, Croteau DL et al (2019) Ageing as a risk factor for neurodegenerative disease. Nat Rev Neurol 15:565–581. https://doi.org/10.1038/s41582-019-0244-7
2. Zuo L, Prather ER, Stetskiv M, Garrison DE, Meade JR, Peace TI et al (2019) Inflammaging and oxidative stress in human diseases: from molecular mechanisms to novel treatments. Int J Mol Sci 20:4472. https://doi.org/10.3390/ijms20184472
3. Aiello A, Farzaneh F, Candore G, Caruso C, Davinelli S, Gambino CM et al (2019) Immunosenescence and its hallmarks: how to oppose aging strategically? A review of potential options for therapeutic intervention. Front Immunol 10:2247. https://doi.org/10.3389/fimmu.2019.02247
4. Heneka MT, Carson MJ, El Khoury J, Landreth GE, Brosseron F, Feinstein DL et al (2015) Neuroinflammation in Alzheimer's disease. Lancet Neurol 14:388–405. https://doi.org/10.1016/S1474-4422(15)70016-5
5. McCauley ME, Baloh RH (2019) Inflammation in ALS/FTD pathogenesis. Acta Neuropathol 137:715–730. https://doi.org/10.1007/s00401-018-1933-9
6. Bright F, Werry EL, Dobson-Stone C, Piguet O, Ittner LM, Halliday GM et al (2019) Neuroinflammation in frontotemporal dementia. Nat Rev Neurol 15:540–555. https://doi.org/10.1038/s41582-019-0231-z
7. Calabrese V, Santoro A, Monti D, Crupi R, Di Paola R, Latteri S et al (2018) Aging and

Parkinson's disease: inflammaging, neuroinflammation and biological remodeling as key factors in pathogenesis. Free Radic Biol Med 115:80–91. https://doi.org/10.1016/j.freeradbiomed.2017.10.379

8. Wyss-Coray T (2016) Ageing, neurodegeneration and brain rejuvenation. Nature 539:180–186. https://doi.org/10.1038/nature20411

9. Wyss-Coray T, Mucke L (2002) Inflammation in neurodegenerative disease—a double-edged sword. Neuron 35:419–432

10. DiSabato DJ, Quan N, Godbout JP (2016) Neuroinflammation: the devil is in the details. J Neurochem 139(Suppl 2):136–153. https://doi.org/10.1111/jnc.13607

11. Sochocka M, Diniz BS, Leszek J (2017) Inflammatory response in the CNS: friend or foe? Mol Neurobiol 54:8071–8089. https://doi.org/10.1007/s12035-016-0297-1

12. Soreq L, Consortium UKBE, North American Brain Expression C, Rose J, Soreq E, Hardy J et al (2017) Major shifts in glial regional identity are a transcriptional hallmark of human brain aging. Cell Rep 18:557–570. https://doi.org/10.1016/j.celrep.2016.12.011

13. von Bernhardi R, Eugenín-von Bernhardi L, Eugenín J (2015) Microglial cell dysregulation in brain aging and neurodegeneration. Front Aging Neurosci 7:124. https://doi.org/10.3389/fnagi.2015.00124

14. Pan J, Ma N, Yu B, Zhang W, Wan J (2020) Transcriptomic profiling of microglia and astrocytes throughout aging. J Neuroinflammation 17:97. https://doi.org/10.1186/s12974-020-01774-9

15. Nimmerjahn A, Kirchhoff F, Helmchen F (2005) Resting microglial cells are highly dynamic surveillants of brain parenchyma in vivo. Science 308:1314–1318. https://doi.org/10.1126/science.1110647

16. Streit WJ, Mrak RE, Griffin WS (2004) Microglia and neuroinflammation: a pathological perspective. J Neuroinflammation 1:14. https://doi.org/10.1186/1742-2094-1-14

17. Kettenmann H, Hanisch UK, Noda M, Verkhratsky A (2011) Physiology of microglia. Physiol Rev 91:461–553. https://doi.org/10.1152/physrev.00011.2010

18. Hong S, Beja-Glasser VF, Nfonoyim BM, Frouin A, Li S, Ramakrishnan S et al (2016) Complement and microglia mediate early synapse loss in Alzheimer mouse models. Science 352:712–716. https://doi.org/10.1126/science.aad8373

19. Ramesh G, MacLean AG, Philipp MT (2013) Cytokines and chemokines at the crossroads of neuroinflammation, neurodegeneration, and neuropathic pain. Mediators Inflamm 2013:480739. https://doi.org/10.1155/2013/480739

20. Bianchi ME (2007) DAMPs, PAMPs and alarmins: all we need to know about danger. J Leukoc Biol 81:1–5. https://doi.org/10.1189/jlb.0306164

21. Rubartelli A, Lotze MT (2007) Inside, outside, upside down: damage-associated molecular-pattern molecules (DAMPs) and redox. Trends Immunol 28:429–436. https://doi.org/10.1016/j.it.2007.08.004

22. Glass CK, Saijo K, Winner B, Marchetto MC, Gage FH (2010) Mechanisms underlying inflammation in neurodegeneration. Cell 140:918–934. https://doi.org/10.1016/j.cell.2010.02.016

23. Liddelow SA, Barres BA (2017) Reactive astrocytes: production, function, and therapeutic potential. Immunity 46:957–967. https://doi.org/10.1016/j.immuni.2017.06.006

24. Grabert K, Michoel T, Karavolos MH, Clohisey S, Baillie JK, Stevens MP et al (2016) Microglial brain region-dependent diversity and selective regional sensitivities to aging. Nat Neurosci 19:504–516. https://doi.org/10.1038/nn.4222

25. Tan YL, Yuan Y, Tian L (2020) Microglial regional heterogeneity and its role in the brain. Mol Psychiatry 25:351–367. https://doi.org/10.1038/s41380-019-0609-8

26. Diaz-Castro B, Gangwani MR, Yu X, Coppola G, Khakh BS (2019) Astrocyte molecular signatures in Huntington's disease. Sci Transl Med 11:eaaw8546. https://doi.org/10.1126/scitranslmed.aaw8546

27. Miller SJ (2018) Astrocyte heterogeneity in the adult central nervous system. Front Cell Neurosci 12:401. https://doi.org/10.3389/fncel.2018.00401

28. Kempuraj D, Thangavel R, Natteru PA, Selvakumar GP, Saeed D, Zahoor H et al (2016) Neuroinflammation induces neurodegeneration. J Neurol Neurosurg Spine 1:1003

29. Sweeney MD, Sagare AP, Zlokovic BV (2018) Blood-brain barrier breakdown in Alzheimer disease and other neurodegenerative disorders. Nat Rev Neurol 14:133–150. https://doi.org/10.1038/nrneurol.2017.188

30. Saito T, Saido TC (2018) Neuroinflammation in mouse models of Alzheimer's disease. Clin Exp Neuroimmunol 9:211–218. https://doi.org/10.1111/cen3.12475

31. Nazem A, Sankowski R, Bacher M, Al-Abed Y (2015) Rodent models of neuroinflammation for Alzheimer's disease. J Neuroinflammation 12:74. https://doi.org/10.1186/s12974-015-0291-y

32. Cebrian C, Loike JD, Sulzer D (2015) Neuroinflammation in Parkinson's disease animal models: a cell stress response or a step in neurodegeneration? Curr Top Behav Neurosci 22:237–270. https://doi.org/10.1007/7854_2014_356

33. Gomez-Nicola D, Perry VH (2015) Microglial dynamics and role in the healthy and diseased brain: a paradigm of functional plasticity. Neuroscientist 21:169–184. https://doi.org/10.1177/1073858414530512

34. Lee Y, Lee S, Chang SC, Lee J (2019) Significant roles of neuroinflammation in Parkinson's disease: therapeutic targets for PD prevention. Arch Pharm Res 42:416–425. https://doi.org/10.1007/s12272-019-01133-0

35. Galimberti D, Venturelli E, Fenoglio C, Guidi I, Villa C, Bergamaschini L et al (2008) Intrathecal levels of IL-6, IL-11 and LIF in Alzheimer's disease and frontotemporal lobar degeneration. J Neurol 255:539–544. https://doi.org/10.1007/s00415-008-0737-6

36. Busse M, Michler E, von Hoff F, Dobrowolny H, Hartig R, Frodl T et al (2017) Alterations in the peripheral immune system in dementia. J Alzheimers Dis 58:1303–1313. https://doi.org/10.3233/JAD-161304

37. Domingues C, da Cruz ESOAB, Henriques AG (2017) Impact of cytokines and chemokines on Alzheimer's disease neuropathological hallmarks. Curr Alzheimer Res 14:870–882. https://doi.org/10.2174/1567205014666170317113606

38. Lee KS, Chung JH, Choi TK, Suh SY, Oh BH, Hong CH (2009) Peripheral cytokines and chemokines in Alzheimer's disease. Dement Geriatr Cogn Disord 28:281–287. https://doi.org/10.1159/000245156

39. Kempuraj D, Thangavel R, Selvakumar GP, Zaheer S, Ahmed ME, Raikwar SP et al (2017) Brain and peripheral atypical inflammatory mediators potentiate neuroinflammation and neurodegeneration. Front Cell Neurosci 11:216. https://doi.org/10.3389/fncel.2017.00216

40. Dorothee G (2018) Neuroinflammation in neurodegeneration: role in pathophysiology, therapeutic opportunities and clinical perspectives. J Neural Transm (Vienna) 125:749–750. https://doi.org/10.1007/s00702-018-1880-6

41. Pasqualetti G, Brooks DJ, Edison P (2015) The role of neuroinflammation in dementias. Curr Neurol Neurosci Rep 15:17. https://doi.org/10.1007/s11910-015-0531-7

42. Mrak RE, Griffin WS (2007) Common inflammatory mechanisms in Lewy body disease and Alzheimer disease. J Neuropathol Exp Neurol 66:683–686. https://doi.org/10.1097/nen.0b013e31812503e1

43. Hamelin L, Lagarde J, Dorothee G, Leroy C, Labit M, Comley RA et al (2016) Early and protective microglial activation in Alzheimer's disease: a prospective study using 18F-DPA-714 PET imaging. Brain 139:1252–1264. https://doi.org/10.1093/brain/aww017

44. Okello A, Edison P, Archer HA, Turkheimer FE, Kennedy J, Bullock R et al (2009) Microglial activation and amyloid deposition in mild cognitive impairment: a PET study. Neurology 72:56–62. https://doi.org/10.1212/01.wnl.0000338622.27876.0d

45. Ouchi Y, Yoshikawa E, Sekine Y, Futatsubashi M, Kanno T, Ogusu T et al (2005) Microglial activation and dopamine terminal loss in early Parkinson's disease. Ann Neurol 57:168–175. https://doi.org/10.1002/ana.20338

46. Hogl B, Stefani A, Videnovic A (2018) Idiopathic REM sleep behaviour disorder and neurodegeneration—an update. Nat Rev Neurol 14:40–55. https://doi.org/10.1038/nrneurol.2017.157

47. Iranzo A, Tolosa E, Gelpi E, Molinuevo JL, Valldeoriola F, Serradell M et al (2013) Neurodegenerative disease status and postmortem pathology in idiopathic rapid-eye-movement sleep behaviour disorder: an observational cohort study. Lancet Neurol 12:443–453. https://doi.org/10.1016/S1474-4422(13)70056-5

48. Stokholm MG, Iranzo A, Ostergaard K, Serradell M, Otto M, Svendsen KB et al (2017) Assessment of neuroinflammation in patients with idiopathic rapid-eye-movement sleep behaviour disorder: a case-control study. Lancet Neurol 16:789–796. https://doi.org/10.1016/S1474-4422(17)30173-4

49. Bevan-Jones WR, Cope TE, Jones PS, Passamonti L, Hong YT, Fryer T et al (2019) In vivo evidence for pre-symptomatic neuroinflammation in a MAPT mutation carrier. Ann Clin Transl Neurol 6:373–378. https://doi.org/10.1002/acn3.683

50. Garcia-Reitboeck P, Phillips A, Piers TM, Villegas-Llerena C, Butler M, Mallach A et al

(2018) Human induced pluripotent stem cell-derived microglia-like cells harboring TREM2 missense mutations show specific deficits in phagocytosis. Cell Rep 24:2300–2311. https://doi.org/10.1016/j.celrep.2018.07.094

51. Birger A, Ben-Dor I, Ottolenghi M, Turetsky T, Gil Y, Sweetat S et al (2019) Human iPSC-derived astrocytes from ALS patients with mutated C9ORF72 show increased oxidative stress and neurotoxicity. EBioMedicine 50:274–289. https://doi.org/10.1016/j.ebiom.2019.11.026

52. Bossu P, Salani F, Alberici A, Archetti S, Bellelli G, Galimberti D et al (2011) Loss of function mutations in the progranulin gene are related to pro-inflammatory cytokine dysregulation in frontotemporal lobar degeneration patients. J Neuroinflammation 8:65. https://doi.org/10.1186/1742-2094-8-65

53. Liu Y, Pattamatta A, Zu T, Reid T, Bardhi O, Borchelt DR et al (2016) C9orf72 BAC mouse model with motor deficits and neurodegenerative features of ALS/FTD. Neuron 90:521–534. https://doi.org/10.1016/j.neuron.2016.04.005

54. Gagliano SA, Pouget JG, Hardy J, Knight J, Barnes MR, Ryten M et al (2016) Genomics implicates adaptive and innate immunity in Alzheimer's and Parkinson's diseases. Ann Clin Transl Neurol 3:924–933. https://doi.org/10.1002/acn3.369

55. Oakes JA, Davies MC, Collins MO (2017) TBK1: a new player in ALS linking autophagy and neuroinflammation. Mol Brain 10:5. https://doi.org/10.1186/s13041-017-0287-x

56. Kokiko-Cochran ON, Saber M, Puntambekar S, Bemiller SM, Katsumoto A, Lee YS et al (2018) Traumatic brain injury in hTau model mice: enhanced acute macrophage response and altered long-term recovery. J Neurotrauma 35:73–84. https://doi.org/10.1089/neu.2017.5203

57. Ferrari R, Hernandez DG, Nalls MA, Rohrer JD, Ramasamy A, Kwok JB et al (2014) Frontotemporal dementia and its subtypes: a genome-wide association study. Lancet Neurol 13:686–699. https://doi.org/10.1016/S1474-4422(14)70065-1

58. Atanasio A, Decman V, White D, Ramos M, Ikiz B, Lee HC et al (2016) C9orf72 ablation causes immune dysregulation characterized by leukocyte expansion, autoantibody production, and glomerulonephropathy in mice. Sci Rep 6:23204. https://doi.org/10.1038/srep23204

59. Umoh ME, Dammer EB, Dai J, Duong DM, Lah JJ, Levey AI, et al. (2018) A proteomic network approach across the ALS-FTD disease spectrum resolves clinical phenotypes and genetic vulnerability in human brain. EMBO Mol Med 10:48-62. https://doi.org/10.15252/emmm.201708202

60. Ahmed RM, Paterson RW, Warren JD, Zetterberg H, O'Brien JT, Fox NC et al (2014) Biomarkers in dementia: clinical utility and new directions. J Neurol Neurosurg Psychiatry 85:1426–1434. https://doi.org/10.1136/jnnp-2014-307662

61. Mora JS, Barbeito L, Hermine O (2017) Masitinib as an add-on therapy to riluzole is beneficial in the treatment of amyotrophic lateral sclerosis (ALS) with acceptable tolerability: results from a randomized controlled phase 3 trial. European Network to Cure ALS (ENCALS). http://videolectures.net/encals2017_barbeito_mora_hermine_therapy/

62. Cerami C, Iaccarino L, Perani D (2017) Molecular imaging of neuroinflammation in neurodegenerative dementias: the role of in vivo PET imaging. Int J Mol Sci 18:993. https://doi.org/10.3390/ijms18050993

63. Werry EL, Bright FM, Piguet O, Ittner LM, Halliday GM, Hodges JR et al (2019) Recent developments in TSPO PET imaging as a biomarker of neuroinflammation in neurodegenerative disorders. Int J Mol Sci 20:3161. https://doi.org/10.3390/ijms20133161

64. Rupprecht R, Papadopoulos V, Rammes G, Baghai TC, Fan J, Akula N et al (2010) Translocator protein (18 kDa) (TSPO) as a therapeutic target for neurological and psychiatric disorders. Nat Rev Drug Discov 9:971–988. https://doi.org/10.1038/nrd3295

65. Guilarte TR, Loth MK, Guariglia SR (2016) TSPO finds NOX2 in microglia for redox homeostasis. Trends Pharmacol Sci 37:334–343. https://doi.org/10.1016/j.tips.2016.02.008

66. Gatliff J, East D, Crosby J, Abeti R, Harvey R, Craigen W et al (2014) TSPO interacts with VDAC1 and triggers a ROS-mediated inhibition of mitochondrial quality control. Autophagy 10:2279–2296. https://doi.org/10.4161/15548627.2014.991665

67. Gatliff J, East DA, Singh A, Alvarez MS, Frison M, Matic I et al (2017) A role for TSPO in mitochondrial ca(2+) homeostasis and redox stress signaling. Cell Death Dis 8: e2896. https://doi.org/10.1038/cddis.2017.186

68. Cosenza-Nashat M, Zhao ML, Suh HS, Morgan J, Natividad R, Morgello S et al

(2009) Expression of the translocator protein of 18 kDa by microglia, macrophages and astrocytes based on immunohistochemical localization in abnormal human brain. Neuropathol Appl Neurobiol 35:306–328. https://doi.org/10.1111/j.1365-2990. 2008.01006.x

69. Abourbeh G, Theze B, Maroy R, Dubois A, Brulon V, Fontyn Y et al (2012) Imaging microglial/macrophage activation in spinal cords of experimental autoimmune encephalomyelitis rats by positron emission tomography using the mitochondrial 18 kDa translocator protein radioligand [(1)(8)F] DPA-714. J Neurosci 32:5728–5736. https://doi.org/10.1523/JNEUROSCI. 2900-11.2012

70. Amhaoul H, Hamaide J, Bertoglio D, Reichel SN, Verhaeghe J, Geerts E et al (2015) Brain inflammation in a chronic epilepsy model: evolving pattern of the translocator protein during epileptogenesis. Neurobiol Dis 82:526–539. https://doi.org/10.1016/j. nbd.2015.09.004

71. Brendel M, Probst F, Jaworska A, Overhoff F, Korzhova V, Albert NL et al (2016) Glial activation and glucose metabolism in a transgenic amyloid mouse model: a triple-tracer PET study. J Nucl Med 57:954–960. https://doi.org/10.2967/jnumed.115. 167858

72. Daugherty DJ, Selvaraj V, Chechneva OV, Liu XB, Pleasure DE, Deng W (2013) A TSPO ligand is protective in a mouse model of multiple sclerosis. EMBO Mol Med 5:891–903. https://doi.org/10.1002/emmm. 201202124

73. Dedeurwaerdere S, Callaghan PD, Pham T, Rahardjo GL, Amhaoul H, Berghofer P et al (2012) PET imaging of brain inflammation during early epileptogenesis in a rat model of temporal lobe epilepsy. EJNMMI Res 2:60. https://doi.org/10.1186/2191-219X-2-60

74. Israel I, Ohsiek A, Al-Momani E, Albert-Weissenberger C, Stetter C, Mencl S et al (2016) Combined [(18)F]DPA-714 micro-positron emission tomography and autoradiography imaging of microglia activation after closed head injury in mice. J Neuroinflammation 13:140. https://doi.org/10.1186/s12974-016-0604-9

75. Martin A, Boisgard R, Theze B, Van Camp N, Kuhnast B, Damont A et al (2010) Evaluation of the PBR/TSPO radioligand [(18)F]DPA-714 in a rat model of focal cerebral ischemia. J Cereb Blood Flow Metab 30:230–241. https://doi.org/10.1038/jcbfm.2009.205

76. Mattner F, Katsifis A, Staykova M, Ballantyne P, Willenborg DO (2005) Evaluation of a radiolabelled peripheral benzodiazepine receptor ligand in the central nervous system inflammation of experimental autoimmune encephalomyelitis: a possible probe for imaging multiple sclerosis. Eur J Nucl Med Mol Imaging 32:557–563. https://doi.org/10.1007/s00259-004-1690-y

77. Mirzaei N, Tang SP, Ashworth S, Coello C, Plisson C, Passchier J et al (2016) In vivo imaging of microglial activation by positron emission tomography with [(11)C]PBR28 in the 5XFAD model of Alzheimer's disease. Glia 64:993–1006. https://doi.org/10.1002/glia.22978

78. Thomas C, Vercouillie J, Domene A, Tauber C, Kassiou M, Guilloteau D et al (2016) Detection of Neuroinflammation in a rat model of subarachnoid hemorrhage using [18F]DPA-714 PET imaging. Mol Imaging 15:1536012116639189. https://doi.org/10.1177/1536012116639189

79. Toth M, Little P, Arnberg F, Haggkvist J, Mulder J, Halldin C et al (2016) Acute neuroinflammation in a clinically relevant focal cortical ischemic stroke model in rat: longitudinal positron emission tomography and immunofluorescent tracking. Brain Struct Funct 221:1279–1290. https://doi.org/10.1007/s00429-014-0970-y

80. Tremoleda JL, Thau-Zuchman O, Davies M, Foster J, Khan I, Vadivelu KC et al (2016) In vivo PET imaging of the neuroinflammatory response in rat spinal cord injury using the TSPO tracer [(18)F]GE-180 and effect of docosahexaenoic acid. Eur J Nucl Med Mol Imaging 43:1710–1722. https://doi.org/10.1007/s00259-016-3391-8

81. Simmons DA, James ML, Belichenko NP, Semaan S, Condon C, Kuan J et al (2018) TSPO-PET imaging using [18F]PBR06 is a potential translatable biomarker for treatment response in Huntington's disease: preclinical evidence with the p75NTR ligand LM11A-31. Hum Mol Genet 27:2893–2912. https://doi.org/10.1093/hmg/ddy202

82. James ML, Belichenko NP, Shuhendler AJ, Hoehne A, Andrews LE, Condon C et al (2017) [(18)F]GE-180 PET detects reduced microglia activation after LM11A-31 therapy in a mouse model of Alzheimer's disease. Theranostics 7:1422–1436. https://doi.org/10.7150/thno.17666

83. Politis M, Lahiri N, Niccolini F, Su P, Wu K, Giannetti P et al (2015) Increased central microglial activation associated with peripheral cytokine levels in premanifest

Huntington's disease gene carriers. Neurobiol Dis 83:115–121. https://doi.org/10.1016/j.nbd.2015.08.011

84. Gulyas B, Toth M, Schain M, Airaksinen A, Vas A, Kostulas K et al (2012) Evolution of microglial activation in ischaemic core and peri-infarct regions after stroke: a PET study with the TSPO molecular imaging biomarker [(((11))C]vinpocetine. J Neurol Sci 320:110–117. https://doi.org/10.1016/j.jns.2012.06.026

85. Yasuno F, Kosaka J, Ota M, Higuchi M, Ito H, Fujimura Y et al (2012) Increased binding of peripheral benzodiazepine receptor in mild cognitive impairment-dementia converters measured by positron emission tomography with [(1)(1)C]DAA1106. Psychiatry Res 203:67–74. https://doi.org/10.1016/j.pscychresns.2011.08.013

86. Corcia P, Tauber C, Vercoullie J, Arlicot N, Prunier C, Praline J et al (2012) Molecular imaging of microglial activation in amyotrophic lateral sclerosis. PLoS One 7:e52941. https://doi.org/10.1371/journal.pone.0052941

87. Cagnin A, Brooks DJ, Kennedy AM, Gunn RN, Myers R, Turkheimer FE et al (2001) In-vivo measurement of activated microglia in dementia. Lancet 358:461–467. https://doi.org/10.1016/S0140-6736(01)05625-2

88. Gerhard A, Pavese N, Hotton G, Turkheimer F, Es M, Hammers A et al (2006) In vivo imaging of microglial activation with [11C](R)-PK11195 PET in idiopathic Parkinson's disease. Neurobiol Dis 21:404–412. https://doi.org/10.1016/j.nbd.2005.08.002

89. Pavese N, Gerhard A, Tai YF, Ho AK, Turkheimer F, Barker RA et al (2006) Microglial activation correlates with severity in Huntington disease: a clinical and PET study. Neurology 66:1638–1643. https://doi.org/10.1212/01.wnl.0000222734.56412.17

90. Gulyas B, Vas A, Toth M, Takano A, Varrone A, Cselenyi Z et al (2011) Age and disease related changes in the translocator protein (TSPO) system in the human brain: positron emission tomography measurements with [11C]vinpocetine. Neuroimage 56:1111–1121. https://doi.org/10.1016/j.neuroimage.2011.02.020

91. Takano A, Piehl F, Hillert J, Varrone A, Nag S, Gulyas B et al (2013) In vivo TSPO imaging in patients with multiple sclerosis: a brain PET study with [18F]FEDAA1106. EJNMMI Res 3:30. https://doi.org/10.1186/2191-219X-3-30

92. Golla SS, Boellaard R, Oikonen V, Hoffmann A, van Berckel BN, Windhorst AD et al (2015) Quantification of [18F]DPA-714 binding in the human brain: initial studies in healthy controls and Alzheimer's disease patients. J Cereb Blood Flow Metab 35:766–772. https://doi.org/10.1038/jcbfm.2014.261

93. Varrone A, Mattsson P, Forsberg A, Takano A, Nag S, Gulyas B et al (2013) In vivo imaging of the 18-kDa translocator protein (TSPO) with [18F]FEDAA1106 and PET does not show increased binding in Alzheimer's disease patients. Eur J Nucl Med Mol Imaging 40:921–931. https://doi.org/10.1007/s00259-013-2359-1

94. Chauveau F, Van Camp N, Dolle F, Kuhnast B, Hinnen F, Damont A et al (2009) Comparative evaluation of the translocator protein radioligands 11C-DPA-713, 18F-DPA-714, and 11C-PK11195 in a rat model of acute neuroinflammation. J Nucl Med 50:468–476. https://doi.org/10.2967/jnumed.108.058669

95. Chauveau F, Boutin H, Van Camp N, Dolle F, Tavitian B (2008) Nuclear imaging of neuroinflammation: a comprehensive review of [11C]PK11195 challengers. Eur J Nucl Med Mol Imaging 35:2304–2319. https://doi.org/10.1007/s00259-008-0908-9

96. Owen DR, Howell OW, Tang SP, Wells LA, Bennacef I, Bergstrom M et al (2010) Two binding sites for [3H]PBR28 in human brain: implications for TSPO PET imaging of neuroinflammation. J Cereb Blood Flow Metab 30:1608–1618. https://doi.org/10.1038/jcbfm.2010.63

97. Owen DR, Gunn RN, Rabiner EA, Bennacef I, Fujita M, Kreisl WC et al (2011) Mixed-affinity binding in humans with 18-kDa translocator protein ligands. J Nucl Med 52:24–32. https://doi.org/10.2967/jnumed.110.079459

98. Mizrahi R, Rusjan PM, Kennedy J, Pollock B, Mulsant B, Suridjan I et al (2012) Translocator protein (18 kDa) polymorphism (rs6971) explains in-vivo brain binding affinity of the PET radioligand [(18)F]-FEPPA. J Cereb Blood Flow Metab 32:968–972. https://doi.org/10.1038/jcbfm.2012.46

99. Owen DR, Yeo AJ, Gunn RN, Song K, Wadsworth G, Lewis A et al (2012) An 18-kDa translocator protein (TSPO) polymorphism explains differences in binding affinity of the PET radioligand PBR28. J Cereb Blood Flow Metab 32:1–5. https://doi.org/10.1038/jcbfm.2011.147

100. Guo Q, Colasanti A, Owen DR, Onega M, Kamalakaran A, Bennacef I et al (2013) Quantification of the specific translocator protein signal of 18F-PBR111 in healthy humans: a genetic polymorphism effect on in vivo binding. J Nucl Med 54:1915–1923. https://doi.org/10.2967/jnumed.113. 121020

101. Project TIH. Hapmap database. http:// hapmap.ncbi.nlm.nih.gov. Accessed

102. Jaremko L, Jaremko M, Giller K, Becker S, Zweckstetter M (2014) Structure of the mitochondrial translocator protein in complex with a diagnostic ligand. Science 343:1363–1366. https://doi.org/10.1126/science.1248725

103. Colasanti A, Guo Q, Muhlert N, Giannetti P, Onega M, Newbould RD et al (2014) In vivo assessment of brain White matter inflammation in multiple sclerosis with (18)F-PBR111 PET. J Nucl Med 55:1112–1118. https://doi.org/10.2967/jnumed.113.135129

104. Zurcher NR, Loggia ML, Lawson R, Chonde DB, Izquierdo-Garcia D, Yasek JE et al (2015) Increased in vivo glial activation in patients with amyotrophic lateral sclerosis: assessed with [(11)C]-PBR28. Neuroimage Clin 7:409–414. https://doi.org/10.1016/j.nicl.2015.01.009

105. Kreisl WC, Lyoo CH, McGwier M, Snow J, Jenko KJ, Kimura N et al (2013) In vivo radioligand binding to translocator protein correlates with severity of Alzheimer's disease. Brain 136:2228–2238. https://doi.org/10.1093/brain/awt145

106. Suridjan I, Pollock BG, Verhoeff NP, Voineskos AN, Chow T, Rusjan PM et al (2015) In-vivo imaging of grey and white matter neuroinflammation in Alzheimer's disease: a positron emission tomography study with a novel radioligand, [18F]-FEPPA. Mol Psychiatry 20:1579–1587. https://doi.org/10.1038/mp.2015.1

107. Kreisl WC, Lyoo CH, Liow JS, Wei M, Snow J, Page E et al (2016) (11)C-PBR28 binding to translocator protein increases with progression of Alzheimer's disease. Neurobiol Aging 44:53–61. https://doi.org/10.1016/j.neurobiolaging.2016.04.011

108. Bradburn S, Murgatroyd C, Ray N (2019) Neuroinflammation in mild cognitive impairment and Alzheimer's disease: a meta-analysis. Ageing Res Rev 50:1–8. https://doi.org/10.1016/j.arr.2019.01.002

109. Datta G, Colasanti A, Kalk N, Owen D, Scott G, Rabiner EA et al (2017) (11)C-PBR28 and (18)F-PBR111 detect White matter inflammatory heterogeneity in multiple sclerosis. J Nucl Med 58:1477–1482. https://doi.org/10.2967/jnumed.116.187161

110. Varrone A, Oikonen V, Forsberg A, Joutsa J, Takano A, Solin O et al (2015) Positron emission tomography imaging of the 18-kDa translocator protein (TSPO) with [18F] FEMPA in Alzheimer's disease patients and control subjects. Eur J Nucl Med Mol Imaging 42:438–446. https://doi.org/10.1007/s00259-014-2955-8

111. Lyoo CH, Ikawa M, Liow JS, Zoghbi SS, Morse CL, Pike VW et al (2015) Cerebellum can serve as a pseudo-reference region in Alzheimer disease to detect neuroinflammation measured with PET radioligand binding to translocator protein. J Nucl Med 56:701–706. https://doi.org/10.2967/jnumed.114.146027

112. Yoder KK, Nho K, Risacher SL, Kim S, Shen L, Saykin AJ (2013) Influence of TSPO genotype on 11C-PBR28 standardized uptake values. J Nucl Med 54:1320–1322. https://doi.org/10.2967/jnumed.112.118885

113. Sucksdorff M, Rissanen E, Tuisku J, Nuutinen S, Paavilainen T, Rokka J et al (2017) Evaluation of the effect of Fingolimod treatment on microglial activation using serial PET imaging in multiple sclerosis. J Nucl Med 58:1646–1651. https://doi.org/10.2967/jnumed.116.183020

114. Zanotti-Fregonara P, Zhang Y, Jenko KJ, Gladding RL, Zoghbi SS, Fujita M et al (2014) Synthesis and evaluation of translocator 18 kDa protein (TSPO) positron emission tomography (PET) radioligands with low binding sensitivity to human single nucleotide polymorphism rs6971. ACS Chem Nerosci 5:963–971. https://doi.org/10.1021/cn500138n

115. Ikawa M, Lohith TG, Shrestha S, Telu S, Zoghbi SS, Castellano S et al (2017) 11C-ER176, a radioligand for 18-kDa translocator protein, has adequate sensitivity to robustly image all three affinity genotypes in human brain. J Nucl Med 58:320–325. https://doi.org/10.2967/jnumed.116.178996

116. Knezevic D, Mizrahi R (2018) Molecular imaging of neuroinflammation in Alzheimer's disease and mild cognitive impairment. Prog Neuropsychopharmacol Biol Psychiatry 80:123–131. https://doi.org/10.1016/j.pnpbp.2017.05.007

117. Stoll HP, Hutchins GD, Winkle WL, Nguyen AT, Appledorn CR, Janzen I et al (2001) Advantages of short-lived positron-emitting

radioisotopes for intracoronary radiation therapy with liquid-filled balloons to prevent restenosis. J Nucl Med 42:1375–1383

118. Qiao L, Fisher E, McMurray L, Milicevic Sephton S, Hird M, Kuzhuppilly-Ramakrishnan N et al (2019) Radiosynthesis of (R,S)-[(18) F]GE387: a potential PET radiotracer for imaging translocator protein 18 kDa (TSPO) with low binding sensitivity to the human gene polymorphism rs6971. ChemMedChem 14(9):982–993. https://doi.org/10.1002/cmdc.201900023

119. Unterrainer M, Mahler C, Vomacka L, Lindner S, Havla J, Brendel M et al (2018) TSPO PET with [(18)F]GE-180 sensitively detects focal neuroinflammation in patients with relapsing-remitting multiple sclerosis. Eur J Nucl Med Mol Imaging 45:1423–1431. https://doi.org/10.1007/s00259-018-3974-7

120. Albert NL, Unterrainer M, Brendel M, Kaiser L, Zweckstetter M, Cumming P et al (2019) In response to: the validity of (18)F-GE180 as a TSPO imaging agent. Eur J Nucl Med Mol Imaging. https://doi.org/10.1007/s00259-019-04294-8

121. Sridharan S, Raffel J, Nandoskar A, Record C, Brooks DJ, Owen D et al (2019) Confirmation of specific binding of the 18-kDa translocator protein (TSPO) Radioligand [(18)F]GE-180: a blocking study using XBD173 in multiple sclerosis normal appearing White and Grey matter. Mol Imaging Biol 21 (5):935–944. https://doi.org/10.1007/s11307-019-01323-8

122. Zanotti-Fregonara P, Veronese M, Pascual B, Rostomily RC, Turkheimer F, Masdeu JC (2019) The validity of (18)F-GE180 as a TSPO imaging agent. Eur J Nucl Med Mol Imaging 46(6):1205–1207. https://doi.org/10.1007/s00259-019-4268-4

123. Zanotti-Fregonara P, Pascual B, Rostomily RC, Rizzo G, Veronese M, Masdeu JC et al (2020) Anatomy of (18)F-GE180, a failed radioligand for the TSPO protein. Eur J Nucl Med Mol Imaging 47(10):2233–2236. https://doi.org/10.1007/s00259-020-04732-y

124. Albert NL, Unterrainer M, Kaiser L, Brendel M, Vettermann FJ, Holzgreve A et al (2020) In response to: anatomy of (18) F-GE180, a failed radioligand for the TSPO protein. Eur J Nucl Med Mol Imaging 47 (10):2237–2241. https://doi.org/10.1007/s00259-020-04885-w

125. Cheng HWA, Sokias R, Werry EL, Ittner LM, Reekie TA, Du J et al (2019) First nondiscriminating translocator protein ligands produced from a Carbazole scaffold. J Med Chem 62:8235–8248. https://doi.org/10.1021/acs.jmedchem.9b00980

126. Sokias R, Werry EL, Chua SW, Reekie TA, Munoz L, Wong ECN et al (2017) Determination and reduction of translocator protein (TSPO) ligand rs6971 discrimination. Med Chem Commun 8:202–210. https://doi.org/10.1039/c6md00523c

127. Chaney A, Cropper HC, Johnson EM, Lechtenberg KJ, Peterson TC, Stevens MY et al (2019) (11)C-DPA-713 versus (18)F-GE-180: a preclinical comparison of translocator protein 18 kDa PET tracers to visualize acute and chronic neuroinflammation in a mouse model of ischemic stroke. J Nucl Med 60:122–128. https://doi.org/10.2967/jnumed.118.209155

128. Tournier BB, Tsartsalis S, Rigaud D, Fossey C, Cailly T, Fabis F et al (2019) TSPO and amyloid deposits in sub-regions of the hippocampus in the 3xTgAD mouse model of Alzheimer's disease. Neurobiol Dis 121:95–105. https://doi.org/10.1016/j.nbd.2018.09.022

129. Stephenson DT, Schober DA, Smalstig EB, Mincy RE, Gehlert DR, Clemens JA (1995) Peripheral benzodiazepine receptors are colocalized with activated microglia following transient global forebrain ischemia in the rat. J Neurosci 15:5263–5274

130. Vowinckel E, Reutens D, Becher B, Verge G, Evans A, Owens T et al (1997) PK11195 binding to the peripheral benzodiazepine receptor as a marker of microglia activation in multiple sclerosis and experimental autoimmune encephalomyelitis. J Neurosci Res 50:345–353. https://doi.org/10.1002/(SICI)1097-4547(19971015)50:2<345::AID-JNR22>3.0.CO;2-5

131. Arlicot N, Katsifis A, Garreau L, Mattner F, Vergote J, Duval S et al (2008) Evaluation of CLINDE as potent translocator protein (18 kDa) SPECT radiotracer reflecting the degree of neuroinflammation in a rat model of microglial activation. Eur J Nucl Med Mol Imaging 35:2203–2211. https://doi.org/10.1007/s00259-008-0834-x

132. Chen MK, Baidoo K, Verina T, Guilarte TR (2004) Peripheral benzodiazepine receptor imaging in CNS demyelination: functional implications of anatomical and cellular localization. Brain 127:1379–1392. https://doi.org/10.1093/brain/awh161

133. Chen MK, Guilarte TR (2006) Imaging the peripheral benzodiazepine receptor response in central nervous system demyelination and

remyelination. Toxicol Sci 91:532–539. https://doi.org/10.1093/toxsci/kfj172

134. Spiller KJ, Restrepo CR, Khan T, Dominique MA, Fang TC, Canter RG et al (2018) Microglia-mediated recovery from ALS-relevant motor neuron degeneration in a mouse model of TDP-43 proteinopathy. Nat Neurosci 21:329–340. https://doi.org/10.1038/s41593-018-0083-7

135. Keren-Shaul H, Spinrad A, Weiner A, Matcovitch-Natan O, Dvir-Szternfeld R, Ulland TK et al (2017) A unique microglia type associated with restricting development of Alzheimer's disease. Cell 169:1276–1290 e1217. https://doi.org/10.1016/j.cell.2017.05.018

136. Geloso MC, Corvino V, Marchese E, Serrano A, Michetti F, D'Ambrosi N (2017) The dual role of microglia in ALS: mechanisms and therapeutic approaches. Front Aging Neurosci 9:242. https://doi.org/10.3389/fnagi.2017.00242

137. Beckers L, Ory D, Geric I, Declercq L, Koole M, Kassiou M et al (2018) Increased expression of translocator protein (TSPO) marks pro-inflammatory microglia but does not predict neurodegeneration. Mol Imaging Biol 20:94–102. https://doi.org/10.1007/s11307-017-1099-1

138. Owen DR, Narayan N, Wells L, Healy L, Smyth E, Rabiner EA et al (2017) Pro-inflammatory activation of primary microglia and macrophages increases 18 kDa translocator protein expression in rodents but not humans. J Cereb Blood Flow Metab 37:2679–2690. https://doi.org/10.1177/0271678X17710182

139. Rustenhoven J, Park TI, Schweder P, Scotter J, Correia J, Smith AM et al (2016) Isolation of highly enriched primary human microglia for functional studies. Sci Rep 6:19371. https://doi.org/10.1038/srep19371

140. Sandiego CM, Gallezot JD, Pittman B, Nabulsi N, Lim K, Lin SF et al (2015) Imaging robust microglial activation after lipopolysaccharide administration in humans with PET. Proc Natl Acad Sci U S A 112:12468–12473. https://doi.org/10.1073/pnas.1511003112

141. Narayan N, Mandhair H, Smyth E, Dakin SG, Kiriakidis S, Wells L et al (2017) The macrophage marker translocator protein (TSPO) is down-regulated on pro-inflammatory 'M1' human macrophages. PLoS One 12: e0185767. https://doi.org/10.1371/journal.pone.0185767

142. Stansley B, Post J, Hensley K (2012) A comparative review of cell culture systems for the study of microglial biology in Alzheimer's disease. J Neuroinflammation 9:115. https://doi.org/10.1186/1742-2094-9-115

143. Chamberlain LM, Holt-Casper D, Gonzalez-Juarrero M, Grainger DW (2015) Extended culture of macrophages from different sources and maturation results in a common M2 phenotype. J Biomed Mater Res A 103:2864–2874. https://doi.org/10.1002/jbm.a.35415

144. Friedman BA, Srinivasan K, Ayalon G, Meilandt WJ, Lin H, Huntley MA et al (2018) Diverse brain myeloid expression profiles reveal distinct microglial activation states and aspects of Alzheimer's disease not evident in mouse models. Cell Rep 22:832–847. https://doi.org/10.1016/j.celrep.2017.12.066

145. Rangaraju S, Dammer EB, Raza SA, Rathakrishnan P, Xiao H, Gao T et al (2018) Identification and therapeutic modulation of a pro-inflammatory subset of disease-associated-microglia in Alzheimer's disease. Mol Neurodegen 13:24. https://doi.org/10.1186/s13024-018-0254-8

146. Krasemann S, Madore C, Cialic R, Baufeld C, Calcagno N, El Fatimy R et al (2017) The TREM2-APOE pathway drives the transcriptional phenotype of dysfunctional microglia in neurodegenerative diseases. Immunity 47:566–581. https://doi.org/10.1016/j.immuni.2017.08.008

147. Conway EL, Gundlach AL, Craven JA (1998) Temporal changes in glial fibrillary acidic protein messenger RNA and [3H]PK11195 binding in relation to imidazoline-I2-receptor and alpha 2-adrenoceptor binding in the hippocampus following transient global forebrain ischaemia in the rat. Neuroscience 82:805–817

148. Myers R, Manjil LG, Cullen BM, Price GW, Frackowiak RS, Cremer JE (1991) Macrophage and astrocyte populations in relation to [3H]PK 11195 binding in rat cerebral cortex following a local ischaemic lesion. J Cereb Blood Flow Metab 11:314–322. https://doi.org/10.1038/jcbfm.1991.64

149. Domene A, Cavanagh C, Page G, Bodard S, Klein C, Delarasse C et al (2016) Expression of phenotypic astrocyte marker is increased in a transgenic mouse model of Alzheimer's disease versus age-matched controls: a presymptomatic stage study. Int J Alzheimers Dis 2016:5696241. https://doi.org/10.1155/2016/5696241

150. Guilarte TR, Kuhlmann AC, O'Callaghan JP, Miceli RC (1995) Enhanced expression of peripheral benzodiazepine receptors in trimethyltin-exposed rat brain: a biomarker of neurotoxicity. Neurotoxicology 16:441–450

151. Lavisse S, Guillermier M, Herard AS, Petit F, Delahaye M, Van Camp N et al (2012) Reactive astrocytes overexpress TSPO and are detected by TSPO positron emission tomography imaging. J Neurosci 32:10809–10818. https://doi.org/10.1523/JNEUROSCI.1487-12.2012

152. Guilarte TR (2019) TSPO in diverse CNS pathologies and psychiatric disease: a critical review and a way forward. Pharmacol Ther 194:44–58. https://doi.org/10.1016/j.pharmthera.2018.09.003

153. Karchewski LA, Bloechlinger S, Woolf CJ (2004) Axonal injury-dependent induction of the peripheral benzodiazepine receptor in small-diameter adult rat primary sensory neurons. Eur J Neurosci 20:671–683. https://doi.org/10.1111/j.1460-9568.2004.03530.x

154. Bonsack F, Alleyne CH Jr, Sukumari-Ramesh S (2016) Augmented expression of TSPO after intracerebral hemorrhage: a role in inflammation? J Neuroinflammation 13:151. https://doi.org/10.1186/s12974-016-0619-2

155. Varga B, Marko K, Hadinger N, Jelitai M, Demeter K, Tihanyi K et al (2009) Translocator protein (TSPO 18kDa) is expressed by neural stem and neuronal precursor cells. Neurosci Lett 462:257–262. https://doi.org/10.1016/j.neulet.2009.06.051

156. Mages K, Grassmann F, Jagle H, Rupprecht R, Weber BHF, Hauck SM et al (2019) The agonistic TSPO ligand XBD173 attenuates the glial response thereby protecting inner retinal neurons in a murine model of retinal ischemia. J Neuroinflammation 16:43. https://doi.org/10.1186/s12974-019-1424-5

157. Notter T, Coughlin JM, Gschwind T, Weber-Stadlbauer U, Wang Y, Kassiou M et al (2018) Translational evaluation of translocator protein as a marker of neuroinflammation in schizophrenia. Mol Psychiatry 23:323–334. https://doi.org/10.1038/mp.2016.248

158. Nutma E, Stephenson JA, Gorter RP, de Bruin J, Boucherie DM, Donat CK et al (2019) A quantitative neuropathological assessment of translocator protein expression in multiple sclerosis. Brain 142:3440–3455. https://doi.org/10.1093/brain/awz287

Chapter 18

Validation of Diffusion Kurtosis Imaging as an Early-Stage Biomarker of Parkinson's Disease in Animal Models

Amit Khairnar, Eva Drazanova, Nikoletta Szabo, and Jana Ruda-Kucerova

Abstract

Diffusion kurtosis imaging (DKI), which is a mathematical extension of diffusion tensor imaging (DTI), assesses non-Gaussian water diffusion in the brain. DKI proved to be effective in supporting the diagnosis of different neurodegenerative disorders. Its sensitively detects microstructural changes in the brain induced by either protein accumulation, glial cell activation or neurodegeneration as observed in mouse models of Parkinson's disease. We applied two experimental models of Parkinson's disease to validate the diagnostic utility of DKI in early and late stage of disease pathology. We present two DKI analysis methods: (1) tract based spatial statistics (TBSS), which is a hypothesis independent data driven approach intended to evaluate white matter changes; and (2) region of interest (ROI) based analysis based on hypothesis of ROIs relevant for Parkinson's disease, which is specifically used for gray matter changes. The main aim of this chapter is to provide detailed information of how to perform the DKI imaging acquisition and analysis in the mouse brain, which can be, to some extent translated to humans.

Keywords Diffusion kurtosis imaging, Diffusion tensor imaging, Magnetic resonance imaging, Methamphetamine, Microstructural changes, Neurodegeneration, α-Synuclein, TNWT-61 Mice

1 Introduction

Neurodegenerative disorders, such as Alzheimer's disease (AD), Parkinson's disease (PD), Huntington's disease (HD), and amyotrophic lateral sclerosis (ALS), represent a major challenge for patients, their families and healthcare providers due to longer life expectancy and aging of population. According to the epidemiological study published in Lancet Neurology in 2017, it has been reported that approximately 46 million people are living with AD or with other dementias, and more than six million people are affected by PD worldwide [1]. It has been predicted that the number of the people suffering from neurodegenerative disorders will double in next 20 years [2]. The economic burden is high; for example, in USA the annual cost of neurodegenerative disorders is

Philip V. Peplow, Bridget Martinez and Thomas A. Gennarelli (eds.), *Neurodegenerative Diseases Biomarkers: Towards Translating Research to Clinical Practice*, Neuromethods, vol. 173, https://doi.org/10.1007/978-1-0716-1712-0_18,
© Springer Science+Business Media, LLC, part of Springer Nature 2022

around $ 800 billion [3], whereas in Europe it is around 800 billion per year (https://ec.europa.eu/).

The currently available pharmacotherapies are symptomatic, and they do not have a significant impact on the inevitable progression of PD or AD. Recent advances in understanding of the disease pathology will hopefully allow development of disease modifying treatment strategies, which may slow down or even halt the progression of the disease. However, such treatments will likely be effective only when the disease is diagnosed at an early stage [4]. Unfortunately, so far, the neuroprotective agents promising at preclinical level failed to show efficacy in clinical studies [5]. Early diagnosis should help identification of patient populations eligible for novel disease modifying therapy [6]. Since there is no single clinical, biochemical or neuroimaging biomarker that allows exact diagnosis at early stage of the particular neurodegenerative disease, it makes clinical development of disease modifying therapies very challenging.

Although neuroimaging biomarkers are available, they do show some intrinsic limitations in early diagnosis or differentiating different neurodegenerative disorders. The diagnosis of these disorders are mainly based on the clinical feature that does not give the certain diagnosis in complicated cases. In the case of AD, postmortem analysis of brain tissue can only provide definitive diagnosis of the disease, while clinical diagnosis of AD is usually made using volumetric magnetic resonance imaging (MRI), which detects the cortical atrophy in the brain [7]. Although volumetric MRI can be used for tracking the progression of brain atrophy, it is of little significance for AD patients without atrophy, where it has low predictive value [8]. Besides volumetric MRI, the FDG-PET (fluorodeoxyglucose positron emission tomography) imaging has been used for decades to assess the alterations in brain glucose metabolism in patients suffering mild cognitive impairment or AD, who show hypometabolism of glucose [9]. Although hypometabolism of glucose has been reported in aging too, there are specific brain regions (e.g. temporal lobe, posterior cingulate cortex), which show very small changes with aging. These regions are severely affected in AD; therefore, the FDG-PET technique is able to capture the changes in this region sensitively [9]. Recently detection of amyloid accumulation was made possible with PET imaging; however, there is a major limitation of this approach in understanding relationship of amyloid burden with cognition [9]. Though neuroimaging techniques are becoming widely available, they do have limitations with respect to early diagnosis of AD.

In the case of PD, clinical diagnosis is fully based on clinical scoring of physical symptoms and psychological and cognitive status of the patient [10]. Early diagnosis of PD is even more challenging because in the initial disease stage it is difficult to differentiate PD from other parkinsonian syndromes, such as multiple system atrophy or progressive supranuclear palsy. Though conventional structural and volumetric MRI approaches exist, they are not regularly used in daily clinical practice since they show changes only at the very late stage of the pathology [11]. Imaging with a radiotracer such as single photon emission computed tomography using the dopamine transporter ligand [^{123}I] FP-CIT (DaTscan) was found to be effective in the early diagnosis of PD [12]. However, physicians still have to rely on clinical diagnosis as it shows a false-positive scan in some cases [13].

2 Diffusion-Weighted Imaging Techniques and Their Translational Validity in PD

Human brain consists of cca 70% of water, therefore motion of water molecules within the brain can provide important insight into the tissue's microstructure. Diffusion of water is associated with structural organization of brain. There are several diffusion-weighted imaging (DWI) MRI techniques. Diffusion tensor imaging (DTI) was developed to assess the motion of water molecules in tissue microstructure noninvasively [14, 15]. DTI is sensitive for detecting the directional diffusion of water molecules. Most importantly, diffusion of water in the white matter is typified by strong directionality. In the white matter, the water diffusion is not restricted parallel to axonal orientation, while it is largely restricted in the perpendicular direction. There are two main metrics in DTI, one is diffusivity, which measures the magnitude of water diffusion; the other is fractional anisotropy (FA) which is a measure of the directionality of diffusion. Diffusivity can provide a measure of axial or parallel diffusion (AD), radial or perpendicular diffusion (RD) or mean diffusion (MD) of AD and RD. Furthermore, FA provides information about directionality of water diffusion in living tissue [16].

Studies with DTI on neurodegenerative disorders have shown potential utility in early detection of white matter changes. Both clinical and preclinical studies on neurotoxin-based animal models of PD showed decrease in FA and increase in diffusivity metrics due to neuronal loss in the soma of substantia nigra (SN) [14, 17–20]. Olfactory dysfunction, which is one of the nonmotor symptoms in PD, develops decades before clinical diagnosis and might be useful for early PD studies [21]. Importantly, one DTI study found a decrease in FA in the olfactory system at an early stage of PD [22]. Conversely, few studies have also reported increase in FA in SN [23, 24]. A DTI imaging by Guimarães et al. on early PD, moderate PD and severe PD patients was able to find the significant DTI metric changes only in severe conditions, thus questioning the sensitivity of DTI in early diagnosis of PD [25]. Of note, three meta-analyses have been published on DTI in PD patients, two of them asserting the results of studies with DTI are encouraging and it can be a promising biomarker for PD [19, 26], whereas the other one questioned the validity of DTI as an imaging biomarker in PD [27]. This controversy may arise from sensitivity of DTI in white matter compared to gray matter. While gray matter allows a relatively isotropic water diffusion, white matter represents a highly anisotropic environment. Thus, sensitivity of DTI in detecting gray matter microstructural changes is probably limited, while it allows a good assessment of the pathological changes in white matter. The major limitation of DTI lies in the fact that it considers the diffusion of water molecules in the brain as Gaussian, which infer the diffusion of water in the brain is free and unrestricted like water in a bucket. Considering the real structural complexity of the brain, it is merely an approximation and it can be expected that significant diffusivity changes may exist among tissue compartments [28, 29].

To overcome this limitation, diffusion kurtosis imaging (DKI) which is a mathematical extension of DTI was developed. One of the most important components is the b value that deciphers the strength and timing of the gradient which is used to generate diffusion images. In the case of DTI only 2 b values are used [16]. However, DKI uses more than 2 b values to understand inherently non-Gaussian water diffusion in the neural tissue [28, 30]. DKI considers the water diffusion to be nonrestricted, and it measures the diffusional hindrance arising from microstructure such as tissue compartments, cell membrane, and cell organelle. Hence, this technique is sensitive in detecting the restriction of water diffusion in both isotropic and anisotropic environments [16]. DKI metrics are mean kurtosis (MK), axial kurtosis (AK), and radial kurtosis (RK). These provide ancillary information in addition to DTI metrics such as FA, MD, AD, and RD and give detailed knowledge about the microstructural characteristics of the tissue. Increase in kurtosis values suggest higher tissue heterogeneity or substantial hindrance to the water diffusion. This might be due to protein accumulation, glial cell activation, or other pathology in the brain [16, 28, 30, 31]. While the validity of DTI is limited to white matter, the information from gray matter is largely controversial [32, 33]. On the other hand, DKI is sensitive in both white matter as well as gray matter [30].

In the case of PD, it was Wang et al. who proposed the importance of DKI as a diagnostic imaging biomarker in PD. Though in their studies he found increases in both kurtosis and FA in SN, MK showed the best diagnostic performance in ipsilateral SN [24]. Similarly, Kamagata et al. found an increase in MK and FA in cingulate fibers. Moreover, the receiver operating characteristic (ROC) curve analysis revealed that sensitivity of MK was higher than FA in this particular region [34]. To support the diagnostic utility of MK in early PD patients, Zhang et al. performed both DTI and DKI in SN and found increase in MK, which was correlated with Hoehn-Yahr (H-Y) staging and UPDRS III staging [35]. UPDRS (Unified

Parkinson's Disease Rating Scale) evaluates the key areas of disability in PD divided in three subscales, while fourth subscale evaluates any complication to treatment. It is often used with H-Y disease rating scale [36]. Considering clinical scales, a recent study on DKI in PD patients also proved the sensitivity of DKI in PD patients [37].

Though there are several clinical studies of DKI with PD patients [24, 34, 37] currently, the exact origin of changes in kurtosis metrics are not yet understood. Because, histological evaluations of human brains are limited to post mortem studies, the DKI experiments using animal models of PD are required to shed more light into the pathophysiological mechanisms responsible for changes in the kurtosis signal. If we validate DKI and explain the source of kurtosis signal in animal models, we will be able to contribute to early diagnosis of PD, when the possibly neuroprotective treatment strategies will likely be useful.

3 Materials

The following text summarizes and explains in detail our methodological approaches and selection of materials and animal models used in our research. Where applicable, we aim to outline additional studies, identify knowledge gaps, and point out potential pitfalls.

3.1 Laboratory Animals and Selection of a Suitable Model

Animal models are an important tool to study pathogenic mechanisms and therapeutic strategies and to develop imaging biomarkers in human diseases. They are crucial for translational science especially in the development of imaging biomarkers. A biomarker can be any quantifiable metric, which can detect a biological process or disease pathology and changes associated with the treatment. Therefore, it is ideal for any imaging biomarker to detect changes consistently in both animal models and humans. Though several animal models were found to be very similar to the human PD [38, 39], the majority of the compounds tested in these animal models fail to show efficacy in the human condition and thus provide low translational value [40]. To reduce this rate of failure and better translate the changes behind imaging biomarker, a number of animal models mimicking the human disease pathology should be used to validate the proposed biomarker.

PD is characterized on the neuropathological level mainly by the loss of 50–70% of the dopaminergic neurons in SN pars compacta, presence of intracytoplasmic inclusions called "Lewy bodies" mainly consisting of α-synuclein. Rodent models of PD-like phenotype match closely some aspects of human pathology, but they fail to reflect the condition in the complexity observed in humans. Therefore, in our research we decided to use two substantially different mouse models of PD—a transgenic one showing mainly accumulation of human α-synuclein with no neurodegeneration and a neurotoxin-based model featuring primarily loss of dopaminergic neurons.

Male mice were used in all studies. The mice were group housed in the Central Animal Facility of Masaryk University, Brno, Czech Republic, and maintained on a normal 12/12-h light/dark cycle (lights on at 6 a.m.) with a constant relative humidity of 50–60% and temperature of 22 ± 1 °C. Water and food were available ad libitum. Later they were transported to the animal house of the Institute of Scientific Instruments, Academy of Sciences of the Czech Republic, Brno, Czech Republic and maintained under the same conditions as in the previous location. All procedures were performed in accordance with EU Directive no. 2010/63/EU and approved by the Animal Care Committee of the Faculty of Medicine, Masaryk University, Czech Republic and the Czech Governmental Animal Care Committee, in compliance with the Czech Animal Protection Act No. 246/1992.

3.1.1 TNWT-61 Mouse Model

The murine Thy-1 promoter turned out to be particularly useful to drive high levels of expression of the human wild-type α-synuclein throughout the brain, a distribution similar to that observed in human PD. TNWT-61 model reproduces many features of sporadic PD, including progressive changes in dopamine release and striatal content, α-synuclein pathology, and early motor and nonmotor deficits, suggesting that this model could be useful for the study of preclinical PD stages. Although TNWT-61 model exhibits progressive loss of tyrosine hydroxylase positive dopaminergic fibers in the striatum and thus worsening behavioral deficits characteristic of PD, the mice do not exhibit dopaminergic neuron loss in the SN. Surprisingly, neuroinflammation has been initially found only in striatum and subsequently in SN, whereas α-synuclein is overexpressed widely in the cortex, cerebellum, hippocampus, olfactory bulb, and brainstem. This makes TNWT-61 particularly well suited for studying the early stages of PD, before dopaminergic neuron loss occurs, and at the same time well suited for developing imaging biomarkers[38].

3.1.2 Methamphetamine (METH) Mouse Model

METH is an addictive psychostimulant, which mainly acts as an indirect sympathomimetic. Several studies have reported that high doses of METH induce chronic dopamine nerve terminal damage in striatum and neuronal body loss in SN pars compacta [41, 42]. It has been consistently reported that repeated administration of very high METH doses induces degeneration of dopaminergic nerve terminals followed by a decrease in dopamine, 3,4-dihydroxyphenylacetic acid (DOPAC) and homovanillic acid (HVA) levels. It is also reported that it selectively induces loss of dopaminergic cell bodies in SN, sparing dopaminergic neurons of mesolimbic pathway [41, 43]. The tenacious loss of dopaminergic nerve terminals is correlated with dopaminergic neuronal loss in substantia nigra pars compacta, which was evident from the Nissl staining and Fluoro-Jade fluorescence [44]. Importantly, METH selectively affects nigrostriatal dopaminergic neurons sparing the mesolimbic dopaminergic pathway simulating the conditions of PD. Additionally, striosomes of striatum compared to matrix are more susceptible to damaging effects of METH which is observed also in the case of MPTP (1-methyl-4-phenyl-1,2,3,6-tetrahydro-pyridine) administration [43]. METH model also shows motor symptoms similar to other neurotoxin-based animal models of PD [45]. Therefore, it is suitable for assessment of neurodegenerative process as found in the late stage PD patients.

However, both models are far from perfect and more rodent models of PD should be used to adequately assess the validity of any neuroimaging biomarker. Ideally, the findings from one model featuring mainly pathological protein accumulation should be replicated in another model with analogous profile and more neurodegenerative models should be employed. Our decision to use these two models was based on their distinct pathological hallmarks, availability, and previous experience with these models. We later performed a comprehensive study using rotenone model, which combines both neurodegenerative process and protein accumulation (data currently in preparation) [46]. However, it would be indeed very interesting to scan eventually all available models, such as MPTP, 6-hydroxydopamine, paraquat, and others.

3.2 Imaging Equipment

3.2.1 MRI System and Coil Setup

Based on our previous pilot experiments we used the MRI coil setup, which rendered the best signal-to-noise ratio and appropriate image quality. MRI measurement was performed on a high-field horizontal 94/30USR scanner (Bruker Biospin MRI, Ettlingen, Germany) equipped with a gradient system with strength up to 660 mT/m. MRI coil setup was composed of a 1H quadrature volume transmitter coil (inner diameter 86 mm) and a 1H four-channel surface mouse phased-array head coil as a receiver.

3.2.2 Data Processing Software

- Paravision 5.1 running in a console Bruker AVANCE III—acquisition commercial software. First of all, Paravision software was used to obtain MRI data for further data processing. Moreover, Paravision was used for conversion of voxel size of acquired data.

Image analyses were carried out using several other tools:

- MATLAB R2010a (The MathWorks Inc., Natick, MA, USA) software—In house Matlab code programmed was used for conversion of MRI DICOM Bruker data format to NIFTI data format.

- ExploreDTI v4.8.4 Software—ExploreDTI was used for calculation of the diffusion maps and diffusivity directions together with parametric maps.

- ImageJ software (NIH, Bethesda, MD, USA)—free software was used for various brain regions delineation according to the mouse brain atlas [47].

- FSL (FMRIB's Software Library) software (FMRIB, Oxford, UK)—FSL software was used for eddy current correction and movement artifacts tract-based spatial statistics analysis together with brain extraction and tract-based spatial statistics (TBSS) [48].

3.2.3 Animal Monitoring and Respiratory Gating Instrumentation

Respiratory gating together with respiratory rate monitoring was used to reduce motion artifacts. In cases where the respiratory curve would not be stable, the respiratory gating would not work properly and the DKI sequence would take longer time. MRI compatible equipment made by Small Animal Instruments (Model 1030, SA Instruments, Inc., Stony Brook, NY, USA) was used during all MRI procedures. Respiratory rate was monitored by small pneumatic pillow sensor and temperature was maintained with a small rectal thermistor probe.

3.2.4 Animal Bed with Warming and Fixation System

The necessity of a proper animal heating system and appropriate fixation system secured normalization of animal vital functions together with reduction of motion artifacts. An animal bed with a water heating system was used to maintain stable animal body temperature. Tooth-bar and ear-bars were required to fix a the head of the animal.

3.3 Animal Anesthesia System

We used easily controllable continuous inhalation isoflurane anesthesia. The selection of anesthetic agent was based on long-term experience in our lab [49] and the evidence that isoflurane seems to be suitable for prolonged MRI scanning compared to other anesthetics such as propofol, dexmedetomidine, or ketamine [50]. Moreover, isoflurane can be used repeatedly and provides short postanesthesia animal recovery. Importantly, the constant isoflurane concentration was maintained at level 1.5–2% to prevent anesthetic agent fluctuation and to secure regular respiratory curve. Isoflurane anesthetic system consisted of an isoflurane vaporizer (G.A.S. Ltd., Keighley, UK), flowmeter with medical grade air carrier gas and induction chamber with scavenger unit with a carbon filter (UNO Actisorb Anaesthetic Gas Filter). Mask for the anesthesia was integrated into the animal bed and tubing.

4 Methods

4.1 Preparation of Animal for Scanning

Animal was anesthetized using a mix of medical air and 0.5% isoflurane at flow rate 1 l/min increasing every minute in the induction chamber until reaching 3% concentration of isoflurane to provide slow anesthesia onset and to prevent stress. In general, stress can alter physiological parameters, deepen interindividual variability and consequently worsen data quality. The mouse was then placed on the animal bed horizontally and teeth were hooked by a tooth bar. Head was immobilized by placing ear bars gently to the ears to maintain motion restriction for brain imaging. Both tooth and ear bars were fixed also on the animal bed. The nose was gently placed into the anesthetic mask for continuous delivery of isoflurane. Pneumatic pillow sensor was gently placed under the abdomen and the respiratory curve was controlled to get a good signal. The sufficient level of anesthesia was controlled according the animal respiration rate. The optimal respiration rate was maintained in the range of 50–60 breaths per minute. Continuous inhalation anesthesia was used during whole MRI protocol to prevent motion artifact and immoderate stress during MRI scanning. The isoflurane was maintained at 1.5–2% level. The thermistor probe was placed intrarectally to control the temperature. The temperature was maintained at around 37 °C to stabilize basal vital functions of animal and to avoid hypothermic stress. Moreover, maintaining temperature of animals was essential for all experiments because temperature instability or fluctuation could potentially alter water spin density and consequently provide false results or results containing errors [51–53]. We switched on the water heating system 1 h before starting measurements. With this procedure, we were able to maintain stable temperature of each animal. The animal did not lose its body heat by being placed on an unheated animal bed. Eye ointment (Vidisic gel) was applied on mouse eyes to prevent corneal ulceration. Finally, the quadrature coil was fixed over the head of animal and the animal was placed in the middle of magnetic field.

4.2 MRI Data Acquisition Protocol

Pilot scan was acquired to check brain position in the center of the coil.

4.2.1 Localizer

- Method: Fast low-angle shot (FLASH) sequence was used to obtain axial, coronal, and sagittal brain images to localize the brain position and to correct brain position inside the magnetic field.

- Parameters: field of view (FOV)—30 × 30 mm, 128 × 128 acquisition matrix, three orthogonal slices of 1 mm thickness, echo time (TE) was 3 ms, and repetition time (TR) was 200 ms with flip angle: 30°.

- Total acquisition time—25 s.

4.2.2 Anatomical Images   T2-weighted brain scans were acquired to obtain anatomical images and localize Bregma 0 slice. The Bregma 0 slice was characterized by anatomical structures—corpus callosum, merging anterior commissures—anterior parts and striatum according to the Paxinos Mouse Brain Atlas [47]. The Bregma 0 was always positioned as eighth slice from 15 adjacent slices to maintain the same brain slices positioning in all experiments. Moreover, this positioning allowed also acquiring all ROI, specifically SN, hippocampus, sensorimotor cortex, striatum, and thalamus. The slice thickness was 0.5 mm to obtain a high-resolution image. The brain scans were obtained to be used as a reference for the DKI.

- Method: 2D rapid acquisition with relaxation enhancement (2D RARE) sequence was used.

- Parameters: FOV—24 × 24 mm, 256 × 256 acquisition matrix, 15 adjacent slices of 0.5 mm slice thickness, RARE factor of 8, TR was 2500 ms with 4 averages.

- Total acquisition time—6 min.

4.2.3 DKI Acquisition   Diffusion-weighted images were acquired. The DKI protocol included the acquisition of six b-values ($b = 0, 500, 1000, 1500, 2000$, and 2500 s/mm^2) along with 30 non-collinear directions, $\delta = 4$ ms, $\Delta = 11$ ms, with seven averages used for $b = 0$ acquisition and four averages for each other b-value.

- Method: spin echo-echo planner imaging (SE-EPI) was used.

- Parameters: FOV—24 × 24 mm, 98 × 128 acquisition matrix, 15 adjacent slices of 0.5 mm slice thickness, TE was 25 ms using 300 kHz bandwidth, TR was ~5 s depending on respiratory rate.

- Total acquisition time—1 h 40 min.

4.2.4 Respiratory Gating Procedure   It was necessary to configure segmented EPI acquisition triggered to the respiration cycle. Three images were acquired in the same time of the breath cycle to prevent imaging artifacts. Nevertheless, due to this fact the scanning time depended on respiratory rate. Therefore, range of 50–60 breaths per minute was maintained which seemed to be optimal.

The animal was removed from the magnetic field immediately after DKI sequence was finished. The animal was placed on a heating pad with the oxygen mask in case it was having breathing difficulty. Otherwise, it was returned to the cage and checked for recovery. Afterward, Nutra-Gel Diet™ (Bio-Serv) was provided to the animal to supply water and nutrition after scanning. Animals were always transport back to the animal facility after complete recovery.

4.3 Processing of Data

The acquisition matrix of DKI images were 98×128 which was reconstructed to 256×256 with the help of Paravision 5.1 software. The raw MRI data comes in Bruker format therefore it was converted to NIfTI format, which can be handled by MRI data analyzing software. A locally modified, freely available MatLab script was used for conversion and the size of the voxels was enlarged. Signal-to-noise ratio was calculated and compared for b-values maps to check data quality. Diffusion data were corrected for eddy currents and movement artifacts to the first non–diffusion-weighted image with FSL [54]. ExploreDTI v4.8.4. Software [55] calculated the diffusion maps using b-values and diffusivity directions. With six b-values we had the possibility to calculate parametric maps in different ways. For DTI parameters, data were processed and calculated on a voxel-by-voxel basis with six b-values to produce parametric maps (MD, AD, RD, FA, MK, AK, and RK), and with two b-values to produce DTI-derived parametric maps (MD, AD, RD, and FA) in a "conventional" manner. From all the possible fitting methods, the robust estimation of tensors by outlier rejection (RESTORE) proved to be the best choice for our data set. Once both DKI and DTI parametric maps are prepared, the ROI and TBSS analysis procedure differs. Please check the respective section for further analysis.

4.4 Results and Interpretation of ROI-Based Analysis

ROI based analysis is hypothesis driven as we select the regions based on the presence of pathology in that particular region. ROI is specific for detecting microstructural changes in gray matter, while TBSS analysis is specific for white matter changes. In our studies, the selected regions were those generally affected in the PD. We have assessed several slices of the brain allowing evaluation of the same ROIs repeatedly and expressed as an average. Specifically, the regions were SN, striatum, sensorimotor cortex, hippocampus, and thalamus.

After getting the maps for different diffusion tensor and diffusion kurtosis metrics, the ROI selection on $b = 0$ images was drawn manually according to the mouse brain atlas [47] with the help of FA maps using Image J software for various brain regions. The ROIs were delineated on FA map, because it gives better contrast and visualization for differentiating the brain regions.

Table 1
Summary of DKI results [45, 56–58]

Model	Time-point	Mean kurtosis					Axial kurtosis					Radial kurtosis				
		SN	STR	HIPP	CTX	THAL	SN	STR	HIPP	CTX	THAL	SN	STR	HIPP	CTX	THAL
METH	5 days	↓	↓		↓							↓				
	1 month	(↑)	↑	↑	↑											
TNWT-61	3 months		↑			↑							↑			
	6 months	↑	↑			↑							↑			↑
	9 months	↑	↑		↑	↑	↑			↑	↑					↑
	14 months	↑	↑	↑	↑	↑				↑	↑				↑	↑

Model	Time-point	Mean diffusivity					Axial diffusivity					Radial diffusivity					Fractional anisotropy				
		SN	STR	HIPP	CTX	THAL	SN	STR	HIPP	CTX	THAL	SN	STR	HIPP	CTX	THAL	SN	STR	HIPP	CTX	THAL
METH	5 days									↑		↑					(↓)				
	1 month		↓	↓											↓		↑	↑	↑	↑	
TNWT-61	3 months		↓																		
	6 months												↓		↓		↑		↑		
	9 months														↓						
	14 months				↓	↓				↓	↓				↓						

↑	increase
↓	decrease
	no change

The table shows an overview of our previously published data. The symbols (↑) and (↓) mean higher or lower value close to significant level, i.e. a trend. The numbers of animals were as follows. METH study: at 5 days METH ($n = 11$) and SAL (saline-treated, $n = 5$) and at 1 month, METH ($n = 9$) and SAL ($n = 6$). TNWT-61 study: 3 and 6 months ($n = 15$ per control and transgenic group), 9 months ($n = 7$ per control and transgenic group), 14 months ($n = 9$ transgenic and $n = 12$ control mice)

Once the ROI was drawn on the FA map, it was saved, and the same ROI was transferred on another metrics to get all diffusivity and kurtosis variables.

DKI results obtained with TNWT-61 mouse model and METH-treated animals are summarized in Table 1. The primary aim of this line of research was to compare the sensitivity of DKI in two different mouse models of PD as described earlier. The changes in the kurtosis, diffusivity, and fractional anisotropy values were compared to wild-type or vehicle-treated mice.

From Table 1 it can be clearly seen that MK is the most sensitive read out as compared to other DKI and DTI metrics. We found significant changes in the MK already in 3-months old TNWT-61 mice in striatum and thalamus, while the zenith of the PD-like phenotype is considered 9 months of age [57]. Importantly, none of the DTI parameters was able to detect these early pathological changes affecting gray matter, which strengthens the importance of DKI in early detection of PD. The observed increase in MK might be due to increased structural complexity or tissue heterogeneity in the respective regions due to α-synuclein accumulation. It also is consistent with previous reports suggesting Lewy body inclusions or α-synuclein accumulation might have influence on change in MK [59]. Interestingly, even though the TNWT-61 mice start to show wide-spread α-synuclein accumulation starting at the age of 10 days, at 3-months almost all the regions we considered for ROI analysis are reported to show α-synuclein accumulation [38, 60]. Surprisingly, although SN, hippocampus and sensorimotor cortex are reported to show similar or even higher expression of α-synuclein accumulation as compare to striatum and thalamus, we observed increase in kurtosis only in striatum and thalamus in 3-months old mice. Based on these results, we can conclude that MK is sensitive in detecting α-synuclein accumulation-induced microstructural changes rather than α-synuclein accumulation *per se* [58, 60].

In contrast to the α-synuclein overexpressing transgenic model, in the case of METH model at the early 5-days' time-point we found decrease in MK in SN, striatum and sensorimotor-cortex. Decrease in MK generally indicates a decrease in hindrance to the diffusion of water molecules. There is a logical explanation of the opposite effect we have observed in TNWT-61 and METH treated mice due to the different nature of the two models. METH model is a neurodegenerative model, while TNWT-61 model is of prodromal type in which degenerative changes occur only when the mice are 14 month old in the striatum [38]. Neurodegenerative changes induced by METH cause lower hindrance to diffusion of water in the brain due to reduction in structural complexity or tissue heterogeneity [16]. Therefore, we can say that the decrease in MK observed in the METH model might be due to degeneration of dopaminergic neurons. Conversely, in the case of TNWT-61 mice either the α-synuclein accumulation or α-synuclein accumulation induced changes have likely increased hindrance to diffusion of water due to increase in structural complexity or tissue heterogeneity.

Taken together, we can conclude that MK is sensitive to detect microstructural changes induced by α-synuclein accumulation as well degenerative changes induced by METH and is having the sensitivity and diagnostic utility in both early as well as late stage of disease pathology.

Importantly at 1-month time-point in METH model we found increase in MK in striatum, hippocampus and cortex which is opposite to the results we obtained with MK at 5-days. This discrepancy can be explained by recovery of dopaminergic nerve terminals by axonal sprouting or protein accumulation in METH treated mice. It has been reported in the preclinical studies there exists a recovery phenomenon for the striatal and SN dopaminergic system [61]. A recent study of DKI in a rat stress model also reported increase in MK in amygdala of stressed rats which was found to be due to an increase in neurite density [62]. This might be the reason we observed increase in MK after 1 month of METH administration due to generation of neurites as recovery phenomenon might have increased non-Gaussian diffusion in striatum and hippocampus.

A study by Fornai et al. reported that administration of METH to mice causes the formation of intracellular inclusions similar to Lewy bodies in PD which mainly consist of α-synuclein

[63]. Hence, a possibility of accumulation of α-synuclein in METH treated mice cannot be ruled out. The common thing we have found in both the METH and TNWT-61 mouse models is at the late stage of disease pathology both the models exhibit increase in MK in SN, striatum and sensorimotor cortex which may be due to increased structural heterogeneity and hence hindrance to diffusion of water molecules (possibly due to α-synuclein accumulation).

It is important to emphasize that we did not find any consistent significant changes of FA in the two models. However, several preclinical and clinical studies with DTI have reported significance of FA metric in early diagnosis of PD [14, 22]. In the METH model at 1-month time-point we have found increase in FA in SN, striatum hippocampus and cortex which agrees with a study performed by Van camp et al. with 6-hydroxydopamine lesioned rats [23], while there is a report showing a decrease of FA in SN due to degeneration of dopaminergic neurons [17]. This discrepancy in DTI results might be due to intrinsic limitation of DTI, as it considers the diffusion of water molecule as gaussian meaning there is no hindrance to diffusion of water molecules in the brain, whereas in actuality it is non-Gaussian and is hindered by cell constituents and cell membrane. Therefore, DTI seems to be sensitive to detecting white matter changes, while DKI is sensitive in detecting gray matter changes. Furthermore, three meta-analyses on DTI in PD have been published and there is so far no consensus on the importance of DTI in early diagnosis of PD [19, 26, 27]. By contrast, DKI by considering non-Gaussian diffusion of water molecule which is close to the situation in the living tissue, is sensitive in detecting both gray as well as white matter microstructural changes.

In PD, there are different post translational modifications in α-synuclein protein, and it has been reported that the α-synuclein, which has been found postmortem in the brains of PD patients, consists of phosphorylated and proteinase K resistant α-synuclein. Similarly, TNWT-61 mice are reported to show large punctate of proteinase k resistant α-synuclein starting from 1 month of age in thalamus, whereas in the SN it showed variable sizes of proteinase k resistant α-synuclein aggregates in 5 months-old mice [64]. This is consistent with our findings in 3-months old TNWT-61 mice, where we observed an increase in MK in thalamus but not in SN. Conversely, in 6-month-old mice we found an increase in MK in both thalamus as well as SN and might be due to presence of proteinase k resistant α-synuclein. These results may have clinical importance as MK seems to be able to detect regional differences in proteinase k resistant α-synuclein induced microstructural changes and a correlation of proteinase K resistant aggregates with Lewy body inclusions was reported, underpinning the importance of these aggregates in synucleopathies [65].

Several studies have reported the increase in MK might be due to an increase in structural complexity or heterogeneity due to presence of protein accumulation or glial cell activation [7, 66–69]. In our DKI study with TNWT-61 mice we also found glial cell activation, so we cannot rule out the changes in kurtosis values due to glial cell activation [64]. Watson et al. reported that TNWT-61 show strong microglial cell activation in striatum as early as 1 month of age, whereas in SN it starts to show at the age of 5–6 months with very similar microglial morphology to that observed in PD patients [64]. There are reports suggesting that the membrane barrier to diffusion of water is altered in the presence of activated microglia. This might be the reason we found significant increase in MK in striatum but not in SN at 3-months' timepoint, whereas at 6-month time-point we were able to see an increase in MK in both SN as well as striatum. We can conclude that MK is probably sensitive also to detecting regional changes with glial cell activation.

There is a debate over the beginning of PD pathology development either from striatum containing dopaminergic nerve terminals or from SN consisting of dopaminergic cell bodies [70]. In our DKI studies with TNWT-61 model we found changes in MK first in the striatum and later in SN, which correlates with the presence of proteinase k resistant α-synuclein and microglial cell activation first in the striatum and later in SN. This suggests that the α-synuclein accumulation-induced changes start in dopaminergic nerve terminals and later the pathology develops in dopaminergic cell bodies as detected by MK.

Interestingly, the most affected region in TNWT-61 mice at all time-points was the thalamus. It might be due to low ratio of endogenous to human α-synuclein protein compared to other brain regions at approximately (1:6.5) [38], which justifies the changes induced by human α-synuclein. Our DKI study with 14-month-old TNWT-61 mice was the first to report significant negative correlation between α-synuclein accumulation and decrease in diffusivity in thalamus and a trend toward positive correlation of α-synuclein accumulation with an increase in kurtosis [58]. In contrast there was no DKI changes in METH model at both time-points in the thalamus. The contradictory results might be due to differences in the induction of PD pathology. TNWT-61 mouse model shows mainly overexpression of α-synuclein without neurodegenerative changes, while METH induces neurodegeneration first and shows presence of α-synuclein after neurodegenerative changes.

On parametric maps brain extraction was performed using Brain Extraction Toolkit from FSL [71]. The default mode is not suitable for mouse brain analysis, since only partial mouse brain was measured. For that reason, center of gravity was determined, and

**4.5 Results
and Interpretation
of Tract-Based Spatial
Statistics (TBSS) Data**

all maps were checked and corrected manually if it was necessary. A non–hypothesis-driven method, a whole-brain analysis of the white matter tracts, TBSS, was chosen to investigate microstructural alterations. TBSS [72] script was modified to fit it to the mouse brain [73]. All of the 3D FA volumes were registered together nonlinearly, and using the registration matrices, the best registration target was chosen with a free-search method. Then a study-specific volume was used as a template for the final transformations. The mean FA map was calculated and a skeleton, the center of the main white matter tracts, was created at the threshold at FA = 0.2. Each mouse brain's aligned FA data was projected on to the skeleton. The process was done for all of the parametric maps. Nonparametric tests were used for statistical analysis: (1) the general linear model (GLM) design with permutation test (10,000 permutations) and (2) cluster-based thresholding were used with the predefined threshold ($t = 2.3$) to compare groups. The altered white matter tracts were identified according to the mouse brain atlas [47].

TBSS data interpretation: TBSS is the analysis technique to measure the DKI changes in white matter. At early stage of both TNWT-61 and METH model, TBSS analysis detected no white matter changes. This is an interesting phenomenon, given the substantially different nature of the two models. Conversely, at later time points such as 6, 9 and 14 months-old TNWT-61 mice and 1-month time point of METH model we detected significant white matter changes by TBSS analysis. This may indicate that the α-synuclein accumulation or neurodegenerative changes are likely to start in gray matter and later spread to white matter. Most importantly in the TBSS analysis both models showed alterations in similar regions. Specifically, we observed changes in tracts coming from or towards the thalamus such as the mammillothalamic tract and lateral thalamic nuclei. This was not the case in gray matter in which only the TNWT-61 mice showed changes in thalamus [56–58]. Thalami play a crucial role in the passage of motor information to the sensory and motor cortex. It is strongly connected to the cerebellum and hippocampus. Impairment in the thalamic network may create difficulties in recalling motor skills and fine tuning of motor movements [74]. It has been reported that thalamic noradrenaline deficiency might be involved in genesis of motor and nonmotor symptoms [75]. So α-synuclein pathology in the thalamus and sensitive detection of α-synuclein accumulation induced changes by DKI may have clinical importance in diagnosis of PD.

The white matter changes previously reported in PD patients were a decrease in MK and FA in cingulate fibers [34, 76]. In contrast, our studies found increases in kurtosis and FA in the cingulate fiber in both models [45, 56, 58]. This contradiction might be due to the intrinsic limitations of animal models being unable to adequately mimic the complex pathology present in human PD.

With TBSS analysis we have found a trend of an increase in kurtosis and a decrease in diffusivity in the anterior and posterior commissure, cingulate fibers, mammillothalamic tract, and thalamic nucleus. Hence, it seems that wherever there is a protein accumulation or glial cell activation we also observe an increase in kurtosis and a decrease in diffusivity due to increased hindrance to diffusion of water molecules. However, this is a hypothesis that remains to be proven.

Taken together, TBSS results further confirm that DKI is sensitive in detecting both gray matter as well as white matter changes induced by PD-like pathology and this imaging technique does have potential to diagnose PD patients both at early stage as well as late stage of disease.

5 Notes

5.1 Stability of Body Temperature

Maintaining body temperatures of animals was essential for all experiments, because any temperature instability or fluctuation could potentially alter water spin density and consequently provide false results or results containing errors [51–53] To ensure optimal conditions, we always switched on the water heating system in the animal bed 1 h before we started scanning. This allowed to maintain a stable temperature of each animal, because the mice did not lose body heat on being anesthetized and placed on a cold bed. Moreover, also in the winter months an air heating system was required for the beginning of measurements. When using an air heating system, a paper blanket was placed over the animal's body to avoid acral body burns (mainly ears and tail).

5.2 Standard Handling Procedures

Standard environmental conditions, housing and handling of animals is an important aspect of any animal experiment. Environmental enrichment may be useful, but not a necessity in group-housed mice. Ideally, all animals should be handled regularly to lower their stress reaction to human contact. Generally, the stress can impair the whole measurement procedure due to instability of animal physiological functions (e.g. irregular breathing or fluctuating temperature) together with higher interindividual variability of results, prolonged anesthesia induction, necessity of higher dosing of anesthetic, or injury of animal due to unpredicted behavior. We place each animal in the preparation room at least 45 min before starting the measurement to habituate. Moreover, we also used red induction chamber for inducing isoflurane anesthesia which seemed to be more effective than using a transparent induction chamber.

5.3 Good Fixation of Animal and Small Instruments During Measurement

We used a tape to fix every mobile component, such as head surface coil, thermistor probe, anesthesia tubing and MRI compatible ERT module used for measuring the respiration rate from a small pneumatic pillow and temperature. The tape was also used for fixation of animal by gently pressing it on to the pneumatic pillow. We used to take an approximately 10 cm long strip of tape and stick an approximately 3 cm long strip of tape in the middle of long strip to acquire nonadhesive part and to avoid the animal hair depilation. The pitfall of using tape occurred only in the case of using an air heating system because the warm air induced the tape glue instability. We solved it by double taping.

5.4 Specific Neuropathological Process of Animal Model of PD

As discussed previously, each animal model mimics certain features of PD-like neuropathology, but none of them is perfect. We have performed a longitudinal DKI study using low-dose intragastric rotenone administration to model PD-like phenotype [46]. The dataset is currently in preparation, but we faced a problem in data interpretation, because we observed the majority of significant results in the gray matter before the mice reached the zenith of the rotenone-induced PD-like behavioral phenotype. It seems, that at early stage of the pathology, increased MK detects pathological protein accumulation. Interestingly, at later stage, the increase of MK was lost, likely due to more pronounced neurodegeneration. Taken together, it is possible, that one pathological phenomenon is able to mask another (i.e. neurodegeneration may mask protein accumulation). This may potentially limit the usefulness of DKI in the diagnosis of PD, but on the other hand it may become a useful tool in tracking the progression of the neuropathology.

5.5 Motion Artifacts

Motion artifacts seem to be crucial in DKI imaging because this sequence is highly sensitive to motion. We experienced very good results with combination of head/ear bars fixation together with respiratory gating as a prevention of motion artifacts and no necessity of ECG gating. Afterward, diffusion data were also corrected for eddy currents and movement artifacts with FSL software.

5.6 Calculation of DTI and DKI Parametric Maps

Beyond eddy currents and movement artifact correction, fitting our DKI model is crucial to have reliable parametric maps. As it was described before, nonparametric tests are required for such MRI data assuming non-Gaussian water diffusion. Models of diffusion signal decay are complex and going deep in mathematical equations it is out of the scope of this chapter. Since in our DKI data we work with six b values ($b = 0, 500, 1000, 1500, 2000,$ and 2500 s/mm^2), the shape of the decay curve is not linear, therefore conventional DTI calculation methods—working with two b values—are not suitable for this data. At the time of our data was analyzed, we voted for REKINDLE in ExploreDTI. The literature provides now more calculation ways, but still, new diffusion models are needed to extract more accurate information about the measured brain tissue.

5.7 Statistical Considerations

Robust statistical analysis requires some kind of correction for multiple comparisons, which represents a golden standard in the majority of preclinical studies. The level of statistical significance was set at $p < 0.05$ in all studies; however, we opted not to correct the statistical results. This kind of studies are exploratory and we analyzed all DKI variables in as many relevant ROIs as possible. This renders a very high number of comparisons. So, the risk of false positive findings is considerable, but after a stringent correction (e.g. Bonferroni), we lose all significant findings. Importantly, our aim was to reveal the most useful metrics and ideally find a pattern and replicate the findings in more animal models. In our later studies, we decided to supplement the data with 95% confidence intervals for mean differences in all cases [45], which appears to be the most transparent way. Clinical relevance of our data will need to be confirmed in future clinical trials.

6   DKI Imaging in PD Patients and Its Comparison with Animal Models

Clinically DKI imaging has gained an attention as a useful diagnostic tool for early diagnosis of neurodegenerative disorders. The first study with DKI imaging on 30 PD patients compared with 30 age matched healthy controls was reported by Wang et al. in 2011. They found higher MK in caudate and putamen and higher MK and FA in SN in PD patients. Importantly the MK in ipsilateral SN has showed the best diagnostic performance with sensitivity of 0.92 and specificity 0.87. Surprisingly, the traditional diffusion tensor metrics including FA have showed little importance in diagnosis of PD in this study [24]. As explained before the increase in tissue heterogeneity induces increase in MK [28] and it was hypothesized that Lewy body inclusions in the SN might be responsible for higher MK values in SN [59]. Similar to Wang et al., 2011, a study involving around 72 untreated PD patients reported also higher MK in SN with very high sensitivity of 0.944 and specificity 0.917 [35]. In this study a correlation with H-Y staging and UPDRS-III scores was performed, indicating positive correlation with MK in SN again further supporting the diagnostic importance of MK in PD. Later a similar author collective published another 2 studies using the DKI imaging in PD patients, which report higher MK in SN especially in patients showing striatal silent lacunar infarction, which might be caused by hyperhomocysteinemia and MK values had positive correlation with disease severity [77, 78]. In contrast to all these clinical trials, a recent DKI study on 26 PD patients found significantly lower MK values bilaterally in SN in both early and advanced PD patients compared to healthy controls [79]. The reason of discrepancy might be due to differences in H and Y staging of the PD patients, difference in scanning protocols and processing of the data [79]. However, all these studies combined do suggest MK may help clinicians in early diagnosis of PD patients and in detection of severity of PD.

Preclinical studies with DKI are scarce and require widespread investigation to detect the molecular mechanism behind MK changes in SN. It is impossible to recapitulate the human PD pathology in animals; therefore, all the features of human pathology in one animal model do not exist. For this reason, we used two models featuring different sets of PD-like hallmarks as explained earlier. In TNWT-61 model we observed higher MK in SN which was related to presence of proteinase k resistant alpha synuclein and microglial cell activation which might have increased the structural heterogeneity in SN [56–58]. This result is in accordance with clinical evidence. Conversely, in the METH model we found lower MK in SN, which was related to acutely induced neurodegenerative changes, whereas in the late stage we found higher MK likely due to accumulation of alpha synuclein or other pathological proteins developed later [45].

Considering the white matter changes in PD patients it was Kamagata and his colleagues in 2013 reported changes in white matter in PD patients using DKI imaging. In a DKI study with 17 PD patients, they reported lower MK and FA in anterior cingulum. MK has showed the best diagnostic performance in anterior cingulum with sensitivity of 0.87 and specificity 0.94 [80]. In a later study using 12 PD patients. The same group reported lower MK values in posterior corona radiata and superior longitudinal fasciculus with TBSS analysis; however, no FA changes were observed. Importantly these regions show presence of crossing fibers and in DTI imaging FA and MD metrics get influenced by presence of crossing fibers [34]. Therefore, we can conclude that MK has a potential to detect changes in both gray and white matter. The reason behind lower MK values observed in white matter is quite difficult to explain. It might have occurred due to neuronal loss or deposition of axonal Lewy neurites.

Conversely, in our preclinical studies with DKI imaging in both animal models we found increase in kurtosis and decrease in diffusivity in anterior and posterior commissure, cingulate fibers, medial longitudinal fasciculus, mammillothalamic tract, and thalamic nucleus which might be related to glial cell activation or deposition of alpha-synuclein aggregates [45, 56–58]. The existence of aggregates of alpha synuclein in the white matter and its potential repercussions has not yet been studied in the TNWT-61 mouse model. So far, there is only one study which described the presence of axonal alpha synuclein aggregates in TNWT-61 mice, especially C-terminal fragments in parallel to human Lewy body disease patients, and found axonal transport deficits in these mice

[81]. Thus, further preclinical studies are needed to determine the mechanism behind changes in kurtosis and diffusivity metrics in the white matter.

Few of the available DKI studies reported changes in putamen and thalamus with DKI imaging in PD patients. DKI study with 105 PD patients found higher MD and lower MK, which were correlated with severe motor and cognitive symptoms, while in the thalamus higher MD and lower FA were correlated negatively with severity of PD symptoms [82]. Later, the similar author collective performed a longitudinal DKI imaging study with 76 PD patients in which gray matter analysis was done by ROI based approach and white matter by TBSS analysis. This 2 year follow up study reported lower FA in the putamen of PD patients as compared to healthy control [37]. Similar to this study, another DKI trial with 35 clinically confirmed PD patients also reported lower FA in the putamen. Along with this, the authors also reported higher radial kurtosis in SN and globus pallidus and lower mean and axial kurtosis in red nucleus and thalamus [83].

In our preclinical studies with TNWT-61 and METH model we also found changes in kurtosis and diffusivity in striatum and thalamus. In TNWT-61 mice we observed significantly higher MK in striatum and thalamus, which might be related to microglial cell activation and presence of proteinase k resistant alpha synuclein, respectively. Conversely, in METH mouse model we found lower MK at the initial time point and higher MK at the later time point. As discussed before, the lower MK values might be related to neurodegenerative changes induced by METH administration, which has decreased the structural heterogeneity, while at later time point we found higher MK likely due to neuronal sprouting as a recovery process which may have increased the tissue heterogeneity [45].

Overall, we can conclude, that our preclinical studies may help to decipher the underlying pathological processes responsible for MK values. We hypothesize that higher MK is exerted by accumulation of alpha synuclein, mainly proteinase k resistant alpha synuclein and microglial cell activation, whereas lower MK suggests the neurodegenerative changes.

7   Conclusion

The present chapter discusses the importance of DKI imaging in diagnosis of neurodegenerative disorders, particularly PD. In DKI imaging, MK was found to be the most sensitive readout to detect both early as well as late stage of disease pathology. This was confirmed by studies in two different animal models of PD with completely different neuropathological hallmarks. Our DKI imaging studies provided indirect evidence of mechanisms underlying changes in kurtosis and diffusivity metrics observed in PD patients such as Lewy body deposits containing α-synuclein and glial cell activation may cause increase in kurtosis while neurodegenerative changes lead to decrease in kurtosis.

We have performed two types of analysis of DKI data, ROI based analysis, which is a hypothesis driven analysis sensitive to detect gray matter changes and a TBSS analysis, which is a data driven approach used for detection of white matter changes. ROI based analysis seems to have importance in early diagnosis of α-synuclein accumulation induced changes while TBSS analysis started to show changes in the later stage indicating gray matter is affected first and later the changes start in white matter.

We found that MK by measuring the hindrance to diffusion of water molecule to be sensitive in detecting microstructural changes in both gray as well as white matter. Hence, we suggest that MK may serve as an early biomarker for detection of PD and it can be explored in other neurodegenerative disorders.

Funding

This study was performed at Masaryk University as part of the project "Pharmacological research in the field of pharmacokinetics, neuropsychopharmacology, and oncology" number MUNI/A/1249/2020 with the support of the Specific University Research Grant, as provided by the Ministry of Education, Youth and Sports of the Czech Republic (MEYS CR) in the year 2021.

This supplement was supported by the seed fund of National Institute of Pharmaceutical Education and Research (NIPER), Ahmedabad, Department of Pharmaceuticals, Ministry of Chemicals and Fertilizers, Government of India. Amit Khairnar gratefully acknowledges the support of Ramalingaswami Fellowship from Department of Biotechology, India. Eva Drazanova was supported by the grant LM2015062 and CZ.02.1.01/0.0/0.0/16_013/0001775 "National Infrastructure for Biological and Medical Imaging (Czech-BioImaging)."

References

1. Nichols E, Szoeke CE, Vollset SE, Abbasi N, Abd-Allah F, Abdela J et al (2019) Global, regional, and national burden of Alzheimer's disease and other dementias, 1990–2016: a systematic analysis for the Global Burden of Disease Study 2016. Lancet Neurol 18 (1):88–106. https://doi.org/10.1016/S1474-4422(18)30403-4

2. Fereshtehnejad SM, Vosoughi K, Heydarpour P, Sepanlou S, Farzadfar F, Tehrani-Banihashemi A et al (2019) Burden of neurodegenerative diseases in the eastern Mediterranean region, 1990–2016: findings from the global burden of disease study 2016. Eur J Neurol 26(10):1252–1265. https://doi.org/10.1111/ene.13972

3. Gooch CL, Pracht E, Borenstein AR (2017) The burden of neurological disease in the United States: a summary report and call to action. Ann Neurol 81(4):479–484

4. Stanzione P, Tropepi D (2011) Drugs and clinical trials in neurodegenerative diseases. Ann Ist Super Sanita 47:49–54. https://doi.org/10.1002/ana.24897

5. Sarkar S, Raymick J, Imam S (2016) Neuroprotective and therapeutic strategies against Parkinson's disease: recent perspectives. Int J Mol Sci 17(6):904. https://doi.org/10.3390/ijms17060904

6. Stoessl AJ (2012) Neuroimaging in the early diagnosis of neurodegenerative disease. Transl Neurodegen 1(1):1–6. (http://www.translationalneurodegeneration.com/content/1/1/5)

7. Vanhoutte G, Pereson S, Delgado y Palacios R, Guns PJ, Asselbergh B, Veraart J et al (2013) Diffusion kurtosis imaging to detect amyloidosis in an APP/PS1 mouse model for Alzheimer's disease. Magn Reson Med 69(4):1115–1121. https://doi.org/10.1002/mrm.24680

8. Praet J, Manyakov NV, Muchene L, Mai Z, Terzopoulos V, de Backer S et al (2018) Diffusion kurtosis imaging allows the early detection and longitudinal follow-up of amyloid-β-induced pathology. Alzheimers Res Ther 10(1):1–16. https://doi.org/10.1186/s13195-017-0329-8

9. Márquez F, Yassa MA (2019) Neuroimaging biomarkers for Alzheimer's disease. Mol Neurodegen 14(1):21. https://doi.org/10.1186/s13024-019-0325-5

10. Marsili L, Rizzo G, Colosimo C (2018) Diagnostic criteria for Parkinson's disease: from James Parkinson to the concept of prodromal disease. Front Neurol 9:156. https://doi.org/10.3389/fneur.2018.00156

11. Pyatigorskaya N, Gallea C, Garcia-Lorenzo D, Vidailhet M, Lehericy S (2014) A review of the use of magnetic resonance imaging in Parkinson's disease. Ther Adv Neurol Disord 7(4):206–220. https://doi.org/10.1177/1756285613511507

12. Pagano G, Niccolini F, Politis M (2016) Imaging in Parkinson's disease. Clin Med 16(4):371. https://doi.org/10.7861/clinmedicine.16-4-371

13. Armstrong MJ, Okun MS (2020) Diagnosis and treatment of Parkinson disease: a review. JAMA 323(6):548–560. https://doi.org/10.1001/jama.2019.22360

14. Lang AE, Mikulis D (2009) A new sensitive imaging biomarker for Parkinson disease? Neurology 72(16):1374–1375. https://doi.org/10.1212/01.wnl.0000343512.36654.41

15. Frederick J, Meijer BG (2014) Brain MRI in Parkinson's disease. Front Biosci 6:360–369. https://doi.org/10.2741/E711

16. Arab A, Wojna-Pelczar A, Khairnar A, Szabó N, Ruda-Kucerova J (2018) Principles of diffusion kurtosis imaging and its role in early diagnosis of neurodegenerative disorders. Brain Res Bull 139:91–98. https://doi.org/10.1016/j.brainresbull.2018.01.015

17. Boska MD, Hasan KM, Kibuule D, Banerjee R, McIntyre E, Nelson JA et al (2007) Quantitative diffusion tensor imaging detects dopaminergic neuronal degeneration in a murine model of Parkinson's disease. Neurobiol Dis 26(3):590–596. https://doi.org/10.1016/j.nbd.2007.02.010

18. Soria G, Aguilar E, Tudela R, Mullol J, Planas AM, Marin C (2011) In vivo magnetic resonance imaging characterization of bilateral structural changes in experimental Parkinson's disease: a T2 relaxometry study combined with longitudinal diffusion tensor imaging and manganese-enhanced magnetic resonance imaging in the 6-hydroxydopamine rat model. Eur J Neurosci 33(8):1551–1560. https://doi.org/10.1111/j.1460-9568.2011.07639.x

19. Cochrane CJ, Ebmeier KP (2013) Diffusion tensor imaging in parkinsonian syndromes: a systematic review and meta-analysis. Neurology 80(9):857–864. https://doi.org/10.1212/WNL.0b013e318284070c

20. Vaillancourt D, Spraker M, Prodoehl J, Abraham I, Corcos D, Zhou X et al (2009) High-resolution diffusion tensor imaging in the substantia nigra of de novo Parkinson disease. Neurology 72(16):1378–1384. https://doi.org/10.1212/01.wnl.0000340982.01727.6e

21. Takeda A, Kikuchi A, Matsuzaki-Kobayashi M, Sugeno N, Itoyama Y (2007) Olfactory dysfunction in Parkinson's disease. J Neurol 254(4):IV2–IV7. https://doi.org/10.1007/s00415-007-4002-1

22. Rolheiser TM, Fulton HG, Good KP, Fisk JD, McKelvey JR, Scherfler C et al (2011) Diffusion tensor imaging and olfactory identification testing in early-stage Parkinson's disease. J Neurol 258(7):1254–1260. https://doi.org/10.1007/s00415-011-5915-2

23. Van Camp N, Blockx I, Verhoye M, Casteels C, Coun F, Leemans A et al (2009) Diffusion tensor imaging in a rat model of Parkinson's disease after lesioning of the Nigrostriatal tract. NMR Biomed 22(7):697–706. https://doi.org/10.1002/nbm.1381

24. Wang J-J, Lin W-Y, Lu C-S, Weng Y-H, Ng S-H, Wang C-H et al (2011) Parkinson disease: diagnostic utility of diffusion kurtosis imaging.

Radiology 261(1):210–217. https://doi.org/10.1148/radiol.11102277

25. Guimarães RP, Campos BM, de Rezende TJ, Piovesana L, Azevedo PC, Amato-Filho AC et al (2018) Is diffusion tensor imaging a good biomarker for early Parkinson's disease? Front Neurol 9:626. https://doi.org/10.3389/fneur.2018.00626

26. Atkinson-Clement C, Pinto S, Eusebio A, Coulon O (2017) Diffusion tensor imaging in Parkinson's disease: review and meta-analysis. Neuroimage Clin 16:98–110. https://doi.org/10.1016/j.nicl.2017.07.011

27. Schwarz ST, Abaei M, Gontu V, Morgan PS, Bajaj N, Auer DP (2013) Diffusion tensor imaging of nigral degeneration in Parkinson's disease: a region-of-interest and voxel-based study at 3 T and systematic review with meta-analysis. NeuroImage Clin 3:481–488. https://doi.org/10.1016/j.nicl.2013.10.006

28. Jensen JH, Helpern JA, Ramani A, Lu H, Kaczynski K (2005) Diffusional kurtosis imaging: the quantification of non-gaussian water diffusion by means of magnetic resonance imaging. Magn Reson Med 53(6):1432–1440. https://doi.org/10.1002/mrm.20508

29. Liu C, Bammer R, Kim D, Moseley ME (2004) Self-navigated interleaved spiral (SNAILS): application to high-resolution diffusion tensor imaging. Magn Reson Med 52(6):1388–1396. https://doi.org/10.1002/mrm.20288

30. Jensen JH, Helpern JA (2010) MRI quantification of non-Gaussian water diffusion by kurtosis analysis. NMR Biomed 23(7):698–710. https://doi.org/10.1002/nbm.1518

31. Steven AJ, Zhuo J, Melhem ER (2014) Diffusion kurtosis imaging: an emerging technique for evaluating the microstructural environment of the brain. Am J Roentgenol 202(1):W26–W33. https://doi.org/10.2214/AJR.13.11365

32. Pierpaoli C, Basser PJ (1996) Toward a quantitative assessment of diffusion anisotropy. Magn Reson Med 36(6):893–906. https://doi.org/10.1002/mrm.1910360612

33. Zhuo J, Xu S, Proctor JL, Mullins RJ, Simon JZ, Fiskum G et al (2012) Diffusion kurtosis as an in vivo imaging marker for reactive astrogliosis in traumatic brain injury. Neuroimage 59(1):467–477. https://doi.org/10.1016/j.neuroimage.2011.07.050

34. Kamagata K, Tomiyama H, Hatano T, Motoi Y, Abe O, Shimoji K et al (2014) A preliminary diffusional kurtosis imaging study of Parkinson disease: comparison with conventional diffusion tensor imaging.

35. Zhang G, Zhang Y, Zhang C, Wang Y, Ma G, Nie K et al (2015) Diffusion kurtosis imaging of substantia nigra is a sensitive method for early diagnosis and disease evaluation in Parkinson's disease. Parkinsons Dis 2015. https://doi.org/10.1155/2015/207624

36. Goetz CG, Fahn S, Martinez-Martin P, Poewe W, Sampaio C, Stebbins GT et al (2007) Movement Disorder Society-sponsored revision of the unified Parkinson's disease rating scale (MDS-UPDRS): process, format, and clinimetric testing plan. Mov Disord 22(1):41–47. https://doi.org/10.1002/mds.21198

37. Surova Y, Nilsson M, Lampinen B, Lätt J, Hall S, Widner H et al (2018) Alteration of putaminal fractional anisotropy in Parkinson's disease: a longitudinal diffusion kurtosis imaging study. Neuroradiology 60(3):247–254. https://doi.org/10.1007/s00234-017-1971-3

38. Chesselet M-F, Richter F, Zhu C, Magen I, Watson MB, Subramaniam SR (2012) A progressive mouse model of Parkinson's disease: the Thy1-aSyn ("line 61") mice. Neurotherapeutics 9(2):297–314. https://doi.org/10.1007/s13311-012-0104-2

39. Gubellini P, Kachidian P (2015) Animal models of Parkinson's disease: an updated overview. Rev Neurol 171(11):750–761. https://doi.org/10.1016/j.neurol.2015.07.011

40. Olanow CW, Kieburtz K, Schapira AH (2008) Why have we failed to achieve neuroprotection in Parkinson's disease? Ann Neurol 64(S2):S101–SS10. https://doi.org/10.1002/ana.21461

41. Ares-Santos S, Granado N, Espadas I, Martinez-Murillo R, Moratalla R (2014) Methamphetamine causes degeneration of dopamine cell bodies and terminals of the nigrostriatal pathway evidenced by silver staining. Neuropsychopharmacology 39(5):1066–1080. https://doi.org/10.1038/npp.2013.307

42. Wilson JM, Kalasinsky KS, Levey AI, Bergeron C, Reiber G, Anthony RM et al (1996) Striatal dopamine nerve terminal markers in human, chronic methamphetamine users. Nat Med 2(6):699–703. https://doi.org/10.1038/nm0696-699

43. Granado N, Ares-Santos S, O'Shea E, Vicario-Abejón C, Colado MI, Moratalla R (2010) Selective vulnerability in striosomes and in the nigrostriatal dopaminergic pathway after methamphetamine administration. Neurotox Res 18

(1):48–58. https://doi.org/10.1007/s12640-009-9106-1

44. Moratalla R, Khairnar A, Simola N, Granado N, García-Montes JR, Porceddu PF et al (2017) Amphetamine-related drugs neurotoxicity in humans and in experimental animals: main mechanisms. Prog Neurobiol 155:149–170. https://doi.org/10.1016/j.pneurobio.2015.09.011

45. Arab A, Ruda-Kucerova J, Minsterova A, Drazanova E, Szabó N, Starcuk Z et al (2019) Diffusion kurtosis imaging detects microstructural changes in a methamphetamine-induced mouse model of Parkinson's disease. Neurotox Res 36(4):724–735. https://doi.org/10.1007/s12640-019-00068-0

46. Pan-Montojo F, Anichtchik O, Dening Y, Knels L, Pursche S, Jung R et al (2010) Progression of Parkinson's disease pathology is reproduced by intragastric administration of rotenone in mice. PLoS One 5(1):e8762. https://doi.org/10.1371/journal.pone.0008762

47. Paxinos G, Franklin KB (2019) Paxinos and Franklin's the mouse brain in stereotaxic coordinates. Academic, San Diego. https://www.elsevier.com/books/paxinos-and-franklins-the-mouse-brain-in-stereotaxic-coordinates-compact/franklin/978-0-12-816159-3

48. Smith SM, Jenkinson M, Woolrich MW, Beckmann CF, Behrens TE, Johansen-Berg H et al (2004) Advances in functional and structural MR image analysis and implementation as FSL. Neuroimage 23:S208–SS19. https://doi.org/10.1016/j.neuroimage.2004.07.051

49. Drazanova E, Ruda-Kucerova J, Kratka L, Horska K, Demlova R, Starcuk Z Jr et al (2018) Poly (I: C) model of schizophrenia in rats induces sex-dependent functional brain changes detected by MRI that are not reversed by aripiprazole treatment. Brain Res Bull 137:146–155. https://doi.org/10.1016/j.brainresbull.2017.11.008

50. Luca C, Salvatore F, Vincenzo DP, Giovanni C, Attilio ILM (2018) Anesthesia protocols in laboratory animals used for scientific purposes. Acta Bio Med Atenei Parmensis 89(3):337. https://doi.org/10.23750/abm.v89i3.5824

51. Young I, Hand J, Oatridge A, Prior M, Forse G (1994) Further observations on the measurement of tissue T1 to monitor temperature in vivo by MRI. Magn Reson Med 31(3):342–345. https://doi.org/10.1002/mrm.1910310317

52. Young IR, Hand JW, Oatridge A, Prior MV (1994) Modeling and observation of temperature changes in vivo using MRI. Magn Reson Med 32(3):358–369. https://doi.org/10.1002/mrm.1910320311

53. Lin W, Venkatesan R, Gurleyik K, He YY, Powers WJ, Hsu CY (2000) An absolute measurement of brain water content using magnetic resonance imaging in two focal cerebral ischemic rat models. J Cereb Blood Flow Metab 20(1):37–44. https://doi.org/10.1097/00004647-200001000-00007

54. Jenkinson M, Smith S (2001) A global optimisation method for robust affine registration of brain images. Med Image Anal 5(2):143–156. https://doi.org/10.1016/S1361-8415(01)00036-6

55. Leemans A, Jeurissen B, Sijbers J, Jones D (2009) ExploreDTI: a graphical toolbox for processing, analyzing, and visualizing diffusion MR data. Proc Intl Soc Mag Reson Med. https://cds.ismrm.org/protected/09MProceedings/files/03537.pdf

56. Khairnar A, Latta P, Drazanova E, Ruda-Kucerova J, Szabó N, Arab A et al (2015) Diffusion kurtosis imaging detects microstructural alterations in brain of α-Synuclein overexpressing transgenic mouse model of Parkinson's disease: a pilot study. Neurotox Res 28(4):281–289. https://doi.org/10.1007/s12640-015-9537-9

57. Khairnar A, Ruda-Kucerova J, Szabó N, Drazanova E, Arab A, Hutter-Paier B et al (2017) Early and progressive microstructural brain changes in mice overexpressing human α-Synuclein detected by diffusion kurtosis imaging. Brain Behav Immun 61:197–208. https://doi.org/10.1016/j.bbi.2016.11.027

58. Khairnar A, Ruda-Kucerova J, Drazanova E, Szabó N, Latta P, Arab A et al (2016) Late-stage α-synuclein accumulation in TNWT-61 mouse model of Parkinson's disease detected by diffusion kurtosis imaging. J Neurochem 136(6):1259–1269. https://doi.org/10.1111/jnc.13500

59. Giannelli M, Toschi N, Passamonti L, Mascalchi M, Diciotti S, Tessa C (2012) Diffusion kurtosis and diffusion-tensor MR imaging in Parkinson disease. Radiology 265(2):645–646. https://doi.org/10.1148/radiol.12121036

60. Delenclos M, Carrascal L, Jensen K, Romero-Ramos M (2014) Immunolocalization of human alpha-synuclein in the Thy1-aSyn ("line 61") transgenic mouse line. Neuroscience 277:647–664

61. Granado N, Ares-Santos S, Moratalla R (2013) Methamphetamine and Parkinson's disease. Parkinson Dis 2013:30805. https://doi.org/10.1016/j.neuroscience.2014.07.042

62. Khan AR, Chuhutin A, Wiborg O, Kroenke CD, Nyengaard JR, Hansen B et al (2016) Biophysical modeling of high field diffusion MRI demonstrates micro-structural aberration in chronic mild stress rat brain. Neuroimage 142:421–430. https://doi.org/10.1016/j.neuroimage.2016.07.001

63. Fornai F, Lenzi P, Ferrucci M, Lazzeri G, Di Poggio AB, Natale G et al (2005) Occurrence of neuronal inclusions combined with increased nigral expression of α-synuclein within dopaminergic neurons following treatment with amphetamine derivatives in mice. Brain Res Bull 65(5):405–413. https://doi.org/10.1016/j.brainresbull.2005.02.022

64. Watson MB, Richter F, Lee SK, Gabby L, Wu J, Masliah E et al (2012) Regionally-specific microglial activation in young mice over-expressing human wildtype alpha-synuclein. Exp Neurol 237(2):318–334. https://doi.org/10.1016/j.expneurol.2012.06.025

65. Neumann M, Müller V, Kretzschmar HA, Haass C, Kahle PJ (2004) Regional distribution of proteinase K-resistant α-synuclein correlates with Lewy body disease stage. J Neuropathol Exp Neurol 63(12):1225–1235. https://doi.org/10.1093/jnen/63.12.1225

66. Hui ES, Du F, Huang S, Shen Q, Duong TQ (2012) Spatiotemporal dynamics of diffusional kurtosis, mean diffusivity and perfusion changes in experimental stroke. Brain Res 1451:100–109. https://doi.org/10.1016/j.brainres.2012.02.044

67. Falangola MF, Jensen JH, Tabesh A, Hu C, Deardorff RL, Babb JS et al (2013) Non-Gaussian diffusion MRI assessment of brain microstructure in mild cognitive impairment and Alzheimer's disease. Magn Reson Imaging 31(6):840–846. https://doi.org/10.1016/j.mri.2013.02.008

68. Rudrapatna SU, Wieloch T, Beirup K, Ruscher K, Mol W, Yanev P et al (2014) Can diffusion kurtosis imaging improve the sensitivity and specificity of detecting microstructural alterations in brain tissue chronically after experimental stroke? Comparisons with diffusion tensor imaging and histology. Neuroimage 97:363–373. https://doi.org/10.1016/j.neuroimage.2014.04.013

69. Guglielmetti C, Veraart J, Roelant E, Mai Z, Daans J, Van Audekerke J et al (2016) Diffusion kurtosis imaging probes cortical alterations and white matter pathology following cuprizone induced demyelination and spontaneous remyelination. Neuroimage 125:363–377. https://doi.org/10.1016/j.neuroimage.2015.10.052

70. Burke RE, O'Malley K (2013) Axon degeneration in Parkinson's disease. Exp Neurol 246:72–83. https://doi.org/10.1016/j.expneurol.2012.01.011

71. Smith SM (2002) Fast robust automated brain extraction. Hum Brain Mapp 17(3):143–155. https://doi.org/10.1002/hbm.10062

72. Smith SM, Jenkinson M, Johansen-Berg H, Rueckert D, Nichols TE, Mackay CE et al (2006) Tract-based spatial statistics: voxelwise analysis of multi-subject diffusion data. Neuroimage 31(4):1487–1505. https://doi.org/10.1016/j.neuroimage.2006.02.024

73. Sierra A, Laitinen T, Lehtimäki K, Rieppo L, Pitkänen A, Gröhn O (2011) Diffusion tensor MRI with tract-based spatial statistics and histology reveals undiscovered lesioned areas in kainate model of epilepsy in rat. Brain Struct Funct 216(2):123–135. https://doi.org/10.1007/s00429-010-0299-0

74. Sakayori N, Kato S, Sugawara M, Setogawa S, Fukushima H, Ishikawa R et al (2019) Motor skills mediated through cerebellothalamic tracts projecting to the central lateral nucleus. Mol Brain 12(1):1–12. https://doi.org/10.1186/s13041-019-0431-x

75. Pifl C, Kish SJ, Hornykiewicz O (2012) Thalamic noradrenaline in Parkinson's disease: deficits suggest role in motor and non-motor symptoms. Mov Disord 27(13):1618–1624. https://doi.org/10.1002/mds.25109

76. Kamagata K, Hatano T, Okuzumi A, Motoi Y, Abe O, Shimoji K et al (2016) Neurite orientation dispersion and density imaging in the substantia nigra in idiopathic Parkinson disease. Eur Radiol 26(8):2567–2577. https://doi.org/10.1007/s00330-015-4066-8

77. Zhang G, Zhang C, Wang Y, Wang L, Zhang Y, Xie H et al (2019) Is hyperhomocysteinemia associated with the structural changes of the substantia nigra in Parkinson's disease? A two-year follow-up study. Parkinsonism Relat Disord 60:46–50. https://doi.org/10.1016/j.parkreldis.2018.10.008

78. Zhang G, Zhang C, Zhang Y, Wang Y, Nie K, Zhang B et al (2017) The effects of striatal silent lacunar infarction on the substantia nigra and movement disorders in Parkinson's disease: a follow-up study. Parkinsonism Relat Disord 43:33–37. https://doi.org/10.1016/j.parkreldis.2017.06.020

79. Guan J, Ma X, Geng Y, Qi D, Shen Y, Shen Z et al (2019) Diffusion kurtosis imaging for detection of early brain changes in Parkinson's disease. Front Neurol 10. https://doi.org/10.3389/fneur.2019.01285

80. Kamagata K, Tomiyama H, Motoi Y, Kano M, Abe O, Ito K et al (2013) Diffusional kurtosis imaging of cingulate fibers in Parkinson disease: comparison with conventional diffusion tensor imaging. Magn Reson Imaging 31 (9):1501–1506. https://doi.org/10.1016/j.mri.2013.06.009

81. Games D, Seubert P, Rockenstein E, Patrick C, Trejo M, Ubhi K et al (2013) Axonopathy in an α-Synuclein transgenic model of Lewy body disease is associated with extensive accumulation of C-terminal–truncated α-Synuclein. Am J Pathol 182(3):940–953. https://doi.org/10.1016/j.ajpath.2012.11.018

82. Surova Y, Lampinen B, Nilsson M, Lätt J, Hall S, Widner H et al (2016) Alterations of diffusion kurtosis and neurite density measures in deep grey matter and white matter in Parkinson's disease. PLoS One 11(6):e0157755. https://doi.org/10.1371/journal.pone.0157755

83. Bingbing G, Yujing Z, Yanwei M, Chunbo D, Weiwei W, Shiyun T et al (2020) Diffusion kurtosis imaging of microstructural changes in gray matter nucleus in Parkinson disease. Front Neurol 11:252. https://doi.org/10.3389/fneur.2020.00252

Chapter 19

Imaging Biomarkers in Huntington's Disease

Edoardo Rosario De Natale, Heather Wilson, and Marios Politis

Abstract

Huntington's disease (HD) is a fatal neurodegenerative disorder caused by an abnormal CAG repeat expansion in the *HTT* gene, that produces a mutant protein thought to directly cause brain cell damage. Premanifest carriers of the CAG expansion represent an ideal population to track the pathophysiological events responsible for the onset of clinical symptoms. Neuroimaging tools, such as positron emission tomography (PET) and magnetic resonance imaging (MRI), are able to investigate in vivo structural, microstructural, and functional brain alterations up to a molecular level and have substantially contributed to understanding the pathophysiology of HD. Neuroimaging techniques have identified potential biomarkers of early detection and progression of disease which can be translated into clinical trials with acceptable sample sizes. This chapter outlines the most important findings from PET and MRI research that has helped understand the pathophysiology of HD in the premanifest and manifest stages, and the translational potential of the neuroimaging biomarkers for the design of future clinical trials with disease-modifying agents.

Keywords Huntington's disease, Positron emission tomography, Magnetic resonance imaging, Neuroimaging, Neurodegeneration

1 Introduction

Huntington's disease (HD) is a progressive neurodegenerative disease clinically characterized by an overlap of a hyperkinetic movement disorder, psychiatric symptoms, and cognitive deterioration, and average life expectancy at phenoconversion of around 15–20 years [1]. HD is caused by a pathological expansion of a CAG trinucleotide repeat in the *HTT* gene, that translates to a polyglutamine (polyQ) chain in the Huntingtin (htt) protein, which is abundantly expressed in striatal medium spiny neurons (MSN), and the cerebral cortex. Mutant htt, both as a whole and as fragments through proteolytic cleavage, accumulates in excess in damaged cells, and has been hypothesized to contribute directly to cell death [2, 3]. The size of the CAG repeat expansion influences the age at onset and the clinical presentation [4], although other

Philip V. Peplow, Bridget Martinez and Thomas A. Gennarelli (eds.), *Neurodegenerative Diseases Biomarkers: Towards Translating Research to Clinical Practice*, Neuromethods, vol. 173, https://doi.org/10.1007/978-1-0716-1712-0_19,
© Springer Science+Business Media, LLC, part of Springer Nature 2022

factors, which are yet to be fully understood, also contribute to the variability of clinical symptoms and severity of manifest HD cases [5]. Nevertheless, it is possible to identify asymptomatic (premanifest) HD gene expansion carriers (HDGECs) and to predict the time of phenoconversion [6]. This window provides a unique opportunity to characterize biomarkers of disease progression that predict phenoconversion and aid in the development and testing of disease modifying treatments.

Neuroimaging techniques such as magnetic resonance imaging (MRI) and positron emission tomography (PET) allow the quantification of structural, microstructural, functional, and molecular alterations taking place in HDGECs with high spatial and temporal resolution, and may represent ideal biomarkers to identify and track disease progression starting in early premanifest stages [7–9]. This chapter will describe the most recent advances in the identification of imaging biomarkers to characterize HD.

2   PET Imaging

PET imaging is a noninvasive technique that measures physiological functions at a molecular level. PET tracers, labeled with isotopes (generally [^{18}F] or [^{11}C]), are specific to the target of interest and, in the case of neuroimaging tracers, can cross the blood–brain barrier following intravenous injection in the bloodstream. Gamma rays emitted from positron decay, of the PET tracer, are detected by the PET scanner to produce a map of the tracer uptake. By this means, it is possible to obtain maps reflecting the biological functions of the target substance, expression of its neurochemistry, to a level that no other technology is able to achieve in vivo [10]. PET imaging is widely employed in the characterization of neurodegenerative diseases and has provided invaluable insights into the events taking place during these diseases [11–13]. The use of PET imaging to quantify functional activity, or the density of specific molecules of interest has attracted attention for the potential of PET to track disease progression and to represent a biomarker for the development of novel treatments. In HD, PET studies have demonstrated changes in brain metabolism, dopaminergic, cannabinoid systems, phosphodiesterase density, and neuroinflammation from early premanifest to manifest disease [9]. Table 1 presents a list of the PET radiotracers used in HD research.

2.1   Brain Metabolism

[^{18}F]Fluorodeoxyglucose ([^{18}F]FDG) is an analogue of glucose that is phosphorylated in the brain to FDG-6-phosphate. It has been used as a surrogate marker of total brain glucose metabolism since the earliest PET studies on HD patients, dating back to the early 80s of last century when genetic confirmation of HD diagnosis was not available yet [14–21]. In these early works, HD

Table 1
Overview of the PET imaging findings and tracers used in Huntington disease research

PET tracer	Target	Main findings	References
Brain metabolism			
[18F]FDG	Brain metabolism	Pre-HDGECs: reduction of uptake in the caudate and putamen, and in frontal and temporal cortex. mHDGECs: progression of uptake reduction in the striatum and, diffusely, in the cerebral cortex.	[14–32]
15H2O	Regional cerebral blood flow	Reduction of activity in the supplementary motor area and in the premotor, parietal, and prefrontal areas, in response to motor tasks. Reduction in the left inferior temporal gyrus in response to cognitive tasks.	[33–35]
Dopaminergic system			
[11C]SCH23390	Postsynaptic D1 receptors	Reduction of tracer binding in the striatum and temporal cortex. Relative reduction of binding potential in the akinetic-rigid form compared with the choreic form.	[36–40]
[11C]Raclopride	Postsynaptic D2 receptors	Pre-HDGECs: striatal reduction of binding potential that precedes in time the reduction of brain metabolism. mHDGECs: reduction of binding potential in the striatum, hypothalamus, and frontal and temporal cortex. Relative reduction of binding potential in the akinetic-rigid form compared with the choreic form. Nonlinear progression of receptor loss in premanifest and manifest disease.	[31, 32, 37, 38, 40–46]
[11C]FLB457	Postsynaptic D2 receptors	No reduction of binding potential in extrastriatal areas of mHDGECs	[47]
[11C]DTBZ	Presynaptic VMAT2 transporter	Significant reduction of binding potential in mHDGECs with akinetic-rigid form compared to the choreic form.	[48]
Neuroinflammation			
[11C]PK11195	Translocator Protein	Increase of tracer uptake in the striatum, cortex, amygdala, hippocampus, hypothalamus, prefrontal cortex from the premanifest stage. Correlation, in pre-HDGECs, between increase of tracer uptake and peripheral levels of cytokines.	[49–52]

(continued)

Table 1
(continued)

PET tracer	Target	Main findings	References
[^{11}C]PBR28	Translocator Protein	Increase, in mHDGECs, of tracer uptake in pallidum, putamen, thalamus, and brainstem.	[53]
Phosphodiesterases			
[^{11}C]JNJ42259152	Phosphodiesterase 10A	Significant reduction of binding potential in the caudate, putamen, and pallidum.	[54]
[^{11}C]MNI-659	Phosphodiesterase 10A	Significant reduction of binding potential in the caudate, putamen, and pallidum. Nonlinear progression of degeneration in premanifest and manifest stage.	[55–57]
[^{11}C]IMA107	Phosphodiesterase 10A	Significant reduction in far from onset pre-HDGECs in caudate and putamen, and in the insular cortex and occipital fusiform gyri. Increase of binding in the motor thalami.	[58, 59]
GABAergic system			
[^{11}C]Flumazenil	Benzodiazepine receptors	Pre-HDGECs: heterogenous reduction of tracer uptake in the caudate. Inverse correlation with [^{11}C]Raclopride uptake mHDGECs: significant reduction of tracer uptake in the caudate.	[60, 61]
Opioidergic system			
[^{11}C] Diprenorphine	Nonspecific ligand of μ, δ, and κ receptors	Reduction of binding potential in the caudate, putamen, and cingulate. Increase in the thalamus and prefrontal areas.	[62]
Cannabinoid system			
[^{18}F]MK-9470	CB1 receptor	Reduction of binding potential in the cortex, cerebellum, and brainstem.	[63, 64]
Adenosinergic system			
[^{18}F]CPFPX	A1A receptor	Far from onset pre-HDGECs have diffuse increased binding. Linear decrease of binding in near onset pre-HDGECs and mHDGECs in the amygdala, frontal, parietal, and temporal cortex.	[65]

A1A adenosine 1 A receptor, *CB1* cannabinoid 1 receptor, *D1* dopamine 1 receptor, *D2* dopamine 2 receptor, *HDGECs* Huntington disease gene expansion carriers, *VMAT2* vesicular monoamine transporter 2

diagnosis was clinical and subjects at risk of developing HD were evaluated on the basis of their kinship ties with the symptomatic HD and by means of DNA polymorphisms [17, 18]. All manifest HD patients showed a marked decrease of caudate and putamen metabolism [14, 15, 19, 21]; about one third of subjects at risk of developing HD showed a similar pattern of striatal hypometabolism, compared with healthy nonrelated controls [17, 18].

More recent studies have benefited from the availability of genetic testing to identify with certainty premanifest HDGECs (pre-HDGECs), and to estimate the predicted time to phenoconversion [66]. Pre-HDGECs display a marked reduction of brain metabolism in the striatum, as well as in the frontal and temporal lobes [22], which progresses on to the manifest stage [23]. Two observational studies have suggested that the degree of reduction in glucose metabolism in the striatum, as seen with $[^{18}F]FDG$ PET, can be an early sign to predict future phenoconversion. In a first study, 22 pre-HDGECs were longitudinally observed for 2 years and then after an additional 8 years. Those who showed abnormal putaminal glucose metabolism at the end of a 2-year observation, all converted to manifest HD within the subsequent 8 years. Whereas, pre-HDGECs who that did not show any significant reduction of $[^{18}F]FDG$ uptake in the putamen, did not show any sign of phenoconversion at the end of the follow-up [24]. In a second study, 43 pre-HDGECs were assessed with $[^{18}F]FDG$ PET and monitored for 5 years. After the follow-up period 26 HDGECs showed signs of clinical conversion. A receiving operating characteristic (ROC) analysis showed that a cutoff of 1.0493 of caudate uptake in the baseline $[^{18}F]FDG$ PET scan was predictive of future phenoconversion with a sensitivity of 81% and a specificity of 100% [25].

Regional impairment of brain metabolism in both striatal and extrastriatal structures has also been associated with clinical characteristics of the disease. Overall disease severity and impairment of motor scores has been associated with greater decreases of $[^{18}F]$ FDG uptake in the striatum [26]. Apathy, in particular, is one of the most common neuropsychiatric symptoms of HD, being present since the premanifest stage and worsening linearly as the disease progresses. Martinez-Horta and colleagues have demonstrated, in 40 manifest HDGECs (mHDGECs) with mild disease, that the severity of apathy was associated with a reduction of $[^{18}F]FDG$ uptake in the supplemental motor area, anterior cingulate, frontopolar prefrontal cortex, and the superior medial prefrontal cortex [27].

Another approach of analyzing the patterns of brain metabolism alterations in HD is to perform network analysis of covariance of regions of hyper- and hypometabolism in premanifest and mHDGECs. Feigin and colleagues applied this analytic method in a population of 18 pre-HDGECs, 13 mHDGECs, and eight healthy controls [28]. They identified, in pre-HDGECs, a recurring pattern of hypometabolism in the striatum (caudate and putamen), and in the mediotemporal cortex, with hypermetabolism in the occipital cortex. It has been hypothesized that this pattern represents the result of an initial disconnection between the striatum and some cortical areas which, in turn, activates local compensatory upregulatory mechanisms, particularly in the mediodorsal thalamus and orbitofrontal cortex [29]. Within this metabolic pattern, higher degrees of regional glucose hypometabolism have been associated with a greater risk of phenoconversion over a 5-year follow-up [30]. The hypothesis of regional hypermetabolism as a possible compensatory mechanism has been also suggested by a more recent study on 60 HDGECs in which striatal and extrastriatal decrease of glucose metabolism coexisted with a regional increase of $[^{18}F]FDG$ uptake in the anterior cingulate, inferior temporal lobule, dentate nucleus, and cerebellar lobules [26]. The authors found that in these patients, regional hypermetabolism was associated with worse motor scores and suggested that compensatory increase of cerebral glucose consumption could have detrimental consequences in some structures [26].

Brain activation can also be measured quantifying the regional cerebral blood flow as surrogate marker, using $^{15}H_2O$ PET imaging. According to this paradigm, subjects are scanned at rest and after an activating motor, or cognitive task. The difference in the uptake of the tracer would reflect the increased blood flow, and hence activation, in a specific brain region. There are a number of studies that have assessed the degree of brain activation in HDGECs. Bartenstein and colleagues have studied 13 mHDGECs and nine controls, with a motor paradigm consisting in finger opposition task at 1.5 Hz with their dominant hand. HD patients showed a significant decrease of blood flow in the striatum at rest, which was more evident after the motor task. In addition, the authors demonstrated a sharp decrease of the activation of the supplementary motor area and of the anterior cingulate to the motor task, that was closely correlated with HD patient's ability to perform their task correctly [33]. Similar results were obtained, in the same year, by Weeks and colleagues. In their study, HD patient showed a reduction of the activation of the contralateral primary motor, medial premotor, bilateral parietal, and bilateral prefrontal areas in response to a joystick movement; accompanied by an increased activation, perhaps compensatory, of the bilateral insula [34]. These results suggest that patients with manifest HD have degeneration of connections between the basal ganglia and their frontal projections that leads to an impaired recruitment of the latter areas for motor ideation and control. More recently,

Lepron and colleagues applied a cognitive (semantic) task, in which the activation task was the generation of words which were semantically appropriate to words previously presented to the subjects. HD patients showed lower activation of the left inferior temporal gyrus compared with controls, that was more marked the more selective the word generation task was. The impaired activation of this area, important for the lexical selection, was partially compensated in HD patients by an activation of the left supramarginal gyrus (implicated in the phonological loop activity) and of the right inferior frontal gyrus (important in effortful retrieval processes) [35].

2.2 Dopaminergic System

HD is characterized by a prominent degeneration of the striatal medium spiny neurons, which compose the vast majority of all cells of the striatum. This explains, at least in part, the prominent cellular atrophy of this basal ganglia region in this neurodegenerative disease [67]. These alterations give rise to the typical HD motor symptoms which encompass hyperkinetic movements in the early stages, followed by a hypokinetic-rigid syndrome in the later stages [68]. These neurons are enriched of postsynaptic D1 and D2 dopaminergic receptors, critical for the regulation of basal ganglia functions across both the direct and indirect dopaminergic pathways. Postmortem studies have demonstrated that advanced HD patients have an over 50% reduction of D1 and D2 receptor density in the striatum [69, 70]. PET imaging using dedicated radiotracers for the study of the dopaminergic system has contributed significantly to the understanding of the pathophysiological alterations underlying HD.

$[^{11}C]SCH23390$ is a noncompetitive, high-affinity D1 receptor antagonist ($K_i = 0.2$–0.3 nM) PET radioligand that has been long used to quantify the density of D1 receptors in striatal and extrastriatal regions in HD. In a first study, by Sedvall and colleagues, five mHDGECs showed a reduction of up to 75% of putaminal $[^{11}C]SCH23390$ nonspecific binding potential. The only pre-HDGEC assessed in that study displayed striatal values in the lower end of controls' mean range [36]. A finding of a reduced D1 receptor density in the striatum of mHDGECs were replicated by successive studies [37, 38], although still with a limited sample size. Ginovart and colleagues found that the reduction of D1 receptor density in the striatum of mHDGECs correlated with disease duration. In addition they reported an additional significant 23.9% reduction of D1 receptor density also in the temporal cortex, suggesting that a reduction of postsynaptic dopaminergic receptors in cortical regions may account for the prominent cognitive deficits of these patients [38]. The hypothesis of a role of D1 receptor loss in the generation of cognitive disturbance in HD has been tested in a successive study from the same research group [39]. In this study, the reduction of caudate $[^{11}C]SCH23390$ binding was correlated with worse performance in verbal fluency, and the reduction of temporal [11C]SCH23390 binding correlated with worse performances in the Tower of Hanoi game, confirming the hypothesis of a dopaminergic role in the generation of cognitive deterioration in HD [39, 40].

The tracers [11C]Raclopride and [11C]FLB457 have been employed as PET imaging probes for the study of the density of D2 receptors. [11C]FLB457 shows better sensitivity compared to [11C]Raclopride for the detection of D2 receptors in the extrastriatal regions, and has found application for the study of alterations of D2 receptors in these areas of premanifest and manifest HDGECs [71]. The reduction of D2 receptor density in the caudate and putamen, as measured with [11C]Raclopride binding potential, has been demonstrated to be similar to that of D1 receptors [37, 38]; and the extent of D1 and D2 receptor striatal loss was significantly correlated [38, 41]. More recent studies have focused on the distribution of D2 pathology in extrastriatal areas. Politis and colleagues have found significant [11C]Raclopride decrease in the hypothalamus of both premanifest and manifest HDGECs, which correlated with increased uptake of [11C]PK11195 in the same region, as sign of microglial activation [42]. A reduction of D2 receptor availability in premanifest and manifest HDGECs is common also in cortical areas. Pavese and colleagues demonstrated that 62.5% of manifest and 54.5% of pre-HDGECs show a reduction of [11C]Raclopride especially in frontal and temporal areas [43]. A more recent study using the more specific extrastriatal D2 tracer [11C]FLB457 on nine mHDGECs did not show any significant reduction of the density of D2 receptors outside the striatum [47].

The potential of D2 receptor density quantification with [11C] Raclopride has been tested against brain glucose metabolism using [18F]FDG PET, for the detection of preclinical alterations in a cohort of 27 pre-HDGECs [31]. In this study, the rate of the 27 pre-HDGECs with striatal reduction of [11C]Raclopride binding potential was higher than that with reduction of glucose metabolism measured with [18F]FDG PET and MRI volumetric measures. Additionally, the reduction of striatal D2 receptor density correlated with CAG repeat length. This suggests that dopamine D2 receptor availability measured by [11C]Raclopride binding potential seems the most sensitive indicator of early neuronal impairment in pre-HDGECs [31].

Advanced cases, and the juvenile form of HD, are characterized by a prevalence of an akinetic-rigid syndrome rather than a hyperkinetic choreic movement disorder. This has been hypothesized to be due to a more advanced and global dopaminergic deterioration in the basal ganglia, that would include also presynaptic damage. These hypotheses have been tested by a few PET imaging studies. Bohnen and collaborators studied six mHDGECs with akinetic-rigid phenotype and 13 with choreic phenotype with the PET tracer [^{11}C]DTBZ, that binds to the presynaptic receptor Vesicular Monoamine Transporter 2 (VMAT2). HD patients with rigid phenotype showed a higher degree of VMAT2 loss compared to choreic HD patients, suggesting that the akinetic-rigid phenotype is associated with presynaptic dopaminergic denervation [48]. HD patients with akinetic-rigid predominant phenotype show similar degrees of postsynaptic D1 and D2 reductions in the striatum compared with those with choreic form. However the former tended to display a trend toward a greater reduction of striatal dopaminergic receptors compared to choreic HD [37].

Postsynaptic dopaminergic deterioration has been also evaluated as a potential marker of disease progression in HDGECs. Both pre-HDGECs and mHDGECs show, as a group, an accelerated progression of D1 and D2 striatal deterioration, compared with healthy controls [32, 44, 45]. HDGECs at different stages, however, may show different degrees of degeneration. Pre-HDGECs have been found to display either a slow [45, 46], or fast rate of progression of both D1 and D2 deterioration [45]. Manifest HDGECs, on the other hand, show average degrees of dopamine receptor loss which suggest a different rate of progression compared with pre-HDGECs [44]. These findings indicate that, in HD neurodegenerative process, the progression of D1 and D2 degeneration is nonlinear and progresses faster in patients with a higher disease burden.

2.3 Neuro-inflammation

Neuroinflammation is thought to represent a common, early, and decisive step in the generation of cellular damage in neurodegenerative diseases. Misfolded proteins, heavy metal accumulation, and other genres of molecular insult are able to activate microglia. In turn, activated microglia, produce cytotoxic substances such as proinflammatory cytokines and reactive oxygen species, that, if above a certain level, feed forward the neuroinflammatory response and promote cellular damage. When microglia are activated, they express the 18 kDA translocator protein TSPO. TSPO is a surface protein that has been used as a sensitive biomarker of reactive gliosis and inflammation associated with a variety of brain insults [72]. PET tracers targeting TSPO have been developed as markers of central microglial activation and have been extensively used in the study of neuroinflammation across the spectrum of neurodegenerative diseases [73].

Early PET studies have employed the first-generation TSPO tracer, [^{11}C]PK11195, in premanifest and manifest HDGECs. [^{11}C]PK11195 PET studies have shown significantly increased levels of microglial activation in the putamen, pallidum, caudate, and various extrastriatal regions starting from early premanifest stages of the disease [42, 49–52]. Second-generation TSPO tracers have a higher signal-to-noise ratio and allow for the investigation of regional heterogeneity at the individual level. However, the TSPO polymorphism Ala147Thr influences the affinity of these tracers' binding to TSPO and the participants need to be stratified according to their genetic profile. To date, only one study has employed the second-generation PET tracer [^{11}C]PBR28 in HD. Seven manifest and one premanifest HDGECs showed increased TSPO binding in the pallidum, putamen, as well as heterogeneous increases in thalamic and brainstem regions, compared to six healthy controls [53]. Further studies using second-generation tracers, and taking into account the TSPO genetic status of participants, are warranted to better define the potential of TSPO as a marker of neuroinflammation in HD.

Some studies have sought to investigate the relationship between microglial activation and depletion of D2 dopaminergic receptors, as a marker of disease severity in HD. Pavese and colleagues have studied a group of premanifest and manifest HDGECs with [^{11}C]PK11195 and [^{11}C]Raclopride PET, and have demonstrated a significant inverse correlation in the striatum between an increase of TSPO microglial binding and a decrease of D2 receptor density [49]. Politis and colleagues have studied dopaminergic and microglial function in ten premanifest and nine manifest HDGECs. They found significant levels of neuroinflammation compared with controls in the hypothalamus, which inversely correlated with the degree of D2 receptor density loss in both premanifest and manifest HDGECs [42]. In a successive study from the same group, eight pre-HDGECs displayed decreased D2 receptor density and increased TSPO activation in the sensorimotor striatum, associative striatum, red nucleus of the stria terminalis, amygdala and hippocampus; mHDGECs showed additional alterations in the globus pallidus, limbic striatum and anterior prefrontal cortex [51].

TSPO PET studies have also investigated whether the degree of microglial activation could be related to an increased risk of phenoconversion or to the presence of peripheral markers of inflammation. In the aforementioned PET study by Politis and colleagues [51], the level of [^{11}C]PK11195 uptake in the associative striatum and in cortical regions associated with cognitive function was correlated with the probability of HD phenoconversion within 5 years [51]. In a more recent study from the same group, involving 12 pre-HDGECs who underwent [^{11}C]PK11195 PET and blood collection, it was found that the peripheral levels of the interleukins IL-1β, IL-6, and IL-8, and of tumor necrosis factor-α, were directly proportional to the level of TSPO binding in the postcentral gyrus [52], therefore highlighting the role of innate immune response in the generation of neuroinflammation in HD.

2.4 Phospho-diesterase Density

Phosphodiesterases (PDE) are a family of 11 intracellular enzymes that hydrolyze a cyclic phosphate ester nucleotide, such as cyclic adenosine monophosphate (cAMP), to produce a nucleoside monophosphate which modulates physiological enzyme cascades for gene expression and cellular responses. The isoform PDE10A is highly expressed in the striatal MSN, influencing cellular response to both the direct and the indirect dopaminergic pathway [74–78]. Mutant htt directly affects the function, and expression, of PDE10A. Therefore, an alteration of PDE10A levels may constitute an early alteration in the HD history. The recent availability of PET tracers, such as [^{11}C]JNJ42259152, [^{11}C]IMA107, and [^{18}F]MNI-659, specifically binding to PDE10A has allowed the in vivo investigation of this enzyme in HDGECs.

Two early cross-sectional studies evaluated small cohorts of manifest and premanifest HDGECs with the PET tracers [^{11}C]JNJ42259152 and [^{18}F]MNI-659 [54, 55]. These studies demonstrated that patients have significantly lower levels of PDE10A in the caudate (71% decrease), putamen (up to 63% decrease) compared to healthy controls. Our group has studied a cohort of 12 early pre-HDGECs, with a mean 90% probability to phenoconversion of 25 years who did not display any volumetric atrophy on MRI, with [^{11}C]IMA107 PET [58]. [^{11}C]IMA107 binding potential was decreased by 33% in the caudate, 31% in the putamen, and by 26% in the pallidum, whereas it was significantly increased by 35% in the motor thalamic nuclei. At a single subject level, 10 out of 12 pre-HDGECs already displayed altered PDE10A levels, in those regions, greater than 2 standard deviations from the control mean. The motor thalamic nuclei/striatopallidal [^{11}C]IMA107 binding potential ratio correlated with the 15 year probability of phenoconversion [58]. Significant depletion of PDE10A density was also demonstrated in extrastriatal regions important for cognitive and limbic functions such as the insular cortex and occipital fusiform gyrus [59]. How the loss of PDE10A in the striatum progresses across the stages of HD has been recently addressed in a study that enrolled 45 HDGECs, divided into early premanifest, late premanifest, manifest HD stage 1 (indicating mild disease) and manifest HD stage 2 (indicating a moderate disease), who underwent [^{18}F]MNI-659 and [^{11}C]Raclopride PET imaging [56]. In this study, the decrease of striatal and pallidal [^{18}F]MNI-659 progressed in the premanifest stages and appeared to stabilize between the manifest stages 1 and 2. Additionally, the extent of PDE10A decline compared to healthy controls, was greater than that of [^{11}C]Raclopride [56]. Taken together, these studies highlight the potential of PDE10A PET to be an early biomarker of pathology in HD which is potentially sensitive to predict phenoconversion.

The potential of PDE10A PET imaging to measure progression of the disease has been recently assessed in two longitudinal studies, employing the tracer [^{18}F]MNI-659. Russell and colleagues studied six manifest and two premanifest HDGECs at baseline and 1-year follow-up. A yearly rate PDE10A loss of 17% in the caudate, 7% in the putamen and 6% in the pallidum was demonstrated, compared with a progression of less than 1% PDE10A loss in healthy controls [57]. Moreover, Fazio and colleagues [56], followed up HDGECs for 18–28 months, illustrating a mean annual PDE10A loss of 6% in the caudate and 4% in the putamen and pallidum [56].

2.5 Other Neurotransmitter Systems

2.5.1 GABAergic System

In humans, the balance between the excitatory glutamatergic system and the inhibitory GABAergic system, along the corticostriatal pathway, is critical for motor and behavior control. This delicate balance is impaired in HD as a consequence of the progressive loss of MSN [79]. Mutant htt interacts indirectly with the trafficking of GABA receptors to the synapses and reduces the total number of available GABA receptors in the membrane, thus reducing the cellular inhibition [80, 81]. [^{11}C]Flumazenil is a non-subtype selective antagonist of benzodiazepines that can provide quantitative measure of the density of GABA receptors in the brain, and as such has been used for the study of the pathology of GABAergic neurons in HD.

Two PET studies using [^{11}C]Flumazenil PET are available in the literature, on a total sample size of 16 manifest and 13 premanifest HDGECs [60, 61]. mHDGECs show a marked decrease of benzodiazepine receptor density in the caudate nucleus, compared with controls, that may reflect the loss of projection neurons [60, 61]. Autoradiographic studies indicate a prominent loss of GABAergic neurons occurs in the caudate as well as in other striatal structures [82, 83]. Together, these findings suggests that loss of GABA receptors in the caudate may take place at an earlier stage of the disease compared to more widespread GABAergic degeneration which may occur at a late stage of HD timeline [60]. Conversely, pre-HDGECs show different patterns of GABA receptors loss in the caudate [61]. Pre-HDGECs show a marked inverse correlation between [^{11}C]Flumazenil caudate uptake and concomitant [^{11}C] Raclopride reduced uptake in the caudate [61]. This finding could be interpreted as a compensatory change that takes place in the premanifest stage before the neuronal loss becomes too severe.

2.5.2 Opioidergic System

Opioid receptors are expressed widely throughout the human brain and are constituted by four categories: δ, κ, μ, and the nociceptor/orphanin FQ peptide receptor. The caudate and putamen contain high densities of μ, δ, and κ receptors [84]. A postmortem study demonstrated that loss of opioid receptors takes place in the premanifest phase of HD pathology [85]. A PET study employing [^{11}C]Diprenorphine, a nonspecific radioligand for all opioid receptors, showed in a cohort of five mHDGECs a 31% decrease in tracer binding in the caudate and a 26% decrease in the putamen, with additional decreases reported in the cingulate cortex, accompanied by an increase in the thalamus and prefrontal areas [62]. The knowledge of the in vivo pathophysiology of the opioidergic system in HD is still limited, more studies employing newer PET tracers specific for the opioid receptors, are warranted.

2.5.3 Cannabinoid System

A number of studies on different animal models of HD have demonstrated a severe loss of cannabinoid 1 (CB1) receptors, which is associated with HD severity and progression [86–88]. This finding has also been replicated in postmortem HD brains [89]. [^{18}F]MK-9470 is a PET radioligand for the CB1 receptor, and has been used in a cohort of 20 mHDGECs and 14 healthy controls [63]. The availability of CB1 receptors was diffusely decreased in both cerebral and cerebellar cortex, and in the brainstem of mHDGECs [63]. Moreover, loss of CB1 receptors in the prefrontal and premotor cortices was associated with higher disease burden. More recently, the same group tested 15 pre-HDGECs with [^{18}F]MK-9470 and clinical assessment, and found that reduced tracer uptake in the frontal and cingulate areas correlated with clinical scores suggestive of apathy, providing a possible molecular basis for the onset of this frequent symptom in pre-HDGECs [64].

2.5.4 Adenosinergic System

The adenosinergic system intervenes in the brain for a number of metabolic functions, including sleep, neuroinflammation, and excitotoxicity. There are a number of adenosine receptors (A1, A2A, A2B, and A3), of which the receptor A1A is the most highly expressed in the human brain. The A1A receptors are thought to protect the cells from excitotoxicity, hypoxia, and inflammation [90]. In HD animal models, the levels of A1A receptors are significantly decreased in the hypothalamus [91]. The PET radiotracer [^{18}F]CPFPX has been developed to quantify in vivo the regional alterations of A1A receptors in neurological diseases [92].

Matusch and colleagues have employed [^{18}F]CPFPX PET imaging in a cross-sectional study to quantify in vivo the alteration of A1A receptor regional distribution in a population of HDGECs, subdivided into far-from onset (mean predicted time of onset of 17.6 years) and near-onset (mean predicted time of onset 8.5 years), as well as in mHDGECs [65]. They detected an initial 13% widespread increase of tracer uptake in far-from onset pre-HDGECs, followed, as predicted time to disease onset approached, by a linear decrease of A1AR availability in mHDGECs in the amygdala and in frontal, parietal, and temporal cortices [65]. There is an accumulating evidence of a possible role of an adenosinergic dysfunction in HD, and the availability of PET tracers for adenosine receptors other than A1A will shed more light in the future about the pathophysiology of this system across all stages of the disease [93].

Overall, PET imaging has provided extremely useful in the understanding of the pathophysiology of HD and in the search of a reliable marker of disease. Changes in brain metabolism of pre-HDGECs and mHDGECs with $[^{18}F]$FDG provide cheap and useful information about the alterations taking place across disease stages and have been used as outcome measure in small randomized clinical trials [94–96]. However, in longitudinal studies, $[^{18}F]$FDG has proven less sensitive as marker of disease progression compared to changes in the D2 receptors ($[^{11}C]$Raclopride) and PDE10A ($[^{18}F]$MNI-659) density in the striatum which, at the moment, could be more suitable as potential markers of disease progression, with outstanding potential in tracking HD pathology in the early premanifest stages. There is a need, on one side, of multitracer PET studies to investigate the potential of combined molecular markers to track disease progression and establish to which degree they contribute to the generation of HD pathology; on the other side, the current blooming of PET molecular tracers targeting more and more metabolic pathways could be exploited to design imaging studies that could improve our understanding of the complex pathophysiology underlying HD.

3 Magnetic Resonance Imaging in Huntington Disease

Magnetic resonance imaging (MRI) is a noninvasive and well-tolerated technique to capture 3D in vivo images of the brain through the application of strong magnetic fields, therefore avoiding the subject to the exposure to potentially dangerous ionizing radiations. Advances in MRI acquisition parameters, as well as techniques for image processing and analysis, over the last decade have enabled researchers to obtain quantitative measures relating to macro- and microstructural alterations, and the functionality of brain white and gray matter connections, as well as measures relating to the concentration of magnetically susceptible elements (e.g., iron). The use of MRI techniques has led to breakthroughs in our understanding of disease mechanisms and the pathological progression of HD. Furthermore, specific MRI sequences have been proposed as disease biomarkers in clinical trials [97]. The following paragraphs will describe recent insights into the pathophysiology of HD, obtained through the use of MRI, and outline the most promising MRI biomarkers of early disease activity and progression. Table 2 summarizes the principal findings obtained with MRI in research in HDGECs.

Table 2
Overview of the MRI imaging findings and sequences used in Huntington disease research

MRI sequence	Findings	References
Structural MRI		
Whole-brain analysis	Pre-HDGECs closer to disease onset show volumetric reductions in the basal ganglia, amygdala, thalamus, insula, and occipital cortex. Pre-HDGECs show reduction in total brain volume. mHDGECs show additional reductions in the premotor, sensorimotor cortex, inferior frontal, frontotemporal, temporoparietal, and cingulate cortex. Near-onset pre-HDGECs show thinning of the sulci mHDGECs show widespread cortical thinning.	[98–119]
Region-of interest analysis	Pre-HDGECs show regional atrophy in the caudate and putamen as far as 15–20 years from predicted phenoconversion. Additional subsequent atrophy in the pallidum, accumbens, and amygdala. Nonlinear rate of progression of atrophy in premanifest and manifest stages.	[120–138]
Diffusion weighted imaging	Pre-HDGECs show alteration of FA and MD in the basal ganglia, thalamus, corpus callous, cortical regions, and corticospinal tract, as well as in frontal corticofugal connections. mHDGECs show additional FA, RD, and MD alterations in connections involving the striatothalamocortical loop. Near onset pre-HDGECs show faster rates of progression of deterioration.	[139–165]
NODDI	Diffuse reduction of axonal density.	[166]
Iron-sensitive MRI		
FDRI, T2, R2 relaxometry	mHDGECs show increase of signal in the basal ganglia, and in areas hit by prominent myelin breakdown.	[140, 145, 167–172]
SWI	Bilateral increases of field map values in the bilateral putamen and pallidum, additional increased values in the caudate, occipital cortex, left pre- and postcentral cortex in mHDGECs.	[173]
QSM	Increase of susceptibility values in the caudate, putamen, and pallidum. Decrease in the hippocampus and substantia nigra in pre-HDGECs.	[174, 175]
Functional MRI task-based		
Motor tasks	Increase of activation of the supplementary motor area.	[176, 177]
Working memory	Reduction, in the pre-HDGECs, of activation of the dorsolateral prefrontal cortex. Alteration of connections of frontostriatal and frontotemporal networks, and in the anterior cingulate. N-BACK tasks cause alterations in the insula, striatum, and inferior frontal gyrus.	[178–183]

(continued)

Table 2
(continued)

MRI sequence	Findings	References
Attention tasks	Increase, in pre-HDGECs of the default mode network and decrease of the connections between the putamen and the motor regions. mHDGECs: increase of the activity of the dorsolateral prefrontal cortex, dorsal cingulate, basal ganglia, and reduction of the anterior cingulate.	[184–186]
Time discrimination tasks	Pre-HDGECs show heterogenous activation of the supplementary motor area and of the anterior cingulate. Near-onset pre-HDGECs show a decrease of caudate connectivity.	[187–189]
Executive function	Pre-HDGECs and mHDGECs show a decrease of left premotor cortex activation.	[190]
Behavioural tasks	Altered connections in areas belonging to the limbic system.	[191–194]
Functional MRI resting state		
Seed based	mHDGECs show a reduction of the default mode network, and increase of supplementary motor area connections. Networks between the primary motor cortex and the insula, between the basal ganglia and the insula, and between the cerebellum and the paracentral nucleus are impaired.	[195–198]
ICA, ALFF	Pre-HDGECs show a decrease of connection between left frontal and right parietal cortices, and between visual cortices. mHDGECs show reduction of connectivity of the visual cortices, and of the calcarine-middle frontal gyri loop. mHDGECs show reduction of ALFF of the precuneus-angular gyrus. Pre-HDGECs and mHDGECs show a reduction of the frontal executive network and of the dorsal attention network, as well as of connections between the putamen and insula. Lack of sensitivity for longitudinal progression.	[176, 199–207]
Connectomics	Pre-HDGECs show weakened frontostriatal connections and loss of hubs in somatosensory and associative networks. mHDGECs show additional reduction of path length, clustering, and betweenness centrality.	[208, 209], [284, 285]
Magnetic resonance spectroscopy		
^{1}H MRS	Pre-HDGECs show decrease of NAA in the striatum; mHDGECs show a general decrease of NAA in striatum, occipital cortex, thalami, and posterior cingulate. Pre-HDGECs and mHDGECs show increased levels of Myo Heterogenous findings, in pre-HDGECs and mHDGECs, of altered levels of Cho in the striatum; decreased Cho levels in the frontal cortex Pre-HDGECs and mHDGECs show increased levels of Lactate and decreased levels of Cr Heterogenous findings of elevated Glu–Cr ratio in pre-HDGECs and mHDGECs; decreased Glu in the posterior cingulate of mHDGECs	[210–230]

(continued)

Table 2
(continued)

MRI sequence	Findings	References
^{31}P MRS	mHDGECs have a reduction of PCr–inorganic phosphate ratio, and fail to increase the levels of PCr in response to activation tasks	[231, 232]

ALFF amplitude of low frequency fluctuations, *Cho* choline, *Cr* creatine, *FA* fractional anisotropy, *HDGECs* Huntington disease gene expansion carriers, *ICA* independent component analysis, *MD* mean diffusivity, *MRI* magnetic resonance imaging, *Myo* myoinositol, *NAA* N-acetylaspartate, *NODDI* neurite orientation dispersion and density imaging, *MRS* magnetic resonance spectroscopy, *QSM* quantitative susceptibility mapping, *PCr* phosphocreatine, *RD* radial diffusivity, *SWI* susceptibility weighted imaging

3.1 Structural MRI

T1 and T2 weighted MRI images both provide important information about the anatomy and structural architecture of the brain. The application of specific T1-weighted sequences, such as the magnetization prepared rapid acquisition gradient echo (MP-RAGE) [233], provides high-resolution structural images, with good contrast between gray and white matter in both cortical and subcortical regions, which can be utilized for structural analysis methods and segmentation of different tissue types (gray matter and white matter) and cerebrospinal fluid. T1-weighted sequences with short inversion recovery to null the white matter signal, such as fast gray matter acquisition T1 inversion recovery (FGATIR) [234], provides enhanced contrast for subcortical regions, such as the internal and external globus pallidus. T2-weighted MRI is commonly used to identify pathological lesions. The main structural MRI information that can be analyzed relates to quantitative volumetric measures of white and gray matter regions, as well as cortical gray matter thickness. Regions of interest (ROIs) can be either manually [235], or automatically [236, 237] delineated. Although manual delineation of ROIs is the gold standard for a subject-specific study of single brain regions, automatic delineations methods allow the assessment of multiple regions according to standardized atlases, which improve consistency across studies, is less labor intensive. Alternative methods for the analysis of structural MRI analyses are the whole-brain analyses, such as voxel-based morphometry (VBM) [98], and cortical thickness [99, 100]. These methods use automated segmentations of the gray and white matter to extract information about the structural integrity of these regions. These methods are used to investigate changes in cortical and subcortical volumes, as well as, in the case of cortical thickness, to obtain measures of cortical thickness taking into account some parts such as cortical folding, for which standard volumetric analyses are not sensitive to.

Whole-brain analysis studies in HD have consistently shown a significant degeneration of cortical and subcortical regions in both mHDGECs, and pre-HDGECs [101]. Far-from onset pre-HDGECs do not seem to display significant volumetric difference compared with healthy controls [102–106]. However, pre-HDGECs, who are less than 10 years from predicted onset, consistently show atrophy in the basal ganglia, amygdala, thalamus, insula, and occipital regions [101, 102, 107]. mHDGECs show additional loss, compared with pre-HDGECs, in the premotor and sensorimotor cortices, as well as in the inferior frontal, frontoparietal, and temporoparietal cortices, and the cingulate [101]. One interesting hypothesis is that HD may represent a neurodevelopmental disease. It has been reported that pre-HDGECs tend to have a 4% smaller brain volume, suggesting that the presence of mutant htt may directly linked with neurodevelopmental dysfunction [108]. Longitudinal studies to assess the rate of progression of whole-brain and gray matter atrophy in pre-HDGECs have yielded conflicting results, with some studies showing an increased rate of atrophy over time [109–111], whereas others did not [104, 112, 113]. It is possible that differences in the characteristics of the pre-HDGECs populations (e.g., time to phenoconversion) could account for these discrepancies. Overall, observational studies have provided evidence for the potential of whole-brain gray matter atrophy to act as a biomarker to predict phenoconversion in pre-HDGECs [111, 113].

Cortical thickness analysis, using Freesurfer imaging package (http://surfer.nmr.mgh.harvard.edu), has shown significant cortical thinning in mHDGECs [114, 115], which correlated with the severity of global clinical deterioration [116], processing speed [117], and perceptual function [238]. Far-onset pre-HDGECs do not show differences in cortical thickness, compared with controls; however, near-onset pre-HDGECs show an altered cortical morphology with enlargement of the gyral sulci and abnormally thin sulci [103, 118]. In a longitudinal study in 22 mHDGECs followed up for 1 year, the highest rates of thinning were detected in the sensorimotor, posterior frontal, and portions of parietal cortex [119].

A number of studies have attempted to find patterns of covariance between regions of atrophy in HD. In these studies, VBM is followed by an independent component analysis (ICA) to identify regions that undergo atrophy at similar stages, also called components [239]. In a large study with 831 pre-HDGECs and 219 healthy controls, regional patterns of co-occurrence of structural alterations, particularly between the parietal and occipital cortex, have been identified with frontostriatal circuits that seem to be affected most prominently and early in the disease course, and other regions, such as the occipital regions, that tend to degenerate latest and to progress more slowly [240]. In another study, pre-HDGECs showed decreased network integrity along the frontostriatal circuits, and limbic-sensorimotor-premotor regions;

whereas mHDGECs additionally showed alterations in networks involving occipital areas [241].

ROI studies have identified the striatum as the region with the highest and earliest volumetric atrophy in HDGECs, and the best potential to track disease progression and predict the development of symptoms of phenoconversion in pre-HDGECs [102, 110, 111, 120]. Atrophy of the caudate and the putamen can be spotted as early as 15–20 years from predicted phenoconversion [120], with a rate of progression that is faster in pre-HDGECs the nearer they are to symptoms onset, and in all mHDGECs stages [102, 110, 111, 121]. Although putamen atrophy seems to progress faster than caudate atrophy in both premanifest and manifest HDGECs [102, 109, 120, 122–124]; caudate volumetric decreases correlate better with indicators of disease burden such as CAG repeat and total functional capacity [111, 124–130] and therefore making caudate atrophy a potential ideal biomarker of disease progression in clinical trials [127].

Other basal ganglia structures that show significant volume loss in HD are the pallidum and the nucleus accumbens. mHDGECs consistently show a significant degeneration of these regions, with around 60% volumetric loss for both regions [117, 125, 131–133], which correlates with measures of total functional capacity [133, 134]. Atrophy of the pallidum and accumbens can be detected from the near-onset stage of pre-HDGECs, but these changes do not seem to correlate with total functional capacity or motor progression [134]. However, the extent of atrophy of pallidum and accumbens, as well as the longitudinal change of pallidal volume over a 1-year follow-up in pre-HDGECs, is able to predict motor onset and could represent valuable biomarkers of disease progression in this population [109, 242].

The amygdala has recently gained attention as a brain region important in HD for its potential association with cognitive and neuropsychiatric problems and its close connectivity with the striatum [135, 136]. VBM studies have identified a degeneration of the amygdala in mHDGECs, mirroring the more widespread degeneration of all other brain regions [132]. This finding has been replicated in pre-HDGECs [137]. More recently, a study involving 35 premanifest, 36 mHDGECs and 35 healthy controls, has demonstrated that the volume of the amygdala was smaller in pre-HDGECs compared with controls, and in mHDGECs compared with pre-HDGECs. Moreover, volume loss in the amygdala was associated with scores of increased deficit in motor and cognitive abilities, as well as with the presence of anxiety [138].

3.2 Diffusion Weighted Imaging

Diffusion weighted imaging (DWI) is an MRI sequence that offers information about the integrity of the white matter and the microstructure of the gray matter. DWI measures the movement of water molecules as free diffusivity along biological tissues. Deviations from the normal diffusivity of water is an indirect marker of micro-structural tissue shape and possibly damage [243]. Diffusion Tensor Imaging (DTI) studies the microstructural integrity and direction of white matter fibers across brain regions through the indirect measurement of the structural orientation and the anisotropy of each voxel. The principal variables extracted from DTI that have been investigated in HD research are fractional anisotropy (FA), a measure of tissular integrity, mean diffusivity (MD), a measure of the directionality of water diffusivity, and axial diffusivity (AD) and radial diffusivity (RD), which measure demyelination of white matter, and axonal damage, respectively [244]. A detailed description of the principles underlying DWI is found in Chapter 18.

Cross-sectional imaging studies in mHDGECs using DTI have evidenced a picture of severe white matter alterations involving the putamen, caudate, pallidum, and the corpus callosum, and an increase of diffusion in the same areas [245], as a likely result of a white matter tracts demyelination [139, 246]. Pre-HDGECs show altered RD and FA in the basal ganglia, with damage not yet involving other structures, such as the corpus callosum and cortical regions [245].

An alteration of FA and an increase of MD and apparent diffusion coefficient (ADC) in the striatum and the thalamus have been consistently detected in pre-HDGECs [140, 141, 247] and m HDGECs [142–144]. In pre-HDGECs, increased MD and FA in the striatum has been associated with striatal volume loss and increased iron accumulation, as measured with T2* relaxation. In early manifest patients, these associations extended to include neighboring regions, such as the thalamus and accumbens [140], as well as the pallidum [145]. Sanchez-Castaneda and colleagues showed that MD of the striatum and thalamus was the best predictor of HD symptomatic onset, making this a potential biomarker for disease progression in pre-HDGECs [140].

DTI studies have also provided evidence that long, white matter tracts connecting distant cortical and subcortical brain areas are particularly affected in HD [141, 144, 146–149, 247]. Douaud and colleagues found, in a group of 14 mHDGECs, an increase of MD and FA in all subcortical gray matter structures involved in the striatothalamocortical loop which was accompanied, on tractography analysis, by a dispersion of corticostriatal connections [150]. This finding can be integrated with the results from a successive study in which DTI connectivity-based parcellation was used to segment the striatum according to its connections to specific cortical areas. It was found that the striatal connections with the largest increases of MD were those between the caudate and the primary motor and somatosensory cortical regions, which also show the highest levels of volumetric atrophy, highlighting a vulnerability of these regions and connections in HD

[151]. Pre-HDGECs also display loss of corticostriatal connections and microstructural alterations of cortical white matter organization [141, 144, 146, 147, 247]. Kloppel and colleagues found a reduction of frontal corticofugal streamlines to the caudate body in a group of 25 pre-HDGECs. Many of these fibers originated from the frontal eye fields, which are important for the control of voluntary saccades and may provide a pathophysiological link for this early symptom of HD [247]. Furthermore, in pre-HDGECs, alterations in MD and FA in the sensorimotor cortex, and in the prefrontal cortex, correlate with clinical cognitive measures [146, 147], as well as with 2- and 5-year probability of onset [144]. Near-onset pre-HDGECs also show a more defined pattern of increased diffusivity in the tract connecting the putamen to the prefrontal and motor cortex [146].

Microstructural alterations of the corpus callosum have been extensively studied in HDGECs. Severe diffusion and anisotropy alterations of all subregions of the corpus callosum have been described in both premanifest and manifest stage [139, 144, 151–153, 247, 248]. Phillips and colleagues analyzed the corpus callosum with DTI and tractography in 25 pre-HDGECs and 25 mHDGECs. They detected a broad reduction of FA and increase of RD across all corpus callosum pathways in pre-HDGECs and, with more severe AD increases in mHDGECs. Furthermore, decreased FA and increased RD correlated with disease burden, CAG repeat length, and both motor and cognitive scores [139]. These findings underline the importance of microstructural alterations in the corpus callosum white matter for the generation of HD-related symptoms. Significant clinical correlations between FA and MD alterations in the corpus callosum have also been found with performances in neuropsychological tests [248]. Importantly, increased diffusivity in the corpus callosum of pre-HDGECs has been found to correlate with 2-year and 5-year probability of phenoconversion [144]. Therefore, loss of integrity in the corpus callosum, similarly to the sensorimotor cortex, could represent a potential biomarker for the progression of HD pathology in premanifest mutation carriers.

Tractography of the corticospinal tract has revealed microstructural connectivity alternations in this tract are bilaterally affected in HD, since the premanifest stage [153, 154]. In pre-HDGECs, alterations in FA, AD, and RD of the corticospinal tract correlate with worse motor scores, assessed using the Unified Huntington's Disease Rating Scale (UHDRS) scale [154]. Rosas and colleagues found that increased diffusivity of the corticospinal tract was associated with thinning of the precentral cortex, and correlated with total motor scores [149]. These results emphasize the role of microstructural alternations in the pyramidal tracts since the premanifest stages of HD.

Microstructural white matter alterations have also been associated with presence of specific clinical features in HD. mHDGECs with high scores on depression scales, compared with those with subthreshold scores, display decreased FA in brain regions associated with major depressive disorder [155], such as the frontal cortex, the anterior cingulate, the insula, and cerebellum [156]. Moreover, the degree of FA decrease in the corpus callosum has been correlated with the intensity of depressive symptoms [152].

The progression of microstructural white matter alterations has been investigated with a number of longitudinal DTI studies [117, 146, 157–163]. While a few studies did not find any significant progression of MD and FA alterations over a follow-up period of up to 24 months [157, 158], other studies have found a significant progression of diffuse axonal injury, particularly in areas already described as affected in HD, such as the corpus callosum, frontal cortex, and striatum [146, 157, 159, 161, 163]. In mHDGECs, the rate of progression of these alterations has been shown to be relatively high. In a short longitudinal study, conducted on 13 mHDGECs, FA decrease in the corpus callosum progressed by 2.19% over 7 months [117]. In another study, in which 48 early-stage HDGECs were followed-up for 2 years, the rate of progression of changes in FA, AD, and RD was highest in the genu and body of the corpus callosum, corona radiata and internal capsule [152]. A pattern of progression can be seen also in pre-HDGECs, although with different characteristics according to their time to disease onset. Far-from onset subjects do not tend to display significant alterations [159, 162, 163], and indeed may show a decrease of white matter diffusivity [160], that has been interpreted as an early indicator of swelling, perhaps due to neuroinflammation [164], before the neurodegenerative process takes place [160]. Near-onset pre-HDGECs show faster increases of MD and decreases of FA, as evidence of the progressive neurodegenerative cascade that leads to clinical onset of disease [159, 160, 162].

Neurite Orientation Dispersion and Density Imaging (NODDI) is a recent diffusion MRI technique that studies the density of the neurite packs and their spatial distribution, intended as orientation dispersion in the brain [165]. NODDI is particularly fit to assess specific aspects of the microstructure of tissues which have a non-Gaussian distribution, whereas DTI relies on a Gaussian distribution of diffusion [249]. In 2018, Zhang and colleagues published a study in which this technique was applied to study a cohort of 38 pre-HDGECs, and 45 healthy controls. They found that pre-HDGECs displayed a reduction in the axonal density in the whole brain, as well as in the corpus callosum, bilateral superior longitudinal fasciculi, posterior limb of internal capsules, external capsules, posterior thalamic radiations, middle cerebellar peduncles, corona radiata, uncinate fasciculi, and the posterior cingulum

[166]. These findings suggest that axonal damage may play a central role in the generation of white matter pathology in the early stages of HD. Moreover, reduction of neurite density in the body and genu of the corpus callosum correlated with UHDRS total motor scores (TMS), indicating that this alteration is critical for the generation of disease-related clinical deterioration [166]. NODDI is a promising technique for the detailed microstructural study of intracellular and extracellular white matter and it is expected that more studies will be published utilizing this technique in the future.

3.3 Iron Deposition

HD, commonly with other neurodegenerative diseases, is characterized by the deposition of iron in the brain, in excess to the physiological age-related iron accumulation [250, 251]. Alterations in iron homeostasis seem to represent an early feature in HD pathophysiology, and mutant huntingtin appears to play a direct role [252]. Brain specimens from patients with advanced HD have showed an excess accumulation of iron in the striatum [253, 254]. A number of recent MRI techniques have been developed to assess and quantify the iron levels in brain tissues. Ferritin, the main storage protein of iron, appears as hyperintense in structural T1, and hypointense in T2, and Gradient Echo Sequences, or T2*, and has been used to assess regional iron overload in HDGECs [145, 167–175, 255]. More recently, variations in the magnetic susceptibility of tissues to iron have been exploited in an attempt to quantify the regional levels of iron accumulation [250]. Susceptibility Weighted Imaging (SWI) and, more recently, Quantitative Susceptibility Mapping (QSM), have been developed as iron-sensitive MRI sequences, and have been employed widely to study neurodegenerative diseases [256]. It has been demonstrated that the susceptibility values of the SWI phase images vary linearly with iron concentrations in postmortem brain tissues, especially in high-iron regions, such as the basal ganglia [250, 257]. However, phase data may be disturbed by the intrinsic geometrical characteristics of tissues, therefore hindering its accuracy and spatial resolution.

QSM directly allows the quantification of iron concentration in brain tissues. QSM gives more accurate measures of magnetic susceptibility, which has been found to correlate with local gray matter tissue iron concentration [258, 259]. Differently from SWI, that generates contrast based on phase images, QSM calculates the underlying susceptibility of each voxel as a scalar quantity. QSM uses multiple echoes to allow the detection of weaker degrees of susceptibility changes, increasing sensitivity, and overcoming the spatial resolution drawbacks encountered with SWI [7].

The first in vivo attempt to assess iron levels in HDGECs with MRI dates back more than 20 years, when Bartzokis and colleagues used field-dependent relaxation rate increase (FDRI), as a measure to quantify iron with MRI [255]. They found that 11 mHDGECs had increased iron levels in the basal ganglia compared with controls, and decreased iron levels in the white matter [255]. In a successive study that reanalyzed the same MRI scans, it was shown that the areas showing higher ferritin content were areas with early myelin breakdown, such as the pallidum, as opposed to other areas, such as the hippocampus and the thalamus, that show slightly lower ferritin levels [169]. The authors postulated that iron accumulation may follow a sequential demyelinating damage in deep brain structures. A link between iron alteration in HD and a breakdown of axonal transport has also been proposed by another study that used T2 relaxation time as a surrogate measure to quantify ferritin levels in mHDGECs [167]. A few successive studies have therefore sought to investigate the possible relationship between white matter damage and iron accumulation in HD. In mHDGECs, the alteration in T2* relaxation rate in the pallidum was associated with increased FA, and correlated with the size of CAG repeat expansion, although not with measures of disease severity [145]. Although indirect signs of iron accumulation in the basal ganglia, and in particular the pallidum are detectable from the premanifest stage [140], myelin breakdown in pre-HDGECs seems to precede iron deposition [170, 173]. In mHDGECs, measures of iron accumulation such as alterations of T2 relaxation rate, as well as magnetic field inhomogeneities have been associated with disease characteristics such as disease burden score, CAG repeat length, and increase of UHDRS motor scores [144, 167, 170]. In pre-HDGECs, it has been shown that T2 hypointensity in the basal ganglia correlates with UHDRS-TMS, longer CAG length, and greater probability of developing symptoms at 5 years [170]. More recently, studies using MR phase imaging have allowed to estimate a quantitative measure which reflects iron content. In a study using ultrahigh-field 7 T MR phase imaging, 13 pre-HDGECs showed increased field shift in the caudate, which correlated with CAG length, and with a trending, nonsignificant correlation with cognitive scores for executive and working memory functions [168]. Rosas and colleagues used SWI phase imaging to assess regional iron accumulation in both pre-HDGECs and mHDGECs [171]. Far-from onset pre-HDGECs showed bilateral increases in field map values in the putamen and pallidum, followed, as diseases stage progressed, by increases in the caudate nucleus, the occipital cortex. In late mHDGECs, increases in field map values were observed in the left precentral and postcentral cortex, the anterior cingulate, and the precuneus. Higher field map values correlated with CAG repeat length and measures of disease severity [171]. More recent cross-sectional studies have used QSM as a quantitative measure to assess iron levels in HDGECs. In these studies, it was shown that an increase of susceptibility values was evident in the caudate, putamen, and pallidum, since the premanifest stage, and significantly associated with volumetric measures of striatal atrophy, the CAG-Age Product score (CAP score), a measure of disease burden [260], and disease severity [172, 175]. One study has also found, in pre-HDGECs, a decrease of iron levels in the hippocampus and the substantia nigra compared with controls, which has been interpreted as a local redistribution of brain iron after the increase of oligodendrocyte density following initial tissue damage [172].

3.4 Functional MRI   Functional MRI (fMRI) is a technique used to investigate complex patterns of brain activation at rest (resting-state fMRI) or while performing a task (task-based fMRI). The assumption underlying fMRI is that an increase in local cellular activity is accompanied by an increase in cerebral blood flow and, on the contrary, local cellular inactivation is characterized by a decrease in regional blood flow. The measure of changes in blood flow (Blood-Oxygen Level-Dependent, BOLD) is thus used as an indirect measure of brain activation or inactivation. BOLD imaging is based on the differential interaction of oxygenated (diamagnetic) and deoxygenated (paramagnetic) blood with the magnetic field. Changes in BOLD signal are measured as a function of time; therefore, fast acquisition techniques are required, such as rapid echo planar imaging (EPI) image acquisition in which multiple echoes are acquired using either gradient echo, spatial echo, or both. The gradients of the spatial magnetic fields help in the localization of the signals. Changes in neuronal activity (fast) and changes in vascular activity (slow) are then temporally coupled to ensure an accurate localization of changes in neuronal activity [199]. Resting-state fMRI studies patterns of temporally correlated functional connectivity between distant brain regions which form specific networks that have been associated with a number of specific functions. Methods of analysis include seed-based connectivity (which requires a priori region(s) of interest as the starting point for the analysis), independent component analysis (ICA), principal component analysis (PCA), clustering, and whole-brain functional connectomics (which does not need a priori region(s) of interest). Effective connectivity (i.e., causality between connections) can be inferred with Dynamic Causal Modeling analysis [261]. Complex networks can be represented according to the graph theory, that represents the brain as an intricated web of nodes and edges, where the nodes are the brain regions and the edges are the connections. A number of functional associations can be calculated according to the functional or effective connectivity [262].

Task-based fMRI studies changes in the spatial distribution of brain activity according to particular motor, cognitive, or behavioral tasks. Task-based fMRI relies on two fundamental types of tests, known as blocked or event-related designs. In blocked designs, stimuli are presented continuously for long intervals of time (blocks). Blocked tests are advantageous due to their robustness and consistency of the design, which can yield large BOLD signal changes [263]. In event-related designs, discrete conditions of short durations are presented to the subject, at different inter-stimulus-intervals. Despite having less statistical power, compared with blocked designs, event-related designs are useful to estimate the hemodynamic impulse response and to minimize the risk of a subject's habituation to the motor or cognitive stimulus presented [264].

Changes in brain activity in response to motor tasks, can be studied with paced tapping tasks. In a first study, it was found that pre-HDGECs tended to activate more the left caudal supplementary motor area in all experimental conditions, whereas the activation of the left superior parietal lobe tended to decrease with more demanding tasks [265]. By contrast, in a large cohort of pre-HDGECs from the TRACK-ON HD study, no activation changes were detected compared with controls [176]. The lack of detectable changes in pre-HDGECs observed in the latter study could be due to a difference in the level of difficulty posed by the motor task.

A number of fMRI studies have been carried out on HDGECs using tasks to test different domains of cognitive function. Working memory has been tested with a number of verbal tasks, including the so-called N-BACK paradigm [177]. Deficits in working memory are consistently present since the premanifest stage of HD [266, 267]. Despite not having any degree of atrophy, pre-HDGECs show less activation of the left dorsolateral prefrontal cortex (DLPFC), the inferior frontal gyrus, anterior insula, and striatum, which was proportional to the difficulty of the task [178, 179]. Longitudinal observation showed a change over time, in pre-HDGECs, of connectivity between the left caudate and the DLPFC in the absence of any volumetric progression of atrophy, which correlated with disease burden and estimated years to disease onset [180, 181]. The inactivation of the DLPFC was correlated with lower scores on the UHDRS cognitive scale, and with presence of neuropsychiatric symptoms, such as obsessive-compulsive behavior, and depression, especially at higher working memory loads and in near-onset pre-HDGECs [178, 179, 182, 183].

Several fMRI studies have investigated the functional alterations underpinning attentional deficits, a constant feature in HD, that can be detected since the premanifest stage [268]. The presence of this alteration can negatively influence other high order cognitive functions, such as working memory, and executive functions. In a first attentional study, a group of 18 pre-HDGECs showed a decrease of connectivity between motor regions and the putamen, and an increase of the Default Motor Network (DMN), more pronounced the closer they were to phenoconversion

[184]. In a 30-month longitudinal study on mHDGECs, they displayed a progressively increased activation of the striatum, and a decreased activation of the anterior cingulate cortex [185]. Alterations of the anterior cingulate connectivity have also been observed in pre-HDGECs after interference tasks [186]. In particular, pre-HDGECs displayed an aberrant activation of a complex network involving the anterior cingulate and the insula, and other frontal and parietal regions [269]. These findings indicate a general state of cortical hyperactivation and highlight the presence of dysfunction in cortical regions involved in attention tasks since the premanifest stage of disease.

The functional disruption of the corticostriatal circuitry in HD has also been studied using a time discrimination task paradigm. This task requires the subject to judge whether the duration of a given time interval is longer or shorter than a reference one [270]. The corticostriatal circuitry has been shown to be altered in mHDGECs and in near-onset pre-HDGECs using this technique [187]. Far-from onset pre-HDGECs show a complex pattern of activation of the anterior cingulate, the thalami, and pre-supplementary motor area function, with a reduction of neural activation in the right anterior cingulate and the right anterior insula; those closer to predicted onset show an hypoactivation of the caudate and thalamus [187–189]. These findings suggest the possibility of an initial compensatory mechanisms that fails when nearing time to onset, and highlight, once more, the complexity and unique upregulatory and downregulatory mechanisms and their role during the development of HD.

Early-stage HDGECs also show subtle degrees of executive function impairment. In a study, the Tower of London task, a paradigm to assess executive function, has been coupled with fMRI in pre-HDGECs and early-stage mHDGECs [190]. Both HD groups showed reduced activation of the left premotor cortex, which correlated with the difficulty of the executive task [190], therefore highlighting the medial prefrontal area as a critical region for the generation of executive dysfunction in HDGECs.

Behavioral disturbance is a typical early feature of HD. Recognition and processing of emotions can be altered in premanifest stages of the disease and constitutes a distinct characteristic from healthy controls [102]. A number of fMRI studies coupling imaging with tasks capable of recognizing or inducing emotions have been performed, especially in pre-HDGECs, to understand functional brain areas most affected. It has been found that tasks aimed at both recognizing and processing emotions show altered connection in critical regions of the limbic system, such as the insula, the amygdala, the cingulate cortex, the orbitofrontal cortex, as well as subcortical structures such as the putamen and the pulvinar [191–193, 265]. Novak and colleagues found, in a group of 16 pre-HDGECs with no detectable alteration in emotion recognition, segregated functional circuits associated with the processing of disgust and happiness emotions [194]. These networks were identified at a group level and after controlling for CAG repeat length and probability of onset at 5 years [194]. Therefore, these results provide evidence for phenotypical variations of HD across single subjects.

Resting-state fMRI studies in HD have surged over the last few years and have significantly helped in identifying signatures of disease since the early stages, when volumetric atrophy of cortical and subcortical regions could not be identified yet. Studies investigating brain function at rest have mainly used seed-to-seed or an independent component analysis (ICA) approach, whereas functional connectomic studies have emerged in more recent years.

The DMN has been studied in a few studies involving both mHDGECs and pre-HDGECs. Two studies have used DMN regions as seeds in cohorts of mHDGECs. An aberrant connection was detected within DMN nuclei that did not show any volumetric atrophy yet, and correlated with performances on the Stroop test [195]; another study showed aberrant increased connectivity between the cingulate cortex and the supplementary motor area, as seed for the somatosensory network, in the absence of relevant atrophy in the DMN. Interestingly, the extent of this aberrant connection was reduced by treatment with an iron-lowering drug, suggesting that this functional alteration could represent a potential biomarker [196]. Connectivity of the motor network has been explored by two studies on mHDGECs. A complex pattern of reduced connectivity within the basal ganglia, between the primary motor cortex and the insula, between the basal ganglia and the insula, and between the cerebellum and the paracentral gyrus have emerged from these studies [197, 198]. Alterations of connectivity between the cerebellum and the paracentral gyrus, and between the motor cortex and the insula correlated with clinical motor scores.

Alterations in functional connectivity between cortical regions has been also investigated using specific whole-brain analysis methods, such as ICA and amplitude of low frequency fluctuations (ALFF). ALFF is an analysis method that evaluates the intrinsic regional brain differences of BOLD signal fluctuation, rather than the differences in region-to-region synchrony. Overall, these studies evidenced two main patterns of functional connectivity alterations. A first pattern consists in a reduced connectivity starting from the visual cortex mHDGECs and pre-HDGECs consistently show an alteration of connectivity in posterior regions [200–203]. Alterations have been detected within the visual cortex [201], as well as between the visual cortex and the frontal regions, which declines with increasing size of CAG repeats [200, 202]. Liu and colleagues also found a reduction of ALFF in posterior areas, namely, the right precuneus and angular gyrus of mHDGECs, which correlated with poor performance on cognitive measures [203].

The other main regions of reduced connectivity are frontoparietal and corticosubcortical connections [202, 204–206]. mHDGECs show alterations of the connectivity of the frontal regions, and between frontal regions and the putamen [202, 204], which correlate with worsening motor and cognitive performances [204, 205]. On the other hand, these patients also show an increase of connections between the supplementary motor area and the inferior and middle frontal gyri, and between motor and parietal cortices, which again correlated with motor impairment [205, 206]. Therefore, these highlight once more the profound levels of connectivity alterations typical of HD.

One longitudinal study is available in the literature for resting-state fMRI. In this study, 22 pre-HDGECs and 17 controls were followed-up for 3 years and analysed with ICA [207]. However, even when divided into far-from onset and near-onset, no connectivity changes were detected over time in pre-HDGECs. These findings suggest that resting-state fMRI may lack of sensitivity to track changes over time in pre-HDGECs.

Whole brain connectivity in HD has also been studied with connectomics, which assumes the brain as composed by a complex network of myriads of nodes (brain nuclei) and edges (white matter connections). The study of the density, functional interaction, and richness of these connections through network science or graph theory allows a detailed map of the intrinsic functional connectivity of the brain. Harrington and colleagues found that pre-HDGECs, as they neared time to predicted onset, and increased disease burden, displayed weakened frontostriatal connections and strengthened frontal-posterior connections, despite the presence of preserved global functional segregation [208]. Another study using graph theory analyzed 182 brain regions in 24 pre-HDGECs and 18 mHDGECs [209]. Pre-HDGECs showed a reduction of hubs in the somatosensory and associative networks; mHDGECs showed additional reductions of path length (an indicator of network randomization), clustering (an indicator of fewer communications between neighboring nodes), and betweenness centrality (which indicates less robustness of the network). These changes in graph theory measures correlated variably with clinical and cognitive measures but, on longitudinal observation, did not show any relevant change [209]. Pre-HDGECs also seem to show an inverse correlation between structural connectivity as measured with DTI, and fMRI functional connectivity which could be explained by compensatory changes to microstructural and functional early alterations taking place in pre-HDGEC [271].

There are two studies that have coupled resting-state and task-based fMRI in HD [176, 272]. In one study, it was found that, as structural atrophy increased in a group of pre-HDGECs, compensatory changes involving functional activation of the right parietal cortex, right dorsolateral prefrontal cortex, and the left hemisphere network occurred during the working memory task. These compensatory phenomena were not detectable in the left hemisphere, and were interpreted as a sign of higher vulnerability of this brain area in pre-HDGECs [176]. A 3-year follow-up of this cohort showed that a maintained global cognition associated with increased effective connectivity between the left and right dorsolateral prefrontal cortex, whereas maintained motor performance was associated with increased connectivity between left and right premotor cortex [272], therefore providing further direct evidence for functional compensatory mechanisms taking place in the premanifest stage of HD.

3.5 Magnetic Resonance Spectroscopy

The principle of magnetic resonance spectroscopy (MRS) resides in the changes that the external magnetic field exerts on protons (^{1}H), as well as other nuclei such as ^{13}C and ^{31}P. Nuclei of the same species, such as ^{1}H in a molecule, can have slightly different resonance frequencies, due to the fact that they are surrounded by electrons, which exert a shielding effect on the magnetic field. The shielding effect of electrons causes a shift of the resonance frequency, that is dependent on the chemical environment and is different according to the organic compound the electrons belong to [273]. Every compound has its own resonance pattern that can be measured, and is an indicator of their concentrations in a certain region of interest. MRS needs larger detection volumes compared to MRI, given the small concentration of metabolites in the tissues. Although the most common MRS modality is ^{1}H MRS, that detects the concentration of hydrogen-containing metabolites, and permits the quantification of a wide range of markers of cellular damage, other modalities have also been used in HD research, such as ^{13}C MRS, that allows the quantification of molecules involved in cellular metabolism, and ^{31}P MRS, that studies the concentration of phosphorylated metabolites [274, 275]. The brain regions most studied with MRS in HD research are the striatum, the frontal cortex, and the occipital cortex. MRS studies in HD generally yield heterogenous results, with large standard deviations and wide variability even in selected populations [274]. This could be explained by the low sample size in the majority of the studies conducted. Nevertheless, some precious results indicative of metabolic changes have been obtained with the use of this MR technique.

3.5.1 Measure of Neuronal Damage

N-acetylaspartate (NAA), found exclusively in the brain, is synthesized by the neuronal mitochondria and, although its exact function is still unclear, it is considered a marker of neuronal function. Because NAA synthesis depends on mitochondrial metabolism, NAA has also been considered a marker of mitochondrial stress [276]. There is a large volume of literature that has established that, in HDGECs, the brain levels of NAA are significantly decreased

[210–221], with only sparse exceptions [222]. Significant decreases of NAA in mHDGECs have been detected in the caudate [210, 218, 219] and putamen [213, 215, 216, 219], but also in other brain regions such as the occipital cortex [219], the thalami [214], and the posterior cingulate [217]. The entity of regional NAA decrease compared with controls is around 17–60% [212, 215], although this high figure has been obtained in a group of more advanced mHDGECs. Correlation studies have demonstrated that loss of striatal NAA was associated with age and CAG repeat length [210], or with volumetric measures [218], suggesting that overall, the reduction of NAA may represent a marker of unspecific, although widespread, neuronal stress in this patient population. ^{1}H MRS studies have also detected regional decreases of NAA in the striata of the pre-HDGECs population [210, 212, 213, 215, 220]. Sanchez-Pernaute and colleagues found, in a small group of 6 pre-HDGECs, a decrease of up to 30% of NAA compared to controls, indicating that severe neuronal loss can occur early in the disease course [212]. Decreases of NAA have been also detected, in the caudate and putamen of the same population [210, 215]. However, in one study conducted on 19 pre-HDGECs, a nonsignificant decrease of putaminal and thalamic NAA was reported compared to controls [222]. In that study, however, to avoid partial volume effect, the researchers used a 2 cm^3 MRS voxel size, which was smaller than those used in other studies, and that may have decreased the spectral signal-to-noise ratio [222].

Two 24-month longitudinal studies have assessed the trajectory of NAA decrease in pre-HDGECs and mHDGECs [221, 223]. In the first study, on a total group of 13 HDGECs, it was found that NAA progressively decreased in the putamen of pre-HDGECs, and in the putamen and caudate of mHDGECs [223]. In contrast, another study on a larger population did not spot any longitudinal changes of NAA in both pre-HDGECs and mHDGECs [221].

3.5.2 Measure of Glial Damage with MRS

Myoinositol (Myo) is produced exclusively in glial cells, particularly astrocytes, and does not cross the blood–brain barrier, and is used as marker of glial damage [277]. The levels of Myo are generally increased in HD [215, 221, 224]. In one large study involving 25 pre-HDGECs and 30 mHDGECs, the putaminal levels of Myo in the mHDGECs have been found to be increased by as much as 50%, compared with pre-HDGECs [215]. Moreover, the putaminal levels of Myo and the Myo–tNAA ratio correlated, in the population of mHDGECs, with UHDRS motor scores. These results suggest that regional glial damage may represent a biological signature of phenoconversion. However, a more recent cross-sectional study, using a 3 T MRS with a semilocalized by adiabatic selective refocusing (semi-LASER) sequence that allows the analysis of a wide number of metabolites, did not find significant changes in the striatal and occipital amount of Myo of ten mHDGECs [225]. Further studies with high field MR and short echo are warranted to further investigate this metabolite in view of the increasing importance that glial dysfunction is getting in HD.

3.5.3 Integrity of Membranes

Choline (Cho) is a metabolic marker of membrane integrity and is measured either as standalone measure, or in relationship with other metabolites (NAA or creatine). The levels of Cho (or of Cho–Cr ratio) in HD have been found to be increased in a number of early studies [210, 211, 220, 224], although there are reports of unchanged levels of striatal Cho [225], and of decreased Cho levels in both pre-HDGEC and mHDGEC subjects [219, 226]. In those two latter studies, the decrease of Cho in the frontal cortex of pre-HDGECs correlated with worse scores on visuomotor tasks [226]. In the other study, mHDGECs showed significantly lower levels of Cho in the caudate compared with pre-HDGECs, suggesting that progression of the disease may be accompanied by a proportional decrease of Cho levels [219]. The latter assumption is also corroborated by the finding of a longitudinal loss of Cho in the putamen of mHDGECs [223].

3.5.4 Energy Metabolism

Two metabolites, Creatine (Cr) and Lactate, have been used to study the levels of energy metabolism in HD. In research, Cr is generally used as an absolute measure or to generate ratios (with NAA and Cho) given its stability in physiological settings. The use of ^{1}H MRS and ^{31}P MRS allows the differentiation, and the separate study, of Cr and phosphocreatine (PCr). Since Cr is not only generated in the brain, its levels may be influenced by concomitant systemic diseases [278]. Lactate is present in minimal quantities in the healthy brain, but increases when the aerobic oxidation mechanisms fails and anaerobic glycolysis takes over [279]. HD is characterized by a severe state of energetic metabolism failure. Several studies have found a decrease of Cr in the caudate and putamen of both pre-HDGECs and mHDGECs [210, 212, 213, 216, 219] and in seven patients with juvenile HD [227]. A decrease of Cr has also been found, in pre-HDGECs and mHDGECs, in the occipital cortex [219]. In a group of pre-HDGECs the decrease of Cr, expressed as higher NAA/Cr and Cho/Cr ratios, correlated with disease progression [210]. In another study, a decrease in Cr levels was associated with poor performances on cognitive tests and CAG length [212]. Longitudinal tracking of the levels of Cr in the caudate have also demonstrated a progressive loss in mHDGECs over time [223]. For these reasons, Cr decrease has been hypothesized to represent a marker of disease progression in HD. Two studies have found increases in Cr levels in extrastriatal areas, particularly in the gray matter [224] and in the visual cortex [225]. Regional concentrations of lactate are consistently increased in mHDGECs and pre-HDGECs in the striatum, frontal, and occipital cortex [210, 220, 228], providing further evidences of aberrant glucose metabolism in HD. Furthermore, the extent of lactate increase correlated with longer disease duration [220].

^{31}P MRS studies can also measure the levels of phosphocreatine (PCr), for example in response to activating tasks [231, 232]. However, HD patients fail to increase the levels of PCr (as inorganic phosphate/PCr ratio) in response to an occipital lobe activating task, compared with controls [231].

3.5.5 Excitotoxicity

Glutamate (Glu) is the main excitatory neurotransmitter in the central nervous system and is used in MRS to estimate the levels of regional excitotoxicity, as well as to quantify the levels of the neurotransmitter in the brain. Early studies used methods which struggled to distinguish the Glu chemical shift from the ones of metabolites with similar spectral properties, such as glutamine and GABA, and quantified these three molecules together as the Glu/Cr ratio. More recent studies employ for sophisticated methods that allow the distinction between these molecules [280]. The Glu/Cr ratio has been found to be consistently elevated in small cohorts of mHDGECs in the striatum [229, 230] and, in one single case, in the thalamus [229]. More recently, increases of Glu were detected in the frontal lobe of mHDGECs [218], as well as in the putamen of pre-HDGECs [213] and in patients with the juvenile form of HD [227]. All these findings suggest that severe excitotoxic damage is taking place diffusely across all stages of HD pathology. Two studies, however, conducted in small cohorts of early mHDGECs, have found a decrease of Glu levels in the putamen and posterior cingulate cortex [216, 217]. In one study, a significant 10.1% decrease of Glu in the posterior cingulate cortex was associated with cognitive impairment [217]. This finding has been interpreted as the effect of a net decrease of glutamatergic neurons in a crucial region for cognitive performances in HD and as a possible signature of disease worsening in mHDGECs [217].

4 Conclusions

Lessons from HD, as well as other neurodegenerative diseases, suggest that neurodegeneration is a complex and lengthy process, with several confounders. There is still a long way to go to understand distinctive tracts that could help identify when and how neurodegeneration begins. Despite significant efforts from the research community, it is difficult to extract definitive findings and to identify a clear biomarker that could explain disease

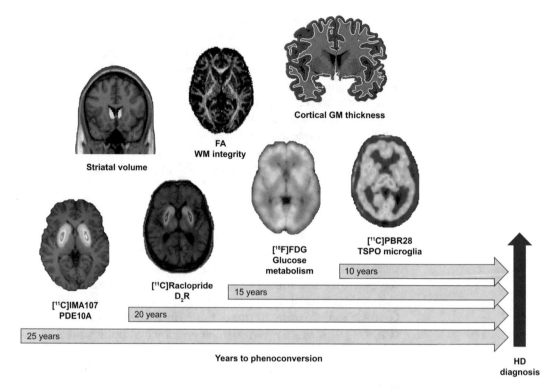

Fig. 1 Schematic representation of the changes in magnetic resonance imaging (MRI), and positron emission tomography (PET) biomarkers in premanifest Huntington's disease (HD) expansion gene carriers (HDGECs). In MRI, structural volumetric changes of the striatum are detected first, followed by changes of fractional anisotropy (FA), indicating white matter (WM) microstructural integrity loss, and by gray matter (GM) loss of cortical thickness (the blue arrows in the image between the yellow line delineating the WM surface, and the red line delineating the pial surface). In PET, loss of phosphodiesterase 10 (PDE10A) concentration in the striatum, as measured with [^{11}C]IMA107, has been spotted in pre-HDGECs as early as 25 years before predicted phenoconversion and represents so far the earliest abnormality in HD pathology identified with neuroimaging. This is followed, in the striatum, by loss of Dopamine 2 receptors (D2R), measured with [^{11}C] Raclopride, and loss of brain glucose metabolism, measured with [^{18}F]FDG. Widespread neuroinflammation, seen with the microglial translocator protein (TSPO) [^{11}C]PBR28, is seen around 10 years before predicted phenoconversion

pathophysiology, progression, and severity. In an era where effort is aimed at the implementation of preventive strategies, including disease-modifying clinical trials, the search of a reliable biomarker is a major challenge in HD (Fig. 1). The contribution by which neuroimaging could help in this quest is threefold. On one side, neuroimaging has helped to gain incredible knowledge about the pathological events underpinning HD in vivo, at premanifest and manifest stage. Molecular and microstructural alterations take place in premanifest HDGECs at a time where no macroscopic and clinical alteration could be detected, and progress in a nonlinear fashion over time. PET tracer studies in these populations have yielded high effect sizes and some of them possess qualities to represent potential ideal biomarkers to be translated into clinical trials with acceptable sample sizes. More studies, including well-designed longitudinal studies are needed to better characterize whether these biomarkers could explain disease progression and be modulated by external intervention. Secondly, neuroimaging is essential to reveal novel pathophysiological mechanisms of disease. PET and MRI techniques are being refined to study microscopic alterations of the brain to an unprecedented level. This would help us understand better the direct causes of disease in living humans and identify novel targets for therapeutic intervention. Finally, for the success of research in HD, a collaborative, multidisciplinary effort is being increasingly required and neuroimaging can play a major contribution in the coupling of anatomical/functional alterations with clinical symptoms and other clinical, digital, or fluid biomarkers, to have a comprehensive picture of HD across all stages.

References

1. Ross CA, Tabrizi SJ (2011) Huntington's disease: from molecular pathogenesis to clinical treatment. Lancet Neurol 10:83–98

2. Chao TK, Hu J, Pringsheim T (2017) Risk factors for the onset and progression of Huntington disease. Neurotoxicology 61:79–99. https://doi.org/10.1016/j.neuro.2017.01.005

3. Halliday GM, McRitchie DA, Macdonald V et al (1998) Regional specificity of brain atrophy in Huntington's disease. Exp Neurol 154:663–672. https://doi.org/10.1006/exnr.1998.6919

4. Andrew SE, Goldberg YP, Kremer B et al (1993) The relationship between trinucleotide (CAG) repeat length and clinical features of Huntington's disease. Nat Genet 4:398–403. https://doi.org/10.1038/ng0893-398

5. Keum JW, Shin A, Gillis T et al (2016) The HTT CAG-expansion mutation determines age at death but not disease duration in Huntington disease. Am J Hum Genet 98:287–298. https://doi.org/10.1016/j.ajhg.2015.12.018

6. Langbehn DR, Hayden MR, Paulsen JS et al (2010) CAG-repeat length and the age of onset in Huntington Disease (HD): a review and validation study of statistical approaches. Am J Med Genet B Neuropsychiatr Genet 153:397–408

7. Wilson H, Dervenoulas G, Politis M (2018) Structural magnetic resonance imaging in Huntington's disease. In: International review of neurobiology. Academic Press Inc., New York, NY, pp 335–380

8. Wilson H, Politis M (2018) Molecular imaging in Huntington's disease. In: International review of neurobiology. Academic Press Inc., New York, NY, pp 289–333

9. Wilson H, De Micco R, Niccolini F, Politis M (2017) Molecular imaging markers to track Huntington's disease pathology. Front Neurol 8:11

10. Rocchi L, Niccolini F, Politis M (2015) Recent imaging advances in neurology. J Neurol 262:2182–2194. https://doi.org/10.1007/s00415-015-7711-x

11. Strafella AP, Bohnen NI, Pavese N et al (2018) Imaging markers of progression in Parkinson's disease. Mov Disord Clin Pract 5:586–596

12. Politis M, Pagano G, Niccolini F (2017) Imaging in Parkinson's disease. Int Rev Neurobiol 132:233–274. https://doi.org/10.1016/bs.irn.2017.02.015

13. Wilson H, Pagano G, Politis M (2019) Dementia spectrum disorders: lessons learnt from decades with PET research. J Neural Transm 126:233–251

14. Kuhl DE, Phelps ME, Markham CH et al (1982) Cerebral metabolism and atrophy in Huntington's disease determined by18FDG and computed tomographic scan. Ann Neurol 12:425–434. https://doi.org/10.1002/ana.410120504

15. Young AB, Penney JB, Starosta-Rubinstein S et al (1986) PET scan investigations of

Huntington's disease: cerebral metabolic correlates of neurological features and functional decline. Ann Neurol 20:296–303. https://doi.org/10.1002/ana.410200305

16. Berent S, Giordani B, Lehtinen S et al (1988) Positron emission tomographic scan investigations of Huntington's disease: cerebral metabolic correlates of cognitive function. Ann Neurol 23:541–546. https://doi.org/10.1002/ana.410230603

17. Hayden MR, Hewitt J, Stoessl AJ et al (1987) The combined use of positron emission tomography and DNA polymorphisms for preclinical detection of Huntington's disease. Neurology 37:1441–1447. https://doi.org/10.1212/wnl.37.9.1441

18. Mazziotta JC, Phelps ME, Pahl JJ et al (1987) Reduced cerebral glucose metabolism in asymptomatic subjects at risk for Huntington's disease. N Engl J Med 316:357–362. https://doi.org/10.1056/NEJM198702123160701

19. Kuwert T, Lange HW, Langen KJ et al (1990) Cortical and subcortical glucose consumption measured by PET in patients with Huntington's disease. Brain 113(Pt 5):1405–1423. https://doi.org/10.1093/brain/113.5.1405

20. Garnett ES, Firnau G, Nahmias C et al (1984) Reduced striatal glucose consumption and prolonged reaction time are early features in Huntington's disease. J Neurol Sci 65:231–237. https://doi.org/10.1016/0022-510x(84)90087-x

21. Leenders KL, Frackowiak RSJ, Quinn N, Marsden CD (1986) Brain energy metabolism and dopaminergic function in Huntington's disease measured in vivo using positron emission tomography. Mov Disord 1:69–77. https://doi.org/10.1002/mds.870010110

22. Ciarmiello A, Cannella M, Lastoria S et al (2006) Brain white-matter volume loss and glucose hypometabolism precede the clinical symptoms of Huntington's disease. J Nucl Med 47:215–222

23. López-Mora DA, Camacho V, Pérez-Pérez J et al (2016) Striatal hypometabolism in premanifest and manifest Huntington's disease patients. Eur J Nucl Med Mol Imaging 43:2183–2189. https://doi.org/10.1007/s00259-016-3445-y

24. Herben-Dekker M, Van Oostrom JCH, Roos RAC et al (2014) Striatal metabolism and psychomotor speed as predictors of motor onset in Huntington's disease. J Neurol 261:1387–1397. https://doi.org/10.1007/s00415-014-7350-7

25. Ciarmiello A, Giovacchini G, Orobello S et al (2012) 18F-FDG PET uptake in the pre-Huntington disease caudate affects the time-to-onset independently of CAG expansion size. Eur J Nucl Med Mol Imaging 39:1030–1036. https://doi.org/10.1007/s00259-012-2114-z

26. Gaura V, Lavisse S, Payoux P et al (2017) Association between motor symptoms and brain metabolism in early Huntington disease. JAMA Neurol 74:1088–1096. https://doi.org/10.1001/jamaneurol.2017.1200

27. Martínez-Horta S, Perez-Perez J, Sampedro F et al (2018) Structural and metabolic brain correlates of apathy in Huntington's disease. Mov Disord 33:1151–1159. https://doi.org/10.1002/mds.27395

28. Feigin A, Leenders KL, Moeller JR et al (2001) Metabolic network abnormalities in early Huntington's disease: an [18F]FDG PET study. J Nucl Med 42:1591–1595

29. Feigin A, Tang C, Ma Y et al (2007) Thalamic metabolism and symptom onset in preclinical Huntington's disease. Brain 130:2858–2867. https://doi.org/10.1093/brain/awm217

30. Tang CC, Feigin A, Ma Y et al (2013) Metabolic network as a progression biomarker of premanifest Huntington's disease. J Clin Invest 123:4076–4088. https://doi.org/10.1172/JCI69411

31. van Oostrom JCH, Maguire RP, Verschuuren-Bemelmans CC et al (2005) Striatal dopamine D2 receptors, metabolism, and volume in preclinical Huntington disease. Neurology 65:941–943. https://doi.org/10.1212/01.wnl.0000176071.08694.cc

32. Antonini A, Leenders KL, Spiegel R et al (1996) Striatal glucose metabolism and dopamine D2 receptor binding in asymptomatic gene carriers and patients with Huntington's disease. Brain 119(Pt 6):2085–2095. https://doi.org/10.1093/brain/119.6.2085

33. Bartenstein P, Weindl A, Spiegel S et al (1997) Central motor processing in Huntington's disease. A PET study. Brain 120:1553–1567. https://doi.org/10.1093/brain/120.9.1553

34. Weeks RA, Ceballos-Baumann A, Piccini P et al (1997) Cortical control of movement in Huntington's disease. A PET activation study. Brain 120(Pt 9):1569–1578. https://doi.org/10.1093/brain/120.9.1569

35. Lepron E, Péran P, Cardebat D, Démonet J-F (2009) A PET study of word generation in Huntington's disease: effects of lexical competition and verb/noun category. Brain Lang

110:49–60.   https://doi.org/10.1016/j.bandl.2009.05.004

36. Sedvall G, Karlsson P, Lundin A et al (1994) Dopamine D1 receptor number--a sensitive PET marker for early brain degeneration in Huntington's disease. Eur Arch Psychiatry Clin Neurosci 243:249–255. https://doi.org/10.1007/BF02191583

37. Turjanski N, Weeks R, Dolan R et al (1995) Striatal D1 and D2 receptor binding in patients with Huntington's disease and other choreas. A PET study. Brain 118 (Pt 3):689–696. https://doi.org/10.1093/brain/118.3.689

38. Ginovart N, Lundin A, Farde L et al (1997) PET study of the pre- and post-synaptic dopaminergic markers for the neurodegenerative process in Huntington's disease. Brain 120 (Pt 3):503–514. https://doi.org/10.1093/brain/120.3.503

39. Bäckman L, Robins-Wahlin TB, Lundin A et al (1997) Cognitive deficits in Huntington's disease are predicted by dopaminergic PET markers and brain volumes. Brain 120 (Pt 1):2207–2217. https://doi.org/10.1093/brain/120.12.2207

40. Lawrence AD, Weeks RA, Brooks DJ et al (1998) The relationship between striatal dopamine receptor binding and cognitive performance in Huntington's disease. Brain 121 (Pt 7):1343–1355. https://doi.org/10.1093/brain/121.7.1343

41. Weeks RA, Piccini P, Harding AE, Brooks DJ (1996) Striatal D1 and D2 dopamine receptor loss in asymptomatic mutation carriers of Huntington's disease. Ann Neurol 40:49–54. https://doi.org/10.1002/ana.410400110

42. Politis M, Pavese N, Tai YF et al (2008) Hypothalamic involvement in Huntington's disease: an in vivo PET study. Brain 131:2860–2869. https://doi.org/10.1093/brain/awn244

43. Pavese N, Politis M, Tai YF et al (2010) Cortical dopamine dysfunction in symptomatic and premanifest Huntington's disease gene carriers. Neurobiol Dis 37:356–361. https://doi.org/10.1016/j.nbd.2009.10.015

44. Antonini A, Leenders KL, Eidelberg D (1998) [11C]raclopride-PET studies of the Huntington's disease rate of progression: relevance of the trinucleotide repeat length. Ann Neurol 43:253–255. https://doi.org/10.1002/ana.410430216

45. Andrews TC, Weeks RA, Turjanski N et al (1999) Huntington's disease progression. PET and clinical observations. Brain 122 (Pt 1):2353–2363. https://doi.org/10.1093/brain/122.12.2353

46. van Oostrom JCH, Dekker M, Willemsen ATM et al (2009) Changes in striatal dopamine D2 receptor binding in pre-clinical Huntington's disease. Eur J Neurol 16:226–231. https://doi.org/10.1111/j.1468-1331.2008.02390.x

47. Esmaeilzadeh M, Farde L, Karlsson P et al (2011) Extrastriatal dopamine D(2) receptor binding in Huntington's disease. Hum Brain Mapp 32:1626–1636. https://doi.org/10.1002/hbm.21134

48. Bohnen NI, Koeppe RA, Meyer P et al (2000) Decreased striatal monoaminergic terminals in Huntington disease. Neurology 54:1753–1759. https://doi.org/10.1212/wnl.54.9.1753

49. Pavese N, Gerhard A, Tai YF et al (2006) Microglial activation correlates with severity in Huntington disease: a clinical and PET study. Neurology 66:1638–1643. https://doi.org/10.1212/01.wnl.0000222734.56412.17

50. Tai YF, Pavese N, Gerhard A et al (2007) Imaging microglial activation in Huntington's disease. Brain Res Bull 72:148–151. https://doi.org/10.1016/j.brainresbull.2006.10.029

51. Politis M, Pavese N, Tai YF et al (2011) Microglial activation in regions related to cognitive function predicts disease onset in Huntington's disease: a multimodal imaging study. Hum Brain Mapp 32:258–270. https://doi.org/10.1002/hbm.21008

52. Politis M, Lahiri N, Niccolini F et al (2015) Increased central microglial activation associated with peripheral cytokine levels in premanifest Huntington's disease gene carriers. Neurobiol Dis 83:115–121. https://doi.org/10.1016/j.nbd.2015.08.011

53. Lois C, González I, Izquierdo-García D et al (2018) Neuroinflammation in Huntington's disease: new insights with 11C-PBR28 PET/MRI. ACS Chem Neurosci 9:2563–2571. https://doi.org/10.1021/acschemneuro.8b00072

54. Ahmad R, Bourgeois S, Postnov A et al (2014) PET imaging shows loss of striatal PDE10A in patients with Huntington disease. Neurology 82:279–281. https://doi.org/10.1212/WNL.0000000000000037

55. Russell DS, Barret O, Jennings DL et al (2014) The phosphodiesterase 10 positron emission tomography tracer, [18F]MNI-659, as a novel biomarker for early

Huntington disease. JAMA Neurol 71:1520–1528. https://doi.org/10.1001/jamaneurol.2014.1954

56. Fazio P, Fitzer-Attas CJ, Mrzljak L et al (2020) PET molecular imaging of phosphodiesterase 10A: an early biomarker of Huntington's disease progression. Mov Disord 35:606. https://doi.org/10.1002/mds.27963

57. Russell DS, Jennings DL, Barret O et al (2016) Change in PDE10 across early Huntington disease assessed by [18F]MNI-659 and PET imaging. Neurology 86:748–754. https://doi.org/10.1212/WNL.0000000000002391

58. Niccolini F, Haider S, Reis Marques T et al (2015) Altered PDE10A expression detectable early before symptomatic onset in Huntington's disease. Brain 138:3016–3029. https://doi.org/10.1093/brain/awv214

59. Wilson H, Niccolini F, Haider S et al (2016) Loss of extra-striatal phosphodiesterase 10A expression in early premanifest Huntington's disease gene carriers. J Neurol Sci 368:243–248. https://doi.org/10.1016/j.jns.2016.07.033

60. Holthoff VA, Koeppe RA, Frey KA et al (1993) Positron emission tomography measures of benzodiazepine receptors in Huntington's disease. Ann Neurol 34:76–81. https://doi.org/10.1002/ana.410340114

61. Künig G, Leenders KL, Sanchez-Pernaute R et al (2000) Benzodiazepine receptor binding in Huntington's disease: [11C]flumazenil uptake measured using positron emission tomography. Ann Neurol 47:644–648

62. Weeks RA, Cunningham VJ, Piccini P et al (1997) 11C-diprenorphine binding in Huntington's disease: a comparison of region of interest analysis with statistical parametric mapping. J Cereb blood flow Metab 17:943–949. https://doi.org/10.1097/00004647-199709000-00003

63. Van Laere K, Casteels C, Dhollander I et al (2010) Widespread decrease of type 1 cannabinoid receptor availability in Huntington disease in vivo. J Nucl Med 51:1413–1417. https://doi.org/10.2967/jnumed.110.077156

64. Ceccarini J, Ahmad R, Van De Vliet L et al (2019) Behavioral symptoms in premanifest Huntington disease correlate with reduced frontal CB 1 R levels. J Nucl Med 60:115–121. https://doi.org/10.2967/jnumed.118.210393

65. Matusch A, Saft C, Elmenhorst D et al (2014) Cross sectional PET study of cerebral adenosine A_1 receptors in premanifest and manifest Huntington's disease. Eur J Nucl Med Mol Imaging 41:1210–1220. https://doi.org/10.1007/s00259-014-2724-8

66. Langbehn DR, Brinkman RR, Falush D et al (2004) A new model for prediction of the age of onset and penetrance for Huntington's disease based on CAG length. Clin Genet 65:267–277. https://doi.org/10.1111/j.1399-0004.2004.00241.x

67. Vonsattel JP, Myers RH, Stevens TJ et al (1985) Neuropathological classification of Huntington's disease. J Neuropathol Exp Neurol 44:559–577. https://doi.org/10.1097/00005072-198511000-00003

68. Phillips W, Shannon KM, Barker RA (2008) The current clinical management of Huntington's disease. Mov Disord 23:1491–1504. https://doi.org/10.1002/mds.21971

69. Cross A, Rossor M (1983) Dopamine D-1 and D-2 receptors in Huntington's disease. Eur J Pharmacol 88:223–229. https://doi.org/10.1016/0014-2999(83)90009-2

70. Joyce JN, Lexow N, Bird E, Winokur A (1988) Organization of dopamine D1 and D2 receptors in human striatum: receptor autoradiographic studies in Huntington's disease and schizophrenia. Synapse 2:546–557. https://doi.org/10.1002/syn.890020511

71. Halldin C, Farde L, Högberg T et al (1995) Carbon-11-FLB 457: a radioligand for extrastriatal D2 dopamine receptors. J Nucl Med 36:1275–1281

72. Chen MK, Guilarte TR (2008) Translocator protein 18 kDa (TSPO): molecular sensor of brain injury and repair. Pharmacol Ther 118:1–17

73. Janssen B, Vugts DJ, Windhorst AD, Mach RH (2018) PET imaging of microglial activation-beyond targeting TSPO. Molecules 23:607. https://doi.org/10.3390/molecules23030607

74. Fujishige K, Kotera J, Omori K (1999) Striatum- and testis-specific phosphodiesterase PDE10A isolation and characterization of a rat PDE10A. Eur J Biochem 266:1118–1127. https://doi.org/10.1046/j.1432-1327.1999.00963.x

75. Coskran TM, Morton D, Menniti FS et al (2006) Immunohistochemical localization of phosphodiesterase 10A in multiple mammalian species. J Histochem Cytochem 54:1205–1213. https://doi.org/10.1369/jhc.6A6930.2006

76. Nishi A, Kuroiwa M, Miller DB et al (2008) Distinct roles of PDE4 and PDE10A in the regulation of cAMP/PKA signaling in the

striatum. J Neurosci 28:10460–10471. https://doi.org/10.1523/JNEUROSCI. 2518-08.2008

77. Roze E, Betuing S, Deyts C et al (2008) Mitogen- and stress-activated protein kinase-1 deficiency is involved in expanded-huntingtin-induced transcriptional dysregulation and striatal death. FASEB J 22:1083–1093. https://doi.org/10.1096/fj.07-9814

78. Girault J-A (2012) Integrating neurotransmission in striatal medium spiny neurons. Adv Exp Med Biol 970:407–429. https://doi.org/10.1007/978-3-7091-0932-8_18

79. Hsu YT, Chang YG, Chern Y (2018) Insights into GABA A ergic system alteration in Huntington's disease. Open Biol 8:180165

80. Twelvetrees AE, Yuen EY, Arancibia-Carcamo IL et al (2010) Delivery of GABAARs to synapses is mediated by HAP1-KIF5 and disrupted by mutant huntingtin. Neuron 65:53–65. https://doi.org/10.1016/j.neuron.2009.12.007

81. Yuen EY, Wei J, Zhong P, Yan Z (2012) Disrupted GABAAR trafficking and synaptic inhibition in a mouse model of Huntington's disease. Neurobiol Dis 46:497–502. https://doi.org/10.1016/j.nbd.2012.02.015

82. Penney JBJ, Young AB (1982) Quantitative autoradiography of neurotransmitter receptors in Huntington disease. Neurology 32:1391–1395. https://doi.org/10.1212/wnl.32.12.1391

83. Walker FO, Young AB, Penney JB et al (1984) Benzodiazepine and GABA receptors in early Huntington's disease. Neurology 34:1237–1240. https://doi.org/10.1212/wnl.34.9.1237

84. Cross AJ, Hille C, Slater P (1987) Subtraction autoradiography of opiate receptor subtypes in human brain. Brain Res 418:343–348. https://doi.org/10.1016/0006-8993(87)90101-6

85. Albin RL, Reiner A, Anderson KD et al (1992) Preferential loss of striato-external pallidal projection neurons in presymptomatic Huntington's disease. Ann Neurol 31:425–430. https://doi.org/10.1002/ana.410310412

86. Denovan-Wright EM, Robertson HA (2000) Cannabinoid receptor messenger RNA levels decrease in a subset of neurons of the lateral striatum, cortex and hippocampus of transgenic Huntington's disease mice. Neuroscience 98:705–713. https://doi.org/10.1016/s0306-4522(00)00157-3

87. Lastres-Becker I, Berrendero F, Lucas JJ et al (2002) Loss of mRNA levels, binding and activation of GTP-binding proteins for cannabinoid CB1 receptors in the basal ganglia of a transgenic model of Huntington's disease. Brain Res 929:236–242. https://doi.org/10.1016/s0006-8993(01)03403-5

88. Dowie MJ, Bradshaw HB, Howard ML et al (2009) Altered CB1 receptor and endocannabinoid levels precede motor symptom onset in a transgenic mouse model of Huntington's disease. Neuroscience 163:456–465. https://doi.org/10.1016/j.neuroscience.2009.06.014

89. Glass M, Faull RL, Dragunow M (1993) Loss of cannabinoid receptors in the substantia nigra in Huntington's disease. Neuroscience 56:523–527. https://doi.org/10.1016/0306-4522(93)90352-g

90. Fredholm BB (2007) Adenosine, an endogenous distress signal, modulates tissue damage and repair. Cell Death Differ 14:1315–1323. https://doi.org/10.1038/sj.cdd.4402132

91. Bauer A, Zilles K, Matusch A et al (2005) Regional and subtype selective changes of neurotransmitter receptor density in a rat transgenic for the Huntington's disease mutation. J Neurochem 94:639–650. https://doi.org/10.1111/j.1471-4159.2005.03169.x

92. Bauer A, Holschbach MH, Cremer M et al (2003) Evaluation of 18F-CPFPX, a novel adenosine A1 receptor ligand: in vitro autoradiography and high-resolution small animal PET. J Nucl Med 44:1682–1689

93. Ishiwata K, Noguchi J, Wakabayashi S et al (2000) 11C-labeled KF18446: a potential central nervous system adenosine A2a receptor ligand. J Nucl Med 41:345–354

94. Squitieri F, Orobello S, Cannella M et al (2009) Riluzole protects Huntington disease patients from brain glucose hypometabolism and grey matter volume loss and increases production of neurotrophins. Eur J Nucl Med Mol Imaging 36:1113–1120. https://doi.org/10.1007/s00259-009-1103-3

95. Esmaeilzadeh M, Kullingsjö J, Ullman H et al (2011) Regional cerebral glucose metabolism after pridopidine (ACR16) treatment in patients with Huntington disease. Clin Neuropharmacol 34:95–100. https://doi.org/10.1097/WNF.0b013e31821c31d8

96. Hjermind LE, Law I, Jønch A et al (2011) Huntington's disease: effect of memantine on FDG-PET brain metabolism? J Neuropsychiatr Clin Neurosci 23:206–210. https://doi.org/10.1176/jnp.23.2.jnp206

97. Fazio P, Paucar M, Svenningsson P, Varrone A (2018) Novel imaging biomarkers for Huntington's disease and other hereditary choreas. Curr Neurol Neurosci Rep 18:85. https://doi.org/10.1007/s11910-018-0890-y

98. Ashburner J, Friston KJ (2000) Voxel-based morphometry--the methods. NeuroImage 11:805–821. https://doi.org/10.1006/nimg.2000.0582

99. Dale AM, Fischl B, Sereno MI (1999) Cortical surface-based analysis. I. Segmentation and surface reconstruction. NeuroImage 9:179–194. https://doi.org/10.1006/nimg.1998.0395

100. Fischl B, Sereno MI, Dale AM (1999) Cortical surface-based analysis. II: inflation, flattening, and a surface-based coordinate system. NeuroImage 9:195–207. https://doi.org/10.1006/nimg.1998.0396

101. Dogan I, Eickhoff SB, Schulz JB et al (2013) Consistent neurodegeneration and its association with clinical progression in Huntington's disease: a coordinate-based meta-analysis. Neurodegener Dis 12:23–35. https://doi.org/10.1159/000339528

102. Tabrizi SJ, Langbehn DR, Leavitt BR et al (2009) Biological and clinical manifestations of Huntington's disease in the longitudinal TRACK-HD study: cross-sectional analysis of baseline data. Lancet Neurol 8:791–801. https://doi.org/10.1016/S1474-4422(09)70170-X

103. Nopoulos PC, Aylward EH, Ross CA et al (2010) Cerebral cortex structure in prodromal Huntington disease. Neurobiol Dis 40:544–554. https://doi.org/10.1016/j.nbd.2010.07.014

104. Wild EJ, Henley SMD, Hobbs NZ et al (2010) Rate and acceleration of whole-brain atrophy in premanifest and early Huntington's disease. Mov Disord 25:888–895. https://doi.org/10.1002/mds.22969

105. Hobbs NZ, Henley SMD, Ridgway GR et al (2010) The progression of regional atrophy in premanifest and early Huntington's disease: a longitudinal voxel-based morphometry study. J Neurol Neurosurg Psychiatry 81:756–763. https://doi.org/10.1136/jnnp.2009.190702

106. Hobbs NZ, Henley SMD, Wild EJ et al (2009) Automated quantification of caudate atrophy by local registration of serial MRI: evaluation and application in Huntington's disease. NeuroImage 47:1659–1665. https://doi.org/10.1016/j.neuroimage.2009.06.003

107. Thieben MJ, Duggins AJ, Good CD et al (2002) The distribution of structural neuropathology in pre-clinical Huntington's disease. Brain 125:1815–1828. https://doi.org/10.1093/brain/awf179

108. Nopoulos PC, Aylward EH, Ross CA et al (2011) Smaller intracranial volume in prodromal Huntington's disease: evidence for abnormal neurodevelopment. Brain 134:137–142. https://doi.org/10.1093/brain/awq280

109. Majid DSA, Stoffers D, Sheldon S et al (2011) Automated structural imaging analysis detects premanifest Huntington's disease neurodegeneration within 1 year. Mov Disord 26:1481–1488. https://doi.org/10.1002/mds.23656

110. Tabrizi SJ, Reilmann R, Roos RAC et al (2012) Potential endpoints for clinical trials in premanifest and early Huntington's disease in the TRACK-HD study: analysis of 24 month observational data. Lancet Neurol 11:42–53. https://doi.org/10.1016/S1474-4422(11)70263-0

111. Tabrizi SJ, Scahill RI, Owen G et al (2013) Predictors of phenotypic progression and disease onset in premanifest and early-stage Huntington's disease in the TRACK-HD study: analysis of 36-month observational data. Lancet Neurol 12:637–649. https://doi.org/10.1016/S1474-4422(13)70088-7

112. Henley SMD, Wild EJ, Hobbs NZ et al (2009) Whole-brain atrophy as a measure of progression in premanifest and early Huntington's disease. Mov Disord 24:932–936. https://doi.org/10.1002/mds.22485

113. Aylward EH, Nopoulos PC, Ross CA et al (2011) Longitudinal change in regional brain volumes in prodromal Huntington disease. J Neurol Neurosurg Psychiatry 82:405–410. https://doi.org/10.1136/jnnp.2010.208264

114. Rosas HD, Liu AK, Hersch S et al (2002) Regional and progressive thinning of the cortical ribbon in Huntington's disease. Neurology 58:695–701. https://doi.org/10.1212/wnl.58.5.695

115. Rosas HD, Hevelone ND, Zaleta AK et al (2005) Regional cortical thinning in preclinical Huntington disease and its relationship to cognition. Neurology 65:745–747. https://doi.org/10.1212/01.wnl.0000174432.87383.87

116. Rosas HD, Salat DH, Lee SY et al (2008) Cerebral cortex and the clinical expression of Huntington's disease: complexity and heterogeneity. Brain 131:1057–1068. https://doi.org/10.1093/brain/awn025

117. Sweidan W, Bao F, Bozorgzad N, George E (2020) White and gray matter abnormalities in manifest Huntington's disease: cross-sectional and longitudinal analysis. J Neuroimaging 30:351. https://doi.org/10.1111/jon.12699

118. Nopoulos P, Magnotta VA, Mikos A et al (2007) Morphology of the cerebral cortex in preclinical Huntington's disease. Am J Psychiatry 164:1428–1434. https://doi.org/10.1176/appi.ajp.2007.06081266

119. Rosas HD, Reuter M, Doros G et al (2011) A tale of two factors: what determines the rate of progression in Huntington's disease? A longitudinal MRI study. Mov Disord 26:1691–1697. https://doi.org/10.1002/mds.23762

120. Paulsen JS, Nopoulos PC, Aylward E et al (2010) Striatal and white matter predictors of estimated diagnosis for Huntington disease. Brain Res Bull 82:201–207. https://doi.org/10.1016/j.brainresbull.2010.04.003

121. Georgiou-Karistianis N, Scahill R, Tabrizi SJ et al (2013) Structural MRI in Huntington's disease and recommendations for its potential use in clinical trials. Neurosci Biobehav Rev 37:480–490. https://doi.org/10.1016/j.neubiorev.2013.01.022

122. Harris GJ, Pearlson GD, Peyser CE et al (1992) Putamen volume reduction on magnetic resonance imaging exceeds caudate changes in mild Huntington's disease. Ann Neurol 31:69–75. https://doi.org/10.1002/ana.410310113

123. Harris GJ, Aylward EH, Peyser CE et al (1996) Single photon emission computed tomographic blood flow and magnetic resonance volume imaging of basal ganglia in Huntington's disease. Arch Neurol 53:316–324. https://doi.org/10.1001/archneur.1996.00550040044013

124. Aylward EH, Sparks BF, Field KM et al (2004) Onset and rate of striatal atrophy in preclinical Huntington disease. Neurology 63:66–72. https://doi.org/10.1212/01.wnl.0000132965.14653.d1

125. Ruocco HH, Lopes-Cendes I, Li LM et al (2006) Striatal and extrastriatal atrophy in Huntington's disease and its relationship with length of the CAG repeat. Braz J Med Biol Res 39:1129–1136. https://doi.org/10.1590/s0100-879x2006000800016

126. Peinemann A, Schuller S, Pohl C et al (2005) Executive dysfunction in early stages of Huntington's disease is associated with striatal and insular atrophy: a neuropsychological and voxel-based morphometric study. J Neurol Sci 239:11–19. https://doi.org/10.1016/j.jns.2005.07.007

127. Hobbs NZ, Farmer RE, Rees EM et al (2015) Short-interval observational data to inform clinical trial design in Huntington's disease. J Neurol Neurosurg Psychiatry 86:1291–1298. https://doi.org/10.1136/jnnp-2014-309768

128. Aylward EH, Anderson NB, Bylsma FW et al (1998) Frontal lobe volume in patients with Huntington's disease. Neurology 50:252–258. https://doi.org/10.1212/wnl.50.1.252

129. Aylward EH, Codori AM, Rosenblatt A et al (2000) Rate of caudate atrophy in presymptomatic and symptomatic stages of Huntington's disease. Mov Disord 15:552–560. https://doi.org/10.1002/1531-8257(200005)15:3<552::AID-MDS1020>3.0.CO;2-P

130. Wijeratne PA, Johnson EB, Eshaghi A et al (2020) Robust Markers and sample sizes for multicenter trials of Huntington disease. Ann Neurol 87:751–762. https://doi.org/10.1002/ana.25709

131. Aylward EH, Brandt J, Codori AM et al (1994) Reduced basal ganglia volume associated with the gene for Huntington's disease in asymptomatic at-risk persons. Neurology 44:823–828. https://doi.org/10.1212/wnl.44.5.823

132. Douaud G, Gaura V, Ribeiro M-J et al (2006) Distribution of grey matter atrophy in Huntington's disease patients: a combined ROI-based and voxel-based morphometric study. NeuroImage 32:1562–1575. https://doi.org/10.1016/j.neuroimage.2006.05.057

133. Rosas HD, Koroshetz WJ, Chen YI et al (2003) Evidence for more widespread cerebral pathology in early HD: an MRI-based morphometric analysis. Neurology 60:1615–1620. https://doi.org/10.1212/01.wnl.0000065888.88988.6e

134. Bogaard SJA, Dumas EM, Acharya TP et al (2011) Early atrophy of pallidum and accumbens nucleus in Huntington's disease. J Neurol 258:412–420. https://doi.org/10.1007/s00415-010-5768-0

135. Roy AK, Shehzad Z, Margulies DS et al (2009) Functional connectivity of the human amygdala using resting state fMRI. NeuroImage 45:614–626. https://doi.org/10.1016/j.neuroimage.2008.11.030

136. Phelps EA, LeDoux JE (2005) Contributions of the amygdala to emotion processing: from animal models to human behavior. Neuron

48:175–187. https://doi.org/10.1016/j.neuron.2005.09.025

137. Kipps CM, Duggins AJ, McCusker EA, Calder AJ (2007) Disgust and happiness recognition correlate with anteroventral insula and amygdala volume respectively in preclinical Huntington's disease. J Cogn Neurosci 19:1206–1217. https://doi.org/10.1162/jocn.2007.19.7.1206

138. Ahveninen LM, Stout JC, Georgiou-Karistianis N et al (2018) Reduced amygdala volumes are related to motor and cognitive signs in Huntington's disease: the IMAGE-HD study. NeuroImage Clin 18:881–887. https://doi.org/10.1016/j.nicl.2018.03.027

139. Phillips O, Sanchez-Castaneda C, Elifani F et al (2013) Tractography of the corpus callosum in Huntington's disease. PLoS One 8: e73280. https://doi.org/10.1371/journal.pone.0073280

140. Sánchez-Castañeda C, Cherubini A, Elifani F et al (2013) Seeking Huntington disease biomarkers by multimodal, cross-sectional basal ganglia imaging. Hum Brain Mapp 34:1625–1635. https://doi.org/10.1002/hbm.22019

141. Reading SAJ, Yassa MA, Bakker A et al (2005) Regional white matter change in pre-symptomatic Huntington's disease: a diffusion tensor imaging study. Psychiatry Res 140:55–62. https://doi.org/10.1016/j.pscychresns.2005.05.011

142. Mascalchi M, Lolli F, Della Nave R et al (2004) Huntington disease: volumetric, diffusion-weighted, and magnetization transfer MR imaging of brain. Radiology 232:867–873. https://doi.org/10.1148/radiol.2322030820

143. Novak MJU, Seunarine KK, Gibbard CR et al (2014) White matter integrity in premanifest and early Huntington's disease is related to caudate loss and disease progression. Cortex 52:98–112. https://doi.org/10.1016/j.cortex.2013.11.009

144. Dumas EM, van den Bogaard SJA, Ruber ME et al (2012) Early changes in white matter pathways of the sensorimotor cortex in premanifest Huntington's disease. Hum Brain Mapp 33:203–212. https://doi.org/10.1002/hbm.21205

145. Syka M, Keller J, Klempíř J et al (2015) Correlation between relaxometry and diffusion tensor imaging in the globus pallidus of Huntington's disease patients. PLoS One 10: e0118907. https://doi.org/10.1371/journal.pone.0118907

146. Poudel GR, Stout JC, Domínguez DJF et al (2015) Longitudinal change in white matter microstructure in Huntington's disease: the IMAGE-HD study. Neurobiol Dis 74:406–412. https://doi.org/10.1016/j.nbd.2014.12.009

147. Matsui JT, Vaidya JG, Johnson HJ et al (2014) Diffusion weighted imaging of prefrontal cortex in prodromal Huntington's disease. Hum Brain Mapp 35:1562–1573. https://doi.org/10.1002/hbm.22273

148. Orth M, Gregory S, Scahill RI et al (2016) Natural variation in sensory-motor white matter organization influences manifestations of Huntington's disease. Hum Brain Mapp 37:4615–4628. https://doi.org/10.1002/hbm.23332

149. Rosas HD, Wilkens P, Salat DH et al (2018) Complex spatial and temporally defined myelin and axonal degeneration in Huntington disease. NeuroImage Clin 20:236–242. https://doi.org/10.1016/j.nicl.2018.01.029

150. Douaud G, Behrens TE, Poupon C et al (2009) In vivo evidence for the selective subcortical degeneration in Huntington's disease. NeuroImage 46:958–966. https://doi.org/10.1016/j.neuroimage.2009.03.044

151. Bohanna I, Georgiou-Karistianis N, Egan GF (2011) Connectivity-based segmentation of the striatum in Huntington's disease: vulnerability of motor pathways. Neurobiol Dis 42:475–481. https://doi.org/10.1016/j.nbd.2011.02.010

152. Gregory S, Cole JH, Farmer RE et al (2015) Longitudinal diffusion tensor imaging shows progressive changes in white matter in Huntington's disease. J Huntingtons Dis 4:333–346. https://doi.org/10.3233/JHD-150173

153. Saba RA, Yared JH, Doring TM et al (2017) Diffusion tensor imaging of brain white matter in Huntington gene mutation individuals. Arq Neuropsiquiatr 75:503–508. https://doi.org/10.1590/0004-282X20170085

154. Phillips O, Squitieri F, Sanchez-Castaneda C et al (2015) The corticospinal tract in Huntington's disease. Cereb Cortex 25:2670–2682. https://doi.org/10.1093/cercor/bhu065

155. Liu Z, Xu C, Xu Y et al (2010) Decreased regional homogeneity in insula and cerebellum: a resting-state fMRI study in patients with major depression and subjects at high risk for major depression. Psychiatry Res 182:211–215. https://doi.org/10.1016/j.pscychresns.2010.03.004

156. Sprengelmeyer R, Orth M, Müller H-P et al (2014) The neuroanatomy of subthreshold depressive symptoms in Huntington's disease: a combined diffusion tensor imaging (DTI) and voxel-based morphometry (VBM) study. Psychol Med 44:1867–1878. https://doi.org/10.1017/S003329171300247X

157. Sritharan A, Egan GF, Johnston L et al (2010) A longitudinal diffusion tensor imaging study in symptomatic Huntington's disease. J Neurol Neurosurg Psychiatry 81:257–262. https://doi.org/10.1136/jnnp.2007.142786

158. Odish OFF, Leemans A, Reijntjes RHAM et al (2015) Microstructural brain abnormalities in Huntington's disease: a two-year follow-up. Hum Brain Mapp 36:2061–2074. https://doi.org/10.1002/hbm.22756

159. Odish OFF, Reijntjes RHAM, van den Bogaard SJA et al (2018) Progressive microstructural changes of the occipital cortex in Huntington's disease. Brain Imaging Behav 12:1786–1794. https://doi.org/10.1007/s11682-018-9849-5

160. Pflanz CP, Charquero-Ballester M, Majid DSA et al (2020) One-year changes in brain microstructure differentiate preclinical Huntington's disease stages. NeuroImage Clin 25:102099. https://doi.org/10.1016/j.nicl.2019.102099

161. Weaver KE, Richards TL, Liang O et al (2009) Longitudinal diffusion tensor imaging in Huntington's Disease. Exp Neurol 216:525–529. https://doi.org/10.1016/j.expneurol.2008.12.026

162. Shaffer JJ, Ghayoor A, Long JD et al (2017) Longitudinal diffusion changes in prodromal and early HD: evidence of white-matter tract deterioration. Hum Brain Mapp 38:1460–1477. https://doi.org/10.1002/hbm.23465

163. Harrington DL, Long JD, Durgerian S et al (2016) Cross-sectional and longitudinal multimodal structural imaging in prodromal Huntington's disease. Mov Disord 31:1664–1675. https://doi.org/10.1002/mds.26803

164. Warach S, Gaa J, Siewert B et al (1995) Acute human stroke studied by whole brain echo planar diffusion-weighted magnetic resonance imaging. Ann Neurol 37:231–241. https://doi.org/10.1002/ana.410370214

165. Sone D (2019) Neurite orientation and dispersion density imaging: clinical utility, efficacy, and role in therapy. Rep Med Imaging 12:17–29. https://doi.org/10.2147/RMI.S194083

166. Zhang J, Gregory S, Scahill RI et al (2018) In vivo characterization of white matter pathology in premanifest Huntington's disease. Ann Neurol 84:497–504. https://doi.org/10.1002/ana.25309

167. Vymazal J, Klempíř J, Jech R et al (2007) MR relaxometry in Huntington's disease: correlation between imaging, genetic and clinical parameters. J Neurol Sci 263:20–25. https://doi.org/10.1016/j.jns.2007.05.018

168. Apple AC, Possin KL, Satris G et al (2014) Quantitative 7T phase imaging in premanifest Huntington disease. AJNR Am J Neuroradiol 35:1707–1713. https://doi.org/10.3174/ajnr.A3932

169. Bartzokis G, Lu PH, Tishler TA et al (2007) Myelin breakdown and iron changes in Huntington's disease: pathogenesis and treatment implications. Neurochem Res 32:1655–1664. https://doi.org/10.1007/s11064-007-9352-7

170. Jurgens CK, Jasinschi R, Ekin A et al (2010) MRI T2 Hypointensities in basal ganglia of premanifest Huntington's disease. PLoS Curr 2:RRN1173. https://doi.org/10.1371/currents.RRN1173

171. Rosas HD, Chen YI, Doros G et al (2012) Alterations in brain transition metals in Huntington disease: an evolving and intricate story. Arch Neurol 69:887–893. https://doi.org/10.1001/archneurol.2011.2945

172. van Bergen JMG, Hua J, Unschuld PG et al (2016) Quantitative susceptibility mapping suggests altered brain iron in premanifest Huntington disease. AJNR Am J Neuroradiol 37:789–796. https://doi.org/10.3174/ajnr.A4617

173. Di Paola M, Phillips OR, Sanchez-Castaneda C et al (2014) MRI measures of corpus callosum iron and myelin in early Huntington's disease. Hum Brain Mapp 35:3143–3151. https://doi.org/10.1002/hbm.22391

174. Ordidge RJ, Gorell JM, Deniau JC et al (1994) Assessment of relative brain iron concentrations using T2-weighted and T2*-weighted MRI at 3 Tesla. Magn Reson Med 32:335–341. https://doi.org/10.1002/mrm.1910320309

175. Domínguez JFD, Ng ACL, Poudel G et al (2016) Iron accumulation in the basal ganglia in Huntington's disease: cross-sectional data from the IMAGE-HD study. J Neurol Neurosurg Psychiatry 87:545–549. https://doi.org/10.1136/jnnp-2014-310183

176. Klöppel S, Gregory S, Scheller E et al (2015) Compensation in preclinical Huntington's disease: evidence from the track-on HD

study. EBioMedicine 2:1420–1429. https://doi.org/10.1016/j.ebiom.2015.08.002

177. Kirchner WK (1958) Age differences in short-term retention of rapidly changing information. J Exp Psychol 55:352–358. https://doi.org/10.1037/h0043688

178. Wolf RC, Vasic N, Schönfeldt-Lecuona C et al (2007) Dorsolateral prefrontal cortex dysfunction in presymptomatic Huntington's disease: evidence from event-related fMRI. Brain 130:2845–2857. https://doi.org/10.1093/brain/awm210

179. Wolf RC, Sambataro F, Vasic N et al (2008) Aberrant connectivity of lateral prefrontal networks in presymptomatic Huntington's disease. Exp Neurol 213:137–144. https://doi.org/10.1016/j.expneurol.2008.05.017

180. Georgiou-Karistianis N, Stout JC, Domínguez DJF et al (2014) Functional magnetic resonance imaging of working memory in Huntington's disease: cross-sectional data from the IMAGE-HD study. Hum Brain Mapp 35:1847–1864. https://doi.org/10.1002/hbm.22296

181. Poudel GR, Stout JC, Domínguez DJF et al (2015) Functional changes during working memory in Huntington's disease: 30-month longitudinal data from the IMAGE-HD study. Brain Struct Funct 220:501–512. https://doi.org/10.1007/s00429-013-0670-z

182. Wolf RC, Sambataro F, Vasic N et al (2008) Altered frontostriatal coupling in pre-manifest Huntington's disease: effects of increasing cognitive load. Eur J Neurol 15:1180–1190. https://doi.org/10.1111/j.1468-1331.2008.02253.x

183. Poudel GR, Driscoll S, Domínguez DJF et al (2015) Functional brain correlates of neuropsychiatric symptoms in presymptomatic Huntington's disease: the IMAGE-HD study. J Huntingtons Dis 4:325–332. https://doi.org/10.3233/JHD-150154

184. Wolf RC, Grön G, Sambataro F et al (2012) Brain activation and functional connectivity in premanifest Huntington's disease during states of intrinsic and phasic alertness. Hum Brain Mapp 33:2161–2173. https://doi.org/10.1002/hbm.21348

185. Gray MA, Egan GF, Ando A et al (2013) Prefrontal activity in Huntington's disease reflects cognitive and neuropsychiatric disturbances: the IMAGE-HD study. Exp Neurol 239:218–228. https://doi.org/10.1016/j.expneurol.2012.10.020

186. Reading SAJ, Dziorny AC, Peroutka LA et al (2004) Functional brain changes in presymptomatic Huntington's disease. Ann Neurol 55:879–883. https://doi.org/10.1002/ana.20121

187. Zimbelman JL, Paulsen JS, Mikos A et al (2007) fMRI detection of early neural dysfunction in preclinical Huntington's disease. J Int Neuropsychol Soc 13:758–769. https://doi.org/10.1017/S1355617707071214

188. Paulsen JS, Zimbelman JL, Hinton SC et al (2004) fMRI biomarker of early neuronal dysfunction in presymptomatic Huntington's Disease. AJNR Am J Neuroradiol 25:1715–1721

189. Saft C, Schüttke A, Beste C et al (2008) fMRI reveals altered auditory processing in manifest and premanifest Huntington's disease. Neuropsychologia 46:1279–1289. https://doi.org/10.1016/j.neuropsychologia.2007.12.002

190. Unschuld PG, Liu X, Shanahan M et al (2013) Prefrontal executive function associated coupling relates to Huntington's disease stage. Cortex 49:2661–2673. https://doi.org/10.1016/j.cortex.2013.05.015

191. Hennenlotter A, Schroeder U, Erhard P et al (2004) Neural correlates associated with impaired disgust processing in pre-symptomatic Huntington's disease. Brain 127:1446–1453. https://doi.org/10.1093/brain/awh165

192. Van den Stock J, De Winter F-L, Ahmad R et al (2015) Functional brain changes underlying irritability in premanifest Huntington's disease. Hum Brain Mapp 36:2681–2690. https://doi.org/10.1002/hbm.22799

193. Malejko K, Weydt P, Süßmuth SD et al (2014) Prodromal Huntington disease as a model for functional compensation of early neurodegeneration. PLoS One 9:e114569. https://doi.org/10.1371/journal.pone.0114569

194. Novak MJU, Warren JD, Henley SMD et al (2012) Altered brain mechanisms of emotion processing in pre-manifest Huntington's disease. Brain 135:1165–1179. https://doi.org/10.1093/brain/aws024

195. Quarantelli M, Salvatore E, Giorgio SMDA et al (2013) Default-mode network changes in Huntington's disease: an integrated MRI study of functional connectivity and morphometry. PLoS One 8:e72159. https://doi.org/10.1371/journal.pone.0072159

196. Sánchez-Castañeda C, de Pasquale F, Caravasso CF et al (2017) Resting-state connectivity and modulated somatomotor and default-mode networks in Huntington disease. CNS

Neurosci Ther 23:488–497. https://doi.org/10.1111/cns.12701

197. Müller H-P, Gorges M, Grön G et al (2016) Motor network structure and function are associated with motor performance in Huntington's disease. J Neurol 263:539–549. https://doi.org/10.1007/s00415-015-8014-y

198. Wolf RC, Thomann PA, Sambataro F et al (2015) Abnormal cerebellar volume and corticocerebellar dysfunction in early manifest Huntington's disease. J Neurol 262:859–869. https://doi.org/10.1007/s00415-015-7642-6

199. Gregory S, Scahill RI (2018) Functional magnetic resonance imaging in Huntington's disease. In: International review of neurobiology. Academic Press Inc., New York, NY, pp 381–408

200. Dumas EM, van den Bogaard SJA, Hart EP et al (2013) Reduced functional brain connectivity prior to and after disease onset in Huntington's disease. NeuroImage Clin 2:377–384. https://doi.org/10.1016/j.nicl.2013.03.001

201. Wolf RC, Sambataro F, Vasic N et al (2014) Abnormal resting-state connectivity of motor and cognitive networks in early manifest Huntington's disease. Psychol Med 44:3341–3356. https://doi.org/10.1017/S0033291714000579

202. Espinoza FA, Turner JA, Vergara VM et al (2018) Whole-brain connectivity in a large study of Huntington's disease gene mutation carriers and healthy controls. Brain Connect 8:166–178. https://doi.org/10.1089/brain.2017.0538

203. Liu W, Yang J, Chen K et al (2016) Resting-state fMRI reveals potential neural correlates of impaired cognition in Huntington's disease. Parkinsonism Relat Disord 27:41–46. https://doi.org/10.1016/j.parkreldis.2016.04.017

204. Poudel GR, Egan GF, Churchyard A et al (2014) Abnormal synchrony of resting state networks in premanifest and symptomatic Huntington disease: the IMAGE-HD study. J Psychiatry Neurosci 39:87–96. https://doi.org/10.1503/jpn.120226

205. Wolf RC, Sambataro F, Vasic N et al (2014) Visual system integrity and cognition in early Huntington's disease. Eur J Neurosci 40:2417–2426. https://doi.org/10.1111/ejn.12575

206. Werner CJ, Dogan I, Saß C et al (2014) Altered resting-state connectivity in Huntington's disease. Hum Brain Mapp 35:2582–2593. https://doi.org/10.1002/hbm.22351

207. Odish OFF, van den Berg-Huysmans AA, van den Bogaard SJA et al (2015) Longitudinal resting state fMRI analysis in healthy controls and premanifest Huntington's disease gene carriers: a three-year follow-up study. Hum Brain Mapp 36:110–119. https://doi.org/10.1002/hbm.22616

208. Harrington DL, Rubinov M, Durgerian S et al (2015) Network topology and functional connectivity disturbances precede the onset of Huntington's disease. Brain 138:2332–2346. https://doi.org/10.1093/brain/awv145

209. Gargouri F, Messé A, Perlbarg V et al (2016) Longitudinal changes in functional connectivity of cortico-basal ganglia networks in manifests and premanifest Huntington's disease. Hum Brain Mapp 37:4112–4128. https://doi.org/10.1002/hbm.23299

210. Jenkins BG, Rosas HD, Chen Y-CI et al (1998) 1H NMR spectroscopy studies of Huntington's disease. Neurology 50:1357–1365. https://doi.org/10.1212/wnl.50.5.1357

211. Clarke CE, Lowry M, Quarrell OWJ (1998) No change in striatal glutamate in Huntington's disease measured by proton magnetic resonance spectroscopy. Parkinsonism Relat Disord 4:123–127. https://doi.org/10.1016/s1353-8020(98)00026-1

212. Sánchez-Pernaute R, García-Segura JM, del Barrio AA et al (1999) Clinical correlation of striatal 1H MRS changes in Huntington's disease. Neurology 53:806–812. https://doi.org/10.1212/wnl.53.4.806

213. Reynolds NC, Prost RW, Mark LP (2005) Heterogeneity in 1H-MRS profiles of presymptomatic and early manifest Huntington's disease. Brain Res 1031:82–89. https://doi.org/10.1016/j.brainres.2004.10.030

214. Ruocco HH, Lopes-Cendes I, Li LM, Cendes F (2007) Evidence of thalamic dysfunction in Huntington disease by proton magnetic resonance spectroscopy. Mov Disord 22:2052–2056. https://doi.org/10.1002/mds.21601

215. Sturrock A, Laule C, Decolongon J et al (2010) Magnetic resonance spectroscopy biomarkers in premanifest and early Huntington disease. Neurology 75:1702–1710. https://doi.org/10.1212/wnl.0b013e3181fc27e4

216. van den Bogaard SJA, Dumas EM, Teeuwisse WM et al (2011) Exploratory 7-Tesla magnetic resonance spectroscopy in Huntington's disease provides in vivo evidence for impaired energy metabolism. J Neurol

258:2230–2239. https://doi.org/10.1007/s00415-011-6099-5

217. Unschuld PG, Edden RAE, Carass A et al (2012) Brain metabolite alterations and cognitive dysfunction in early Huntington's disease. Mov Disord 27:895–902. https://doi.org/10.1002/mds.25010

218. Padowski JM, Weaver KE, Richards TL et al (2014) Neurochemical correlates of caudate atrophy in Huntington's disease. Mov Disord 29:327–335. https://doi.org/10.1002/mds.25801

219. Alcauter-Solórzano S, Pasaye-Alcaraz EH, Alvarado-Alanis P et al (2010) Hydrogen magnetic resonance quantitative spectroscopy at 3 T in symptomatic and asymptomatic Huntington's disease patients. Rev Neurol 51:208–212. https://doi.org/10.33588/rn.5104.2009173

220. Jenkins BG, Koroshetz WJ, Beal MF, Rosen BR (1993) Evidence for impairment of energy metabolism in vivo in Huntington's disease using localized 1H NMR spectroscopy. Neurology 43:2689–2695. https://doi.org/10.1212/wnl.43.12.2689

221. Sturrock A, Laule C, Wyper K et al (2015) A longitudinal study of magnetic resonance spectroscopy Huntington's disease biomarkers. Mov Disord 30:393–401. https://doi.org/10.1002/mds.26118

222. van Oostrom JCH, Sijens PE, Roos RAC, Leenders KL (2007) 1H magnetic resonance spectroscopy in preclinical Huntington disease. Brain Res 1168:67–71. https://doi.org/10.1016/j.brainres.2007.05.082

223. van den Bogaard SJA, Dumas EM, Teeuwisse WM et al (2014) Longitudinal metabolite changes in Huntington's disease during disease onset. J Huntingtons Dis 3:377–386. https://doi.org/10.3233/JHD-140117

224. Hoang TQ, Bluml S, Dubowitz DJ et al (1998) Quantitative proton-decoupled 31P MRS and 1H MRS in the evaluation of Huntington's and Parkinson's diseases. Neurology 50:1033–1040. https://doi.org/10.1212/wnl.50.4.1033

225. Adanyeguh IM, Monin M, Rinaldi D et al (2018) Expanded neurochemical profile in the early stage of Huntington disease using proton magnetic resonance spectroscopy. NMR Biomed 31:e3880. https://doi.org/10.1002/nbm.3880

226. Gómez-Ansón B, Alegret M, Muñoz E et al (2007) Decreased frontal choline and neuropsychological performance in preclinical Huntington disease. Neurology

68:906–910. https://doi.org/10.1212/01.wnl.0000257090.01107.2f

227. Reynolds NC, Prost RW, Mark LP, Joseph SA (2008) MR-spectroscopic findings in juvenile-onset Huntington's disease. Mov Disord 23:1931–1935. https://doi.org/10.1002/mds.22245

228. Harms L, Meierkord H, Timm G et al (1997) Decreased N-acetyl-aspartate/choline ratio and increased lactate in the frontal lobe of patients with Huntington's disease: a proton magnetic resonance spectroscopy study. J Neurol Neurosurg Psychiatry 62:27. https://doi.org/10.1136/jnnp.62.1.27

229. Taylor-Robinson SD, Weeks RA, Bryant DJ et al (1996) Proton magnetic resonance spectroscopy in Huntington's disease: evidence in favour of the glutamate excitotoxic theory? Mov Disord 11:167–173. https://doi.org/10.1002/mds.870110209

230. Taylor-Robinson S, Weeks R, Sargentoni J et al (1994) Evidence for glutamate excitotoxicity in Huntington's disease with proton magnetic resonance spectroscopy. Lancet 343:1170. https://doi.org/10.1016/s0140-6736(94)90280-1

231. Mochel F, N'Guyen T, Deelchand D et al (2012) Abnormal response to cortical activation in early stages of Huntington disease. Mov Disord 27:907–910. https://doi.org/10.1002/mds.25009

232. Koroshetz WJ, Jenkins BG, Rosen BR, Beal MF (1997) Energy metabolism defects in Huntington's disease and effects of coenzyme Q10. Ann Neurol 41:160–165. https://doi.org/10.1002/ana.410410206

233. Mugler JP III, Brookeman JR (1990) Three-dimensional magnetization-prepared rapid gradient-echo imaging (3D MP RAGE). Magn Reson Med 15:152–157. https://doi.org/10.1002/mrm.1910150117

234. Sudhyadhom A, Haq IU, Foote KD et al (2009) A high resolution and high contrast MRI for differentiation of subcortical structures for DBS targeting: the Fast Gray Matter Acquisition T1 Inversion Recovery (FGATIR). NeuroImage 47(Suppl 2):T44–T52. https://doi.org/10.1016/j.neuroimage.2009.04.018

235. Tziortzi AC, Searle GE, Tzimopoulou S et al (2011) Imaging dopamine receptors in humans with [11C]-(+)-PHNO: dissection of D3 signal and anatomy. NeuroImage 54:264–277. https://doi.org/10.1016/j.neuroimage.2010.06.044

236. Heckemann RA, Keihaninejad S, Aljabar P et al (2010) Improving intersubject image

registration using tissue-class information benefits robustness and accuracy of multi-atlas based anatomical segmentation. NeuroImage 51:221–227. https://doi.org/10.1016/j.neuroimage.2010.01.072

237. Khan AR, Wang L, Beg MF (2008) FreeSurfer-initiated fully-automated subcortical brain segmentation in MRI using Large Deformation Diffeomorphic Metric Mapping. NeuroImage 41:735–746. https://doi.org/10.1016/j.neuroimage.2008.03.024

238. Coppen EM, Jacobs M, van den Berg-Huysmans AA et al (2018) Grey matter volume loss is associated with specific clinical motor signs in Huntington's disease. Parkinsonism Relat Disord 46:56–61. https://doi.org/10.1016/j.parkreldis.2017.11.001

239. Xu L, Groth KM, Pearlson G et al (2009) Source-based morphometry: the use of independent component analysis to identify gray matter differences with application to schizophrenia. Hum Brain Mapp 30:711–724. https://doi.org/10.1002/hbm.20540

240. Ciarochi JA, Calhoun VD, Lourens S et al (2016) Patterns of co-occurring gray matter concentration loss across the Huntington disease prodrome. Front Neurol 7:147. https://doi.org/10.3389/fneur.2016.00147

241. Coppen EM, van der Grond J, Hafkemeijer A et al (2016) Early grey matter changes in structural covariance networks in Huntington's disease. NeuroImage Clin 12:806–814. https://doi.org/10.1016/j.nicl.2016.10.009

242. Paulsen JS, Long JD, Ross CA et al (2014) Prediction of manifest Huntington's disease with clinical and imaging measures: a prospective observational study. Lancet Neurol 13:1193–1201. https://doi.org/10.1016/S1474-4422(14)70238-8

243. Le Bihan D, Mangin JF, Poupon C et al (2001) Diffusion tensor imaging: concepts and applications. J Magn Reson Imaging 13:534–546. https://doi.org/10.1002/jmri.1076

244. Assaf Y, Pasternak O (2008) Diffusion tensor imaging (DTI)-based white matter mapping in brain research: a review. J Mol Neurosci 34:51–61. https://doi.org/10.1007/s12031-007-0029-0

245. Liu W, Yang J, Burgunder J et al (2016) Diffusion imaging studies of Huntington's disease: a meta-analysis. Parkinsonism Relat Disord 32:94–101. https://doi.org/10.1016/j.parkreldis.2016.09.005

246. Stoffers D, Sheldon S, Kuperman JM et al (2010) Contrasting gray and white matter changes in preclinical Huntington disease: an MRI study. Neurology 74:1208–1216. https://doi.org/10.1212/WNL.0b013e3181d8c20a

247. Klöppel S, Draganski B, Golding CV et al (2008) White matter connections reflect changes in voluntary-guided saccades in pre-symptomatic Huntington's disease. Brain 131:196–204. https://doi.org/10.1093/brain/awm275

248. Rosas HD, Lee SY, Bender AC et al (2010) Altered white matter microstructure in the corpus callosum in Huntington's disease: implications for cortical "disconnection". NeuroImage 49:2995–3004. https://doi.org/10.1016/j.neuroimage.2009.10.015

249. De Santis S, Gabrielli A, Palombo M et al (2011) Non-Gaussian diffusion imaging: a brief practical review. Magn Reson Imaging 29:1410–1416. https://doi.org/10.1016/j.mri.2011.04.006

250. Haacke EM, Cheng NYC, House MJ et al (2005) Imaging iron stores in the brain using magnetic resonance imaging. Magn Reson Imaging 23:1–25. https://doi.org/10.1016/j.mri.2004.10.001

251. Muller M, Leavitt BR (2014) Iron dysregulation in Huntington's disease. J Neurochem 130:328–350. https://doi.org/10.1111/jnc.12739

252. Niu L, Ye C, Sun Y et al (2018) Mutant huntingtin induces iron overload via up-regulating IRP1 in Huntington's disease. Cell Biosci 8:41. https://doi.org/10.1186/s13578-018-0239-x

253. Chen JC, Hardy PA, Kucharczyk W et al (1993) MR of human postmortem brain tissue: correlative study between T2 and assays of iron and ferritin in Parkinson and Huntington disease. AJNR Am J Neuroradiol 14:275–281

254. Dexter DT, Carayon A, Javoy-Agid F et al (1991) Alterations in the levels of iron, ferritin and other trace metals in Parkinson's disease and other neurodegenerative diseases affecting the basal ganglia. Brain 114 (Pt 4):1953–1975. https://doi.org/10.1093/brain/114.4.1953

255. Bartzokis G, Cummings J, Perlman S et al (1999) Increased basal ganglia iron levels in Huntington disease. Arch Neurol 56:569–574. https://doi.org/10.1001/archneur.56.5.569

256. Yousaf T, Dervenoulas G, Politis M (2018) Advances in MRI methodology. In:

International review of neurobiology. Academic Press Inc., New York, NY, pp 31–76

257. Hopp K, Popescu BFG, McCrea RPE et al (2010) Brain iron detected by SWI high pass filtered phase calibrated with synchrotron X-ray fluorescence. J Magn Reson Imaging 31:1346–1354. https://doi.org/10.1002/jmri.22201

258. Deistung A, Schäfer A, Schweser F et al (2013) Toward in vivo histology: a comparison of quantitative susceptibility mapping (QSM) with magnitude-, phase-, and R2*-imaging at ultra-high magnetic field strength. NeuroImage 65:299–314. https://doi.org/10.1016/j.neuroimage.2012.09.055

259. Lim IAL, Faria AV, Li X et al (2013) Human brain atlas for automated region of interest selection in quantitative susceptibility mapping: application to determine iron content in deep gray matter structures. NeuroImage 82:449–469. https://doi.org/10.1016/j.neuroimage.2013.05.127

260. Ross CA, Aylward EH, Wild EJ et al (2014) Huntington disease: natural history, biomarkers and prospects for therapeutics. Nat Rev Neurol 10:204–216. https://doi.org/10.1038/nrneurol.2014.24

261. Friston KJ, Harrison L, Penny W (2003) Dynamic causal modelling. NeuroImage 19:1273–1302. https://doi.org/10.1016/s1053-8119(03)00202-7

262. Friston KJ, Frith CD, Liddle PF, Frackowiak RS (1993) Functional connectivity: the principal-component analysis of large (PET) data sets. J Cereb Blood Flow Metab 13:5–14. https://doi.org/10.1038/jcbfm.1993.4

263. Brockway JP (2000) Two functional magnetic resonance imaging f(MRI) tasks that may replace the gold standard, Wada testing, for language lateralization while giving additional localization information. Brain Cogn 43:57–59

264. Birn RM, Cox RW, Bandettini PA (2002) Detection versus estimation in event-related fMRI: choosing the optimal stimulus timing. NeuroImage 15:252–264. https://doi.org/10.1006/nimg.2001.0964

265. Klöppel S, Draganski B, Siebner HR et al (2009) Functional compensation of motor function in pre-symptomatic Huntington's disease. Brain 132:1624–1632. https://doi.org/10.1093/brain/awp081

266. Lemiere J, Decruyenaere M, Evers-Kiebooms G et al (2002) Longitudinal study evaluating neuropsychological changes in so-called asymptomatic carriers of the Huntington's disease mutation after 1 year. Acta Neurol Scand 106:131–141. https://doi.org/10.1034/j.1600-0404.2002.01192.x

267. Lemiere J, Decruyenaere M, Evers-Kiebooms G et al (2004) Cognitive changes in patients with Huntington's disease (HD) and asymptomatic carriers of the HD mutation--a longitudinal follow-up study. J Neurol 251:935–942. https://doi.org/10.1007/s00415-004-0461-9

268. Sprengelmeyer R, Lange H, Hömberg V (1995) The pattern of attentional deficits in Huntington's disease. Brain 118 (Pt 1):145–152. https://doi.org/10.1093/brain/118.1.145

269. Georgiou-Karistianis N, Sritharan A, Farrow M et al (2007) Increased cortical recruitment in Huntington's disease using a Simon task. Neuropsychologia 45:1791–1800. https://doi.org/10.1016/j.neuropsychologia.2006.12.023

270. Rao SM, Mayer AR, Harrington DL (2001) The evolution of brain activation during temporal processing. Nat Neurosci 4:317–323. https://doi.org/10.1038/85191

271. McColgan P, Gregory S, Razi A et al (2017) White matter predicts functional connectivity in premanifest Huntington's disease. Ann Clin Transl Neurol 4:106–118. https://doi.org/10.1002/acn3.384

272. Gregory S, Long JD, Klöppel S et al (2018) Testing a longitudinal compensation model in premanifest Huntington's disease. Brain 141:2156–2166. https://doi.org/10.1093/brain/awy122

273. Mlynárik V (2017) Introduction to nuclear magnetic resonance. Anal Biochem 529:4–9. https://doi.org/10.1016/j.ab.2016.05.006

274. Mochel F, Dubinsky JM, Henry P-G (2016) Magnetic resonance spectroscopy in Huntington's disease. In: Öz G (ed) Magnetic resonance spectroscopy of degenerative brain diseases. Springer International Publishing, Cham, pp 103–120

275. Buonocore MH, Maddock RJ (2015) Magnetic resonance spectroscopy of the brain: a review of physical principles and technical methods. Rev Neurosci 26:609–632. https://doi.org/10.1515/revneuro-2015-0010

276. Xu S, Yang J, Shen J (2008) Measuring N-acetylaspartate synthesis in vivo using proton magnetic resonance spectroscopy. J Neurosci Methods 172:8–12. https://doi.org/10.1016/j.jneumeth.2008.04.001

277. Soares DP, Law M (2009) Magnetic resonance spectroscopy of the brain: review of

metabolites and clinical applications. Clin Radiol 64:12–21. https://doi.org/10. 1016/j.crad.2008.07.002

278. Danielsen ER, Ross B (1999) Magnetic resonance spectroscopy diagnosis of neurological diseases. CRC Press, Boca Raton, FL

279. López-Villegas D, Lenkinski RE, Wehrli SL et al (1995) Lactate production by human monocytes/macrophages determined by proton MR spectroscopy. Magn Reson Med 34:32–38

280. Srinivasan R, Cunningham C, Chen A et al (2006) TE-averaged two-dimensional proton spectroscopic imaging of glutamate at 3 T. NeuroImage 30:1171–1178

Chapter 20

Imaging Biomarkers in Amyotrophic Lateral Sclerosis

Leonor Cerdá Alberich, Juan Francisco Vázquez-Costa, Amadeo Ten-Esteve, Miguel Mazón, and Luis Martí-Bonmatí

Abstract

Amyotrophic lateral sclerosis (ALS) is a group of heterogeneous disorders characterized by degenerative changes in upper motor neurons within the motor cortex and/or lower motor neurons within the brain stem and spinal cord. The histologic hallmark is brain, spinal cord, nerves and muscles atrophy with reactive astrogliosis and iron accumulation in the motor cortex. This chapter will summarize the available imaging diagnostic biomarkers to understand the disease pathophysiology, to differentiate between phenotypes and to allow an early diagnosis. Moreover, progression and prognostic biomarkers will be also described to quantify the loss of upper and lower motor neurons and to stratify patients. ALS complexity and appropriate clinical endpoints are important considerations when defining the role of new imaging biomarkers, mainly from brain MR and neuromuscular ultrasound studies.

Keywords Neurodegenerative disorders, Amyotrophic lateral sclerosis, Brain MR image analysis, Imaging biomarkers, Neuromuscular ultrasound, Structured report

1 Introduction and Clinical Needs

Amyotrophic lateral sclerosis (ALS), also known as Lou Gehrig's disease, was first described by Jean-Martin Charcot more than 140 years ago [1]. ALS is a group of heterogenous disorders characterized by degenerative changes in upper motor neurons (UMN) within the motor cortex and/or lower motor neurons (LMN) within the brain stem and spinal cord. Although motor neurons are primarily affected in an earlier and more severe way, ALS is considered a multisystem disorder with nonmotor regions additional involvement, particularly in the prefrontal regions and temporal lobe [2–4].

The median age of onset is 55 years [5]; the disease usually having a fatal course. The median survival from onset to death is less than 3 years [6], although there are large individual variations. A small proportion of patients has a slower rate of progression with a median survival time longer than 10 years [7].

Philip V. Peplow, Bridget Martinez and Thomas A. Gennarelli (eds.), *Neurodegenerative Diseases Biomarkers: Towards Translating Research to Clinical Practice*, Neuromethods, vol. 173, https://doi.org/10.1007/978-1-0716-1712-0_20,
© Springer Science+Business Media, LLC, part of Springer Nature 2022

The etiology of ALS is not fully understood, with most cases (90%) being sporadic. The remaining cases have a family history of the disease, usually the result of dominantly inherited autosomal mutations [8, 9], with SOD1, C9ORF72, FUS, and TARDBP being the most frequently involved genes [9]. While sporadic cases remain mostly unexplained, a large genetic component with high heritability can be found [10]. The pathogenic mechanisms and cellular pathways implicated in ALS are diverse, including disturbances in RNA metabolism, oxidative stress, mitochondrial dysfunction, neuroinflammation, excitotoxicity, impaired DNA repair, protein misfolding and aggregation, defects in nucleocytoplasmic and axonal transport, and oligodendrocyte dysfunction [9, 11, 12].

ALS has a particular and intricate neuropathology. The histologic hallmark is the loss of UMN and LMN with associated corticospinal degeneration [13, 14]. Gross pathologic abnormalities are variable, including spinal cord atrophy of the anterior nerve roots and minor brain atrophy of the frontal and/or temporal cortex with white matter reduction in the corticospinal tract [11, 13, 15]. Microscopic changes include loss of the motor neurons in the spinal cord and pyramidal neurons in the motor cortex (Betz cells), reactive astrogliosis surrounding degenerated motor neurons, microglial activation in the affected areas, and the presence of ubiquitinated inclusion bodies with thread-like, skein-like, or compact morphology in the cytoplasm of surviving neurons and occasionally in glial cells [11, 13, 16, 17]. The transactive response DNA binding protein 43 (TDP-43) is the major component of these ubiquitinated inclusions in ALS, but is not invariably present [13]. Interestingly, iron accumulates in the motor cortex as ferritin, which is sequestered in activated microglia in the middle and deep layers. This is a relevant microscopic feature due to the iron detection capacity of magnetic resonance (MR) imaging [18].

ALS frequently begins in a single region, typically the limb or bulbar regions, usually spreading to contiguous body regions [19]. The site of onset, together with the relative UMN and LMN deficits, region of onset, rate of progression, presence of nonmotor deficits (mainly cognitive and behavioral impairments), and the age of onset, results in a striking heterogeneous phenotypic expression [7, 20–22] A minority of patients show an exclusive impairment of UMN and are diagnosed as Primary Lateral Sclerosis (PLS), while others with an exclusive LMN impairment are diagnosed as Progressive Muscular Atrophy (PMA).

The diagnosis of ALS remains mainly clinical [21], relying on the presence of UMN and LMN signs and symptoms, clinical progression, and exclusion of ALS mimics. While the electromyography is sensitive to detect LMN impairment, clinical UMN biomarkers are lacking. Moreover, the current diagnostic criteria (revised El Escorial criteria and the Awaji criteria) are primarily designed to be used in clinical trials and, although specific, show poor sensitivity [21]. The absence of an UMN objective biomarker and disease heterogeneity result in a long diagnostic delay, preventing early neuroprotective treatments [23, 24] and hindering the recruitment and results' interpretation in clinical trials [25].

Current options are based on symptomatic management, with disease-modifying treatments, such as riluzole and edaravone, increasing survival by a few months in a variable percentage of patients [26, 27]. Other clinical trials have failed to show efficacy, perhaps because of the clinical outcome measures and analysis [28].

Summarizing, there is an urgent need of diagnostic biomarkers to understand the disease pathophysiology, to differentiate between phenotypes and to allow an early diagnosis. Moreover, progression and prognostic biomarkers are also needed in clinical trials to quantify the loss of UMN and LMN and to stratify patients [29].

2   Clinical Endpoints in ALS

As a result of disease heterogeneity, clinical endpoints in ALS are multiple and complex.

2.1   Clinical Endpoints of Disease Progression

The most widely used endpoint in both research and clinical practice is the disability ALS Functional Rating Scale-Revised ALSFRS-R [30], and ordinal rating that scores 12 items from 0 to 4 (from more to less disability) for a total maximum score of 48. The ALSFRS-R is reliable and reproducible, can be generated online and correlates with other important measures such as muscle strength or survival [31–34]. The rate of disease progression measures the number of ALSFRS-R points that an individual loses per month [35], a useful parameter to quantify disease progression and to predict prognosis [35].

Some ALSFRS-R scale pitfalls have to be considered. Namely, it is multidimensional, measuring function but also compliance of noninvasive ventilation; not linearly weighted, meaning that a 1-point change is not a consistent measurable unit across the scale; and contains questions that show improvement due to changes in symptoms management and behavior but not on the disease hallmark [36]. All these aspects reduce its ability to detect changes and limits its utility in clinical trials. For example 25% of patients failed to show decline and 14% experienced an apparent improvement on the ALSFRS-R over a 6-months period in an ALS pooled clinical trial database (PRO-ACT) [37], despite the well-known progressive course of the disease. Using Rasch analysis, a new questionnaire measuring disability and improving item targeting of ALSFRS-R was developed and validated, but it has not been transferred yet into research or clinical practice [36].

Disability in ALS is a result of UMN and LMN impairment. Biomarkers targeting one of these motor neurons will be only moderately associated with disability, fostering the independent clinical measures of UMN and LMN impairment.

Staging criteria are simple clinical milestones in the course of a disease that reflect severity, prognosis and options for treatment [38]. ALSFRS-R measures the disability as reported by the patient, providing an easy, universal and objective measure of disease progression. Two main staging systems have been validated: the Milano Torino Staging system (MiToS) and the King's [39, 40]. The MiToS system uses six stages, from 0 to 5, based on functional ability assessed by the ALSFRS-R, with stage 0 being normal function and stage 5, death. The King's system uses five stages, from 1 to 5, based on the number of the body regions affected (from stage 0 to 3), feeding and respiratory impairment (stage 4), with stage 5 being death. The King's College staging system has shown a higher homogeneity with smaller differences in survival among patients in the same stage and a higher discriminatory ability with greater differences in survival among patients in different stages. This system seems more suitable for individualized prognosis and for measuring efficacy of therapeutic interventions [40, 41].

The UMN are all those fibers with motor functions that descend through the pyramids in the lower brainstem on each side. They include crossed and uncrossed corticospinal tracts and corticobulbar, tectospinal, rubrospinal, vestibulospinal, and reticulospinal tracts, as well as short internuncial and cerebellar connections [42]. UMN symptoms include numbness, loss of distal dexterity, slowness of movements, while weakness is often only mild and predominates in hip and knee flexors [42]. The nine-hole peg test and the finger/foot tapping velocity are widely used to measure these features. UMN signs include spasticity, hyperreflexia, and extensor reflexes (Babinski and Hoffman). Spasticity can be measured with the Ashworth scale [43], while the UMN score measures the number of pathological reflexes [44]. However, in the UMN signs and symptoms are frequently hidden or modified by the usually more prominent LMN signs and symptoms.

The LMN impairment causes prominent weakness, hypotonia, and areflexia and is the result of the degeneration of the alfa-motor neurons in the anterior horn of the spinal cords. Moreover, it is the main cause of respiratory impairment and dysphagia.

Muscle strength is usually measured with the medical research council (MRC) scale [45], having considerable inter- and intraobserver variability. More reliable and reproducible is the strength quantitative measure using hand-held dynamometry [31]. As mentioned above, the co-occurrence of UMN impairment hinders a direct correlation between LMN impairment and muscle strength.

Both respiratory impairment and dysphagia are relevant clinical endpoints in ALS since they constitute the main cause of death. The respiratory impairment is mainly the result of the LMN degeneration at C4 and thoracic levels, causing weakness of the diaphragm and intercostal muscles. Similarly, dysphagia is the result of the degeneration of LMN at the brainstem. The most common clinical endpoint is the time from symptoms onset to noninvasive ventilation or gastrostomy. A patient requiring any of them is considered at the last stage of the disease according to the King's staging [38]. Other clinical endpoints are the need of continuous ventilation (>20 h/day) and to tracheostomy. The respiratory decline in ALS is frequently measured with either the forced vital capacity or the slow vital capacity [46], although the sniff nasal inspiration pressure can be useful in patients with bulbar symptoms [47].

Both cognitive and behavioral impairment are common in ALS, their frequency increases with disease progression and are probably the result of the extension of the neurodegeneration to prefrontal and temporal areas of the brain [48, 49]. About 15% of ALS patients will meet criteria of frontotemporal dementia [50]. Criteria for the diagnosis of mild cognitive and/or behavioral impairment, as well as for frontotemporal dementia in ALS patients have been defined [50]. Although several questionnaires have been developed for the screening and quantification of cognitive and behavioral impairment, the Edinburgh Cognitive and Behavioral ALS Screen is probably the most widely used tool [51].

2.2 Clinical Endpoints of Disease Prognosis

ALS is a severe disease leading to death a few years after disease onset. Consequently, survival is the most relevant clinical endpoint of disease prognosis. At diagnosis, the following factors are independent predictors of survival [52]: age at onset, diagnostic delay, progression rate, forced vital capacity, the presence of frontotemporal dementia, and *C9ORF72* mutation expansion. Consequently, all these factors must be considered when analyzing the role of prognostic biomarkers.

3 Brain Image Biomarkers

Digital medical images allow the qualitative evaluation and computational extraction of derived quantitative data that might define disease stage and estimate clinical endpoints and prognosis. The different features, pathological correlate, potential application, clinical surrogates, and main limitations are summarized in Table 1.

Table 1
Summary of the most relevant imaging biomarkers studied in ALS, their pathologic correlates and clinical relevance. Brain biomarkers have been more thoroughly studied than nerve and muscle ones. However, the clinical heterogeneity of the disease together with several studies limitations (mainly the methodological differences and the limited study populations) hinder the incorporation of most biomarkers into the clinical practice or clinical trials

Biomarker	Pathological correlate	Potential application	Clinical endpoints	Limitations	Ref.
BRAIN					
Qualitative/semiquantitative brain MR imaging biomarkers					
FLAIR/T2 images	Reduced axonal density	Diagnostic biomarker	Differentiates between patients and controls	Low sensitivity and specificity	[29, 53]
T2/T2*/SW images	Hypointensity of the motor cortex	Diagnostic biomarker	Differentiates between patients and controls	Inconsistent results due to clinical and methodological heterogeneity	[29, 53]
	Iron deposition (ferritin) in activated microglia	Progression's biomarker	Associates with ALSFRS-R, phenotype, site of onset and UMN score	Few studies	[29–44]
		Prognostic biomarker	Associates with progression rate	Few studies. No survival studies	[44, 54]
Quantitative MR imaging biomarkers					
Voxel based morphometry (VBM)	Reduction of gray and white matter density	Biomarker	Patients vs. controls differences: Inconsistent results	Insensitive biomarker	[29, 53, 55]
Surface based morphometry (SBM)	Reduction of cortical thickness and brain volumes	Diagnostic biomarker	Thickness and volume reductions in motor and nonmotor cortical areas differentiates patients vs. controls	Few studies. Limited accuracy	[29, 53, 55]
	Gray and white matter atrophy	Diagnostic biomarker	Thickness and volume reductions in motor and nonmotor cortical areas differentiates patients with and without cognitive impairment	Few studies with methodological differences	[29, 53, 55]

		Progression's biomarker	Progressive atrophy of gray and white matter volumes. Correlation with: ALSFRS-R, UMN score, Disease duration, Clinical staging, Site of onset, Phenotype	Inconsistent results due to clinical and methodological heterogeneity	[29, 53, 55]
		Prognostic biomarker	Atrophy of motor and nonmotor cortical areas correlates with progression rate. Survival prediction	Few studies. Only one survival study	[56–59]
Quantitative iron mapping	Relaxometry (R2*)	Iron deposition (ferritin) in activated microglia	Diagnostic biomarker	Differentiates patients vs. controls	Few studies. Limited accuracy	[29, 53]
	Quantitative susceptibility mapping (QSM)	Iron deposition (ferritin) in activated microglia	Diagnostic biomarker	Differentiates patients vs. controls	Few studies. Methodological heterogeneity	[29, 53]
			Progression's biomarker	Correlates with ALSFRS-R and UMN impairment	Few studies. Methodological heterogeneity	[29, 53]
Diffusion tensor imaging (DTI)	Reduction of fractional anisotropy (FA)	Wallerian degeneration of white matter tracts	Diagnostic biomarker	FA reduction in CST differentiates patients from controls	Variable accuracy. Methodological heterogeneity	[29, 53, 55, 60]
			Progression's biomarker	Correlation with: ALSFRS-R, UMN score/spasticity, Muscle strength, Disease duration, Clinical staging, Site of onset, Phenotype (UMN vs. LMN predominant)	Methodological heterogeneity	[29, 53, 55]
			Prognostic biomarker	Correlation with progression rate	Few studies. No survival studies	[29, 53, 55]

(continued)

Table 1
(continued)

Biomarker	Pathological correlate	Potential application	Clinical endpoints	Limitations	Ref.
		Progression's biomarker	Correlation with: Spasticity Disease duration Site of onset ALSFRS-R	Methodological heterogeneity	[29, 53, 55]
Increase of mean diffusivity (MD), axial diffusivity (AD) and radial diffusivity (RD)	Wallerian degeneration of white matter tracts	Prognostic biomarker	Correlation with progression rate	Few studies. No survival studies	[29, 53, 55]
		Pathophysiologic biomarker	Direct connections of the motor cortex are more affected	Few studies	[29, 53, 55]
Structural connectivity	Wallerian degeneration of white matter tracts	Progression's biomarker	Evidence of disease spread, resembling the pathological stages	Few studies. Methodological heterogeneity	[29, 53, 55]
		Pathophysiologic biomarker	Direct connections of the motor cortex are more affected and there is evidence of disease spread, resembling the pathological stages. Functional connectivity increases at early stages, but reductions when the WM tracts are affected by DTI.	Few studies with inconsistent results. Methodological heterogeneity	[29, 53, 55]

Functional MR (fMR)	Resting state fMR (functional connectivity)	Functional connectivity changes	Progression's biomarker	Evidence of disease spread, resembling the pathological stages. Correlation with: ALSFRS-R Muscle strength	Few studies. Methodological heterogeneity	[29, 53, 55]
			Prognostic biomarker	Correlation with progression rate	Few studies	[29, 53, 55]
			Progression's biomarker	Correlation with: ALSFRS-R UMN impairment Disease duration Site of onset Phenotype (UMN vs. LMN predominant)	Methodological heterogeneity	[29, 53, 55]
Multiparametric MR imaging	Most frequently, combinations of cortical thickness and FA, but also combinations of structural and functional biomarkers	Gray and white matter degeneration	Diagnostic, prognostic and progression's biomarker	The multimodal approach has been shown superior for the diagnosis, progression's monitoring and prognostic stratification	Few studies. Methodological heterogeneity	[61–65]
Roots, peripheral nerves and muscles						
Ultrasound						
Roots and nerves	Cross-sectional area (CSA)	Root/nerve Wallerian gdegeneration vs. inflammation	Diagnostic biomarker	Differentiates between ALS patients and multifocal motor neuropathy patients	Considerable overlap between patients and controls. Limited accuracy	[66]
			Progression's biomarker	No longitudinal changes nor relevant correlation with clinical variables	Combination of neurodegeneration and nerve inflammation	[66]

(continued)

Table 1
(continued)

Biomarker	Pathological correlate	Potential application	Clinical endpoints	Limitations	Ref.
				probably affects the negative results	
Muscle	Fasciculations as defined by electromyography	Pathophysiological biomarker	Origin, rise and fall of fasciculations	Few studies	[67]
		Diagnostic biomarker	Increases detection of fasciculations	Needs additional electromyography (for chronic denervation)	[68]
Muscle thickness	Muscle atrophy	Diagnostic biomarker	Reduced in patients vs. controls	Considerable overlap between patients and controls. Values Influenced by age, sex and BMI	[68]
		Progression's biomarker	Thickness reduction with disease progression. Weak correlation with muscle strength	Few longitudinal studies. Values influenced by age, sex, and BMI	[68]
Echogenicity	Muscle fat/fibrous infiltration	Diagnostic biomarker	Reduced in patients vs. controls	Considerable overlap between patients and controls. Values Influenced by age, sex, and BMI	[68]
		Progression's biomarker	Little association with muscle strength or disability. Discrepancy about longitudinal changes	Few longitudinal studies. Values influenced by age, sex, and BMI	[68]
		Prognostic biomarker	Changes in signal predict survival	Only one study	[69]
Echovariation	Muscle fat/fibrous infiltration	Diagnostic biomarker	Differentiates between patients and controls	Only one study	[69]

	Biomarker	Pathologic correlate	Biomarker type	Findings	Limitations	References
Magnetic resonance	Textural changes	Muscle fat/fibrous infiltration	Progression's biomarker	Correlates with muscle strength and disability. Discrepancy about longitudinal changes	Few studies	[70, 71]
			Diagnostic biomarker	Differentiate between patients and controls	Few studies. More complex post-processing	[68]
Roots and nerves — Qualitative biomarkers	T2 hyperintensities, gadolinium enhancement	Nerve inflammation	Diagnostic and pathophysiologic biomarker	Characteristic of rapidly progressing patients	Few studies. Selection bias. Limited accuracy	[66]
Quantitative biomarkers	Cross-sectional area	Root/nerve Wallerian degeneration vs. inflammation	Diagnostic and pathophysiologic biomarker	Differentiate between patients and controls, and detection of "inflammatory" patients	Few studies. Selection bias. Limited accuracy	[66]
Muscle — Qualitative/semiquantitative biomarkers	T2 relaxation time increase and STIR hyperintensity	Muscle denervation and fat/fibrous infiltration	Progression's biomarker	Correlate with neurophysiological and clinical endpoints. Longitudinal changes detected	Few studies	[72]
Quantitative biomarkers	Muscle thickness	Muscle atrophy	Progression's biomarker	Longitudinal changes detected	Few studies	[72]

3.1 Qualitative Biomarkers

Qualitative brain MR imaging features, such as hyperintensity of the corticospinal tract on FLAIR/T2 and a low signal intensity rim in the precentral cortex on the T2 weighted images, are considered as classical MR imaging findings in ALS patients. Corticospinal tract hyperintensities might reflect areas of reduced axonal and myelin density [73] and have been associated with UMN signs and bulbar onset [44]. However, they show limited sensitivity and specificity [29]. Hypointensities of the motor cortex correspond to iron deposition, in the form of ferritin, in activated microglia in the middle and deep layers of the motor cortex, as proved in histologic analysis and postmortem examinations [18]. Iron causes a concentration-dependent irregularity in the local magnetic field and a signal darkening on T2 and T2* weighted images that is greater with increasing magnetic field strength. While initial studies noted these hypointensities in T2 and T2* weighted images in a subset of ALS patients, such changes were neither sensitive nor specific [29].

Susceptibility weight (SW) imaging is a three-dimensional gradient echo sequence with full flow compensation and high resolution that enhances image contrast by using the susceptibility differences between tissues, which aids in the identification of paramagnetic nonheme iron. SW images are created by combining both magnitude and phase information in the gradient echo data, iron appearing hypointense and easily assessed. The main advantage of SW as opposed to T2 weighted images is to provide a higher contrast image with more sensitivity to the presence of iron in the tissue. Consequently, SW images have been proved superior to T2 and T2* weighted images to detect the iron deposition in the precentral cortex in ALS patients [74]. In our experience, a careful semiquantitative assessment of the SW images in 3 T MR images shows good sensitivity and very good specificity for ALS diagnosis and could therefore act as a diagnostic biomarker. Moreover, these hypointensities are a marker of UMN degeneration, more frequently found in bulbar onset patients [44]. Interestingly, their intensity and extent in the different motor homunculus regions (lower limbs, upper limbs, and bulbar) are linked to the symptom's onset site, suggesting that the regional semiquantitative measurement of iron-related hypointensities following the motor homunculus can be used as a measure of disease progression [44].

Summarizing, the semiquantitative assessment of the motor cortex on SW images is an easy to implement tool that could act as a diagnostic and progression's biomarker of UMN degeneration. However, further studies are warranted before it can be implemented in both clinical trials and clinical practice.

3.2 Quantitative Biomarkers

Computational quantitative imaging evaluates the properties and behavior of tissues through their acquired images following a well-define pipeline, including image acquisition and preparation, brain parcellation, specific areas segmentation, data extraction, data analysis, and the correlation with the relevant outcomes (Fig. 1). To be considered as imaging biomarkers, the extracted features have to be subrogated to biological processes or pathological changes, providing their regional distribution and magnitude. These parametric computational images are resolved in space (parametric distribution analysis) and time (longitudinal studies) [75].

To be used clinically as imaging biomarkers, the different extracted features and parameters have to pass a technical validation of variability, including the exclusion of redundant data, the analysis of repeatability (stability on test-retest data sets) and reproducibility (stability on similar but independent data sets). Even more, the extracted biomarkers have to be accurate on the clinical validation with a high correlation with the endpoints to which they are surrogated [76].

If proven useful, the selected biomarkers distribution and values should be incorporated within the structured radiological report as coded objective quantitative entries and controlled ontology. The structured report constitutes an indexed and traceable database, allowing large-scale data extraction and exploitation, improving consistency and completeness, and fostering clinical innovation (Fig. 2).

To be reproducible and robust, imaging biomarkers should be extracted from normalized images. The following sections will deal with the pipeline of image preparation and biomarkers extraction.

4 Brain Image Preparation

Image preparation is an important and challenging factor in the quantitative medical imaging field to reduce technical variability across images and to remove unwanted artifacts and transform the data into a standard format.

There are several neuroimaging data processing tools that are widely used, such as SPM [77], AFNI [78], FSL [79], FreeSurfer [80], ANTs [81], Workbench [82], fMRIPrep [83], and CONN [84]. Different data preprocessing pipelines yield slightly differing results. In this section, we describe a preprocessing pipeline for structural and functional MRI data by combining components of well-known software packages to fully incorporate recent developments in MRI preprocessing into a single coherent image preparation strategy for ALS patients.

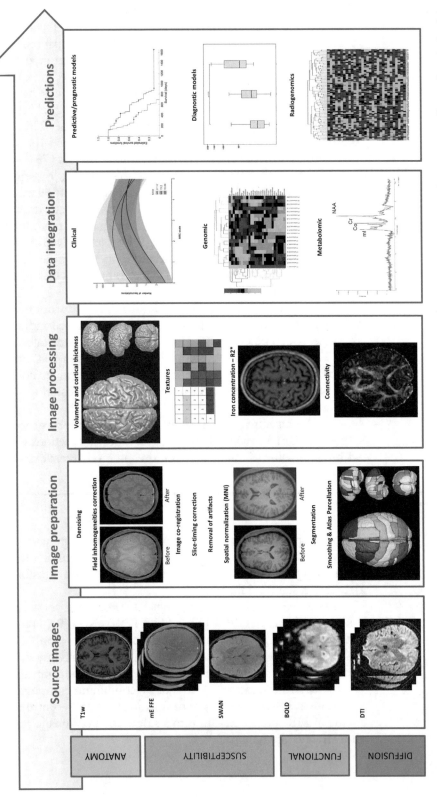

Fig. 1 Image processing pipeline for the extraction of brain imaging biomarkers in ALS patients, including source images, image preparation and analysis, data integration, radiogenomics and clinical outcomes modelling

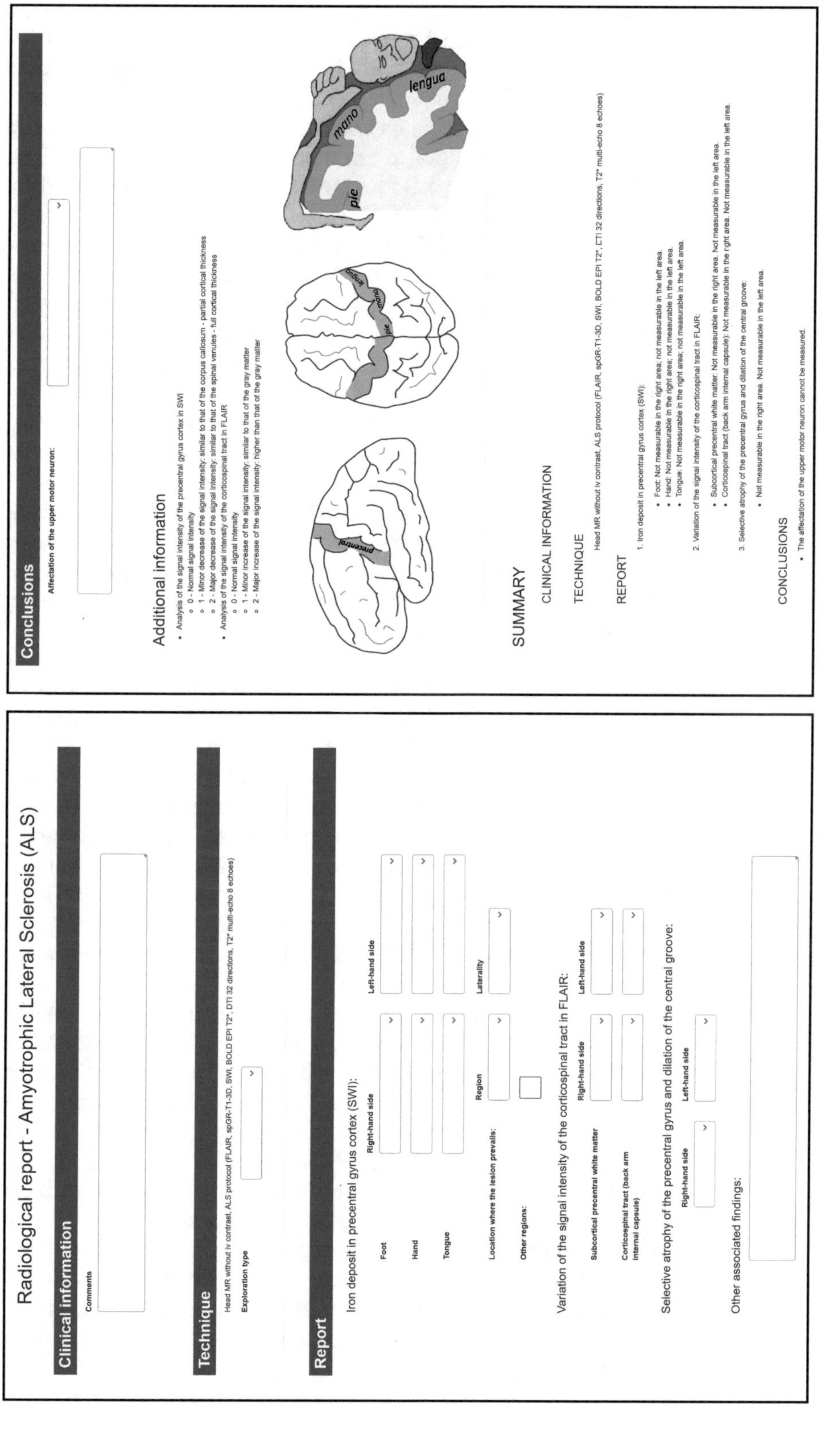
Radiological report - Amyotrophic Lateral Sclerosis (ALS)

Clinical information

Comments

Technique

Head MR without iv contrast, ALS protocol (FLAIR, spGR-T1-3D, SWI, BOLD EPI T2*, DTI 32 directions, T2* multi-echo 8 echoes)

Exploration type

Report

Iron deposit in precentral gyrus cortex (SWI):

Right-hand side

Left-hand side

Foot

Hand

Tongue

Location where the lesion prevails:

Region

Laterality

Other regions:

Variation of the signal intensity of the corticospinal tract in FLAIR:

Subcortical precentral white matter

Corticospinal tract (back arm internal capsule)

Selective atrophy of the precentral gyrus and dilation of the central groove:

Other associated findings:

Conclusions

Affectation of the upper motor neuron:

Additional information

- Analysis of the signal intensity of the precentral gyrus cortex in SWI
  - 0 - Normal signal intensity
  - 1 - Minor decrease of the signal intensity: similar to that of the corpus callosum - partial cortical thickness
  - 2 - Major decrease of the signal intensity: similar to that of the spinal venules - full cortical thickness
- Analysis of the signal intensity of the corticospinal tract in FLAIR
  - 0 - Normal signal intensity
  - 1 - Minor increase of the signal intensity: similar to that of the gray matter
  - 2 - Major increase of the signal intensity: higher than that of the gray matter

SUMMARY

CLINICAL INFORMATION

TECHNIQUE

Head MR without iv contrast, ALS protocol (FLAIR, spGR-T1-3D, SWI, BOLD EPI T2*, DTI 32 directions, T2* multi-echo 8 echoes)

REPORT

1. Iron deposit in precentral gyrus cortex (SWI):
   - Foot: Not measurable in the right area; not measurable in the left area.
   - Hand: Not measurable in the right area; not measurable in the left area.
   - Tongue: Not measurable in the right area; not measurable in the left area.
2. Variation of the signal intensity of the corticospinal tract in FLAIR:
   - Subcortical precentral white matter: Not measurable in the right area. Not measurable in the left area.
   - Corticospinal tract (back arm internal capsule): Not measurable in the right area. Not measurable in the left area.
3. Selective atrophy of the precentral gyrus and dilation of the central groove:
   - Not measurable in the right area. Not measurable in the left area.

CONCLUSIONS

- The affectation of the upper motor neuron cannot be measured.

Fig. 2 Example of our Structured Report in ALS patients, including quantitative biomarkers analysis and graphic display of main findings

4.1 Denoising and Field Inhomogeneities Correction

Denoising is an important preprocessing step in medical image analysis, whose main goal is to reconstruct the original image with a higher quality while preserving important detail features such as edges and textures. Various denoising techniques and algorithms have been proposed, such as the anisotropic diffusion filter, the bilateral filter, the nonlocal mean filter, and the Wiener filter. The performance of these methods can be quantitatively evaluated by calculating objective Image Quality Metrics, such as the peak signal-to-noise ratio and the structural similarity index values.

Field inhomogeneities correction avoids artefactual nonuniform image intensities that might affect the extracted images features. Correcting for such variance produces tissue classes that have similar intensity signatures. In the proposed image preprocessing pipeline, the obtained denoised images can be used to produce N4-bias corrected scans by employing the ANTs-based toolkit, where care must be taken to select the most optimal parameters and arguments.

These steps improve the image registration and segmentation processes, image visualization, and reliability of associative quantitative analysis. The impact of the denoising and field inhomogeneities correction algorithms on the performances of post-processing algorithms has to be further evaluated.

4.2 Image Registration

Image registration automatically align and transform images into a common reference coordinate framework, so corresponding pixels represent homologous biological points. Multisequence image registration constitutes an essential step aiming to extract reliable imaging biomarkers and build computational models for prediction of clinical outcomes. The development of an image registration pipeline enables comparisons between subjects, image sequences, and longitudinal in time. The reference frame can either be a common template space to which every subject's image is registered or a region-labeling system obtained with multiatlas segmentation.

Intrasequence registration implies registration of images acquired using the same scanning sequence, as in multiecho GRE or BOLD MR series, being internally registered to the reference highest quality image by linear transformations. Intersequence registration allows the registration of different sequences within the same modality, such as DTI or SW images to T1 weighted images, by combining linear and nonlinear transformations. Intrasubject spatial registration of brain images usually makes use of rigid or affine transformations to remove small movements that take place during scanning. In the cases of intrasubject anatomical variability across time, and intersubject registration tasks, nonrigid, elastic, or symmetric diffeomorphic normalization transformations using B-splines are recommended.

The definition of accurate and robust similarity measures to evaluate the spatial correspondence between images has become a challenging research area. Among the different proposed image registration measures, mutual information and normalized mutual information are the most commonly used due to their accuracy, robustness, and universality. Mutual information is based on the statistical relationship between images and normalized mutual information improves the robustness of mutual information minimizing misregistration. Two of the most popular state-of-the-art image registration tools are ANTs [81] and Elastix [85].

4.3 Removal of Artifacts

Image analysis can lead to erroneous conclusions when original data is of low quality. MR images are unlikely to be artifact-free, and assessing their quality has long been a challenge. Traditionally, MR images are visually inspected by experts, excluding those showing an insufficient level of quality. Visual assessment is time consuming and prone to variability. An additional concern is that some artifacts evade human detection. Automated quality control (QC) protocols for fully automated, robust, and minimally biased checks are mandatory. The absence of a "gold standard" impedes the definition of relevant quality metrics, and human experts introduce biases with their visual assessment. In addition, cross-study and intersite acquisition differences introduce uncharacterized variability. Machine-specific artifacts have generally been tracked down in a quantitative manner using phantoms. However, many forms of image degradation are participant-specific or arise from practical settings. Artifact detection for MR raw data can be performed using Artifact Detection Tools [86]. This post-processing tool facilitates the identification of outlier values in MR time series that likely reflect artifacts and provides diagnostics tools which assist in appropriate design specification and model estimation. It offers an automatic and manual detection of global mean intensity and motion outliers in MRI data that may be omitted from subsequent statistical analyses.

4.4 Slice-Timing Correction

Exact timing is essential for functional MR imaging (fMRI) data analysis. Datasets are commonly measured using repeated 2D imaging methods, resulting in a temporal offset between slices. To compensate for this timing difference, slice-timing correction [87] has been used as a preprocessing step for more than 15 years and is currently included in all major fMRI software packages, such as SPM, AFNI, BrainVoyager [88], or FSL. In slice-timing correction, the individual slice is temporally realigned to a reference slice based on its relative timing using an appropriate resampling method. Different data interpolation methods have been proposed, including *linear*, *sinc*, and *cubic spline* interpolation, increasing the robustness of the data analysis.

4.5 Brain Spatial Normalization and Segmentation

Normalization methods transform individual brain images into a standard anatomical coordinate space. They require a suitable template image, which is already in the standard space, and an elastic matching procedure for appropriately warping images. Available templates (atlases) use *atlas* images representing a "standard" anatomy imaged with a certain modality. The MNI (Montreal Neurological Institute) templates are commonly used in this setting [89]. Image warping standardized a coordinate space using a flexible matching technique, such as a nonlinear diffeomorphic normalization algorithm supplied by ANTs.

Segmentation is the process of assigning a label to each image voxel, allowing differentiating those that share a certain characteristic. This step attempts to differentiate the previously normalized brain into its different tissues of interest (gray matter [GM], white matter [WM], and cerebrospinal fluid [CSF]). Before starting the segmentation process, a routine is applied to remove the brain tissue from nonrelevant tissues, such as skull, vessels, scalp, eyes, fat, and muscles. This algorithm first performs a process of erosion of the WM until obtaining a homogeneous set of voxels. Then, conditional dilations are performed during several iterations by requiring the presence of GM or WM. This skull-stripping process can be achieved using the Brain Extraction Tool (BET) algorithm from FSL.

The SPM software includes one of the most popular strategies to perform GM, WM, and CSF segmentation, providing probability maps of each tissue. In these maps, voxels have an intensity value normalized between 0 and 1, indicating the probability of belonging to a specific tissue. In order to generate these probability maps, a priori probability atlases representing the usual probabilities of belonging to a given tissue for a group of healthy adults are used. Iteratively, the algorithm calculates the mean and variance of voxels contained in each tissue. Taking these values as a reference, the probability of membership is recalculated following a Gaussian probability function. The algorithm continues iterating until the convergence criterion is reached or the number of maximum iterations specified is exceeded.

4.6 Smoothing and Atlas Parcellation

The smoothing process increases the signal-to-noise ratio by reducing the influence of high frequencies in the image. In this procedure, the images are convoluted with an isotropic gaussian kernel, each image voxel becoming the weighted average of its neighboring voxels, as defined by the shape of the kernel used. Through smoothing, a normal distribution of the information is also achieved, following the central limit theorem. These features provide greater reliability and robustness to the subsequent statistical results. Furthermore, smoothing compensates those defects or inaccuracies that have been generated during the spatial normalization of the image. This also benefits the statistical analysis, where homologous regions are compared with a voxel-by-voxel strategy to find variations or significant local differences.

In order to study the brain by regions and to establish functional relationships between different regions of the GM/WM/CSF and known functional networks, it is necessary to previously parcellate the brain by applying an anatomical atlas. The Harvard-Oxford atlas [90] is one of the most commonly used templates to segment the GM in its different regions. To be able to correlate the brain functional activity with the already known networks, the brain can be segmented by another specific atlas, called the Human Connectome Project (HCP) *brain atlas* [91].

5 Brain Image Processing

To decipher the relationship between the motor manifestation and UMN/LMN deficit, the obtention of an objective and reliable UMN and LMN deficit marker in ALS is highly desirable.

5.1 Brain Volumetric Analysis

Structural studies aim to detect or locate morphological abnormalities. MR imaging allows a great visualization of the different structures of the brain and the inclusion of those alterations that are not directly related to physical trauma but to pathology (such as ALS, schizophrenia or Alzheimer's disease). Measurements of brain volume and morphometry (Fig. 3) become effective tools to assess disease classification and progression, offering reliable quantitative information on the underlying central nervous system disorders [92].

Voxel-based morphometry (VBM) is one of the most commonly used methods to analyze anatomical abnormalities. VBM conducts a study of GM volume variations in the brain by analyzing the image intensity in order to identify potential local structural alterations. By using a General Linear Model [93], a specific brain is compared to a group of healthy subjects in order to infer the presence of tissue atrophy or expansion. The main advantage of VBM is that it allows the evaluation of anatomical changes throughout the brain with great regional specificity in a voxel-by-voxel way. In addition, results are invariant with respect to different users and do not depend on a particular hypothesis [94]. VBM studies have shown volume changes and atrophy in classic motor areas, including the precentral gyrus, and in white matter areas, particularly along the corticospinal tracts [95].

Sub-cortical atrophy, related with cognitive and behavioral impairment can be assessed by the measurement of the width of the third ventricle. This easily obtained biomarker needs to be validated [96].

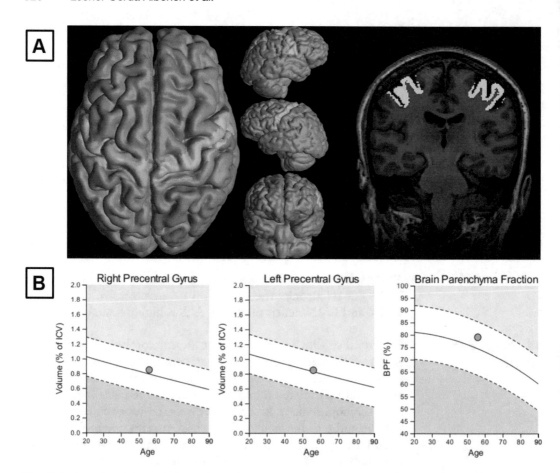

Fig. 3 Brain volumetric analysis performed on a 56 years old patient with predominant LMN ALS (**a**). Percentage of right (0.855%) and left (0.852%) precentral gyrus volume, with respect to the intracranial volume (ICV), and the corresponding brain parenchyma fraction (79.1%) showed no significant atrophy in this case (**b**)

5.2 Cortical Thickness

The measurement of motor cortical thickness appears to be a useful and sensitive method for UMN impairment in ALS [58]. Several surface-based MR imaging morphometry studies have demonstrated significant thinning of the motor cortex in ALS patients with the potential to track ALS spread along the motor cortex [15, 97]. Bulbar-onset patients were reported to show cortical thinning in the bilateral bulbar-segment of motor homunculus. In contrast, patients with lower limb-onset demonstrated thinning in the bilateral segments of the motor homunculus representing arms and legs [98]. The different region onset was an aspect of clinical heterogeneity in ALS [20]. It has been suggested that different body onsets correspond to different motor homunculus partitions [98]. Therefore, when localizing the anatomic distribution of UMN loss, more detailed motor partitions are bound to help

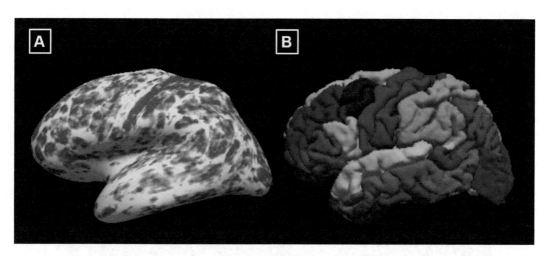

Fig. 4 Representation of cortical thickness map (**a**) and corresponding Desikan-Killiany brain parcellation (**b**) in a 56 years old patient with predominant LMN ALS showing results from general linear models compared to a normal adjusted population

characterize UMN damages. Recently, a more detailed and refined division of the motor homunculus cortex was proposed, which comprises head–face, tongue–larynx, upper-limb, trunk, and lower-limb areas [99].

The FreeSurfer software is one of the most popular tools to measure per subject cortical thickness (Fig. 4). The tissue classification is allowed by the definition of the boundaries between WM and GM (white matter surface) and GM and CSF (pial surface). Cortical thickness is then defined as the distance from the white matter surface to the nearest point on the pial surface. It is then recommended to perform a visual evaluation of cortical thickness by an experienced radiologist with several years of MR imaging reporting experience. Cortical thickness should be included in the structured report of brain MR exams in ALS patients as a relevant phenotypic, progression and prognostic biomarker (Table 1).

5.3 Iron Concetration-R2*

MR susceptibility is an intrinsic characteristic of the elements and, if handled properly, can be of great utility [100]. The ferritin/iron deposits observed in patients with ALS have a strong ferromagnetic susceptibility, which can be detected, staged and quantified. The presence of these deposits produces a local inhomogeneity in the magnetic field that affects the decay of the T2* transverse relaxation time, caused by the spin-spin interaction and the field inhomogeneity [29]. However, for magnetic fields larger than 1.5 T the signal-to-noise ratio of the phase images of GRE sequences are shown to be higher than those of modulus [101].

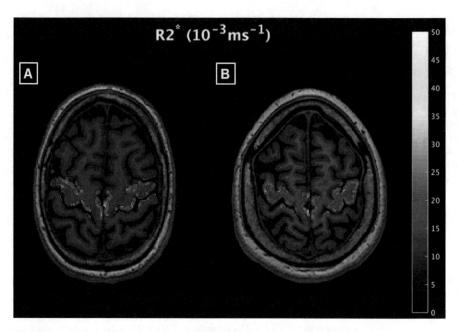

Fig. 5 R2* mapping of the primary motor cortex performed on a patient with predominant LMN ALS (**a**) and a paired healthy subject (**b**). The median R2* values calculated in the right and left motor cortices of the ALS patient were higher (15.1 and 13.9 × 10⁻³ ms⁻¹) than in the control subject (12.7 and 11.8 × 10⁻³ ms⁻¹, respectively) due to higher iron deposition in these regions

Different approaches based on GRE sequences allow us to quantify susceptibility effects in MR imaging: T2*/R2* mapping, Susceptibility Weighted Imaging (SWI), Quantitative Susceptibility Mapping (QSM) and Susceptibility Tensor Imaging (STI). The first two have been observed to be relevant in the evaluation of suspected ALS because abnormal hypointensity in the precentral gyrus and GM has been noted to be highly sensitive and specific for ALS. In addition, they are the easiest to implement and are widely used in current diagnostic equipment, thus being the ones that have the fastest translation to the clinic.

The T2*/R2* mapping is used to obtain the T2*/R2* value of each voxel from the decay of the transverse magnetization in each voxel. The multiple number of echoes allows to sample in time this decay to obtain the T2*/R2* value by solving the signal decay equations. The T2*/R2* values are related to the amount of iron or ferritin deposit present (Fig. 5).

SW images take into account both magnitude and filtered phase images, combined in a multiplied way, adding information to the contrast between tissues of different susceptibility. In this scenario, the discontinuity due to the phase periodicity $[-\pi, \pi]$ has to be considered. The contribution of the background noise generated by the receiving coils, the disturbance of the patient's own body to the static magnetic field, as well as the large differences in susceptibility of the tissues to air constitute sources of uncertainty.

It is necessary to use unwrapping techniques and a high-pass filter, since the contributions of the background to the phase variation are generally smooth. The elimination of low frequencies leads to a loss of information at the tissues of interest. More complex methods try to minimize this loss of information, such as the Sophisticated Harmonic Artifact Reduction for Phase data (SHARP), the Projection onto Dipole Field (PDF) and the Harmonic Phase Removal using the Laplacian operator (HARPERELLA) [100]. Defining the extent and quantity of iron deposits should be included in the MR structured report as relevant diagnostic and progression biomarkers.

5.4   Texture-Radiomics

Radiomics is an emerging translational field of research characterized by the high-throughput extraction, storage, and analysis of a large number of quantitative imaging features (that might constitute imaging biomarkers) providing quantitative information (virtual biopsies) for early disease diagnosis, disease phenotyping, disease grading, targeting therapies, and evaluation of disease response to treatment [75, 102].

As an example, hyperechogenicity of nigrosome 1 in the substantia nigra has been found to be a marker of neuronal vulnerability in ALS patients [103]. Characterization of this nigrosome alteration can also be performed by MR susceptibility weighted imaging. The use of textures analysis in this setting has to be defined within the data driven domain to allow the obtention of a huge amount of picture features attributes which describe the signal intensity relationships within the processed images. These quantitative descriptors exhibit different levels of complexity and express properties of the spatial arrangement of the intensity values at voxel level. They can be extracted by developing and implementing mathematical algorithms either directly on the original images or after applying different filters or transforms (e.g., fractal transform).

One of the most popular tools for feature extraction is the Pyradiomics software [104], which allows for the extraction of thousands of quantitative radiomics features. Such features can be derived, for instance, from the gray-level co-occurrence matrix, quantifying the incidence of voxels with same intensities at a predetermined distance along a fixed direction, from the gray-level run-length matrix, quantifying consecutive voxels with the same intensity along fixed directions, or from the neighborhood-gray-tone-difference matrix, quantifying the difference between a gray value and the average gray value of its neighbors within a predetermined distance.

Other higher-order statistics features are obtained by statistical methods after applying filters or mathematical transforms to the images. For instance, with the aim of suppressing noise, identifying repetitive or nonrepetitive patterns, or highlighting details. These include methods such as fractal analysis, Minkowski functionals, wavelet transform, and Laplacian transforms of Gaussian-filtered images, which can extract areas with increasingly coarse texture patterns.

The innovation brought by radiomics lies in the use of higher-level data extracted from a single region, which is mathematically processed with advanced statistical methods and big data analytics under the hypothesis that when combined with clinical data, a tissue can be characterized accurately. Radiomics can provide a platform for a more personalized, higher-quality, and cost-effective diagnosis, prognosis, and outcome for a patient. Although radiomics research shows great potential, efforts are still required for the development of a translational radiomics pipeline from methodology to implementation in real-world clinical applications. Reproducibility and clinical value of radiomic features are required to be tested with an internal cross-validation and then validated on independent external cohorts to become a viable tool for medical decision-making. The potential application of gray matter texture analysis is still a matter of research in ALS patients.

5.5 Functional Connectivity

Changes produced in brain structure, such as those described in previous sections, have an implication on different brain function, including cognitive function and deficit of nonmotor component, and vice versa [105].

Brain function is identified by the activation of neurons, which, due to the lack of energy/glucose reserves, require additional oxygen to be supplied by the bloodstream on demand. This creates a close relationship between neuronal activation, functional activity, and the hemodynamic response of the brain. This hemodynamic response is characterized by the greater presence of oxyhemoglobin in the presence of activation than of deoxyhemoglobin. As deoxyhemoglobin is more paramagnetic than oxyhemoglobin, $T2^*$-sensitive GRE sequences can detect these temporal variations. This process is known as Blood Oxygen Level Dependent (BOLD) weighting.

BOLD sequences must be sensitive to small changes in $T2^*$ and have sufficient temporal and spatial resolution to cover the entire brain while sampling the temporal variations in the hemodynamic response. The temporal resolution below 4 s is almost certainly the most important parameter for this type of analysis. The echo time should be adjusted to be close to the $T2^*$ of the tissue, between 30 and 35 ms in 3 T equipment. Regarding the spatial resolution, 2–4 mm slice thickness is usually obtained with a 2–3 mm in-plane resolution. In all cases, the acquisition must be made in the transversal plane and parallel to the anterior-posterior commissure.

Image preparation follows standardized pipelines, like the one suggested by CONN toolbox [106]. Firstly, an intrasequential registration of all the time moments acquired is made. This allows to have all the temporal instants aligned, correcting possible movements of the patient during the acquisition and providing a registration to be used as a covariate to detect potential outliers. This preprocessing of the images is possible with the realign and unwrap algorithms provided by SPM. The next step is to perform a slice-timing correction that allows to correct the temporary displacement obtained due to the nature of the acquisition. After the spatial and temporal realignment performed in the previous steps, the identification of outliers is performed based on the delimitation of thresholds. Some recommended thresholds are: >1 mm of displacement and >5 SD of variation of the BOLD signal. One specific toolbox is the Artifact Detection Tools [86], which supports SPM and is adaptable to the FSL environment.

Then, an intersequence registration is made, where 3D T1 weighted anatomical and BOLD sequences are registered to the MNI standard space, and the GM, WM, and CSF are segmented. Finally, the BOLD signal is filtered with a convolutional Gaussian kernel filter to improve the signal-to-noise ratio. Denoising takes into account the noise components obtained from the WM and CSF signal [107], motion parameters, and other possible confounders extracted during image preparation and, finally, a bandpass filtering [108] which allows to eliminate those frequencies that are not expected in the range of the hemodynamic response (between 0.008 and 0.09 Hz).

The variation of the signal in time for each voxel/region of the brain of each subject allows to generate the region-to-region connectivity matrices, in which each element of the matrix represents the bivariate Fisher-transformed correlation coefficient for each pair of regions (Fig. 6).

The information provided by these matrices can be used to find differences between patients with ALS by performing an analysis of covariance (ANCOVA), in which it is possible to consider the influence of other factors, such as age, sex, schooling and laterality that might also have an effect on connectivity. Due to the multiple tests performed, it is recommended to use p-value correction such as the False Discovery Rate technique [109].

Resting state fMR information might be incorporated in the structured report as prognostic and progression biomarkers, although the level of evidence is still weak.

5.6 Tractography

Water molecule movement within the brain is limited by the intracellular, extracellular and intravascular spaces. In addition, diffusion might have a preferential direction of movement, which is characterized by the fractional anisotropy (FA). The anisotropy fraction can take values between 0 and 1, being 0 the absence of preferential direction and 1 the total restriction to a single preferential direction.

Currently, diffusion sequences based on pulsed gradient spin echo (PGSE) with symmetrical gradients on each side of the 180°

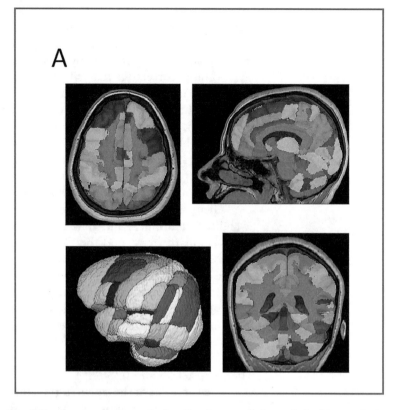

Fig. 6 The three main steps of a functional connectivity analysis: (**a**) parcellation of brain cortical regions from a predefined atlas

pulse are used. In this way, stationary molecules are not affected since the phase increase produced by the first gradient pulse is counteracted by the second, while molecules that show movement, being in different positions between the first and second pulse, are out of phase and lose their signal. To carry out the diffusion tensor analysis, 32–128 directions are recommended. This sensitivity to the direction of movement is achieved by means of controlled activation of the gradients, which allows to quantify the movement in the number of noncollinear desired directions. As chemical displacement artifacts may appear, the use of selective fat saturation pulses or nonselective inversion pulses are recommended. Geometrical distortion may also occur due to the high Echo Planar Imaging (EPI) factors used.

For DTI image analysis there are different toolboxes to obtain the maps and the WM tracts. There are semiautomatic toolboxes that provides the tracts by means of predefined atlases, as is the case of ENIGMA DTI pipeline [110], and more manual processes with delimitation of regions and inclusion conditions as Medinria [111]. Recent studies show that FA is postulated as a promising imaging biomarker for the diagnosis and follow-up of patients with

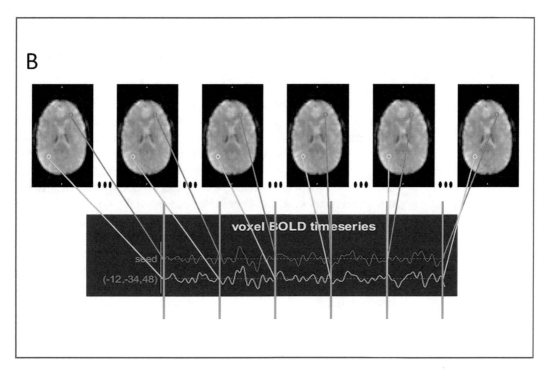

Fig. 6 The three main steps of a functional connectivity analysis: (**b**) in-time changes within a voxel

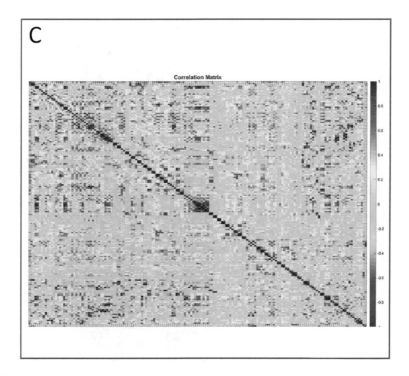

Fig. 6 The three main steps of a functional connectivity analysis: (**c**) correlation matrix between the voxels

ALS, due to the observed FA decrease, especially in the corticospinal and corpus callosum tracts. Similarly, changes in the motor and nonmotor regions related tracts are useful progression biomarkers [29] (Fig. 7).

The use of proton MR spectroscopy (MRS) provides a measure of cerebral metabolites relevant to neurodegeneration in vivo. Changes in motor gyri and CST reflecting neuronal impairment (*N*-acetylaspartate, NAA), inhibitory neurotransmission (gamma aminobutyric acid, GABA), glial destruction (myoinositol, mIns), energy metabolism (creatine, Cr; and phosphocreatine, PCr), cell membrane turnover (Choline, Cho), and excitatory neurotransmitter (glutamate, Glu) markers have been related to diagnosis, prognostic, and progression biomarkers [29, 53, 55, 61–65, 112]. Despite MRS high expectations, data analysis remains challenging, most studies had few cases, there were no validation analysis, and no main clinical endpoints have been subrogated. In the upcoming years, MRS might play a role in addressing phenotypic heterogeneity.

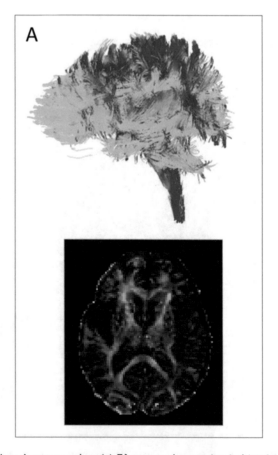

Fig. 7 DW imaging processing. (a) FA map and reconstructed tracts

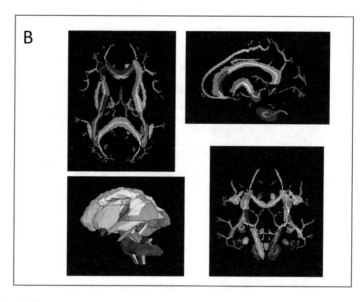

Fig. 7 DW imaging processing. (**b**) Atlas based on tracts

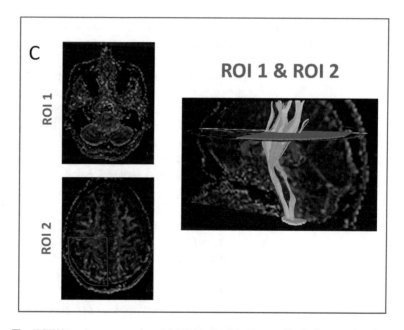

Fig. 7 DW imaging processing. (**c**) DTI tract extraction method after seed regions delimitation

D

Volume (mm³)	Number of fibers	min length (mm)	max length (mm)	mean length (mm)	std length (mm)
4014.38	15	138.373	181.194	162.851	14.5784

Label	Min	Max	Mean	Std
FA	0.200	0.966	0.535	0.130
ADC (mm²/s)	0.850	12.650	2.331	0.662
Lambda 1 (mm²)	0.463	5.865	1.275	0.330
Lambda 2 (mm²)	0.059	3.580	0.661	0.230
Lambda 3 (mm²)	0.004	3.204	0.395	0.219
Cl	0.004	0.889	0.268	0.143
Cp	0.001	0.661	0.239	0.146
Cs	0.009	0.858	0.493	0.134
RA	0.116	0.873	0.433	0.126
VR	0.000	0.036	0.025	0.006

Fig. 7 DW imaging processing. (**d**) Extracted imaging biomarkers from selected tract

6   Neuromuscular Image Biomarkers

The quantification of LMN degeneration in ALS is of upmost importance for the diagnosis (ALS cannot be diagnosed without LMN impairment), for the monitoring of the disease progression and for the prognosis (respiratory insufficiency, dysphagia and survival largely depend on the LMN impairment) [113]. Until recently, neurophysiology (electromyography and electroneurography) have provided most biomarkers of LMN degeneration in ALS [113]. However, imaging biomarkers have several advantages over neurophysiological ones. They are easy to obtain, painless and not explorer-dependent.

The anterior horn denervation causes secondary atrophy and fat infiltration in muscles and nerves which can be assessed, and potentially employed as biomarkers. Thus, in the last few years ultrasonography, MRI and PET of the peripheral nerves, and muscles are being evaluated as possible relevant biomarkers of LMN impairment.

6.1   Ultrasound

Several ultrasound biomarkers of nerves and muscles have emerged as candidates for diagnostic, progression's monitoring and prognostic purposes.

In ALS, pathological studies have shown a marked reduction of large myelinated fibers in ventral roots and peripheral nerves and these changes correlate with the muscle strength in the corresponding myotomes [114]. Since ALS is a neuronopathy, and given that muscle weakness and atrophy in ALS are not evident until one third of the LMN have degenerated [113], it could be expected that nerve changes are detectable previous to muscle changes. Indeed, the cross sectional area of the median and ulnar nerves as measured by ultrasound has been consistently found to be smaller in ALS patients than in controls [68, 115–118]. Images of the nerves should be obtained proximal (Fig. 8) or even at the nerve roots, since previous studies have assessed the median nerve at different levels, all favoring the study of more proximal ones [116, 119, 120]. This proximal level has also the advantage of having a greater number of axons and of avoiding other nerve lesions or entrapments that usually occur more distally. Despite this, there is considerable overlap in the cross sectional area of nerves and roots in ALS patients and healthy controls, and therefore its use as a diagnostic biomarker is limited.

It has been suggested that the combination of the cross sectional area of the median nerve and progranulin in CSF could indicate axonal damage [66, 121]. However, previous transversal studies have failed to find a correlation between the cross sectional area of the median nerve and several clinical variables such as disease duration, ALSFRS-R, forced vital capacity or strength of the abductor pollicis brevis and wrist flexors muscles [66]. Moreover, a previous longitudinal study also failed to detect longitudinal

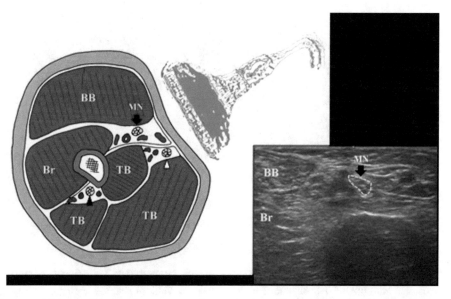

Fig. 8 Ultrasound image of the median nerve at the proximal position of the upper arm. The cross sectional area of the median nerve is delimited with a dotted yellow line. BB: biceps brachii muscle; Br: brachialis muscle; MN: median nerve; TB: triceps brachii muscle; black arrowhead, radial nerve; white arrowhead, ulnar nerve. (From [118] with authors' permission)

changes in the cross sectional area of the median nerve of ALS patients, albeit the median nerve was assessed distally [122].

Fasciculations, muscle atrophy and replacement by fat and fibrous tissue all are hallmarks of the LMN impairment detectable by means of muscle ultrasound [113]. Several studies have shown greater sensitivity for the detection of fasciculations with ultrasound than with electromyography [67, 123] and the detection of fasciculations by ultrasound is easy and fast. Since fasciculations are one of the hallmarks of LMN impairment in the new diagnostic criteria of ALS [124], this exam could provide additional benefit in the diagnostic work-up. Moreover, the combination of fasciculations and other muscle ultrasound biomarkers could add information about the origin, rise and fall of fasciculations (Fig. 9).

Muscle thickness and muscle cross sectional area, as measured by ultrasound, are reduced in ALS patients compared with healthy controls [70, 125–127]. However, muscle strength and disability show only weak correlation with muscle thickness [70, 128]. Most longitudinal studies [71, 129, 130], but not all [128], have detected changes in muscle thickness with disease progression, suggesting it could be useful as a progression biomarker (Fig. 10).

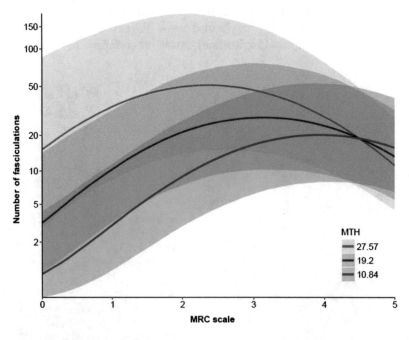

Fig. 9 Graphic representation of fasciculations number, muscle strength (MRC) and thickness (MTh). Greater MTh is associated with greater fasciculations number (colours) with an inverted U-shape correlation of the fasciculations' number with muscle strength. The number of fasciculations initially increases with the loss of muscle strength until a particular point, determined by the MTh, where it progressively decreases. Thicker muscles need to be severely weak to show a decrease in fasciculations, whereas thinner muscles show an early decrease in fasciculations with the onset of weakness. From [67] with the permission of the authors

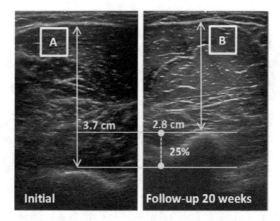

Fig. 10 Ultrasound analysis of thickness in the biceps/brachialis muscle group in an ALS patients (**a**) and after a follow-up of 20 weeks (**b**). A loss of 25% of muscle thickness is depicted. (From [71] with permission of authors)

Together with the atrophy, muscles of ALS patients suffer progressive replacement by fibrous and fatty tissue, which is detectable by ultrasound. This results in a progressive increase in muscle echogenicity [70, 125, 131], a loss of muscle homogeneity [68], together with textural and elasticity changes [132, 133]. Echogenicity shows little association with muscle strength or disability [70, 127, 128] and there is discrepancy about its ability to detect changes with disease progression [71, 128, 129]. Interestingly, it was found to be a relevant prognostic biomarker [69]. Signal homogeneity has been scarcely studied. However, compared with echogenicity and muscle thickness, it differentiated better between patients and controls and correlated better with muscle strength and disability [70]. Moreover, unlike echogenicity and muscle thickness, it was independent of age, sex and body mass index [70]. A longitudinal study detected echovariation changes in some, but not all, muscles of ALS patients [71] (Fig. 11). Textural muscle changes have shown potential to differentiate ALS from controls [133], but further studies are warranted to determine its utility in the clinical practice or clinical trials. Interestingly, the combination of nerve and muscle ultrasound biomarkers could increase their role for diagnostic or progression's monitoring purposes [68].

Diaphragmatic ultrasound can also be useful to identify or predict respiratory insufficiency in ALS patients. Diaphragmatic thickness and the thickening ratio (difference in thickness between maximal inhalation/expiration) are reduced in ALS patients compared with controls and correlate with several measures of functional pulmonary testing such as forced vital capacity or sniff nasal inspiratory pressure [134–136]. This relationship is lost in patients with bulbar symptoms where the functional pulmonary testing

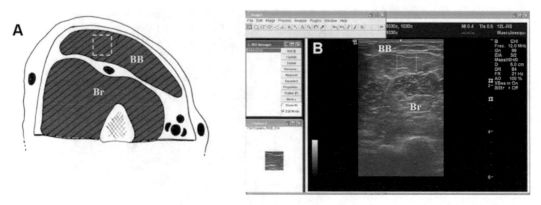

Fig. 11 Biceps brachii muscle ultrasound quantitative analysis (ImageJ, v.1.48) (**a, b**) of echogenicity, echovariation, and gray level co-occurrence matrix (From [132, 133] with the permission of the authors)

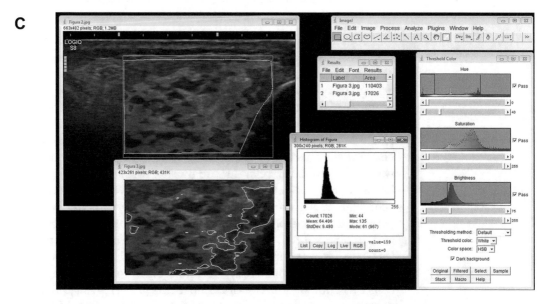

Fig. 11 Elastography of the tibialis anterior muscle (**c**). (From [132, 133] with the permission of the authors)

shows limitations; however, an attempt to measure longitudinal changes in diaphragmatic thickness or the thickening ratio failed [137].

6.2 MR Imaging

Peripheral nerves have been scarcely studied by means of MR in ALS patients. One initial study found changes in 85% of ALS patients [138]. In a longitudinal study, the fractional anisotropy of tibial and peroneal nerve was found to decrease with disease progression and to correlate with ALSFRS-R, suggesting its potential as a progression's biomarker [139]. More recent studies have revealed T2 hyperintensities in peripheral nerves and nerve roots, associated with rapid disease progression [140, 141].

Muscle volume reduction, T2 relaxation time increase, and Short Time Inversion Recovery (STIR) MR images hyperintensity are common findings among ALS patients, being related with neurophysiological biomarkers and clinical endpoints [140–145]. Moreover, longitudinal T2 and volume changes have been detected in these patients [140–145].

Few studies have assessed metabolites changes related to cellular bioenergetics using phosphorus-31 spectroscopy (31P-MRS) [72] and results showed discrepancies.

6.3 PET Imaging

To the best of our knowledge, only one study has performed FDG-PET in spinal cord and muscle in ALS patients. This study found hypermetabolism in both spinal cord and psoas muscle. This is not surprising since hypermetabolism is linked to the existence of fasciculations [67]. The spinal cord hypermetabolism was found to be an independent prognostic biomarker [146].

7 Conclusion

ALS is a complex disease with multiple pathophysiological processes and, consequently, multiple clinical endpoints. Diagnostic, progression, and prognostic imaging biomarkers are being evaluated and are urgently needed. ALS complexity and appropriate clinical endpoints are important considerations when defining the role of new biomarkers in ALS, mainly from brain MR and neuromuscular ultrasound studies.

References

1. Kumar DR, Aslinia F, Yale SH et al (2011) Jean-Martin Charcot: the father of neurology. Clin Med Res 9:46–49. https://doi.org/10.3121/cmr.2009.883
2. Lloyd CM, Richardson MP, Brooks DJ et al (2000) Extramotor involvement in ALS: PET studies with the GABA(A) ligand [11C]flumazenil. Brain 123:2289–2296. https://doi.org/10.1093/brain/123.11.2289
3. Murphy J, Henry R, Lomen-Hoerth C (2007) Establishing subtypes of the continuum of frontal lobe impairment in amyotrophic lateral sclerosis. Arch Neurol 64:330–334. https://doi.org/10.1001/archneur.64.3.330
4. Phukan J, Pender NP, Hardiman O (2007) Cognitive impairment in amyotrophic lateral sclerosis. Lancet Neurol 6:994–1003. https://doi.org/10.1016/S1474-4422(07)70265-X
5. Shaw PJ, Wood-Allum C (2010) Motor neurone disease: a practical update on diagnosis and management. Clin Med (Northfield IL) 10:252–258. https://doi.org/10.7861/clinmedicine.10-3-252
6. Kiernan MC, Vucic S, Cheah BC et al (2011) Amyotrophic lateral sclerosis. Lancet 377:942–955. https://doi.org/10.1016/S0140-6736(10)61156-7
7. Swinnen B, Robberecht W (2014) The phenotypic variability of amyotrophic lateral sclerosis. Nat Rev Neurol 10:661–670. https://doi.org/10.1038/nrneurol.2014.184
8. Chen S, Sayana P, Zhang X et al (2013) Genetics of amyotrophic lateral sclerosis: an update. Mol Neurodegener 8:28. https://doi.org/10.1186/1750-1326-8-28

9. Mejzini R, Flynn LL, Pitout IL et al (2019) ALS genetics, mechanisms, and therapeutics: where are we now? Front Neurosci 13:1310. https://doi.org/10.3389/fnins.2019.01310

10. Zou Z-Y, Zhou Z-R, Che C-H et al (2017) Genetic epidemiology of amyotrophic lateral sclerosis: a systematic review and meta-analysis. J Neurol Neurosurg Psychiatry 88:540–549. https://doi.org/10.1136/jnnp-2016-315018

11. Ferraiuolo L, Kirby J, Grierson AJ et al (2011) Molecular pathways of motor neuron injury in amyotrophic lateral sclerosis. Nat Rev Neurol 7:616–630. https://doi.org/10.1038/nrneurol.2011.152

12. Cluskey S, Ramsden DB (2001) Mechanisms of neurodegeneration in amyotrophic lateral sclerosis. Mol Pathol 54:386–392. https://doi.org/10.1136/mp.54.6.386

13. Saberi S, Stauffer JE, Schulte DJ et al (2015) Neuropathology of amyotrophic lateral sclerosis and its variants. Neurol Clin 33:855–876. https://doi.org/10.1016/j.ncl.2015.07.012

14. Ince PG, Evans J, Knopp M et al (2003) Corticospinal tract degeneration in the progressive muscular atrophy variant of ALS. Neurology 60:1252–1258. https://doi.org/10.1212/01.WNL.0000058901.75728.4E

15. Roccatagliata L, Bonzano L, Mancardi G et al (2009) Detection of motor cortex thinning and corticospinal tract involvement by quantitative MRI in amyotrophic lateral sclerosis. Amyotroph Lateral Scler 10:47–52. https://doi.org/10.1080/17482960802267530

16. Boillée S, Vande Velde C, Cleveland DW (2006) ALS: a disease of motor neurons and their nonneuronal neighbors. Neuron 52:39–59. https://doi.org/10.1016/j.neuron.2006.09.018

17. McGeer PL, McGeer EG (2002) Inflammatory processes in amyotrophic lateral sclerosis. Muscle Nerve 26:459–470. https://doi.org/10.1002/mus.10191

18. Kwan JY, Jeong SY, van Gelderen P et al (2012) Iron accumulation in deep cortical layers accounts for MRI signal abnormalities in ALS: correlating 7 tesla MRI and pathology. PLoS One 7:e35241. https://doi.org/10.1371/journal.pone.0035241

19. Turner MR, Kiernan MC (2012) Does interneuronal dysfunction contribute to neurodegeneration in amyotrophic lateral sclerosis? Amyotroph Lateral Scler 13:245–250. https://doi.org/10.3109/17482968.2011.636050

20. Ravits JM, La Spada AR (2009) ALS motor phenotype heterogeneity, focality, and spread: deconstructing motor neuron degeneration. Neurology 73:805–811. https://doi.org/10.1212/WNL.0b013e3181b6bbbd

21. Al-Chalabi A, Hardiman O, Kiernan MC et al (2016) Amyotrophic lateral sclerosis: moving towards a new classification system. Lancet Neurol 15:1182–1194. https://doi.org/10.1016/S1474-4422(16)30199-5

22. Arlandis S, Vázquez-Costa JF, Martínez-Cuenca E, Sevilla T, Boronat F, Broseta E (2017) Urodynamic findings in amyotrophic lateral sclerosis patients with lower urinary tract symptoms: results from a pilot study. Neurourol Urodyn 36(3):626–631. https://doi.org/10.1002/nau.22976

23. Zoccolella S, Beghi E, Palagano G et al (2006) Predictors of delay in the diagnosis and clinical trial entry of amyotrophic lateral sclerosis patients: a population-based study. J Neurol Sci 250:45–49. https://doi.org/10.1016/j.jns.2006.06.027

24. Vázquez-Costa J, Martínez-Molina M, Fernández-Polo M et al (2018) Analysis of the pathway and diagnostic delay of amyotrophic lateral sclerosis patients in Valencian Community. Neurologia

25. Agosta F, Chiò A, Cosottini M et al (2010) The present and the future of neuroimaging in amyotrophic lateral sclerosis. AJNR Am J Neuroradiol 31:1769–1777. https://doi.org/10.3174/ajnr.A2043

26. Petrov D, Mansfield C, Moussy A et al (2017) ALS clinical trials review: 20 years of failure. are we any closer to registering a new treatment? Front Aging Neurosci 9:68

27. Sawada H (2017) Clinical efficacy of edaravone for the treatment of amyotrophic lateral sclerosis. Expert Opin Pharmacother 18:735–738. https://doi.org/10.1080/14656566.2017.1319937

28. Berry JD, Miller R, Moore DH et al (2013) The Combined Assessment of Function and Survival (CAFS): a new endpoint for ALS clinical trials. Amyotroph Lateral Scler Front Degener 14:162–168. https://doi.org/10.3109/21678421.2012.762930

29. Mazón M, Costa JFV, Ten-Esteve A et al (2018) Imaging biomarkers for the diagnosis and prognosis of neurodegenerative diseases. The example of amyotrophic lateral sclerosis. Front Neurosci 12:784. https://doi.org/10.3389/fnins.2018.00784

30. Cedarbaum JM, Stambler N, Malta E et al (1999) The ALSFRS-R: a revised ALS functional rating scale that incorporates

assessments of respiratory function. BDNF ALS Study Group (Phase III). J Neurol Sci 169:13–21

31. Shefner JM, Liu D, Leitner ML et al (2016) Quantitative strength testing in ALS clinical trials. Neurology 87:617–624. https://doi.org/10.1212/WNL.0000000000002941

32. Kaufmann P, Levy G, Montes J et al (2007) Excellent inter-rater, intra-rater, and telephone-administered reliability of the ALSFRS-R in a multicenter clinical trial. Amyotroph Lateral Scler 8:42–46. https://doi.org/10.1080/17482960600888156

33. Maier A, Holm T, Wicks P et al (2012) Online assessment of ALS functional rating scale compares well to in-clinic evaluation: a prospective trial. Amyotroph Lateral Scler 13:210–216. https://doi.org/10.3109/17482968.2011.633268

34. Kollewe K, Mauss U, Krampfl K et al (2008) ALSFRS-R score and its ratio: a useful predictor for ALS-progression. J Neurol Sci 275:69–73. https://doi.org/10.1016/j.jns.2008.07.016

35. Labra J, Menon P, Byth K et al (2016) Rate of disease progression: a prognostic biomarker in ALS. J Neurol Neurosurg Psychiatry 87:628–632. https://doi.org/10.1136/jnnp-2015-310998

36. Fournier CN, Bedlack R, Quinn C et al (2020) Development and validation of the rasch-built overall amyotrophic lateral sclerosis disability scale (ROADS). JAMA Neurol 77:480. https://doi.org/10.1001/jamaneurol.2019.4490

37. Bedlack RS, Vaughan T, Wicks P et al (2016) How common are ALS plateaus and reversals? Neurology 86:808–812. https://doi.org/10.1212/WNL.0000000000002251

38. Roche JC, Rojas-Garcia R, Scott KM et al (2012) A proposed staging system for amyotrophic lateral sclerosis. Brain 135:847–852. https://doi.org/10.1093/brain/awr351

39. Ferraro D, Consonni D, Fini N et al (2016) Amyotrophic lateral sclerosis: a comparison of two staging systems in a population-based study. Eur J Neurol 23:1426–1432. https://doi.org/10.1111/ene.13053

40. Fang T, Al Khleifat A, Stahl DR et al (2017) Comparison of the King's and MiToS staging systems for ALS. Amyotroph Lateral Scler Front Degener 18:227–232. https://doi.org/10.1080/21678421.2016.1265565

41. Thakore NJ, Lapin BR, Pioro EP (2020) Stage-specific riluzole effect in amyotrophic lateral sclerosis: a retrospective study. Amyotroph Lateral Scler Front Degener 21:140–143. https://doi.org/10.1080/21678421.2019.1655060

42. Swash M, Burke D, Turner MR et al (2020) Upper motor neuron syndrome in amyotrophic lateral sclerosis. J Neurol Neurosurg Psychiatry 91:227–234. https://doi.org/10.1136/jnnp-2019-321938

43. Vázquez-Costa JF, Máñez I, Alabajos A et al (2016) Safety and efficacy of botulinum toxin A for the treatment of spasticity in amyotrophic lateral sclerosis: results of a pilot study. J Neurol 263:1954–1960. https://doi.org/10.1007/s00415-016-8223-z

44. Vázquez-Costa JF, Mazón M, Carreres-Polo J et al (2017) Brain signal intensity changes as biomarkers in amyotrophic lateral sclerosis. Acta Neurol Scand 137:262–271. https://doi.org/10.1111/ane.12863

45. Florence JM, Pandya S, King WM et al (1992) Intrarater reliability of manual muscle test (Medical Research Council scale) grades in Duchenne's muscular dystrophy. Phys Ther 72:115–122

46. Pinto S, De Carvalho M (2019) SVC is a marker of respiratory decline function, similar to FVC, in patients with ALS. Front Neurol 10:109. https://doi.org/10.3389/fneur.2019.00109

47. Rafiq MK, Proctor AR, McDermott CJ et al (2012) Respiratory management of motor neurone disease: a review of current practice and new developments. Pract Neurol 12:166–176. https://doi.org/10.1136/practneurol-2011-000199

48. Menke RAL, Körner S, Filippini N et al (2014) Widespread grey matter pathology dominates the longitudinal cerebral MRI and clinical landscape of amyotrophic lateral sclerosis. Brain 137:2546–2555. https://doi.org/10.1093/brain/awu162

49. Crockford C, Newton J, Lonergan K et al (2018) ALS-specific cognitive and behavior changes associated with advancing disease stage in ALS. Neurology 91:e1370–e1380. https://doi.org/10.1212/WNL.0000000000006317

50. Strong MJ, Abrahams S, Goldstein LH et al (2017) Amyotrophic lateral sclerosis - frontotemporal spectrum disorder (ALS-FTSD): revised diagnostic criteria. Amyotroph Lateral Scler Frontotemporal Degener 18:153–174. https://doi.org/10.1080/21678421.2016.1267768

51. Abrahams S, Newton J, Niven E et al (2014) Screening for cognition and behaviour changes in ALS. Amyotroph Lateral Scler

Frontotemporal Degener 15:9–14. https://doi.org/10.3109/21678421.2013.805784

52. Westeneng HJ, Debray TPA, Visser AE et al (2018) Prognosis for patients with amyotrophic lateral sclerosis: development and validation of a personalised prediction model. Lancet Neurol 17:423–433. https://doi.org/10.1016/S1474-4422(18)30089-9

53. Grolez G, Moreau C, Danel-Brunaud V, Delmaire C, Lopes R, Pradat PF, El Mendili MM, Defebvre L, Devos D (2016) The value of magnetic resonance imaging as a biomarker for amyotrophic lateral sclerosis: a systematic review. BMC Neurol 16(1):155. https://doi.org/10.1186/s12883-016-0672-6. PMID: 27567641; PMCID: PMC5002331

54. Cosottini M, Donatelli G, Costagli M et al (2016) High-resolution 7T MR imaging of the motor cortex in amyotrophic lateral sclerosis. Am J Neuroradiol 37:455–461. https://doi.org/10.3174/ajnr.A4562

55. Menke RAL, Agosta F, Grosskreutz J et al (2017) Neuroimaging endpoints in amyotrophic lateral sclerosis. Neurotherapeutics 14:11–23. https://doi.org/10.1007/s13311-016-0484-9

56. Verstraete E, Veldink JH, Hendrikse J et al (2012) Structural MRI reveals cortical thinning in amyotrophic lateral sclerosis. J Neurol Neurosurg Psychiatry 83:383–388. https://doi.org/10.1136/jnnp-2011-300909

57. Agosta F, Valsasina P, Riva N et al (2012) The cortical signature of amyotrophic lateral sclerosis. PLoS One 7:e42816. https://doi.org/10.1371/journal.pone.0042816

58. Walhout R et al (2015) Cortical thickness in ALS: towards a marker for upper motor neuron involvement. J Neurol Neurosurg Psychiatry 86(3):288–294. https://doi.org/10.1136/jnnp-2013-306839

59. Westeneng H-J, Verstraete E, Walhout R et al (2015) Subcortical structures in amyotrophic lateral sclerosis. Neurobiol Aging 36:1075–1082. https://doi.org/10.1016/j.neurobiolaging.2014.09.002

60. Foerster BR, Dwamena BA, Petrou M et al (2013) Diagnostic accuracy of diffusion tensor imaging in amyotrophic lateral sclerosis: a systematic review and individual patient data meta-analysis. Acad Radiol 20:1099–1106. https://doi.org/10.1016/j.acra.2013.03.017

61. Spinelli EG, Agosta F, Ferraro PM et al (2016) Brain MR imaging in patients with lower motor neuron–predominant disease. Radiology 280:545–556. https://doi.org/10.1148/radiol.2016151846

62. Van der Burgh HK, Westeneng HJ, Walhout R et al (2020) Multimodal longitudinal study of structural brain involvement in amyotrophic lateral sclerosis. Neurology 94:e2592–e2604. https://doi.org/10.1212/WNL.0000000000009498

63. Spinelli EG, Riva N, Rancoita PMV et al (2020) Structural MRI outcomes and predictors of disease progression in amyotrophic lateral sclerosis. NeuroImage Clin 27:102315. https://doi.org/10.1016/j.nicl.2020.102315

64. Schuster C, Hardiman O, Bede P (2017) Survival prediction in Amyotrophic lateral sclerosis based on MRI measures and clinical characteristics. BMC Neurol 17:1–10. https://doi.org/10.1186/s12883-017-0854-x

65. Ferraro PM, Agosta F, Riva N et al (2017) Multimodal structural MRI in the diagnosis of motor neuron diseases. NeuroImage Clin 16:240–247. https://doi.org/10.1016/j.nicl.2017.08.002

66. Schreiber S, Vielhaber S, Schreiber F et al (2020) Peripheral nerve imaging in amyotrophic lateral sclerosis. Clin Neurophysiol 131:2315. https://doi.org/10.1016/j.clinph.2020.03.026

67. Vázquez-Costa JF, Campins-Romeu M, Martínez-Payá JJ et al (2018) New insights into the pathophysiology of fasciculations in amyotrophic lateral sclerosis: an ultrasound study. Clin Neurophysiol 129:2650–2657. https://doi.org/10.1016/j.clinph.2018.09.014

68. Hobson-Webb LD, Simmons Z (2019) Ultrasound in the diagnosis and monitoring of amyotrophic lateral sclerosis: a review. Muscle Nerve 60:114–123. https://doi.org/10.1002/mus.26487

69. Arts IMP, Overeem S, Pillen S et al (2011) Muscle ultrasonography to predict survival in amyotrophic lateral sclerosis. J Neurol Neurosurg Psychiatry 82:552–554. https://doi.org/10.1136/jnnp.2009.200519

70. Martínez-Payá JJ, del Baño-Aledo ME, Ríos-Díaz J et al (2017) Muscular echovariation: a new biomarker in amyotrophic lateral sclerosis. Ultrasound Med Biol 43:1153–1162. https://doi.org/10.1016/j.ultrasmedbio.2017.02.002

71. Martínez-Payá JJ, Ríos-Díaz J, Medina-Mirapeix F et al (2018) Monitoring progression of amyotrophic lateral sclerosis using ultrasound morpho-textural muscle biomarkers: a pilot study. Ultrasound Med Biol 44:102–109. https://doi.org/10.1016/j.ultrasmedbio.2017.09.013

72. Verber NS, Shepheard SR, Sassani M et al (2019) Biomarkers in motor neuron disease: a state of the art review. Front Neurol 10:291. https://doi.org/10.3389/fneur.2019.00291

73. Yagishita A, Nakano I, Oda M et al (1994) Location of the corticospinal tract in the internal capsule at MR imaging. Radiology 191:455–460. https://doi.org/10.1148/radiology.191.2.8153321

74. Adachi Y, Sato N, Saito Y et al (2015) Usefulness of SWI for the detection of iron in the motor cortex in amyotrophic lateral sclerosis. J Neuroimaging 25:443–451. https://doi.org/10.1111/jon.12127

75. Alberich-Bayarri Á, Hernández-Navarro R, Ruiz-Martínez E, García-Castro F, García-Juan D, Martí-Bonmatí L (2017) Development of imaging biomarkers and generation of big data. Radiol Med 122(6):444–448

76. Martí-Bonmatí L, Ruiz-Martínez E, Ten A, Alberich-Bayarri A. (2018) How to integrate quantitative information into imaging reports for oncologic patients [Cómo integrar la información cuantitativa en el informe radiológico del paciente oncológico]. Radiologia 60(Suppl 1):43–52. https://doi.org/10.1016/j.rx.2018.02.005

77. Penny W, Friston K, Ashburner J, Kiebel S, Nichols T (2006) Statistical parametric mapping: the analysis of functional brain images, 1st edn. Academic Press, New York, NY. Hardcover ISBN: 9780123725608. Paperback ISBN: 9781493300952. eBook ISBN: 9780080466507

78. Cox RW (1996) AFNI: software for analysis and visualization of functional magnetic resonance neuroimages. Comput Biomed Res 29(3):162–173. https://doi.org/10.1006/cbmr.1996.0014

79. Woolrich MW, Jbabdi S, Patenaude B, Chappell M, Makni S, Behrens T, Beckmann C, Jenkinson M, Smith SM (2009) Bayesian analysis of neuroimaging data in FSL. NeuroImage 45:S173–S186

80. Fischl B (2012) FreeSurfer. NeuroImage 62(2):774–781

81. Avants BB, Tustison NJ, Song G, Cook PA, Klein A, Gee JC (2011) A reproducible evaluation of ANTs similarity metric performance in brain image registration. NeuroImage 54(3):2033–2044. https://doi.org/10.1016/j.neuroimage.2010.09.025

82. Elam JS, Van Essen D (2013) Human connectome project. In: Jaeger D, Jung R (eds) Encyclopedia of computational neuroscience. Springer, New York, NY. https://doi.org/10.1007/978-1-4614-7320-6_592-1

83. Esteban O, Markiewicz CJ, Goncalves M, DuPre E, Kent JD, Ciric R, et al (2020) fMRIPrep: a robust preprocessing pipeline for functional MRI. Zenodo. https://zenodo.org/record/852659

84. Whitfield-Gabrieli S, Nieto-Castanon A (2012) Conn: a functional connectivity toolbox for correlated and anticorrelated brain networks. Brain Connect 2:125. https://doi.org/10.1089/brain.2012.0073

85. Klein S, Staring M, Murphy K, Viergever MA, Pluim JPW (2010) Elastix: a toolbox for intensity based medical image registration. IEEE Trans Med Imaging 29(1):196–205

86. Mazaika P, Whitfield S, Cooper JC (2005) Detection and repair of transient artifacts in fMRI data. NeuroImage 26:S36

87. Sladky R, Friston K, Tröstl J, Cunnington R, Moser E, Windischberger C (2011) Slice-timing effects and their correction in functional MRI. NeuroImage 58:588–594

88. Goebel R, Esposito F, Formisano E (2006) Analysis of FIAC data with BrainVoyager QX: from single-subject to cortically aligned group GLM analysis and self-organizing group ICA. Hum Brain Mapp 27(5):392–401

89. Mazziotta JC, Toga AW, Evans A, Fox P, Lancaster J (1995) A probabilistic atlas of the human brain: theory and rationale for its development. The International Consortium for Brain Mapping (ICBM). NeuroImage 2(2):89–101. https://doi.org/10.1006/nimg.1995.1012

90. Makris N, Goldstein JM, Kennedy D, Hodge SM, Caviness VS, Faraone SV, Tsuang MT, Seidman LJ (2006) Decreased volume of left and total anterior insular lobule in schizophrenia. Schizophr Res 83(2–3):155–171. https://doi.org/10.1016/j.schres.2005.11.020

91. Van Essen DC, Glasser MF, Dierker DL, Harwell J, Coalson T (2012) Parcellations and hemispheric asymmetries of human cerebral cortex analyzed on surface-based atlases. Cereb Cortex 22:2241–2262

92. Giorgio A, De Stefano N (2013) Clinical use of brain volumetry. J Magn Reson Imaging 37(1):1–14. https://doi.org/10.1002/jmri.23671

93. Friston KJ, Holmes AP, Poline JB et al (1995) Analysis of fMRI time-series revisited. NeuroImage 2(1):45–53. https://doi.org/10.1006/nimg.1995.1007

94. Pletson JE (2007) Psychology and schizophrenia. Nova Science Publishers, Hauppauge, NY. ISBN-10: 1594548676

95. Kassubek J, Ludolph AC, Müller HP (2012) Neuroimaging of motor neuron diseases. Ther Adv Neurol Disord 5(2):119–127. https://doi.org/10.1177/1756285612437562. PMID: 22435076; PMCID: PMC3302203

96. Vázquez-Costa JF, Carratalà-Boscà S, Tembl JI, Fornés-Ferrer V, Pérez-Tur J, Martí-Bonmatí L, Sevilla T (2019) The width of the third ventricle associates with cognition and behaviour in motor neuron disease. Acta Neurol Scand 139(2):118–127. https://doi.org/10.1111/ane.13022. PMID: 30183086

97. Thorns J et al (2013) Extent of cortical involvement in amyotrophic lateral sclerosis - an analysis based on cortical thickness. BMC Neurol 13:148–158. https://doi.org/10.1186/1471-2377-13-148

98. Schuster C et al (2013) Focal thinning of the motor cortex mirrors clinical features of amyotrophic lateral sclerosis and their phenotypes: a neuroimaging study. J Neurol 260(11):2856–2864. https://doi.org/10.1007/s00415-013-7083-z

99. Li H et al (2016) The Human brainnetome atlas: a new brain atlas based on connectional architecture. Cereb Cortex 26(8):3508–3526. https://doi.org/10.1093/cercor/bhw157

100. Liu C, Li W, Tong KA, Yeom KW, Kuzminski S (2015) Susceptibility-weighted imaging and quantitative susceptibility mapping in the brain. J Magn Reson Imaging 42(1):23–41. https://doi.org/10.1002/jmri.24768

101. Langkammer C, Schweser F, Krebs N, Deistung A, Goessler W, Scheurer E et al (2012) Quantitative susceptibility mapping (QSM) as a means to measure brain iron? A post mortem validation study. NeuroImage 62(3):1593–1599. https://doi.org/10.1016/j.neuroimage.2012.05.049

102. Gillies R, Kinahan P, Hricak H (2015) Radiomics: images are more than pictures, they are data. Radiology 278:563. https://doi.org/10.1148/radiol.2015151169

103. Vázquez-Costa JF, Tembl JI, Fornés-Ferrer V et al (2017) Genetic and constitutional factors are major contributors to substantia nigra hyperechogenicity. Sci Rep 7:7119. https://doi.org/10.1038/s41598-017-07835-z

104. Van Griethuysen JJM et al (2017) Computational radiomics system to decode the radiographic phenotype. Cancer Res 77(21): e104–e107. https://doi.org/10.1158/0008-5472.CAN-17-0339

105. Calhoun V (2018) Data-driven approaches for identifying links between brain structure and function in health and disease. Dialogues Clin Neurosci 20(2):87–99. PMID: 30250386; PMCID: PMC6136124

106. Nieto-Castanon A (2020) Handbook of functional connectivity Magnetic Resonance Imaging methods in CONN. Hilbert Press, Boston, MA. ISBN: 978-0-578-64400-4

107. Behzadi Y, Restom K, Liau J, Liu TT (2007) A component based noise correction method (CompCor) for BOLD and perfusion based fMRI. NeuroImage 37(1):90–101. https://doi.org/10.1016/j.neuroimage.2007.04.042

108. Hallquist M, Hwang K, Luna B (2013) The nuisance of nuisance regression: spectral misspecification in a common approach to resting-state fMRI preprocessing reintroduces noise and obscures functional connectivity. NeuroImage 82:208. https://doi.org/10.1016/j.neuroimage.2013.05.116

109. Ioannidis JPA (2018) The proposal to lower P value thresholds to .005. JAMA 319(14):1429–1430. https://doi.org/10.1001/jama.2018.1536

110. Stein JL, Medland SE, Vasquez AA et al (2012) Identification of common variants associated with human hippocampal and intracranial volumes. Nat Genet 44(5):552–561. https://doi.org/10.1038/ng.2250

111. Vichot F, Cochet H, Bleuzé B, Toussaint N, Jaïs P, Sermesant M (2012) Cardiac interventional guidance using multimodal data processing and visualisation: medinria as an interoperability platform. Midas J

112. Kalra S (2019) Magnetic resonance spectroscopy in ALS. Front Neurol 10:482. https://doi.org/10.3389/fneur.2019.00482

113. De Carvalho M, Swash M (2016) Lower motor neuron dysfunction in ALS. Clin Neurophysiol 127:2670–2681. https://doi.org/10.1016/j.clinph.2016.03.024

114. Tandan R, Bradley WG (1985) Amyotrophic lateral sclerosis: Part 1. Clinical features, pathology, and ethical issues in management. Ann Neurol 18:271–280. https://doi.org/10.1002/ana.410180302

115. Cartwright MS, Walker FO, Griffin LP et al (2011) Peripheral nerve and muscle ultrasound in amyotrophic lateral sclerosis. Muscle Nerve 44:346–351. https://doi.org/10.1002/mus.22035

116. Nodera H, Takamatsu N, Shimatani Y et al (2014) Thinning of cervical nerve roots and peripheral nerves in ALS as measured by sonography. Clin Neurophysiol 125:1906–1911. https://doi.org/10.1016/j.clinph.2014.01.033

117. Grimm A, Décard BF, Athanasopoulou I et al (2015) Nerve ultrasound for differentiation between amyotrophic lateral sclerosis and multifocal motor neuropathy. J Neurol 262:870–880. https://doi.org/10.1007/s00415-015-7648-0

118. Rios-Diaz J, Del Bano-Aledo ME, Tembl-Ferrairo JI et al (2019) Quantitative neuromuscular ultrasound analysis as biomarkers in amyotrophic lateral sclerosis. Eur Radiol 29:4266. https://doi.org/10.1007/s00330-018-5943-8

119. Schreiber S, Abdulla S, Debska-Vielhaber G et al (2015) Peripheral nerve ultrasound in amyotrophic lateral sclerosis phenotypes. Muscle Nerve 51:669–675. https://doi.org/10.1002/mus.24431

120. Noto Y-I, Garg N, Li T et al (2018) Comparison of cross-sectional areas and distal-proximal nerve ratios in amyotrophic lateral sclerosis. Muscle Nerve 58:777–783. https://doi.org/10.1002/mus.26301

121. Schreiber S, Schreiber F, Garz C et al (2019) Toward in vivo determination of peripheral nervous system immune activity in amyotrophic lateral sclerosis. Muscle Nerve 59:567–576. https://doi.org/10.1002/mus.26444

122. Schreiber S, Dannhardt-Stieger V, Henkel D et al (2016) Quantifying disease progression in amyotrophic lateral sclerosis using peripheral nerve sonography. Muscle Nerve 54:391–397. https://doi.org/10.1002/mus.25066

123. Misawa S, Noto Y, Shibuya K et al (2011) Ultrasonographic detection of fasciculations markedly increases diagnostic sensitivity of ALS. Neurology 77:1532–1537. https://doi.org/10.1212/WNL.0b013e318233b36a

124. De Carvalho M, Dengler R, Eisen A et al (2008) Electrodiagnostic criteria for diagnosis of ALS. Clin Neurophysiol 119:497–503. https://doi.org/10.1016/j.clinph.2007.09.143

125. Arts IMP, van Rooij FG, Overeem S et al (2008) Quantitative Muscle Ultrasonography in Amyotrophic Lateral Sclerosis. Ultrasound Med Biol 34:354–361. https://doi.org/10.1016/j.ultrasmedbio.2007.08.013

126. Arts IMP, Overeem S, Pillen S et al (2012) Muscle ultrasonography: a diagnostic tool for amyotrophic lateral sclerosis. Clin Neurophysiol 123:1662–1667. https://doi.org/10.1016/j.clinph.2011.11.262

127. Grimm A, Prell T, Décard BF et al (2015) Muscle ultrasonography as an additional diagnostic tool for the diagnosis of amyotrophic lateral sclerosis. Clin Neurophysiol 126:820–827. https://doi.org/10.1016/j.clinph.2014.06.052

128. Arts IM, Overeem S, Pillen S et al (2011) Muscle changes in amyotrophic lateral sclerosis: a longitudinal ultrasonography study. Clin Neurophysiol 122:623–628. https://doi.org/10.1016/j.clinph.2010.07.023

129. Lee CD, Song Y, Peltier AC et al (2010) Muscle ultrasound quantifies the rate of reduction of muscle thickness in amyotrophic lateral sclerosis. Muscle Nerve 42:814–819. https://doi.org/10.1002/mus.21779

130. Pathak S, Caress JB, Wosiski-Kuhn M et al (2019) A pilot study of neuromuscular ultrasound as a biomarker for amyotrophic lateral sclerosis. Muscle Nerve 59:181–186. https://doi.org/10.1002/mus.26360

131. Pillen S, Tak RO, Zwarts MJ et al (2009) Skeletal muscle ultrasound: correlation between fibrous tissue and echo intensity. Ultrasound Med Biol 35:443–446. https://doi.org/10.1016/j.ultrasmedbio.2008.09.016

132. Martínez-Payá JJ, del Baño-Aledo ME, Ríos-Díaz J et al (2018) Sonoelastography for the assessment of muscle changes in amyotrophic lateral sclerosis: results of a pilot study. Ultrasound Med Biol 44:2540–2547. https://doi.org/10.1016/j.ultrasmedbio.2018.08.009

133. Martínez-Payá JJ, Ríos-Díaz J, Del Baño-Aledo ME et al (2017) Quantitative Muscle Ultrasonography Using Textural Analysis in Amyotrophic Lateral Sclerosis. Ultrason Imaging 39:357–368. https://doi.org/10.1177/0161734617711370

134. Pinto S, Alves P, Pimentel B et al (2016) Ultrasound for assessment of diaphragm in ALS. Clin Neurophysiol 127:892–897. https://doi.org/10.1016/j.clinph.2015.03.024

135. Fantini R, Mandrioli J, Zona S et al (2016) Ultrasound assessment of diaphragmatic function in patients with amyotrophic lateral sclerosis. Respirology 21:932–938. https://doi.org/10.1111/resp.12759

136. Hiwatani Y, Sakata M, Miwa H (2013) Ultrasonography of the diaphragm in amyotrophic lateral sclerosis: clinical significance in

assessment of respiratory functions. Amyotroph Lateral Scler Front Degener 14:127–131. https://doi.org/10.3109/17482968.2012.729595

137. Pinto S, Alves P, Swash M et al (2017) La stimulation du nerf phrénique est plus sensible que la mesure échographique de l'épaisseur du diaphragme dans l'évaluation du début de la progression de la SLA. Neurophysiol Clin 47:69–73. https://doi.org/10.1016/j.neucli.2016.08.001

138. Gerevini S, Agosta F, Riva N et al (2016) MR imaging of Brachial Plexus and limb-girdle muscles in patients with amyotrophic lateral sclerosis. Radiology 279:553–561. https://doi.org/10.1148/radiol.2015150559

139. Simon NG, Lagopoulos J, Paling S et al (2017) Peripheral nerve diffusion tensor imaging as a measure of disease progression in ALS. J Neurol 264:882–890. https://doi.org/10.1007/s00415-017-8443-x

140. Gerevini S, Agosta F, Riva N et al (2015) MR imaging of brachial plexus and limb-girdle muscles in patients with amyotrophic lateral sclerosis. Radiology 279:553. https://doi.org/10.1148/radiol.2015150559

141. Staff NP, Amrami KK, Howe BM (2015) Magnetic resonance imaging abnormalities of peripheral nerve and muscle are common in amyotrophic lateral sclerosis and share features with multifocal motor neuropathy.

Muscle Nerve 52:137–139. https://doi.org/10.1002/mus.24630

142. Jenkins TM, Alix JJP, David C et al (2018) Imaging muscle as a potential biomarker of denervation in motor neuron disease. J Neurol Neurosurg Psychiatry 89:248–255. https://doi.org/10.1136/jnnp-2017-316744

143. Jenkins TM, Alix JJP, Fingret J et al (2020) Longitudinal multi-modal muscle-based biomarker assessment in motor neuron disease. J Neurol 267:244–256. https://doi.org/10.1007/s00415-019-09580-x

144. Bryan WW, Reisch JS, McDonald G et al (1998) Magnetic resonance imaging of muscle in amyotrophic lateral sclerosis. Neurology 51:110–113. https://doi.org/10.1212/WNL.51.1.110

145. Klickovic U, Zampedri L, Sinclair CDJ et al (2019) Skeletal muscle MRI differentiates SBMA and ALS and correlates with disease severity. Neurology 93:E895–E907. https://doi.org/10.1212/WNL.0000000000008009

146. Baucknent M, Lai R, Miceli A et al (2020) Spinal cord hypermetabolism extends to skeletal muscle in amyotrophic lateral sclerosis: a computational approach to [18F]-fluorodeoxyglucose PET/CT images. EJNMMI Res 10:23. https://doi.org/10.1186/s13550-020-0607-5

Part IV

Conclusion

Chapter 21

Trends in Biomarkers of Neurodegenerative Diseases

Philip V. Peplow, Bridget Martinez, and Thomas A. Gennarelli

Abstract

An agreed-upon action plan is necessary by national health funding and research agencies over the next decade to lower the number of new cases of neurodegenerative disease. Many of the biomarkers currently used clinically for many neurodegenerative diseases are inadequate, and have inherent limitations in regard to sensitivity and specificity. There is a need for newer and better biomarkers of neurodegenerative disease that can be accurately quantitated in cerebrospinal fluid, blood, saliva, or urine. Biomarkers hold promise for enabling more effective drug development in neurodegenerative diseases and a more personalized medicine approach. There is a need for biomarkers to enable presymptomatic diagnosis of neurodegenerative disorders that would allow early treatment strategies to be instituted and evaluated in slowing or preventing development of the disease. In clinical trials of new potential disease-modifying drugs, careful selection of patients with defined inclusion and exclusion criteria, and recruiting patients and assigning them to treatment groups to avoid possible bias, is paramount. This book presents advances in the medical application of neurofunctional, structural, and molecular biomarkers through innovative neuromethods development.

Key words Biomarkers development, Neurodegenerative disease, Sensitivity, Specificity, Experimental animal models, Clinical trials

1 Introduction

An agreed-upon action plan is necessary by national health funding and research agencies over the next decade to make some inroads into lowering the number of new cases of neurodegenerative disease. As noted in Chapter 1 to stem the increase in numbers of patients and costs of neurodegenerative disorders, a new treatment that delays the onset of Alzheimer's disease by only 5 years would eliminate 50% of cases, and to delay onset by 10 years would eliminate 75% of patients, with a potential saving of >$175 billion annually (in US dollars) [1].

Philip V. Peplow, Bridget Martinez and Thomas A. Gennarelli (eds.), *Neurodegenerative Diseases Biomarkers: Towards Translating Research to Clinical Practice*, Neuromethods, vol. 173, https://doi.org/10.1007/978-1-0716-1712-0_21,
© Springer Science+Business Media, LLC, part of Springer Nature 2022

2   Newer and Better Biomarkers of Neurodegenerative Diseases

The biomarkers currently used clinically for many neurodegenerative diseases are inadequate. Three tests are currently used in the clinical diagnosis and management of multiple sclerosis [2], namely the oligoclonal bands in the cerebrospinal fluid [3], which are attributable mostly to immunoglobulins; the white matter/Gadolinium (Gad)-enhancing lesions detected by magnetic resonance imaging [4, 5], which correspond to active lesions with inflammation; and the John Cunningham (JC) virus antibody titers [6, 7], which demonstrate exposure of the patient to this virus and the likely risk of developing progresive multifocal leukoencephalopathy following immunosuppressive therapy. While these tests have been used consistently in the clinic, all have inherent limitations in regard to sensitivity and specificity [8, 9]. Furthermore, the current health systems have limited capacity to select those individuals likely to have Alzheimer's disease pathology in order to confirm the diagnosis with available cerebrospinal fluid and imaging biomarkers at memory clinics. Significant gender and race disparities in CSF biomarkers amyloid beta 42 and tau levels have been reported both in healthy subjects and patients with Alzheimer's disease [10, 11] This poses a barrier to the effective conduct of clinical trials of new drug candidates against Alzheimer's disease and to identifying patients for receiving disease-modifying treatments [12].

There is a need for newer and better biomarkers of neurodegenerative disease that can be accurately quantitated in cerebrospinal fluid, blood, saliva, or urine. Brain-derived biomarkers are usually present at relatively low concentrations in the blood because of the blood-brain barrier preventing free passage of molecules between the CNS and blood compartments. Blood biomarkers able to detect neurodegenerative disease in its earliest stages (prodromal, e.g., MCI stages of Alzheimer's disease) are predicted to have the most impact for use as a screening tool [12].

A biomarker is defined as a characteristic that is objectively measured and evaluated as an indicator of normal biological processes, pathogenic processes, or pharmacological responses to a therapeutic intervention [13]. Biomarkers hold promise for enabling more effective drug development in neurodegenerative diseases and a more personalized medicine approach. The following criteria should be met before acceptance as a valid biomarker for Alzheimer's disease.

- Sensitivity (>85%; 100% indicates all patients are identified with the disease),

- Specificity (>85%; 100% identifies all individuals free of the disease),

- Prior probability (the background prevalence of the diseases in the population tested),

- Positive predicted value (>80%; refers to the percentage of people who are positive for the biomarker and have the disease at autopsy),

- Negative predicted value (percentage of people with a negative test, and no disease at autopsy) [14].

MicroRNAs have emerged as promising biomarkers for Parkinson's disease, Alzheimer's disease, diabetic retinopathy, and multiple sclerosis [15–18] and sensitivity and specificity values for discriminating these diseases from normal healthy controls are shown in Table 1. Biomarkers should be able to differentiate neurodegenerative diseases from other related diseases or subtypes.

3   Biomarkers to Detect Neurodegenerative Diseases at Presymptomatic Stage, to Distinguish Stages of Disease Progression, and Response to Therapy

There is a need for biomarkers to enable presymptomatic diagnosis of neurodegenerative disorders that would allow early treatment strategies to be instituted and evaluated in slowing or preventing development of the disease. MicroRNAs have been shown to be biomarkers for mild cognitive impairment and could be used to assess the possible progression to Alzheimer's disease [16]. Interestingly, a high percentage of patients with diabetic retinopathy subsequently develop Alzheimer's disease, and microRNAs are biomarkers for diabetic retinopathy [17]. Also, microRNAs allow distinguishing different degrees of severity in patients with multiple sclerosis [18].

4   New Drugs to Be Approved for Treating Neurodegenerative Diseases

No new treatments for Alzheimer's disease have been approved in the last 16 years [35]. An application is being made to the US Food and Drug Administration (FDA) by the pharmaceutical company Biogen for approval for a new drug, an antibody called aducanumab, which was shown to slow cognitive decline in the early stages of Alzheimer's disease. The drug offers the best hope for the first significant treatment for the disease since 2003, when the FDA approved memantine, which is commonly marketed as Namenda and relieves some of the symptoms of dementia [36, 37].

Table 1
Examples of candidate microRNA biomarkers in human studies of neurodegenerative diseases

Candidate biomarker(s)	Sample	AUC value	Sensitivity	Specificity	Reference
Alzheimer's disease					
miR-223	Blood serum	0.786			[19]
miR-501-3p	Blood serum	0.82	53%	100%	[20]
miR-455-3p	Blood serum	0.79			[21]
miR-34c	Blood plasma	0.82			[22]
miR-483-5p, miR-502-3p	Blood plasma	>0.9	>80%	>80%	[23]
miR-29a	CSF		78%	60%	[24]
miR-455-3p	Frontal cortex	0.792	89%	67%	[21]
Parkinson's disease					
Set of 29 microRNAs	Prefrontal cortex		97%	94%	[25]
Multiple sclerosis					
miR-145	Blood serum	0.670	73%	61%	[26]
miR-223	Blood serum	0.702	73%	61%	[26]
miR-128-3p	Blood serum	0.727	65%	71%	[27]
miR-191-5p	Blood serum	0.808	74%	81%	[27]
miR-320a	Blood serum	0.707			[28]
miR-150	CSF	0.744	89%	50%	[29]
Diabetic retinopathy					
miR-211	Blood serum	0.864			[30]
let-7a-5p/miR-28-3p/miR-novel-chr5_15976	Blood serum	0.937	92%	95%	[31]
miR-126	Blood serum	0.976	81%	90%	[32]
miR-93	Blood plasma	0.866	73%	89%	[33]
miR-21	Blood plasma	0.825	66%	90%	[34]

AUC (area under curve) by ROC (receiver operating characteristics curve) analysis

5 Improved Experimental Animal Models of Neurodegenerative Diseases

In the JPND Research and Innovation Strategy document under Theme Two is included "Develop novel animal models that are relevant to neurodegenerative disease and take into account factors such as the progressive nature of neurodegenerative disease, comorbidities, sex differences and ageing." [38]. In many of the experimental animal studies on elucidating possible biomarkers in neurodegenerative diseases, the animals have been young adults and of mainly one sex. As neurodegenerative diseases are increased in the elderly, studies should be performed using aged animals of both sexes.

6 Clinical Trials of New Potential Disease-Modifying Agents

Careful selection of patients with defined inclusion and exclusion criteria, as well as recruiting patients and assigning them to treatment groups to avoid possible bias, is paramount. Combinations of drugs are likely to be more effective than administering a single drug. For example, a combination of an immunomodulatory agent tuftsin and a drug promoting remyelination benztropine improved multiple sclerosis-like pathologies in an experimental autoimmune encephalomyelitis (EAE) animal model [39]. Also a combination of the immunomodulator glatiramer acetate and an antioxidant drug epigallocatechin-3-gallate delayed disease onset and decreased clinical symptoms in an EAE relapsing-remitting mouse model [40]. MicroRNAs are also therapeutic targets and microRNA mimics and inhibitors (antagomirs) were effective in alleviating disease in EAE models [41] and have the potential to be disease-modifying agents in clinical trials.

7 Conclusion

New biomarkers with high sensitivity and specificity are critical for identifying prodromal and early stages of neurodegenerative diseases and thereby being able to initiate early treatment of the disease to reduce its severity and slow its progression. The purpose of this issue is to provide readers—neuroscience graduate students, clinical chemists, physicians, and medical residents—with the latest advances in neurodegenerative disease pathophysiological mechanisms, biotechnology, neuroimaging, and emergent care. This book presents advances in the medical application of neurofunctional, structural, and molecular biomarkers through innovative neuromethods development.

References

1. Brookmeyer R, Gray S, Kawas C (1988) Projections of Alzheimer's disease in the United States and the public health impact of delaying disease onset. Am J Public Health 88 (9):1337–1342. https://doi.org/10.2105/ajph.88.9.1337

2. Housley WJ, Pitt D, Hafler DA (2015) Biomarkers in multiple sclerosis. Clin Immunol 161(1):51–58. https://doi.org/10.1016/j.clim.2015.06.015

3. Stangel M, Fredrikson S, Meinl E, Petzold A, Stüve O, Tumani H (2013) The utility of cerebrospinal fluid analysis in patients with multiple

sclerosis. Nat Rev Neurol 9(5):267–276. https://doi.org/10.1038/nrneurol.2013.41

4. Zivadinov R, Leist TP (2005) Clinical-magnetic resonance imaging corelations in multiple sclerosis. J Neuroimaging 15 (4 Suppl):10S–21S. https://doi.org/10.1177/1051228405283291

5. Fisniku LK, Brex PA, Altmann DR, Miszkiel KA, Benton CE, Lanyon R, Thompson AJ, Miller DH (2008) Disability and T2 MRI lesions: a 20-year follow-up of patients with relapse onset of multiple sclerosis. Brain 131:808–817. https://doi.org/10.1093/brain/awm329

6. Antoniol C, Stankoff B (2014) Immunological markers for PML prediction in MS patients treated with natalizumab. Front Immunol 5:668. https://doi.org/10.3389/fimmu.2014.00668

7. Outteryck O, Zéphir H, Salleron J, Ongagna JC, Etxeberria A, Collongues N, Lacour A, Fleury MC, Blanc F, Giroux M, de Seze J, Vermersch P (2014) JC-virus conversion in multiple sclerosis patients receiving natalizumab. Mult Scler 20(7):822–829. https://doi.org/10.1177/1352458513505353

8. Li DK, Held U, Petkau J, Daumer M, Barkhof F, Fazekas F, Frank JA, Kappos L, Miller DH, Simon JH, Wolinsky JS, Filippi M (2006) MRI T2 lesion burden in multiple sclerosis: a plateauing relationship with clinical disability. Neurology 66(9):1384–1389. https://doi.org/10.1212/01.wnl.0000210506.00078.5c

9. Plavina T, Subramanyam M, Bloomgren G, Richman S, Pace A, Lee S, Schlain B, Campagnolo D, Belachew S, Ticho B (2014) Anti-JC virus antibody levels in serum or plasma further define risk of natalizumab-associated progressive multifocal leukoencephalopathy. Ann Neurol 76(6):802–812. https://doi.org/10.1002/ana.24286

10. Koran ME, Wagener M, Hohman TJ (2017) Sex differences in the association between AD markers and cognitive decline. Brain Imaging Behav 11(1):205–213. https://doi.org/10.1007/s11682-016-9523-8

11. Morris JC, Schindler SE, McCue LM, Moulder KL, Benzinger TL, Cruchaga C, Fagan AM, Grant E, Gordon BA, Holtzman DM, Xiong C (2019) Assessment of racial disparities in biomarkers for Alzheimer disease. JAMA Neurol 76(3):264–273. https://doi.org/10.1001/jamaneurol.2018.4249

12. Zetterberg H, Burnham SC (2019) Blood-based biomarkers for Alzheimer's disease. Mol Brain 12:26. https://doi.org/10.1186/s13041-019-0448-1

13. Biomarkers Definitions Working Group (2001) NIH biomarkers definitions Working Group biomarkers and surrogate endpoints: preferred definitions and conceptual framework. Clin Pharmacol Ther 69(3):89–95. https://doi.org/10.1067/mcp.2001.113989

14. Lewczuk P, Riederer P, O'Bryant SE, Verbeek MM, Dubois B, Visser PJ et al (2018) Cerebrospinal fluid and blood biomarkers for neurodegenerative dementias: an update of the consensus of the task force on biological markers in psychiatry of the world federation of societies of biological psychiatry. World J Biol Psychiatry 19(4):244–328. https://doi.org/10.1080/15622975.2017.1375556

15. Martinez B, Peplow PV (2017) MicroRNAs in Parkinson's disease and emerging therapeutic targets. Neural Regen Res 12(12):1945–1959. https://doi.org/10.4103/1673-5374.221147

16. Martinez B, Peplow PV (2019) MicroRNAs as diagnostic and therapeutic tools for Alzheimer's disease: advances and limitations. Neural Regen Res 14(2):242–255. https://doi.org/10.4103/1673-5374.244784

17. Martinez B, Peplow PV (2019) MicroRNAs as biomarkers of diabetic retinopathy and disease progression. Neural Regen Res 14(11):1858–1869. https://doi.org/10.4103/1673-5374.259602

18. Martinez B, Peplow PV (2020) MicroRNAs in blood and cerebrospinal fluid as diagnostic markers of multiple sclerosis and to monitor disease progression. Neural Regen Res 15:606–619. https://doi.org/10.4103/1673-5374.266905

19. Jia LH, Liu YN (2016) Downregulated serum miR-223 serves as biomarker in Alzheimer's disease. Cell Biochem Funct 34(4):233–237. https://doi.org/10.1002/cbf.3184

20. Hara N, Kikuchi M, Miyashita A, Hatsuta H, Saito Y, Kasuga K, Murayama S, Ikeuchi T, Kuwano R (2017) Serum microRNA miR-501-3p as a potential biomarker related to the progression of Alzheimer's disease. Acta Neuropathol Commun 5(1):10. https://doi.org/10.1186/s40478-017-0414-z

21. Kumar S, Vijayan M, Reddy PH (2017) MicroRNA-455-3p as a potential peripheral biomarker for Alzheimer's disease. Hum Mol Genet 26(19):3808–3822. https://doi.org/10.1093/hmg/ddx26

22. Zirnheld AL, Shetty V, Chertkow H, Schipper HM, Wang E (2016) Distinguishing mild cognitive impairment from Alzheimer's disease by increased expression of key circulating microRNAs. Curr Neurobiol 7(2):38–50

23. Najaraj S, Laskowska-Kaszub K, Dębski KJ, Wojsiat J, Dąbrowski M, Gabryelewicz T, Kuźnicki J, Wojda U (2017) Profile of 6 micro-RNA in blood plasma distinguish early stage Alzheimer's disease patients from non-demented subjects. Oncotarget 8 (10):16122–16143. https://doi.org/10.18632/oncotarget.15109

24. Müller M, Jäkel L, Bruinsma IB, Claassen JA, Kuiperij HB, Verbeek MM (2016) MicroRNA-29a is a candidate biomarker for Alzheimer's disease in cell-free cerebrospinal fluid. Mol Neurobiol 53(5):2894–2899. https://doi.org/10.1007/s12035-015-9156-8

25. Hoss AG, Labadorf A, Beach TG, Latourelle JC, Myers RH (2016) microRNA profiles in Parkinson's disease prefrontal cortex. Front Aging Neurosci 8:36. https://doi.org/10.3389/fnagi.2016.00036

26. Sharaf-Eldin WE, Kishk NA, Gad YZ, Hassan H, Ali MAM, Zaki MS, Mohamed MR, Essawi ML (2017) Extracellular miR-145, miR-223 and miR-326 expression signature allow for differential diagnosis of immune-mediated neuroinflammatory diseases. J Neurol Sci 383:188–198. https://doi.org/10.1016/j.jns.2017.11.014

27. Vistbakka J, Elovaara I, Lehtimäki T, Hagman S (2017) Circulating microRNAs as biomarkers in progressive multiple sclerosis. Mult Scler 23(3):403–412. https://doi.org/10.1177/1352458516651141

28. Regev K, Paul A, Healy B, von Glenn F, Diaz-Cruz C, Gholipour T, Mazzola MA, Raheja R, Nejad P, Glanz BI, Kivisakk P, Chitnis T, Weiner HL, Gandhi R (2016) Comprehensive evaluation of serum microRNAs as biomarkers in multiple sclerosis. Neurol Neuroimmunol Neuroinflamm 3(5):e267. https://doi.org/10.1212/NXI.0000000000000267

29. Bergman P, Piket E, Khademi M, James T, Brundin L, Olsson T, Piehl F, Jagodic M (2016) Circulating miR-150 in CSF is a novel candidate biomarker for multiple sclerosis. Neurol Neuroimmunol Neuroinflamm 3(3):e219. https://doi.org/10.1212/NXI.0000000000000219

30. Liu HN, Cao NJ, Li X, Qian W, Chen XL (2018) Serum microRNA-211 as a biomarker for diabetic retinopathy via modulating Sirtuin 1. Biochem Biophys Res Commun 505 (4):1236–1243. https://doi.org/10.1016/j.bbrc.2018.10.052

31. Liang Z, Gao KP, Wang YX, Liu ZC, Tian L, Yang XZ, Ding JY, Wu WT, Yang WH, Li YL, Zhang ZB, Zhai RH (2018) RNA sequencing identified specific circulating miRNA biomarkers for early detection of diabetes retinopathy. Am J Physiol Endocrinol Metab 315(3):E374–E385. https://doi.org/10.1152/ajpendo.00021.2018

32. Qin LL, An MX, Liu YL, Xu HC, Lu ZQ (2017) MicroRNA-126: a promising novel biomarker in peripheral blood for diabetic retinopathy. Int J Ophthalmol 10(4):530–534. https://doi.org/10.18240/ijo.2017.04.05

33. Zou HL, Wang Y, Gang Q, Zhang Y, Sun Y (2017) Plasma level of miR-93 is associated with higher risk to develop type 2 diabetic retinopathy. Graefes Arch Clin Exp Ophthalmol 255(6):1159–1166. https://doi.org/10.1007/s00417-017-3638-5

34. Jiang Q, Lyu XM, Yuan Y, Wang L (2017) Plasma miR-21 expression: an indicator for the severity of Type 2 diabetes with diabetic retinopathy. Biosci Rep 37(2):BSR20160589. https://doi.org/10.1042/BSR20160589

35. Fillit HM. Scientific American. https://blogs.scientificamerican.com/observations/we-need-new-biomarkers-for-alzheimers-disease/

36. https://www.sciencemag.org/news/2019/12/skepticism-persists-about-revived-alzheimer-s-drug-after-conference-presentation?utm_campaign=news_weekly_2019-12-06&et_rid=544227157&et_cid=3113276

37. https://www.washingtonpost.com/dc-md-va/2019/12/05/biogens-potential-new-drug-alzheimers-disease-gets-cautiously-optimistic-review-following-presentation/

38. https://www.neurodegenerationresearch.eu/wp-content/uploads/2019/04/Full-JPND-Research-and-Innovation-Strategy-3.04.pdf

39. Thompson KK, Nissen JC, Pretory A, Tsirka SE (2018) Tuftsin combines with remyelinating therapy and improves outcomes in models of CNS demyelinating disease. Front Immunol 9:2784. https://doi.org/10.3389/fimmu.2018.02784

40. Herges K, Millward JM, Hentschel N, Infante-Duarte C, Aktas O, Zipp F (2011) Neuroprotective effect of combination therapy of glatiramer acetate and epigallocatechin-3-gallate in neuroinflammation. PLoS One 6(10):e25456. https://doi.org/10.1371/journal.pone.0025456

41. Martinez B, Peplow PV (2020) MicroRNAs as disease progression biomarkers and therapeutic targets in experimental autoimmune encephalomyelitis model of multiple sclerosis. Neural Regen Res 15:1831–1837. https://doi.org/10.4103/1673-5374.280307

Addendum

In the course of putting together this book we came across several recent publications relevant to the content of the book. They are listed below.

Aguzzi A, De Cecco E (2020) Shifts and drifts in prion science. Science 370:32–34. https://doi.org/10.1126/science.abb8577

Ahmad A, Patel V, Xiao J, Khan MM (2020) The role of the neurovascular system in neurodegenerative diseases. Mol Neurobiol 57:4373–4393. https://doi.org/10.1007/s12035-020-02023-z

Bartels T, De Schepper S, Hong S (2020) Microglia modulate neurodegeneration in Alzheimer's and Parkinson's diseases. Science 370:66–69. https://doi.org/10.1126/science.abb8587

Dawson TE, Golde TE, Lagier-Tourenne C (2018) Animal models of neurodegenerative diseases. Nat Neurosci 21:1370–1379. https://doi.org/10.1038/s41593-018-0236-8

De Jager PL, Yang HS, Bennett DA (2018) Deconstructing and targeting the genomic architecture of human neurodegeneration. Nat Neurosci 21:1310–1317. https://doi.org/10.1038/s41593-018-0240-z

Editorial (2018) Focus on neurodegenerative disease. Nat Neurosci 21:1293. https://doi.org/10.1038/s41593-018-0250-x

Ehrenberg AJ, Khatun A, Coomans E, Betts MJ, Capraro F, Thijssen EH, Senkevich K et al (2020) Relevance of biomarkers across different neurodegenerative diseases. Alzheimers Res Ther 12:71. https://doi.org/10.1186/s13195-020-00637-y

Fu H, Hardy J, Duff KE (2018) Selective vulnerability in neurodegenerative diseases. Nat Neurosci 21:1350–1358. https://doi.org/10.1038/s41593-018-0221-2

Gan L, Cookson MR, Petrucelli L, La Spada AR (2018) Converging pathways in neurodegeneration, from genetics to mechanisms. Nat Neurosci 21:1300–1309. https://doi.org/10.1038/s41593-018-0237-7

Philip V. Peplow, Bridget Martinez and Thomas A. Gennarelli (eds.), *Neurodegenerative Diseases Biomarkers: Towards Translating Research to Clinical Practice*, Neuromethods, vol. 173, https://doi.org/10.1007/978-1-0716-1712-0,
© Springer Science+Business Media, LLC, part of Springer Nature 2022

Guedes VA, Devoto C, Leete J, Sass D, Acott JD, Mithani S, Gill JM (2020) Extracellular vesicle proteins and microRNAs as biomarkers for traumatic brain injury. Front Neurol 11:663. https://doi.org/10.3389/fneur.2020.00663

Hickman S, Izzy S, Sen P, Morsett L, El Khoury J (2018) Microglia in neurodegeneration. Nat Neurosci 21:1359–1369. https://doi.org/10.1038/s41593-018-0242-x

Hrelia P, Sita G, Ziche M, Ristori E, Marino A, Cordaro M, Molteni R et al (2020) Common protective strategies in neurodegenerative disease: focusing on risk factors to target the cellular redox system. Oxidative Med Cell Longev 2020:8363245. https://doi.org/10.1155/2020/8363245

Jiménez-Jiménez FJ, Alonso-Navarro H, García-Martín E, Agúndez JA (2020) Cerebrospinal and blood levels of amino acids as potential biomarkers for Parkinson's disease: review and meta-analysis. Eur J Neurol 27:2336–2347. https://doi.org/10.1111/ene.14470

Jucker M, Walker LC (2018) Propagation and spread of pathogenic protein assemblies in neurodegenerative diseases. Nat Neurosci 21:1341–1349. https://doi.org/10.1038/s41593-018-0238-6

Keihani S, Kluever V, Mandad S, Bansal V, Rahman R, Fritsch E, Caldi Gomes L et al (2019) The long noncoding RNA*neuroLNC* regulates presynaptic activity by interacting with the neurodegeneration-associated protein TDP-43. Sci Adv 5: eaay2670. https://doi.org/10.1126/sciadv.aay2670

Liu YH, Wang J, Li QX, Fowler CJ, Zeng F, Deng J, Zu ZQ et al (2021) Association of naturally occurring antibodies to β-amyloid with cognitive decline and cerebral amyloidosis in Alzheimer's disease. Sci Adv 7:eabb0457. https://doi.org/10.1126/sciadv.abb0457

Martinez B, Peplow PV (2019) MicroRNAs as biomarkers of diabetic retinopathy and disease progression. Neural Regen Res 14:1858–1869. https://doi.org/10.4103/1673-5374.259602

Martinez B, Peplow PV (2020a) MicroRNAs as disease progression biomarkers and therapeutic targets in experimental autoimmune encephalomyelitis model of multiple sclerosis. Neural Regen Res 15:1831–1837. https://doi.org/10.4103/1673-5374.280307

Martinez B, Peplow PV (2020b) MicroRNAs in blood and cerebrospinal fluid as diagnostic markers of multiple sclerosis and to monitor disease progression. Neural Regen Res 15:606–619. https://doi.org/10.4103/1673-5374.266905

Martinez B, Peplow PV (2021a) MicroRNAs in laser-induced choroidal neovascularization in mice and rats: their expression and potential therapeutic targets. Neural Regen Res 16:621–627. https://doi.org/10.4103/1673-5374.295271

Martinez B, Peplow PV (2021b) MicroRNAs as diagnostic and prognostic biomarkers of age-related macular degeneration: advances and limitations. Neural Regen Res 16:440–447. https://doi.org/10.4103/1673-5374.293131

Mathieu C, Pappu RV, Taylor JP (2020) Beyond aggregation: pathological phase transitions in neurodegenerative disease. Science 370:56–60. https://doi.org/10.1126/science.abb8032

Nedergaard M, Goldman SA (2020) Glymphatic failure as a final common pathway to dementia. Science 370:50–56. https://doi.org/10.1126/science.abb8739

Palmqvist S, Janelidze S, Quiroz YT, Zetterberg H, Lopera F, Stomrud E, Su Y et al (2020) Discriminative accuracy of plasma phospho-tau217 for Alzheimer disease vs other neurodegenerative disorders. JAMA 324:772–781. https://doi.org/10.1001/jama.2020.12134

Sbodio JI, Snyder SH, Paul BD (2019) Redox mechanisms in neurodegeneration: from disease outcomes to therapeutic opportunities. Antioxid Redox Signal 30:1450–1499. https://doi.org/10.1089/ars.2017.7321

Sell SL, Widen SG, Prough DS, Hellmich HL (2020) Principal component analysis of blood microRNA datasets facilitates diagnosis of diverse diseases. PLoS One 15:e0234185. https://doi.org/10.1371/journal.pone.0234185

Sierksma A, Escott-Price V, De Strooper B (2020) Translating genetic risk of Alzheimer's disease into mechanistic insight and drug targets. Science 370:61–66. https://doi.org/10.1126/science.abb8575

Soto C, Pritzkow S (2018) Protein misfolding, aggregation, and conformational strains in neurodegenerative diseases. Nat Neurosci 21:1332–1340. https://doi.org/10.1038/s41593-018-0235-9

Sweeney MD, Kissler K, Montagne A, Toga AW, Zlokovic BV (2018) The role of brain vasculature in neurodegenerative disorders. Nat Neurosci 21:1318–1331. https://doi.org/10.1038/s41593-018-0234-x

Tanaka M, Toldi J, Vécsei L (2020) Exploring the etiological links behind neurodegenerative diseases: inflammatory cytokines and bioactive kynurenines. Int J Mol Sci 21:2431. https://doi.org/10.3390/ijms21072431

Wang W, Zhao F, Ma X, Perry G, Zhu X (2020) Mitochondrial dysfunction in the pathogenesis of Alzheimer's disease: recent advances. Mol Neurodegener 15:30. https://doi.org/10.1186/s13024-020-00376-6

Yu X, Ji C, Shao A (2020) Neurovascular unit dysfunction and neurodegenerative disorders. Front Neurosci 14:334. https://doi.org/10.3389/fnins.2020.00334

Zeng HM, Han HB, Zhang QF, Bai H (2020) Application of modern neuroimaging technology in the diagnosis and study of Alzheimer's disease. Neural Regen Res 16:73–79. https://doi.org/10.4103/1673-5374.286957

Zhou Y, Zhu F, Liu Y, Zheng M, Wang Y, Zhang D, Anraku Y et al (2020) Blood-brain barrier–penetrating siRNA nanomedicine for Alzheimer's disease therapy. Sci Adv 6:eabc7031. https://doi.org/10.1126/sciadv.abc7031

Philip V. Peplow
Bridget Martinez
Thomas A. Gennarelli

INDEX

Philip V. Peplow, Bridget Martinez and Thomas A. Gennarelli (eds.), *Neurodegenerative Diseases Biomarkers: Towards Translating Research to Clinical Practice*, Neuromethods, vol. 173, https://doi.org/10.1007/978-1-0716-1712-0,
© Springer Science+Business Media, LLC, part of Springer Nature 2022

Printed in the United States
by Baker & Taylor Publisher Services